Biotechnology of Human Reproduction

Biotechnology of Human Reproduction

Edited by

Alberto Revelli
University of Turin, Italy

Ilan Tur-Kaspa
Barzilai Medical Center, Ben Gurion University, Ashkelon, Israel

Jan Gunnar Holte
Carl von Linné Clinic, Uppsala, Sweden

Marco Massobrio
University of Turin, Italy

The Parthenon Publishing Group
International Publishers in Medicine, Science & Technology

A CRC PRESS COMPANY
BOCA RATON   LONDON   NEW YORK   WASHINGTON, D.C.

Cover illustrations

Laser-assisted intracytoplasmic sperm injection. Reproduced with the kind permission of Filippo Ubaldi and Laura Rienzi, Center for Reproductive Medicine, European Hospital, Rome, Italy

Library of Congress Cataloging-in-Publication Data
Data available on application

British Library Cataloguing in Publication Data
Biotechnology of human reproduction
  1. Human reproductive technology
  I. Revelli, Alberto
616.6'9206

ISBN   1842141325

Published in the USA by
The Parthenon Publishing Group
345 Park Avenue South, 10th Floor
New York, NY 10010, USA

Published in the UK and Europe by
The Parthenon Publishing Group
23–25 Blades Court
Deodar Road
London SW15 2NU, UK

Copyright © 2003 The Parthenon Publishing Group

First published 2003

Typeset by AMA DataSet, Preston, UK
Printed and bound by Butler & Tanner Ltd.,
Frome and London, UK

Contents

List of Contributors

Gautam N. Allahbadia
The Center for Human Reproduction
672 Kalpak Gulistan
Perry Cross Road
Bandra (West)
Mumbai 400050
India

Michal Amit
Department of Obstetrics and Gynecology
Rambam Medical Center
Haifa 31096
Israel

Michael W. Berns
Beckman Laser Institute
Department of Obstetrics and Gynecology
University of California
Irvine
CA 92612
USA

Andrea Bettella
Department of Medical and Surgical Sciences
Medical Clinic 3
Via Ospedale 105
University of Padova
Padova 35128
Italy

Graziella Bracone
Institute of Obstetrics and Gynecology
University of Bologna
Via Massarenti 13
Bologna 40138
Italy

Kathleen H. Burns
Departments of Pathology and Molecular and Human
    Genetics
Baylor College of Medicine
One Baylor Plaza
Houston
TX 77030
USA

Cilar Celik-Ozenci
The Sperm Physiology Laboratory
Department of Obstetrics and Gynecology
Yale University School of Medicine
333 Cedar Street
New Haven
CT 06510
USA

Patrizia M. Ciotti
Infertility and IVF Center
University of Bologna
Via Massarenti 13
Bologna 40138
Italy

Giuseppe Damiano
Infertility and IVF Center
University of Bologna
Via Massarenti 13
Bologna 40138
Italy

Franco Dondero
The University Laboratory of Seminology and
    Immunology of Reproduction
Department of Physiopathology
University of Rome 'La Sapienza'
Rome 00161
Italy

Raffaella Fabbri
Reproductive Medicine Unit
Institute of Obstetrics and Gynecology
University of Bologna
Via Massarenti 13
Bologna 40138
Italy

Anna Pia Ferraretti
S.I.S.M.E.R.
Reproductive Medicine Unit
Via Mazzini 12
Bologna 40138
Italy

Carlo Foresta
Department of Medical and Surgical Sciences
Medical Clinic 3
Via Ospedale 105
University of Padova
Padova 35128
Italy

Carlo Galeazzi
Department of Medical and Surgical Sciences
Medical Clinic 3
Center for Male Gamete Cryopreservation
Via Giustiniani 2
University of Padova
Padova 35128
Italy

Sandro Gambino
Department of Obstetrics and Gynecology
University of Turin
OIRM-S Anna Hospital
Turin
Italy

Loredana Gandini
The University Laboratory of Seminology and
    Immunology of Reproduction
Department of Physiopathology
University of Rome 'La Sapienza'
Rome 00161
Italy

David K. Gardner
Colorado Center for Reproductive Medicine
799 E. Hampden Avenue
Suite 520
Englewood
CO 80110
USA

Gianluca Gennarelli
Department of Obstetrics and Gynecology
St Anna Hospital
University of Turin
Via Ventimiglia 3
Turin 10126
Italy

Luca Gianaroli
S.I.S.M.E.R.
Reproductive Medicine Unit
Via Mazzini 12
Bologna 40138
Italy

Norbert Gleicher
The Center for Human Reproduction
1585 N. Barrington Road
Suite 406
Hoffman Estates
Illinois 60194
USA

Ermanno Greco
Center for Reproductive Biology and Medicine
European Hospital
Via Portuense 700
Rome 00149
Italy

Alessandro Di Gregorio
A.R.T.E.S.
Via Lamarmora 33
Turin 10128
Italy

Geraldine M. Hartshorne
Department of Biological Sciences
University of Warwick
Coventry CV4 7AL
UK

Jan Gunnar Holte
Carl von Linné Clinic
Uppsala University
Uppsala S-75183
Sweden

Outi Hovatta
Karolinska Institute
Department of Obstetrics and Gynecology
Huddinge University Hospital
Stockholm 14186
Sweden

Gabor Huszar
The Sperm Physiology Laboratory
Department of Obstetrics and Gynecology
Yale University School of Medicine
333 Cedar Street
New Haven
CT 06510
USA

Cosetta Iannascoli
Institute of Obstetrics and Gynecology
University of Bologna
Via Massarenti 13
Bologna 40138
Italy

Joseph Itskovitz-Eldor
Department of Obstetrics and Gynecology
Rambam Medical Center
Haifa 31096
Israel

Attila Jakab
The Sperm Physiology Laboratory
Department of Obstetrics and Gynecology
Yale University School of Medicine
333 Cedar Street
New Haven
CT 06510
USA

Vishvanath Karande
The Center for Human Reproduction
1585 N. Barrington Road
Suite 406
Hoffman Estates
Illinois 60194
USA

Marijo Kent-First
Department of Obstetrics and Gynecology
University of Wisconsin
Madison and Promega Corp.
2800 Woods Hollow Road
Madison
WI 53711
USA

Tamas Kovacs
The Sperm Physiology Laboratory
Department of Obstetrics and Gynecology
Yale University School of Medicine
333 Cedar Street
New Haven
CT 06510
USA

Lewis C. Krey
Program for IVF, Reproductive Surgery and Infertility
New York University School of Medicine
550 First Avenue
New York
NY 10016
USA

Michelle Lane
Colorado Center for Reproductive Medicine
799 E. Hampden Avenue
Suite 520
Englewood
CO 80110
USA

Giovanni B. La Sala
Department of Obstetrics and Gynecology
Gynecological Endoscopy and Reproductive Medicine
Reggio Emilia
Italy

Dan Leivin
Reproductive Medicine and Infertility Unit
Department of Obstetrics and Gynecology
Kaplan Medical Center
Rehovot 76100
Israel

Andrea Lenzi
The University Laboratory of Seminology and
    Immunology of Reproduction
Department of Medical Physiopathology
University of Rome 'La Sapienza'
Rome 00161
Italy

Francesco Lombardo
The University Laboratory of Seminology and
    Immunology of Reproduction
Department of Physiopathology
University of Rome 'La Sapienza'
Rome 00161
Italy

Maria Cristina Magli
S.I.S.M.E.R.
Reproductive Medicine Unit
Via Mazzini 12
Bologna 40138
Italy

Amnon Makler
Male Infertility Institute
Rambam Medical Center
Technion-Israel Institute of Technology
Haifa
Israel

Dorit Manor
Department of Obstetrics and Gynecology
Rambam Medical Center
Haifa 31096
Israel

Marco Massobrio
Department of Obstetrics and Gynecology
University of Turin
Via Ventimiglia 3
Turin 10126
Italy

Martin M. Matzuk
Department of Pathology
Baylor College of Medicine
One Baylor Plaza
Houston
TX 77030
USA

Carmen Mendoza
Department of Biochemistry and Molecular Biology
University of Granada
University Campus Fuentenueva
18008 Granada
Spain

Federica Moffa
Department of Obstetrical and Gynecological Sciences
IVF Unit
University of Turin
Turin
Italy

Henk J. Out
NV Organon
PO Box 20
5340 HB Oss
The Netherlands

Avinash Phadnis
Shree IVF Clinic
Pune
India

Eleonora Porcu
Infertility and IVF Center
University of Bologna
Via Massarenti 13
Bologna 40138
Italy

Francesca Poso
Department of Obstetrical and Gynecological Sciences
University of Turin
Via Ventimiglia 3
Turin 10126
Italy

Alberto Revelli
Department of Obstetrics and Gynecology
University of Turin
Via Ventimiglia 3
Turin 10126
Italy

Laura Rienzi
Center for Reproductive Medicine
European Hospital
Via Portuense 700
Rome 00149
Italy

Marco Rossato
Department of Medical and Surgical Sciences
Medical Clinic 3
Via Ospedale 105
University of Padova
Padova 35128
Italy

Denny Sakkas
The Sperm Physiology Laboratory
Department of Obstetrics and Gynecology
Yale University School of Medicine
333 Cedar Street
New Haven
CT 06510
USA

Angela Scarano
Infertility and IVF Center
University of Bologna
Via Massarenti 13
Bologna 40138
Italy

Stefan Schlatt
Institute of Reproductive Medicine
University of Münster
Domagkstr. 11
Münster 48149
Germany

Shmuel Segal
Department of Obstetrics and Gynecology
Barzilai Medical Center
Ben-Gurion University
Ashkelon
Israel

Zeev Shoham
Reproductive Medicine and Infertility Unit
Department of Obstetrics and Gynecology
Kaplan Medical Center
Rehovot 76100
Israel

Ruslan V. Sobolev
Department of Human Reproduction
Odessa State Medical University
Odessa 65026
Ukraine

Davide Spolaore
Department of Medical and Surgical Sciences
Medical Clinic 3
Via Ospedale 105
University of Padova
Padova 35128
Italy

Edith Suss-Toby
Department of Obstetrics and Gynecology
Rambam Medical Center
Haifa 31096
Israel

Yona Tadir
Ramat Marpe Hospital
3 Spiegel St.
Petach-Tikva 49361
Israel

Jan Tesarik
Molecular Assisted Reproduction & Genetics
   (MAR&Gen)
Gracia 36
18002 Granada
Spain

Bruce J. Tromberg
Beckman Laser Institute
Department of Obstetrics and Gynecology
University of California
Irvine
CA 92612
USA

Ilan Tur-Kaspa
Department of Obstetrics and Gynecology
Barzilai Medical Center
Ben Gurion University
Ashkelon 78306
Israel

Filippo Ubaldi
Center for Reproductive Medicine
European Hospital
Via Portuense 700
Rome 00149
Italy

Stefano Venturoli
Infertility and IVF Center
University of Bologna
Via Massarenti 13
Bologna 40138
Italy

Lynne Vigue
The Sperm Physiology Laboratory
Department of Obstetrics and Gynecology
Yale University School of Medicine
333 Cedar Street
New Haven
CT 06510
USA

Maria T. Villani
Department of Obstetrics and Gynecology
University of Udine
Udine
Italy

Ariel Weissman
IVF Department
Department of Obstetrics and Gynecology
Wolfson Medical Center
Holon
Israel

Valeriy N. Zaporozhan
Department of Obstetrics and Gynecology
Odessa State Medical University
Valikhovskiy per. 2
Odessa 65026
Ukraine

John Zhang
IVF Unit
Medical Center
New York University
New York
USA

Efraim Zohav
Department of Obstetrics and Gynecology
Barzilai Medical Center
Ben-Gurion University
Askkelon 78306
Israel

Preface

The application of biotechnology to human reproduction offers new strategies to increase our basic knowledge on reproduction and to improve our chances of diagnosing and treating male and female infertility. We felt the need for a book describing the most innovative tools that biotechnology offers to reproductive medicine and that are likely to change our approach to the diagnosis and treatment of human infertility.

Recombinant DNA technology for the synthesis of new drugs, transgenic animal models to investigate the intimate mechanisms of reproductive genetics, fluorescence *in situ* hybridization (FISH) and polymerase chain reaction (PCR) techniques to study genes and chromosomes in the human embryo all represent direct applications of biotechnology to reproductive medicine. Cryopreservation and micromanipulation of human embryos and gametes, *in vitro* maturation of human gametes, embryonic stem cells transplantation and nuclear transfer techniques are other currently developed options in human reproduction. Technologies such as laser and three-dimensional ultrasound will probably make important contributions to the diagnosis and treatment of infertile patients. Many of these technologies are already available for clinical application and some of them are currently under clinical investigation. In human reproduction, the gap between basic research and clinical science is narrowing.

Aim of this book is to provide reproductive scientists, physicians, fellows and students with a complete review of the most advanced acquisitions in the field of biotechnology of human reproduction.

We wish to express our deepest gratitude to all authors that contributed to increase the value of this book with their competent and outstanding work.

Alberto Revelli, Ilan Tur-Kaspa,
Jan Gunnar Holte, Marco Massobrio
October 2002

Physiology of human reproduction

New insights into gametogenesis, sperm migration, fertilization and implantation in humans

1

Alberto Revelli, Francesca Poso and Ilan Tur-Kaspa

INTRODUCTION

Our knowledge of human reproduction has noticeably grown in the last years, and several new technologies have been introduced as novel tools in reproductive medicine as well as in reproductive biology research. The efforts to understand the physiologic mechanisms of animal and human reproduction are at the basis of such progress and of future advances in reproductive biotechnologies. This chapter provides an updated synthesis of the known processes that regulate reproduction in humans, from gamete production to fertilization, embryo development and implantation.

GAMETOGENESIS

The embryonic cells establishing the germ cell line separate from somatic cell lineages very early in development, in the peri-implantation stage[1], and first appear during the third week post-fertilization in the endoderm of the yolk sac[2]. They are termed primordial germ cells (PGC) and have a round shape with 15–20 μm diameter, an eccentric nucleus with fine granular chromatin and 1–2 nucleoli. Their cytoplasm, stained with alkaline phosphatase, is relatively devoid of organelles, with a few mitochondria around the nucleus, glycogen and lipid droplets that form an energy reserve to be used during migration to the genital ridges[3].

Male and female primordial germ cells are phenotypically indistinguishable and reach their intra-embryonic position migrating from the yolk sac to the gonadal anlage both by ameboid self-propulsion[4] and by virtue of the remodeling of the growing embryo layout.

Indeed, when the embryo loses its discoid shape and acquires a tubular aspect, it incorporates the yolk sac endoderm in the posterior primitive intestine area, where the gonads will develop. Although genetically programed, PGC migration seems to be partially dependent on the production of chemotactic substances from the gonad-forming tissues. Fibronectin is suspected to be one of these factors[5], while locally synthesized cytokines (Kit ligand and transforming growth factor β1 (TGFβ1)) appear to promote germ cell survival and multiplication during migration[1].

Oogenesis

At the end of the fifth week of development, PGCs reach the gonadal anlage and colonize the developing ovary, actively migrating through the ovarian tissue[6]. From the ninth week post-fertilization, PGCs begin to differentiate into oogonia, characterized by a large, central nucleus and a high mitotic activity. From the 12–13th week post-fertilization, oogonia start to differentiate into oocytes that soon begin meiosis, increase in volume and reorganize the distribution of intracellular organelles. In the middle part of gestation (17–22 weeks), oocytes have progressed to the diplotene stage of meiosis and associate with cords of somatic cells, which will become the pregranulosa cells of primordial follicles[7]. These pregranulosa cells are devoid of follicle stimulating hormone (FSH) receptors and are endocrinally silent[8].

The number of germ cells reaches a peak of 6–7 million during the fifth month of gestation, but an impressive reduction in number occurs during prenatal development, producing a population of about 1–2 million primordial follicles at birth. Such a restriction of the gamete pool is due to a combination of two processes: (1) germ cell degeneration resulting from genetic errors, metabolic and/or vascular disturbances; and (2) germ cell exfoliation into the coelomic cavity, due to massive desquamation of the most superficial ovarian areas[4]. No new follicles and oocytes are formed after birth; they are reduced to approximately 250 000 by puberty and progressively disappear during the fertile age, until menopause[9]. The genetically driven rate of follicle recruitment, maturation and atresia is the major factor affecting the number of primordial follicles present at any given age.

After several rounds of mitotic amplification in the fetal gonads, oogonia begin meiosis around the fourth month of pregnancy. They undergo an extended prophase I that progresses for some weeks and finally stops at the early diplotene stage, when oocytes are in intimate association with somatic pregranulosa cells, forming quiescent primordial follicles (Figure 1). The oocyte is now diploid (2n), with a 4C DNA content: it is in the G2 stage of the cell cycle, following DNA replication that occurs before the beginning of meiosis. At this stage, the oocyte nucleus is pale and large, and is called 'germinal vesicle' (GV).

Growth and maturation of the human oocyte are very slow, requiring some months from primordial to mature follicles. During these weeks, the oocyte enlarges from 35 to 120 μm in diameter and increases 100-fold

in volume, becoming one of the largest cells in the human body[10]. Cytoplasmic organelles, ribosomes and mitochondria increase in number; the Golgi apparatus expands and migrates to the periphery of the cytoplasm, starting production and exocytosis of the zona pellucida (ZP) glycoproteins[11]. Crystalline bodies, fat droplets and glycogen granules accumulate in the oocyte cytoplasm, reflecting active synthesis and storage of materials.

Primordial follicles containing a GV-stage (prophase I) oocyte are unpredictably recruited into the growing pool throughout all reproductive life and the oocyte remains at the diplotene stage until it is stimulated to restart meiotic maturation by the luteinizing hormone (LH) preovulatory surge, that precedes meiotic maturation by approximately 36 h. Morphologically, when meiosis reinitiates, the nuclear membrane breaks and disappears (GV breakdown, GVBD). During progression to metaphase I, a meiotic spindle forms at the periphery of the oocyte cytoplasm and guides the chromosomal alignment on an equatorial plane. The eccentric location of the meiotic spindle has the consequence that at the end of the first meiotic division, a small polar body containing a set of chromosomes is expelled into the perivitelline space. The oocyte progresses to the second meiotic division without a new S phase and proceeds until metaphase II, when it arrests unless fertilization occurs. The segregation of chromatids during the second meiotic division reduces the number of chromosomes to a haploid set (n). The mature oocyte has a very limited life-span, approximately 24 h.

Oocyte entrance into meiosis is controlled by complex mechanisms, most of which are still poorly understood:

(1) A factor responsible for the oocyte's transition from G2 to M phase has been identified in the mouse and named M-phase promoting factor (MPF)[12]. Its expression rises at the time of GVBD and continues growing until metaphase I, following which MPF activity disappears. During GVBD, MPF probably promotes dissolution of the nuclear membrane, chromosomal condensation and microtubule reorganization to form the meiotic spindle. The combination of two components, cyclin B and p34[cdc2] (homologous to the *cdc2* gene), form an inactive pre-MPF, which is activated by specific tyrosine phosphatases via dephosphorylation of tyrosine residues[13].

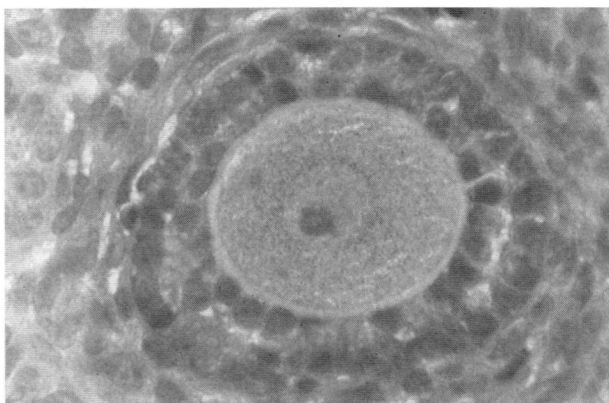

Figure 1 Human follicle with two layers of cuboidal granulosa cells surrounding an immature oocyte at the germinal vesicle (GV) stage

(2) Mitogen-activating protein (MAP) kinase is likely to be the factor that prevents the oocyte from entering interphase between metaphase I and II by phosphorylating substrates modified by MPF[14].

(3) The cAMP/protein kinase A (PKA) pathway seems to prevent GVBD by inhibiting MPF activation[15]. In fact, GVBD spontaneously occurs when intracellular cAMP levels decrease following granulosa cell removal. The synthesis of hyaluronic acid in response to the LH midcycle peak elicits mucification of the granulosa cells with loss of gap junctions and cell dispersion. The subsequent fall of cAMP levels in the oocyte below the threshold that maintains GV integrity triggers GVBD[15].

(4) Hypoxanthine and adenosine prevent GVBD acting, respectively, via a block of cAMP hydrolysis and being a substrate of adenylate cyclase in the cAMP synthesis pathway[16].

(5) The protein p39[mos], product of the *c-mos* proto-oncogene, activates MFP by interacting with cyclin B. It is a serine–threonine kinase degraded by a Ca^{2+}-dependent protease. When intracellular Ca^{2+} increases at fertilization, the p39[mos] activity decreases[17] and phosphodiesterase is activated, reducing cAMP intracellular levels.

Besides the acquisition of meiotical competence (nuclear maturation), the oocyte must undergo changes in the pattern of protein synthesis and in the distribution of cytoplasmic organelles (cytoplasmic maturation). Cytoplasmic maturation should be co-ordinated to nuclear maturation in order to obtain a fully mature, fertilizable oocyte. The acquisition of cytoplasmic maturation is a stepwise event that takes place during the final growth phase of the oocyte, in parallel to the progression of meiosis. It seems to be promoted by estrogens and counteracted by androgens[18]. Interestingly, a prevalently estrogenic environment is a characteristic of the dominant follicle, that will ovulate and provide the egg for fertilization, whereas an androgen prevalence is observed in non-dominant follicles, that will undergo atresia.

Active RNA synthesis, storage and translation take place in the developing oocyte. Transcription is highly active in oogonia during mitosis and less active in the oocyte during meiosis. It is reduced by the association of pregranulosa cells with the egg, probably as a consequence of the transfer of nucleotides and other small molecules from these cells to the oocyte[19]. The RNA polymerase transcriptional activity of the growing oocyte decreases sharply after full size is attained and is practically absent when GVBD starts[20]. After GVBD, no new RNA molecules are transcribed until the embryonic genome starts to be expressed[21]. The oocyte, as well as the developing embryo, utilizes the abundant pool of stored RNA that is accumulated and packaged in the growth phase. Many RNAs transcribed during oocyte growth are stored and expressed only during embryo development[22]. Activation of stored RNAs occurs at specific stages of oocyte maturation or embryo development.

The translation of mRNA into a protein is linked to mRNA polyadenylation[23]: the addition of long poly(A)* tails promotes translation, and may occur during oocyte growth or only later, following ovulation[24]. A further control of RNA translation in the oocyte is due to RNA association with RNA-binding proteins that package the molecule in a way that prevents the ribosomes from starting translation[24]. In summary, during oocyte growth protein synthesis occurs for use by the oocyte itself, for communication with granulosa cells or for use during the early stages of embryo development. Protein synthesis is particularly active in the small, developing oocytes, whereas in larger oocytes an active uptake of exogenous proteins takes place, as revealed by the appearance of numerous membrane-bound vesicles in the ooplasm[19].

Oocyte–granulosa cells interplay during follicle growth

A detailed description of the endocrine and paracrine mechanisms leading to follicle recruitment, follicle growth and ovulation is beyond the scope of this chapter. Suffice to say, oocyte growth and maturation are strictly linked to follicle growth and the interaction between granulosa cells and the oocyte deeply influences the oocyte's development.

In brief, clusters of primordial follicles are continuously drawn into a growth phase, during which pregranulosa cells become cuboidal and actively proliferate around the oocyte. When granulosa cell numbers have risen to some thousands, as a consequence of the stimulation provided by gonadotropins and local factors, a fluid-filled cavity (antrum) becomes evident and the growing follicle is termed the 'antral follicle' (Figure 2). At this time the oocyte has considerably increased its volume and has the competence to resume meiosis (GVBD).

In the antral follicle, two populations of granulosa cells can be distinguished: (1) cumulus–granulosa cells (CGCs), surrounding the oocyte; and (2) mural cells, delimiting the follicle wall. Much experimental evidence has pointed out that oocyte and CGCs are clearly interdependent. In fact, gap junctions link CGCs to each other and to the oocyte, mediating the transfer of metabolites and small molecules[25]. CGCs provide both a physical support and a nutritive as well as regulatory activity to the oocyte, whose *in vitro* development without CGCs is abnormal[26], leading to a reduced competence to achieve a normal preimplantation embryo development[27].

The presence of gap junctions between CGCs and the oocyte is essential for oocyte growth. The c-Kit ligand KL, produced by CGCs, significantly stimulates oocyte growth when added *in vitro* to growing follicles[28]. KL acts at all stages of follicle development, being probably involved in the adhesiveness of the oocyte to CGCs[29]. The c-Kit KL receptor is expressed in oocytes[29] and its activation exerts a negative control over oocyte meiosis via the production of a meiosis-inhibiting substance[30]; this negative control, that can be overcome only by the preovulatory LH surge, is important to prevent the premature onset of meiosis.

On the other hand, differentiation and survival of the granulosa cells are positively affected by the oocyte[31], that actively influences the normal development of its own follicle. The oocyte sustains estradiol production and inhibits progesterone (P) secretion by CGCs[32],

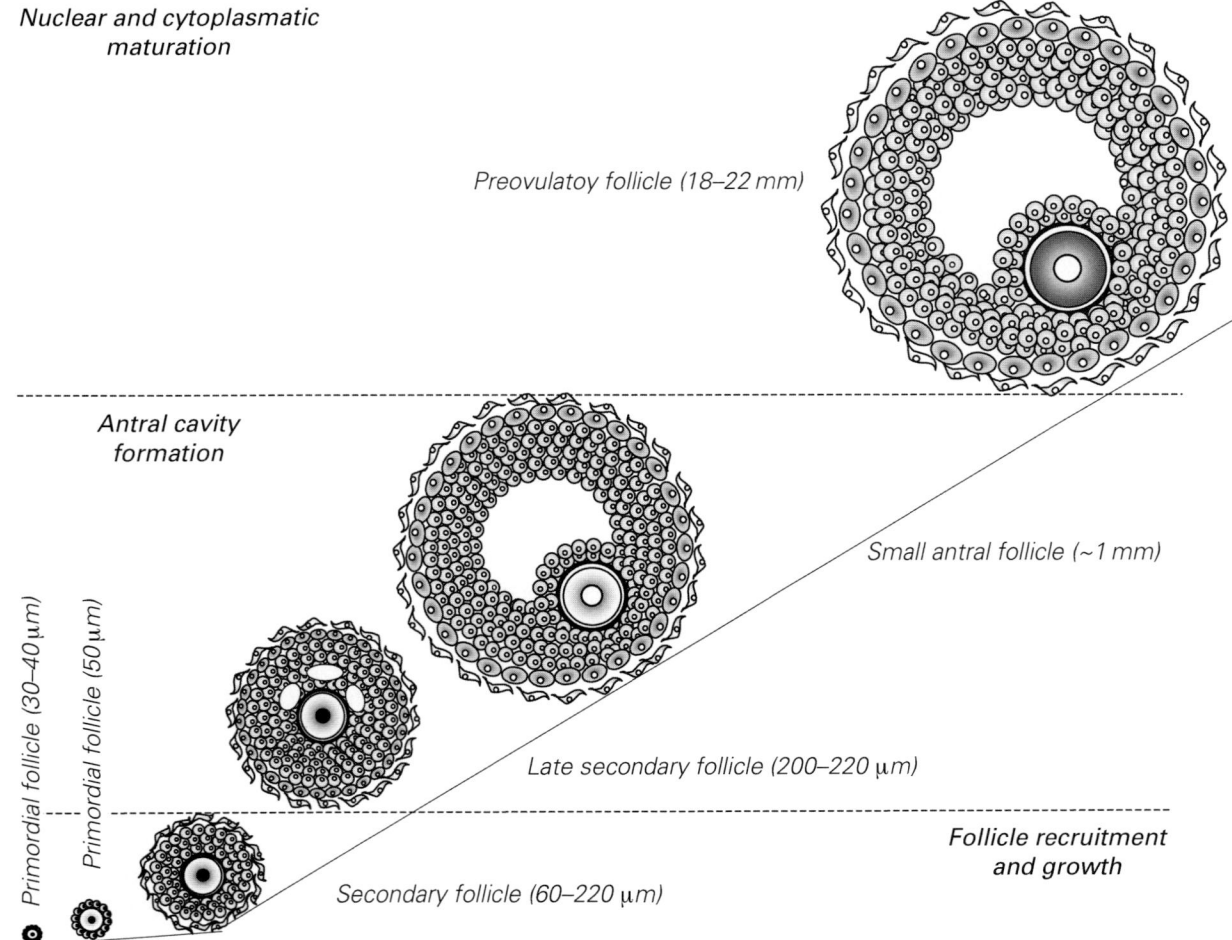

Figure 2   Growth and maturation of the human ovarian follicle from the primordial to the preovulatory stage

induces CGC proliferation[33] and blocks the expression of mRNA for LH receptors in CGCs[34].

The oocyte also stimulates the FSH- and cAMP-induced hyaluronic acid secretion by CGCs, promoting the formation of a viscoelastic mucinous mass that compacts the cumulus facilitating its pick-up by the oviductal fimbriae at ovulation[35]. If the oocyte is removed, CGCs remain unresponsive to FSH and cAMP and no hyaluronic acid is secreted[36]. The oocyte even prevents the dissociation of this viscoelastic matrix by blocking the gonadotropin-stimulated secretion of urokinase plasminogen activator (uPA) by CGCs[37].

A rather recently studied oocyte factor, belonging to the TGFβ superfamily and exclusively expressed by the oocyte, is growth differentiation factor 9 (GDF-9), that is involved in the control of granulosa cell proliferation and follicle development[38]. GDF-9 mRNA appears in oocytes included in primary follicles and continues to be transcribed throughout oocyte maturation until ovulation. Mutant mice not expressing the *gdf-9* gene show smaller ovaries with no granulosa cell proliferation and a follicle development arrested at the primary follicle stage[39]. Oocytes in these follicles have an abnormal distribution of the intracellular organelles, tend to activate parthenogenetically and have impaired meiotic competence[39]. The presence of GDF-9 has also been demonstrated in human oocytes[38].

Spermatogenesis

Differently from oogenesis, spermatogenesis is a continuous process occurring in the male gonads from puberty through adulthood, until old age. Its efficiency in humans is much lower than in other mammalian species, and many morphologically abnormal and non-functional spermatozoa are present in all human semen samples, indicating a highly defective gamete-producing machinery.

After the arrival of PGCs in the undifferentiated gonad, spermatogenesis is determined by the expression of the *sry* gene (the sex-determining region of the Y chromosome), that drives the primordial gonad to become a testis[40]. At birth, the spermatogenic epithelium is composed of immature Sertoli cells enclosing a relatively limited number of undifferentiated spermatogonia. The number of Sertoli cells is a major rate-limiting factor for the final length of the testicular tubules: it depends on their mitotic activity, which is submitted to the control of genetic[41] and endocrine factors (among which there are FSH and thyroid hormones)[42].

From puberty onwards, spermatogenesis takes place in the testicular tubules, where the spermatogenic cells are complexed with Sertoli cells, that support a number of different processes. The interaction among Sertoli cells is important for establishing the so-called blood–testis barrier, that allows the creation of a specific, protected microenvironment in which spermatogenesis may take place[43]. At the onset of puberty, a marked and progressive rise in circulating gonadotropins induces the activation of testicular Leydig and Sertoli cells, leading to initiation of spermatogenesis. Clusters of germ cells enter spermatogenesis at regular time intervals and every single step of spermatogenesis has a precise duration. As a consequence, the cellular composition of the spermatogenic epithelium is highly ordered (spermatogenic cycle), a single Sertoli cell being always in contact with a defined set of spermatogenic cell types[44].

When spermatogenesis begins, spermatogonia make contact with the basal membrane and become mitotically active, whereas the mitotic activity of Sertoli cells ceases. In the human, the mitotically active population of spermatogonia is named 'A pale', as opposed to the more silent 'A dark' spermatogonia[45]. After a series of mitotic divisions, undifferentiated A pale spermatogonia become A1, A2, A3, A4, intermediate and, finally, differentiated B spermatogonia, whose mitotic division results in the formation of primary spermatocytes[45] (Figure 3). This mitotic step represents the point of entry into meiosis. Primary spermatocytes are transported through the Sertoli cell into the adluminal compartment of the tubule, leaving space for new spermatogonia in the basal compartment. Spermatocytes form syncytia in which cells are connected by cytoplasmic bridges[43]; proteolytic enzymes produced by the Sertoli cells (plasminogen activator and cyclic protein-2, the proenzyme form of cathepsin L) promote the Sertoli cell crossing[46]. Primary spermatocytes perform the last DNA duplication, resulting in diploid cells with a 4C DNA content. They go through leptotene, zygotene, pachytene and diplotene stages, after which two meiotic divisions rapidly occur, leading to the formation of two

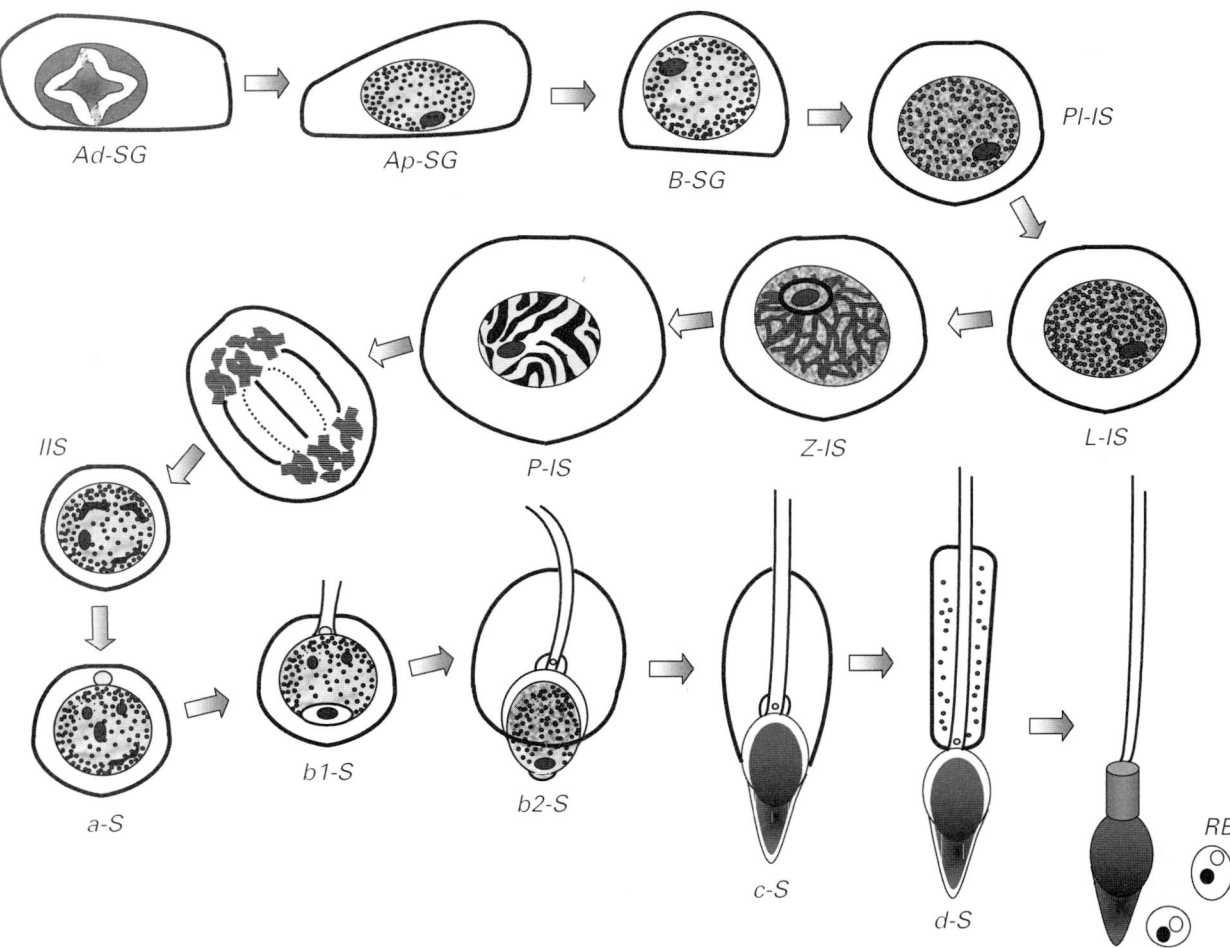

Figure 3 Stages of human spermatogenesis. Mature spermatozoa are obtained following mitotic, meiotic and post-meiotic phases. The cell types shown are spermatogonia type A dark (Ad-SG) and type A pale (Ap-SG), spermatogonia type B (B-SG), pre-leptotene (Pl), leptotene (L), zigotene (Z) and pachytene (P) primary spermatocytes (IS), secondary spermatocytes (IIS), spermatids (S) at different stages of maturation (a, b1, b2, c, d) and mature spermatozoa with residual bodies (RB)

haploid secondary spermatocytes with a 2C DNA content, and then four haploid round spermatids with a 1C DNA content. In spermatocytes, the diplotene stage is immediately followed by the two meiotic divisions, with no sign of a temporary metaphase II arrest, as observed in oocytes[47].

Haploid spermatids appear as round cells, but soon they begin differentiation through a process named spermiogenesis, that takes approximately 20 days (Figure 3). During spermiogenesis, the spermatid elongates, being firmly attached to Sertoli cells by junctions whose formation requires cell adhesion molecules such as integrins and cadherins[48]. Spermiogenesis leads to

spermatid elongation, acrosome development, nuclear condensation, formation of midpiece and flagellum and reduction of cytoplasmic volume. At the end of the process, the cytoplasmic droplet containing the remains of the Golgi apparatus (residual body) is taken up by the Sertoli cell and the mature spermatids (now called testicular spermatozoa) are released into the tubular lumen (spermiation). It is believed that the residual body gives a signal to the Sertoli cell, leading on one side to the release of proteases able to release spermatids and, on the other side, to the transport of a new cohort of early spermatocytes across the Sertoli cell[46].

Testosterone and FSH are two major regulators of spermatogenesis. They do not act on the germ cells directly, but interact with Sertoli cells stimulating the latter to become mature and fully active in supporting spermatogenesis. FSHβ knockout mice[49] and men with inactive FSH receptors[50] show a greatly impaired spermatogenesis. FSH acts on Sertoli cells via stimulation of a G-protein-coupled surface receptor inducing the activation of adenylate cyclase, cAMP synthesis, PKA activation and protein phosphorylation[51]. Testosterone stimulates the Sertoli cells both directly, through interaction with nuclear androgen receptors[52], and via a signaling molecule that is produced by the peritubular cells and travels across the basal membrane reaching Sertoli cells[53]. Several other growth factors and cytokines regulate the mitotic activity of spermatogonia: among them are activin[54] and the so-called steel factor (SLF) that acts through activation of the c-Kit receptor system[55].

During meiosis in spermatogenesis, pairing of the X and Y chromosomes is limited to a short region and complex mechanisms prevent the unpaired regions of the X and Y chromosomes from undergoing non-homologous recombination events[56]. This avoids the synthesis of gene products encoded by the X chromosome, possibly incompatible with further progression through meiosis[57]. Since the expression of genes on the X chromosome is impaired, autosomal genes with testis-specific expression take over their function, compensating the need of the corresponding proteins: for instance, the pyruvate dehydrogenase E1 α-subunit is encoded by genes on the X chromosome in somatic cells but by an autosomal gene in spermatids[58]. Many of the proteins encoded by testis-specific autosomal genes are enzymes of the energy-producing machinery, others are testis-specific histones or heat shock proteins (HSP) expressed during spermatogenesis. Among HSPs, HSP70-2 plays a role in meiosis[59] and cell cycle regulation[60]: in HSP70-2 knock-out mice, spermatocytes arrest in meiosis I and undergo apoptosis. Testis-specific transcripts of autosomes are generated through alternative gene splicing or via activation of an alternative promoter[61].

Several testis-specific genes are expressed during spermatogenesis. Some of them encode proteins essential for sperm function rather than for spermatogenesis itself, e.g., the proteolytic enzyme acrosin or fertilin, a surface protein involved in fertilization. Specific transcription-regulatory proteins interact with regulatory DNA elements in the promoter region of each gene, determining the rate of gene transcription. Transcription regulatory factors that show highly specific expression during spermatogenesis are DNA-binding proteins such as homeobox proteins (Hox-1.4, Sperm-1, Esx1) and zinc-finger proteins (Zfy, Ret)[62]. The factor CREMτ (cAMP response element modulator τ), generated through a testis-specific alternative splicing of the CREM gene, is an important regulatory factor of spermatogenesis, particularly when testis-specific gene promoters containing a cAMP-responsive element are involved[63]. Interestingly, in somatic cells and in the early steps of spermatogenesis, the CREM gene expression yields a repressor CREM protein; in contrast, CREMτ functions as an activator of gene transcription in post-meiotic cells[63]. During spermatogenesis, the 'on' or 'off' status of a gene depends also on the chromatin structure: in fact, inactive chromatin (heterochromatin) shows a typical pattern of cytosine methylation in DNA areas rich in cytosine and guanine (CpG islands)[64]. Chromatin structure and gene expression are influenced also by histone acetylation or phosphorylation[65]. Some testis-specific histones are exclusively expressed in spermatogonia, whose chromatin is packaged in DNA–histone complexes[66]. During development from spermatogonia into meiotic spermatocytes, the nucleosomal histone-based structure of the chromatin is replaced by a tightly packed protamine-based structure. Elongation and condensation of the spermatid nucleus requires the replacement of histones first by transition proteins (TP1 and TP2) and then by protamines (P1 and P2)[67], preceded by post-translational histone modifications, such as H4 acetylation[68]. Testis-specific proteins of the high-mobility group (tsHMG) are also expressed during nuclear elongation, and contribute to the chromatin rearrangement through modulation of topoisomerase activity[69].

The DNA–protamine complexing results in the formation of a tightly compacted DNA. The genome is already transcriptionally inactivated in spermatids, with the shut-off of protein synthesis; mRNAs in spermatids are stored and undergo translational regulation; gene transcription and mRNA synthesis are practically absent in spermatozoa[70].

SPERM TRANSPORT

Sperm transport along the upper female genital tract includes complex, co-ordinated events that ensure that some spermatozoa reach the fertilization site in the upper part of the salpinx and express their fertilization potential. Three main steps occur during the sperm's trip along the female genital tract: sperm separation from seminal plasma, sperm chemotaxis and capacitation.

Sperm separation from seminal plasma

The first step in the series of events leading to fertilization competence in spermatozoa is separation from seminal plasma, that begins when the sperm-coagulating proteins of the ejaculate are dissolved and spermatozoa are allowed to move unrestrictedly. This process, termed sperm liquefaction, takes place in a few seconds in the upper part of the vagina and is temperature-dependent, relying on the activity of specific enzymes of prostatic origin. Soon after liquefaction, the spermatozoa actively enter into cervical mucus, where the proteinic coat adsorbed to sperm membrane in seminal plasma ('sperm coat') is totally removed. As a consequence, previously hidden surface antigens are exposed, giving new antigenic properties to spermatozoa and allowing proteins with receptor function to be exposed to their ligands.

Seminal plasma has the important function of maintaining sperm in a biologically restricted state, to prevent fruitless sperm activation far from the fertilization site. As spermatozoa progress through the cervical mucus, seminal plasma components are removed: this allows the beginning of a complex series of processes, collectively called capacitation[71].

Sperm chemotaxis

Sperm chemotaxis is defined as an oriented movement in response to a chemical gradient, resulting in approaching the chemical attractant, or in retreating from a chemical repellent. In some invertebrate species, whose gametes are spawned into water before fertilization, chemotaxis is a key event in reproduction as the spermatozoa must be guided towards the eggs over long distances. Egg-derived, species-specific peptides able to attract spermatozoa have been identified and purified in echinoderms, whose spermatozoa have specific surface receptors for these chemotactic substances[72]. Activation of such receptors, which belong to the membrane guanylate cyclase family, causes an increase in intracellular cGMP (reviewed in reference 73) and mediates ion fluxes across the sperm membrane[74], that in turn conditions flagellar motion and determines the direction of movement[75].

In species with internal fertilization, and particularly in mammals, chemotaxis may appear to be less crucial, as millions of spermatozoa are ejaculated directly into the female genital tract and sperm can be stored in the cervix and/or in the oviduct[76], not far from the fertilization site. It was calculated, however, that only about 1 in 25 000 spermatozoa inseminated into the vagina reaches the fallopian tubes, and the chance of a successful random collision between a spermatozoon and the egg is statistically minimal[77,78]. Interestingly, it has been observed that the tubal ampulla where the egg resides contains a significantly larger number of sperm than the contralateral ampulla[79].

Sperm chemotaxis in humans was demonstrated *in vitro*, and recent studies suggest that its physiologic role is probably to select the fertilizing capacitated spermatozoa and to ensure their continuous availability for an extended period of time[77].

In order to exert a chemical attraction to human spermatozoa, a substance (or substances) should have an increasing concentration from the lower to the upper part of the salpinx. Such a substance should also be present in the fluid surrounding the oocyte or secreted by the cells surrounding the egg, so as to guide spermatozoa towards the oocyte. Follicular fluid (FF) factors have been considered ideal candidates, as they are also able to increase sperm motility and velocity *in vitro*[77,80,81]. Sperm chemotaxis to FF factors has been reported[82] and has been distinguished from other processes (e.g., chemokinesis, the enhancement of sperm motility) which might cause sperm accumulation in the accumulation assays used to assess chemical attraction[83–85].

Despite active research, the follicular factors responsible for sperm chemotaxis have not yet been identified. Human FF contains the guanylate cyclase activator atrial natriuretic factor (ANP)[86,87], whose specific receptors have been identified on the surface of human spermatozoa[88]. Experiments accomplished by accumulation[89] and choice assays[90] led to the hypothesis of sperm chemotaxis by ANP towards human spermatozoa. However, no correlation was found between the chemotactic activity of a given FF and its ANP content[89], and sperm chemotaxis to ANP at physiologic concentrations is observed only when phosphoramidon, a neutral endopeptidase inhibitor probably absent *in vivo*, is added to the system[90]. Thus, ANP might simply be a substance capable of activating the guanylate cyclase system *in vitro* in a way similar to the activation caused by the physiologic attractant *in vivo*. As to other putative sperm chemoattractants contained in FF, such as P[91] and N-formylated peptides[92], there is no evidence for the involvement of guanylate cyclase in their signaling.

Capacitation

Capacitation is a unique preparatory event for fertilization. Despite its pivotal importance, its intimate mechanisms are not yet completely understood, and a definitive marker to identify it has not been defined. The onset of sperm hyperactivated motility and the exocytosis of acrosomal content (acrosome reaction) represent the endpoints of capacitation, but several intermediate steps occur during its time course. The following changes have been identified in capacitating spermatozoa:

(1) The cholesterol removal from the sperm plasma membrane, with a decrease in the cholesterol-to-phospholipid ratio and an increase in the membrane fluidity[93];

(2) A facilitated movement of proteins within the plasma membrane, with the subsequent expression of mannose binding sites and the activation of some membrane receptors, such as progesterone receptors (PR), that become active after dimerization[94];

(3) An increase in the phosphorylation of tyrosine residues in membrane proteins[95];

(4) An influx of Ca^{2+} ions from the extracellular environment[96];

(5) An increase in intracellular reactive oxygen species (ROS), oxidative processes and cAMP production[97].

The time needed to induce capacitation of human spermatozoa *in vitro* is 2–4 h during incubation at 37°C

in balanced saline containing 0.3–3% albumin, HCO^{3-} and Ca^{2+} ions[98].

In vitro studies have shown that albumin supports capacitation, as it facilitates the removal of cholesterol from the sperm membranes[99]. As a consequence of the increased membrane fluidity, the expression of membrane proteins with a receptor function is enhanced: these receptors, upon interaction with their ligands, will initiate the signaling sequence that leads to the acrosome reaction. Some proteins with a receptor function need to undergo phosphorylation on tyrosine residues to get a fully activated state: tyrosine phosphorylation is increased during capacitation and appears to be upregulated by ROS (e.g., hydrogen peroxide)[100], and through the activation of the bicarbonate- and Ca^{2+}-sensitive adenylate cyclase/cAMP/PKA pathway[95]. When sufficient time has elapsed to complete the capacitation process, spermatozoa become susceptible to stimuli able to induce the acrosome reaction.

ACROSOME REACTION

The acrosome reaction (AR) is an exocytotic event that includes the dissolution and release of the acrosomal content. It is absolutely required for successful penetration through the oocyte vestments, the cumulus oophorus and the ZP (Figure 4).

The AR initiates with point fusions between the plasma and the outer acrosomal membrane, immediately followed by the formation of wider fenestrations[101]. As a consequence, the acrosomal matrix becomes exposed and solubilized, and several lytic enzymes are released. Among the acrosomal enzymes, a serine glycoproteinase called acrosin seems to play a major role in sperm–oocyte fusion, specifically at the level of ZP binding and penetration[102]. Acrosin exists in a proenzyme form called proacrosin, both in the acrosomal membranes and in the acrosomal matrix. Proacrosin conversion into acrosin (activation) involves a change in acrosomal pH[103] and occurs immediately before membrane vesiculation. After the AR, residual acrosin remains associated with the non-released, posterior part of the acrosome to facilitate sperm binding with the oolemma[103].

Some spermatozoa in the ejaculate may prematurely undergo acrosomal exocytosis, becoming temporally and/or spatially inadequate for fertilization. In this

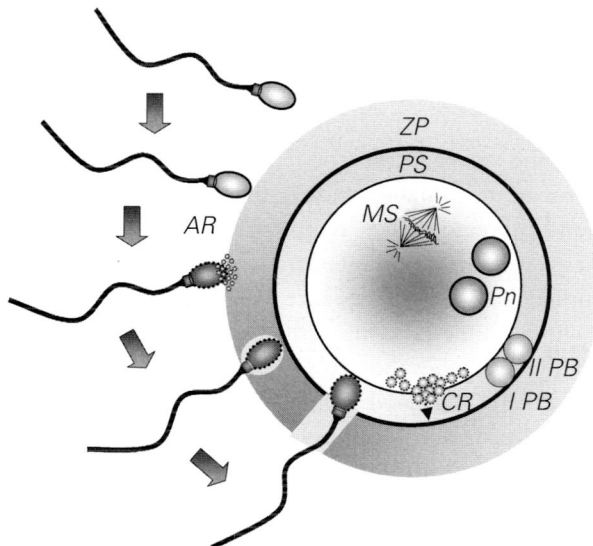

Figure 4 Fertilization in humans. The interaction between the protein ZP3 of the zona pellucida (ZP) and its receptor on the sperm head leads to the acrosomal reaction (AR) and to the subsequent enzymatic attack of the ZP. The sperm penetrates the ZP and reaches the perivitelline space (PS). The oocyte and sperm membranes fuse and fertilization takes place. The cortical reaction (CR) prevents polyspermia. The second meiotic division of the oocyte is completed and the second polar body extruded (II PB). Upon fusion, the two pronuclei (Pn) give rise to the formation of the mitotic spindle (MS) and to the first cleavage. I PB, first polar body

respect, seminal plasma proteins play a key role in stabilizing sperm membranes. On the other hand, the ability of the spermatozoon to respond to proper stimuli and to activate the right intracellular signaling cascade is of utmost importance. Although in human spermatozoa the AR can be biochemically induced without a previous capacitation by substances that bypass membrane-mediated events[104], in physiologic human fertilization the only spermatozoa that undergo the AR are believed to be those that fully complete capacitation.

In humans, the physiologic inducers of the AR appear to be the following: the steroid hormone P and a protein of the ZP.

FF mixed with the oviductal fluid at the time of ovulation or trapped in the extracellular matrix of the cumulus represents a source of P acting on spermatozoa at the fertilization site. P has been shown to stimulate the motility and acrosomal exocytosis of human sperm *in vitro*[105], to increase the percentage of spermatozoa exhibiting hyperactivated motility[106] and to increase the velocity of spermatozoa when added to peritoneal fluid[107]. A significant correlation between P levels in FF and the ability of FF to induce the acrosomal exocytosis of human sperm has been reported[108] and preincubation of FF with P-specific antibodies prevents the FF-induced Ca^{2+} influx in human spermatozoa, a phenomenon which is strictly related to acrosomal exocytosis[109].

The action of P on human spermatozoa is enhanced by acrosin, that cleaves the P-binding protein CBG, finally producing a high local P concentration[110,111]. Specific binding sites for P have been demonstrated on the part of the sperm plasma membrane overlying the acrosome[112], where three types of surface PR have been identified: (1) a plasma membrane Ca^{2+} channel (PR1); (2) a membrane-associated protein tyrosine kinase (PTK; PR2); and (3) a plasma membrane chloride channel (PR3)[113]. PR2, that appears to be sensitized during the capacitation process, is probably the one responsible for the effect of P on the AR and on hyperactivated motility[114]. PR1 is responsible for the rapid opening of the Ca^{2+} channel, but is not capable of initiating the AR. The third receptor (PR3) is likely to be a $GABA_A$-receptor/chloride channel complex, and probably mediates the Cl^- fluxes occurring during acrosomal exocytosis[115]. After binding to the PTK-associated surface receptor (PR2), P produces receptor aggregation on the sperm surface, which is facilitated by the increased membrane fluidity acquired during capacitation[116]. The aggregation of PR2 stimulates the tyrosine phosphorylation of a 94 kDa phosphoprotein, in turn able to activate the opening of voltage-dependent Ca^{2+} channels. In parallel, voltage-independent Ca^{2+} channels are opened via activation of PR1. The result is a biphasic Ca^{2+} influx into the sperm, coupled to a Cl^- efflux arising from the activation of PR3[117].

Thus, P effects on spermatozoa appear to be mediated by a multireceptor system. When one member of this system is not operational, the biological response may fail even if the other types of receptor function normally. PR in spermatozoa are probably relevant to the sperm's fertilizing ability: in fact, an impaired P-binding capacity with subsequent failure generate the typical Ca^{2+} influx wave and the AR has been observed in some cases of unexplained infertility[118], and in some oligozoospermic men[119].

The ZP is a specific coat (15 μm-thick) composed of glycoproteins secreted by the oocyte into the perivitelline space. Three of these glycoproteins are termed ZP1, ZP2 and ZP3. The genes encoding for ZP proteins are present as single copies on chromosomes 19 (ZP1), 7 (ZP2) and 5 (ZP3); they are expressed during the growth phase of the oocyte, but their transcripts are rapidly degraded when meiosis is resumed[120]. ZP3 is a surface protein that acts as a tyrosine kinase-coupled receptor and binds the sperm head when the acrosome is still intact[121]. ZP2 acts as a secondary sperm receptor anchoring the sperm to the zona after the AR and helps the sperm's penetration through the zona. Chains of ZP2 and ZP3 are linked together by ZP1, to form the zona matrix[122].

The ZP, whether solubilized or intact, stimulates the AR in a dose-dependent manner[123]. ZP3 is considered the primary ligand for sperm–zona binding and AR induction[124]. ZP3 is coupled to a PTK, which acts as a sperm receptor and is stimulated by both ZP3 and P (PR2)[125]. In fact, in the mouse P enhances the acrosomal exocytotic response to the subsequent exposition to solubilized ZP3[126]. The receptor for ZP3 on the sperm plasma membrane is probably a lectin that binds mannose-containing ligands[94], whose expression increases during capacitation.

Both ZP3 and P stimulate multiple signal transduction pathways involving second messengers, which interact via convergent cross-talk mechanisms[127]. The effect of ZP3 and P on the AR is linked to their capacity to induce a rapid increase in intracellular Ca^{2+}, which is almost entirely due to the influx of extracellular Ca^{2+}, as it may be blocked by Ca^{2+} channel blockers or by the addition of Ca^{2+} chelators to the incubation medium[109]. Ca^{2+} acts as a second messenger molecule, in turn activating the following major pathways:

(1) *The adenylate cyclase/cAMP/PKA pathway* cAMP analogs (dibutyryl cAMP) that artificially increase intracellular cAMP levels exert a dose-dependent stimulation of the AR in capacitated human spermatozoa[127]. Furthermore, inhibition of sperm cAMP phosphodiesterase by pentoxifylline or activation of adenylate cyclase and cAMP synthesis by forskolin[127] induces the AR in capacitated human spermatozoa. Inhibition of PKA, which phosphorylates proteins as a downstream step of the activation cascade, prevents AR induction by dibutyryl cAMP or forskolin[128].

(2) *The PLC/PIP2/PKC pathway* Phospholipase C (PLC) activation increases diacylglycerol (DAG) and inositol 1,4,5-triphosphate (InsP3) levels. DAG is responsible for the activation of the Ca^{2+} and phospholipid-dependent kinase (PKC), whereas InsP3 actively stimulates the release of intracellular Ca^{2+} stores that, although very scarce in the spermatozoon, are represented by the acrosome itself. PLC inhibitors have been shown to prevent the AR[129], whereas stimulators of PKC induce a dose-dependent AR that can be prevented by PKC blockers[130].

(3) *G-protein pathway* Activation of G-proteins after sperm–ZP3 binding or P–PR complex activation has been hypothesized as one of the effector mechanisms involved in AR; however, a clear demonstration of the involvement of a particular class of G-protein in the human AR is still lacking[131].

Other molecules have been recently proposed as part of the complex mechanisms leading to the occurrence of AR. Studies *in vitro* have shown that low concentrations of the guanylate cyclase agonists ANP and nitric oxide (NO) are able to induce the AR in human capacitated spermatozoa via stimulation of guanylate cyclase, cGMP synthesis and Ca^{2+} influx[132,133]. Activation of the constitutive isoform of NO synthase (eNOS), an enzyme that is expressed by human spermatozoa, is implicated in the FF-induced AR, and could work as a subsidiary physiologic mechanism through an autocrine pathway[132].

FERTILIZATION

Fertilization is the process that leads to the union of the sperm and oocyte nuclei within the activated oocyte cytoplasm. Part of the knowledge about fertilization has been obtained during *in vitro* fertilization (IVF) procedures in humans, but the majority of information has been derived from murine or non-human primate (rhesus monkey) models. As a consequence, not all of the information can be extrapolated to humans.

The spermatozoon contributes to fertilization with three important components: (1) the paternal haploid genome; (2) the oocyte-activating factor(s); and (3) the centrosome, which organizes microtubule assembly and directs the formation of the mitotic spindle.

The process of fertilization begins when the exocytosis of hydrolytic enzymes from the sperm acrosome, together with the vigorous hyperactivated motility, allows the sperm to penetrate into the oocyte overcoming its outer vestments (Figure 4). The first stimulus leading to AR is provided by P, which is highly concentrated in the FF surrounding the cumulus oophorus and in the intercellular matrix within the cumulus cells. Some sperm promptly respond to P undergoing the AR. Among the acrosomal enzymes, hyaluronidase plays a key role in breaking intercellular bridges within the cumulus and promotes the dispersion of cumulus cells. Several spermatozoa can now reach the ZP passing through the dispersed cumulus cells, and start interacting with the zona surface proteins, among them the powerful AR inducer ZP3. Sperm–ZP3 interaction leads to a further stimulation of the AR, that amplifies the enzymatic aggression to the zona. Acrosome-reacted spermatozoa, that, losing their acrosome, have lost the ZP3 receptors on their head plasma membrane, remain bound to the zona by binding with another zona protein, ZP2[134]. They achieve zona penetration by a combination of vigorous sperm motility and enzymatic hydrolysis. The latter appear to be important, but not crucial, since spermatozoa belonging to acrosin-null mice can successfully penetrate the ZP, although with some delay[135].

After zona crossing, acrosome-reacted sperm reach the perivitelline space, binding to and fusing with the oocyte plasmalemma by using their plasma membrane at the post-acrosomal region. In humans, the spermatozoon in the perivitelline space rests tangentially with the egg surface. During sperm incorporation, microfilaments assembled from maternal actin form the so-called 'incorporation cone' at the oocyte's cortex. In the mouse, binding of the spermatozoon to the egg plasma membrane is mediated by the interaction between the oocyte integrin $\alpha6\beta1$ receptors and two members of the ADAM protein family, fertilin-β and cyristein, on the sperm[136]. While fertilin-β supports binding of the sperm to the oocyte plasma membrane, fertilin-α is probably implicated in the subsequent step, the fusion of sperm and oocyte membranes[137]. Another plasma membrane protein involved in sperm–egg fusion is the oocyte integrin-associated protein CD9: in mice, homozygous-

null CD9 females have decreased fertility due to impaired sperm–egg fusion[138]. Since in the egg plasma membrane CD9 protein is intimately associated with integrin $\alpha6\beta1$, to which fertilin-β binds, it is likely that CD9 regulates the interaction between integrins and fertilin, ultimately responsible for sperm–egg fusion[139]. However, it appears probable that other mechanisms and other egg and sperm proteins are involved in the sperm–egg fusion process.

Sperm–egg binding and fusion rapidly leads to egg activation, which is strictly linked to the abrupt rise in intracellular Ca^{2+}. Ca^{2+} represents the central messenger of the activation signal, as the artificial increase in egg intracellular Ca^{2+} levels is able to elicit activation and parthenogenetic development of the oocyte[140]. According to the most accredited hypothesis, the spermatozoon activates the egg by introducing a proteinic factor named oscillin, which is responsible for the occurrence of a specific pattern of Ca^{2+} oscillations, that in the human egg can only be observed as a consequence of sperm entry[141]. A correct pattern of Ca^{2+} oscillations is supposed to be a prerequisite for normal embryo development.

The eventuality that more than one spermatozoon could enter the egg, giving rise to a polyspermic, abnormal fertilization, is prevented by a complex mechanism linked to egg activation called 'cortical reaction'. Cortical reaction consists of the rapid exocytosis of cortical granules from the oocyte, that brings high concentrations of proteolytic enzymes into the perivitelline space (Figure 4). These enzymes digest the sites for sperm binding on the zona (ZP3, ZP2) and modify the zona structure to toughen and harden it[142]. In some vertebrates, the cortical reaction is paralleled by a fast electrical block due to the sudden opening of ion channels by sperm factors, resulting in the reversal of the electrical membrane potential that prevents further sperm binding and fusion[143].

At the time of ovulation in humans, the oocyte is arrested at the metaphase of the second meiotic division (MII). When the sperm enters, the oocyte cell cycle is still arrested at MII, with condensed meiotic chromosomes. This cell cycle arrest is due to a cytostatic mechanism that involves both the *c-mos* gene product and the MAP kinase pathway[144]. When the sperm crosses the perivitelline space and fuses with the oocyte membrane inducing oocyte activation, the oocyte resumes meiosis, proceeds to anaphase, then to telophase, and finally completes meiosis by dividing into two unequal cells: the fertilized oocyte and the second polar body (Figure 4). The latter is extruded into the perivitelline space, while the maternal chromosomes are enclosed by a nuclear envelope and form the female pronucleus. Most of the molecules stored during oogenesis are retained in the cytoplasm of the fertilized oocyte and play a key role in further embryo development.

On the other hand, the chromatin of the fertilizing spermatozoon is highly condensed, tightly packaged with nuclear protamines and transcriptionally inactive. After incorporation into the oocyte, the sperm nucleus loses the nuclear envelope and undergoes chromatin decondensation and replacement of sperm-specific protamines by histones. The paternal DNA binds maternal histones and is surrounded by a new nuclear envelope of maternal origin, forming the male pronucleus. The fertilized oocyte with two pronuclei is termed 'zygote'.

Fertilization leads to the decondensation and remodeling of both parental genomes. Active DNA synthesis takes place in both the male and the female pronuclei. Contemporaneously to pronuclear formation, a new microtubule-formed structure termed 'sperm aster' assembles within the egg, being organized by the sperm-derived centrosome[145]. The centrosome is paternally inherited in humans: it attracts and binds several egg proteins, e.g., γ-tubulin and the nuclear mitotic apparatus protein (NuMA), that transform the inactive precursor sperm centrosome into the microtubule-nucleating, zygote centrosome. The sperm aster is a radially arrayed, three-dimensional structure adjacent to and affixed to the sperm nucleus. It pushes the two pronuclei until they appose in the centre of the oocyte, where they subsequently fuse.

Following migration of the pronuclei to the center of the fertilized egg, the breakdown of the nuclear envelopes occurs and the pronuclei fuse together (singamy). In parallel, the centrosome duplicates and separates, originating the two poles of the first mitotic spindle. In mammals, pronuclear fusion does not occur during interphase, but at first mitosis, when parental chromosomes are mixed as they align at the metaphase plate of the first mitotic spindle (Figures 4 and 5). At this stage cleavage begins, resulting in the production of a zygotic nucleus that contains both sets of parental chromosomes.

The duration of the full mitotic cell cycle of the human zygote is 20–22 h, but the length of each phase shows marked differences among zygotes. *In vitro* the G1 phase, that begins with the completion of the second meiotic division, is accomplished by the majority of zygotes between 3 and 14 h post-insemination; the S phase begins between 8 and 14 h post-insemination and lasts until 14–24 h post-insemination; the G2 phase starts between 12 and 30 h post-insemination; the M phase lasts 3–4 h, and the majority of zygotes undergo pronuclear breakdown and first-cleavage division 24–30 h and 27–33 h post-insemination[146].

The most common fertilization abnormalities observed *in vitro* are the appearance of one or three pronuclei, and the precocious cleavage arrest. Oocytes with only one pronucleus represent about 1% of oocytes fertilized *in vitro*; they may result from a parthenogenetic activation of the oocyte (without a sperm contribution), from asynchrony in pronuclei formation (with retarded appearance of the second pronucleus) or from precocious singamy. Tripronuclear zygotes, that represent about 5% of the embryos fertilized *in vitro*, may derive from the fertilization of a diploid oocyte by a single sperm (digynic triploids), or from the fertilization of a normal egg by two spermatozoa (diandric triploids). The cleavage arrest at the pronuclear stage may be linked to failed oocyte activation or by structural defects or incorrect location of the sperm centrosome[147].

PREIMPLANTATION DEVELOPMENT

Preimplantation development in humans is characterized by reductive cleavage divisions, i.e., cell proliferation in the absence of significant cell growth. The overall volume of the preimplantation embryo is practically constant during all preimplantation development and the embryo remains surrounded by the ZP (Figure 5). The first embryo cleavage, leading to the two-cell stage, occurs meridionally in the human embryo, and is the only division that is synchronous, resulting in two equal cells (blastomeres) which contain similar allocations of the animal and vegetal poles of the oocyte. The second-cleavage division is slightly asynchronous and associated with rotation, resulting in the four-cell stage.

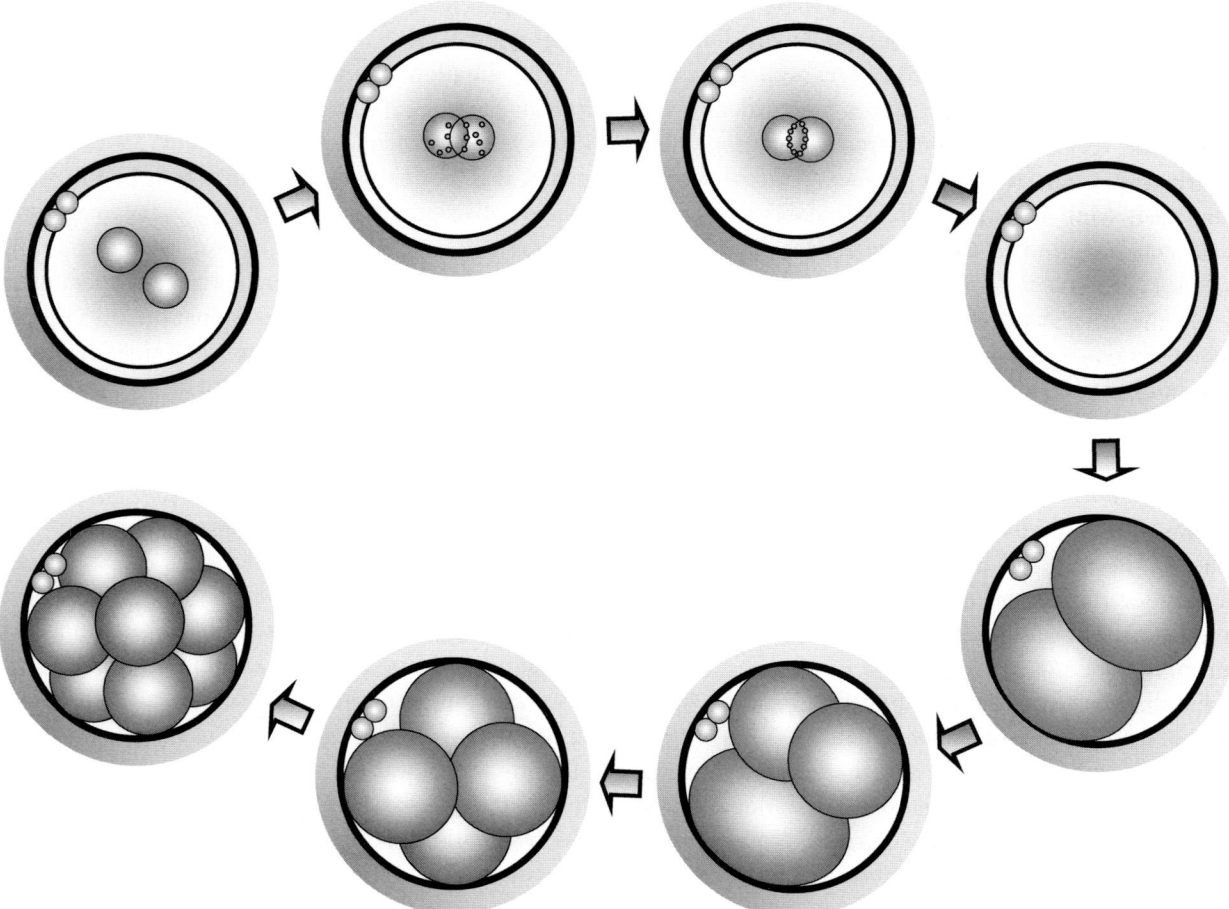

Figure 5 Schematic representation of pronuclear syngamy and fusion, of the first cleavage rounds and of preimplantation embryo development up to the eight-cell stage. The entire process is accomplished without an increase in volume, with the embryo remaining within the confines of the zona pellucida

All following cleavages, that still take place in the space delimited by the ZP, are asynchronous, leading to embryos with an odd number of cells.

During preimplantation development, the embryo undergoes a critical process termed maternal-to-zygotic transition, during which the genetic program of early development is shifted from maternally derived proteins and mRNAs to the products of the embryonic genome. Prior to zygotic gene activation (ZGA), embryo development is supported by organelles, proteins and mRNAs previously stored in the oocyte cytoplasm. The process of ZGA generates novel transcripts and begins a drastic reprograming of gene expression, finally leading to the full expression of the totipotent nucleus of the embryo.

In the mouse, ZGA is completed by the two-cell stage, when many mRNAs that are not expressed in the egg appear[21]. However, the mouse embryo is still transcriptionally active at the one-cell stage when nucleosomes (functional units of DNA and histones that block the binding of transcription factors to promoters) are displaced, allowing the access of maternally derived transcription factors to their DNA-binding sequences on the embryo genome[148]. The newly expressed mRNAs are in general poorly translated in the one-cell mouse embryo: this allows the embryo to avoid the synthesis of functional proteins whose expression could be inappropriate before the formation of the zygotic nucleus[150]. Interestingly, in the mouse, the male pronucleus supports a higher level of transcription than the female one[151]. This is probably due to the fact that following fertilization, sperm-bound protamines are replaced by maternal histones and maternal transcription factors gain access to their DNA-binding sequences; in the female pronucleus, on the contrary, DNA is already complexed with histones and no replacement occurs: as a consequence, DNA-binding sequences remain inaccessible.

In humans, ZGA occurs at about the four- to eight-cell stage[152] because human eggs contain a 3–4 times higher content of maternal histones than mouse eggs, and one to two additional cell cycles are required for maternal transcription factors to form transcription complexes. However, certain genes (about 5–10% of the genome) are expressed already from the pronuclear stage[153]. Among them are some 'imprinted' genes, which are expressed differently depending on whether they are inherited from the mother or the father (genomic imprinting)[154]. When genomic imprinting occurs, only one of the alleles for a particular gene is expressed in the developing zygote, depending on its parental source[155]. The paternal genome is mainly involved in the development of the extra-embryonic tissues, whereas the maternal genome is more closely associated with post-implantation embryonic development. Genomic imprinting is established during gametogenesis and is mediated through methylation of cytidines located within dinucleotide CG repeating sequences termed CpG islands, outside the gene's coding region. Several embryo abnormalities as well as human diseases appear to be linked to defects in genomic imprinting[156].

Overall, the occurrence of abnormal embryos in human IVF is common and it probably reflects what happens *in vivo*. Major chromosomal abnormalities (e.g., aneuploidy) are often reflected by gross morphologic abnormalities or by an arrested or heavily delayed development. Nevertheless, chromosomal abnormalities are frequently observed even in morphologically normal, early human embryos[157]. Sometimes these abnormal genomic patterns are already present in the zygote and are simply amplified by the subsequent cleavage divisions; in other embryos, the rapidity of mitotic divisions may cause occasional errors, often compatible with embryo development until the blastocyst stage.

Fragmentation of blastomeres occurs in many cleaving embryos and is considered the result of apoptotic processes within the embryo itself. When structural defects are present, developmental arrest occurs, caused by chromosomal abnormality, failure of embryonic genome activation or cytoplasmic deterioration[158].

Up to the eight-cell stage the blastomeres are quite distinct, whereas from this stage onwards compaction begins, with the formation of junctions between neighboring blastomeres (Figure 6). An interior embryo compartment is formed, bounded by impermeable tight junctions. Before compaction, the ZP prevents the blastomeres from dissociating, as well as from attaching to the oviduct epithelial cells. During compaction, the blastomeres flatten and form different intercellular junctions: gap junctions, adherens junctions and tight junctions. Gap junctions have the function of providing channels for the direct passage of small molecules, adherens junctions are involved in cadherin-linked intracellular associations, while tight junctions, composed of several specific proteins (e.g., cingulin, occludin), form a permeability seal between cells.

During compaction, the preimplantation embryo cells become highly polarized: microvilli appear at the apical surface, but not at the basolateral surface (surface polarity); actin microfilaments concentrate in the apical part of the cytoplasm, microtubuli in the basolateral part (cytoplasmic polarity)[159]. The protein E-cadherin (uvomorulin) is thought to be the main responsible for compaction[160]. It is of maternal origin, and can be found in the egg's cytoplasm; it appears at the egg's surface after fertilization, remains uniformly distributed on the blastomere's surface, concentrating particularly in the regions of blastomere contact[161]. The timing of compaction could be regulated by phosphorylation of E-cadherin[162].

Following compaction at the eight-cell stage, cleavage divisions occur on a random plane and the deriving daughter cells are allocated randomly to the inside or the outside of the developing morula. Interestingly, the cleavage plane of an outer blastomere of a 16-cell morula is not random, but produces 80% daughter cells that remain on the outside part of the embryo. As a result of this, in the early 32-cell blastocyst the majority of cells remain on the outside of the embryo and differentiate into trophoectoderm cells, whereas only 10–12 cells

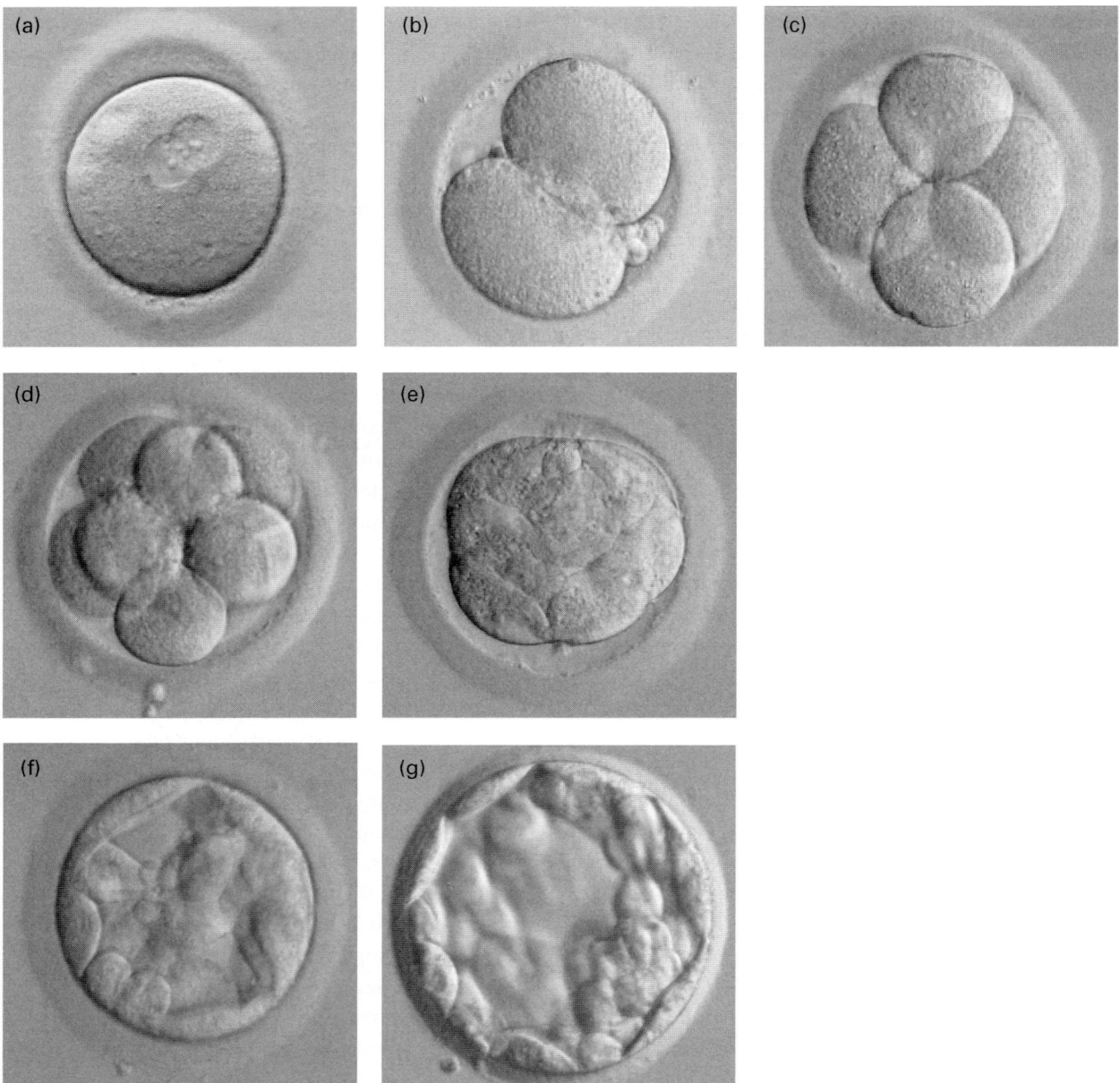

Figure 6 Preimplantation embryo development with pronuclear fusion, cleavage, compaction and formation of the blastocyst. (a) Pronuclear stage; (b) two-cell cleavage stage; (c) four-cell embryo; (d) eight-cell embryo; (e) compacting morula; (f) cavitating blastocyst; (g) cavitated blastocyst

comprise the inner cell mass (ICM) and remain fully totipotent[163]. ICM cells preferentially communicate via gap junctions with each other and not with trophoectoderm cells[164]. Gene expression is very different in the two cell types that form the blastocyst, some genes being preferentially expressed in the ICM, others in the trophoectoderm[165]. Apoptosis is restricted to the ICM, where cells are thought to require survival signals, among which is TGFα in the mouse embryo[166]. Some of these survival factors may be secreted by cells of the upper female genital tract or even by the embryo itself: in fact, the incidence of apoptosis is reduced if several embryos are cultivated *in vitro* in a single medium drop of low volume.

Between the 16-cell and the 32-cell stages, cells from the outer part of the embryo start to transport water into the interior of the embryo, forming the blastocoele cavity. Fluid accumulation is mainly due to a Na^+–K^+-ATPase activity that pumps Na^+ into the blastocoele, while water diffuses osmotically[167]. Ca^{2+}, Cl^-, bicarbonate and cAMP also contribute to blastocyst cavitation. Tight junctions form a continuous belt around the trophoectoderm cells, and, being impermeable, prevent leakage of ions from the blastocoele. A correct ion composition in blastocoele fluid is fundamental for ICM cell proliferation and differentiation and probably contributes to the initial events of implantation.

Early cleavage divisions take place in the oviduct and the embryo enters into the uterus around day 4, at the morula stage (16 cells). The last event of preimplantation development is the escape from the ZP (hatching), that occurs in the uterus 5 days after fertilization. In the mouse, hatching is due to the action of a trypsin-like protease named strypsin and appears as an extended dissolution of the zona[168]. In rodents, the powerful oxidant molecule superoxide anion, generated by the blastocyst itself, seems to contribute to hatching[169]. In humans, hatching is limited to a small area of the blastocyst surface: the blastocyst gradually escapes from the zona evaginating towards a narrow hole. Very little is known about the molecular basis of hatching, but the transcriptional events that underlie hatching are thought to be completed just before hatching starts[170].

IMPLANTATION

The human endometrium is structured as a scaffold of extracellular matrix (including collagen, proteoglycans and basement membranes), to which cells are fixed via adhesion molecules (e.g., laminin and fibronectin). Luminal endometrial cells express cell surface receptors for adhesion molecules, mainly membrane glycoproteins referred to as integrins.

Implantation is a three-step process in which the blastocyst is guided towards a specific area in the uterine cavity (apposition), binding to endometrial cell integrins (adhesion) and, finally, activating proteases that destruct the extracellular matrix and allow migration into the endometrial stroma (invasion) (Figure 7). Most of our

knowledge about implantation is derived from animal models as well as from *in vitro* studies with human endometrial cells and explant cultures.

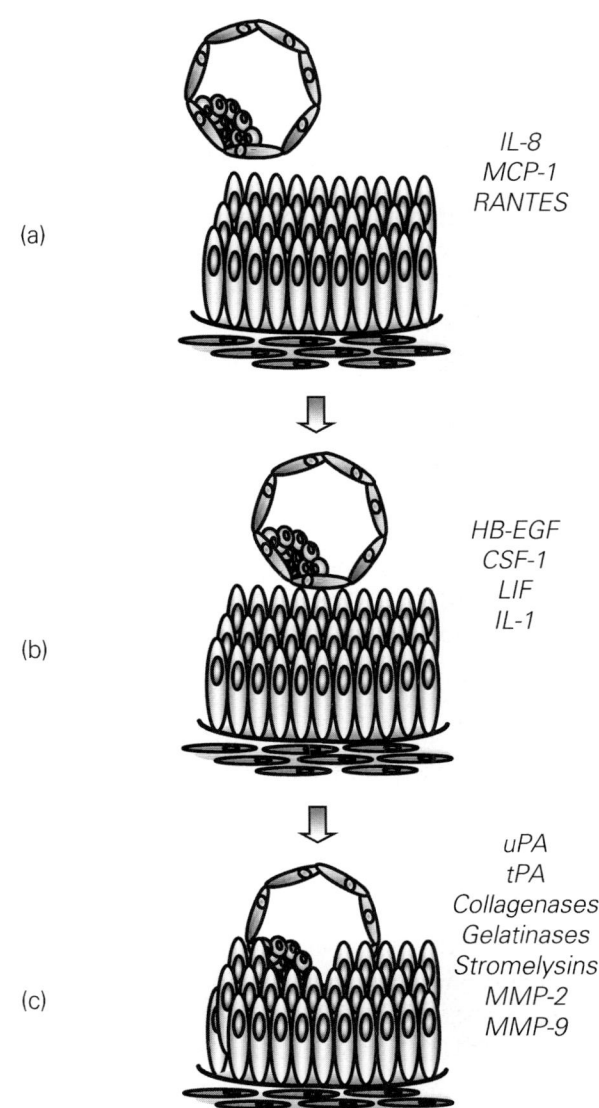

Figure 7 The three stages of blastocyst implantation in humans, apposition (a) and adhesion (b) to the endometrium as well as endometrial invasion (c), are highly regulated by the coordinated action of chemokines and cytokines. IL-8, interleukin 8; MCP-1, monocyte chemotactic protein-1; HB-EGF, heparin-binding epidermal growth factor; CSF-1, colony-stimulating factor-1; LIF, leukemia inhibitory factor; IL-1, interleukin-1; uPA, urokinase-type plasminogen activator; tPA, tissue-type plasminogen activator; MMP-1 and –9, matrix metalloproteinases 2 and 9

The embryo generally apposes near the posterior wall of the uterus, with its embryonic pole directed towards the endometrial epithelium. Subsequently, the embryo attaches to the endometrial epithelium and passes through the epithelial cell lines, reaching and subsequently invading the epithelial stroma. A complex molecular dialog occurs between the blastocyst and the endometrium, involving cell-to-cell and cell-to-extracellular matrix interactions, that are mediated by several growth factors and cytokines. The time period during which this dialog takes place is called the 'implantation window'.

The most relevant morphologic marker of the implantation window is the appearance of pinopodes, P-dependent apical protrusions involved in endocytosis and pinocytosis that appear on the apical surface of endometrial luminal epithelium[171]. The long, thin microvilli of the luminal epithelial cells are gradually converted into irregular, flattened pinopodes that last about 24–48 h[172] and are involved in the uptake of macromolecules and in the expression of specific adhesion molecules mediating embryo attachment. In the implantation window, the surface endometrial cells even decrease their negative surface charge by reducing sialic acid moieties and express adhesion molecules. Among them, integrin subunits β3, α4 and α1 are considered markers of uterine receptivity, as well as integrin α5β3, which appears to play a major role during implantation in humans[173]. Interestingly, the human blastocyst selectively promotes the expression of β3 integrin subunit via activation of the embryonic interleukin-1 (IL-1) system and the binding of IL-1α and -β to specific receptors (IL-1Rt1) on the endometrial cell surface[174]. Also the expression of some mucins (MAG and MUC-1) and of the membrane protein trophinin is enhanced in the peri-implantation period: endometrial areas in which these proteins are particularly concentrated may function as efficient adhesion sites[175].

Ovarian steroid hormones play a fundamental role in preparing the endometrium for successful embryonic implantation. However, ovarian steroids are not the final effectors, but rather activate a series of co-ordinated molecular events through local autocrine/paracrine factors. Several growth factors and cytokines secreted by all cells involved in implantation (embryo, endometrium, local T lymphocytes and macrophages) mediate the effect of estradiol and P in the endometrium during implantation.

In humans, the apposition phase appears to occur under the control of chemokines such as interleukin-8 (IL-8), monocyte chemotactic protein-1 (MCP-1) and RANTES, produced and secreted by the endometrial epithelium[176,177]. Abnormal secretion and/or activity of these factors is likely to be involved in the pathogenesis of ectopic pregnancy as well as of placental site defects such as placenta previa.

In the adhesion phase, heparin-binding epidermal growth factor (EGF)-like growth factor (HB-EGF), colony-stimulating factor-1 (CSF-1), leukemia inhibitory factor (LIF) and the IL-1 system appear to be involved in the complex mechanisms that provide the physical contact between the blastocyst and the endometrial surface[178–181]. The pathology of this phase is probably at the basis of some cases of unexplained infertility, repeated implantation failures in IVF and precocious pregnancy loss.

After attachment, the trophoblast invasion phase occurs via a finely regulated local balance between extracellular matrix proteolysis, cell migration and regulatory inhibition of these processes. The blastocyst is embedded in the endometrial stroma and the site of entry is covered by fibrin, soon colonized by epithelial cells. The embryo trophoblast expresses mainly integrins α5β1 and α1β1, that enable the blastocyst to anchor and migrate through the maternal decidua[182]. After breaking of the basement membrane, the trophoblast penetrates the stromal compartment reaching maternal blood vessels. Urokinase-type (uPA) and tissue-type (tPA) plasminogen activators catalyze the conversion of plasminogen into plasmin, a protease with a broad proteolytic action on the extracellular matrix[183]. Collagenases digest collagen types I, II, III, IV and X, gelatinases digest collagen type IV, stromelysins degrade fibronectin, laminin, proteoglycans and collagen types IV, V and VII[183]. Specific matrix metalloproteinases (MMP-2 and MMP-9) are expressed by the trophoblast under the autocrine control of IL-1[184].

A tight regulation of the invasion phase is needed to prevent the onset of pathologic conditions such as placenta accreta. The endometrium limits invasion by three major mechanisms: (1) inhibition of the trophoblast-derived matrix-degrading enzymes and synthesis of matrix proteins restraining invasion; (2) inhibition of the autocrine/paracrine stimulation of invasion; and (3) inhibition of the trophoblast cell differentiation into the invasive phenotype. Decidualized endometrial stromal cells secrete the extracellular matrix protein fibronectin, that plays an important role in restraining trophoblast invasiveness[182]. The insulin-like growth factor/insulin-like growth factor binding protein (IGF/IGFBP) system is involved in the regulation of decidual invasion: IGF-II is stimulatory to trophoblast migration, while IGFBP-1 is able to modulate invasion via binding of $\alpha5\beta1$ integrin[185]. The complexity of the blastocyst/endometrial epithelium interplay is evident in the following example: endometrial TGFβ1 induces the synthesis of plasminogen activator inhibitor (PAI), that with its antiproteolytic effect counteracts invasion; the latent form of TGFβ1 in maternal decidua is activated by plasmin, in turn secreted by the invading trophoblast[186]. All is further complicated by the fact that the trophoblast enhances its own invasiveness by modulating endometrial restraint on invasion, as suggested by the observation that the synthesis of metalloprotease inhibitor TIMP-3 in human endometrium is downregulated by IL-1[187].

ACKNOWLEDGEMENTS

The authors wish to thank Dr. Cristina Garello (LIVET Center, Turin, Italy) for providing photographs of embryos and follicles.

References

1. Gosden RG. Ovulation 1: oocyte development throughout life. In Grudzinskas JG, Yovich JL, eds. *Cambridge Reviews in Human Reproduction, Gametes – The Oocyte.* Cambridge: Cambridge University Press, 1995:119–49
2. Witschi E. Migration of germ cells of human embryos from the yolk sac to the primitive gonadal folds. *Contrib Embryol Carnegie Inst* 1948;32:67–80
3. Fukuda O. Ultrastructure of primordial germ cells in human embryo. *Virchows Arch Abt B Cell Pathol* 1976;20: 85–9
4. Makabe S, Motta PM. Migration of human germ cells and their relationship with the developing ovary: ultrastructural aspects. *Prog Clin Biol Res* 1989;296:41–54
5. Fujimoto T, Yoshinaga K, Kono I. Distribution of fibronectin on the migratory pathway of primordial germ cells in mice. *Anat Rec* 1985;211:271–8
6. Motta PM, Makabe S. Development of the ovarian surface and associated germ cells in the human fetus. A correlated study by scanning and transmission electron microscopy. *Cell Tissue Res* 1982;226:493–510
7. Byskov AG. Differentiation of mammalian embryonic gonad. *Physiol Rev* 1986;66:71–117
8. Oktay K, Briggs D, Gosden RG. Ontogeny of FSH receptor gene expression in isolated human ovarian follicles demonstrated by the nested polymerase chain reaction. *J Clin Endocrinol Metab* 1997;82:3748–51
9. Faddy MJ, Gosden RG, Gougeon A, *et al.* Accelerated disappearance of ovarian follicles in mid-life – implications for forecasting menopause. *Hum Reprod* 1992;7:1342–6
10. Gosden RG, Bownes M. Cellular and molecular aspects of oocyte development. In Grudzinskas JG, Yovich JL, eds. *Cambridge Reviews in Human Reproduction, Gametes – The Oocyte.* Cambridge: Cambridge University Press, 1995:23–53
11. Mehlmann LM, Terasaki M, Jaffe LA, Kline D. Reorganization of the endoplasmic reticulum during meiotic maturation of the mouse oocyte. *Dev Biol* 1995;170: 607–15
12. Sorenson RA, Cyert MS, Pedersen RA. Active maturation-promoting factor is present in mature mouse oocyte. *J Cell Biol* 1985;100:1637–40
13. Lohka MJ, Hayes MK, Maller JL. Purification of maturation-promoting factor, an intracellular regulator of early mitotic events. *Proc Natl Acad Sci USA* 1988;85: 3009–13
14. Dekel N. Molecular control of meiosis. *Trends Endocrinol Metab* 1995;6:165–9
15. Schultz RM, Montgomery RR, Belanoff JR. Regulation of mouse oocyte maturation: implications of a decrease in oocyte cAMP and protein dephosphorylation in commitment to resume meiosis. *Dev Biol* 1983;97: 267–73
16. Eppig JJ, Ward-Bailey PF, Coleman DL. Hypoxanthine and adenosine in murine ovarian follicular fluid: concentrations and activity in maintaining oocyte meiotic arrest. *Biol Reprod* 1985;33:1041–9
17. Tombes RM, Simerly C, Borisy G, Schatten G. Meiosis, egg activation, nuclear envelope breakdown are differentially reliant on Ca^{2+}, whereas germinal vesicle breakdown is Ca^{2+}-independent in the mouse oocyte. *J Cell Biol* 1992; 117:799–811
18. Revelli A, Massobrio M, Tesarik J. Nongenomic actions of steroid hormones in reproductive tissues. *Endocr Rev* 1998; 19:3–17
19. Schultz RM, LaMarca MJ, Wassarman PM. Absolute rates of protein synthesis during meiotic maturation of mammalian oocyte *in vitro. Proc Natl Acad Sci USA* 1978; 75:4160–4

20. Moore GPM, Linten-Moore S. Transcription of the mouse oocyte genome. *Biol Reprod* 1978;18:865–70

21. Schultz RM. Regulation of zygotic gene activation in the mouse. *BioEssays* 1993;15:531–8

22. Taylor KD, Piko L. Expression of the *rig*-gene in mouse oocytes and early embryos. *Mol Reprod Dev* 1991;28:319–24

23. Bachvarova RF. A maternal tail of poly(a) – the long and the short of it. *Cell* 1992;69:895–7

24. Meric F, Searfoss AN, Wormington M, Wolffe AP. Masking and unmasking maternal mRNA: the role of polyadenylation, transcription, splicing and nuclear history. *J Biol Chem* 1996;271:30804–10

25. Heller DT, Schultz RM. Ribonucleoside metabolism by mouse oocytes: metabolic cooperativity between fully grown oocytes and cumulus cells. *J Exp Zool* 1980;214:355–64

26. Cecconi S, Rossi G, De Felici M, Colonna R. Mammalian oocyte growth *in vitro* is stimulated by soluble factor(s) produced by preantral granulosa cells and by Sertoli cells. *Mol Reprod Dev* 1996;44:540–6

27. Eppig JJ, O'Brien MJ, Pendola FL, *et al*. Factors affecting the developmental competence of mouse oocytes grown *in vitro*: follicle-stimulating hormone and insulin. *Biol Reprod* 1998;59:1445–53

28. Yoshida H, Takakura N, Kataoka H, *et al*. Stepwise requirement of c-kit tyrosine kinase in mouse ovarian follicle development. *Dev Biol* 1997;184:122–37

29. Packer AI, Hsu YC, Besmer P, *et al*. The ligand of the c-kit receptor promotes oocyte growth. *Dev Biol* 1994;161:194–205

30. Racowsky C, Baldwin KV. *In vitro* and *in vivo* studies reveal that hamster oocyte meiotic arrest is maintained only transiently by follicular fluid, but persistently by membrana/cumulus granulosa cell contact. *Dev Biol* 1989;134:297–306

31. Buccione R, Schroeder AC, Eppig JJ. Interactions between somatic cells and germ cells throughout mammalian oogenesis. *Biol Reprod* 1990;43:543–7

32. Vanderhyden BC, Cohen JN, Morley P. Mouse oocytes regulate granulosa cell steroidogenesis. *Endocrinology* 1993;133:423–6

33. Vanderhyden BC, Telfer EE, Eppig JJ. Mouse oocytes promote proliferation of granulosa cells from preantral and antral follicles *in vitro*. *Biol Reprod* 1992;46:1196–204

34. Eppig JJ, Pendola FL, Wigglesworth K. Mouse oocytes suppress cAMP-induced expression of LH receptor mRNA by granulosa cells *in vitro*. *Mol Reprod Dev* 1998;49:327–32

35. Meizel S. Molecules that initiate or help stimulate the acrosome reaction by their interaction with the mammalian sperm surface. *Am J Anat* 1985;17:285–302

36. Buccione R, Vanderhyden BC, Caron PJ, Eppig JJ. FSH-induced expansion of the mouse cumulus oophorus *in vitro* is dependent upon a specific factor(s) secreted by the oocyte. *Dev Biol* 1990;138:16–25

37. Canipari R, Epifano O, Siracusa G, *et al*. Mouse oocytes inhibit plasminogen activator production by ovarian cumulus and granulosa cells. *Dev Biol* 1995;167:371–8

38. McGrath SA, Esquela AF, Lee S-J. Oocyte-specific expression of growth/differentiation factor 9. *Mol Endocrinol* 1995;9:131–6

39. Dong J, Albertini DF, Nishimori K, *et al*. Growth differentiation factor-9 is required during early ovarian folliculogenesis. *Nature* 1996;383:531–5

40. Koopman P, Gubbay J, Vivian N, *et al*. Male development of chromosomally female mice transgenic for *Sry*. *Nature* 1991;351:117–21

41. Chubb C. Genes regulating testis size. *Biol Reprod* 1992;47:29–36

42. van Haaster LH, de Jong FH, Docter R, de Rooij DG. The effect of hypothyroidism on Sertoli cell proliferation and differentiation and hormone levels during testicular development in the rat. *Endocrinology* 1992;131:1574–6

43. de Kretser DM, Kerr JB. The cytology of the testis. In Knobil E, Neill JD, eds. *The Physiology of Reproduction*. New York: Raven Press, 1994:1177–290

44. Clermont Y. The cycle of the seminiferous epithelium in man. *Am J Anat* 1963;112:35–51

45. Sharpe RM. Regulation of spermatogenesis. In Knobil E, Neill JD, eds. *The Physiology of Reproduction*. New York: Raven Press, 1994:1363–434

46. Fritz IB, Tung PS, Ailenberg M. Proteases and anti-proteases in the seminiferous tubule. In Russel LD, Griswold MD, eds. *The Sertoli Cell*. Clearwater, FL: Cache River Press, 1993:217–35

47. Chandley AC. Meiosis in man. *Trends Genet* 1988;4:79–84

48. Palombi F, Salanova M, Tarone G, *et al*. Distribution of β1 integrin subunit in rat seminiferous epithelium. *Biol Reprod* 1992;47:1173–82

49. Kumar TR, Wang Y, Naifang L, Matzuk MM. Follicle stimulating hormone is required for ovarian follicle maturation but not male fertility. *Nature Genet* 1997;15:201–4

50. Tapanainen JS, Aittomaki K, Min J, *et al*. Men homozygous for an inactivating mutation of the follicle-stimulating hormone (FSH) receptor gene present variable suppression of spermatogenesis and fertility. *Nature Genet* 1997;15:205–6

51. Simoni M, Gromoll J, Nieschlag E. The follicle-stimulating hormone receptor: biochemistry, molecular biology, physiology, and pathophysiology. *Endocr Rev* 1997;18:739–73

52. van Roijen JH, van Assen S, van der Kwast T, *et al*. Androgen receptor immunoexpression in the testes of subfertile men. *J Androl* 1995;16:510–6

53. Skinner MK. Cell–cell interactions in the testis. *Endocr Rev* 1991;12:45–77

54. Vale W, Bilezikjian LM, Rivier C. Reproductive and other roles of inhibins and activins. In Knobil E, Neill JD, eds. *The Physiology of Reproduction*. New York: Raven Press, 1994:1861–78

55. Rossi P, Dolci S, Albanesi C, *et al*. Follicle-stimulating hormone induction of steel factor (SLF) mRNA in mouse Sertoli cells and stimulation of DNA synthesis in spermatogonia by soluble SLF. *Dev Biol* 1993;155:68–74

56. McKee BD, Handel MA. Sex chromosomes, recombination, and chromatin conformation. *Chromosoma* 1993;102:71–80

57. Lifschytz E, Lindsley DL. The role of X chromosome inactivation during spermatogenesis. *Proc Natl Acad Sci USA* 1972;69:182–6

58. Eddy EM. 'Chauvinist genes' of male germ cells: gene expression during mouse spermatogenesis. *Reprod Fertil Dev* 1995;7:695–704

59. Dix DJ, Allen JW, Collins BW, *et al*. HSP70-2 is required for desynapsis of synaptonemal complexes during meiotic prophase in juvenile and adult mouse spermatocytes. *Development* 1997;124:4595–603

60. Zhu D, Dix DJ, Eddy EM. HSP70-2 is required for CDC2 kinase activity in meiosis I of mouse spermatocytes. *Development* 1997;124:3007–14

61. Albanesi C, Geremia R, Giorgio M, *et al*. A cell- and developmental stage-specific promoter drives the expression of a truncated c-kit protein during mouse spermatid elongation. *Development* 1996;122:1291–302

62. Wolgemuth DJ, Viviano CM, Gizang-Ginsberg E, *et al*. Differential expression of the mouse homeobox-containing gene *Hox-1.4* during male germ cell differentiation and embryonic development. *Proc Natl Acad Sci USA* 1987;84: 5813–7

63. Delmas V, Sassone-Corsi P. The key role of CREM in the cAMP signaling pathway in the testis. *Mol Cell Endocrinol* 1994;100:121–4

64. Felsenfeld G. Chromatin unfolds. *Cell* 1996;86:13–19

65. Davie JR. Histone modifications, chromatin structure, and the nuclear matrix. *J Cell Biochem* 1996;62:149–57

66. Bhatnagar YM, Romrell LJ, Bellve AR. Biosynthesis of specific histones during meiotic prophase of mouse spermatogenesis. *Biol Reprod* 1985;32:599–609

67. Oliva R, Dixon GH. Vertebrate protamine genes and the histone-to-protamine replacement reaction. *Prog Nucl Acid Res Mol Biol* 1991;40:25–94

68. Meistrich ML, Trostle-Weige PK, Lin R, *et al*. Highly acetylated: H4 is associated with histone displacement in rat spermatids. *Mol Reprod Dev* 1992;31:170–81

69. Alami-Ouahabi B, Veilluex S, Meistrich ML, Boissonneault G. The testis-specific high-mobility-group protein, a phosphorylation-dependent DNA-packaging factor of elongating and condensing spermatids. *Mol Cell Biol* 1996;16:3720–9

70. Kleene KC. Patterns of translational regulation in the mammalian testis. *Mol Reprod Dev* 1996;43:268–81

71. Zaneveld LJD, De Jonge CJ, Anderson RA, Mack SR. Human sperm capacitation and the acrosome reaction. *Hum Reprod* 1991;6:1265–74

72. Shimomura H, Dangott LJ, Garbers DL. Covalent coupling of a resact analogue to guanylate cyclase. *J Biol Chem* 1986;261:15778–82

73. Revelli A, Ghigo D, Moffa F, *et al*. Guanylate cyclase activity and sperm function. *Endocr Rev* 2002;23:484–94

74. Repaske DR, Garbers DL. A hydrogen ion flux mediates stimulation of respiratory activity by speract in sea urchin spermatozoa. *J Biol Chem* 1983;258:6025–9

75. Shapiro BM, Cook S, Quest AF, *et al*. Molecular mechanisms of sea-urchin sperm activation before fertilization. *J Reprod Fertil Suppl* 1990;42:3–8

76. Pacey AA, Hill CJ, Scudamore IW, *et al*. The interaction *in vitro* of human spermatozoa with epithelial cells from the human uterine (fallopian) tube. *Hum Reprod* 1995;10: 360–6

77. Eisenbach M, Tur-Kaspa I. Do human eggs attract spermatozoa? *BioEssays* 1999;21:203–10

78. Tur-Kaspa I. Pathophysiology of the fallopian tube. In Gleicher N, ed. *Tubal Catheterization*. New York: Wiley-Liss Inc., 1992:5–14

79. Williams M, Hill CJ, Scudamore I, *et al*. Sperm numbers and distribution within the human fallopian tube around ovulation. *Hum Reprod* 1993;8:2019–26

80. Revelli A, Soldati G, Stamm J, *et al*. Effect of volumetric mixtures of peritoneal and follicular fluid from the same woman on sperm motility and acrosomal reactivity *in vitro*. *Fertil Steril* 1992;57:654–60

81. Eisenbach M, Tur-Kaspa I. Human sperm chemotaxis is not enigmatic anymore. *Fertil Steril* 1994;62:233–5

82. Tacconis P, Revelli A, Massobrio M, *et al*. Chemotactic responsiveness of human spermatozoa to follicular fluid is enhanced by capacitation but is impaired in dyspermic semen. *J Assist Reprod Genet* 2001; 18:36–44

83. Ralt D, Manor M, Cohen-Dayag A, *et al*. Chemotaxis and chemokinesis of human spermatozoa to follicular factors. *Biol Reprod* 1994;50:774–85

84. Cohen-Dayag A, Tur-Kaspa I, Dor J, *et al*. Sperm capacitation in humans is transient and correlates with chemotactic responsiveness to follicular factors. *Proc Natl Acad Sci USA* 1995;92:11039–43

85. Cohen-Dayag A, Ralt D, Tur-Kaspa I, *et al*. Sequential acquisition of chemotactic responsiveness by human spermatozoa. *Biol Reprod* 1994;50:786–90

86. Sundsfjord JA, Forsdahl F, Thibault G. Physiological levels of immunoreactive ANH-like peptides in human follicular fluid. *Acta Endocrinol* 1989;121:578–80

87. Steegers EA, Hollanders JM, Jongsma HW, Hein PR. Atrial natriuretic peptide and progesterone in ovarian follicular fluid. *Gynecol Obstet Invest* 1990;29:185–7

88. Silvestroni L, Palleschi S, Guglielmi R, Tosti Croce C. Identification and localization of atrial natriuretic factor receptors in human spermatozoa. *Arch Androl* 1992;28: 75–82

89. Anderson RA, Feathergill KA, Rawlins RG, *et al*. Atrial natriuretic peptide: a chemoattractant of human spermatozoa by a guanylate cyclase-dependent pathway. *Mol Reprod Dev* 1995;40:371–8

90. Zamir N, Riven-Kreitman R, Manor M, *et al*. Atrial natriuretic peptide attracts human spermatozoa *in vitro*. *Biochem Biophys Res Commun* 1993;197:116–22

91. Jaiswal BS, Tur-Kaspa I, Dor J, *et al*. Human sperm chemotaxis: is progesterone a chemoattractant? *Biol Reprod* 1999;60:1314–19

92. Gnessi L, Fabbri A, Silvestroni L, *et al*. Evidence for the presence of specific receptors for *N*-formyl chemotactic peptides on human spermatozoa. *J Clin Endocrinol Metab* 1986;63:841–6

93. O'Rand MG. Modification of the sperm membrane during capacitation. *Ann N Y Acad Sci* 1982;383:392–404

94. Benoff S. Carbohydrates and fertilization: an overview. *Mol Hum Reprod* 1997;3:599–637

95. Carrera A, Moos J, Gerton GL, *et al*. Regulation of protein tyrosine phosphorylation in human sperm by a calcium/

calmodulin dependent mechanism: identification of A kinase anchor proteins as a major substrates for tyrosine phosphorylation. *Dev Biol* 1996;180:284–96

96. Baldi E, Casano R, Flasetti C, *et al*. Intracellular calcium accumulation and responsiveness to progesterone in capacitating human spermatozoa. *J Androl* 1991;12:323–30

97. de Lamirande E, Leclerc P, Gagnon C. Capacitation as a regulatory event that primes spermatozoa for the acrosome reaction and fertilization. *Mol Hum Reprod* 1997;3:175–94

98. Fraser LR. Mechanisms regulating capacitation and the acrosome reaction. In Fénichel P, Parinaud J, eds. *Human Sperm Acrosome Reaction*. Colloque INSERM: John Libbey Eurotext, Ltd., 1995:17–33

99. Hamamah S, Grizard G, Gadella BM, *et al*. Lipid composition of sperm plasma membrane: alteration during the fertilization process. In Hamamah S, Mieusset R, eds. *Male Gametes, Production and Quality*. Paris: INSERM, 1996:187–202

100. Aitken RJ. Molecular mechanisms regulating human sperm function. *Mol Hum Reprod* 1997;3:169–73

101. Zaneveld LJD, De Jonge CJ. Mammalian sperm acrosomal enzymes and the acrosome reaction. In Dunbar BS, O'Rand MG, eds. *A Comparative Overview of Mammalian Fertilization*. New York: Plenum Press, 1991:63–79

102. Tesarik J. Appropriate timing of the acrosome reaction is a major requirement for the fertilizing spermatozoon. *Hum Reprod* 1989;4:957–61

103. Tesarik J, Drahorad J, Peknicova J. Subcellular immunochemical localization of acrosin in human spermatozoa during the acrosome reaction and the zona pellucida penetration. *Fertil Steril* 1988;50:133–41

104. Anderson RA, Feathergill KA, De Jonge CJ, *et al*. Facilitative effect of pulsed addition of dibutyryl cAMP on the acrosome reaction of uncapacitated human spermatozoa. *J Androl* 1992;13:398–408

105. Parinaud J, Labal B, Vieitez G. High progesterone concentrations induce acrosome reaction with a low cytotoxic effect. *Fertil Steril* 1992;58:599–602

106. Uhler ML, Leung A, Chan SY, Wang C. Direct effects of progesterone and antiprogesterone on human sperm hyperactivated motility and acrosome reaction. *Fertil Steril* 1992;58:1191–8

107. Modotti M, Togni G, Medici G, *et al*. Effect of peritoneal fluid supplemented with exogenous progesterone on sperm motility *in vitro*. *Hum Reprod* 1994;9:303–9

108. Morales P, Llanos M, Gutierrez G, *et al*. The acrosome reaction-inducing activity of individual human follicular fluid samples is highly variable and is related to the steroid content. *Hum Reprod* 1992;7:646–51

109. Blackmore PF, Beebe SJ, Danforth DR, Alexander N. Progesterone and 17α-hydroxyprogesterone. Novel stimulators of calcium influx in human sperm. *J Biol Chem* 1990;265:1376–80

110. Miska W, Fehl P, Henkel R. Biochemical and immunological characterization of the acrosome reaction-inducing substance (ARIS) of hFF. *Biochem Biophys Res Commun* 1994;199:125–9

111. Tesarik J. The role of proteases in the mammalian sperm acrosome reaction. In Fénichel P, Parinaud J, eds. *Human Sperm Acrosome Reaction*. Colloque INSERM: John Libbey Eurotext, Ltd., 1995:123–32

112. Tesarik J, Mendoza C, Moos J, Carreras A. Selective expression of a progesterone receptor on the human sperm surface. *Fertil Steril* 1992;58:784–92

113. Revelli A, Massobrio M, Tesarik J. Nongenomic effects of steroid hormones in reproductive tissues. *Endocrine Rev* 1998;19:3–18

114. Parinaud J, Milhet P. Progesterone induces Ca^{++}-dependent 3′,5′-cyclic adenosine monophosphate increase in human sperm. *J Clin Endocrinol Metab* 1996;81:1357–60

115. Wistrom CA, Meizel S. Evidence suggesting involvement of a unique human sperm steroid receptor/Cl$^-$ channel complex in the progesterone-initiated acrosome reaction. *Dev Biol* 1993;159:679–90

116. Tesarik J, Mendoza C. Insights into the function of a sperm-surface progesterone receptor: evidence of ligand-induced receptor aggregation and the implication of proteolysis. *Exp Cell Res* 1993;205:111–17

117. Turner KO, Garcia MA, Meizel S. Progesterone initiation of the human sperm acrosome reaction: the obligatory increase in intracellular calcium is independent of the chloride requirement. *Mol Cell Endocrinol* 1994;101:221–5

118. Tesarik J, Mendoza C. Defective function of a nongenomic progesterone receptor as a sole sperm anomaly in infertile patients. *Fertil Steril* 1992;58:793–7

119. Falsetti C, Baldi E, Krausz C, *et al*. Decreased responsiveness to progesterone of spermatozoa in oligozoospermic patients. *J Androl* 1993;14:17–22

120. Gosden RG, Krapez J, Briggs D. Growth and development of the mammalian oocyte. *BioEssays* 1997;19:857–82

121. Miller DJ, Macek MB, Shur BD. Complementarity between sperm surface β-1,4-galactosyl-transferase and egg-coat ZP3 mediates sperm–egg binding. *Nature* 1992;357:589–93

122. Wassarman PM, Liu C, Litscher ES. Constructing the mammalian egg zona pellucida: some new pieces of an old puzzle. *J Cell Sci* 1996;109:2001–4

123. Cross NL, Morales P, Overstreet JW, Hanson FW. Induction of acrosome reactions by the human zona pellucida. *Biol Reprod* 1988;38:235–44

124. Barrat CLR, Hornby DP. Induction of the human acrosome reaction by rhuZP3. In Fénichel P, Parinaud J, eds. *Human Sperm Acrosome Reaction*. Colloque INSERM: John Libbey Eurotext, Ltd., 1995:105–22

125. Tesarik J, Moos J, Mendoza C. Stimulation of protein tyrosine phosphorylation by a progesterone receptor on the cell surface of human sperm. *Endocrinology* 1993;133:328–35

126. Roldan ER, Murase T, Shi QX. Exocytosis in spermatozoa in response to progesterone and zona pellucida. *Science* 1994;266:1578–81

127. De Jonge CJ. Role of cAMP pathways: cross-talk mechanisms for the acrosome reaction. In Fénichel P, Parinaud J, eds. *Human Sperm Acrosome Reaction*. Colloque INSERM: John Libbey Eurotext, Ltd., 1995:257–76

128. Doherty CM, Tarchala SM, Radwanska E, De Jonge CJ. Characterization of two second messenger pathways and

their interactions in eliciting the human sperm acrosome reaction. *J Androl* 1995;16:36–46

129. Spungin B, Margalit I, Breitbart H. Sperm exocytosis reconstructed in a cell-free system. Evidence for the involvement of phospholipase C and actin filaments in membrane fusion. *J Cell Sci* 1995;108:2525–35

130. Breitbart H, Lax Y, Rotem R, Naor Z. Role of protein kinase C in the acrosome reaction of mammalian spermatozoa. *Biochem J* 1992;281:473–6

131. Lee MA, Check LH, Kopf GS. Guanine nucleotide-binding regulatory protein in human sperm mediates acrosomal exocytosis induced by the human zona pellucida. *Mol Reprod Dev* 1992;31:78–86

132. Revelli A, Soldati G, Costamagna C, *et al*. Follicular fluid proteins stimulate nitric oxide (NO) synthesis in human sperm: a possible role for NO in acrosomal reaction. *J Cell Physiol* 1999;178:85–92

133. Anderson RA, Feathergill KA, Drisdel RC, *et al*. Atrial natriuretic peptide (ANP) as a stimulus of the human acrosome reaction and a component of ovarian follicular fluid: correlation of follicular ANP content with *in vitro* fertilization outcome. *J Androl* 1994;15:61–70

134. Bleil JD, Greve JM, Wassarman PM. Identification of a secondary sperm receptor in the mouse egg zona pellucida: role in maintenance of binding of acrosome-reacted sperm to eggs. *Dev Biol* 1988;128:376–85

135. Adham IM, Nayernia K, Engel W. Spermatozoa lacking acrosin protein show delayed fertilization. *Mol Reprod Dev* 1997;46;370–6

136. Primakoff P, Myles DG. The ADAM gene family: surface proteins with adhesion and protease activity. *Trends Genet* 2000;16:83–7

137. Bigler D, Chen M, Waters S, White JM. A model for sperm–egg binding and fusion based on ADAMs and integrins. *Trends Cell Biol* 1997;7:220–5

138. Miyado K, Yamada G, Yamada S, *et al*. Requirement of CD9 on the egg plasma membrane for fertilization. *Science* 2000;287:321–4

139. Chen MS, Tung KS, Coonrod SA, *et al*. Role of the integrin-associated protein CD9 in binding between sperm ADAM 2 and the egg integrin α6β1: implications for murine fertilization. *Proc Natl Acad Sci USA* 1999;96:11830–5

140. Schultz RM, Kopf GS. Molecular basis of mammalian egg activation. *Curr Topics Dev Biol* 1995;30:21–62

141. Parrington J, Swann K, Shevchenko VI, *et al*. Calcium oscillations in mammalian eggs triggered by a soluble sperm factor. *Nature* 1996;379:364–8

142. Yanagimachi R. Mammalian fertilization. In Knobil E, Neill JD, eds. *The Physiology of Reproduction*. New York: Raven Press, 1994:189–317

143. Jaffe LA, Cross NL. Electrical regulation of sperm–egg fusion. *Annu Rev Physiol* 1986;48:191–200

144. Masui Y, Markert CL. Cytoplasmic control of nuclear behavior during meiotic maturation of frog oocytes. *J Exp Zool* 1971;117:129–46

145. Schatten G. The centrosome and its mode of inheritance: the reduction of the centrosome during gametogenesis and its restoration during fertilization. *Dev Biol* 1994; 165:299–335

146. Nagy ZP, Liu J, Joris H, *et al*. Time-course of oocyte activation, pronuclear formation and cleavage in human oocytes fertilized by intracytoplasmic sperm injection. *Hum Reprod* 1995;9:1743–8

147. Van Blerkom J, Davis P, Merriam J, *et al*. Nuclear and cytoplasmic dynamics of sperm penetration, pronuclear formation and microtubule organization during fertilization and early preimplantation development in the human. *Hum Reprod Update* 1995;1:429–61

148. Wolffe AP. The transcription of chromatin templates. *Curr Opin Genet Dev* 1994;4:245–54

149. Pazin MJ, Kadonaga JT. SWI2/SNF2 and related proteins: ATP-driven motors that disrupt protein–DNA interactions. *Cell* 1997;88:737–40

150. Majumder S, Miranda M, DePamphilis ML. Analysis of gene expression in mouse preimplantation embryos demonstrates that the primary role of enhancers is to relieve repression of promoters. *EMBO J* 1993;12:1131–40

151. Wiekowski M, Miranda M, DePamphilis ML. Requirements for promoter activity in mouse oocytes and embryos distinguish paternal pronuclei from maternal and zygotic nuclei. *Dev Biol* 1993;159:366–78

152. Telford NA, Watson AJ, Schultz GA. Transition from maternal to embryonic control in early mammalian development: a comparison of several species. *Mol Reprod Dev* 1990;26:90–100

153. Edwards RG, Beard H. Oocyte polarity and cell determination in early mammalian embryos. *Mol Hum Reprod* 1997;13:863–905

154. Huntriss J, Daniels R, Bolton V, *et al*. Imprinted expression of *SNRPN* in human preimplantation embryos. *Am J Hum Genet* 1998;63:1009–14

155. Surani MAH. Differential roles of paternal and maternal genomes during embryogenesis in the mouse. *BioEssays* 1984;1:224–7

156. Lyle R. Gamete imprinting in development and disease. *J Endocrinol* 1997;155:1–12

157. Kuo HC, Ogilvie CM, Handyside AH. Chromosomal mosaicism in cleavage-stage human embryos and the accuracy of single-cell genetic analysis. *J Assist Reprod Genet* 1998;15:276–80

158. Hartshorne G. The embryo. *Hum Reprod* 2000;15 (Suppl. 4):31–41

159. Johnson MH, Maro B. Time and space in the mouse early embryo: a cell biological approach to cell diversification. In Rossant J, Pedersen RA, eds. *Experimental Approaches to Mammalian Embryonic Development*. Cambridge: Cambridge University Press, 1986:35–65

160. Hyafil F, Morello D, Babinet C, Jacob F. A cell surface glycoprotein involved in the compaction of embryonal carcinoma cells and cleavage stage embryos. *Cell* 1980; 21:927–34

161. Clayton L, Stinchcombe SV, Johnson MH. Cell surface localization and stability of uvomorulin during early mouse development. *Zygote* 1993;1:333–44

162. Sefton M, Johnson MH, Clayton L. Synthesis and phosphorylation of uvomorulin during mouse early development. *Development* 1992;115:313–8

163. Pedersen RA. Potency, lineage, and allocation in preimplantation mouse embryos. In Rossant J, Pedersen

RA, eds. *Experimental Approaches to Mammalian Embryonic Development*. Cambridge: Cambridge University Press, 1986:3–33

164. Lo CW, Gilula NB. Gap junctional communication in the preimplantation mouse embryo. *Cell* 1979;18:399–409

165. Van Blerkom J, Barton SSC, Johnson MH. Molecular differentiation of the preimplantation mouse embryo. *Nature* 1976;259:319–21

166. Brison DR, Schultz RM. Apoptosis during mouse blastocyst formation: evidence for a role for survival factors including transforming growth factor α. *Biol Reprod* 1997; 56:1088–96

167. Watson AJ, Pape C, Emanuel JR, *et al.* Expression of Na,K-ATPase α- and β-subunit genes during preimplantation development of the mouse. *Dev Genet* 1990;11:41–8

168. Yamazaki K, Suzuki R, Hojo E, *et al.* Trypsin-like hatching enzyme of mouse blastocysts: evidence for its participation in hatching process before zona shedding of embryos. *Dev Growth Diff* 1994;36:149–54

169. Thomas M, Jain S, Kumar GP, Laloray M. A programmed oxyradical burst causes hatching of mouse blastocysts. *J Cell Sci* 1997;110:1597–602

170. Kidder GM, McLachlin JR. Timing of transcription and protein synthesis underlying morphogenesis in preimplantation mouse embryos. *Dev Biol* 1985;112: 265–75

171. Psychoyos A, Nikas G. Uterine pinopodes as markers of uterine receptivity. *Assist Reprod Rev* 1994;4:26–32

172. Murphy CR. The cytoskeleton of uterine epithelial cells: a new player in uterine receptivity and the plasma membrane transformation. *Hum Reprod Update* 1995;1: 567–80

173. Lessey BA, Damjanovich L, Coutifaris C, *et al.* Integrin adhesion molecules in the human endometrium: correlation with the normal and abnormal menstrual cycle. *J Clin Invest* 1992;90:188–95

174. Simón C, Gimeno MJ, Mercader A, *et al.* Embryonic regulation of integrins β3, α4 and α1 in human endometrial epithelial cells *in vitro*. *J Clin Endocrinol Metab* 1997;82:2607–16

175. Fukuda MN, Sato T, Nakayama J, *et al.* Trophinin and tastin, a novel adhesion molecule complex with potential involvement in embryo implantation. *Genes Dev* 1995;9: 1199–210

176. Arici A, Head JR, MacDonald PC, Casey ML. Regulation of interleukin-8 gene expression in human endometrial cells in culture. *Mol Cell Endocrinol* 1993;94:195–204

177. Wood G, Hausmann E, Cloudluri R. Relative role of CSF-1, MCP-1/JE and RANTES in macrophage recruitment during successful pregnancy. *Mol Reprod Dev* 1997;46:62–70

178. Simón C, Frances A, Piquette GN, *et al.* Embryonic implantation in mice is blocked by interleukin-1 receptor antagonist. *Endocrinology* 1994;134:521–8

179. Stewart CL. Leukemia inhibitory factor and the regulation of pre-implantation development of the mammalian embryo. *Mol Reprod Dev* 1994;39:233–8

180. Das SK, Wang X, Paria BC, *et al.* Heparin-binding EGF-like growth factor gene is induced in the mouse uterus temporally by the blastocyst solely at the site of its apposition: a possible ligand for interaction with blastocyst EGF-receptor in implantation. *Development* 1994;120: 1071–83

181. Pollard JW, Hunt JS, Wiktor-Jedrzejczak W, *et al.* A pregnancy defect in the osteopetrotic (op/op) mouse demonstrates the requirement for CSF-1 in female fertility. *Dev Biol* 1991;148:273–83

182. Damsky CH, Librach C, Lim KH, *et al.* Integrin switching regulates normal trophoblast invasion. *Development* 1994; 120:3657–66

183. Alexander CM, Werb Z. Extracellular matrix degradation. In Hay ED, ed. *Cell Biology of Extracellular Matrix*. New York: Plenum Press, 1991:255–302

184. Librach CL, Feigenbaum SL, Bass KE, *et al.* Interleukin-1β regulates human cytotrophoblast metalloproteinase activity and invasion *in vitro*. *J Biol Chem* 1994;269: 17125–31

185. Irwin JC, Giudice LC. IGFBP-1 binds to the α5β1 integrin in human cytotrophoblasts and inhibits their invasion into decidualized endometrial stromal cells *in vitro*. *Growth Horm IGF Res* 1998;8:21–31

186. Graham CH, Lysiak JJ, McCrae KR, *et al.* Localization of transforming growth factor beta at the human fetal–maternal interface: role in trophoblast growth and differentiation. *Biol Reprod* 1992;46:561–72

187. Huang HY, Wen Y, Irwin JC, *et al.* Cytokine mediated regulation of tissue inhibitor of metalloproteinase-1 (TIMP-1), TIMP-3, and 92 kD type IV collagenase mRNA expression in human endometrial stromal cells. *J Clin Endocrinol Metab* 1998;83:1721–9

Ovarian reserve and managing poor responders in assisted reproductive technologies

2

Jan Gunnar Holte

OVARIAN AGING, DECLINE IN FERTILITY POTENTIAL AND DIMINISHING 'OVARIAN RESERVE'

The need to estimate a woman's ovarian capacity or 'reserve' during infertility investigations has become increasingly more evident during the last decade. There are several reasons for this. Of all female and male factors involved in the reproductive process, the single most important and ultimately most commonly limiting factor is the quality of the egg. The experience gained from intracytoplasmic sperm injection (ICSI) has produced firm evidence that suboptimal sperm quality *in vivo* in most cases can be overcome *in vitro*, provided the sperms fertilize eggs of a high quality. Similarly, the experience gained from egg donation procedures shows that the receptivity of the uterus is rarely the limiting factor, since simple hormonal priming of the endometrium is sufficient to result in pregnancy rates in aging women equally as high as in young women undergoing *in vitro* fertilization (IVF) with their own eggs[1–3].

Importantly, all eggs that will eventually be involved in conception and gametogenesis are already present from early fetal life. From a peak at the fifth month of gestation the number of oocytes shows a continuous decline, dropping to approximately 300 000 at menarche. At menopause about 1000 eggs remain[4]. In parallel with decreasing numbers, egg quality becomes increasingly compromised. The percentage of chromosomally normal oocytes and, secondly, of embryos – already, in a young healthy woman, presumably fairly low[5] – becomes lower still, a factor which with increasing age also results in increasing miscarriage rates[1].

The process of ovarian aging follows a reasonably strict schedule in parallel with chronologic and biological age. This is the reason for the fairly uniform cessation of natural fecundity in ethnically homogenous populations who practice no birth control program, as shown in the classic studies of the Hutterites[6]. The natural decrease in fertility potential is illustrated by the longer intervals between deliveries with increasing age, depending primarily on the increasingly compromised gamete quality in the final years before the last conception at 42 to 43 years.

However, age itself is too crude a measurement when estimating the fertility potential in an individual couple with infertility problems, since there is a fairly wide variation in ovarian aging within the infertile population, especially in the 37 to 43 age group. During this transitional period, an increasing proportion of women lose their fertility potential. The mere finding of regular ovulatory menstrual cycles discloses nothing of oocyte integrity, since more or less ovulatory regular cycles, with essentially unchanged sex steroid levels, will continue in most women long after cessation of the fertile period[7,8]. Clinically, the final loss of fertility potential usually precedes the loss of ovulatory cycles, overt estrogen deficiency and the menopause by a decade. The variation in age at menopause in the general female population can therefore be seen as reflecting the variation in age at which the fertile period ends. Although the mean age at menopause is 51, the normal variation spans between

40 and 60 years, and about 10% of women enter menopause before the age of 46[9]. Given this high figure, it is hardly surprising that a significant proportion of women over 35 years will exhibit a definite decrease in their fertility potential due to compromised gametogenesis.

The term 'ovarian reserve' is vaguely defined, but summarizes intuitively the remaining fertility potential in a woman, and thus provides an integrated estimate of the quantity and quality of the oocytes. Ideally, ovarian reserve should be described by a continuous variable from 100 to 0%, but the methods are as yet too inexact to achieve such an accurate prediction of the individual patient's pregnancy chances. Up to today, most methods have simply provided information in two distinct categories – either 'normal' or 'diminished' ovarian reserve – whereas the variation of fertility potential within the large 'normal' group appears more difficult to predict.

Finding means to evaluate the ovarian reserve in women with an age approaching the end of the fertile period would serve two purposes. One of them is to extend the infertility investigation, since an occult ovarian or egg factor could in many cases be the cause of the infertility. The other, and most widely accepted purpose, is to estimate the couple's chances of succeeding during treatment with assisted reproductive technologies (ART). If prediction models with high accuracy could be constructed, the couple could be counseled either to proceed to ART or to avoid undertaking costly and strenuous treatment attempts.

Estimations of ovarian reserve should not, some recent findings suggest, be confined to women approaching the end of their fourth decade, but should also include younger women, some of whom will already present with a diminished reserve. This suggests that the ovarian factor could be an underestimated cause of infertility and that it could in other cases limit a couple's chances of success during ART procedures, irrespective of the woman's age[10].

Although the evaluation of a couple's fertility potential overall could be regarded as the ultimate object of an accurate estimate of the woman's ovarian reserve, more pragmatic aspects have often been the incitation for studies in this field. The purpose may then be to predict ovarian response to human menopausal gonadotropin (hMG) or follicle stimulating hormone (FSH) stimulation in IVF treatments. Thus, findings of a reduced ovarian reserve usually predict a low response to exogenous FSH, and the clinician might therefore administer a higher starting dose, warning the couple before starting the ovarian stimulation of the higher risk of cancellation.

In the following review of the literature, biochemical and morphologic methods for estimating ovarian reserve are discussed separately.

BIOCHEMICAL TESTS

The physiologic basis for measuring serum levels of circulating hormones in the basal state as an estimate of the ovarian reserve is the knowledge of endocrine changes taking place in the aging ovary. From being originally isolated to levels of FSH in the early follicular phase, other endocrine markers have gradually been introduced in parallel with increasing knowledge on the physiology of the failing ovarian function. Furthermore, a number of dynamic tests have been conceived, with the aim of augmenting the diagnostic accuracy of basal hormone levels.

Basal levels of hormones

The use of basal FSH levels as a marker for ovarian reserve originates from studies showing that subtle elevations of this gonadotropin are originally seen in the early follicular phase, with an increasing incidence in women from their mid-30s being the earliest sign of incipient ovarian failure[11]. They were first used as a measurement of fertility potential by Muasher and co-workers in 1988, who, in a small series of IVF treatments, noted poor results in patients who prior to the treatment had exhibited relatively high FSH levels[12]. In two subsequent reports with increasingly larger material the association between basal FSH concentrations and the outcome of IVF treatments was further evaluated[13,14]. Regression analyses showed that FSH and age independently contributed to predict cancellation rates, peak estradiol levels, the number of oocytes retrieved, fertilized and transferred, as well as total and ongoing pregnancy rates. For all outcome variables, FSH was a better predictor than age, and FSH remained significant after accounting for etiology of infertility and semen quality[14]. In a study comprising 637 IVF cycles an FSH level above 13 U/l was found to predict cycle cancellation with a sensitivity and specificity of 52% and 91%, respectively[15]. In a later British study comprising 344 first IVF attempts, FSH was found to be a stronger determinant than age of the number of oocytes collected, whereas age was the only independent factor associated with pregnancy rate[16].

More recently, the authors of a paper from the Netherlands[17] argue against basal FSH as a powerful prognostic indicator. Using FSH levels above 15 U/l as a cutoff level for treatment failure, only 5% of 435 women undergoing their first IVF treatment showed such an elevated FSH concentration. In a multivariate analysis, FSH, age and infertility diagnosis were selected as predictors for the ongoing pregnancy rate. The inclusion of FSH as a basal test increased the number of patients with an identified extremely low chance of success from one to 22, but the change in prognosis of achieving a viable pregnancy was of low clinical value, from 5–12% to below 5%. Together with the findings that FSH as a prognosticator was insufficient at cutoff levels lower than 15 U/l and that only very few patients were identified, the authors concluded that the clinical value of a single basal FSH concentration is questionable.

A recent retrospective analysis of a large number of treatments further underlines the poor prognosis for women with clearly high FSH levels, adding increased miscarriage rates to high cancellation and low implantation rates. Among the 9802 patients with basal FSH levels analyzed, 1034 had values above 14.2 U/l. Although 28 (2.7%) of these conceived, 20 of the pregnancies were lost in the first trimester, resulting in an ongoing pregnancy rate of only 1.7%. Age had an additional impact on the risk of an early miscarriage[18].

The conclusion from the vast amount of collected data is that IVF treaments should not be performed on women with basal FSH levels above 14–15 U/l, since the results are so strikingly poor, whereas a single basal FSH sample giving a value below that cutoff level is too crude a measurement to add any significant predictive information on a couple's chances in ART procedures.

An important issue when determining the accuracy of a single laboratory test as a prognosticator is the intraindividual variation. When this was investigated in 81 women, the mean variation was 4.2 ± 0.4 U/l, with the highest variation among those with levels above 15 U/l[19]. Although the authors state that such a variation in FSH values would not generally mean a patient in the 'normal' group being reclassified to a group with a poor prognosis or vice versa, this size of the variation certainly brings into question the value of a single sample.

In attempts to improve the predictive power of basal FSH concentrations, luteinizing hormone (LH) and/or estradiol levels analyzed concomitantly with FSH have been introduced by some investigators. When this was done in the first study from the Norfolk group[12], neither hormone added significant information to that acquired by use of FSH alone[13]. In a smaller study from New York, however, a low basal LH/FSH ratio identified women with a poor prognosis in IVF. These patients did not differ in FSH levels from the controls[20], suggesting that a lowering of the LH/FSH ratio might be an earlier sign of diminished ovarian reserve than increases in absolute FSH levels. Furthermore, the addition of estradiol levels, analyzed together with FSH, could help differentiate between patients who will subsequently exhibit a low response to ovarian stimulation and those whose response will be normal. Women who exhibited relatively high estradiol levels basally subsequently had a lower oocyte yield and a higher risk of cancellation than those with similarly normal FSH levels but low levels of estradiol[21]. Conversely, in a retrospective study comprising 2634 IVF treatments, basal estradiol concentrations did not correlate with the general ovarian response or pregnancy chances, except for women 40 years of age or older, where an association to stimulation variables was seen[22].

The role of inhibin B is still not clear, but, being produced by antral follicles, it is potentially an interesting indicator of ovarian reserve. A 1996 study suggested, as expected, that inhibin B correlated inversely with age and basal FSH, but added no independent prognostic information prior to IVF[23]. Conversely, another study comprising a limited number of patients concluded that inhibin B had an independent value together with FSH and the parity to predict oocyte recovery[24]. Furthermore, inhibin B levels were lower among 47 low responders whose FSH concentrations were similar to those of normal responders, suggesting that inhibin B might be an earlier marker for declining ovarian reserve than FSH[25]. However, further and larger studies are warranted before the role of inhibin B, *theoretically* a good marker of ovarian reserve, is fully established.

Hormones after various types of stimulation

The unifying concept behind various hormonal provocation tests for ovarian capacity is that subtle or incipient decline in ovarian reserve, not obvious in the basal state, will be disclosed when an additional 'stress' is imposed on the ovary and the pituitary–ovarian axis. Most published studies in the field suggest that, indeed, additional information on the ovarian function can be achieved in this way.

The clomiphene citrate (CC) challenge test is the most widely used dynamic test for ovarian reserve. Usually, 100 mg of CC is given for five consecutive days during the follicular phase (normally cycle days 5–9), and FSH alone or in combination with LH and/or estradiol is analyzed on cycle days 3 and 10. The test was introduced in 1987, before the first publication on basal FSH levels as a predictor[26]. Among 51 women with normal basal FSH levels, 18 responded to CC with elevated levels, and only one of them conceived compared to 14 of 33 in the group who had normal FSH levels after CC. A later study comprised 114 patients in whom IVF outcome variables were evaluated as a function of the sum of FSH before and after CC provocation. Patients who exhibited summed FSH levels above 26 U/l performed poorly in all aspects, including a cancellation rate of 25% compared with 1% among those with levels below the cutoff value[27]. While this study did not answer the question of the additional value of the post-challenge analysis, this point was clearly highlighted in a later study comprising 91 women in the 35 years and above age group. Out of 21 women who had an initially raised level of FSH, 20 also had an excessive response to CC. More interestingly, another 17 patients who had normal basal levels responded to CC with FSH levels above the cutoff level. The stimulated FSH values correlated better than the basal levels with the response variables during the subsequent controlled ovarian hyperstimulation, and could predict a poor response with 85% accuracy[28].

Thus, the CC challenge test seems to be more sensitive than single basal FSH levels, and abnormal tests clearly predict a poor fertility potential. This is also the case when a cohort of infertile couples is followed with life table analysis. While the couples with normal tests showed age-dependent cumulative pregnancy rates, the pregnancy rates for those with abnormal CC tests were uniformly poor independent of age[29].

However, within the much larger group of patients with normal CC tests, FSH levels do not seem to provide stratified information on a patient's fertility potential. This is best illustrated by the strong age dependence of pregnancy rates within this group, with very low chances for patients over 40 years of age to conceive, including those with a normal CC test (see Scott and co-workers for a life table analysis[29], Pearlstone and co-workers, and Watt and co-workers, for treatment data[30,31]). Given the high success rates with donor eggs in this age group, which underlines the relatively low impact of factors other than ovarian, it can be concluded that CC tests have a fairly low sensitivity as a predictor of oocyte quality in the aging woman. Increasing FSH levels after CC provocation may be a late phenomenon in the declining ovarian function. It is not clear whether the accuracy of the test could be increased by applying lower cutoff levels in the higher age groups.

As mentioned above, an interesting application of the CC test is its use in basal infertility investigations. When 236 couples in an unselected general infertility population were assessed with the CC test, 10% of them showed FSH levels in the abnormal range, with an increasing prevalence from 3% below the age of 30 years, 7% between 30 and 34 years, 10% between 35 and 39 years, to 26% among women aged 40 years or older. Unexplained infertility was a more prevalent diagnosis in couples with an abnormal test (12 of 23) than in the group with normal tests (20 of 213), suggesting that this method might disclose a proportion of occult ovarian factors in the large group of patients without an obvious cause of their infertility according to the conventional investigation[32].

In analogy with basal FSH measurements, the intercycle variability for CC-provoked FSH levels is considerable, with a significant variation being seen in 75% of 62 tested patients, changing prognostic values in as many as 40% of the patients[33]. Other concerns about the reliability of the CC test, as well as of the gonadotropin releasing hormone (GnRH) analog stimulation test (see below), are raised by a study showing poor correlations between these test results and the number of follicles histologically proven after oophorectomy[34].

Responses of FSH, estradiol or inhibins to hMG, FSH or GnRH analog stimulations

Various models to evaluate ovarian reserve using the early ovarian response to either hMG/FSH or the flare-up effect of GnRH analogs (GnRH-a) have been described, with several groups finding the early estradiol increase after such stimulation valuable. It has been suggested that the dose–response nature of the correlations thus attained between estradiol increases and pregnancy rates would rank a model higher than basal FSH measurements or CC tests, which generally limit prediction to the crude binomial categories of normal and abnormal chances to conceive[35]. Clearly, providing more stratified information on a couple's chances of success would greatly increase a model's clinical value, since the vast majority of the patients, though classified as normal with basal or stimulated FSH measurements, will nevertheless perform with great variation in all important endpoints in ART procedures. In the first published studies[35,36], peak estradiol levels after 1 mg of leuprolide correlated better to various main outcome variables than did either age or basal FSH levels, and there was a suggestion of a continuous association between peak estradiol and pregnancy rates. Later studies corroborated the findings of a greater predictive value of the estradiol surge after GnRH-a[37], hMG[38] or FSH[39] than with basal FSH. The combination of basal and stimulated measurements seems to improve the power of the model further. In a couple of studies, the responses of inhibin A or B were included, with findings either of a stronger predictive power[40,41] or else complementary[42] to those of estradiol increases and basal FSH.

MORPHOLOGIC ASSESSMENT OF OVARIAN RESERVE

The increased resolution of modern ultrasound machines, and their more widespread use in endocrine investigations – such as those concerning the diagnostics of polycystic ovaries – as well as their employment in ovarian stimulation, form the basis for a more detailed morphologic evaluation of the ovaries. Ovarian volume decreases with increasing age, and this is clearly visible on ultrasound examination after 40 years of age[43]. In parallel, the number of visible antral follicles, 2–5 mm large, decreases. These events, easily visualized on ultrasound, constitute an apparently valuable guide to a woman's declining fertility potential.

In a retrospective review of 188 first-treatment cycles, the total pretreatment ovarian volume was positively associated with peak estradiol levels and the number of oocytes recovered. A low total ovarian volume was a predictor of cycle cancellation, and the volume of the smallest ovary was negatively associated with the pregnancy chance[44]. In another study, the treatment records of women with an ovarian volume below minus one standard deviation from the mean volume of 6.3 cm^3, i.e., below 3 cm^3, were compared with those of the women with volumes above 3 cm^3. The group with small ovaries had higher FSH levels, although mainly within the normal range, than the group with larger ovaries (9.5 versus 7.0 U/l). The women with small ovaries had higher cancellation rates, required higher doses of hMG and had fewer eggs retrieved than the women with larger ovaries[45].

The number of antral follicles, calculated before the start of the treatment, were shown to correlate with the number of oocytes recovered[46–48], with peak estradiol concentration, the number of follicles and inversely with the total doses of hMG/FSH administered[47], and with cancellation rates[47–49]. Furthermore, there was also an increased risk of ovarian hyperstimulation syndrome (OHSS) in patients with a high number of antral follicles[48].

A generally low pregnancy rate in women with a low antral follicle count has been observed in several studies[47,49,50]. Furthermore, when patients are divided into three or four groups determined by their follicle count, higher pregnancy rates have been reported in groups with higher numbers of antral follicles[50,51]. This finding suggests that antral follicle counting could be a more valuable means of predicting outcome than what has been achieved so far by basal or stimulated FSH, since no correlation to pregnancy rates has been reported for FSH levels within the normal range. When tested together in a recent study, antral follicle count was indeed a better predictor for IVF treatments outcome, including pregnancy chance, than basal FSH and age[49]. Similarly, antral follicle count was a stronger predictor of oocyte recovery than ovarian volume, FSH, age and body mass index in another study[48].

In a recent prospective study of all patients scheduled to receive their first IVF treatment at our clinic during an eight month period, the ovaries were carefully scanned in the basal state and the total number of antral and growing follicles were counted. The ovaries were arbitrarily classified as oligofollicular (OFO; less than five follicles/ovary), normal (five to nine follicles/ovary), multifollicular (MFO; more than nine follicles/ovary, normal stroma) and polycystic (PCO; more than nine follicles/ovary, increased stroma). These variables, the number of follicles and the ovarian type, were related to variables reflecting ovarian response, embryo quality and pregnancy. The 291 cycles leading to oocyte pickup (OPU) (IVF/ICSI) were included. The number of antral follicles (the sum of both ovaries) ranged from three to 60, with a mean of 17, and with 22% OFO, 42% normal ovaries, 11% MFO and 25% PCO of the entire study group. As with most variables, the differences between MFO and PCO were small; these groups were analyzed together (M/P). The number of antral follicles correlated inversely with age ($r = -0.38$, $p < 0.0001$) and the mean age for the OFO group was slightly higher than for the other two groups. The antral follicle number correlated with the number of eggs at OPU ($r = 0.65$, $p < 0.0001$; Figure 1) and inversely with the total dose of FSH administered ($r = -0.58$, $p < 0.0001$). Embryo quality scores of the transferred embryos were higher in M/P versus the other groups, independent of age. The number of clinical pregnancies was higher when the number of antral follicles increased (OFO: 19%; N: 32%; M/P: 47%; $p < 0.01$ for M/P versus OFO or N, independent of age and number of embryos transferred; Figure 2). The number of antral follicles was an independent predictor of pregnancy, with a similar power as age[51].

This study confirms that follicle counting provides a means for basal pregnancy prediction of similar and independent power as age. Again, it suggests that antral follicle counting enables stratified prediction of outcome of IVF treatments also in the large group of patients who do not have a clearly diminished ovarian reserve.

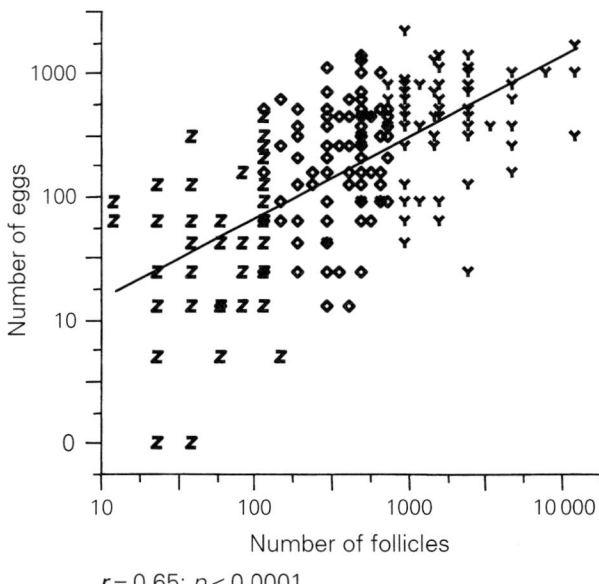

$r = 0.65$; $p < 0.0001$

Figure 1 The number of eggs retrieved as a function of basal antral follicle counts in 291 cycles leading to oocyte pickup.

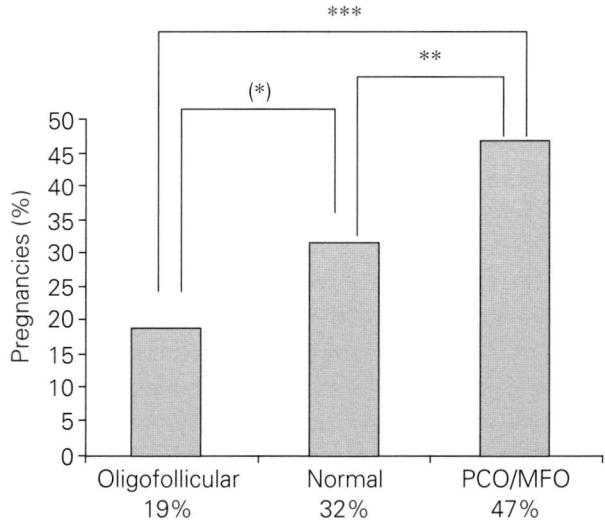

Figure 2 Pregnancy rates in groups of patients classified according to basal ovarian morphology: oligofollicular ovaries (less than five antral follicles/ovary); normofollicular ovaries (five to nine antral follicles/ovary); multifollicular ovaries (more than nine antral follicles/ovary and normal stroma); and polycystic ovaries (more than nine antral follicles/ovary and increased stroma). The two latter groups were analyzed together. *** denotes $p < 0.001$, ** denotes $p < 0.01$ and (*) denotes $p = 0.06$

SUMMARY – OVARIAN RESERVE: A REAPPRAISAL OF THE CURRENT CONCEPT OF OVARIAN MORPHOLOGIC TYPES

Since the reports in the mid-1980s which introduced the ultrasound diagnosis of polycystic ovaries, surprisingly few attempts at further extending our concept of ovarian types have been published, in spite of the improved resolution of ultrasound machines and the introduction of the transvaginal mode of investigation.

An approach based on the visible number of antral follicles, possibly assisted by ovarian volume, may be useful in improving our understanding of the endocrine and the clinical spectrum in a woman's fertile period. Although much research remains to be done in this field, a few major conclusions can be drawn from the studies cited above. The number of antral follicles appears to be closely associated with ovarian reserve, an increasing number seeming to correspond with a linear increase in fertility potential. On the endocrine level, an increasing number of antral follicles is probably paralleled by increasing levels of inhibin B and slightly lower early follicular phase levels of FSH, though with increasing LH/FSH ratios. Clinically, the length of the follicular phase and hence the menstrual cycle increases with the number of antral follicles, from short cycles of 25 days or lower in women with very few antral follicles, to 30–35 days in women with multifollicular or polycystic ovaries, and an increasing number of anovulatory cycles in the polycystic ovaries. Theoretically, the spectrum could also involve clinical differences related to the overall increase in estrogen and androgen levels, with an increase in the antral follicle number.

In the ART situation, there appears to be a generally increased sensitivity to exogenous FSH stimulation with an increasing number of antral follicles. Thus, low responders are mainly found in the oligofollicular group of patients, whereas a high response and a high risk of OHSS are essentially restricted to those with multifollicular and polycystic ovaries. With a low antral follicle count, there is an increased risk of cycle cancellation, lower embryo quality scores, lower implantation figures and increased risks of miscarriage. Since implantation rates are higher in women with many antral follicles, multiple pregnancy risks are increased in this group, but this can be counteracted by transfering fewer embryos.

Could, then, polycystic ovaries really form part of such a normal spectrum of ovarian types? Although the polycystic ovary was originally described as containing ten or more follicles visible in the ovary (using an abdominal scan)[52], in clinical practice this was later changed to 'ten or more follicles per scanning plane'. The figure is arbitrarily chosen, the total number of antral follicles in a typical PCO being usually at least 20, and ranging often up to 30–40 follicles, when the ovary is carefully scanned from pole to pole. We also know that all visible antral follicles are viable when exposed to increased levels of FSH, hence the term 'cyst' is misleading. Given the high prevalance of the ovarian type and the heterogeneous clinical and endocrine spectrum of women with PCO, with generally more subtle endocrine diversions in women with PCO who regularly or sometimes ovulate, it seems reasonable to regard the PCO per se as a variant of the normal, with a somewhat diffuse transitional zone to the normal ovary, and especially to the multifollicular ovary with its many antral follicles but normal stroma. This notion is supported by the finding that when polycystic ovaries (without taking into account other factors such as the presence or absence of ovulatory cycles, the presence of obesity and insulin resistance, increased androgens and the presence of hirsutism) are placed at the extreme end of a spectrum based simply on number of antral follicles counted by means of transvaginal ultrasound, they fit the place remarkably well in terms of ovarian reserve and fertility potential, i.e., they exhibit the highest implantation rates. Indeed, our results (unpublished) suggest that a higher number of antral follicles is related to higher implantation rates in the PCO and multifollicular ovaries groups (although obesity and insulin resistance may secondarily affect the take-home baby rate through increased obstetric risks).

The normal spectrum of ovarian types, using the approach of antral follicle counting, needs to be defined via large-scale population studies centering on healthy women of different age groups. It is not known if antral follicle counts thus obtained would follow a Gaussian distribution, but the spectrum of women with proven fertility would possibly differ from that found in the infertility population. Although it is not clear what the cutoff level for antral follicle counting should be when seeking to define a 'normal' ovary, we have tentatively

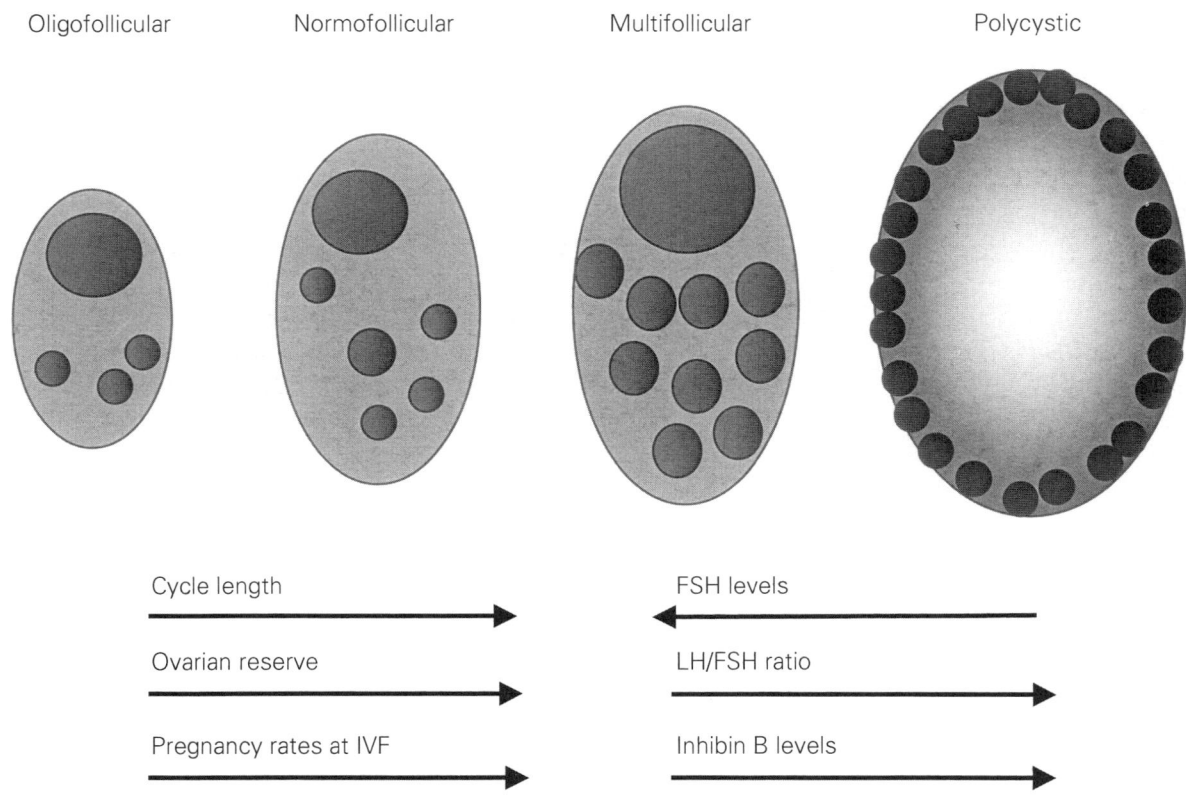

Oligofollicular Normofollicular Multifollicular Polycystic

Cycle length

Ovarian reserve

Pregnancy rates at IVF

FSH levels

LH/FSH ratio

Inhibin B levels

Figure 3 The tentative classification of ovarian types, seen as a function of antral follicle counts, and the relationship with cycle length, gonadotropins (FSH, follicle stimulating hormone; LH, luteinizing hormone), inhibin B, ovarian reserve and pregnancy rates at *in vitro* fertilization (IVF)

opted for between five and nine visible follicles, the term 'oligofollicular' being reserved for an ovary with less than five follicles. When more than nine follicles are visible, we use the term 'multifollicular' or 'polycystic', depending on the presence of a normal or an increased stroma (Figure 3). Needless to say, the distinction between these two latter ovarian types is not a clear one, since an ovarian stroma is notoriously difficult to grade. It could also be argued that the cutoff level for a polycystic ovary should be higher than ten follicles, since the majority of such ovaries exhibit far more, but studies are again lacking. However, accepting that the polycystic ovaries are included in the normal spectrum, a seamless approach in which the antral follicle number is a continuous variable is preferable to a concept of rigidly defined entities. Thus, independent from what terminology is used, antral follicle counting appears a promising simple and non-invasive approach to improving our understanding of female endocrinology and fertility.

TREATMENT OPTIONS FOR POOR RESPONDERS AND PATIENTS WITH A DIMINISHED OVARIAN RESERVE

The poor response to exogenous FSH stimulation that is encountered in some patients continues to be a major challenge in assisted reproduction techniques. Although numerous different approaches have been suggested to increase the ovarian response and improve pregnancy rates, these patients remain a large subgroup among couples undergoing IVF treatments, with a generally reduced fertility prognosis.

Definitions and problems interpreting published data

No universal definition for a poor response exists, although most clinicians working in the field of ART have an intuitive feeling for what the term represents, i.e., a less than normal ovarian response to hMG/FSH. The problems encountered in defining the term center around the choice of endpoint and what cutoff level to apply when evaluating that endpoint, whether it is the number of large follicles, or the level of estradiol or the presentation of the retrieved oocytes. The most commonly applied and most logical option is to take the number of large follicles after controlled ovarian stimulation, mainly because of the shortcomings of the other two approaches: variations between different laboratories in estradiol measurements and, in the case of retrieved oocytes, the problem of categorizing the interrupted cycles.

An important second factor in the definition of a poor response, unfortunately not always clearly accounted for in the literature, is the total number of units of hMG/FSH administered. Often, the only ovarian stimulation a patient receives in many centers before being classified as a poor responder is 150 units of hMG/FSH a day, increasing to 225 or 300 units after 5–8 days in case of suboptimal response. Some of these patients would show a normal number of growing follicles if the stimulation was instead initiated on an hMG/FSH level above the threshold for recruiting a dominant follicle. The problem can be offset by involving the total hMG/FSH dose in the measurement of the response, in effect producing an 'ovarian sensitivity index' between the total number of large follicles and the total dose of hMG/FSH administered. This would result in a continuous variable where cutoff values could be put at appropriate levels. Thus, a severe poor response could be represented by less than four large follicles after a total dose of 5400 units of hMG/FSH (12 days of 450 units), whereas the same number of follicles at half the total dose of hMG/FSH would not mean a proven poor response and result in an index twice as large.

Another important point when classifying a poor response is whether or not any test of ovarian reserve has been performed in the patient. Although the fertility prognosis is clearly worse in patients with increased FSH levels (either basally or after stimulation) than in those with normal levels, FSH levels are nevertheless not investigated in many studies, or else the study group contains a mix of patients, some with increased levels, some with normal levels – either way interfering with the interpretation of the results.

Age is another important basic factor, a reduced ovarian response being normal in an older patient. The results are therefore stratified, usually in a dicotomic way, with women aged 40 years or more being treated as a separate group.

Finally, the literature on poor responders is to a large extent characterized by statistical shortcomings. Many papers are first reports on a new therapeutic protocol, and the material is then often limited to a few cases, reducing its statistical power. Commonly, the diagnosis of a 'poor response' is based on a patient's first treatment failure, with this failed cycle being used as a historical control. To say the least, statistics based on such a strategy should be interpreted with great care. Other reports are retrospective in character. In both of the last two instances, papers may be influenced by publication bias, i.e., the likelihood that the author or authors would have submitted the report, or that it would have been published, would be much lower if the results had been negative. Controlled prospective, randomized studies with study groups large enough to obtain statistical power on pregnancy rates are rare. Even when such studies have been performed, they have generally been unable to corroborate the promising results from a previous pilot study on a new protocol.

Pathophysiology and clinical features

A poor response is presumably in most cases a sign of diminishing ovarian reserve, i.e., an early stage of ovarian failure due to follicular depletion. In that sense, it is a normal reaction to the aging process in the ovary, depending primarily on egg depletion (see above). Histopathologically, a reduction in number of both primordial and antral follicles is seen, the latter also visible on transvaginal ultrasound (see above). The morphologic changes of a young poor responder are likely to be similar to those normally seen in a woman at the end of her normal fertile period.

Typically, the poor responder's menstrual cycle is shorter than 28 days, reflecting a shortened follicular phase. The process of follicle selection to dominance is faster than normal. Endocrinologically, this rapid growth and recruitment of a dominant follicle is, in a natural cycle, reflected by an unusually high estradiol level early in the cycle. Inhibin B levels are low, determined by the reduced number of antral follicles. Often an LH/FSH ratio below 1 is seen in the early follicular phase. This is determined by slighty higher FSH levels than in normal responders, and perhaps also to some degree by lower LH levels. Only later, when ovarian reserve is more diminished, will FSH concentrations in the early follicle phase rise to pathologic levels (above 15 U/l).

Few studies have attempted to describe the follicular growth characteristics, fertilization and cleavage process in poor responders during IVF as distinct from the results achieved with normal responders. However, in a British study, 61 women who demonstrated a poor response to a conventional hMG stimulation were in a subsequent cycle stimulated with higher doses to a satisfactory response. When compared to normal responders, the poor responders, in spite of producing almost as many oocytes (8.9 versus 11.8; $p < 0.01$), took longer to achieve follicles of a mature size, showed a lower fertilization rate, a lower rate of cell division and a lower implantation rate (all $p < 0.01$). Thus, even when essentially overcoming the problem of oocyte underproduction by increasing the dose of hMG/FSH, a series of more or less subtle divergences remain, presumably intrinsic to the state of reduced ovarian capacity[53].

Prediction

The visualization on ultrasound of a low number of antral follicles and lower ovarian volumes indicates to the clinician that the patient will exhibit a low response, with a generally reduced prognosis, even with the appropriate ovarian stimulation. It appears reasonable to analyze the basal FSH and perhaps also to perform a CC test in women with a low antral follicle count (less than five to six follicles per ovary) and/or small ovaries (less than 3–3.5 ml). An increased alertness to the possibility of a poor response should also be kept in patients with short menstrual cycles.

While increased FSH levels indicate a poor fertility prognosis, the overall clinical value of these measurements is limited in the overall infertility population, as only a relatively small group of women below the age of 40 will fall into this category, among them some poor responders. However, the majority of young poor responders will show FSH levels within the normal range. In most cases their poor response is probably a sign of early diminished ovarian reserve, although it remains a possibility that there are other pathophysiologic mechanisms. It seems clear, though, that a basal or stimulated FSH should be examined in patients with a proven or anticipated poor response, since the prognosis is much poorer in those with an elevated FSH level than in those in which it is normal. In fact, a few studies have suggested that in younger women (below the age of 40) with normal FSH levels, the implantation rates in poor responders may be comparable to, or only slightly lower than, those encountered in normal responders[54–56].

Treatment

Increased hMG/FSH dose

The basis of ovarian stimulation for poor responders is, traditionally, the administration of higher doses of FSH, which generally decreases the risks of cancellation and increases the chances of recruiting more eggs, which in turn should increase the chance of achieving a pregnancy. Although the latter conclusion is more or less generally accepted as a clinical fact, the scientific evidence is not overwhelming. Indeed, a recent retrospective study showed a reasonable pregnancy rate, also per cycle, in poor responders with normal FSH, who elected not to cancel the stimulation, but proceeded to oocyte retrieval with one to four eggs[56]. Similarly, the doubling of hMG doses from 225 to 450 U/day in the second cycle in a group of 126 poor responders with normal FSH did not result in improved pregnancy rates, in spite of increased numbers of follicles and oocytes[57]. Solid conclusions are difficult to draw from these two studies, however, because of their inherrent limitations (the first study being retrospective with no controls, and the second a comparison with the previous cycle), but they do question the utility of increased doses and underline the need for prospective randomized studies in this field.

Downregulation

As the process of follicle selection to dominance is faster than normal in poor responders, ovarian stimulation initiated in the early follicular phase often results in a single dominant follicle or a few dominant follicles.

The strategy to overcome this problem is to start the stimulation as early as possible in the follicular phase. However, in more severe cases, even introducing FSH on the first day of the menstrual cycle is leaving it late, since recruitment of the growing follicles is initiated in the previous late luteal phase. One way of circumventing this is to downregulate the patient prior to FSH stimulation. This has been the routine stimulation protocol for a decade now, after it was found that more eggs could be recruited with this method, and it seems likely that especially the large group of poor responders has benefited from this strategy.

Among the different downregulation regimes, the accumulated data suggest no advantages in the main outcome measures are to obtained with either flare-up or long protocols[58]. Although not restricted to poor responders, a meta-analysis by Bhattacharya and Templeton[59] concluded that other factors like drug costs (lower for flare-up) and planning acceptability (better for long protocols) should guide the clinician. Among the long protocols, a less physiologic approach to the initiation of GnRH-a on the first day of the menstrual cycle results in a high number of cyst formations especially in poor responders, given the already often advanced follicular growth early in the cycle.

More recently, it has been observed that GnRH-a may have detrimental effects on follicular growth[60], effects presumably mediated through direct action on ovarian receptors. It was therefore suggested that the analog should be withdrawn after the downregulation period, at the start of hMG/FSH stimulation. Although this is now a frequently used protocol in many centers, most published studies are not conducted in a randomized, prospective way, and previous cycles are often used as control cycles. Apart from this obvious statistical shortcoming, all studies have reported an improved response with generally more oocytes retrieved with the 'stop protocol'. Two recent studies were randomized, however, both with a low number of included cycles[61,62], and showed higher or similar cancellation rates and a similar (greater when restricted to patients with normal FSH levels) or greater number of oocytes retrieved, with the stop protocol. The protocols vary, but theoretically the most promising is the mini-dose protocol, in which a short period of GnRH-a treatment in the mid-luteal phase is followed by the initiation of a high dose of hMG/FSH directly at the start of the menstrual bleeding. By these means, the chance of recruiting a coordinated cohort of follicles would be increased. Another approach would be to administer a single high dose of a GnRH antagonist, instead of an analog, in the mid-luteal phase, and then to initiate FSH treatment directly at the start of the bleeding. So far, only pilot studies have been performed with this protocol (L. Nilsson, Sahlgrenska Hospital, Gothenborg, Sweden, personal communication), but it has theoretical advantages in its direct withdrawal of endogenous gonadotropins at the time in the cycle when the initiation of follicle growth and, presumably often in poor responders, selection could take place.

Other protocols

Pretreatment with gestagens or oral contraceptive pills (OC) in the cycle before starting ovarian stimulation serves the same theoretical purpose, i.e., to ensure a coordinated cohort of growing follicles. Two retrospective studies (gestagen versus OC[63] and GnRH versus OC[64]) concluded, however, that neither of these protocols increased the amount of oocytes retrieved or improved pregnancy rates.

Other attempts at improving pregnancy rates in poor responders include the natural cycle[65,66] and clomiphene citrate[67] alternative stimulation protocols. Although no improvements have been found compared with the conventional GnRH-a protocol, it is of interest that the results did not indicate a clear advantage with downregulation.

The addition of LH in completely downregulated cycles has been suggested as a means to improve the ovarian response. When recombinant LH was added to a highly purified FSH in a small, randomized study of normal responders, the tendency was towards better results in those who did not have additional LH[68]. Conversely, when 150 units of hMG were added in the latter part of the stimulation, instead of an increase in recombinant FSH (rFSH) in women who initially showed a low response on 300 units of FSH, there was a higher final estradiol level and a larger quantity of retrieved oocytes, compared with those who were randomly allocated increased rFSH[69]. A recent study from Denmark suggested that low LH levels during ovarian stimulation for

IVF might have a detrimental effect[70]. Although such a finding may reflect endocrine differences in the patients, low LH levels signaling an early stage of diminished ovarian reserve, the follow-up randomized study suggested that stimulation protocols leading to lower LH levels could result in lower implantation rates[71]. Similar results can be extracted from the dose-finding studies with antagonist protocols. Clearly, the role for LH during ovarian stimulation deserves further evaluation in large-scale studies, with reference to various degrees of ovarian reserve and responsiveness.

Growth hormone (GH) has been tried as an adjuvant agent with hMG/FSH in a series of investigations, but a recent Cochrane analysis of six fairly small studies showed it to have no significant effect[72]. However, the authors concluded that in patients with a previously proven poor response, as opposed to those with a normal response, the results did indicate a tendency to an improved outcome with GH treatment. Although non-significant, the findings suggest that adjuvant GH therapy in the poor-responder group may deserve further investigation in larger studies.

Dexamethasone has also been mooted as a suitable candidate for the role of adjuvant drug to be used during ovarian stimulation. However, two recent, randomized, controlled studies showed it to have no beneficial effect on pregnancy rates in ART cycles[73,74], although one did reveal a lower frequency of canceled cycles because of a poor response in the dexamethasone-treated group[75].

SUMMARY – POOR RESPONDERS

The basis of ovarian stimulation for poor responders is the administration of higher doses of FSH, which generally increases the chances of recruiting more eggs, which in turn should increase the chance of achieving a pregnancy. However, the number of follicles that it is possible to recruit remains a limiting factor, and an increased miscarriage rate further supports the notion that egg quality, apart from egg number, is often diminished in this group of patients. To date, no specific protocol has been shown to possess any additional advantages in the ovarian stimulation of poor responders, when compared with traditional downregulation/ hMG/FSH regimes. Although numerous uncontrolled studies, in which new protocols have suggested improved results, have been published, either the studies have not been followed up by controlled randomized studies, or, when this has been done, the results have been disappointing. Measurements of ovarian reserve should be taken, since a diminished ovarian reserve with FSH levels above 15 U/l is consistent with very low chances of success, whereas a poor response at normal FSH levels in young women may still be compatible with a reasonable pregnancy chance.

References

1. Abdalla HI, Burton G, Kirkland A, *et al*. Age, pregnancy and miscarriage: uterine versus ovarian factors. *Hum Reprod* 1993; 8:1512–17
2. Navot D, Drews MR, Bergh PA, *et al*. Age-related decline in female fertility is not due to diminished capacity of the uterus to sustain embryo implantation. *Fertil Steril* 1994;61:97–101
3. Rosenwaks Z. Donor eggs: their application in modern reproductive technologies. *Fertil Steril* 1987;47:895–909
4. Faddy MJ, Gosden RG. A model conforming the decline in follicle numbers to the age of menopause in women. *Hum Reprod* 1996;11:1484–6
5. Wramsby H, Hansson A, Liedholm P. Chromosome preparations from *in vitro* matured human oocytes using a simple air-drying technique. *Clin Reprod Fertil* 1982;1:323–6
6. Tietze C. Reproductive span and rate of reproduction among Hutterite women. *Fertil Steril* 1957;8:89
7. Cameron IT, O'Shea FC, Rolland JM, *et al*. Occult ovarian failure: a syndrome of infertility, regular menses, and elevated follicle-stimulating hormone concentrations. *J Clin Endocrinol Metab* 1988;67:1190–4
8. Lee SJ, Lenton EA, Sexton L, Cooke ID. The effect of age on the cyclical patterns of plasma LH, FSH, oestradiol and progesterone in women with regular menstrual cycles. *Hum Reprod* 1988;3:851–5
9. Gindoff PR, Jewelewicz R. Reproductive potential in the older woman. *Fertil Steril* 1986;46:989–1001
10. Scott RT Jr, Hofmann GE. Prognostic assessment of ovarian reserve. *Fertil Steril* 1995;63:1–11
11. Sherman BM, West JH, Korenman SG. The menopausal transition: analysis of LH, FSH, estradiol, and progesterone concentrations during menstrual cycles of older women. *J Clin Endocrinol Metab* 1976;42:629–36

12. Muasher SJ, Oehninger S, Simonetti S, *et al.* The value of basal and/or stimulated serum gonadotropin levels in prediction of stimulation response and *in vitro* fertilization outcome. *Fertil Steril* 1988;50:298–307

13. Scott RT, Toner JP, Muasher SJ, *et al.* Follicle-stimulating hormone levels on cycle day 3 are predictive of *in vitro* fertilization outcome. *Fertil Steril* 1989;51:651–4

14. Toner JP, Philput CB, Jones GS, Muasher SJ. Basal follicle-stimulating hormone level is a better predictor of *in vitro* fertilization performance than age. *Fertil Steril* 1991;55:784–91

15. Gurgan T, Urman B, Yarali H, Duran HE. Follicle-stimulating hormone levels on cycle day 3 to predict ovarian response in women undergoing controlled ovarian hyperstimulation for *in vitro* fertilization using a flare-up protocol. *Fertil Steril* 1997;68:483–7

16. Sharif K, Elgendy M, Lashen H, Afnan M. Age and basal follicle stimulating hormone as predictors of *in vitro* fertilisation outcome. *Br J Obstet Gynaecol* 1998;105:107–12

17. Bancsi LF, Huijs AM, den Ouden CT, *et al.* Basal follicle-stimulating hormone levels are of limited value in predicting ongoing pregnancy rates after *in vitro* fertilization. *Fertil Steril* 2000;73:552–7

18. Levi AJ, Raynault MF, Bergh PA, *et al.* Reproductive outcome in patients with diminished ovarian reserve. *Fertil Steril* 2001; 76:666–9

19. Scott RT Jr, Hofmann GE, Oehninger S, Muasher SJ. Intercycle variability of day 3 follicle-stimulating hormone levels and its effect on stimulation quality in *in vitro* fertilization. *Fertil Steril* 1990;54:297–302

20. Mukherjee T, Copperman AB, Lapinski R, *et al.* An elevated day three follicle-stimulating hormone:luteinizing hormone ratio (FSH:LH) in the presence of a normal day 3 FSH predicts a poor response to controlled ovarian hyperstimulation. *Fertil Steril* 1996;65:588–93

21. Evers JL, Slaats P, Land JA, *et al.* Elevated levels of basal estradiol-17beta predict poor response in patients with normal basal levels of follicle-stimulating hormone undergoing *in vitro* fertilization. *Fertil Steril* 1998;69: 1010–4

22. Frattarelli JL, Bergh PA, Drews MR, *et al.* Evaluation of basal estradiol levels in assisted reproductive technology cycles. *Fertil Steril* 2000;74:518–24

23. Balasch J, Creus M, Fabregues F, *et al.* Inhibin, follicle-stimulating hormone, and age as predictors of ovarian response in *in vitro* fertilization cycles stimulated with gonadotropin-releasing hormone agonist–gonadotropin treatment. *Am J Obstet Gynecol* 1996;175:1226–30

24. Tinkanen H, Blauer M, Laippala P, *et al.* Prognostic factors in controlled ovarian hyperstimulation. *Fertil Steril* 1999;72: 932–6

25. Seifer DB, Scott RT Jr, Bergh PA, *et al.* Women with declining ovarian reserve may demonstrate a decrease in day 3 serum inhibin B before a rise in day 3 follicle-stimulating hormone. *Fertil Steril* 1999;72:63–5

26. Navot D, Rosenwaks Z, Margalioth EJ. Prognostic assessment of female fecundity. *Lancet* 1987;2:645–7

27. Loumaye E, Billion JM, Mine JM, *et al.* Prediction of individual response to controlled ovarian hyperstimulation by means of a clomiphene citrate challenge test. *Fertil Steril* 1990;53:295–301

28. Tanbo T, Dale PO, Lunde O, *et al.* Prediction of response to controlled ovarian hyperstimulation: a comparison of basal and clomiphene citrate-stimulated follicle-stimulating hormone levels. *Fertil Steril* 1992;57:819–24

29. Scott RT, Opsahl MS, Leonardi MR, *et al.* Life table analysis of pregnancy rates in a general infertility population relative to ovarian reserve and patient age. *Hum Reprod* 1995;10: 1706–10

30. Pearlstone AC, Fournet N, Gambone JC, *et al.* Ovulation induction in women aged 40 and older: the importance of basal follicle-stimulating hormone level and chronological age. *Fertil Steril* 1992;58:674–9

31. Watt AH, Legedza AT, Ginsburg ES, *et al.* The prognostic value of age and follicle-stimulating hormone levels in women over forty years of age undergoing *in vitro* fertilization. *J Assist Reprod Genet* 2000;17:264–8

32. Scott RT, Leonardi MR, Hofmann GE, *et al.* A prospective evaluation of clomiphene citrate challenge test screening of the general infertility population. *Obstet Gynecol* 1993;82: 539–44

33. Hannoun A, Abu Musa A, Awwad J, *et al.* Clomiphene citrate challenge test: cycle to cycle variability of cycle day 10 follicle stimulating hormone level. *Clin Exp Obstet Gynecol* 1998;25: 155–6

34. Gulekli B, Bulbul Y, Onvural A, *et al.* Accuracy of ovarian reserve tests. *Hum Reprod* 1999;14:2822–6

35. Winslow KL, Toner JP, Brzyski RG, *et al.* The gonadotropin-releasing hormone agonist stimulation test – a sensitive predictor of performance in the flare-up *in vitro* fertilization cycle. *Fertil Steril* 1991;56:711–17

36. Padilla SL, Bayati J, Garcia JE. Prognostic value of the early serum estradiol response to leuprolide acetate in *in vitro* fertilization. *Fertil Steril* 1990;53:288–94

37. Ranieri DM, Quinn F, Makhlouf A, *et al.* Simultaneous evaluation of basal follicle-stimulating hormone and 17 beta-estradiol response to gonadotropin-releasing hormone analogue stimulation: an improved predictor of ovarian reserve. *Fertil Steril* 1998;70:227–33

38. Fabregues F, Balasch J, Creus M, *et al.* Ovarian reserve test with human menopausal gonadotropin as a predictor of *in vitro* fertilization outcome. *J Assist Reprod Genet* 2000;17: 13–19

39. Fanchin R, de Ziegler D, Olivennes F, *et al.* Exogenous follicle stimulating hormone ovarian reserve test (EFORT): a simple and reliable screening test for detecting 'poor responders' in *in-vitro* fertilization. *Hum Reprod* 1994;9:1607–11

40. Penarrubia J, Balasch J, Fabregues F, *et al.* Day 5 inhibin B serum concentrations as predictors of assisted reproductive technology outcome in cycles stimulated with gonadotrophin-releasing hormone agonist–gonadotrophin treatment. *Hum Reprod* 2000;15:1499–504

41. Dzik A, Lambert-Messerlian G, Izzo VM, *et al.* Inhibin B response to EFORT is associated with the outcome of oocyte retrieval in the subsequent *in vitro* fertilization cycle. *Fertil Steril* 2000;74:1114–17

42. Hall JE, Welt CK, Cramer DW. Inhibin A and inhibin B reflect ovarian function in assisted reproduction but are less useful at predicting outcome. *Hum Reprod* 1999;14:409–15

43. Andolf E, Jorgensen C, Svalenius E, Sunden B. Ultrasound measurement of the ovarian volume. *Acta Obstet Gynecol Scand* 1987;66:387–9

44. Syrop CH, Willhoite A, Van Voorhis BJ. Ovarian volume: a novel outcome predictor for assisted reproduction. *Fertil Steril* 1995;64:1167–71

45. Lass A, Skull J, McVeigh E, *et al*. Measurement of ovarian volume by transvaginal sonography before ovulation induction with human menopausal gonadotrophin for *in-vitro* fertilization can predict poor response. *Hum Reprod* 1997;12:294–7

46. Tomas C, Nuojua-Huttunen S, Martikainen H. Pretreatment transvaginal ultrasound examination predicts ovarian responsiveness to gonadotrophins in *in-vitro* fertilization. *Hum Reprod* 1997;12:220–3

47. Chang MY, Chiang CH, Hsieh TT, *et al*. Use of the antral follicle count to predict the outcome of assisted reproductive technologies. *Fertil Steril* 1998;69:505–10

48. Ng EH, Tang OS, Ho PC. The significance of the number of antral follicles prior to stimulation in predicting ovarian responses in an IVF programme. *Hum Reprod* 2000;15: 1937–42

49. Nahum R, Shifren JL, Chang Y, *et al*. Antral follicle assessment as a tool for predicting outcome in IVF – is it a better predictor than age and FSH? *J Assist Reprod Genet* 2001;18:151–5

50. Huang FJ, Chang SY, Tsai MY, *et al*. Determination of the efficiency of controlled ovarian hyperstimulation in the gonadotropin-releasing hormone agonist-suppression cycle using the initial follicle count during gonadotropin stimulation. *J Assist Reprod Genet* 2001;18:91–6

51. Holte J, Isaksson J, Bergh T. Antral follicle count – strong predictive power for pregnancy chance at IVF treatments. In: Oskarsson T, ed. *XIII Nordic IVF Congress*, Reykjavik, 2000

52. Adams J, Polson DW, Franks S. Prevalence of polycystic ovaries in women with anovulation and idiopathic hirsutism. *Br Med J (Clin Res Ed)* 1986;293:355–9

53. Jenkins JM, Davies DW, Devonport H, *et al*. Comparison of 'poor' responders with 'good' responders using a standard buserelin/human menopausal gonadotrophin regime for *in-vitro* fertilization. *Hum Reprod* 1991;6:918–21

54. Biljan MM, Buckett WM, Dean N, *et al*. The outcome of IVF-embryo transfer treatment in patients who develop three follicles or less. *Hum Reprod* 2000;15:2140–4

55. Hanoch J, Lavy Y, Holzer H, *et al*. Young low responders protected from untoward effects of reduced ovarian response. *Fertil Steril* 1998;69:1001–4

56. Lashen H, Ledger W, Lopez-Bernal A, Barlow D. Poor responders to ovulation induction: is proceeding to *in-vitro* fertilization worthwhile? *Hum Reprod* 1999;14:964–9

57. Land JA, Yarmolinskaya MI, Dumoulin JC, Evers JL. High-dose human menopausal gonadotropin stimulation in poor responders does not improve *in vitro* fertilization outcome. *Fertil Steril* 1996;65:961–5

58. Karacan M, Erkan H, Karabulut O, *et al*. Clinical pregnancy rates in an IVF program. Use of the flare-up protocol after failure with long regimens of GnRH-a. *J Reprod Med* 2001; 46:485–9

59. Bhattacharya S, Templeton A. A systematic review of the long protocol versus the short protocol of GnRH administration for *in-vitro* fertilization cycles. In: Filicori M, Flamigni C, eds. *Ovulation Induction. Update '98. Proceedings of the 2nd World Conference on Ovulation Induction*. Carnforth, UK: Parthenon Publishing, 1998:83–9

60. Hugues JN, Cedrin Durnerin IC. Revisiting gonadotrophin-releasing hormone agonist protocols and management of poor ovarian responses to gonadotrophins. *Hum Reprod Update* 1998;4:83–101

61. Dirnfeld M, Fruchter O, Yshai D, *et al*. Cessation of gonadotropin-releasing hormone analogue (GnRH-a) upon down-regulation versus conventional long GnRH-a protocol in poor responders undergoing *in vitro* fertilization. *Fertil Steril* 1999;72:406–11

62. Garcia-Velasco JA, Isaza V, Requena A, *et al*. High doses of gonadotrophins combined with stop versus non-stop protocol of GnRH analogue administration in low responder IVF patients: a prospective, randomized, controlled trial. *Hum Reprod* 2000;15:2292–6

63. al-Mizyen E, Sabatini L, Lower AM, *et al*. Does pretreatment with progestogen or oral contraceptive pills in low responders followed by the GnRHa flare protocol improve the outcome of IVF-ET? *J Assist Reprod Genet* 2000;17:140–6

64. Kovacs P, Barg PE, Witt BR. Hypothalamic–pituitary suppression with oral contraceptive pills does not improve outcome in poor responder patients undergoing *in vitro* fertilization-embryo transfer cycles. *J Assist Reprod Genet* 2001;18:391–4

65. Bar-Hava I, Ferber A, Ashkenazi J, *et al*. Natural-cycle *in vitro* fertilization in women aged over 44 years. *Gynecol Endocrinol* 2000;14:248–52

66. Bassil S, Godin PA, Donnez J. Outcome of *in-vitro* fertilization through natural cycles in poor responders. *Hum Reprod* 1999;14:1262–5

67. Awonuga AO, Nabi A. *In vitro* fertilization with low-dose clomiphene citrate stimulation in women who respond poorly to superovulation. *J Assist Reprod Genet* 1997;14:503–7

68. Sills ES, Levy DP, Moomjy M, *et al*. A prospective, random-ized comparison of ovulation induction using highly purified follicle-stimulating hormone alone and with recombinant human luteinizing hormone in *in-vitro* fertilization. *Hum Reprod* 1999;14:2230–5

69. De Placido G, Mollo A, Alviggi C, *et al*. Rescue of IVF cycles by HMG in pituitary down-regulated normogonadotrophic young women characterized by a poor initial response to recombinant FSH. *Hum Reprod* 2001;16:1875–9

70. Westergaard LG, Erb K, Laursen S, *et al*. The effect of human menopausal gonadotrophin and highly purified, urine-derived follicle stimulating hormone on the outcome of *in-vitro* fertilization in down-regulated normogonadotrophic women. *Hum Reprod* 1996;11:1209–13

71. Westergaard LG, Erb K, Laursen SB, *et al*. Human menopausal gonadotropin versus recombinant follicle-stimulating hormone in normogonadotropic women down-regulated with a gonadotropin-releasing hormone agonist who were undergoing *in vitro* fertilization and intracytoplasmic sperm injection: a prospective randomized study. *Fertil Steril* 2001; 76:543–9

72. Kotarba D, Kotarba J, Hughes E. Growth hormone for *in vitro* fertilization. *Cochrane Database Syst Rev* 2000: CD000099

73. Bider D, Blankstein J, Levron J, Tur-Kaspa I. Gonadotropins and glucocorticoid therapy for 'low responders' – a controlled study. *J Assist Reprod Genet* 1997;14:328–31

74. Keay SD, Liversedge NH, Mathur RS, Jenkins JM. Assisted conception following poor ovarian response to gonado-trophin stimulation. *Br J Obstet Gynaecol* 1997;104:521–7

Pharmacology of assisted reproduction

Synthesis and pharmacology of recombinant follicle stimulating hormone

3

Henk J. Out

INTRODUCTION

The use of gonadotropins for the treatment of various infertility conditions has grown tremendously during the last decade. Particularly the clinical application of follicle stimulating hormone (FSH) for ovarian stimulation prior to *in vitro* fertilization (IVF) or intracytoplasmic sperm injection (ICSI) has become an essential part of assisted reproductive technologies (ART). The use of luteinizing hormone (LH) activity is essential for treatment of hypogonadotropic hypogonadal women to provide the necessary androgen precursors for FSH-induced steroidogenesis[1]. Human chorionic gonadotropin (hCG) is employed for final maturation of oocytes and triggering of ovulation. Until recently, these three gonadotropins could only be obtained by extraction from urine either from postmenopausal (human menopausal gonadotropin (hMG) which contains FSH and LH activities) or pregnant women (hCG). hMG has been available since the beginning of the 1960s and hCG was already on the market in the 1930s. To obtain these products, commercial manufacturers had to build up a urine-collecting program, gathering millions of liters of postmenopausal and pregnancy urine per year. Urinary extracts of gonadotropins were subsequently produced as freeze-dried cakes in ampules.

Because of the increased demand for ART during the early 1990s, patients and doctors were confronted with shortages of hMG and urinary FSH (uFSH) products. The lead time between increasing the pool of urine donors and having hMG/uFSH in the pharmacy is approximately 1 year. Other disadvantages of urinary gonadotropins include batch-to-batch inconsistencies because of the unknown and changing nature of the urine pool. In addition, most urinary preparations contain urinary proteins of unknown origin[2]. As a result of these contaminations, local intolerance with intramuscular (IM) or subcutaneous (SC) injections has been described[3].

It is highly likely that any urine pool obtained from thousands of donors is contaminated by viruses and remnants of medicines. Also prions can be transmitted via the kidney to the urine[4]. It is impossible to prevent these contaminations even with very selected populations of donors. The safety of medicinal products derived from human urine is almost exclusively ensured by the reproducibility and efficacy of antiviral steps of the manufacturing process. Although no infectivity in humans has been described for prion disease with urine, the severity of it, its long incubation period and the emergence of variants of the disease, justify extreme caution in the use of biological sources. It can therefore be expected that (inter)national regulatory bodies in the future will re-examine safety regulations for medicines extracted from urine, in line with guidelines for other biological products of human origin, such as blood plasma for which the microbiological quality of the starting material is crucial. Since the late 1990s, nearly 100% pure gonadotropins made by recombinant DNA technology have become available, overcoming most of the problems seen with gonadotropins from urinary origin. In 1995, recombinant FSH was introduced (rFSH, follitropin-α, Gonal-F®, Serono, Switzerland; follitropin-β, Puregon®/Follistim®, Organon, The

Netherlands). Recently, recombinant LH (luteotropin, Luveris®, Serono, Switzerland) and recombinant hCG (choriotropin, Ovidrel®, Serono, Switzerland) have also become available, making the continued use of urinary products unnecessary and obsolete.

This chapter will describe the synthesis and pharmacology of recombinant FSH (especially follitropin-β), being the most frequently used gonadotropin in anovulatory infertility and ART.

SYNTHESIS

FSH is produced by the gonadotropic cells of the anterior pituitary and released into the circulation. FSH and LH belong to a family of glycoproteins that are heterodimers, containing two non-covalently linked α- and β-subunits. The subunits are encoded by separate genes. Other members of the glycoprotein hormone family are thyroid stimulating hormone (TSH) and hCG. Within an animal species the amino acid sequence of the α-subunits is identical, whereas the β-subunits differ and confer biological specificity on the individual gonadotropins. Both the α- and β-subunits are glycosylated. The α- and β-subunit of FSH both have two potential asparagin-linked glycosylation sites, characterized by the consensus sequence Asn-X-Ser, on positions α52, α78 and β7, β24, respectively.

After initial glycosylation in the endoplasmic reticulum further processing involves trimming by glucosidases and mannosidases, and remodeling of the carbohydrates in a complex series of biochemical reactions. The final structure of the Asn-linked carbohydrates on glycoproteins is dependent on the protein itself and the tissue in which it is produced. In particular the terminal residues on the carbohydrate antennas may differ: LH carbohydrates terminate with sulfate-4-*N*-acetyl-galactosamine, whereas FSH bears more highly branched sialylated structures. As a result of the extensive biochemical processing each of the glycan chains demonstrates considerable microheterogeneity resulting in numerous glycoforms, which can be resolved by isoelectric focusing. The carbohydrates on the gonadotropins serve many important functions. They are required for proper folding, assembly and secretion of the gonadotropins. Furthermore, carbohydrates are also highly relevant for the biological activity. It is well established that glycosylation determines the halflife of the gonadotropins. Furthermore, alterations in the carbohydrate structures may result in molecules with decreased ability to stimulate adenylate cyclase and steroidogenesis, but with unaffected receptor affinity. Thus, post-translational modification, such as glycosylation, is an absolute requirement for proper expression and full biological activity of the glycoprotein hormones.

The Chinese hamster ovary (CHO) K1 cell line – which had been created in 1957 from the ovarian biopsy of an adult hamster – was selected as the 'protein factory' for the production of rFSH. This cell line proved to be easily transfected with both genes encoding for the α- and β-subunits, with the assurance of proper glycoprotein expression and safety. Analyses showed that the selected CHO clone was genetically stable and capable of secreting biologically active human rFSH for a prolonged period[5].

The CHO cell line is an anchorage-dependent cell line, which means that a proper surface must be provided for growth of the cells. In order to obtain a favorable surface/volume ratio cells are grown on small beads with a diameter of approximately 0.2 mm. The use of microcarriers in cell culture also provides an opportunity for easy physical separation of the cells from the culture supernatant. This enabled the development of a perfusion-type continuous culture of CHO cells. CHO cells are grown to high cell density in a continuously stirred bioreactor. rFSH is isolated from pooled culture supernatant by a series of chromatographic steps including anion and cation exchange chromatography, hydrophobic interaction chromatography and size exclusion chromatography[5].

PHARMACOLOGY

With respect to pharmacokinetics, FSH differs from classical chemical substances in two aspects:

(1) A given dose is expressed in terms of *in vivo* bioactivity as determined in the rat Steelman–Pohley assay[6].

(2) On the basis of their heterogeneous isohormone character, gonadotropins cannot be considered as single-component drugs.

Natural gonadotropins display so-called microheterogeneity because they occur in various isoforms. This is due to differences in carbohydrate chain structure especially in the degree of sialylation[7–9].

Isohormones can be separated by chromatofocusing or electrofocusing techniques on the basis of differences in isoelectric points (p*I*). Acidic isohormones combine relatively low receptor-binding affinity and intrinsic bioactivity with a long plasma halflife, whereas the basic isoforms display relatively high receptor binding and intrinsic bioactivity together with a short plasma halflife[7]. It was shown that human pituitary FSH could be separated into at least 20 isohormone fractions, which displayed seven discrete levels of FSH receptor-binding activities[9].

In a comparative study of 13 batches of rFSH (follitropin-β) and 10 batches of uFSH it was shown that rFSH contains an approximately two-fold higher proportion of basic (p*I* > 4.7) isoforms (32.0% versus 17.0%) and a two-fold lower proportion of acidic (p*I* < 4.1) isoforms (14.7% versus 31.4%). In *in vitro* assays it was shown that rFSH had a higher potency than uFSH[10].

Differences in isohormone composition, especially in sialylation, have a direct effect on the kinetics of gonadotropin preparations and may therefore influence their bioactivity.

Gonadotropins can be quantified with four essentially different types of assays, all having their own specific merits, i.e., immunoassays, receptor-binding assays, *in vitro* bioassays and *in vivo* bioassays. These assays measure four different basic characteristics of gonadotropin molecules.

(1) Immunoassays measure a structural feature of a gonadotropin molecule. It is general belief that immunoassays provide a 'relative' measure for the mass of gonadotropins. In other words, immunoassays measure the number of molecules present.

(2) Receptor-binding assays provide information on the proper conformation for receptor binding.

(3) *In vitro* bioassays measure, in contrast to the two previous assays, a functional aspect of gonadotropins, namely their intrinsic biological activity in terms of second messenger activation and subsequent steroid biosynthesis.

(4) *In vivo* assays measure the overall bioactivity of gonadotropins. This *in vivo* bioactivity is determined by the number of molecules injected, the pharmacokinetic behavior of these molecules, their receptor-binding affinity and intrinsic bioactivity.

A single-dose of 300 IU rFSH was administered intramuscularly (buttock) to female gonadotropin-deficient, but otherwise healthy subjects[11]. After a wash-out period of at least 2 weeks, patients from this group received a single IM injection of 300 IU uFSH.

The extent of absorption of immunoreactive FSH was significantly higher for uFSH than for rFSH (C_{max} and AUC both about 65% of those after uFSH). However, serum bioactive FSH (as determined in the *in vitro* bioassay at 6, 24 and 72 h after drug administration) indicated that the circulating intrinsic FSH bioactivity was higher after rFSH than after uFSH injection[12]. Apparently, administration of rFSH leads to low FSH immunolevels combined with high FSH bioactive levels. Since basic isohormones are known to have a higher intrinsic bioactivity (and a shorter plasma elimination halflife) than acidic isohormones, the described difference might be related to the fact that rFSH contains more basic isohormones than uFSH.

CLINICAL EFFICACY AND SAFETY

Assisted reproduction

The clinical effectiveness and safety of rFSH in controlled ovarian hyperstimulation has been established in many trials. The largest trial was published in 1995 and encompassed approximately 1000 cycles[13]. This study was set up in order to detect even small differences between both groups. The primary endpoints as defined prior to starting the study were the number of oocytes retrieved and the ongoing pregnancy rates, defined as the presence of a vital pregnancy at least 12 weeks after embryo transfer. With a randomization in a 3:2 ratio between rFSH and uFSH, 80% power and a two-sided significance level of 5%, a difference of 1.2 oocytes retrieved (standard deviation, 6) and 6% in pregnancy rates could be detected. The trial was designed as a prospective, randomized, assessor-blind, multicenter study. Eighteen centers from 11 European countries participated.

Selection criteria were age between 18 and 39 years, good physical health, normal weight, at least 1 year of infertility, no male factor, no endocrine abnormalities and normal regular ovulatory cycles. The protocol included (1) intranasal buserelin downregulation in a long protocol; (2) 150 or 225 IU for the first 4 days after which the dose was adapted according to ovarian response; (3) hCG administration when at least three follicles ≥ 17 mm were seen; and (4) a maximum replacement per transfer of three embryos. The results are given in Table 1.

The main efficacy parameter, i.e., the number of oocytes retrieved, was consistently higher after rFSH treatment in all 18 participating centers (Figure 1). Ongoing pregnancy rates including frozen–thawed embryo replacements in subsequent natural cycles were significantly in favor of rFSH ($p = 0.05$). The incidence of ovarian hyperstimulation syndrome (OHSS) leading to hospitalization was seen in 19 out of 585 rFSH-treated subjects (3.2%) versus eight out of 396 uFSH-treated subjects (2%), which was not significantly different.

In summary, this study demonstrated a significantly higher number of oocytes, embryos and ongoing pregnancies (efficacy endpoints), using a lower total dose during a shorter treatment period (efficiency endpoints) with a similar incidence of OHSS (safety endpoint) for rFSH when compared with uFSH.

Numerous studies have confirmed the higher potency of both follitropin-α and -β[14]. As a result more oocytes are retrieved as compared with uFSH and more embryos are available to select for transfer and to freeze. It has been shown that the higher availability of frozen embryos increases the chance for a pregnancy per stimulated cycle using rFSH by frozen–thawed embryo replacements in natural cycles[15]. In addition, a recent Cochrane systematic review, meta-analyzing all comparisons between rFSH and uFSH, indicated a significantly higher chance for a pregnancy after fresh embryo

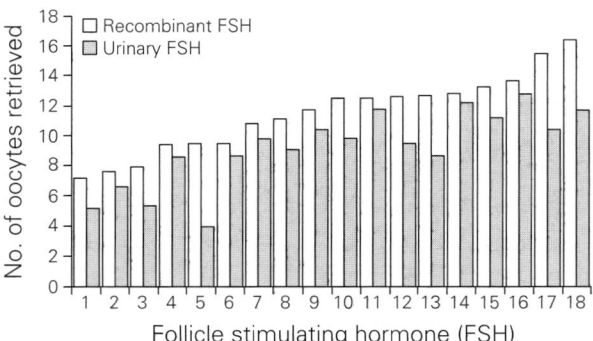

Figure 1 Number of oocytes retrieved in 18 centers participating in a randomized trial comparing recombinant and urinary follicle stimulating hormone[13]. Reproduced with permission of Oxford University Press/Human Reproduction, from Out HJ, Mannaerts BM, Driessen SG, Bennink HJ. A prospective, randomized, assessor-blind, multicentre study comparing recombinant and urinary follicle stimulating hormone (Puregon versus Metrodin) in *in vitro* fertilization. *Hum Reprod* 1995;10:2534–40, ©European Society of Human Reproduction and Embryology

Table 1 Results of recombinant versus urinary follicle stimulating hormone (FSH) in *in vitro* fertilization[13]

Parameter	rFSH	uFSH	95% CI of treatment difference	p Value
Number of subjects treated	585	396	na	
Total number of oocytes retrieved	10.8	9.0	1.2, 2.6	< 0.0001
Number of mature oocytes retrieved	8.6	6.8	1.1, 2.4	< 0.0001
Total FSH dose (IU)	2138	2385	−338, −158	< 0.0001
Duration of treatment (days)	10.7	11.3	−0.9, −0.3	< 0.0001
Number of follicles ≥ 17 mm	4.6	4.4	−0.0, 0.5	0.09
Number of follicles ≥ 15 mm	7.5	6.7	0.4, 1.2	0.0002
Maximum serum estradiol (pmol/l)	6084	5179	494, 1317	< 0.0001
Number of high-quality embryos	3.1	2.6	0.2, 0.8	0.003
Ongoing pregnancy rate per attempt	22.2%	18.2%	−1.1, 9.0	0.13
Ongoing pregnancy rate per transfer	26.0%	22.0%	−1.9, 9.8	0.19
Ongoing pregnancy rate per attempt including frozen embryo cycles	25.6%	20.4%	0.0, 10.6	0.05

rFSH, recombinant FSH; uFSH, urinary FSH; CI, confidence interval; na, not applicable

transfer[14]. The reason for this higher pregnancy potential is not known. A number of mechanisms can be hypothesized. First, the higher pregnancy rates directly following the fresh transfer suggests a higher embryo quality after follitropin-β treatment. This may be related to an increased number of available embryos enabling the embryologist to replace the morphologically best embryos. It is believed that embryo morphology correlates well with the chance for implantation and therefore pregnancy. However, one cannot exclude the possibility that oocyte quality and therefore embryo quality are influenced by the type of gonadotropin preparation used. It has been shown in *in vitro* experiments with mammalian oocytes that meiotic progression, polar body emission, cumulus–oocyte interactions and oocyte cytoskeletal organization are influenced by the presence or absence of gonadotropins in the culture medium[16]. One can speculate that the different nature of rFSH compared to traditional urinary preparations as manifested in the absence of impurities and a relatively more basic isohormone profile may influence the ability of the embryo to implant. Further research is needed to elucidate these issues.

Ovulation induction

In monofollicular approaches as used in clomiphene-resistant anovulatory women, the higher potency of rFSH has been confirmed in a large prospective, randomized trial comparing it with uFSH[17]. In this study, selection criteria included age between 18 and 39 years, no male factor, at least one open fallopian tube, and a body mass index (BMI) between 19 and 32 kg/m². At cycle day 3, a low-dose step-up dose regimen was started (initial dose 75 IU/day IM) including a fixed-dose for the first 2 weeks in the first treatment cycle and, when needed, weekly upward adjustments by half an ampule. In the second and third cycles, upward adjustments were already allowed in the second treatment week. Randomization was in a 3:2 ratio between rFSH and uFSH. hCG (10 000 IU) was given when one follicle ≥ 18 mm, or two or three follicles ≥ 15 mm were seen on ultrasound.

A total of 178 subjects (rFSH: $n = 109$; uFSH: $n = 69$) were randomized. One hundred and seventy-two subjects (rFSH: $n = 105$; uFSH: $n = 67$) were treated

in the first treatment cycle, 111 (rFSH: $n = 69$; uFSH: $n = 42$) subjects in the second and 78 subjects (rFSH: $n = 49$; uFSH: $n = 29$) in the third treatment cycle.

The cumulative ovulation rates did not differ significantly between both treatment groups and were 95% and 96% for rFSH and uFSH, respectively. Taking all cycles together, 155 out of 223 rFSH cycles were ovulatory (69.5%) compared to 92 out of 138 uFSH cycles (66.7%). Cumulative pregnancy rates were 27% in the rFSH group, compared to 24% in the uFSH group. The miscarriage rates were 31% and 32%, respectively, in the rFSH and uFSH groups. The higher efficiency of rFSH was demonstrated by a significantly shorter treatment duration to reach ovulation: a median of 10 days (range 4–27) compared to 13 days (range 4–49) after uFSH treatment ($p < 0.001$). Correspondingly, the total dose used in the rFSH group was 750 IU (range 300–2738) and 1035 IU (range 300–7350) in the uFSH group ($p < 0.001$).

The higher activity of rFSH was also demonstrated by a significantly increased number of follicles ≥ 12 mm and serum estradiol levels. Serum FSH levels were significantly lower after rFSH treatment as compared with uFSH. The higher potency of rFSH was not correlated with an increased incidence of cycle cancellation due to too many follicles and (or) too high serum estradiol levels. Also, multiple gestation rates were low and similar when compared to uFSH. In both groups one twin (4.2% and 7.7% in the rFSH and uFSH group, respectively) was seen, and in the rFSH group only one triplet (4.2%). All cases of OHSS were mild or moderate in severity. In only one case (rFSH) did the occurrence of OHSS lead to hospitalization.

Given the higher potency of rFSH, ampules/vials containing 50 IU have been made available which seem ideal for anovulatory women who are especially prone to the detrimental side-effects of gonadotropins.

This study confirmed the current impression that a low-dose step-up protocol is correlated with an acceptable efficacy and safety in patients with anovulatory infertility. Starting doses in such protocols range from 37.5 IU to 75 IU. Good results have been obtained using 52.5 IU (0.7 ampule of 75 IU) as first dose[18]. Therefore, given the higher efficacy of rFSH, it seems best to start treatment with 50 IU daily and not to change the dose for the first 7 to 14 days. This will further diminish the total dose of gonadotropins to be administered in these patients with, at the same time, an equal efficacy[19,20].

DOSING STRATEGIES

A practice of starting doses of 150–300 IU has emerged, with the possibility to increase the dose after 4–6 days when an inadequate response is seen or to decrease the dose when estradiol levels rise too rapidly and the risk of OHSS is considered to be unacceptably high. The basis for this practice is empirical rather than evidence-based.

It is interesting to note that in more than 20 years of IVF, hardly any randomized comparisons have been made to look at what the appropriate gonadotropin (starting) dose is. In addition, the relevance of FSH or hMG dose adaptations in downregulated cycles has hardly been studied. In fact, the only prospective randomized study conducted in this field showed that doubling the hMG dose in patients with a low response after 5 days of 225 IU hMG did not result in a higher number of oocytes retrieved compared to patients treated with a fixed dose[21]. It may be that follicular recruitment takes places in the late luteal and early follicular phase and is insensitive to later stimulation. Furthermore, pharmacokinetic/dynamic studies with rFSH have shown that follicular growth continues for 4 days after cessation of rFSH administration[22,23]. Because of the lag between drug administration and follicular growth,

Table 2 Combined results of five randomized clinical trials[24–28] comparing 100 and 200 IU or 150 and 250 IU of recombinant follicle stimulating hormone (FSH) in assisted reproduction

Studies		1	2	3	4	5
Outcome variable		100 IU	150 IU	200 IU	250 IU	p Value
Number of women		355	268	353	273	na
Oocytes retrieved (n)	Mean	5.7	9.1	12.0	10.5	< 0.001
	SD	5.41	6.69	8.38	8.74	
Treatment duration (days)	Mean	12.0	10.9	10.3	10.2	< 0.001
	SD	3.51	2.62	2.50	2.26	
Total FSH dose (IU)	Mean	1196	1632	2047	2554	na
	SD	352.5	391.1	498.7	563.8	
Serum FSH day hCG (IU/l)	Mean	7.9	10.9	12.6	15.2	< 0.001
	SD	4.53	4.36	6.19	5.79	
Serum LH day hCG (IU/l)	Mean	2.9	2.5	2.9	2.2	0.071
	SD	3.43	4.98	3.31	2.47	
Serum estradiol day hCG (pmol/l)	Mean	5066	7260	7691	7607	<0.001
	SD	3742.7	7401.1	5186.4	5797.2	
Serum progesterone day hCG (pmol/l)	Mean	3.5	4.2	4.2	5.1	0.053
	SD	6.55	9.08	5.14	6.65	
Follicles ≥ 13 mm day hCG	Mean	7	8.5	10	9	< 0.001
	SD	0.43	0.34	0.32	0.42	
Number of transferable embryos	Mean	3.8	4.2	6.5	4.2	< 0.001
	SD	3.07	3.02	4.54	3.13	
Number of embryos transferred	Mean	2.4	3.0	2.7	3.0	< 0.001
	SD	0.89	1.29	0.87	1.31	
Vital pregnancy rate per started cycle	n (%)	61 (17)	51 (19)	64 (18)	50 (18)	0.96
Vital pregnancy rate per embryo transfer	n (%)	61 (18)	51 (20)	64 (19)	50 (19)	0.90
Incidence of OHSS	n (%)	7 (2)	5 (2)	14 (4)	8 (3)	0.34

na, not applicable; SD, standard deviation; hCG, human chorionic gonadotropin; LH, luteinizing hormone; OHSS, ovarian hyperstimulation syndrome

the follicular consequences of any change in the FSH dose are not seen for at least 4 days. As the average treatment period of controlled ovarian hyperstimulation regimens in downregulated women lasts for 10–11 days, the usefulness of dose changes may therefore be questioned.

Recently, five trials have been published addressing FSH dosing in a randomized, double-blind way[24–28]. All women had (1) a cause of infertility potentially solvable by IVF or ICSI; (2) normal ovulatory cycles with a mean length of between 24 and 35 days; (3) good physical and mental health; (4) a BMI between 18 and 29 kg/m^2; (5) no endocrine abnormalities; and (6) absence of previous low response. In three studies, all comparing 100 and 200 IU[24–26], the age was between 18 and 39 years. In the two trials comparing 150 and 250 IU[27,28], the age was between 30 and 39 years.

rFSH (follitropin-β) was started and kept fixed throughout a maximum treatment period of 3 weeks after downregulation with gonadotropin releasing hormone (GnRH) agonists was achieved (serum estradiol < 200 pmol/l). All GnRH agonists were allowed except the IM depot preparations. Administration of hCG took place when at least three follicles ≥ 17 mm[24–27] or at least two follicles ≥ 20 mm[28] were seen on ultrasound. Three[24–27] or four[28] embryos could maximally be transferred. All other procedures were performed as was routine in the participating center. In total, 1260 women from 41 IVF clinics in 18 countries were randomized, and 1249 were subsequently treated with rFSH. Combined results have been given in Table 2.

More oocytes could be retrieved using increasing dosages of rFSH. However, that effect seemed to diminish with increasing age. Fresh pregnancy rates were similar between groups. rFSH consumption was nearly 1400 IU higher in the 250 IU-treated women as compared to the 100 IU-treated women.

CONCLUSIONS

The availability of pure recombinant human FSH has been welcomed as a major step forward in the treatment of infertility[29]. The highly controlled manufacturing conditions ensure production of batches of FSH consistent in biochemical characteristics and quality. The possibility of using a nearly 100% pure FSH preparation with an FSH isohormone profile very much resembling the profile of natural FSH in fertile women has created new impulses to research the role of gonadotropins in influencing pregnancy. Randomized clinical trials in IVF have shown that pregnancy rates are significantly higher using rFSH as compared to urinary gonadotropins. The higher potency of rFSH has enabled the introduction of presentation forms lower than the traditional 75 IU ampules, for example 50 IU, which are specifically suitable for the monofollicular approach in anovulatory women. It has become clear that a substantial proportion of women undergoing assisted reproduction may benefit from daily dosages as low as 100 IU. Increasing the daily dose will lead to a higher number of retrievable eggs particularly in younger patients, but it will not increase the rate of new pregnancies. Cumulative pregnancy rates including frozen– thawed embryo replacements per single stimulation cycle, however, might be higher using this high-dose approach. Hopefully, further research into the pharmacologic mechanisms of rFSH treatment will contribute to a better understanding of infertility and to a higher chance of pregnancy.

References

1. Schoot DC, Coelingh Bennink HJT, Mannaerts BMJL, *et al.* Human recombinant follicle-stimulating hormone induces growth of preovulatory follicles without concomitant increase in androgen and estrogen biosynthesis in a woman with isolated gonadotropin deficiency. *J Clin Endocrinol Metab* 1992;74:1471–3

2. Giudice E, Crisci C, Eshkol A, Papoian R. Composition of commercial gonadotrophin preparations extracted from human post-menopausal urine: characterization of non-gonadotrophin proteins. *Hum Reprod* 1994;9:2291–9

3. Li TC, Hindle JE. Adverse local reaction to intramuscular injections of urinary-derived gonadotrophins. *Hum Reprod* 1993;8:1835–6

4. Shaked GM, Shaked Y, Kariv Z, *et al.* A protease resistant PrP isoform is present in urine of animals and humans affected with prion diseases. *J Biol Chem* 2001;276:31479–82

5. Olijve W, De Boer W, Mulders JWM, Van Wezenbeek PMGF. Molecular biology and biochemistry of human recombinant follicle stimulating hormone (Puregon). *Mol Hum Reprod* 1996;2:371–82

6. Steelman SL, Pohley FM. Assay of the follicle-stimulating hormone based on the augmentation with human chorionic gonadotropin. *Endocrinology* 1953;53: 604–16

7. Ulloa-Aguirre A, Espinoza R, Damian-Matsumura P, Chappel SC. Immunological and biological potencies of the different molecular species of gonadotrophins. *Hum Reprod* 1988;3:491–501

8. Ulloa-Aguirre A, Cravioto A, Damian-Matsumura P, *et al.* Biological characterization of the naturally occurring analogues of intrapituitary human follicle-stimulating hormone. *Hum Reprod* 1992;7:23–30

9. Ulloa-Aguirre A, Damian-Matsumura P, Jimenez M, *et al.* Biological characterization of the isoforms of urinary human follicle-stimulating hormone contained in a purified commercial preparation. *Hum Reprod* 1992;7:1371–8

10. Lambert A, Rodgers M, Mitchell R, *et al. In vitro* potency and glycoform distribution of recombinant human follicle stimulating hormone (Org 32489), Metrodin and Metrodin-HP. *Mol Hum Reprod* 1995;10:1928–35

11. Mannaerts B, Shoham Z, Schoot D, *et al.* Single-dose pharmacokinetics and pharmacodynamics of recombinant human follicle-stimulating hormone (Org 32489) in gonadotropin-deficient volunteers. *Fertil Steril* 1993;59: 108–14

12. Matikainen T, Leeuw R de, Mannaerts B, Huhtaniemi I. Circulating bioactive and immunoreactive recombinant follicle stimulating hormone (Org 32489) after administration to gonadotropin-deficient subjects. *Fertil Steril* 1994;61: 62–9

13. Out HJ, Mannaerts BMJL, Driessen SGAJ, Coelingh Bennink HJT. A prospective, randomized, assessor-blind, multicentre study comparing recombinant and urinary follicle-stimulating hormone (Puregon vs Metrodin) in *in-vitro* fertilization. *Hum Reprod* 1995;10:2534–40

14. Daya S, Gunby J. Recombinant versus urinary follicle stimulating hormone for ovarian stimulation in assisted reproduction cycles (Cochrane Review). In: *The Cochrane Library*, Issue 4. Oxford: Update Software, 2000. www.update-software.com/ccweb/cochrane/rebabstr/mainindex.htm

15. Jones HW Jr, Out HJ, Hoomans EHM, *et al.* Cryopreservation: the practicalities of evaluation. *Hum Reprod* 1997;12: 1522–4

16. Plancha CE, Albertini DF. Hormonal regulation of meiotic maturation in the hamster oocyte involves a cytoskeleton-mediated process. *Biol Reprod* 1994;51:852–64

17. Coelingh Bennink HJT, Fauser BCJM, Out HJ, for the European Puregon Collaborative Anovulation Study Group. Recombinant FSH (Puregon) is more efficient than urinary FSH (Metrodin) in clomiphene-resistant normogonado-

tropic chronic anovulatory women: a prospective, multi-center, assessor-blind, randomised, clinical trial. *Fertil Steril* 1998;69:19–25

18. White DM, Polson DW, Kiddy D, *et al.* Induction of ovulation with low-dose gonadotropins in polycystic ovary syndrome: an analysis of 109 pregnancies in 225 women. *J Clin Endocrinol Metab* 1996;81:3821–4

19. Hayden CJ, Rutherford AJ, Balen AH. Induction of ovulation using a starting dose of 50 units of recombinant follicle stimulating hormone (Puregon). *Fertil Steril* 1999;71:106–8

20. Hoomans E, Voortman G. Efficacy and efficiency of ovulation induction in anovulatory women using a low-dose step-up scheme with recombinant FSH (Puregon). *Hum Reprod* 1999;14:327

21. Van Hooff MH, Alberda AT, Huisman GJ, *et al.* Doubling the human menopausal gonadotrophin dose in the course of an *in vitro* fertilization treatment cycle in low responders: a randomized study. *Hum Reprod* 1993;8:369–73

22. Porchet HC, le Cotonnec JY, Loumaye E. Clinical pharmacology of recombinant human follicle-stimulating hormone. III. Pharmacokinetic–pharmacodynamic modeling after repeated subcutaneous administration. *Fertil Steril* 1994; 61:687–95

23. Mannaerts BMJL, Rombout F, Out HJ, Coelingh Bennink HJT. Clinical profiling of recombinant follicle stimulating hormone (rFSH; Puregon): relationship between serum FSH and efficacy. *Hum Reprod Update* 1996:2:153–61

24. Out HJ, Lindenberg S, Mikkelsen AL, *et al.* A prospective, randomised, double-blind clinical trial to study the efficacy and efficiency of a fixed dose of recombinant follicle stimulating hormone (Puregon®) in women undergoing controlled ovarian hyperstimulation. *Hum Reprod* 1999;14:622–7

25. Out HJ, David I, Ron-El R, *et al.* A randomised double-blind clinical trial using fixed daily doses of 100 or 200 IU of recombinant FSH in ICSI cycles. *Hum Reprod* 2001;16: 1104–9

26. Out HJ, Hoomans EHM. New stimulation regimens with recombinant follicle stimulating hormone. In Filicori M, ed. *Proceedings of the Third World Congress of Ovulation Induction*. Rome: CIC Edizioni Internazionali, 2001

27. Out HJ, Braat DDM, Lintsen BME, *et al.* Increasing the daily dose of recombinant follicle-stimulating hormone (Puregon®) does not compensate for the age-related decline in retrievable oocytes after controlled ovarian hyperstimulation. *Hum Reprod* 2000;15:29–35

28. The Latin-American Puregon IVF study group. A double-blind clinical trial comparing a fixed daily dose of 150 and 250 IU recombinant follicle stimulating hormone in *in vitro* fertilisation. *Fertil Steril* 2001;76:950–6

29. McDonough P. The coming of wonders [Letter]. *Fertil Steril* 1997;67:411–13

Gonadotropin releasing hormone antagonists: synthesis, pharmacology and clinical applications in human reproduction

4

Dan Leivin, Ariel Weissman and Zeev Shoham

INTRODUCTION

Gonadotropin releasing hormone (GnRH) is the primary hypothalamic regulator of reproductive function. It was first isolated, characterized and synthesized independently in 1971 by Andrew Schally and Roger Guillemin, who were subsequently awarded the Nobel prize for their achievement. GnRH is a decapeptide which, like several other brain peptides, is synthesized as part of a much larger precursor peptide, the GnRH-associated peptide (GAP), which is made up of a sequence of 56 amino acids. The structure of GnRH is common to all mammals, including humans, and its action is similar in both the male and the female. GnRH is produced and released from a group of loosely connected neurons located in the medial basal hypothalamus, primarily within the arcuate nucleus, and in the preoptic area of the ventral hypothalamus. It is released by axonal transport and in a pulsatile fashion into the complex capillary net of the portal system, and binds to specific receptors in the plasma membrane of the anterior pituitary gonadotrophs, where it stimulates synthesis, storage and release of luteinizing hormone (LH) and follicle stimulating hormone (FSH).

Antagonist analogs of GnRH have a direct inhibitory, reversible suppression effect on gonadotropin secretion. Antagonistic molecules compete for, and occupy, pituitary GnRH receptors, thus competitively blocking the access of endogenous GnRH and precluding substantial receptor occupation and stimulation. Suppression attained by GnRH antagonists is immediate (no flare-up effect) and, as receptor loss does not occur, it requires a constant supply of antagonist to the gonadotroph so that all GnRH receptors are continuously occupied. Consequently, compared with agonistic analogs, a higher dose range of antagonists is required for effective pituitary suppression.

SYNTHESIS OF GnRH ANTAGONISTS

The elucidation of the structure, function and metabolic pathways of native GnRH has prompted an intensive effort by research laboratories and the pharmaceutical industry to synthesize potent and long-acting agonist and antagonist analogs[1]. Over the past three decades, thousands of analogs of GnRH, both agonists and antagonists, have been synthesized. The first generation of antagonistic analogs were hydrophilic and contained replacements for His at position 2 and for Trp at position 3. Inhibitory activity increased after incorporation of a D-amino acid at position 6 but histamine release was also increased, resulting in anaphylactic reactions which prevented their clinical use. In the third generation, the undesirable risk of anaphylaxis and edema was eliminated by replacing the D-Arg at position 6 by neutral D-ureidoalkyl amino acids, to produce compounds such

Table 1 Structure of native gonadotropin releasing hormone (GnRH) and the antagonist analogs

Amino acid	Native GnRH	Abarelix	Antarelix	Cetrorelix	Ganirelix	Iturelix	Nal-Glu
1	pGlu	D-Ala	D-Nal	D-Nal	D-Nal	D-Nal	D-Nal
2	His	D-Phe	D-Phe	D-Phe	D-Phe	D-Phe	D-Phe
3	Trp	D-Ala	D-Pal	D-Pal	D-Pal	D-Pal	D-Pal
4	Ser	Ser	Ser	Ser	Ser	Ser	Ser
5	Tyr	Tyr	Phe	Tyr	Tyr	NicLys	D-Glu
6	Gly	D-Asp	D-Hcit	D-Cit	D-Arg	D-NicLys	D-Glu
7	Leu	Leu	Leu	Leu	Leu	Leu	Leu
8	Arg	Lys(iPr)	Lys(iPr)	Arg	hArg	Lys(iPr)	Arg
9	Pro	Pro	Pro	Pro	Pro	Pro	Pro
10	Gly-NH$_2$	D-Ala	D-Ala	D-Ala	D-Ala	D-Ala	D-Ala

as cetrorelix, iturelix, azaline B, ganirelix, abarelix and antarelix (Table 1)[2–8].

GnRH ANTAGONISTS AND THE ASSISTED REPRODUCTIVE TECHNOLOGIES

Over the past 15 years, GnRH agonists have been extensively used in *in vitro* fertilization (IVF) treatment protocols. The need for this treatment arose from the high rate of spontaneous premature LH surges and ovulation resulting in cancellation of oocyte retrieval in approximately 25% of cycles where gonadotropins were administered alone[9]. Premature high LH levels and consequently elevated progesterone levels after luteinization may also have adverse effects on oocyte maturation, fertilization and implantation rates of resulting embryos[10].

Various GnRH agonist protocols are utilized in IVF cycles. The most commonly used and most successful is the so-called long protocol. In the long protocol, the agonist is administered around 14 days before commencing gonadotropin therapy (either upon menstruation or in the mid-luteal phase of the previous cycle). An initial flare-up response is followed by desensitization of the pituitary via downregulation of GnRH receptors, resulting in an almost complete suppression of ovarian steroid production within about 10 days. Although this method was found to be successful in reducing premature LH surges to < 2% of cycles[10] and in improving oocyte quantity, quality, fertilization and implantation rates[11], this was achieved at the cost of substantial side-effects. The disadvantages of using GnRH agonists in the long protocol include side-effects of estrogen depletion, such as hot flushes, profuse sweating, headaches and vaginal dryness. The initial stimulatory (flare-up) effect on hormone secretion has no medical advantage but may be associated with ovarian cyst formation. The relatively longer treatment period (14–21 days to achieve suppression) and the increase in total exogenous gonadotropins required for controlled ovarian stimulation increase both the cost and the patients' inconvenience during the treatment cycle[12] and may in turn increase the risk for the ovarian hyperstimulation syndrome (OHSS)[13]. Complete suppression through downregulation of GnRH receptors ensues a depletion of pituitary gonadotropin storage vesicles. These changes necessitate a pituitary recovery period, which substantiates the need for luteal phase support and precludes the use of GnRH agonists for ovulation triggering, thus requiring administration of human chorionic gonadotropin (hCG) that may also increase the risk of OHSS. As a result of these practical and medical disadvantages, the potential for immediate inhibition of gonadotropin release via competitive blocking of GnRH receptors by GnRH antagonists emerges as an attractive and elegant alternative.

SAFETY AND TOLERABILITY STUDIES

The introduction of GnRH antagonists into clinical use was delayed because of the property of the first generation of antagonists to induce systemic histamine release and ensue a general edematogenic state. Studies in rat mast cells confirmed that incorporation of D-Cit at position 6 of the antagonists results in reduced histamine release[14,15]. This characteristic of cetrorelix was first assessed in *in vitro* assay systems that found effective plasma concentrations to be significantly smaller ($< 10^{-3}$) than the effective dose (ED_{50}) for systemic histamine secretion and therefore it can confidently be regarded as insignificant. Due to large disparities in such assays cetrorelix safety was further tested in *in vivo* settings. Cetrorelix injected at doses of 1.5 mg/kg (subcutaneous; SC), or 1 and 4 mg/kg (intravenous; IV) into rats caused no systemic adverse effects such as edema, respiratory dysfunction or cardiovascular compromise. In these animal studies no teratogenic effects or detrimental influence on implantation rates or on embryonic development were noted when administration was in the peri-conceptional period. Several thousand human patients have now been treated with third-generation GnRH antagonists (i.e., ganirelix, cetrorelix or abarelix) without evidence of systemic or major local skin reactions and no cessation of therapy was warranted due to side-effects[14,16–20]. The common side-effects observed were injection site reactions and possibly nausea, headache, fatigue and malaise. Local skin reactions seem to be less common with antagonists than they are with agonists.

No drug interactions were demonstrated *in vitro* with medications metabolized through the P450 cytochrome pathway and no significant changes in serum chemistry or hematologic parameters were noted after administration, although substantial liver or kidney dysfunctions are contraindications to GnRH antagonist use. To date, clinical data on GnRH antagonists show a high level of safety and tolerability. Since the discovery of extrapituitary human GnRH receptors the safety of GnRH agonists and antagonists with respect to various structures has become a cardinal question. Effects of GnRH antagonists on the ovary, oocyte and granulosa cells, the endometrium and the embryo in relation to fertility and implantation rates are being investigated. Direct effects of GnRH antagonists on human ovarian steroidogenesis *in vitro* have not been demonstrated. In preliminary *in vitro* and animal studies recent data reveal that some adverse effects on oocyte maturation and on preimplantation development of embryos may be inflicted by GnRH antagonists through inhibition of GnRH receptors in these structures[21–23].

The hypothesis that GnRH may play a role in the preimplantation development of embryos was examined by Raga and co-workers[24]. This group of investigators assessed GnRH and GnRH receptor mRNA by reverse transcription–polymerase chain reaction and protein expression by immunohistochemistry in preimplantation murine embryos at various developmental stages. In addition, preimplantation murine embryos were cultured with GnRH agonist and antagonist *in vitro* to assess the influence of GnRH analogs on embryo development. GnRH is expressed in the developing mouse embryo from morula to hatching blastocyst stages at the mRNA and protein levels. GnRH receptor mRNA was also present in the developing embryos studied. Preimplantation embryonic development was significantly enhanced by incubation with increasing concentrations of GnRH agonists and is significantly decreased by GnRH antagonists compared with that of the control group. Moreover, GnRH antagonist (5 and 10 μmol/l) was able to completely block embryo development. The deleterious effect of GnRH antagonist on embryo development was reversed by increasing concentrations of the agonist, as determined by the number of embryos reaching the blastocyst stage.

PHASE I AND II CLINICAL TRIALS

Several studies of antagonist dose and treatment schedules have been conducted, and two general approaches have emerged. The first is a single SC injection of a large dose on about the eighth day of gonadotropin stimulation. The alternative is five or six daily injections of a smaller dose from about day 6 of stimulation until the day of hCG administration.

In 15 phase I studies[25], cetrorelix was administered in single and multiple SC injections as well as single IV injections. The dose range tested for single doses was 0.25–20.0 mg SC . Following single-dose administration of 3 and 5 mg of cetrorelix between days 6 and 10 of the menstrual cycle to healthy premenopausal female subjects, serum LH, FSH and estradiol decreased immediately. The LH surge was postponed in all women without differences in the two dose groups and occurred between 1 and 2.5 weeks after the GnRH antagonist injection. When cetrorelix was administered during the late follicular phase spontaneous LH surges were also postponed in all women.

Further investigations were performed in order to determine pharmacokinetics, absolute bioavailability and pharmacodynamic effects. Subjects were evaluated during the menstrual cycle before receiving cetrorelix, during a cycle in which 3 mg cetrorelix was administered SC daily for 1 week, and during the cycle following the one with administration of cetrorelix. The first injection was administered on day 8 of the individual cycle. By 24 h after the first application, LH was strongly suppressed and estradiol levels decreased to menopausal values. The mean duration of the suppressive effects of cetrorelix after its last injection compared to baseline values was 13.0 days for LH[26,27]. In the post-treatment cycle LH surges and a post-ovulatory rise in progesterone were noticed in all women, implying normal ovarian function and ovulation.

In an additional study, females receiving daily doses of 0.25, 0.5 and 1 mg cetrorelix SC from cycle days 3 to 16 revealed a linear dose–response and a single dose of 3 mg resulted in a plasma halflife of ~8 h after IV injection and 2.5 h after SC injection with a bioavailability of 92%[28]. The amount of daily cetrorelix injections was further evaluated in a total of 90 patients. Doses of 0.5, 0.25 and 0.1 mg were administered daily. The study concluded that the minimal effective daily dose needed to postpone LH surge and to receive good quality oocytes was 0.25 mg[29].

Further looking at the influence of the different doses of GnRH antagonists in women, a multicenter, double-blind, randomized dose-finding study of Org 37462 (ganirelix) was conducted in 333 women to establish the minimal effective dose preventing premature LH surges during ovarian stimulation[30]. From cycle day 7 onwards, up to and including the day of hCG administration, ganirelix (dosages 0.0625, 0.125, 0.25, 0.5, 1.0 and 2.0 mg) was administered once daily by SC injection, and the recombinant FSH dose was adjusted depending on ovarian response. The lowest (0.0625 mg) and highest (2.0 mg) dose groups were terminated prematurely on the advice of an external independent advisory committee. Serum ganirelix concentrations increased in a linear, dose-proportional manner, whereas serum LH and increases of estradiol fell with increasing ganirelix

dose. The implantation rate was highest in the 0.25-mg group (21.9%) and lowest in the 2-mg group (1.5%). The early miscarriage rates (first 6 weeks after embryo transfer) were 8.5 and 13% in the 1- and 2-mg group respectively, whereas in the other dose groups this incidence was zero (0.0625%) up to a maximum of 3.7% (0.5-mg group). These results, which are summarized in Table 2, further support the notion that the administration of high doses of GnRH antagonists might have some deleterious effects on the embryos and the endometrium.

Pituitary suppression by GnRH agonists usually results in residual LH secretion sufficient to cover the needs of a successful cycle of assisted reproductive technologies (ART). In contrast, with antagonist administration cessation of LH secretion is abrupt, in a dose-dependent manner[30]. This may be of clinical relevance with potential implications on embryo implantation. With ganirelix, for example, implantation rates decreased from 20% to 1% when 0.25 mg or 2 mg of the antagonist were administered daily, respectively. The extent of pituitary response to antagonists may vary between patients, as demonstrated at the other extreme by the sporadic occurrence of spontaneous LH rises. Thus, in some patients, a profound effect on the pituitary may interfere with implantation, and perhaps provide a partial explanation for the somewhat lower pregnancy rates reported with antagonists in phase II studies where recombinant human FSH was used.

One of the goals in introducing GnRH antagonists into clinical practice was to shorten the duration of

Table 2 Outcome of the six dose groups of ganirelix examined by The Ganirelix Dose-Finding Study Group. Data derived from reference 30

Dose (mg)	Egg collection	Embryos	Implantation (%)	Miscarriage (%)	Pregnancy (%)
0.125	9.5	5.9	16.6	3.3	25
0.25	10.0	5.4	21.0	1.6	37.1
0.5	8.8	4.6	9.0	3.7	13.0
1.0	9.3	5.3	8.8	8.5	15.3
2.0	8.8	4.9	1.5	13.0	0

Results are presented as a mean number.

treatment and to reduce the inconvenience of multiple injections. Additional studies were therefore carried out in order to verify whether a single dose given during the follicular phase was also effective in delaying LH surges. In patients who were injected with single doses of 5 mg of cetrorelix in the mid-late follicular phase LH levels were suppressed and no LH surges were noted until administration of hCG[31]. In further studies the minimal effective dose for single administration was found to be 3 mg and when final ovulation triggering was to be further postponed by 48–72 h another similar dose was advocated[32,33].

PHASE III TRIALS

The next step was to explore whether a GnRH antagonist is at least as effective as a GnRH agonist as a reference medication. So far, four such multicenter, multinational studies of repeated daily antagonist injections have been reported, one of them using the single-dose regimen[18–20,34,35].

In multiple-dose studies 0.25 mg/day of cetrorelix starting on stimulation day 5 or 6 were compared to intranasal buserelin (0.6 mg/day) starting on day 20 of the previous cycle. In the single-dose studies, the use of 3 mg cetrorelix administered on stimulation day 7 was evaluated in comparison with a single dose of 3.75 mg depot triptorelin in the control group. Results revealed that with multiple-dose injections the duration of the gonadotropin stimulation is shortened by 1–2 days. In the antagonist study group, slightly fewer follicles were noticed at the time of hCG injection and therefore fewer oocytes were recovered. A likely explanation is that the longer stimulation phase required in the control group increases the cohort of mature follicles. Percentages of post-ovulatory oocytes, fertilization rates and number of good quality embryos did not differ between the study and control group. In all studies published to date, the pregnancy rates with antagonists were relatively high but still somewhat lower than with agonist regimens. This finding does not correlate with the preclinical findings showing that cetrorelix does not affect steroid biosynthesis of granulosa-lutein cells or growth factor-induced granulosa cell proliferation[36]. A direct detrimental effect on the developing embryo is probably not the cause for this difference because of the minimal

probability of embryo exposure to an antagonist in IVF protocols. Cetrorelix plasma and follicular fluid levels fall significantly after hCG administration, and at minimal effective doses cetrorelix was not detectable during ovum retrieval and embryo transfer. This finding needs to be enforced through long-term follow-up studies of pregnancy and childhood developmental outcomes.

The results of IVF–embryo transfer with GnRH antagonist protocols have been analyzed recently in comparison with the GnRH agonist long protocol in a Cochrane review[37]. On the question of prevention of premature LH surges no significant statistical difference was found. The number of oocytes retrieved was consistently smaller in patients treated with antagonists, which correlated with a smaller cohort of affected follicles, smaller amount of gonadotropins used and a shorter stimulation period. When the results of all five clinical trials included in the meta-analysis were pooled, the clinical pregnancy rate was significantly lower in those treated with antagonists with an odds ratio (OR) of 0.79 (95% confidence interval (CI) 0.63–0.99), and there was no significant difference in the rate of severe OHSS (OR 0.79; 95% CI 0.22–1.18) between the antagonist- and agonist-treated patients. In the different studies a trend was noticed in favor of the agonists in terms of number of oocytes retrieved and number of embryos obtained. The reduction in pregnancy rates may be explained by this trend and by yet unrevealed adverse effects of antagonists on the endometrium interfering with implantation. To date, no study has incorporated endometrial sampling. Another field warranting more investigation is optimal dose and timing of administration of the antagonists. An interesting statistic arising from this review is that for every 20 couples treated with IVF/intracytoplasmic sperm injection one more pregnancy will be achieved in the agonist group compared to the antagonist group.

Many believe that this finding reflects the learning curve of contemporary treatment regimens, and that after further investigation and more experience gained, the learning curve will be accomplished and results will be similar.

Follow-up of neonatal outcome after pregnancies established in large phase II/III trials of around 700 pregnancies after GnRH antagonists and 134 after a long agonist protocol did not show any negative effect[38,39].

Utilizing GnRH antagonists to prevent premature LH surges offers the possibility of using GnRH agonists as an option to induce final oocyte maturation and ovulation due to the unaffected pituitary responsiveness to GnRH with this treatment. In addition to the lower number of follicles and the lower concentration of estradiol on day of hCG administration, this may play a role in reducing the incidence of severe OHSS, thought in part to be affected by exogenous hCG administration.

Another potential benefit associated with GnRH antagonists is the reduced requirement for exogenous gonadotropins. For regimens using a single administration of cetrorelix[40] the mean number of human menopausal gonadotropin (hMG) ampoules (27.7 ± 4.2) was lower than the mean number of vials administered in a similar population treated with GnRH agonists (38.3 ± 15.4). In a trial of a single 3 mg dose of cetrorelix compared with the administration of a depot form of triptorelin, the duration of stimulation, number of ampoules administered, serum estradiol levels on the day of hCG administration and number of developing follicles were significantly lower in the cetrorelix group[32,33,35].

In protocols using daily administration of cetrorelix, the mean number of hMG ampoules varied between a mean of 26 and 36 in different studies. When results of cetrorelix cycles (0.25 mg/day, 188 patients) were compared with buserelin treatment in the long protocol (600 μg/day intranasally, 85 patients), the mean number of hMG ampoules, duration of hMG treatment and serum estradiol level on the day of hCG injection were again significantly lower in the cetrorelix group[34]. The difference in estradiol levels in various studies was attributed to fewer medium-sized follicles whereas the number of larger follicles was comparable in the agonist and antagonist protocols.

Another potential benefit attributed to antagonist therapy may be the lack of requirement for luteal phase supplementation. It was postulated that the swift recovery of pituitary function following cessation of antagonist administration may obviate the need for progesterone supplementation in the luteal phase as required in agonist protocols due to early luteolysis[41]. However, most studies cited already incorporated luteal phase support as part of their regimen. In one study that commenced without luteal phase support, bleeding appeared in the mid-luteal period in all six subjects and all subsequent patients received support. None of the first six patients achieved pregnancy[29,42]. The authors concluded that corpus luteum function remains impaired in treatment regimens incorporating cetrorelix, hMG and hCG. In a study focused on luteal phase effects in patients treated with antagonists and hMG as compared to patients stimulated with hMG alone, the luteal phase was supported in both groups by hCG. LH levels in both groups were low but without significant difference, which raised the theory that these low levels of LH may result from an hCG effect[43]. Theoretically, it is possible that substitution of hCG by recombinant LH or a GnRH agonist will not disrupt the hypothalamic–pituitary–ovarian axis in the same manner as a large dose of hCG, and this may obviate the need for luteal phase supplementation in GnRH antagonist cycles.

SUMMARY

GnRH antagonists are now available for clinical use in ART cycles. Their main disadvantage at the moment is the lower pregnancy rate reported in the preliminary phase II and III studies, as reflected by a recent meta-analysis[37]. The main advantages are in the area of patient convenience, i.e., shorter treatment regimens and less discomfort, as well as the lower cost associated with the above. It still has to be proven whether antagonist regimens are more cost-effective per pregnancy than the agonist regimens in direct and indirect costs. The various regimens seem to be safe and with time, it is possible that once the learning curve is completed, clinical pregnancy rates may be comparable with agonist protocols which are currently the 'gold standard' in ART.

References

1. Nestor JJ Jr. Development of agonistic LHRH analogs. In Vickery BH, Nestor JJ Jr, Hafez ESE, eds. *LHRH and its Analogs.* Lancaster, UK: MTP Press, 1984:3–15

2. Bajusz S, Kovacs M, Gazdag M, *et al.* Highly potent antagonists of luteinizing hormone-releasing hormone free of edematogenic effects. *Proc Natl Acad Sci USA* 1988;85: 1637–41

3. Ljungqvist A, Feng DM, Hook W, *et al.* Antide and related antagonists of luteinizing hormone release with long action and oral activity. *Proc Natl Acad Sci USA* 1988;85:8236–40

4. Rivier J, Porter J, Hoeger C, *et al.* Gonadotropin-releasing hormone antagonists with *N* omega-triazolylornithine, -lysine, or -*p*-aminophenylalanine residues at positions 5 and 6. *J Med Chem* 1992;35:4270–8

5. Nestor JJ Jr, Tahilramani R, Ho TL, *et al.* Potent gonadotropin releasing hormone antagonists with low histamine-releasing activity. *J Med Chem* 1992;35:3942–8

6. Garnick MB, Campion M. Abarelix depot, a GnRH antagonist, v LHRH superagonists in prostate cancer: differential effects on follicle-stimulating hormone. *Mol Urol* 2000;4:275–7

7. Cook T, Sheridan WP. Development of GnRH antagonists for prostate cancer: new approaches to treatment. *Oncologist* 2000;5:162–8

8. Deghenghi R, Boutignon F, Wuthrich P, Lenaerts V. Antarelix (EP 24332), a novel water soluble LHRH antagonist. *Biomed Pharmacother* 1993;47:107–10

9. Schmutzler RK, Diedrich K. Basic and clinical aspects of GnRH-agonists in reproduction. *Int J Gynecol Obstet* 1990; 32:311–24

10. Stanger JD, Yovich JL. Reduced *in-vitro* fertilization of human oocytes from patients with raised basal luteinizing hormone levels during the follicular phase. *Br J Obstet Gynaecol* 1985;92:335–93

11. Hughes EG, Fedorkow DM, Daya S, *et al.* The routine use of gonadotropin-releasing hormone agonists prior to *in vitro* fertilization and gamete intrafallopian transfer: a meta-analysis of randomized controlled trials. *Fertil Steril* 1992; 58:888–96

12. Ben-Rafael Z, Lipitz S, Bider D, *et al.* Ovarian hyporesponsiveness in combined gonadotropin-releasing hormone agonist and menotropin therapy is associated with low serum follicle stimulating hormone levels. *Fertil Steril* 1991;55:272–5

13. Rizk B, Smitz J. Ovarian hyperstimulation syndrome after superovulation using GnRH agonists for IVF and related procedures. *Hum Reprod* 1992;7:320–7

14. Felberbaum R, Diedrich K. Ovarian stimulation for *in-vitro* fertilization/intracytoplasmic sperm injection with gonadotrophins and gonadotrophin-releasing hormone analogues: agonists and antagonists. *Hum Reprod* 1999;14:207–21

15. Bajusz S, Csernus VJ, Janaky T, *et al.* New antagonists of LHRH. II. Inhibition and potentiation of LHRH by closely related analogues. *Int J Peptide Protein Rec* 1988;32: 425–35

16. Diedrich K, Diedrich C, Santos E, *et al.* Suppression of the endogenous luteinizing hormone surge by the gonadotrophin-releasing hormone antagonist Cetrorelix during ovarian stimulation. *Hum Reprod* 1994;9:788–91

17. Felberbaum RE, Albano C, Ludwig M, *et al.* Ovarian stimulation for assisted reproduction with HMG and concomitant mid-cycle administration of the GnRH antagonist cetrorelix according to the multiple dose protocol: a prospective uncontrolled phase III study. *Hum Reprod* 2000;15:1015–20

18. Borm G, Mannaerts B. Treatment with the gonadotrophin-releasing hormone antagonist ganirelix in women undergoing ovarian stimulation with recombinant follicle stimulating hormone is effective, safe and convenient: results of a controlled, randomized, multicentre trial. The European Orgalutran Study Group. *Hum Reprod* 2000;15:1490–8

19. The European and Middle East Orgalutran Study Group, Comparable clinical outcome using the GnRH antagonist ganirelix or a long protocol of the GnRH agonist triptorelin for the prevention of premature LH surges in women undergoing controlled ovarian hyperstimulation. *Hum Reprod* 2001;16:644–51

20. Fluker M, Crifo J, Leader A, *et al.* Efficacy and safety of ganirelix acetate (Antagon/Orgalutran) versus leuprolide acetate in women undergoing controlled ovarian hyperstimulation. *Fertil Steril* 2001;75:38–45

21. Mannaerts B, Gordon K. Embryo implantation and GnRH antagonists: GnRH antagonists do not activate the GnRH receptor. *Hum Reprod* 2000;15:1882–3

22. Demirel LC, Weiss JM, Polack S, *et al.* Effect of the gonadotropin-releasing hormone antagonist ganirelix on cyclic adenosine monophosphate accumulation of human granulosa-lutein cells. *Fertil Steril* 2000;74:1001–7

23. Ortmann O, Weiss JM, Diedrich K. Embryo implantation and GnRH antagonists: ovarian action of GnRH antagonists. *Hum Reprod* 2001;16:608–11

24. Raga F, Casan EM, Kruessel J, *et al.* The role of gonadotropin-releasing hormone in murine preimplantation embryonic development. *Endocrinology* 1999;140:3705–12

25. Leroy I, d'Acremont M, Brailly-Tabard S, *et al.* A single injection of a gonadotropin-releasing hormone (GnRH) antagonist (Cetrorelix) postpones the luteinizing hormone (LH) surge: further evidence for the role of GnRH during the LH surge. *Fertil Steril* 1994;62:461–7

26. Gonzalez-Barcena D, Vadillo-Buenfil M, Garcia-Procel E, *et al.* Inhibition of luteinizing hormone, follicle-stimulating hormone and sex-steroid levels in men and women with a potent antagonist analog of luteinizing hormone-releasing hormone, Cetrorelix (SB-75). *Eur J Endocrinol* 1994;131: 286–92

27. Sommer L, Zanger K, Dyong T. Seven-day administration of the gonadotropin-releasing hormone antagonist Cetrorelix in normal cycling women. *Eur J Endocrinol* 1994;131:280–5

28. Herman R, Schneider E, Romeis P. Pharmacokinetics and pharmacodynamics of the LHRH antagonist Cetrorelix in healthy male and female subjects. *Gynecol Endocrinol* 1996; 10:18

29. Albano C, Smitz J, Camus M, *et al.* Comparison of different doses of gonadotropin-releasing hormone antagonist

cetrorelix during controlled ovarian hyperstimulation. *Fertil Steril* 1997:67:917–22

30. The Ganirelix Dose-Finding Study Group. A double-blind, randomized, dose-finding study to assess the efficacy of the gonadotrophin-releasing hormone antagonist ganirelix (Org 37462) to prevent premature luteinizing hormone surges in women undergoing ovarian stimulation with recombinant follicle stimulating hormone (Puregon). *Hum Reprod* 1998; 13:3023–31

31. Duijkers IJ, Klipping C, Willemsen WN, *et al.* Single and multiple dose pharmacokinetics and pharmacodynamics of the gonadotrophin-releasing hormone antagonist cetrorelix in healthy female volunteers. *Hum Reprod* 1998;13: 2392–8

32. Olivennes F, Fanchin R, Bouchard P, *et al.* Scheduled administration of a gonadotrophin-releasing hormone antagonist (Cetrorelix) on day 8 of *in vitro* fertilization cycles: a pilot study. *Hum Reprod* 1995;10:1382–6

33. Olivennes F, Álvarez S, Bouchard P. The use of a GnRH antagonist (cetrorelix) in a single dose protocol in IVF-embryo transfer: a dose finding study of 3 versus 2 mg. *Hum Reprod* 1998:13:2411–14

34. Albano C, Felberbaum RE, Smitz J, *et al.* European Cetrorelix Study Group. Ovarian stimulation with HMG: results of a prospective randomized phase III European study comparing the luteinizing hormone-releasing hormone (LHRH)-antagonist cetrorelix and the LHRH-agonist buserelin. *Hum Reprod* 2000;15:526–31

35. Olivennes F, Belaisch-Allart J, Emperaire JC, *et al.* Prospective, randomized, controlled study of *in vitro* fertilization-embryo transfer with a single dose of a luteinizing hormone-releasing hormone (LH-RH) antagonist (cetrorelix) or a depot formula of an LH-RH agonist (triptorelin). *Fertil Steril* 2000;73:314–20

36. Yano T, Yano N, Matsumi H, *et al.* Effect of luteinizing hormone-releasing hormone analogs on the rat ovarian follicle development. *Horm Res* 1997;48 (Suppl.3):35–41

37. Al-Inany H, Aboulghar M. GnRH antagonist in assisted reproduction: a Cochrane review. *Hum Reprod* 2002;17: 874–85

38. Ludwig M, Riethmuller-Winzen H, Felberbaum RE, *et al.* Health of 227 children born after controlled ovarian stimulation for *in vitro* fertilization using the luteinizing hormone releasing hormone antagonist cetrorelix. *Fertil Steril* 2001;75: 18–22

39. Bonduelle M. Neonatal outcome of pregnancies established after treatment with recombinant FSH and Ganirelix for ART. Presented at the *6th International Symposium on GnRH Analogues in Cancer and Human Reproduction*, Geneva, 2001; abstr.0056

40. Olivennes P, Fanchin R, Bouchard P, *et al.* The single or dual administration of the gonadotropin-releasing hormone antagonist Cetrorelix in an in vitro fertilization-embryo transfer program. *Fertil Steril* 1994:62:468–76

41. Ditkoff EC, Cassidenti DL, Paulson RJ, *et al.* The gonadotropin-releasing hormone antagonist (Nal-Glu) acutely blocks the luteinizing hormone surge but allows for resumption of folliculogenesis in normal women. *Am J Obstet Gynecol* 1991;165:1811–17

42. Albano C, Smitz J, Camus M, *et al.* Hormonal profile during the follicular phase in cycles stimulated with a combination of human menopausal gonadotrophin and gonadotrophin releasing hormone antagonist (cetrorelix). *Hum Reprod* 1996; 11:2114–18

43. Tavaniotou A, Albano C, Smitz J, Devroey P. Comparison of LH concentrations in the early and mid-luteal phase in IVF cycles after treatment with HMG alone or in association with the GnRH antagonist cetrorelix. *Hum Reprod* 2001:16:663–7

The role of luteinizing hormone in ovulation induction for assisted reproductive technologies

Gautam N. Allahbadia and Avinash Phadnis

5

INTRODUCTION

While initial attempts of *in vitro* fertilization (IVF) procedures were conducted in spontaneous menstrual cycles or by using mild ovarian stimulation with clomiphene citrate (CC), current assisted reproductive technologies (ART) include ovarian stimulation with exogenous gonadotropins combined with gonadotropin-releasing hormone (GnRH) analogs. Another critical development that occurred in the last decade is the progressive abandonment of luteinizing hormone (LH)-containing menotropins in favor of purified, highly purified or recombinant follicle stimulating hormone (FSH). The combination of these LH-deprived gonadotropin preparations and GnRH analogs has resulted in follicular stimulation virtually devoid of LH activity. Nevertheless, growing evidence suggests that LH plays an important physiologic and clinical role in this setting. Moderate amounts of LH activity appear to optimize ovulation induction, reduce gonadotropin dose requirements, shorten follicular stimulation, and improve overall outcome of ART, possibly also through estrogen-mediated actions on the oocyte.

PRINCIPLES OF USE OF CONTROLLED OVARIAN HYPERSTIMULATION IN ART

Injections of gonadotropins achieve supraphysiologic levels of FSH in the circulation, which override the normal ovarian hormone feedback mechanisms that control the pituitary secretion of gonadotropins. The high levels of LH and FSH in the circulation result in the growth and development of multiple follicles in both ovaries, instead of the single, dominant follicle characteristic of an unstimulated cycle[1,2]. Controlled ovarian hyperstimulation (COH) with gonadotropins is started in the early follicular phase, around day 2 of the cycle. This procedure is believed to rescue the cohort of follicles that would have degenerated under normal conditions without hyperstimulation. Response to COH is monitored over time with (1) serial measurements of the size of the growing follicles with ultrasound examination, and (2) serial determinations of the concentration of estradiol in the serum, which increases proportionally with follicular maturation. The gonadotropin injections are continued until the follicles reach the appropriate size and the estradiol level confirms follicular maturation. At this point the gonadotropin therapy is stopped and the patient receives an injection of human chorionic gonadotropin (hCG), a surrogate LH surge needed for the final maturation of the oocytes, and the stimulus for ovulation. COH is used to increase the number of oocytes available at the time of oocyte retrieval for IVF. Retrievals for IVF are planned 34–36 h following the surrogate LH surge.

EVOLUTION OF CONTROLLED OVARIAN HYPERSTIMULATION IN ART

The first successful IVF attempt, and most treatment cycles for a while thereafter, was conducted in spontaneous menstrual cycles. Nevertheless, realization that availability of a crop of mature oocytes markedly increased chances of success in this therapy prompted most centers to adopt some form of COH. At the outset CC alone or in combination with human menopausal gonadotropins (hMG) was used but eventually exogenous gonadotropins emerged as the sole stimulatory drug for COH. Initially, hMG were used for COH, but poor reproductive outcome was ascribed to high levels of circulating LH associated with hMG therapy. Lunenfeld hypothesized that 'a high concentration of LH through the follicular phase allows the developing oocyte to mature prematurely, producing at ovulation an oocyte that is physiologically aged. Such oocytes may have a decreased capacity to fertilize; if they fertilize, they are unlikely to implant; and if they implant, their survival rate is decreased, resulting in early abortion'[3]. This LH hypothesis changed worldwide thinking and the direction of stimulation protocols in ART. 'Pure' FSH preparations with reduced LH content such as purified FSH (PoFSH), highly purified FSH (PoFSH-HP) and recombinant FSH (rFSH) were introduced.

Soon after the introduction of COH to ART it became evident that in the course of ovulation induction the midcycle LH surge could be unexpectedly triggered by rising ovarian steroid levels; as a result premature ovulation or follicle luteinization could occur and cause cycle cancellations in up to 30% of cases. In order to identify patients on the verge of spontaneous ovulation, close monitoring of LH and progesterone plasma levels with multiple daily determination used to be the rule in ART. Thus, in the mid-1980s GnRH agonist supplementation was tested in ovulation induction[4] and then successfully applied in most cycles. The pituitary response to GnRH agonist administration is biphasic: an initial agonistic phase during which circulating levels of LH and FSH rise followed by pituitary desensitization and downregulation with shutdown of gonadotropin secretion. The major advantage of the use of GnRH agonist is the complete elimination of the preovulatory LH surge; in addition to preventing cycle cancellations due to premature ovulation, the use of GnRH agonist has permitted a dramatic reduction in endocrine monitoring and its related costs. Very recently, GnRH antagonists were introduced in stimulation protocols for ART. GnRH antagonists offer a new and powerful tool for ovarian stimulation. Administered at pharmacologic dosages,

they allow an immediate yet completely reversible competitive blockade of the GnRH receptors at the level of the pituitary. The use of GnRH analogs in conjunction with gonadotropins for COH has afforded better control of the cycle, and has provided the versatility to tailor specific COH protocols to specific groups of patients[5–8].

WHAT IS THE IDEAL STIMULATION PROTOCOL?

The ideal gonadotropin stimulation regimen should result in mature oocytes capable of developing into embryos with high implantation potential, while being both patient-friendly and safe. A meta-analysis of clinical trials comparing urinary (u)-hMG and u-PoFSH in IVF showed better results with PoFSH than hMG[9]. A larger prospective randomized trial comparing hMG and PoFSH in GnRH agonist downregulated cycles concluded that the LH contained in hMG may be detrimental to the outcome of IVF, and that rFSH should replace hMG in COH for IVF[10]. On the other hand a large prospective, randomized trial showed improved IVF outcome with hMG as compared to PoFSH-HP[11]. Fleming and co-workers reported that a combination of GnRH agonist and PoFSH-HP results in such profound suppression of mid-follicular endogenous LH secretion (< 0.5 IU/l) that the estradiol biosynthesis is significantly decreased and oocyte maturation and fertilization capacity may be compromised[12]. Subsequently Westergaard and co-workers, in a retrospective IVF study in GnRH agonist and rFSH cycles, reported a significantly increased risk of early pregnancy loss in women with a mid-follicular serum concentration of LH below 0.5 IU/l[13].

IS LUTEINIZING HORMONE REQUIRED AT ALL IN THE FOLLICULAR PHASE IN CONTROLLED OVARIAN HYPERSTIMULATION CYCLES FOR ART?

Recent scientific and clinical evidence indicates that LH has a key role in the stimulation of multiple folliculogenesis. There are four different scientific propositions that we will discuss in detail to emphasize the necessity of LH in ovulation induction for ART:

(1) The fertilization potential of the oocyte appears to be compromised in a reduced LH environment.

(2) Patients undergoing ovulation induction benefit from the administration of exogenous LH (hMG), since FSH only regimens are suboptimal.

(3) hMG and FSH are comparable in effectiveness in the treatment of patients with polycystic ovary syndrome (PCOS) and in ART.

(4) LH accelerates FSH stimulation of folliculogenesis, reduces monitoring time and gonadotropin dose, and optimizes COH, thereby markedly reducing the cost.

The fertilization potential of the oocyte appears to be compromised in a reduced luteinizing hormone environment

It is becoming increasingly apparent that hormonal events before the mid-cycle LH peak, specifically the levels of LH and FSH during the early follicular phase, may significantly influence the development capacity of the oocyte. Earlier studies have shown a detrimental effect of high levels of LH in the follicular phase on oocyte maturation, fertilization and conception[14–16]. However, these studies were referring to concentrations of LH > 10 IU/l. With the advent of GnRH analogs, pituitary suppression is now routine in most clinics and those high levels of LH are unlikely to be achieved.

Maturity of the oocyte is a complex issue and the majority of studies use morphological parameters for assessment, which are subjective. Tornell and co-workers[17] showed that while both rFSH and rLH were independently capable of causing germinal vesicle breakdown in isolated preovulatory rat follicles, a combination of these gonadotropins in submaximal concentrations resulted in an additive effect on oocyte meiosis. In a randomized prospective study in GnRH-agonist-suppressed IVF patients, Gordon and co-workers compared rFSH treatment with urinary gonadotropin preparations containing 1, 25 and 75 IU of LH per ampoule, keeping the FSH dose constant[18]. A lower percentage of mature oocytes and a higher percentage of immature oocytes were noted with rFSH treatment compared to the other three groups[18].

Westergaard and co-workers[13] showed the mean fertilization rate to be significantly higher with hMG (56%) compared to PoFSH (50%) in their randomized trial of downregulated normogonadotropic women. Another study compared PoFSH with hMG in their oocyte donation program and showed higher fertilization with hMG (36 versus 48%, respectively)[19]. In a woman with isolated congenital gonadotropin deficiency, oocytes obtained in an LH-depleted environment showed considerably lower fertilization rates with rFSH or PoFSH (27–28%) than with hMG (93%)[20]. Fleming and co-workers[12] have identified a subgroup of patients on GnRH agonists and PoFSH treatment who show profound suppression of endogenous LH. These cases, with mean serum follicular phase LH levels of less than 0.5 IU/l, showed significantly lower estradiol levels in the circulation and lower oocyte fertilization rates[21]. Concern has also been raised about a higher incidence of failed fertilization in IVF cycles with low LH. Westergaard and co-workers, in their randomized trial comparing PoFSH-HP with hMG, reported a significantly higher percentage of complete failure of fertilization with PoFSH-HP (16 versus 6%, respectively)[11].

Results with regard to the role of LH in embryo development vary. A few clinical studies examined the effects of different urinary gonadotropin preparations on embryo cleavage rates following IVF, and reported no significant differences[22–25]. In contrast, others reported better cleavage rates[26] and more transferable embryos[27] with hMG treatment compared to rFSH. Embryo viability and preimplantation development were also shown to be optimal in GnRH-antagonist-treated macaques when folliculogenesis occurred in the presence of exogenous LH[28]. A significant difference in embryo cryosurvival (56% after rFSH versus 78% after rFSH + rLH) and developmental rate to the hatched blastocyst stage (12 versus 10 days, respectively) was noted. The rescue of the corpus luteum also occurred later (16 versus 12–13 days, respectively). Fleming and co-workers[21] showed that while the degree of LH suppression in the follicular phase influences the number of embryos available, due to a lower oocyte yield and fertilization rates, no impact was noted upon the fertilized embryos to undergo cleavage and expanded blastocyst formation.

Rekha and co-workers[29] showed no correlation of circulating LH to implantation and pregnancy rates in GnRH-agonist-suppressed IVF patients. In contrast, Weston and co-workers[28] showed that embryos derived from rFSH alone without any added LH undergo delayed development *in vitro* leading to a decreased frequency of implantation in macaques. While two pregnancies resulted from three embryo transfer attempts with rFSH plus rLH treatment, one twin pregnancy resulted from eight embryo transfer attempts in the rFSH-only group. In addition, while pregnancies have been reported in hypogonadal women treated with rFSH and rLH[30,31] no pregnancies to date have been documented with rFSH alone.

Patients undergoing ovulation induction benefit from the administration of exogenous luteinizing hormone (human menopausal gonadotropins)

Currently, many protocols for gonadotropin administration have eliminated LH bioactivity. However, several recent primate studies have added to our understanding of the ovarian actions of androgens, estrogens and gonadotropins, allowing a less speculative approach[32,33]. Indeed, if androgens do not promote follicular atresia, if the presence of LH directly influences oocyte quality or embryo growth, and if excessive LH concentrations are detrimental only to PCOS granulosa cells, the addition of LH to ovarian stimulation protocols might have beneficial effects.

Thecal cells express LH receptors constitutionally, and LH stimulates androgen substrate production by the thecal cells from fetal life until after the menopause[34,35]. Granulosa cells acquire their own LH receptors in the mid- to late follicular phase under the influence of FSH[36]. At that stage, FSH and LH act in synergy to support follicular development, increase granulosa cell aromatase activity and inhibin production, and prepare the follicle and its oocyte for the mid-cycle LH surge.

Androgens serve as a substrate for granulosa cell estrogen production through aromatization, but may also have an ovarian autocrine/paracrine role. Androgens were thought to be involved in follicular atresia and granulosa cell apoptosis because of their actions in the rodent ovary[37]. However, recent evidence suggests that they may act as growth factors in primate preantral and small antral follicles[33]. Locally produced androgens are thought to enhance the FSH responsiveness of immature follicles, thereby promoting the selection of the follicles that will achieve full preovulatory maturity under FSH stimulation[38].

Several clinical situations, where LH is either absent or completely inactive, provide important clues to our understanding of the roles of LH in follicular development. In Kallmann's syndrome, women are profoundly hypogonadotropic, and follicle development may be induced by the exogenous administration of gonadotropins. Treatment of these patients with purified or recombinant FSH alone allows multiple follicle development, but produces inadequate estradiol concentrations. In some studies, fewer preovulatory follicles developed compared with patients treated with a combination of FSH and LH[39], while others have observed no differences[40]. Administration of FSH without LH to hypogonadotropic hypogonadal patients results in lower serum and follicular fluid estradiol concentrations, normal inhibin concentrations, decreased endometrial thickness, reduced occurrence of ovulation, reduced oocyte fertilization rates and lower embryo cryosurvival rates, compared with hMG treatment[31,39–41]. More importantly, no pregnancies were observed in these women when they received FSH alone for ovarian stimulation, despite estradiol replacement[30,40]. In the same way, women with primary amenorrhea and infertility attributable to a homozygous inactivating mutation in the LH receptor gene exhibit low concentrations of estradiol, although ovarian histology reveals all stages of follicular development up to large antral follicles[42,43].

According to Filicori and co-workers,[44] low-dose hCG therapy can improve sensitivity to exogenous FSH in patients with secondary amenorrhea. They conducted a study to assess the effect of supplementing an ovulation induction regimen of PoFSH-HP with LH activity in the form of low-dose hCG therapy. A woman with weight-related secondary hypogonadotropic amenorrhea was treated first with PoFSH-HP alone and then received PoFSH-HP in combination with low-dose hCG therapy (50 IU/day). The concomitant administration of low-dose hCG and PoFSH-HP markedly reduced the duration of treatment and the dose of PoFSH-HP, and resulted in a quadruplet pregnancy in a patient in whom several previous ovulation induction procedures had been unsuccessful. The authors concluded that supplementation of an ovulation induction regimen with an agent that has LH activity can enhance FSH-induced folliculogenesis and markedly reduce costs in women with hypogonadotropic hypogonadism[44].

GnRH antagonist treatment at high doses can mimic naturally occurring LH deficiencies[45]. In the European Ganirelix Multi-Centre Dose-Finding Study[45], the use of GnRH antagonist in the highest dose group (2 mg daily) during ovarian stimulation cycles with rFSH resulted in profound gonadotropin suppression, decreased estradiol concentrations, and shortening of the follicular phase. These observations were accompanied by low implantation (1.5%) and pregnancy rates (3.8%), and higher early miscarriage rates (13%)[45]. Interestingly, these disparate outcomes occurred even though the six different dose groups exhibited a similar number of antral follicles, oocytes recovered, fertilization rates, and numbers of transferable embryos[45].

The aim of a recent study by De Placido and co-workers[46] was to investigate the effects of adding hMG during controlled ovarian stimulation in normo-ovulatory, normogonadotropic patients showing an initial suboptimal response to a standardized long protocol therapy with rFSH (300 IU/day). The data suggested that LH supplementation improves the ovarian outcome in patients characterized by an inadequate initial response to rFSH therapy in a long protocol[46].

Based on these observations, it can be postulated that: (1) follicular development, at least until the preovulatory stage, can occur in the absence of high estradiol concentrations, and in the absence of significant follicular phase LH bioactivity; and (2) profound LH suppression, and the consequently decreased steroid concentrations, may interfere with optimal oocyte maturation and/or endometrial development. Patients undergoing ovulation induction will benefit from the administration of exogenous LH (hMG), since FSH-only regimens are suboptimal.

Human menopausal gonadotropins and follicle stimulating hormone are comparable in effectiveness in the treatment of patients with polycystic ovary syndrome and in ART

Chronically administered hMG or FSH in PCOS leads to a progressive decline of endogenous LH with reduction of peripheral LH concentrations. In a study by Gordon and co-workers[47], the 75 IU of LH present in an ampoule of hMG administered daily from the early follicular phase led to progressive decrease of endogenous LH levels in the preovulatory period. Stimulation by hMG in PCOS subjects for ovulation induction was as effective as PoFSH-HP therapy in these subjects. Although the excessive LH secretion that is present in some disorders is detrimental to reproductive function, this is not applicable to ovulation induction with hMG because this menotropin does not increase daily plasma LH levels[48].

A randomized clinical trial was performed comparing rFSH ($n = 54$) with hMG (Humegon, Organon, The Netherlands, $n = 35$) in infertile women undergoing IVF without the use of a GnRH agonist[49]. In 15 subjects (rFSH: $n = 9$, 16.7%; hMG: $n = 6$, 17.1%) LH concentrations higher than 10 IU/l were seen during stimulation. In two of them, both from the rFSH group, ongoing pregnancies were achieved. The results indicate that rFSH is at least as efficacious as hMG and that acceptable pregnancy rates can be achieved without the use of a GnRH agonist[49].

Jacob and co-workers[50] compared the clinical results during the cross-over from menopausal urinary products (hMG) with rFSH in order to determine whether the manufacturer's recommendation for equivalence of ampoule to ampoule (50 IU rFSH:75 IU hMG) would prove clinically correct. A total of 353 consecutive IVF/intracytoplasmic sperm injection (ICSI) treatment cycles were studied. Greater efficacy was seen in the hMG group in terms of days of stimulation required, need to increase dosage, cycle discontinuation, number of follicles punctured, the numbers of oocytes retrieved and their quality. The hormonal response to stimulation assessed by estradiol concentrations on days 5, 8 and day of hCG was significantly lower in the rFSH group. Theratio of estradiol per follicle to per oocyte was significantly lower in the rFSH group. There was a highly significant increase in cost with rFSH therapy. Clinical pregnancy rates were 14% per cycle with rFSH and 20% per cycle with hMG[50].

Luteinizing hormone accelerates follicle stimulating hormone stimulation of folliculogenesis, reduces monitoring time and gonadotropin dose, and optimizes controlled ovarian hyperstimulation

Studies in humans and non-human primates have shown that the follicular development potential is influenced by

LH. The presence of LH appears to facilitate a shortening of the follicular phase and/or a reduction in the amount of gonadotropin ampoules required. Zelinski-Wooten and co-workers[51] treated macaques with GnRH antagonist (Antide, Organon, The Netherlands) for a period of 90 days to achieve suppression of endogenous LH comparable to hypophysectomized animals. They showed that an average of 12 days was needed to achieve follicular growth with rFSH compared with 9 days with rFSH plus rLH treatment. No difference was noted in the number of follicles recruited or the oocytes collected. Shoham and co-workers[39] compared hMG treatment with PoFSH in subsequent cycles in patients with isolated hypogonadotropic hypogonadism and found that the former required significantly fewer ampoules of gonadotropin. Balasch and co-workers[40] compared rFSH, hMG and PoFSH plus estradiol valerate in a patient with isolated congenital gonadotropin deficiency and showed that the number of gonadotropin ampoules used was lower in the hMG cycle. Studies in immature female rats rendered hypogonadotropic by hypophysectomy[52] or LH-releasing hormone antagonist[53] have also shown that while FSH alone can induce multiple follicular development, the addition of a small amount of hCG increased the percentage of healthy follicles and reduction in atresia in various follicle size classes. The European Recombinant Human LH Study Group[54] investigated the efficacy of rLH for supporting rFSH-induced follicular development in hypogonadotropic hypogonadal women. Supplementation with 75 IU rLH resulted in optimal follicular development and luteinization in the majority of patients.

To investigate the role of LH activity during COH, Filicori and co-workers[55] supplemented PoFSH-HP with low-dose hCG in GnRH-agonist-suppressed women. Twenty normo-ovulatory women were pretreated with a GnRH agonist and after 2 weeks were randomly assigned to receive PoFSH-HP (150 IU/day) alone (group A) or combined with hCG (50 IU/day; group B). The PoFSH-HP dose was increased after 14 days only in cases of inadequate response. Treatment was monitored with pelvic ultrasound and daily hormone determinations. None of the ten patients in group B and eight of ten in group A required more than 14 days of treatment and increments of the FSH dose. Folliculogenesis and 17β-estradiol secretion progressed more rapidly and evenly in group B. Although preovulatory follicle number and 17β-estradiol concentrations were comparable, patients in group B required a shorter stimulation time (12.50.6 versus 17.30.7 days in group A; $p < 0.0001$) and a lower PoFSH-HP dose (1725 versus 2670 IU in group A; $p < 0.0001$). The authors concluded that LH activity promotes folliculogenesis in synergy with FSH in the mid- to late follicular phase and that low-dose hCG co-administration optimizes COH by: (1) enhancing FSH action; (2) accelerating ovarian follicle development; (3) shortening COH duration; (4) lowering PoFSH-HP requirements; and (5) reducing COH cost[55]. Thus, moderate LH activity in the follicular phase plays a positive physiological and clinical role in folliculogenesis and ovulation induction.

INDUCTION PROTOCOLS FOR THE NEW MILLENIUM USING RECOMBINANT LUTEINIZING HORMONE

It is likely that treatment with gonadotropins derived from human urine for induction of ovulation and ovarian stimulation in IVF will soon be replaced by the use of wholly biosynthetic preparations. The availability of a large amount of a highly pure rLH preparation with potential uses in humans allows for the first time a precise assessment of the potential use of LH as a pharmaceutical agent in reproduction and development.

The first pregnancies with rFSH used to stimulate ovulation in women with adequate endogenous LH were reported in 1992[56]. The aim of giving FSH is to stimulate preovulatory maturation. Selective modulation of the several follicles that start maturing is not possible. By contrast, new understanding of the actions of rLH suggests that the dose of rLH could be individually tailored during follicular stimulation (by rFSH) to promote monoovulation by inducing atresia of secondary follicles[57]. Furthermore, a large enough dose by single injection could mimic the LH surge more closely than by use of hCG, which lasts much longer[58]. These applications are now being explored clinically, as is the need to define the minimum LH required for steroidogenesis to synergise with FSH to induce ovulation in gonadotropin-deficient women. Only small amounts are needed and many such women produce enough LH to act with exogenous FSH to stimulate full functional maturation of follicles[59]. Women with profound LH deficiency develop follicles in response to administered FSH but estrogen production by the proliferating granulosa cells is lacking[39].

Rationale for use of recombinant luteinizing hormone as an ovulation trigger

In natural menstrual cycles, final follicular maturation, ovulation and corpus luteum formation are induced by the mid-cycle LH surge. Infertility therapy includes stimulation of follicular development by administration of FSH followed by an injection of urine-derived hCG to mimic the endogenous LH surge. Urinary hCG is used because it shares biological properties with LH and is easier to obtain than LH. But hCG has longer pharmacodynamic activity than LH, which could be disadvantageous in patients at high risk of ovarian hyperstimulation syndrome (OHSS)[60–61]. Recombinant DNA technology has made it possible to develop a form of LH that is structurally and functionally very similar to the natural hormone. Recombinant LH in appropriate high dosage can mimic the preovulatory surge required to induce final follicular and oocyte maturation and a recent report showed that a single administration of rLH can induce adequate final follicular maturation and corpus luteum formation, and lead to a viable pregnancy without clinical and ultrasound signs of OHSS in a patient with a high response to FSH[62].

Rationale for use of recombinant luteinizing hormone in ovulation induction and superovulation

Since the introduction of gonadotropin therapy, multiple pregnancy has been an almost inevitable complication as a result of ovulation induction with or without various procedures of assisted reproduction. The higher order of maternal and neonatal complications that resulted iatrogenically from ovulation induction should therefore raise the awareness of the treating physicians to take any available steps to avoid multiple pregnancy under almost all circumstances. Careful monitoring during treatment by ultrasound, assessment of serum estradiol, and the use of low-dose stimulation protocols, have reduced these risks but have not been able to prevent them in all patients. These problems are directly related to the difficulty of obtaining the growth of a single dominant follicle leading to non-physiologic multifollicular development.

It is well known that follicular development is FSH-dependent. It was Brown[63] in 1978 who was the first to raise the theory of the FSH 'threshold', which indicated the level of FSH required to prevent atresia and promote further growth of a follicle. The FSH threshold level varies with time, and at a given time-point the follicles that are currently in a growth phase have different FSH threshold levels.

A second parameter contributing to the phenomenon of follicle dominance and mono-ovulation is LH. Indeed, although some LH is essential for estrogen synthesis during folliculogenesis, there is evidence indicating that excessive exposure to LH will trigger follicular atresia[44]. Depending on the stage of development, follicles exposed to inappropriately high concentrations of LH may enter atresia or be prematurely luteinized, and oocyte development may be compromised. Thus developing follicles appear to have finite requirements for stimulation by LH, beyond which normal development ceases. Whereas each follicle is believed to have a 'threshold' beyond which it must be stimulated by FSH to initiate preovulatory development[63], it may also have a 'ceiling' within which it should be stimulated by LH if termination of preovulatory development is to be avoided[57]. This basic knowledge is likely to have practical implications for the development of improved ovarian stimulation regimens where the desired aims are mono-ovulation and conception *in vivo*, e.g., in group I-type infertility patients as defined by the World Health Organization (WHO). Once a sufficiently advanced (i.e., LH-responsive) stage of follicular development has been achieved in response to treatment with FSH, there are grounds for reducing or completely withdrawing FSH and maintaining tonic (subceiling) stimulation of the dominant follicle with exogenous LH. Such a low-dose LH 'coast', for no more than 1 or 2 days, could have the dual advantage of promoting the terminal maturation of a single preovulatory follicle and simultaneously arresting the development of multiple, less mature follicles that would otherwise occur in response to treatment with FSH. A multicenter clinical trial to evaluate these possibilities is now in progress.

The rationale for adding rLH to rFSH stimulation protocols takes advantage of the FSH threshold and LH ceiling theories, in which by simultaneously reducing FSH levels and enhancing LH levels it might be possible to induce atresia of secondary follicles, allowing further promotion of the dominance of one or a few leading follicles.

In a recent study[64], a retrospective analysis of the role of serum LH on IVF and embryo transfer outcome in patients treated with a GnRH agonist and stimulated with rFSH or urinary PoFSH alone (*n* = 323) was conducted to identify a putative IVF patient population that might benefit from rLH administration, i.e., patients with very low LH concentrations. However, a small population (6% of patients characterized by a high ovarian response) displayed a very low estradiol/oocyte ratio (0–70 pg/ml per oocyte) and a significantly lower pregnancy rate (5.3 versus 31.3% in patients with a ratio of 70–140 pg/ml per oocyte). Because a low estradiol/oocyte ratio may indicate insufficient exposure to LH (because of deep downregulation), which is not measured by a single serum-level assay, one option to be investigated for improving the outcome in this small population could be to enhance exposure to LH during the follicular phase[64].

CLINICAL APPLICATIONS

Induction of final follicular maturation and early luteinization in women undergoing superovulation with rFSH using rLH

The administration of GnRH analogs instead of hCG, to produce an endogenous LH surge, has been proposed to prevent OHSS in women at risk[65]. The use of rLH may provide another physiological route to stimulate follicular rupture and final oocyte maturation, while at the same time helping to prevent OHSS in women at risk; this notion is based on its short halflife which is similar to that of human LH[66,67].

A future potential use of rLH will be its replacement of hCG as the ovulatory trigger, as has been demonstrated in non-human primates[68], and the rLH may mimic the physiological endogenous LH surge more closely than hCG[69]. This may help prevent OHSS[60],

possibly as a result of very different halflives of natural and rLH (measured in hours)[70], compared with hCG (measured in days). Also, in a study by Romeu and co-workers[69], the implantation rate was significantly higher with rLH versus hCG possibly because of the better embryo quality produced by rLH.

A recent clinical study[58] tested four doses of rLH: 5000, 15 000, 30 000 and 15 000 + 10 000 IU. Two hundred and fifty-nine patients were enrolled, of whom 129 received rLH and 121 were administered 5000 IU u-hCG. All 250 patients were included in the safety and efficacy analysis. The primary efficacy endpoint was to compare the number of oocytes recovered 34–38 h after u-hCG/rLH administration. The results revealed that 5000, 15 000, 30 000 and 15 000 + 10 000 IU of rLH are equivalent to 5000 IU u-hCG for inducing final follicular maturation, allowing oocyte recovery.

In the safety assessment, there were no significant differences in the incidence or type of adverse events between the u-hCG and rLH treatments[58]. Regarding the evaluation of OHSS, a clear-dose response relationship was observed, a single dose of 5000, 15 000 or 30 000 IU appeared safer than 15 000 + 10 000 IU of rFSH and 5000 IU of u-hCG. In the 5000, 15 000 and 30 000 IU rLH groups the size of the ovaries, the proportion of patients developing ascites and the total renin levels were statistically significantly lower than in patients treated with 5000 IU u-hCG. The safety profile of the 15 000 + 10 000 IU rLH group is clearly poorer, and closer to that of u-hCG.

The study demonstrated the clinical efficacy of rLH in inducing final follicular maturation and early luteinization in IVF patients[58]. These findings support the hypothesis that the use of rLH as a shorter-lasting and therefore more physiologic surrogate surge would be beneficial in terms of reducing the risk of OHSS. A single dose between 15 000 and 30 000 IU provides the best efficacy/safety ratio[58].

In another study[64] where ovulatory patients were pretreated with GnRH agonist (*n* = 250) and stimulated with rFSH, doses ranging between 5000 and 30 000 IU rLH were used to trigger final follicular maturation and luteinization prior to IVF and embryo transfer. All doses tested were shown to be as effective as 5000 IU hCG to trigger cumulus and oocyte maturation and early luteinization. Mid-luteal phase ovarian volume increase, serum renin concentration and fluid accumulation in the abdominal cavity were positively correlated with the dose of rLH used to trigger final follicular maturation[64].

Induction of mono-folliculogenesis using recombinant luteinizing hormone

A multicenter, randomized, double-blind, placebo-controlled study[71] evaluated the safety of two doses of rLH administered alone during the late follicular phase of anovulatory patients (WHO group II) who over-responded to treatment with rFSH. In this study 17 patients were identified in the population undergoing routine ovulation induction with FSH. The distinguishing eligibility criteria were the presence of four or more follicles ≥ 8 mm and ≤ 13 mm in diameter, no larger follicles and endometrium ≥ 8 mm in thickness. At this point according to the randomization, five patients received a daily injection of placebo, four received a daily injection of 225 IU of rLH and eight received 450 IU of rLH only until the final follicular response was judged to be adequate (at least one follicle > 18 mm and fewer than three > 11 mm), for triggering ovulation by the administration of hCG. No luteal support was given.

The results showed that at baseline when the patients were randomized for the different modes of treatment the median number of follicles of 8–13 mm were: for placebo, 12 (range 8–22); for 225 IU/day of rLH, 8.50 (range 4–18); and for 450 IU/day of rLH, 9 (range 4–17). At the time of hCG administration or following 7 days of treatment, the median number of follicles > 14 mm in diameter were: for placebo, 2 (range 1–10); for 225 IU/day of rLH, 1 (range 0–3); and for 450 IU/day of rLH, 1 (range 0–4). This pilot study clearly showed that by increasing serum LH concentration during the late follicular phase it might be possible to minimize the number of large follicles.

Looking at the concepts of the 'FSH threshold' and the 'LH ceiling' for minimizing the number of pre-ovulatory follicles from another point of view, a second double-blind, placebo-controlled, randomized, parallel group study comparing rFSH alone, rFSH and rLH, and rLH alone, administered during the late follicular phase to patients with hypogonadotropic hypogonadism, was

performed[71]. In this study 24 patients received initial treatment with 112.5 IU rFSH/day and 225 IU rLH/day. When a follicle reached a mean diameter between 10 and 15 mm, 20 patients were randomly allocated to treatments with either: rFSH and rLH (eight patients); rFSH/stop rLH (six patients); or rLH/stop rFSH (six patients). The three options are believed to be relevant for dissecting out the respective roles of stopping FSH and maintaining LH in promoting the dominance of the minimum number of follicles. The median number of follicles of more than 10 mm at the time when the patients were randomized to the different study groups were: rFSH and placebo, 1.80 (range 1–4); rLH and placebo, 1.50 (range 1–2); and rFSH and rLH, 2 (range 1–4). Following 7 days of study drug administration or on the day of hCG the following results in term of number of follicles of more than 14 mm in diameter were obtained: rFSH and placebo, 2.5 (range 1–4); rLH and placebo, 0 (range 0–2); and rFSH and rLH, 1 (range 0–8).

This study[71] gives initial support for the 'LH ceiling' hypothesis: increasing serum LH concentrations in the mid/late follicular phase may facilitate monofollicular development without premature luteinization. However, to further test these preliminary findings, larger clinical studies are required to examine this phenomenon and to optimize the protocol for therapeutic use. Also, this study raises the hope that by separately administering both gonadotropins we will be able to control the ovarian response in such a way that only one dominant follicle will reach the final stage of ovulation, thus minimizing the complications of this treatment.

Induction of ovulation in women with hypogonadotropic hypogonadism

The majority of women with hypogonadotropic hypogonadism (WHO group I anovulation)[72] do not have the threshold level of endogenous LH required to achieve optimal follicular development and steroidogenesis during therapy with FSH alone. Among these women, urinary and rFSH have been shown to stimulate considerably lower estradiol levels than those obtained with an hMG preparation containing both FSH and LH[39,59,73]. It also appears that in this population, the follicles stimulated by FSH alone do not consistently rupture after hCG administration, they luteinize poorly[39,59,73] and oocytes may have a lower fertilization rate[39,40,59,73]. An exogenous supply of LH is required if an adequate follicular response is to be achieved.

The successful induction of ovulation in women with hypogonadotropic hypogonadism and intact pituitary function has been achieved with pulsatile GnRH therapy or hMG therapy. Pulsatile GnRH therapy has been the treatment of choice because it restores the pulsatile release of both gonadotropins from the pituitary. This results in predominantly unifollicular cycles and satisfactory pregnancy rates. The treatment is associated with low rates of multiple pregnancy and is not complicated by OHSS[74,75].

However, for patients who do not respond adequately to pulsatile GnRH or those with pituitary disease, hMG therapy given by once-daily injection has been the only alternative treatment for ovulation induction. Urine-derived hMG preparations contain a fixed dose of FSH and LH. Until recently, hMG was the only source of exogenous LH for women with hypogonadotropic hypogonadism. However, rLH is now available for clinical use, providing a new treatment option. By titration of both gonadotropin doses for individual patients, the potential complications seen with hMG therapy (i.e., multiple pregnancy and OHSS) may be reduced with rLH and rFSH therapy. Because the two preparations are given separately, the dose of each gonadotrophin can be tailored to the individual's requirements and achieve the goal of unifollicular cycles. Recombinant LH has indeed been shown to have pharmacokinetic characteristics comparable to those of human pituitary LH[70], and to promote estradiol secretion by FSH-stimulated follicles in monkeys treated with a GnRH antagonist[51]. Moreover, a few case reports have suggested that rLH is effective and safe in humans[30,31,76,77].

Compared to hMG treatment for hypogonadotropic hypogonadism, the use of rLH offers a number of differences. It is the first preparation of LH devoid of FSH activity that is suitable and available for extensive clinical use. It has a high specific activity suitable for subcutaneous injection, allowing self-administration by the patient. By comparison, hMG preparations are given intramuscularly and contain a large proportion (about 95%) of non-specific co-purified urinary proteins, which can cause hypersensitivity reactions[78]. Furthermore, only a once-daily injection of rFSH and rLH is required in comparison with GnRH, which has to be administered every 60–120 min.

Preliminary studies in the treatment of hypogonadotropic hypogonadal women indicate that it promotes normal FSH-induced estradiol and follicular development[30,79]. In WHO group I anovulatory patients ($n = 38$), when co-administered with a fixed dose of rFSH (150 IU), rLH promoted in a dose-related manner estradiol secretion and luteinization of the follicles[64], and enhanced ovarian sensitivity to FSH. Although individual requirements for rLH varied, a daily dose of 75 IU rLH was effective in the majority of patients for obtaining a preovulatory serum estradiol concentration > 400 pmol/l and a mid-luteal phase progesterone concentration > 25 nmol/l after hCG administration. The mean endometrial thickness was 8.8 mm in this treatment group. This study demonstrated that rLH is safe at doses up to 225 IU/day and is not immunogenic[64]. The use of this preparation may also help define more closely the role of LH in assisted reproduction programs, particularly as there is evidence to suggest that too much LH may be associated with poor therapeutic outcome[16,57,80] and that LH of a different quality is present in patients with PCOS[81]. Furthermore, although rFSH treatment alone during the pre-ovulatory period has been demonstrated to be adequate to promote multiple follicular growth in macaque monkeys[51], concomitant treatment with rLH improves embryo viability[28].

CONCLUSION

With the advent of recombinant FSH and LH, free from contamination with each other, it is possible to dissect out the individual contributions of FSH and LH to the regulation of ovarian function and to define optimal gonadotropin usage in COH. Basic and clinical research evidence is accumulating to propose that the future use of these new pharmaceuticals will allow systematic improvements in the treatment regimes used to achieve COH.

References

1. Corson GH, Kemmann E. The role of superovulation with menotropins in ovulatory infertility: a review. *Fertil Steril* 1991;55:468–70

2. Ginsburg J, Hardiman P. Ovulation induction with human menopausal gonadotropins – a changing scene. *Gynecol Endocrinol* 1991;5:57–62

3. Lunenfeld B, Lunenfeld E. Gonadotropic preparations: lessons learnt. *Fertil Steril* 1997;67:812–14

4. Dodson WC, Hughes CL, Whitesides DB, *et al*. The effect of leuprolide acetate on ovulation induction with human menopausal gonadotropins in polycystic ovary syndrome. *J Clin Endocrinol Metab* 1987;65:95–100

5. Meldrum D. GnRH agonists as adjuncts for *in vitro* fertilization. *Obstet Gynecol Surv* 1989;44:314–17

6. Meldrum DR, Wisot A, Hamilton F, *et al*. Routine pituitary suppression with leuprolide before ovarian stimulation for oocyte retrieval. *Fertil Steril* 1989;51:455–9

7. Hughes FG, Fedorkow DM, Daya S, *et al*. The routine use of gonadotropin releasing hormone agonists prior to *in vitro* fertilization and gamete intrafallopian transfer: a meta-analysis of randomized controlled trials. *Fertil Steril* 1992;58:888–90

8. Cedars MI, Surey E, Hamilton E, *et al*. Leuprolide acetate lowers circulating, bioactive luteinizing hormone and testosterone concentrations during ovarian stimulation for oocyte retrieval. *Fertil Steril* 1990;53:627–31

9. Daya S, Gunby J, Hughes EG, *et al*. Randomized controlled trial of follicle stimulating hormone versus human menopausal gonadotrophin in *in-vitro* fertilization. *Hum Reprod* 1995;10:1392–6

10. Daya S. In hMG versus FSH: is there any difference? Filicori M, Flamigni C, eds. *Ovulation Induction Update 1988*. London: Parthenon Publishing, 1998:183–92

11. Westergaard LG, Erb K, Laursen S, *et al*. The effect of human menopausal gonadotrophin and highly purified, urine-derived follicle stimulating hormone on the outcome of *in-vitro* fertilization in down-regulated normogonadotropic women. *Hum Reprod* 1996;11:1209–13

12. Fleming R, Chung CC, Yates RWS, *et al*. Purified urinary follicle stimulating hormone induces different hormone profiles compared with menotropins, dependent upon the route of administration and endogenous luteinizing hormone activity. *Hum Reprod* 1996;11:1854–8

13. Westergaard LG. Presented at the *11th World Congress on IVF*, Sydney, November 1999;279–82

14. Stanger JD, Yovich JL. Reduced *in-vitro* fertilization of human oocytes from patients with raised basal luteinising hormone levels during the follicular phase. *Br J Obstet Gynaecol* 1985;92:385–93

15. Homburg R, Armar NA, Eshel A, *et al*. Influence of serum luteinising hormone concentrations on ovulation, conception and early pregnancy loss in polycystic ovary syndrome. *Br Med J* 1988;297:1024–6

16. Regan L, Owen EJ, Jacobs HS. Hypersecretion of luteinising hormone, infertility and miscarriage. *Lancet* 1990;336:1141–4

17. Tornell J, Bergh C, Selleskog U, *et al*. Effect of recombinant human gonadotropins on oocyte meiosis and steroidogenesis in isolated pre-ovulatory rat follicles. *Hum Reprod* 1955;10:1619–22

18. Gordon UD, Gordon AC, Bonnar J, *et al*. Effect of differing doses of LH on oocyte maturation and fertilization: abstracts of the 13th annual meeting of the ESHRE. *Hum Reprod* 1997;53

19. Soderstrom-Anttila V, Foudila T, Hovatta O. A randomized comparative study of highly purified follicle stimulating hormone and human menopausal gonadotrophin for ovarian hyperstimulation in an oocyte donation programme. *Hum Reprod* 1996;11:1864–70

20. Balasch J, Vidal E, Penarrubia J. Suppression of LH during ovarian stimulation: analysing threshold values and effects on ovarian response and the outcome of assisted reproduction in down-regulated women stimulated with recombinant FSH. *Hum Reprod* 2001;16:1636–43

21. Fleming R, Lloyd F, Herbert M, *et al*. Effects of profound suppression of LH during controlled ovarian stimulation on follicular activity, oocyte and embryo function in cycles stimulated with purified FSH. *Hum Reprod* 1988;13:1788–92

22. McNatty KP, Smith MD, Makris A, *et al*. The micro-environment of the human antral follicle: interrelationship among the steroid levels in antral fluid, and population of granulose cells, and the status of the oocyte *in vivo*. *J Clin Endocrinol Metab* 1979;49:851–60

23. Meirow D, Schenker JG, Rosler A. Ovarian hyperstimulation syndrome with low oestradiol in non-classical 17alpha-hydroxylase 17, 20-lyase deficiency: what is the role of estrogens? *Hum Reprod* 1996;11:2119–21

24. Meldrum DR. Blastocyst transfer – a natural evolution. *Fertil Steril* 1999;72:216–17

25. Catt KJ, Dufau ML. Spare receptors in rat testes. *Nature* 1977;244:219–222

26. Tanbo T, Hhaug E, Dale PO, *et al*. Stimulation with human menopausal gonadotropin versus follicle stimulating hormone after pituitary suppression in polycystic ovarian syndrome. *Fertil Steril* 1990;53:798–803

27. Imthurn B, Macas E, Rosselli M, *et al*. Nuclear maturity and oocyte morphology after stimulation with highly purified follicle stimulating hormone compared to human menopausal gonadotrophin. *Hum Reprod* 1995;11:2387–91

28. Weston AM, Zelinski-Wooten MB, Stouffer RL, *et al*. Developmental potential of embryos produced by *in vitro* fertilization from gonadotropin-releasing hormone antagonist treated macaques stimulated with recombinant human follicle stimulating hormone alone or in combination with luteinizing hormone. *Hum Reprod* 1996;11:608–13

29. Rekha P, Mowat L, Jamieson ME, *et al*. Effect of profound suppression of luteinizing hormone during treatment with gonadotropin releasing hormone analogue and purified follicle stimulating hormone upon development of cryopreserved embryos. *Hum Reprod* 1998;13:696–8

30. Hull MGR, Corrigan E, Piazzi A, *et al.* Recombinant human luteinising hormone: an effective new gonadotropin preparation. *Lancet* 1994;344:334–5

31. Kousta E, White DM, Piazzi A, *et al.* Successful induction of ovulation and completed pregnancy using recombinant human luteinizing hormone and follicle-stimulating hormone in a woman with Kallmann's syndrome. *Hum Reprod* 1996;11:70–1

32. Zelinski-Wooten MB, Hess DL, Wolf DP, *et al.* Steroid reduction during ovarian stimulation impairs oocyte fertilization, but not folliculogenesis, in rhesus monkeys. *Fertil Steril* 1994;61:1147–55

33. Vendola KA, Zhou J, Adesanya O, *et al.* Androgens stimulate early stages of follicular growth in the primate ovary. *J Clin Invest* 1998;101:2622–9

34. Gougeon A. Regulation of ovarian follicular development in primates – facts and hypothesis. *Endocr Rev* 1996;17:121–55

35. Adashi EY. The climacteric ovary: an androgen-producing gland. In Adashi EY, Rock JA, Rosenwaks Z, eds. *Reproductive Endocrinology, Surgery and Technology*, Vol. 2. Philadelphia: Lippincott-Raven Publishers, 1996:1745–57

36. Erickson GF, Wang C, Hsueh AJW. FSH induction of functional LH receptors in granulose cells cultured in a chemically defined medium. *Nature* 1979;279:336–8

37. Billig H, Furuta I, Hsueh AJW. Estrogens inhibit and androgens enhance ovarian granulose cell apoptosis. *Endocrinology* 1993;133:2204–12

38. Hillier SG. Role of androgens in ovarian folliculogenesis. In Filicori M., ed. *The Role of Luteinizing Hormone in Folliculogenesis and Ovulation Induction*. Bologna, Italy: Monduzzi Editore, 1999:69–78

39. Shoham Z, Balen A, Patel A, *et al.* Results of ovulation induction using human menopausal gonadotropin or purified follicle-stimulating hormone in hypogonadotropic hypogonadism patients. *Fertil Steril* 1991;56:1048–53

40. Balasch J, Miro F, Burzaco I, *et al.* The role of luteinizing hormone in human follicle development and oocyte fertility: evidence from *in-vitro* fertilization in a women with long-standing hypogonadotropic hypogonadism and using recombinant human follicle stimulating hormone. *Hum Reprod* 1995;10:1678–83

41. Schoot DC, Harlin J, Shoham Z, *et al.* Recombinant human follicle-stimulating hormone and ovarian response in gonadotrophin-deficient women. *Hum Reprod* 1994;9: 1237–42

42. Latronico AA, Anasti J, Arnhold IJP, *et al.* Brief report: testicular and ovarian resistance to luteinizing hormone caused by inactivating mutations of the luteinizing hormone-receptor gene. *N Engl J Med* 1996;334:507–12

43. Toledo SPA, Brunner HG, Kraaij R, *et al.* An inactivating mutation of the luteinizing hormone receptor causes amenorrhea in a 46, XX female. *J Clin Endocrinol Metab* 1996;81:3850–4

44. Filicori M, Cognigni GE, Taraborrelli S, *et al.* Low-dose human chorionic gonadotropin therapy can improve sensitivity to exogenous follicle-stimulating hormone in patients with secondary amenorrhea. *Fertil Steril* 1999;72:1118–20

45. European Ganirelix Multi-Centre Dose-Finding Study Group. A double-blind, randomized, dose-finding study to assess the efficacy of the GnRH antagonist Ganirelix (Org 37462) to prevent premature LH surges in women undergoing controlled ovarian hyperstimulation with recombinant FSH (Puregon). *Hum Reprod* 1998;13:3023–31

46. De Placido G, Mollo A, Alviggi C, *et al.* Rescue of IVF cycles by HMG in pituitary down-regulated normogonadotrophic young women characterized by a poor initial response to recombinant FSH. *Hum Reprod* 2001;16:1875–9

47. Gordon UD, Gordon AC, Bonnar J, *et al.* Chronically administered hMG or FSH in PCOS leads to a progressive decline of endogenous LH with reduction of peripheral LH concentrations. *Hum Reprod* 1997;12(Suppl.1):53–7

48. Filicori M. The role of luteinizing hormone in folliculogenesis and ovulation induction. *Fertil Steril* 1999;71:405–14

49. Jansen CA, van Os HC, Out HJ, *et al.* A prospective randomized clinical trial comparing recombinant follicle stimulating hormone (Puregon) and human menopausal gonadotrophins (Humegon) in non-down-regulated *in-vitro* fertilization patients. *Hum Reprod* 1998;13:2995–9

50. Jacob S, Drudy L, Conroy R, *et al.* Outcome from consecutive *in-vitro* fertilization/intracytoplasmic sperm injection attempts in the final group treated with urinary gonadotrophins and the first group treated with recombinant follicle stimulating hormone. *Hum Reprod* 1998;13:1783–7

51. Zelinski-Wooten MB, Hutchison JS, Hess DL, *et al.* Follicle stimulating hormone alone supports follicle growth and oocyte development in gonadotrophin-releasing hormone antagonist-treated monkeys. *Hum Reprod* 1995;10:1658–66

52. Mannaerts B, Uilenbroek J, Schoot P, *et al.* Folliculogenesis in hypophysectomised rats after treatment with recombinant human follicle-stimulating hormone. *Biol Reprod* 1994;51: 72–81

53. Uilenbroek JT, Kramer P, Karels B, *et al.* Significance of oestradiol for follicular development in hypogonadotropic immature rats treated with FSH and hCG. *J Reprod Fertil* 1997;110:231–6

54. The European Recombinant Human LH Study Group. Recombinant human luteinizing hormone (LH) to support recombinant human follicle stimulating hormone (FSH)-induced follicular development in LH- and FSH-deficient anovulatory women: a dose-finding study. *J Clin Endocrinol Metab* 1998;83:1507–14

55. Filicori M, Cognigni GE, Taraborrelli S, *et al.* Luteinizing hormone activity supplementation enhances follicle-stimulating hormone efficacy and improves ovulation induction outcome. *J Clin Endocrinol Metab* 1999;84:2659–63

56. O'Dea L, Currie K, Chang P, *et al.* Recombinant hCG: optimizing follicular maturation with r-hCG. In abstract book of recombinant LH & hCG for the new millennium: new solutions for old problems. Presented at the *11th World Congress on IVF and Human Reproductive Genetics*, Vancouver, DC, October 1999:12–13

57. Hillier SG. Current concepts of the roles of follicle stimulating hormone and luteinizing hormone in folliculogenesis. *Hum Reprod* 1994;9:188–91

58. Loumaye E, Engrand P, Piazzi A, *et al.* Use of recombinant human LH to reduce the risk of OHSS. In abstract book of recombinant LH & hCG for the new millennium: new solutions for old problems. Presented at the *11th World Congress on IVF and Human Reproductive Genetics*, 1999:4–5

59. Couzinet B, Lestrat N, Brailly S, *et al*. Stimulation of ovarian follicle-stimulating hormone in women with gonadotropin deficiency. *J Clin Endocrinol Metab* 1988;66:552–6

60. Emperaire JC, Ruffie A. Triggering ovulation with endogenous hormone may prevent ovarian hyperstimulation syndrome. *Hum Reprod* 1991;6:506–10

61. Shoham Z, Schachter M, Loumaye E, *et al*. The luteinizing hormone surge – the final stage in ovulation induction: modern aspects of ovulation triggering. *Fertil Steril* 1995; 64:237–51

62. Imthurn B, Piazzi A, Loumaye E. Recombinant human luteinizing hormone to mimic mid-cycle LH surge. *Lancet* 1996;348:332–3

63. Brown J. Pituitary control of ovarian function: concepts derived from gonadotropin therapy. *Aust N Z J Obstet Gynaecol* 1978;18:47–55

64. Loumaye E, Piazzi A, Warne D, *et al*. Clinical use of recombinant human LH. Presented at *13th Annual Meeting of the ESHRE*, Edinburgh, 1997;50

65. Allahbadia GN, Gandhi GN, Phadke A, *et al*. Stimulation of endogenous surge of luteinizing hormone with subcutaneous leuprolide acetate (Lupride) after ovarian stimulation for intrauterine insemination. Presented at *Second World Congress of APART*, Budapest, Hungary, September 2000

66. Simon JA, Danforth DR, Hutchinson JS, *et al*. Characterization of recombinant DNA derived-human LH *in-vitro* and *in-vivo*. Efficacy in ovulation induction and corpus luteum support. *J Am Med Assoc* 1988;259:3290–5

67. Marshall JC, Anderson DC, Russel-Fraser T, *et al*. Human LH in man: studies of metabolism and biological action. *J Endocrinol* 1973;56:431–9

68. Chandrasekher YA, Hutchison JS, Zelinski-Wooten MB. Initiation of preovulatory events in primate follicles using recombinant and native human LH to mimic gonadotropin surge. *J Clin Endocrinol Metab* 1994;79:298–306

69. Romeu A, Molina I, Tresguerres JAF, *et al*. Effect of r-hLH versus hCG: effects on ovulation, embryo quality and transport, steroid balance and implantation in rabbits. *Hum Reprod* 1995;10:1290–6

70. Porchet HC, Le Cotonnec JY, Neuteboom S. Pharmacokinetics of recombinant human LH after IV, IM and SC administration in monkeys and comparison with IV administration of pituitary human LH. *J Clin Endocrinol Metab* 1995;80:667–73

71. Shoham Z. Use of recombinant LH to facilitate monofollicular development. In abstract book of recombinant LH & hCG for the new millennium:new solutions for old problems. Presented at the *11th World Congress on IVF and Human Reproductive Genetics*, Vancouver, 1999:7–10

72. WHO Scientific Group. Agents stimulating gonadal function in the human. WHO Press, Geneva, Switzerland, 1973;514: 1–28

73. Schoot DC, Harlin J, Shoham Z, *et al*. Recombinant human follicle stimulating hormone and ovarian response in gonadotropin-deficient women. *Hum Reprod* 1994;9:1237–42

74. Balen AH, Braat DDM, *et al*. Cumulative conception rates after the treatment of anovulatory infertility: safety and efficacy of ovulation induction in 200 patients. *Hum Reprod* 1994;9:1563–70

75. Homburg R, Eshel A, Armar NA, *et al*. One hundred pregnancies after treatment with pulsatile luteinising hormone releasing hormone to induce ovulation. *Br Med J* 1989;298:809–12

76. Agarwal R, West C, Conway GS, *et al*. Pregnancy after treatment with three recombinant gonadotropins. *Lancet* 1997;349:29–30

77. Shoham Z, Loumaye E, Piazzi A. A dose finding study to determine the effective dose of recombinant human luteinizing hormone to support FSH-induced follicular development in hypogonadotropin hypogonadal (HH) women. *Proceedings of the 51st annual meeting of the Am Soc Reprod Med, Seattle, WA* 1995;(abstr.S69–S70)

78. Giudice E, Crisci C, Eshkol A, *et al*. Composition of commercial gonadotrophin preparations extracted from human post-menopausal urine characterization of non-gonadotrophin proteins. *Hum Reprod* 1994;9:2291–9

79. Baird DT. Recent developments in gonadotropins for clinical therapy. *J Endocrinol* 1996;148:S38–39

80. Chappel SC, Howles C. Re-evaluation of the roles of LH and FSH in the ovulatory process. *Hum Reprod* 1991;6:1206–12

81. Ding Y, Huhtanieimi I. Preponderance of basic isoforms of serum LH is associated with the high bio/immuno ratio of LH in healthy women and in women with PCOS. *Hum Reprod* 1991;6:346–50

Ovulation induction strategies in women with polycystic ovary syndrome

6

Gianluca Gennarelli, Marco Massobrio and Jan Gunnar Holte

INTRODUCTION

Polycystic ovary syndrome (PCOS) is probably the most common endocrine abnormality in women, and the main cause of menstrual disturbances and anovulatory infertility. A large number of studies have been performed on women with PCOS, in particular during the last 20 years (more than 3300 papers dating back to 1982 are available on MEDLINE), but most aspects of this syndrome remain only partially understood. What we know today is that PCOS encompasses a wide range of clinical pictures where signs and symptoms of hyperandrogenism, increased concentrations of circulating androgens and disturbed ovulatory function combine in various proportions. We also know that important metabolic aberrations, such as obesity, insulin resistance/hyperinsulinemia, abnormal lipid and clotting profiles, and a tendency to hypertension, are associated with PCOS. These features of the so-called metabolic syndrome may have profound implications for the long-term health of women with PCOS. Importantly, insulin resistance and hyperinsulinemia seem closely linked to the pathogenetic mechanisms of chronic anovulation and infertility. In recent years, in fact, strategies aimed at lowering hyperinsulinemia have been successfully added to the traditional therapies for treating anovulatory infertility in women with PCOS. The pathophysiology of PCOS is, to date, unknown. A family association for the syndrome has been demonstrated, and it is suggested that several genes involved in the control of hormone metabolism and carbohydrate balance could be part of a complex model of inheritance. The heterogeneous clinical presentation could be the net result of the interaction between a genetic background and environmental/nutritional factors. Such phenotypic heterogeneity could in part determine the somehow unpredictable response to ovulation induction in women with PCOS, whereas it suggests that different therapeutic strategies, targeted to different subgroups of these patients, should be used.

On the following pages, the definition, the potential mechanisms of anovulation, the hormonal profile and the problems of obesity and insulin resistance in women with PCOS will be briefly reviewed. Thereafter, the issue of medical and surgical methods of ovulation induction will be discussed.

DEFINING POLYCYSTIC OVARY SYNDROME

There is no consensus on how PCOS should be defined. Whereas North American investigators use a somehow limited definition which includes the presence of ovulatory dysfunction (oligo/amenorrhea) in combination with clinical (hirsutism, acne, male pattern baldness) and/or laboratory evidence of hyperandrogenism, most European authors consider the presence of polycystic ovaries essential to the diagnosis. Indeed, ovarian morphology at ultrasound (polycystic ovary, PCO) is the most sensitive marker of PCOS[1]; it is not subjected to sudden changes, in contrast to the physiologic variations of circulating hormones[2], nor is it influenced by recent ovulation, which occurs unpredictably in women with PCOS[3].

One further advantage of ultrasound diagnosis is the possibility to include a wide range of clinical presentations, overcoming the problem of the extreme heterogeneity of PCOS. Since the introduction of ultrasound imaging, it has been clear that PCOS should be considered as a spectrum of anomalies[1], ranging from the oligosymptomatic normal-weight woman with only minor disturbances of ovarian cyclicity, to the complex picture originally described as the Stein–Leventhal syndrome, characterized by amenorrhea, obesity, hirsutism, diabetes and infertility[4].

Ultrasound criteria for PCOS were initially introduced in 1985 by Adams, who described enlarged ovaries, with ten or more follicles per ovary 2–8 mm in diameter, disposed peripherally or scattered throughout an increased amount of stroma[5]. Adams' criteria, with minor variations, are still used by most investigators, despite various efforts having been made to improve the sensitivity and specificity of the method. Whereas the specificity of parameters such as the number of cystic follicles[6] and ovarian volume[7] has been questioned, the amount (absolute or in relation to total ovarian volume) and the density of the stroma are considered the strongest predictors of PCOS[7–10].

Despite high sensitivity, the presence of polycystic ovaries alone is not specific for full-blown PCOS, as polycystic ovaries are found in about 20% of normally cycling women[11–13]. However, when their presence is combined with clinical signs of chronic anovulation (oligo/amenorrhea), the diagnosis will usually be consistent with the *classic*[14] endocrine profile of PCOS. In such cases, the prevalence of PCOS among women of reproductive age ranges from 6 to 10%.

Anovulation

The mechanisms of anovulation in women with PCOS are not yet understood. In the polycystic ovary, the growth of antral follicles is typically arrested, with a consequent failure to enter the preovulatory phase. Accelerated atresia does not seem involved in the pathogenesis[15,16]. Instead, these follicles are viable, produce androgens and estradiol and respond vigorously to stimulation with exogenous follicle stimulating hormone (FSH)[15], as proven by the higher incidence of ovarian hyperstimulation syndrome (OHSS) during induction of ovulation in women with PCOS compared to normally cycling women. Many hypotheses, not mutually exclusive, have been suggested to explain such clinical observations. An altered pattern of pulsatile gonadotropin secretion could derange follicular maturation, whereas an altered follicular milieu could decrease the follicular growth mediated by endogenous FSH. Finally, as insulin resistance and hyperinsulinemia are found in a large proportion of women with PCOS, it is likely that potential interactions between elevated insulin concentrations and gonadotropins at the ovarian level could contribute significantly to altered folliculogenesis and steroidogenesis in genetically predisposed ovaries.

The main neuro-endocrine abnormality reported in women with PCOS is an increased secretion of gonadotropin releasing hormone (GnRH)/luteinizing hormone (LH). Indeed, increased frequency and/or amplitude of LH pulses have for many years been considered central features of the syndrome[17]. An analysis of the complex mechanisms involved in the neuromodulation of PCOS is beyond the scope of this chapter, and the interested reader could refer to recent reviews[18]. Studies on the pattern of gonadotropin secretion during adolescence[19] support the notion that hypothalamic abnormalities could be an intrinsic defect in at least some of the women with PCOS. However, studies in adult women with PCOS have shown that progesterone administration[20], ovarian wedge resection[21] or treatment with anti-androgens[22] restores normal gonadotropin secretion. These findings seem to exclude abnormalities of the hypothalamic–pituitary axis as a *primum movens* in the complex array of disturbances in PCOS.

Conversely, there is accumulating evidence from both *in vivo* and *in vitro* studies that defects of steroidogenesis and cell proliferation within the ovary could be the primary mechanism of chronic anovulation in these women[23–25]. It is suggested that intrinsic ovarian defects interact with abnormally high LH and insulin levels, frequently found in women with PCOS, to determine the arrest of follicular growth[26].

Granulosa cells: premature luteinization

Granulosa cells in normally ovulating women acquire responsiveness to LH at the 9–10 mm stage[27], when LH receptors are exposed on the cell surface. This is an important signal for follicle selection and maturation, since only two more cell divisions occur before the proliferative capacity of granulosa cells is lost and the dominant follicle reaches the preovulatory stage[28]. Granulosa cells from anovulatory women with PCOS respond to LH stimulation already at the 4 mm stage[29], and steroidogenesis *in vitro* is increased in granulosa cells from women with PCOS in comparison with granulosa cells from size-matched follicles in normally cycling women[15,29]. These findings have led to the hypothesis that the granulosa cells in PCOS are more differentiated than is appropriate for the stage of follicle development[29]. In line with this notion is the overexpression of both LH receptors and the mRNA for the cholesterol side-chain cleavage protein (P450ssc), involved in the rate-limiting step of all steroidogenic tissues, in follicles from polycystic ovaries[30].

Several mechanisms could be involved in the pathogenesis of accelerated maturation of granulosa cells in PCOS. Concentrations of LH constantly above a hypothetical physiologic threshold[31] could cause premature luteinization and disruption of normal follicular development[32]. This hypothesis represents the rationale for suppressing pituitary function by GnRH analogs in some protocols of ovulation induction in these patients (see below). However, increased LH concentrations are not found in all women with PCOS[33,34]. Raised insulin concentrations could be a co-factor causing premature maturation. Previous studies have shown how insulin is able to increase the ability of granulosa cells to respond to LH *in vitro*[26]. This gonadotropic activity of insulin could be effective *in vivo*, since ovarian sensitivity to insulin in terms of steroid production is preserved in women with PCOS, despite the presence of peripheral insulin resistance[26,35]. The most probable scenario is that of a positive interaction between increased LH and insulin concentrations, where insulin increases the activity of LH on granulosa cells' function and maturation, in a steady-state fashion, to levels usually seen only at the onset of the LH surge in normally ovulating women.

Granulosa cells: impaired estrogen production

It is a common experience that exogenous FSH, even in small doses, is able to restore the mechanism of follicle selection in a large proportion of women with PCOS, a fact which suggests functional integrity[16] but low activation of granulosa cells. There are indications of lower FSH plasma concentrations in women with PCOS compared to the levels observed during the early follicular phase in ovulatory women[36], although this difference is not a constant finding[37]. However, follicles from polycystic ovaries do not contain low concentrations of bioactive FSH[38], and granulosa cells from the same follicles express high levels of FSH receptors[16]. Furthermore, several findings support the notion of an increased *in vitro* aromatase activity of granulosa cells from women with PCOS[15,38–40], both in basal conditions and under FSH or insulin stimulation. All these findings suggest a potentially increased steroidogenic activity as a primary abnormality of granulosa cells. However, *in vivo*, follicular concentrations of estradiol in PCOS do not reach levels able to maintain the estrogenic microenvironment typical of the dominant follicle[41]. One logical explanation for this apparent paradox would be the presence of specific aromatase inhibitor(s) exerting paracrine activities within the follicle. In an elegant experiment, Andreani and co-workers[42] showed that aromatase activity, evaluated by measuring estradiol production and granulosa cell proliferation, expressed as the rate of thymidine incorporation, were significantly reduced in granulosa cells from women with PCOS when follicular fluid from the same women was added to the culture medium. This effect was not observed when follicular fluid from normo-ovulatory women was added to the culture. Obviously, follicular fluid plays a pivotal role in both steroidal response and follicle selection and maturation. However, so far, no conclusive data are available on which factors or mechanisms could be responsible for the abnormalities observed in follicles from polycystic ovaries, despite several explanations having been proposed.

The presence of specific FSH receptor inhibitors of unknown origin acting within the follicle has been postulated[43]. However, the level of inhibition of FSH receptor activation, expressed as cAMP production, in both serum and follicular fluid, does not differ between women with PCOS and normo-ovulatory women[44]. These findings argue against an increased FSH threshold in women with PCOS, and are compatible with the notion of a normal or even increased responsiveness to exogenous FSH.

A number of intrafollicular growth factors, with either stimulating or inhibiting activities on granulosa cells, are potential candidates responsible for disturbed follicular growth in women with PCOS. The intraovarian insulin-like growth factor (IGF) system shows some alterations in PCOS[45]. IGF-I in physiologic concentrations stimulates aromatase activity *in vitro* in women with PCOS to the same extent as FSH, acting synergistically with FSH in controlling the level of estradiol production[46]. Intrafollicular concentrations of IGF-I, the main stimulating factor of the IGF system, seem to be normal in women with PCOS[47]. However, IGF-I activity could be impaired due to a decrease of its unbound fraction within the follicle. The ovarian production of IGF-binding proteins (IGF-BP) in PCOS is increased for IGF-BP2 and IGF-BP4, similarly to what happens in small (atretic) androgen-dominant follicles from normally cycling women, and in marked contrast to estrogen-dominant follicles[48–51]. These observations suggest that the physiologic changes occurring in the IGF system during follicle maturation, consisting of a decrease of the bound fraction in favor of the unbound fraction of IGFs, are disrupted in PCOS follicles. Furthermore, a recent *in vitro* study demonstrated a potent and direct inhibitory activity of steroidogenesis exerted by IGF-BP4 on granulosa cells from both normal and PCOS ovaries[52]. The physiologic meaning of this finding is not clear, but the observation that IGF-BP4 localization within the ovary is a function of insulin resistance is intriguing[53]. Hyperinsulinemia could be involved in one more aspect of follicular abnormalities in the PCOS.

Epidermal growth factor (EGF) and transforming growth factor α (TGFα), an EGF analog acting through EGF receptors, have been shown to inhibit estradiol production in granulosa cells, in both normal and PCOS ovaries[54]. Whereas TGFα intrafollicular concentrations are not abnormal in PCOS ovaries[55], EGF levels are higher than normal[56], and EGF receptors in the same cells are overexpressed[57], thus increasing the sensitivity to the blocking activity of these factors.

The results of recent studies suggest that raised concentrations of intrafollicular androgens could also be responsible for impaired activity of aromatase in unstimulated PCOS ovaries. Magoffin and co-workers[41] demonstrated that 5α-reduced metabolites of andro-stenedione are found in higher concentrations in follicular fluid from PCOS ovaries than from normal ovaries, and that these hormones are able to inhibit estrogen production from granulosa cells, probably via competitive mechanisms for the aromatase enzyme. These results have been corroborated by the finding of increased 5α-reductase activity in granulosa cells from PCOS ovaries compared to normal ovaries[58]. The reasons for this finding are not known, but the authors speculate that elevated androgen concentrations could play a role in stimulating ovarian 5α-reductase. On the other hand, it has recently been shown that high androstenedione concentrations, within the ranges observed in follicular fluid from PCOS ovaries, are able to inhibit *in vitro* the combined stimulatory effect of LH and insulin on granulosa cell steroidogenesis[59].

Theca cells: excessive androgen production

Theca cells from polycystic ovaries show an increased synthesis of androgens, irrespective of the ovulatory status[60]. The results of clinical studies suggests that increased theca cell androgen production in the polycystic ovary could result from a dysregulation of cytochrome P450c17 enzyme activities, 17-hydroxylase and 17,20-lyase[23,61,62]. Genetic studies seem to exclude a major causative role for the gene encoding the P450c17 in the pathogenesis of PCOS[63–65]. Conversely, the observation that progesterone production is also increased in the same cells[24] suggests an overall enhancement of theca cells' bioactivity in PCOS. This notion has been corroborated by *in vivo* studies of ovarian androgen secretion[66]. Genetic investigations have shown that the gene *cyp11*, encoding the cholesterol side-chain cleavage, the rate-limiting step in the steroid biosynthesis, might be a major genetic susceptibility locus for PCOS[67].

However, this genetic constitution seems strongly influenced by hyperinsulinemia, frequently associated with PCOS. Theca cells from human ovaries increase androgen biosynthesis when incubated with insulin[68,69]. Despite the presence of peripheral insulin resistance, the effects of insulin on theca cells seem to be preserved in women with PCOS. The mechanism behind this paradox is not fully understood, but one proposed explanation is the activation of alternative post-receptor pathways of insulin activity, such as inositolglycan mediators[70].

Whatever the underlying mechanisms, several studies have shown that androgen production in women with PCOS correlates with the level of insulin resistance/hyperinsulinemia[60], and that partial normalization of ovarian androgen secretion is obtained after reduction of hyperinsulinemia[71,72]. This effect is not observed in normal-weight healthy women[73], suggesting that insulin effects on androgen biosynthesis are amplified in polycystic ovaries compared to normal ovaries[70].

Finally, a further consequence of hyperinsulinemia, which could contribute indirectly to hyperandrogenism, derives from the trophic activity of insulin, which leads to hyperplasia of the thecal/stromal compartment of polycystic ovaries[74].

OBESITY, INSULIN RESISTANCE AND HYPERINSULINEMIA

Women with PCOS show an abnormally high prevalence of signs and symptoms of the metabolic syndrome[75], such as impaired glucose tolerance (IGT) and non-insulin-dependent diabetes mellitus (NIDDM)[76,77], obesity[78], an adverse lipid profile[79,80], decreased fibrinolytic capacity[81] and a labile control of blood pressure[82]. Insulin resistance and compensatory hyperinsulinemia, which play a pivotal role in the metabolic syndrome, are commonly considered central features of PCOS[83]. The pathogenesis of insulin resistance in these patients is not fully elucidated, despite several studies having been performed in recent years[84]. Some lines of investigation (in particular, North American studies) point to an intrinsic defect of insulin sensitivity, which is worsened by, but not dependent on, the presence of obesity[85–87]. Specific defects of the insulin receptor are reported only in rare cases of severe hyperandrogenism[88], and do not seem to be a common cause of insulin resistance in women with PCOS. Post-receptor defects in insulin signaling have been suggested by *in vitro* studies[89], which showed excessive serine phosphorylation activity in cultured fibroblasts obtained from women with PCOS. However, studies on the tyrosine kinase domain of the insulin receptor gene have failed to find any abnormality in these women[90]. Very recently, Dunaif and co-workers found another kind of post-receptor defect in a limited group of women with PCOS, who showed decreased activity of phosphatidylinositol 3-kinase[91]. Further studies on larger groups of patients are needed to evaluate the prevalence and the importance of such abnormalities in the general population of women with PCOS.

In contrast to the studies by Dunaif and co-workers, European studies usually fail to find insulin resistance in normal-weight women with PCOS[84]. Such heterogeneous results could partly be explained by differences in study methods, ethnic composition or dietary habits of the study groups. Therefore, in order to elucidate the mechanisms of insulin resistance in PCOS, our group performed extensive studies in Scandinavian women with PCOS[77,92,93], using the euglycemic hyperinsulinemic clamp, considered the most reliable method for measuring insulin sensitivity *in vivo*.

The first questions we had were whether all women with PCOS have some degree of insulin resistance, and how excessive body weight affects insulin sensitivity in women with PCOS compared with normally ovulating women. As expected, in both groups of women the body mass index (BMI) showed a strong negative correlation with insulin sensitivity, expressed as glucose disposal during the euglycemic clamp. However, the decrease in insulin sensitivity was much more profound per unit of BMI in women with PCOS than for normal women, resulting in wide differences in insulin sensitivity between the groups at a BMI between 25 and 30 kg/m^2, but no differences at all for a BMI below 25 kg/m^2[77] (Figure 1). These data suggest a negative synergism between obesity and PCOS on insulin sensitivity, and give support to the notion of preserved insulin sensitivity in normal-weight women with PCOS, a result not in line with the hypothesis of insulin resistance as an intrinsic defect of the PCOS.

A further question was whether differences in hormonal, clinical or anthropometric variables could at least in part explain the different impact of BMI on insulin sensitivity. The variable which showed the stronget association (inverse) with insulin sensitivity was truncal-abdominal body fat, as measured by skinfolds and waist-to-hip ratio (WHR). This association was similar in both women with PCOS and controls, reflecting what is observed in other populations[94], and suggesting a cause–effect relationship, probably involving the

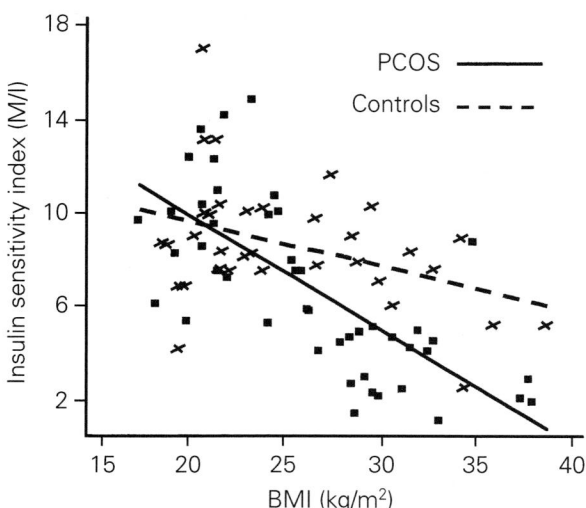

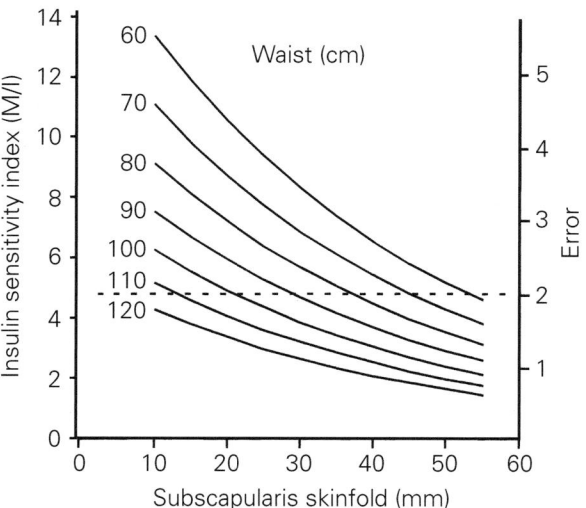

Figure 1 Relationship between insulin sensitivity index and body mass index (BMI) in women with polycystic ovary syndrome (PCOS) (squares, $p = 0.001$) and control women (crosses, $p < 0.01$). The lines intersect at a BMI of 21 kg/m^2. Reproduced with permission from Holte J, Bergh T, Berne C, *et al.* Enhanced early insulin response to glucose in relation to insulin resistance in women with polycystic ovary syndrome and normal glucose tolerance. *J Clin Endocrinol Metab* 1994;78:1052–8

Figure 2 Nomogram indicating the relationship between waist girth and the insulin sensitivity index (M/I) for different values of subscapularis skinfold in women with polycystic ovary syndrome. Left axis, predicted value of M/I; right axis, prediction error at that level of M/I. Reproduced with permission of Oxford University Press/Human Reproduction, from Gennarelli G, Holte J, Berglund I., *et al.* Prediction models for insulin resistance in the polycystic ovary syndrome. *Hum Reprod* 2000;15:2098–102, ©European Society of Human Reproduction and Embryology

metabolism of free fatty acids[77,92], released in large amounts by truncal-abdominal fat[95]. Importantly, with increasing BMI the amount of truncal-abdominal fat increased much more in women with PCOS than in normally cycling women[77]. Mirroring what happened for insulin sensitivity, the differences in truncal-abdominal fat became clinically important starting from a BMI of 25 kg/m^2, whereas similar amounts of truncal-abdominal fat were observed in normal-weight women with PCOS and normal-weight control women. The increasing differences between the groups in insulin resistance with increasing BMI were completely abolished after adjusting for indices of upper-body fat distribution[77], further supporting the hypothesis of a cause–effect relationship. These results were confirmed by extended studies[93]. Construction of prediction models derived from clamp investigations on 72 women with PCOS resulted in a model, containing only waist circumference and the subscapularis skinfold, which can explain a large part of the variation in insulin sensitivity in women with PCOS, better than any model containing hormonal and other clinical variables[96] (Figure 2). Intervention studies showed that weight loss following

dietary treatment in women with PCOS was characterized by a preferential reduction of body fat at the truncal-abdominal sites, which was paralleled by improvement of insulin sensitivity to levels comparable to those found in normal women with comparable BMI[92]. These results and those from other studies (see following section on weight reduction) further question the hypothesis that a unique defect of insulin activity is a common cause of insulin resistance in PCOS. Moreover, they could partly explain the varying results for insulin resistance in non-obese women with PCOS, since varying proportions of women with increased truncal-abdominal fat within non-obese groups could be enroled in different studies[97,98].

Whereas the synergism between obesity and PCOS seems to be strongly associated with insulin resistance, the reasons why women with PCOS tend to be overweight in a proportion significantly larger than normal women, and what determines their preferential accumulation of upper-body fat, remain matters of debate. However, the results of our studies point to β-cell function and adrenal response to stress as potential key factors in these metabolic abnormalities.

As previously mentioned, a large proportion of women with PCOS have glucose intolerance or diabetes mellitus already at a young age[76]. Obviously, these women have reduced β-cell function, as shown by lower acute responses to intravenously administered glucose. Given the high proportion of these patients among women with PCOS (for the majority obese), some studies support the notion of defective insulin secretion as a general feature of PCOS[99,100]. However, Holte and co-workers found that after excluding from the analysis all women with glucose intolerance, those with PCOS had a significantly higher acute insulin response to glucose than those in the control group. The exaggerated insulin secretion was independent from obesity, and was out of proportion for the degree of insulin resistance, nor was it affected by weight reduction and improvement of insulin resistance[77,92]. For the first time it was suggested that hyperinsulinemia in women with PCOS may not only depend on a compensatory mechanism to insulin resistance. These results were confirmed in a larger series of women with PCOS[93] and are in line with other studies of the same kind[101]. In contrast to insulin resistance, exaggerated insulin secretion could be a constitutional feature of PCOS. Consistent with this notion is the finding of an association between PCOS and allelic variation at the INS VNTR locus[102]. A logical consequence of β-cell activity exaggerated for the degree of insulin resistance would be a more rapid uptake and storage of circulating glucose. Indeed, there is evidence for lower mean concentrations and lower postprandial rises of blood glucose in women with PCOS[77,103]. Furthermore, obese women with PCOS show reduced hormonal and clinical reponses to insulin-induced hypoglycemia, a fact which suggests possible adaptation to hypoglycemia[104]. It is theoretically possible that repeated episodes of slight hypoglycemia could induce carbohydrate cravings or a subnormal feeling of satiety. The finding of bulimic eating behavior in a large proportion of women with PCOS[105,106], and the association between polycystic ovary morphology and bulimia nervosa[107], fit well in this hypothetical chain of events.

The negative impact of an increased tendency to energy intake seems further aggravated by lower energy expenditure, as reduced postprandial thermogenesis[108]

and low sympathetic activity[104] have been reported in these women. The net result would be overweight/obesity in women with PCOS. What determines the preferential accumulation of truncal-abdominal fat is so far not known. However, a long line of investigation suggests that the balance between sex steroids and glucocorticoids could be a key factor in humans[94]. In women, estrogens and progesterone seem to play in concert to favor the accumulation of gluteo-femural fat, which is metabolically inert except during lactation[95]. On the other hand, cortisol is likely to play an important role in accumulating truncal-abdominal fat, a view supported by the presence of large amounts of cortisol receptors in this type of fat[95], and by the clinical evidence of Cushing syndrome. In women with PCOS the activity of cortisol in terms of upper-body fat accumulation could be enhanced, due to the relative lack of progesterone, which has proven anti-corticoid activity. More importantly, there is accumulating evidence for abnormal activity of the hypothalamic–pituitary–adrenal axis and increased cortisol production in women with PCOS[109–113]. In a recent study by our group, cortisol response to hypoglycemia was more rapid and reached similar incremental levels, in spite of lower adrenocorticotropic hormone (ACTH) secretion, in women with PCOS compared with controls[114], suggesting increased adrenal sensitivity in PCOS. The same study showed a more rapid decline of circulating cortisol concentrations, suggesting a more rapid turnover of this steroid in women with PCOS.

In conclusion, any situation causing a reduction of sex steroids, in particular progesterone, and/or an increase in cortisol production would favor accumulation of truncal-abdominal fat, thus increasing the metabolic risk. Lifestyle habits, such as smoking and alcohol consumption, may contribute to such endocrine changes, through increased activation of the hypothalamic–pituitary–adrenal axis[115]. Modifications in these habits have recently been shown to improve significantly menstrual cyclicity in women with PCOS[116].

INDUCTION OF OVULATION IN WOMEN WITH PCOS

The proximate mechanism behind anovulation in women with PCOS consists of an impairment of the selection of a dominant follicle, while the number of smaller follicles is increased. Notably, any intervention able to raise circulating FSH concentrations, or to decrease the threshold for FSH stimulation in PCOS[117], can unblock the process of follicle selection in the majority of cases.

Weight reduction

There is clear evidence that dietary treatment and weight loss in obese women with PCOS have positive effects on insulin resistance, sex hormone concentrations, ovulatory function and fertility rates. This cost-effective approach[118] should be a first-line treatment to induce ovulation in overweight women. The exact mechanisms behind the beneficial effects of weight reduction are not known. However, given the important impact of hyperinsulinemia on steroidogenesis and on ovulatory function, the improvement of insulin resistance[92,116,119–126] probably plays a primary role in a chain of events leading to decreased androgen production[92,121,124–126], increased synthesis of sex hormone-binding globulin (SHBG)[92,119,123,126] and partial normalization of gonadotropin pulsatile secretion[116,120,124]. This view is also supported by the results of recent trials on the use of insulin-lowering drugs for induction of ovulation in women with PCOS (see below).

Interestingly, weight loss does not need to be dramatic, nor does it need to restore body weight to within the normal range, in order to obtain a good clinical and hormonal response. In the study by Kiddy and co-workers[122], a weight loss as little as 5% was observed in some women with PCOS who improved their ovarian function, whereas in the study by Holte and co-workers[92], amelioration of hormonal variables and of menstrual cyclicity was obtained at a post-diet BMI of 28 kg/m^2. As shown by our group[92] and by others[116,120], the increase of insulin sensitivity and the amelioration of menstrual cyclicity correlate strongly with the reduction of the indices of central fat distribution. As mentioned above, one recent study showed how lifestyle modifications (exercise counseling), with only minimal calorie restriction and a weight loss ranging from 2 to 5%, were able to restore ovulation in nine out of 15 obese anovulatory women[116]. These

findings underline the great metabolic impact of abdominal fat, and are encouraging for women candidates for this therapy, since the efforts required should be aimed only at reducing this specific area of body fat. Additionally, these observations suggest that lifestyle and dietary modifications could also be a therapeutic option in the subgroup of non-obese women with PCOS who nevertheless have an increased amount of abdominal fat[60,98,127].

Strong clinical evidence suggests that weight reduction should be encouraged at least in all overweight patients with PCOS, before any attempt to conceive is made, whether by induction of ovulation or by assisted reproduction. A significantly higher rate of spontaneous abortions is reported in the (even moderately) obese than in lean women with PCOS[128]. Furthermore, it is well known that obesity and insulin resistance are associated with complications of pregnancy such as pre-eclampsia and gestational diabetes mellitus (GDM)[129,130]. Women with PCOS at any level of clinical expression of the syndrome (even women with only the ovarian morphology of PCOS) bear an intrinsic risk of developing GDM and/or pre-eclampsia. Holte and co-workers reported an increased prevalence of women with polycystic ovarian morphology among patients with previous GDM[131]. These findings have been confirmed by several subsequent studies in women with PCOS, whether full-blown[132] or not[133–135]. As intimated above, PCOS is also associated with an abnormally high incidence of pre-eclampsia, irrespective of body mass[136,137]. Therefore, the combination of excess weight and PCO/PCOS would result in a sum of risk factors for metabolic aberrations during pregnancy.

Two problems need consideration. First of all, none of the mentioned studies addressed the long-term efficacy of weight reduction, nor the stability of the weight attained after diet. It is a common clinical experience that women with PCOS encounter great difficulties in maintaining a stable body weight. There are indications that these patients could suffer from carbohydrate cravings and disturbed eating behavior[105–107]. This is probably one feature of a particular constitution, characterized by a strong anabolic drive[138], which could be expressed more markedly in subgroups of women with PCOS. Obviously, a therapeutic approach based on diet only would not be enough for these patients, who are therefore potential candidates for therapies with insulin-lowering drugs.

Second, ovulation is not restored in all obese women with PCOS, despite significant weight loss. In the study by our group, six out of 13 women with PCOS remained anovulatory after diet (the clinical, metabolic and hormonal variables did not differ between the subgroups of women who did and did not restore ovulation)[92]. It therefore seems impossible to predict the response for each single patient. It is possible that in those women with PCOS who do not respond to diet treatment and weight reduction, the genetic predisposition to abnormal ovarian function might play a stronger role in determining anovulation than the associated metabolic aberrations, which stresses once again the heterogeneity of the syndrome.

Clomiphene citrate

Clomiphene citrate (CC) represents the most common and simple method to induce ovulation in women with normogonadotropic anovulatory infertility (World Health Organization group 2)[139]. CC is traditionally administered for five days during the follicular phase (following spontaneous or progesterone-induced menstruations), with a starting dose of 50 mg daily, increasing to 100, 150 or even 200 mg daily during consecutive cycles, in cases of no response. Doses higher than 200 mg are in general of no additional benefit, whereas the adverse effects, in particular on cervical mucus thickness and the endometrium, are increased[140]. CC therapy has been considered for years the first-choice treatment for infertility in PCOS. Approximately 70–80% of PCOS patients ovulate[141], and up to 60–70% of ovulatory patients conceive during the first six cycles of CC administration[142,143]. However, a substantial subgroup of women with PCOS, defined as clomiphene-resistant, remain anovulatory, even at high doses of CC. Despite extensive use of CC over the last four decades, defined clinical and/or hormonal characteristics able to identify beforehand CC-resistant women are lacking. It is a common experience that obese women with PCOS require higher doses of CC and are more likely to remain anovulatory[144,145]. Furthermore, in a series of 128 anovulatory women, Kousta and co-workers[143] recently reported that CC-resistant women had a higher BMI

than responsive patients, whereas no differences could be identified among other clinical and hormonal variables. Larger ovarian volume and a higher number of antral follicles[146], as well as decreased IGF-BP1 circulating concentrations[147], have also been associated with CC resistance in PCOS. More recently, a multivariate analysis of clinical and hormonal variables in 201 oligo/amenorrheic women treated with CC resulted in a prediction model of CC unresponsiveness, which included free androgen index (FAI), BMI, cycle history (oligo- or amenorrhea) and mean ovarian volume[141]. These results are in line with the notion that obese, highly hyperandrogenic women are less likely to ovulate during CC therapy. A following study from the same group showed that indirect indices of insulin resistance, such as elevated fasting insulin or insulin/glucose ratio, low IGF-BP1 and high leptin concentrations, could predict a poor response to clomiphene[148]. These observations provide a rationale for the use of insulin-lowering strategies (weight reduction and/or hypoglycemic drugs – see below) in combination with CC, in overweight/hyperinsulinemic women with PCOS.

Several other treatment protocols (with CC given alone or in combination with other drugs) have been used in CC-resistant women with PCOS, but none has shown advantages over the others. Improved results have occasionally been reported with extended use of CC for up to seven to 10 days in the follicular phase[149–151], but no randomized studies are available. The association of CC with corticosteroids has been proposed as an alternative treatment. The only randomized study showed a higher rate of ovulation and conception in women with PCOS treated with 0.5 mg of dexamethasone daily in combination with CC[152]. The elevated adrenal androgens found in a high proportion of women with PCOS provide the rationale for this therapy. There are, however, suggestions that even CC-resistant patients with normal adrenal androgens could benefit from such a schedule[153]. Attempts to partially suppress the hypothalamic–pituitary–ovarian axis with oral contraceptives[154] before CC treatment seem to improve the response in CC-resistant women with PCOS.

Very recently, Mitwally and Casper[155] reported interesting results in a limited number of CC-resistant PCOS patients, using letrozole, an aromatase inhibitor, administered in a dose of 2.5 mg daily, from days 3 to 7 of the cycle. An advantage of letrozole over CC suggested by the authors is a lack of anti-estrogenic effects on the endometrium, and a potential stimulatory effect on follicular maturation, mediated by an increase of FSH receptors.

An interesting alternative CC protocol has been studied in 200 women who failed to ovulate after three consecutive cycles of conventional CC therapy with incremental doses up to 150 mg per day. By adding a single injection of 75 UI of urinary FSH, the authors of this study reported ovulation in almost 80% of the women under investigation, and pregnancy in 34% of those who ovulated, with no occurrence of OHSS[156].

Whatever the strategy adopted, the number of CC cycles which should be attempted before switching to other methods of ovulation induction is still a controversial issue. Recent reports agree on the fact that the majority of women who conceive during CC therapy will do so within the first six ovulatory cycles[142,143]. In light of these results and of the putative increased risk of ovarian cancer associated with prolonged use of CC[157], it is generally not recommended to treat patients with CC for more than six cycles, once an ovulatory reponse is obtained.

Hypoglycemic drugs

Since the first report of resumption of ovulation in women with PCOS following treatment with metformin[158], there has been an extensive use of insulin-sensitizing drugs in these patients. The rationale for the employment of such agents in the context of infertility treatment is the fact that insulin resistance and compensatory hyperinsulinemia, observed in a large pro-portion of women with PCOS, are not only associated with increased metabolic risk, but also involved at multi-ple levels in the derangements of ovarian steroidogenesis and follicular maturation (see above). Therefore, it is believed that the positive effects on ovulation induction are principally mediated by a substantial decrease of hyperinsulinemia.

Metformin, a biguanide with potent glucose-lowering effects in patients with NIDDM, is the most widely used hypoglycemic drug in women with PCOS, but it is not able to induce hypoglycemia, nor to stimulate β-cell function, and is rarely associated with serious complica-tions such as lactate acidosis. The exact mechanisms of action of metformin are not fully understood, but it appears that the positive effects on carbohydrate and insulin metabolism are exerted at different levels[159]. At least two studies in women with PCOS reported a significant improvement of peripheral insulin sensitivity during metformin therapy, as shown by glucose disposal during euglycemic clamps[71,160]. In women with PCOS, metformin is usually administered in doses ranging from 500 mg TID to 1000 mg BID, the only contraindication being impaired kidney function (serum creatinine > 1.5 mg/dl), which is associated with a higher potential for metformin toxicity. An incremental dosage protocol starting with 500 mg once a day for one week and increasing by 500 mg in the second and third weeks is recommended to avoid minor side-effects[161].

After the uncontrolled study by Velazquez and co-workers[158], several authors have demonstrated a reduction of circulating androgens and resumption of ovulation either spontaneously or in response to CC in obese women with PCOS[71,121,160,162–167]. In the first placebo-controlled trial, Nestler and Jakubowicz[168] showed that only four to eight weeks of treatment resulted in significant decreases in fasting and glucose-stimulated insulin concentrations, paralleled by decreases in both basal and GnRH-stimulated androgen concen-trations in obese, severely hyperinsulinemic women with PCOS. These results were confirmed in non-obese women with PCOS in a following study by the same authors[72], who concluded that the effects of metformin on ovarian androgen production (ovulation was not assessed) are independent from changes in body mass. However, it is of note that in both obese and normal weight groups of women with PCOS a partial redistrib-ution of body fat was observed, as shown by significant reductions of the WHR, despite the limited period of metformin administration. In a third short-term study by Nestler and co-workers[165], 31 of 35 obese women with PCOS treated with metformin ovulated either spontane-ously or in response to clomiphene, as compared with only three of 26 women treated with a placebo.

More recently, Moghetti and co-workers[71] performed the first long-term (six months), randomized, double-blind, placebo-controlled study of 32 obese women with PCOS. In a subgroup of these women the study was continued in an open-label fashion for a further six months. After the first six months, the women treated with metformin showed significant reductions of ovarian androgens and fasting insulin, and increased insulin sensitivity, in comparison with the women in the placebo group. Improvement of menstrual cyclicity was observed in about 50% of the treated women, who showed an increased ovulation rate already in the first month, confirming an early reversal of reproductive abnormalities by metformin therapy. All clinical, hormonal and metabolic changes occurred without significant variations in body weight or WHR. In the follow-up study, the beneficial changes obtained in the treated women were sustained, with even further normalization of menstrual cycles, whereas only one woman discontinued the therapy, due to severe gastrointestinal discomfort, demonstrating an overall good tolerability and the possibility of chronic administration of this therapy. The women who improved their menstrual function in respose to metformin had higher pretreatment plasma insulin and lower serum androgens than those who did not respond. The authors concluded that metformin therapy could be effective, for reproductive functions, in subgroups of hyperinsulinemic women in whom metabolic aberrations play a predominant role in altering the function of the pituitary–ovarian axis.

In another observational long-term study of obese women with PCOS, circulating insulin and androgen concentrations were reduced, with only a slight (and not significant) improvement of insulin resistance[163]. Menstrual cyclicity improved in six of eight subjects. Body weight was unaltered by metformin, whereas WHR was strongly reduced.

In most of the mentioned studies, the beneficial effects of metformin are not associated with changes in body weight, in contrast to what is observed during a hypocaloric diet (see above). To clarify the independent effect of diet and metformin therapy further, Pasquali and co-workers[121] performed a randomized, placebo-controlled trial in 20 obese women with PCOS. Each woman underwent a standardized hypocaloric diet and, after one month, was placed randomly on either metformin or a placebo as an adjunction to diet. Both groups experienced a reduction in body weight, truncal-abdominal fat mass, serum insulin and C-peptide concentrations, and serum testosterone concentrations, and an improvement of menstrual

cyclicity, but these changes were significantly more pronounced in the women treated with diet plus metformin than in those on diet plus placebo. On the basis of their results, the authors concluded that diet and metformin could preferentially address one of two different aspects, abdominal obesity and hyperinsulinemia, respectively, having complementary effects in the pathogenesis of PCOS.

Unfortunately, so far only one study on metformin in PCOS has considered pregnancy as an endpoint[164]. In 11 obese clomiphene-resistant women with PCOS, metformin increased the ovulatory rate by three-fold and the pregnancy rate by eight-fold in comparison with a group of 15 women on a placebo, during a course of six cycles of CC-induced ovulation. This is the only study suggesting a positive effect of metformin plus CC in women who do not respond to CC alone, possibly explained by the limited number of patients involved.

At variance with these positive results, some authors failed to find beneficial effects of metformin in PCOS[169–172]. The reason for this discrepancy is not clear. In two studies racial differences could be in part responsible[171,172], whereas in the study by Ehrmann and co-workers[170] morbidly obese women with PCOS were included (BMI up to 52.5 kg/m^2), suggesting that metformin alone is not sufficient to treat individuals with obesity of this magnitude.

Despite these varying results, most evidence seems to favor the use of metformin, alone or in combination with a hypocaloric diet, to induce ovulation either spontaneously or by CC or gonadotropin[173] administration in all overweight, hyperinsulinemic women with PCOS. Furthermore, there are suggestions of a beneficial effect of metformin in clomiphene-resistant women with PCOS during IVF treatment[174]. The mean number of mature oocytes, embryos cleaved, fertilization rates and clinical pregnancy rates was increased in women treated with metformin and FSH, in comparison to women treated with FSH alone[174]. Interestingly, during gonadotropin administration for either ovulation induction[173] or IVF[174], metformin reduced the circulating concentrations of estradiol and the total number of developing follicles, minimizing the risk of OHSS. The most likely mechanism behind such findings is a change in follicular environment mediated by reductions of insulin resistance and hyperinsulinemia. It is also possible that metformin exerts some direct effects within the ovary, at the level of specific enzyme activities, a phenomenon recently observed for thiazolidinediones, another family of insulin-lowering drugs[175].

The results of recent studies suggest that the use of metformin, extended during early pregnancy, could decrease the incidence of early pregnancy loss, which is a major complication observed in women with PCOS[176,177]. Although the exact mechanisms are not known, hyperinsulinemia is believed to alter the endometrial milieu and disrupt normal embryo implantation. A North American study[178] reported a significant improvement of pregnancy outcome in a limited number of women with PCOS, compared to the incidence of spontaneous abortion in their reproductive history. These results are encouraging, but it should be stressed that although in none of the occasional pregnancies occurring during metformin treatment were any fetal malformations reported, safety of such therapy during gestation has not yet been definitively established. Until official consensus opinions are available, therefore, metformin should not be considered a therapeutic option during pregnancy.

The only other hypoglycemic drug extensively studied in women with PCOS is troglitazone, the first of a recent class of antidiabetic drugs defined as thiazolidinediones. The main mechanism of action is an increase of peripheral insulin sensitivity, without stimulating insulin secretion[179]. Several investigations have reported a significant decrease, independent of changes in body weight, of insulin and androgen concentrations in obese women with PCOS[180–184]. These changes are accompanied by important increases in the ovulation rate. In a recent long-term, multicenter, placebo-controlled trial performed with 410 patients of different ethnic backgrounds, Azziz and co-workers[182] showed a significant increase of ovulations, which was dose-dependent (from 300 to 600 mg per day) with respect to placebo. As in the case of metformin, women who remained anovulatory during therapy were more obese, hyperandrogenic and hyperinsulinemic than those who responded to troglitazone. Thus, it seems that no real advantage over metformin could be offered by this molecule, although no studies comparing metformin versus troglitazone are available. Furthermore, the use of troglitazone has been associated with rare, but serious, acute hepatic failure in patients with NIDDM[185]. For this reason, troglitazone should not be recommended for the treatment of anovulation in PCOS. New molecules, presumably with fewer side-effects, such as pioglitazone and rosiglitazone need to be studied in women with PCOS.

Gonadotropins

Injectable gonadotropin administration represents the conventional approach to infertility in women with PCOS who do not get pregnant after CC therapy, but the main problem is the very narrow therapeutic window. Due to the large cohort of recruitable follicles[186] and to the particular sensitivity of polycystic ovaries to exogenous stimulation, traditional regimens of gonadotropin stimulation, consisting of incremental doses of 75 UI every five to seven days, have yielded an unacceptable incidence of multiple pregnancies and OHSS[187]. For this reason, protocols employing low doses of gonadotropins have become widely used. Different stimulation schedules, using urinary or recombinant FSH, with or without various adjunct therapies, have been combined, but to date no clear evidence is available to establish which is the best option.

Step-up protocol

The low-dose step-up protocol consists of the administration of a starting dose of FSH, which is maintained for 14 days, beginning on day 3 of the menstrual cycle. The dose is then increased weekly in steps of a third to a half of the initial dose, until selection of a dominant follicle occurs[188]. The rationale for such a protocol is that the threshold for FSH stimulation has wide individual variations, a fact which requires a dose-finding approach. Obviously, this often results in prolonged stimulations, with negative consequences in terms of patient compliance and total cost-per-treatment. Rates as high as 70% of uni-ovulatory cycles, defined as a single follicle ≥ 16 mm on the day of human chorionic gonadotropin (hCG) administration, and 20% of pregnancies have been reported by several investigators, using either urinary or recombinant FSH, whereas the incidence of multiple gestation is 5–6% and that of OHSS is very rare[189–195]. A starting dose of 75 UI was initially used. However, there is evidence that lower starting doses can be equally effective in these patients[195,196]. The largest study so far (225 women with PCOS) reported the results of ovulation induction by traditional low-dose protocols, using a starting dose of either 75 UI or 52.5 UI[197]. The lower starting dose allowed a substantial decrease of cycle cancellation due to multiple follicle responses, without affecting the pregnancy rate. Given the obvious problem of patient compliance, attempts aimed at reducing the duration of the initial phase from 14 to seven days have been made. The change does not seem to affect pregnancy rate, whereas it reduces the amount of FSH administered and the mean duration of treatment. However, the risk for multiple pregnancy seems significantly increased[198], an effect which strongly reduces the advantage of such approach.

Step-down protocols

Step-up protocols raise FSH concentrations throughout the follicular phase, whereas in naturally occurring cycles endogenous FSH concentrations decrease progressively during this phase. Based on the idea that elevated FSH during the late follicular phase could interfere with monofollicular selection, and on the observation that low-dose step-up gonadotropin therapy is still associated with a percentage of hyperstimulated cycles[199], the step-down principle in gonadotropin stimulation was introduced[117]. This type of stimulation protocol is based on a large initial dose of 150 UI of FSH daily, starting in the early follicular phase. The dose is reduced by 37.5 UI, when at least one follicle of 10 mm is visualized by ultrasound. The dose is further reduced to 75 UI after three more days, and maintained at 75 UI until hCG administration. A prospective, randomized study of 37 clomiphene-resistant women with PCOS showed a monofollicular response in 88% of step-down stimulation cycles, compared with 56% of the low-dose, step-up protocol[192]. Furthermore, a shorter stimulation schedule and fewer FSH units were needed with the step-down regimen[192].

Based on the same principle, an alternative protocol recently proposed is the sequential step-up/step-down regimen[200], in which the FSH dose is reduced by 50% once the leading follicle has reached 14 mm in diameter.

In a randomized study, this protocol resulted in a lower number of midsized follicles compared with the conventional step-up protocol[200].

Lately, a randomized study compared the conventional step-up protocol with a modified step-down protocol[201]. The latter started with 300 UI of recombinant FSH (rFSH) on cycle day 3, followed by three days with no gonadotropin administration, which was resumed on day 7 with 75 UI rFSH until day 9, under ultrasound monitoring. From day 9 on, the protocol was the same as in the step-up protocol. The results of the trial showed a better outcome from the modified step-down regimen compared to the traditional step-up, in terms of single follicle recruitment and number of cancelled cycles, but the study was too limited to detect significant differences in pregnancy rate.

Gonadotropin preparations

A recent meta-analysis[202] failed to find any advantage of urinary FSH (uFSH) over human menopausal gonadotropin (hMG) in terms of pregnancy rate, although FSH appeared to be associated with a reduction in moderate to severe OHSS. In an *in vitro* fertilization program no differences were detected between rFSH and hMG in terms of oocyte maturity and fecundity[203]. It seems, therefore, that the low LH content in hMG preparations does not impose a negative effect in women with PCOS.

Several studies have been performed to evaluate the efficacy of uFSH compared to rFSH. In a recent study of 50 women with PCOS treated with a chronic low-dose protocol, cycles with rFSH performed better than those with uFSH in terms of ovulation rate, duration of the stimulation, number of ampules administered and incidence of OHSS[204]. A previous randomized investigation involving 172 women[193] did show rFSH to be biologically more active than uFSH, resulting in shorter treatment protocols, a lower number of FSH units and comparable ovulation and pregnancy rates. On the other hand, women treated with rFSH were more prone to develop OHSS, suggesting the need for closer monitoring during FSH administration. Others have confirmed the higher bioactivity of rFSH with respect to uFSH[205]. However, to date there are no sufficient data to determine whether rFSH or uFSH is preferable for ovulation induction in women with PCOS[206].

Use of GnRH analogs

The deleterious effects of elevated LH concentrations associated with PCOS[207], and the premature luteinization of granulosa cells observed in some cases (see above), have been the rationale for the use of pituitary desensitization by GnRH analogs (GnRH-a) before ovulation induction. Furthermore, retrospective studies suggest that pretreatment with GnRH-a could improve the high miscarriage rate observed in these patients[177,208]. The use of GnRH-a in combination with gonadotropins seems to avoid premature luteinization[209], but the results of several investigations failed to confirm any advantage over the use of gonadotropins alone, in terms of ovulation rate or pregnancy rate[210,211]. Indeed, the results of a randomized trial on women with a history of recurrent miscarriage, polycystic ovaries and elevated LH concentrations failed to confirm any positive effect of GnRH-a therapy during ovulation induction[212]. Conversely, a higher incidence of OHSS has been reported with the use of GnRH-a[213].

Negative impact of obesity and insulin resistance on ovulation induction with gonadotropins

A large proportion of clomiphene-resistant women with PCOS are obese. Conventionally, these patients are candidates for gonadotropin treatment. It has long been known that obese women with PCOS require higher doses of gonadotropins during ovulation induction, and that ovulation rates are lower than for lean counterparts[188,190]. Even when ovulation occurs, these patients have an increased risk of multiple follicle recruitment, a lower pregnancy rate and a higher incidence of early miscarriage[128,197]. Results from multiple regression analyses in a group of Scandinavian women with PCOS suggest that these negative outcomes do not depend on obesity per se, whereas they are strongly associated with the degree of insulin resistance[214]. The results of a recent randomized Italian study lend support to this notion[173]. Women with PCOS treated with metformin, an insulin sensitizer, during ovulation induction with FSH showed a lower number of follicles > 15 mm in diameter on the day of hCG administration, and a lower number of canceled cycles. These results suggest that any attempt to induce ovulation in overweight women with PCOS should be preceded by metabolic investigations and improvement of insulin resistance either by diet alone or by pharmacologic treatment (see above).

OVARIAN SURGERY

Ovarian electrocautery and laser vaporization have in recent years replaced the more invasive ovarian wedge resection as surgical methods for inducing ovulation in clomiphene-resistant women with PCOS. A large body of literature supports the efficacy of such treatment, which exerts results similar to gonadotropin stimulation.

The physiologic mechanisms behind restored ovulation or improved response to gonadotropin stimulation are not clear. Destruction of a small part of the ovary reduces circulating androgens and probably contributes to altering the function of paracrine factors within the ovary. There are, however, indications that the main mechanism would act at the level of the pituitary–ovarian relationship, since monolateral diathermy is able to restore ovulation also in the contralateral ovary[215].

Which patients should be candidates for laparoscopic ovarian surgery is an open question. The impressive results reported by several authors[216] have led some to argue that ovarian surgery should be considered as an option for all women with PCOS who fail to respond to clomiphene *before* stimulation with gonadotropins. However, some key points need consideration. Patients who are more likely to respond to surgery are those with elevated LH concentrations[217]. Clearly, these patients represent only a subgroup of women with PCOS, who are less likely to be obese and insulin-resistant[36,218]. Ovarian surgery might not be appropriate as a first-line approach in obese, hyperinsulinemic women with PCOS, since it does not seem to affect insulin resistance[219], and therefore cannot compensate for the negative impact of metabolic aberrations on the reproductive outcome.

Finally, the risks associated with the surgical procedure and the possibility of ovarian adhesions are variables that must also be taken into account.

IN VITRO FERTILIZATION

Traditionally, the use of *in vitro* fertilization (IVF) has been limited to women with PCOS who either present with associated causes of infertility or fail to conceive after repeated courses of ovulation induction. Conventional pituitary desensitization with GnRH-a, followed by stimulation with either rFSH or uFSH[220], seems to be the most advantageous protocol for superovulation induction in PCOS[221]. However, two specific problems need consideration: first, the high incidence of OHSS observed in these patients during ovarian stimulation; and second, the question of whether the results of IVF are comparable between women with PCOS and women with normal ovaries. There is evidence of lower fertilization rates in oocytes obtained from women with PCOS with respect to patients with other causes of infertility[222–224]. On the other hand, the number of oocytes retrieved is usually higher in women with PCOS, resulting in a comparable number of good-quality embryos available for transfer and similar pregnancy rates between women with PCOS and women with normal ovaries[224,225]. The high number of follicles could, however, be one of several factors behind the 'explosive' response of polycystic ovaries to exogenous FSH, and the associated high incidence of OHSS. It has been shown that a substantial quantity of the follicles recruited during ovarian stimulation in women with PCOS are of small diameter (8–13 mm) and contain oocytes not meiotically competent, in contrast to larger follicles (> 14 mm), which contain meiotically competent oocytes[226]. Theoretically, minimizing the number of small follicles in favor of larger ones would result in both a lower risk of OHSS and a larger number of mature oocytes. Recent studies in clomiphene-resistant women with PCOS report interesting results with the use of metformin during both ovulation[173] and superovulation[174] induction, with a significant decrease of small follicles and lower levels of circulating estradiol, whereas the number of large follicles is unchanged, more mature oocytes are recovered and more cleaved embryos are obtained. While these findings once more point to the negative impact of hyperinsulinemia on the intraovarian milieu, they suggest that the quality of mature oocytes is not compromised in women with PCOS. Indeed, oocytes obtained from polycystic ovaries in women with no hormonal or metabolic signs of the PCO *syndrome* perform better than oocytes from normal ovaries in terms of IVF outcome[227]. Should these results be confirmed by further studies, IVF could potentially be considered as an alternative to ovulation induction in clomiphene-resistant women with PCOS. A recent randomized study from Sweden has shown how ovulation induction in this specific group of patients is not cost-effective, costs per pregnancy being lower in clomiphene-resistant women who switched directly to an IVF program than in those randomized to receive ovulation induction with gonadotropins[228].

CONCLUSIONS

As shown, a vast range of therapeutic options are today available for infertility treatment in women with PCOS. Since many aspects of PCOS are as yet poorly understood, the response to different methods of ovulation induction is often unpredictable. It is therefore an open question as to which is the more appropriate step in treating infertility in these women. The results of recent clinical and laboratory investigations have highlighted the central role of insulin resistance and hyperinsulinemia in the pathogenesis of anovulation and reproductive failure in a large proportion of women with PCOS. A logical conclusion of these results is that diet and insulin-sensitizing agents should be the first-line treatment for overweight/obese women with PCOS, and should precede CC or FSH stimulation. Whether this approach could also be helpful in normal-weight women with PCOS remains a question for future investigations. Finally, it should be considered that the conventional approach of repeated courses of CC followed, in case of failure to conceive, by repeated cycles of gonadotropin administration, in turn followed by ovarian surgery, might no longer be appropriate, whereas IVF might be confirmed as a more convenient approach in terms of clinical outcome and economic impact.

References

1. Balen AH, Conway GS, Kaltsas G, *et al.* Polycystic ovary syndrome: the spectrum of the disorder in 1741 patients. *Hum Reprod* 1995;10:2107–11
2. Franks S. Polycystic ovary syndrome: a changing perspective. *Clin Endocrinol (Oxf)* 1989;31:87–120
3. van Hooff M, van der Meer M, Lambalk CB, Schoemaker J. Variation of luteinizing hormone and androgens in oligomenorrhoea and its implications for the study of polycystic ovary syndrome. *Hum Reprod* 1999;14:1684–9
4. Stein IF, Leventhal ML. Amenorrhea associated with bilateral polycystic ovaries. *Am J Obstet Gynecol* 1935;29:181–91
5. Adams J, Franks S, Polson DW, *et al.* Multifollicular ovaries: clinical and endocrine features and response to pulsatile gonadotropin releasing hormone. *Lancet* 1985;2:1375–9
6. Fox R, Hull M. Ultrasound diagnosis of polycystic ovaries. *Ann N Y Acad Sci* 1993;687:217–23
7. Pache TD, Wladimiroff JW, Hop WC, Fauser BC. How to discriminate between normal and polycystic ovaries: transvaginal US study. *Radiology* 1992;183:421–3
8. Dewailly D, Robert Y, Helin I, *et al.* Ovarian stromal hypertrophy in hyperandrogenic women. *Clin Endocrinol (Oxf)* 1994;41:557–62
9. Kyei MA, LinTan S, Zaidi J, Jacobs HS. Relationship of ovarian stromal volume to serum androgen concentrations in patients with polycystic ovary syndrome. *Hum Reprod* 1998;13:1437–41
10. Fulghesu AM, Ciampelli M, Belosi C, *et al.* A new ultrasound criterion for the diagnosis of polycystic ovary syndrome: the ovarian stroma/total area ratio. *Fertil Steril* 2001;76:326–31
11. Polson DW, Adams J, Wadsworth J, Franks S. Polycystic ovaries – a common finding in normal women. *Lancet* 1988;1:870–2
12. Clayton RN, Ogden V, Hodgkinson J, *et al.* How common are polycystic ovaries in normal women and what is their significance for the fertility of the population? *Clin Endocrinol (Oxf)* 1992;37:127–34
13. Koivunen R, Laatikainen T, Tomas C, *et al.* The prevalence of polycystic ovaries in healthy women. *Acta Obstet Gynecol Scand* 1999;78:137–41
14. Yen SSC. The polycystic ovary syndrome. *Clin Endocrinol* 1980;12:177–208
15. Mason HD, Willis DS, Beard RW, *et al.* Estradiol production by granulosa cells of normal and polycystic ovaries: relationship to menstrual cycle history and concentrations of gonadotropins and sex steroids in follicular fluid. *J Clin Endocrinol Metab* 1994;79:1355–60
16. Almahbobi G, Anderiesz C, Hutchinson P, *et al.* Functional integrity of granulosa cells from polycystic ovaries. *Clin Endocrinol* 1996;44:571–80
17. Burger CW, Korsen T, van Kessel H, *et al.* Pulsatile luteinizing hormone patterns in the follicular phase of the menstrual cycle, polycystic ovarian disease (PCOD) and non-PCOD secondary amenorrhea. *J Clin Endocrinol Metab* 1985;61:1126–32
18. Kalro BN, Loucks TL, Berga SL. Neuromodulation in polycystic ovary syndrome. *Obstet Gynecol Clin N Am* 2001;28:35–62
19. Apter D, Butzow T, Laughlin GA, Yen SS. Accelerated 24-hour luteinizing hormone pulsatile activity in adolescent girls with ovarian hyperandrogenism: relevance to the developmental phase of polycystic ovarian syndrome. *J Clin Endocrinol Metab* 1994;79:119–25
20. Buckler HM, Bangah M, Healy DL, Burger HG. Vaginal progesterone administration in physiological doses normalizes raised luteinizing hormone levels in patients with polycystic ovarian syndrome. *Gynecol Endocrinol* 1992;6:275–82
21. Graf MA, Bielfeld P, Graf C, Distler W. Pattern of gonadotrophin secretion in patients with hyperandrogenaemic amenorrhoea before and after ovarian wedge resection. *Hum Reprod* 1994;9:1022–6
22. Eagleson CA, Gingrich MB, Pastor CL, *et al.* Polycystic ovarian syndrome: evidence that flutamide restores sensitivity of the gonadotropin-releasing hormone pulse generator to inhibition by estradiol and progesterone. *J Clin Endocrinol Metab* 2000;85:4047–52
23. White D, Leigh A, Wilson C, *et al.* Gonadotrophin and gonadal steroid response to a single dose of a long-acting agonist of gonadotrophin-releasing hormone in ovulatory and anovulatory women with polycystic ovary syndrome. *Clin Endocrinol (Oxf)* 1995;42:475–81
24. Gilling SC, Willis DS, Beard RW, Franks S. Hypersecretion of androstenedione by isolated thecal cells from polycystic ovaries. *J Clin Endocrinol Metab* 1994;79:1158–65
25. Franks S, Gharani N, Gilling SC. Polycystic ovary syndrome: evidence for a primary disorder of ovarian steroidogenesis. *J Steroid Biochem Mol Biol* 1999;69:269–72
26. Willis D, Mason H, Gilling SC, Franks S. Modulation by insulin of follicle-stimulating hormone and luteinizing hormone actions in human granulosa cells of normal and polycystic ovaries. *J Clin Endocrinol Metab* 1996;81:302–9
27. McGee EA, Hsueh AJW. Initial and cyclic recruitment of ovarian follicles. *Endocr Rev* 2000;21:200–14
28. McNatty KP, Smith DM, Makris A, *et al.* The microenvironment of the human antral follicle: interrelationships among the steroid levels in antral fluid, the population of granulosa cells, and the status of the oocyte *in vivo* and *in vitro. J Clin Endocrinol Metab* 1979;49:851–60
29. Willis DS, Watson H, Mason HD, *et al.* Premature response to luteinizing hormone of granulosa cells from anovulatory women with polycystic ovary syndrome: relevance to mechanism of anovulation. *J Clin Endocrinol Metab* 1998;83:3984–91
30. Jakimiuk AJ, Weitsman SR, Navab A, Magoffin DA. Luteinizing hormone receptor, steroidogenesis acute regulatory protein, and steroidogenic enzyme messenger

ribonucleic acids are overexpressed in thecal and granulosa cells from polycystic ovaries. *J Clin Endocrinol Metab* 2001;86:1318–23

31. Hillier SG. Current concepts of the roles of follicle stimulating hormone and luteinizing hormone in folliculogenesis. *Hum Reprod* 1994;9:188–91

32. Shoham Z, Jacobs HS, Insler V. Luteinizing hormone: its role, mechanism of action, and detrimental effects when hypersecreted during the follicular phase. *Fertil Steril* 1993;59:1153–61

33. Grulet H, Hecart AC, Delemer B, *et al.* Roles of LH and insulin resistance in lean and obese polycystic ovary syndrome. *Clin Endocrinol (Oxf)* 1993;38:621–6

34. Anttila L, Ding YQ, Ruutiainen K, *et al.* Clinical features and circulating gonadotropin, insulin, and androgen interactions in women with polycystic ovarian disease. *Fertil Steril* 1991;55:1057–61

35. Willis D, Franks S. Insulin action in human granulosa cells from normal and polycystic ovaries is mediated by the insulin receptor and not the type-I insulin-like growth factor receptor. *J Clin Endocrinol Metab* 1995;80:3788–90

36. Holte J, Bergh T, Gennarelli G, Wide L. The independent effect of polycystic ovary syndrome and obesity on serum concentrations of gonadotrophins and sex steroids in premenopausal women. *Clin Endocrinol* 1994;41:473–81

37. Fauser BC, Pache TD, Lamberts SW, *et al.* Serum bioactive and immunoreactive luteinizing hormone and follicle-stimulating hormone levels in women with cycle abnormalities, with or without polycystic ovarian disease. *J Clin Endocrinol Metab* 1991;73:811–17

38. Erickson GF, Magoffin DA, Garzo VG, *et al.* Granulosa cells of polycystic ovaries: are they normal or abnormal? *Hum Reprod* 1992;7:293–9

39. Pierro E, Andreani CL, Lazzarin N, *et al.* Further evidence of increased aromatase activity in granulosa luteal cells from polycystic ovary. *Hum Reprod* 1997;12:1890–6

40. Andreani CL, Pierro E, Lanzone A, *et al.* Effect of gonadotropins, insulin and IGF I on granulosa luteal cells from polycystic ovaries. *Mol Cell Endocrinol* 1994;106:91–7

41. Agarwal SK, Judd HL, Magoffin DA. A mechanism for the suppression of estrogen production in polycystic ovary syndrome. *J Clin Endocrinol Metab* 1996;81:3686–91

42. Andreani CL, Pierro E, Lazzarin N, *et al.* Effect of follicular fluid on granulosa luteal cells from polycystic ovary. *Hum Reprod* 1996;11:2107–13

43. Fauser BC. Interference with follicle stimulating hormone regulation of human ovarian function. *Mol Hum Reprod* 1996;2:227–34

44. Schipper I, Rommerts FF, Ten HP, Fauser BC. Low levels of follicle-stimulating hormone receptor-activation inhibitors in serum and follicular fluid from normal controls and anovulatory patients with or without polycystic ovary syndrome. *J Clin Endocrinol Metab* 1997;82:1325–31

45. Giudice LC. Growth factor action on ovarian function in polycystic ovary syndrome. *Endocrinol Metab Clin N Am* 1999;28:325–39

46. Erickson GF, Magoffin DA, Cragun JR, Chang RJ. The effects of insulin and insulin-like growth factors-I and -II on estradiol production by granulosa cells of polycystic ovaries. *J Clin Endocrinol Metab* 1990;70:894–902

47. Homburg R, Pariente C, Lunenfeld B, Jacobs HS. The role of insulin-like growth factor-1 (IGF-1) and IGF binding protein-1 (IGFBP-1) in the pathogenesis of polycystic ovary syndrome. *Hum Reprod* 1992;7:1379–83

48. San Roman G, Magoffin DA. Insulin-like growth factor binding proteins in ovarian follicles from women with polycystic ovarian disease: cellular source and levels in follicular fluid. *J Clin Endocrinol Metab* 1992;75:1010–16

49. Cataldo NA, Giudice LC. Follicular fluid insulin-like growth factor binding protein profiles in polycystic ovary syndrome. *J Clin Endocrinol Metab* 1992;74:695–7

50. Schuller AG, Lindenbergh KD, Pache TD, *et al.* Insulin-like growth factor binding protein-2, 28 kDa and 24 kDa insulin-like growth factor binding protein levels are decreased in fluid of dominant follicles, obtained from normal and polycystic ovaries. *Regul Pept* 1993;48:157–63

51. el Roeiy A, Chen X, Roberts VJ, *et al.* Expression of the genes encoding the insulin-like growth factors (IGF-I and -II), the IGF and insulin receptors, and IGF-binding proteins-1–6 and the localization of their gene products in normal and polycystic ovary syndrome ovaries. *J Clin Endocrinol Metab* 1994;78:1488–96

52. Mason HD, Cwyfan-Hughes S, Holly JMP, Franks S. Potent inhibition of human ovarian steroidogenesis by insulin-like growth factor binding protein-4 (IGFBP-4). *J Clin Endocrinol Metab* 1998;83:284–7

53. Peng X, Maruo T, Samoto T, Mochizuki M. Comparison of immunocytologic localization of insulin-like growth factor binding protein-4 in normal and polycystic ovary syndrome human ovaries. *Endocr J* 1996;43:269–78

54. Mason HD, Margara R, Winston RM, *et al.* Inhibition of oestradiol production by epidermal growth factor in human granulosa cells of normal and polycystic ovaries. *Clin Endocrinol (Oxf)* 1990;33:511–17

55. Mason HD, Carr L, Leake R, Franks S. Production of transforming growth factor-alpha by normal and polycystic ovaries. *J Clin Endocrinol Metab* 1995;80:2053–6

56. Volpe A, Coukos G, D'Ambrogio G, *et al.* Follicular fluid steroid and epidermal growth factor content, and in vitro estrogen release by granulosa-luteal cells from patients with polycystic ovaries in an IVF/ET program. *Eur J Obstet Gynecol Reprod Biol* 1991;42:195–9

57. Almahbobi G, Misajon A, Hutchinson P, *et al.* Hyperexpression of epidermal growth factor receptors in granulosa cells from women with polycystic ovary syndrome. *Fertil Steril* 1998;70:750–8

58. Jakimiuk AJ, Weitsman SR, Magoffin DA. 5alpha-Reductase activity in women with polycystic ovary syndrome. *J Clin Endocrinol Metab* 1999;84:2414–18

59. Greisen S, Ledet T, Ovesen P. Effects of androstenedione, insulin and luteinizing hormone on steroidogenesis in human granulosa luteal cells. *Hum Reprod* 2001;16:2061–5

60. Chang PL, Lindheim SR, Lowre C, *et al.* Normal ovulatory women with polycystic ovaries have hyperandrogenic pituitary–ovarian responses to gonadotropin-releasing hormone-agonist testing. *J Clin Endocrinol Metab* 2000; 85:995–1000

61. Rosenfield RL, Barnes RB, Cara JF, Lucky AW. Dysregulation of cytochrome P450c 17 alpha as the cause of polycystic ovarian syndrome. *Fertil Steril* 1990;53: 785–91

62. Ehrmann DA, Barnes RB, Rosenfield RL. Polycystic ovary syndrome as a form of functional ovarian hyperandrogenism due to dysregulation of androgen secretion. *Endocr Rev* 1995;16:322–53

63. Techatraisak K, Conway GS, Rumsby G. Frequency of a polymorphism in the regulatory region of the 17 alpha-hydroxylase-17,20-lyase (*cyp17*) gene in hyperandrogenic states. *Clin Endocrinol (Oxf)* 1997;46:131–4

64. Gharani N, Waterworth DM, Williamson R, Franks S. 5′ Polymorphism of the CYP17 gene is not associated with serum testosterone levels in women with polycystic ovaries. *J Clin Endocrinol Metab* 1996;81:4174

65. Witchel SF, Lee PA, Suda HM, *et al.* 17 alpha-Hydroxylase/17,20-lyase dysregulation is not caused by mutations in the coding regions of *cyp17*. *J Pediatr Adolesc Gynecol* 1998;11:133–7

66. Gilling-Smith C, Story H, Rogers V, Franks S. Evidence for a primary abnormality of thecal cell steroidogenesis in the polycystic ovary syndrome. *Clin Endocrinol (Oxf)* 1997;47:93–9

67. Gharani N, Waterworth DM, Batty S, *et al.* Association of the steroid synthesis gene CYP11a with polycystic ovary syndrome and hyperandrogenism. *Hum Mol Genet* 1997;6:397–402

68. Barbieri RL, Makris A, Randall RW, *et al.* Insulin stimulates androgen accumulation in incubations of ovarian stroma obtained from women with hyperandrogenism. *J Clin Endocrinol Metab* 1986;62:904–10

69. Nahum R, Thong KJ, Hillier SG. Metabolic regulation of androgen production by human thecal cells *in vitro*. *Hum Reprod* 1995;10:75–81

70. Nestler JE. Inositolphosphoglycans (IPGs) as mediators of insulin's steroidogenic actions. *J Basic Clin Physiol Pharmacol* 1998;9:197–204

71. Moghetti P, Castello R, Negri C, *et al.* Metformin effects on clinical features, endocrine and metabolic profiles, and insulin sensitivity in polycystic ovary syndrome: a randomized, double-blind, placebo-controlled 6-month trial, followed by open, long-term clinical evaluation. *J Clin Endocrinol Metab* 2000;85:139–46

72. Nestler JE, Jakubowicz DJ. Lean women with polycystic ovary syndrome respond to insulin reduction with decreases in ovarian P450c17 alpha activity and serum androgens. *J Clin Endocrinol Metab* 1997; 82:4075–9

73. Nestler JE, Singh R, Matt DW, *et al.* Suppression of serum insulin level by diazoxide does not alter serum testosterone or sex hormone-binding globulin levels in healthy, nonobese women. *Am J Obstet Gynecol* 1990; 163:1243–6

74. Duleba AJ, Spaczynski RZ, Olive DL. Insulin and insulin-like growth factor I stimulate the proliferation of human ovarian theca-interstitial cells. *Fertil Steril* 1998; 69:335–40

75. Reaven GM. Syndrome X. *Blood Pressure* 1992;1 (Suppl.4):13–16

76. Ehrmann DA, Barnes RB, Rosenfield RL, *et al.* Prevalence of impaired glucose tolerance and diabetes in women with polycystic ovary syndrome. *Diabetes Care* 1999;22:141–6

77. Holte J, Bergh T, Berne C, *et al.* Enhanced early insulin response to glucose in relation to insulin resistance in women with polycystic ovary syndrome and normal glucose tolerance. *J Clin Endocrinol Metab* 1994;78: 1052–8

78. Pasquali R, Casimirri F. The impact of obesity on hyperandrogenism and polycystic ovary syndrome in pre-menopausal women. *Clin Endocrinol* 1993;39:1–16

79. Holte J, Bergh T, Berne C, Lithell H. Serum lipoprotein lipid profile in women with the polycystic ovary syndrome: relation to anthropometric, endocrine and metabolic variables. *Clin Endocrinol* 1994;41:463–71

80. Talbott E, Clerici A, Berga SL, *et al.* Adverse lipid and coronary heart disease risk profiles in young women with polycystic ovary syndrome: results of a case–control study. *J Clin Epidemiol* 1998;51:415–22

81. Sampson M, Kong C, Patel A, *et al.* Ambulatory blood pressure profiles and plasminogen activator inhibitor (PAI-1) activity in lean women with and without the polycystic ovary syndrome. *Clin Endocrinol (Oxf)* 1996; 45:623–9

82. Holte J, Gennarelli G, Berne C, *et al.* Elevated ambulatory day-time blood pressure in women with polycystic ovary syndrome: a sign of a pre-hypertensive state? *Hum Reprod* 1996;11:23–8

83. Dunaif A. Insulin resistance and the polycystic ovary syndrome: mechanisms and implications for pathogenesis. *Endocr Rev* 1997;18:774–800

84. Holte J. Disturbances in insulin secretion and sensitivity in women with the polycystic ovary syndrome. *Baillières Clin Endocrinol Metab* 1996;10:221–47

85. Dunaif A, Segal KR, Futterweit W, Dobrjansky A. Profound peripheral insulin resistance, independent of obesity, in polycystic ovary syndrome. *Diabetes* 1989;38: 1165–74

86. Dunaif A, Segal KR, Shelley DR, *et al.* Evidence for distinctive and intrinsic defects in insulin action in polycystic ovary syndrome. *Diabetes* 1992;41:1257–66

87. Morales AJ, Laughlin GA, Butzow T, *et al.* Insulin, somatotropic, and luteinizing hormone axes in lean and obese women with polycystic ovary syndrome: common and distinct features. *J Clin Endocrinol Metab* 1996; 81:2854–64

88. Kahn BB, Flier JS. The syndromes of insulin resistance and acanthosis nigricans: insulin-receptor disorders in man. *N Engl J Med* 1976;294:739

89. Dunaif A, Xia J, Book CB, *et al.* Excessive insulin receptor serine phosphorylation in cultured fibroblasts and in skeletal muscle. A potential mechanism for insulin resistance in the polycystic ovary syndrome. *J Clin Invest* 1995;96:801–10

90. Conway GS, Avey C, Rumsby G. The tyrosine kinase domain of the insulin receptor gene is normal in women with hyperinsulinaemia and polycystic ovary syndrome. *Hum Reprod* 1994;9:1681–3

91. Dunaif A, Wu X, Lee A, Diamanti-Kandarakis E. Defects in insulin receptor signaling *in vivo* in the polycystic ovary

syndrome (PCOS). *Am J Physiol Endocrinol Metab* 2001;281:E392–9

92. Holte J, Bergh T, Berne C, *et al*. Restored insulin sensitivity but persistently increased early insulin secretion after weight loss in obese women with polycystic ovary syndrome. *J Clin Endocrinol Metab* 1995;80:2586–93

93. Gennarelli G. *Polycystic ovary syndrome. A study on factors of potential impact on body fat and carbohydrate metabolism.* PhD thesis, Uppsala University, 1998

94. Bjorntorp P. Classification of obese patients and complications related to the distribution of surplus fat. *Nutrition* 1990;6:131–7

95. Bjorntorp P. Adipose tissue distribution and function. *Int J Obes* 1991;15(Suppl.2):67–81

96. Gennarelli G, Holte J, Berglund L, *et al*. Prediction models for insulin resistance in the polycystic ovary syndrome. *Hum Reprod* 2000;15:2098–102

97. Good C, Tulchinsky M, Mauger D, *et al*. Bone mineral density and body composition in lean women with polycystic ovary syndrome. *Fertil Steril* 1999;72:21–5

98. Kirchengast S, Huber J. Body composition characteristics and body fat distribution in lean women with polycystic ovary syndrome. *Hum Reprod* 2001;16:1255–60

99. Ehrmann DA, Sturis J, Byrne MM, *et al*. Insulin secretory defects in polycystic ovary syndrome. Relationship to insulin sensitivity and family history of non-insulin-dependent diabetes mellitus. *J Clin Invest* 1995;96:520–7

100. Dunaif A, Finegood DT. Beta-cell dysfunction independent of obesity and glucose intolerance in the polycystic ovary syndrome. *J Clin Endocrinol Metab* 1996;81:942–7

101. Ciampelli M, Fulghesu AM, Cucinelli F, *et al*. Heterogeneity in beta cell activity, hepatic insulin clearance and peripheral insulin sensitivity in women with polycystic ovary syndrome. *Hum Reprod* 1997;12:1897–901

102. Franks S, Gharani N, McCarthy M. Candidate genes in polycystic ovary syndrome. *Hum Reprod Update* 2001;7:405–10

103. Prelevic GM, Wurzburger MI, Balint PL, Ginsburg J. Twenty-four-hour serum growth hormone, insulin, C-peptide and blood glucose profiles and serum insulin-like growth factor-I concentrations in women with polycystic ovaries. *Horm Res* 1992;37:125–31

104. Gennarelli G, Holte J, Stridsberg M, *et al*. The counterregulatory response to hypoglycaemia in women with the polycystic ovary syndrome. *Clin Endocrinol (Oxf)* 1997;46:167–74

105. Jahanfar S, Eden JA, Nguyent TV. Bulimia nervosa and polycystic ovary syndrome. *Gynecol Endocrinol* 1995;9:113–17

106. McCluskey S, Evans C, Lacey JH, *et al*. Polycystic ovary syndrome and bulimia. *Fertil Steril* 1991;55:287–91

107. Raphael FJ, Rodin DA, Peattie A, *et al*. Ovarian morphology and insulin sensitivity in women with bulimia nervosa. *Clin Endocrinol (Oxf)* 1995;43:451–5

108. Robinson S, Chan SP, Spacey S, *et al*. Postprandial thermogenesis is reduced in polycystic ovary syndrome and is associated with increased insulin resistance. *Clin Endocrinol (Oxf)* 1992;36:537–43

109. Ciampelli M, Guido M, Cucinelli F, *et al*. Hypothalamic–pituitary–adrenal axis sensitivity to opioids in women with polycystic ovary syndrome. *Fertil Steril* 2000;73:712–17

110. Lanzone A, Petraglia F, Fulghesu AM, *et al*. Corticotropin-releasing hormone induces an exaggerated response of adrenocorticotropic hormone and cortisol in polycystic ovary syndrome. *Fertil Steril* 1995;63:1195–9

111. Rodin A, Thakkar H, Taylor N, Clayton R. Hyperandrogenism in polycystic ovary syndrome. Evidence of dysregulation of 11 beta-hydroxysteroid dehydrogenase. *N Engl J Med* 1994;330:460–5

112. Modell E, Goldstein D, Reyes FI. Endocrine and behavioral responses to psychological stress in hyperandrogenic women. *Fertil Steril* 1990;53:454–9

113. Marin P, Darin N, Amemiya T, *et al*. Cortisol secretion in relation to body fat distribution in obese premenopausal women. *Metabolism* 1992;41:882–6

114. Gennarelli G, Holte J, Stridsberg M, *et al*. Response of the pituitary–adrenal axis to hypoglycemic stress in women with the polycystic ovary syndrome. *J Clin Endocrinol Metab* 1999;84:76–81

115. Bjorntorp PA. Overweight is risking fate. *Baillières Best Pract Res Clin Endocrinol Metab* 1999;13:47–69

116. Huber-Buchholz MM, Carey DG, Norman RJ. Restoration of reproductive potential by lifestyle modification in obese polycystic ovary syndrome: role of insulin sensitivity and luteinizing hormone. *J Clin Endocrinol Metab* 1999;84:1470–4

117. Fauser BC, Donderwinkel P, Schoot DC. The step-down principle in gonadotrophin treatment and the role of GnRH analogues. *Baillières Clin Obstet Gynaecol* 1993;7:309–30

118. Clark AM, Thornley B, Tomlinson L, *et al*. Weight loss in obese infertile women results in improvement in reproductive outcome for all forms of fertility treatment. *Hum Reprod* 1998;13:1502–5

119. Guzick DS, Wing R, Smith D, *et al*. Endocrine consequences of weight loss in obese, hyperandrogenic, anovulatory women. *Fertil Steril* 1994;61:598–604

120. Pasquali R, Antenucci D, Casimirri F, *et al*. Clinical and hormonal characteristics of obese amenorrheic hyperandrogenic women before and after weight loss. *J Clin Endocrinol Metab* 1989;68:173–9

121. Pasquali R, Gambineri A, Biscotti D, *et al*. Effect of long-term treatment with metformin added to hypocaloric diet on body composition, fat distribution, and androgen and insulin levels in abdominally obese women with and without the polycystic ovary syndrome. *J Clin Endocrinol Metab* 2000;85:2767–74

122. Kiddy DS, Hamilton FD, Bush A, *et al*. Improvement in endocrine and ovarian function during dietary treatment of obese women with polycystic ovary syndrome. *Clin Endocrinol (Oxf)* 1992;36:105–11

123. Hamilton-Fairley D, Kiddy D, Anyaoku V, *et al*. Response of sex hormone binding globulin and insulin-like growth factor binding protein-1 to an oral glucose tolerance test in obese women with polycystic ovary syndrome before and after calorie restriction. *Clin Endocrinol (Oxf)* 1993;39:363–7

124. Harlass FE, Plymate SR, Fariss BL, Belts RP. Weight loss is associated with correction of gonadotropin and sex steroid

abnormalities in the obese anovulatory female. *Fertil Steril* 1984;42:649–52

125. Jakubowicz DJ, Nestler JE. 17 alpha-Hydroxyprogesterone responses to leuprolide and serum androgens in obese women with and without polycystic ovary syndrome after dietary weight loss. *J Clin Endocrinol Metab* 1997;82:556–60

126. Wahrenberg H, Ek I, Reynisdottir S, *et al.* Divergent effects of weight reduction and oral anticonception treatment on adrenergic lipolysis regulation in obese women with the polycystic ovary syndrome. *J Clin Endocrinol Metab* 1999;84:2182–7

127. Rebuffe SM, Cullberg G, Lundberg PA, *et al.* Anthropometric variables and metabolism in polycystic ovarian disease. *Horm Metab Res* 1989;21:391–7

128. Hamilton FD, Kiddy D, Watson H, *et al.* Association of moderate obesity with a poor pregnancy outcome in women with polycystic ovary syndrome treated with low dose gonadotrophin. *Br J Obstet Gynaecol* 1992;99:128–31

129. Solomon CG, Seely EW. Brief review: hypertension in pregnancy: a manifestation of the insulin resistance syndrome? *Hypertension* 2001;37:232–9

130. Xiong X, Saunders LD, Wang FL, Demianczuk NN. Gestational diabetes mellitus: prevalence, risk factors, maternal and infant outcomes. *Int J Gynaecol Obstet* 2001;75:221–8

131. Holte J, Gennarelli G, Wide L, *et al.* High prevalence of polycystic ovaries and associated clinical, endocrine, and metabolic features in women with previous gestational diabetes mellitus. *J Clin Endocrinol Metab* 1998;83:1143–50

132. Radon PA, McMahon MJ, Meyer WR. Impaired glucose tolerance in pregnant women with polycystic ovary syndrome. *Obstet Gynecol* 1999;94:194–7

133. Anttila L, Karjala K, Penttila RA, *et al.* Polycystic ovaries in women with gestational diabetes. *Obstet Gynecol* 1998;92:13–16

134. Kousta E, Cela E, Lawrence N, *et al.* The prevalence of polycystic ovaries in women with a history of gestational diabetes. *Clin Endocrinol (Oxf)* 2000;53:501–7

135. Koivunen RM, Juutinen J, Vauhkonen I, *et al.* Metabolic and steroidogenic alterations related to increased frequency of polycystic ovaries in women with a history of gestational diabetes. *J Clin Endocrinol Metab* 2001;86:2591–9

136. de Vries M, Dekker GA, Schoemaker J. Higher risk of preeclampsia in the polycystic ovary syndrome. A case control study. *Eur J Obstet Gynecol Reprod Biol* 1998;76:91–5

137. Fridstrom M, Nisell H, Sjoblom P, Hillensjo T. Are women with polycystic ovary syndrome at an increased risk of pregnancy-induced hypertension and/or preeclampsia? *Hypertens Pregnancy* 1999;18:73–80

138. Holte J. Polycystic ovary syndrome and insulin resistance: thrifty genes struggling with over–feeding and sedentary life style? *J Endocrinol Invest* 1998;21:589–601

139. Hughes E, Collins J, Vandekerckhove P. Clomiphene citrate for ovulation induction in women with oligo-amenorrhoea. *Cochrane Database Syst Rev* 2000;2:CD000056

140. Hardiman P, Ginsburg J. Induction of ovulation. In Ginsburg J, Prelevic JM, eds. *Drug Therapy in Reproductive Endocrinology*. London: Arnold, 1996:86–106

141. Imani B, Eijkemans MJ, te Velde ER, *et al.* Predictors of patients remaining anovulatory during clomiphene citrate induction of ovulation in normogonadotropic oligo-amenorrheic infertility. *J Clin Endocrinol Metab* 1998;83:2361–5

142. Imani B, Eijkemans MJ, te Velde ER, *et al.* Predictors of chances to conceive in ovulatory patients during clomiphene citrate induction of ovulation in normogonadotropic oligomenorrheic infertility. *J Clin Endocrinol Metab* 1999;84:1617–22

143. Kousta E, White DM, Franks S. Modern use of clomiphene citrate in induction of ovulation. *Hum Reprod Update* 1997;3:359–65

144. Lobo RA, Gysler M, March CM, *et al.* Clinical and laboratory predictors of clomiphene response. *Fertil Steril* 1982;37:168–74

145. Dickey RP, Taylor SN, Curole DN, *et al.* Relationship of clomiphene dose and patient weight to successful treatment. *Hum Reprod* 1997;12:449–53

146. Takahashi K, Uchida A, Yamasaki H, *et al.* Transvaginal ultrasonic assessment of the response to clomiphene citrate in polycystic ovarian syndrome. *Fertil Steril* 1994;62:48–53

147. Tiitinen AE, Laatikainen TJ, Seppala MT. Serum levels of insulin-like growth factor binding protein-1 and ovulatory responses to clomiphene citrate in women with polycystic ovarian disease. *Fertil Steril* 1993;60:58–62

148. Imani B, Eijkemans MJ, de Jong FH, *et al.* Free androgen index and leptin are the most prominent endocrine predictors of ovarian response during clomiphene citrate induction of ovulation in normogonadotropic oligo-amenorrheic infertility. *J Clin Endocrinol Metab* 2000;85:676–82

149. Lobo RA, Granger LR, Davajan V, Mishell DR Jr. An extended regimen of clomiphene citrate in women unresponsive to standard therapy. *Fertil Steril* 1982;37:762–6

150. Fluker MR, Wang IY, Rowe TC. An extended 10-day course of clomiphene citrate (CC) in women with CC-resistant ovulatory disorders. *Fertil Steril* 1996;66:761–4

151. Isaacs JD Jr, Lincoln SR, Cowan BD. Extended clomiphene citrate (CC) and prednisone for the treatment of chronic anovulation resistant to CC alone. *Fertil Steril* 1997;67:641–3

152. Daly DC, Walters CA, Soto-Albors CE, *et al.* A randomized study of dexamethasone in ovulation induction with clomiphene citrate. *Fertil Steril* 1984;41:844–8

153. Trott EA, Plouffe L Jr, Hansen K, *et al.* Ovulation induction in clomiphene-resistant anovulatory women with normal dehydroepiandrosterone sulfate levels: beneficial effects of the addition of dexamethasone during the follicular phase. *Fertil Steril* 1996;66:484–6

154. Branigan EF, Estes MA. Treatment of chronic anovulation resistant to clomiphene citrate (CC) by using oral contraceptive ovarian suppression followed by repeat CC treatment. *Fertil Steril* 1999;71:544–6

155. Mitwally MF, Casper RF. Use of an aromatase inhibitor for induction of ovulation in patients with an inadequate response to clomiphene citrate. *Fertil Steril* 2001;75:305–9

156. Ghosh S, Goswami S, Chakravarty BN. Induction of ovulation with a single dose of Metrodin-HP in PCOS. *Fertil Steril* 2000;74:S33

157. Rossing MA, Daling JR, Weiss NS, *et al.* Ovarian tumors in a cohort of infertile women. *N Engl J Med* 1994;331:771–6

158. Velazquez EM, Mendoza S, Hamer T, *et al.* Metformin therapy in polycystic ovary syndrome reduces hyperinsulinemia, insulin resistance, hyperandrogenemia, and systolic blood pressure, while facilitating normal menses and pregnancy. *Metabolism* 1994;43:647–54

159. Wood AJJ. Metformin. *N Engl J Med* 1996;334:574–9

160. Diamanti KE, Kouli C, Tsianateli T, Bergiele A. Therapeutic effects of metformin on insulin resistance and hyperandrogenism in polycystic ovary syndrome. *Eur J Endocrinol* 1998;138:269–74

161. Sills ES, Perloe M, Palermo GD. Correction of hyperinsulinemia in oligoovulatory women with clomipheneresistant polycystic ovary syndrome: a review of therapeutic rationale and reproductive outcomes. *Eur J Obstet Gynecol Reprod Biol* 2000;91:135–41

162. Morin PL, Koivunen RM, Ruokonen A, Martikainen HK. Metformin therapy improves the menstrual pattern with minimal endocrine and metabolic effects in women with polycystic ovary syndrome. *Fertil Steril* 1998;69:691–6

163. Morin-Papunen LC, Vauhkonen I, Koivunen RM, *et al.* Endocrine and metabolic effects of metformin versus ethinyl estradiol–cyproterone acetate in obese women with polycystic ovary syndrome: a randomized study. *J Clin Endocrinol Metab* 2000;85:3161–8

164. Vandermolen DT, Ratts VS, Evans WS, *et al.* Metformin increases the ovulatory rate and pregnancy rate from clomiphene citrate in patients with polycystic ovary syndrome who are resistant to clomiphene citrate alone. *Fertil Steril* 2001;75:310–15

165. Nestler JE, Jakubowicz DJ, Evans WS, Pasquali R. Effects of metformin on spontaneous and clomiphene-induced ovulation in the polycystic ovary syndrome. *N Engl J Med* 1998;338:1876–80

166. Glueck CJ, Wang P, Fontaine R, *et al.* Metformin to restore normal menses in oligo-amenorrheic teenage girls with polycystic ovary syndrome (PCOS). *J Adolesc Health* 2001;29:160–9

167. Glueck CJ, Wang P, Fontaine R, *et al.* Metformin-induced resumption of normal menses in 39 of 43 (91%) previously amenorrheic women with the polycystic ovary syndrome. *Metabolism* 1999;48:511–19

168. Nestler JE, Jakubowicz DJ. Decreases in ovarian cytochrome P450c17 alpha activity and serum free testosterone after reduction of insulin secretion in polycystic ovary syndrome. *N Engl J Med* 1996;335:617–23

169. Crave JC, Fimbel S, Lejeune H, *et al.* Effects of diet and metformin administration on sex hormone-binding globulin, androgens, and insulin in hirsute and obese women. *J Clin Endocrinol Metab* 1995;80:2057–62

170. Ehrmann DA, Cavaghan MK, Imperial J, *et al.* Effects of metformin on insulin secretion, insulin action, and ovarian steroidogenesis in women with polycystic ovary syndrome. *J Clin Endocrinol Metab* 1997;82:524–30

171. Acbay O, Gundogdu S. Can metformin reduce insulin resistance in polycystic ovary syndrome? *Fertil Steril* 1996;65:946–9

172. Ng EH, Wat NM, Ho PC. Effects of metformin on ovulation rate, hormonal and metabolic profiles in women with clomiphene-resistant polycystic ovaries: a randomized, double-blinded placebo-controlled trial. *Hum Reprod* 2001;16:1625–31

173. De Leo V, la Marca A, Ditto A, *et al.* Effects of metformin on gonadotropin-induced ovulation in women with polycystic ovary syndrome. *Fertil Steril* 1999;72:282–5

174. Stadtmauer LA, Toma SK, Riehl RM, Talbert LM. Metformin treatment of patients with polycystic ovary syndrome undergoing *in vitro* fertilization improves outcomes and is associated with modulation of the insulinlike growth factors. *Fertil Steril* 2001;75:505–9

175. Arlt W, Auchus RJ, Miller WL. Thiazolidinediones but not metformin directly inhibit the steroidogenic enzymes P450c17 and 3beta-hydroxysteroid dehydrogenase. *J Biol Chem* 2001;276:16767–71

176. Sagle M, Bishop K, Ridley N, *et al.* Recurrent early miscarriage and polycystic ovaries. *Br Med J* 1988;297:1027–8

177. Balen AH, Tan SL, MacDougall J, Jacobs HS. Miscarriage rates following *in-vitro* fertilization are increased in women with polycystic ovaries and reduced by pituitary desensitization with buserelin. *Hum Reprod* 1993;8:959–64

178. Glueck CJ, Phillips H, Cameron D, *et al.* Continuing metformin throughout pregnancy in women with polycystic ovary syndrome appears to safely reduce firsttrimester spontaneous abortion: a pilot study. *Fertil Steril* 2001;75:46–52

179. Day C. Thiazolidinediones: a new class of antidiabetic drugs. *Diabet Med* 1999;16:179–92

180. Mitwally MF, Kuscu NK, Yalcinkaya TM. High ovulatory rates with use of troglitazone in clomiphene-resistant women with polycystic ovary syndrome. *Hum Reprod* 1999;14:2700–3

181. Hasegawa I, Murakawa H, Suzuki M, *et al.* Effect of troglitazone on endocrine and ovulatory performance in women with insulin resistance-related polycystic ovary syndrome. *Fertil Steril* 1999;71:323–7

182. Azziz R, Ehrmann D, Legro RS, *et al.* Troglitazone improves ovulation and hirsutism in the polycystic ovary syndrome: a multicenter, double blind, placebo-controlled trial. *J Clin Endocrinol Metab* 2001;86:1626–32

183. Ehrmann DA, Schneider DJ, Sobel BE, *et al.* Troglitazone improves defects in insulin action, insulin secretion, ovarian steroidogenesis, and fibrinolysis in women with polycystic ovary syndrome. *J Clin Endocrinol Metab* 1997;82:2108–16

184. Dunaif A, Scott D, Finegood D, *et al.* The insulinsensitizing agent troglitazone improves metabolic and reproductive abnormalities in the polycystic ovary syndrome. *J Clin Endocrinol Metab* 1996;81:3299–306

185. Watkins PB, Whitcomb RW. Hepatic dysfunction associated with troglitazone. *N Engl J Med* 1998;338: 916–17

186. Van Der Meer M, Hompes PG, De Boer J, *et al*. Cohort size rather than follicle-stimulating hormone threshold level determines ovarian sensitivity in polycystic ovary syndrome. *J Clin Endocrinol Metab* 1998;83:423–6

187. Hamilton FD, Franks S. Common problems in induction of ovulation. *Baillières Clin Obstet Gynaecol* 1990;4:609–25

188. Hamilton FD, Kiddy D, Watson H, *et al*. Low-dose gonadotrophin therapy for induction of ovulation in 100 women with polycystic ovary syndrome. *Hum Reprod* 1991;6:1095–9

189. Shoham Z, Patel A, Jacobs HS. Polycystic ovarian syndrome: safety and effectiveness of stepwise and low-dose administration of purified follicle-stimulating hormone. *Fertil Steril* 1991;55:1051–6

190. Dale O, Tanbo T, Lunde O, Abyholm T. Ovulation induction with low-dose follicle-stimulating hormone in women with the polycystic ovary syndrome. *Acta Obstet Gynecol Scand* 1993;72:43–6

191. Homburg R, Levy T, Ben RZ. A comparative prospective study of conventional regimen with chronic low-dose administration of follicle-stimulating hormone for anovulation associated with polycystic ovary syndrome. *Fertil Steril* 1995;63:729–33

192. van Santbrink EJ, Fauser BC. Urinary follicle-stimulating hormone for normogonadotropic clomiphene-resistant anovulatory infertility: prospective, randomized comparison between low dose step-up and step-down dose regimens. *J Clin Endocrinol Metab* 1997;82:3597–602

193. Coelingh Bennink HJ, Fauser BC, Out HJ. Recombinant follicle-stimulating hormone (FSH; Puregon) is more efficient than urinary FSH (Metrodin) in women with clomiphene citrate-resistant, normogonadotropic, chronic anovulation: a prospective, multicenter, assessor-blind, randomized, clinical trial. European Puregon Collaborative Anovulation Study Group. *Fertil Steril* 1998;69:19–25

194. Hedon B, Hugues JN, Emperaire JC, *et al*. A comparative prospective study of a chronic low dose versus a conventional ovulation stimulation regimen using recombinant human follicle stimulating hormone in anovulatory infertile women. *Hum Reprod* 1998;13:2688–92

195. Balasch J, Fabregues F, Creus M, *et al*. Recombinant human follicle-stimulating hormone for ovulation induction in polycystic ovary syndrome: a prospective, randomized trial of two starting doses in a chronic low-dose step-up protocol. *J Assist Reprod Genet* 2000;17:561–5

196. Hayden CJ, Rutherford AJ, Balen AH. Induction of ovulation with the use of a starting dose of 50 units of recombinant human follicle-stimulating hormone (Puregon). *Fertil Steril* 1999;71:106–8

197. White DM, Polson DW, Kiddy D, *et al*. Induction of ovulation with low-dose gonadotropins in polycystic ovary syndrome: an analysis of 109 pregnancies in 225 women. *J Clin Endocrinol Metab* 1996;81:3821–4

198. Homburg R, Howles CM. Low-dose FSH therapy for anovulatory infertility associated with polycystic ovary syndrome: rationale, results, reflections and refinements. *Hum Reprod Update* 1999;5:493–9

199. Herman A, Ron ER, Golan A, *et al*. Overstimulated cycles under low-dose gonadotrophins in patients with polycystic ovary syndrome: characterization and management. *Hum Reprod* 1993;8:30–4

200. Hugues JN, Cedrin DI, Avril C, *et al*. Sequential step-up and step-down dose regimen: an alternative method for ovulation induction with follicle-stimulating hormone in polycystic ovarian syndrome. *Hum Reprod* 1996;11: 2581–4

201. Balasch J, Fabregues F, Creus M, *et al*. Follicular development and hormone concentrations following recombinant FSH administration for anovulation associated with polycystic ovarian syndrome: prospective, randomized comparison between low-dose step-up and modified step-down regimens. *Hum Reprod* 2001;16:652–6

202. Hughes E, Collins J, Vandekerckhove P. Ovulation induction with urinary follicle stimulating hormone versus human menopausal gonadotropin for clomiphene-resistant polycystic ovary syndrome. *Cochrane Database Syst Rev* 2000;2:CD000087

203. Teissier MP, Chable H, Paulhac S, Aubard Y. Recombinant human follicle stimulating hormone versus human menopausal gonadotrophin induction: effects in mature follicle endocrinology. *Hum Reprod* 1999;14: 2236–41

204. Fulghesu AM, Apa R, Belosi C, *et al*. Recombinant versus urinary follicle-stimulating hormone in the low-dose regimen in anovulatory patients with polycystic ovary syndrome: a safer and more effective treatment. *Horm Res* 2001;55:224–8

205. Balasch J, Fabregues F, Penarrubia J, *et al*. Follicular development and hormonal levels following highly purified or recombinant follicle-stimulating hormone administration in ovulatory women and WHO group II anovulatory infertile patients. *J Assist Reprod Genet* 1998;15:552–9

206. Bayram N, van Wely M, van Der Veen F. Recombinant FSH versus urinary gonadotrophins or recombinant FSH for ovulation induction in subfertility associated with polycystic ovary syndrome (Cochrane Review). *Cochrane Database Syst Rev* 2001;2:CD002121

207. Homburg R. Adverse effects of luteinizing hormone on fertility: fact or fantasy. *Baillières Clin Obstet Gynaecol* 1998;12:555–63

208. Homburg R, Levy T, Berkovitz D, *et al*. Gonadotropin-releasing hormone agonist reduces the miscarriage rate for pregnancies achieved in women with polycystic ovarian syndrome. *Fertil Steril* 1993;59:527–31

209. Lidor AL, Goldenberg M, Cohen SB, *et al*. Management of women with polycystic ovary syndrome who experienced premature luteinization during clomiphene citrate treatment. *Fertil Steril* 2000;74:749–52

210. Schoot DC, Pijlman B, Stijnen T, Fauser BC. Effects of gonadotropin releasing hormone agonist addition to gonadotropin induction of ovulation in polycystic ovary syndrome patients. *Eur J Obstet Gynecol Reprod Biol* 1992; 45:53–8

211. Hughes E, Collins J, Vandekerckhove P. Gonadotrophin-releasing hormone analogue as an adjunct to gonadotropin therapy for clomiphene-resistant polycystic ovarian syndrome. *Cochrane Database Syst Rev* 2000;2:CD000097

212. Clifford K, Rai R, Watson H, *et al*. Does suppressing luteinising hormone secretion reduce the miscarriage rate? Results of a randomised controlled trial. *Br Med J* 1996;312:1508–11

213. Jacobs HS, Agrawal R. Complications of ovarian stimulation. *Baillières Clin Obstet Gynaecol* 1998;12:565–79

214. Dale PO, Tanbo T, Haug E, Abyholm T. The impact of insulin resistance on the outcome of ovulation induction with low-dose follicle stimulating hormone in women with polycystic ovary syndrome. *Hum Reprod* 1998;13:567–70

215. Balen AH, Jacobs HS. A prospective study comparing unilateral and bilateral laparoscopic ovarian diathermy in women with the polycystic ovary syndrome. *Fertil Steril* 1994;62:921–5

216. Cohen J. Laparoscopic surgical treatment of infertility related to polycystic ovary syndrome. In Kovacs GT, ed. *Polycystic Ovary Syndrome*. Cambridge: Cambridge University Press, 2000:144–58

217. Abdel GA, Khatim MS, Alnaser HM, *et al*. Ovarian electrocautery: responders versus non-responders. *Gynecol Endocrinol* 1993;7:43–8

218. Dale PO, Tanbo T, Vaaler S, Abyholm T. Body weight, hyperinsulinemia, and gonadotropin levels in the polycystic ovarian syndrome: evidence of two distinct populations. *Fertil Steril* 1992;58:487–91

219. Lemieux S, Lewis GF, Ben CA, *et al*. Correction of hyperandrogenemia by laparoscopic ovarian cautery in women with polycystic ovarian syndrome is not accompanied by improved insulin sensitivity or lipid-lipoprotein levels. *J Clin Endocrinol Metab* 1999;84:4278–82

220. Nugent D, Vandekerckhove P, Hughes E, *et al*. Gonadotrophin therapy for ovulation induction in subfertility associated with polycystic ovary syndrome. *Cochrane Database Syst Rev* 2000;4:CD000410

221. Buckett WM, Tan SL. Use of luteinizing hormone releasing hormone agonists in polycystic ovary syndrome. *Baillières Clin Obstet Gynaecol* 1998;12:593–606

222. Dor J, Shulman A, Levran D, *et al*. The treatment of patients with polycystic ovarian syndrome by *in-vitro* fertilization and embryo transfer: a comparison of results with those of patients with tubal infertility. *Hum Reprod* 1990;5:816–18

223. Urman B, Fluker MR, Yuen BH, *et al*. The outcome of *in vitro* fertilization and embryo transfer in women with polycystic ovary syndrome failing to conceive after ovulation induction with exogenous gonadotropins. *Fertil Steril* 1992;57:1269–73

224. MacDougall MJ, Tan SL, Balen A, Jacobs HS. A controlled study comparing patients with and without polycystic ovaries undergoing *in-vitro* fertilization. *Hum Reprod* 1993;8:233–7

225. Buyalos RP, Lee CT. Polycystic ovary syndrome: pathophysiology and outcome with *in vitro* fertilization. *Fertil Steril* 1996;65:1–10

226. Teissier MP, Chable H, Paulhac S, Aubard Y. Comparison of follicle steroidogenesis from normal and polycystic ovaries in women undergoing IVF: relationship between steroid concentrations, follicle size, oocyte quality and fecundability. *Hum Reprod* 2000;15:2471–7

227. Engmann L, Maconochie N, Sladkevicius P, *et al*. The outcome of *in-vitro* fertilization treatment in women with sonographic evidence of polycystic ovarian morphology. *Hum Reprod* 1999;14:167–71

228. Fridstrom M, Sjoblom P, Granberg M, Hillensjo T. A cost comparison of infertility treatment for clomiphene resistant polycystic ovary syndrome. *Acta Obstet Gynecol Scand* 1999;78:212–16

Human seminology

Human seminology: semen examination and *in vitro* evaluation of human seminal cells

7

Amnon Makler

INTRODUCTION

Human seminology has undergone remarkable changes since the middle of the last century. From the simple microscopic semen analysis in small laboratories that was the main methodology at that time, its progress has been steadily growing throughout the last 50 years by ample sophisticated means. The computer-assisted semen analyzers (CASA) heralded a new era in seminology. The success of assisted reproductive technologies (ART) and *in vitro* fertilization (IVF) is the result of our understanding of the sequence of events from the moment sperm are ejaculated until one of them penetrates and fertilizes the egg. Thus, it promoted the development of sperm function tests that were devoted to explain the success or failure of these treatments. This progress culminated shortly after the report by Palermo and co-workers on the first successful intracytoplasmic sperm injection (ICSI) treatment. Consequently, the inevitable question arose: how far should the male partner be investigated before deciding on ICSI, the ultimate promising choice?

Whatever the answer to this question, it is still essential that the male partner should undergo a careful evaluation of his semen, including, if necessary, the more profound tests described in this chapter.

SEMEN ANALYSIS

Semen analysis is indispensable in the evaluation of male infertility, and at present there seems to be no substitute for the information provided by this procedure. Semen quality tends to vary unpredictably from one day to another. These variations are caused by duration of abstinence, how the specimen was provided, fear of results, ambient temperature, transportation time, etc. For screening purposes, a basic analysis may be sufficient. It includes the assessment of semen volume, viscosity, sperm concentration, motility and morphology. The extended semen analysis should be performed only when specially required, and includes motility and movement characteristics assessment by CASA, profound morphology assessment, biochemical assays of certain biological substances, microbiology studies and some of the sperm function tests.

Preliminary procedures

Collection and transportation

As noted above, semen quality partly depends on the time gap since the previous ejaculation. For the purposes of clinical analysis, it is recommended that this time approximately match the typical frequency of sexual activity by the couple, between two and four days but not more than seven. Too short an interval may impose difficulties on the patient, decrease semen volume and sperm concentration, while an extended abstinence period may artificially increase sperm count and the percentage of non-motile and abnormal forms.

The preferred mode of semen collection is by masturbation into a clean container with a wide opening. No lubricants should be used during this act. As an alternative, semen may be provided by interrupted coitus. However, the concentrated first fraction of the ejaculate may be lost. A risk of contamination by harmful vaginal contents also exists. Specimens transported from the patient's home should arrive within 60–90 min and exposure to cold weather below 18°C should be avoided by keeping the well-sealed container in intimate contact with the body of the carrier.

ROUTINE SEMEN ANALYSIS

Macroscopic evaluation of the analyzed semen

This procedure includes assessment of semen liquefaction, viscosity, volume, color and pH. Macroscopic evaluation should be initiated only after liquefaction of the naturally coagulated semen has been completed.

Semen liquefaction

Liquefaction starts immediately after ejaculation. Its mechanism is still not entirely clear. It is probably induced by fibrinogen-like factors (seminoglobin, fibronectin, lactoferrin[2]) of the seminal vesicles which combine with the prostatic enzyme vesiculase to form a clot. Normally, within 10–30 min, this clot undergoes proteolysis by a series of chymotrypsin-like prostatic enzymes that has been identified as the prostatic specific antigen (PSA)[3].

Non-liquefaction of more than 30–60 min is considered abnormal. To ensure that liquefaction has been completed, a drop of the seminal fluid is deposited between a slide and a cover slip. With a phase contrast microscope, a typical mesh containing trapped, non-progressive shaking sperm can be seen. Normally, this mesh slowly disappears while the slide remains untouched.

Semen viscosity

Seminal fluid viscosity should be checked only after the complete liquefaction process has taken place, since it is an entirely different entity. Although viscosity can be determined by special instruments and expressed by international units (centipoise), for daily clinical purposes it is enough to describe it verbally. An applicator is inserted into the seminal fluid and immediately withdrawn. Normal semen shows threading of less than 1 cm above the surface, whereas up to 10–12 cm is observed in viscous semen. Non-liquefaction and viscous semen may

impair sperm free movement. In severe cases, the captured sperm are unable to get access to the cervical mucus and achieve penetration. There are several suggested means to overcome non-liquefaction, most involving the addition of various lytic enzymes such as amylase, chymotrypsin and hyalorunidase. However, such substances can cause damage to the sperm membrane. Viscous semen can be treated by being streamed several times with a syringe through a rigid plastic cannula, thereby breaking the architectural network of the fluid.

Seminal volume

Though this is simply assessed using a graduated test tube, the measurement should not be ignored, since it is the only means to determine the total number of sperm in the ejaculate. Reduced volume (hypospermia) of less than 1.5 ml is responsible for failure of the ejaculated sperm to reach the cervix, while excessive volume (hyperspermia), of 6.0 ml and above, may dilute an otherwise adequate sperm concentration and decrease their number below the normal level.

The patient should always be asked if the container holds his entire specimen, since sperm are not uniformly spread along the stream of the ejaculate. The first one-third is combined from the prostatic fluid and 70–80% of sperm that were stored within the ampulla of the vas deferens. The other two-thirds are of seminal vesicle origin and contain a low number of sperm. In fact, this fraction is responsible for the dilution effect in hyperspermic specimens. Hence, results of analysis where part of the specimen is missing should be considered invalid.

Total lack of seminal fluid (aspermia) generally indicates failure to ejaculate due to inability to produce a specimen 'on demand', fear of results, diabetic neuropathy, postoperative 'retrograde ejaculation' into the bladder and more.

Color and pH

Normal semen appears milky white. A yellow color is due to contents of flavoproteins, while red-brown raises the suspicion of red blood cells (hematospermia). The pH of fresh semen varies between narrow limits, 7.2–7.8, that do not exceed the levels that can affect sperm motility and viability[4]. Although many laboratories measure semen pH routinely, either by litmus paper or electronic pH meters, such tests in fresh ejaculates are of limited importance.

MICROSCOPIC SPERM ANALYSIS

This is the core of routine semen analysis and no adequate information about the male factor can be obtained without it. Three parameters are assessed microscopically: sperm concentration, sperm motility and sperm morphology. Many laboratories also add the supravital staining (sperm live–dead ratio) and the sperm longevity test (sperm survival assessment).

Sperm concentration

This parameter has always been expressed as millions per ml. In addition, the total number of sperm in the entire ejaculate should be calculated. This is simply done by multiplying concentration by volume. Usually, out of the two, only sperm concentration is mentioned. It is considered the more important because it determines how many sperm would penetrate the cervical mucus before the rest are immobilized by the hostile vaginal fluid. However, just mentioning one and excluding the other will provide insufficient information.

Determination of sperm concentration

It is recommended to start the microscopic analysis by scanning a sample from a well-mixed semen spread on a slide and secured by a cover slip. This can help to eliminate technical errors later on and to validate final results when sperm counts are performed by non-microscopic methods. Obviously, it cannot replace the need to determine sperm parameters by accurate methods, however.

Quantitative assessment of sperm concentration is based on counting the number of sperm in a sample of a known volume, picked from well-mixed liquefied semen. The sample should be small enough not to exceed 200 counted sperm. For this purpose several kinds of counting chambers are employed. The applied sample is observed microscopically under $\times 200$ magnification ($\times 20$ objective and $\times 10$ eyepiece), preferably with a phase-contrast microscope.

Chambers for performance of sperm counts

Counting chambers are divided into two groups: deep chambers by which sperm counts are performed only with diluted semen, and shallow chambers where sperm can be counted directly from an undiluted specimen.

The hemocytometer, a deep chamber, is the oldest device used for counting sperm. The best known is 'The Improved Neubauer', originally developed for blood-cells counts. Since the chamber is 100 μm deep, both blood cells and sperm in the crude semen look multi-layered, blurred and not in one focal plane. As such, they cannot be counted unless diluted 1:10 or 1:50, depending on the concentration level of the original specimen. The dilution is made with isotonic saline containing 3.5% formaldehyde to immobilize sperm so that all of them, motile and immotile, can be counted. Each compartment of the chamber contains a single drop from the well-mixed suspension. However, counting can be started only after 20–30 min has elapsed, the time needed for sperm to sink and spread in one layer upon the bottom. Sperm heads in five out of 25 double-lined squares, located at the center of the hemocytometer, are counted (Figure 1). Each of these squares is $0.2 \text{ mm} \times 0.2 \text{ mm} = 0.04 \text{ mm}^2$, subdivided into 16 tiny squares just to facilitate counting. Heads that touch the upper horizontal and the left double lines are counted while those that touch the lower horizontal and the right double lines are ignored. The total number of sperm counted in the five double-lined squares indicates their number in a space bounded within $5 \times 0.04 \text{ mm}^2 \times 0.1 \text{ mm}$ (chamber depth), which is 0.02 mm^3. After dividing by two and multiplying by the dilution factor, five zeroes are added to obtain the concentration in millions/ml. The basic formula for the calculation is:

$$\text{Sperm concentration in } 10^6/\text{ml} = \text{N}/2 \times \text{D} \times 100\ 000,$$

where N = number counted in five double-lined squares and D = dilution factor. The counts in both compartments are compared and, unless the difference does not exceed 20%, an average is made. When the difference is higher, the results should be ignored and another count on a new sample started from scratch. The accuracy of the sperm counts depends on several factors: how well the semen is mixed and whether the sperm are uniformly

(a)

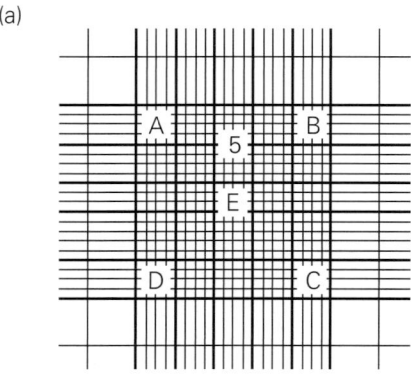

(b)

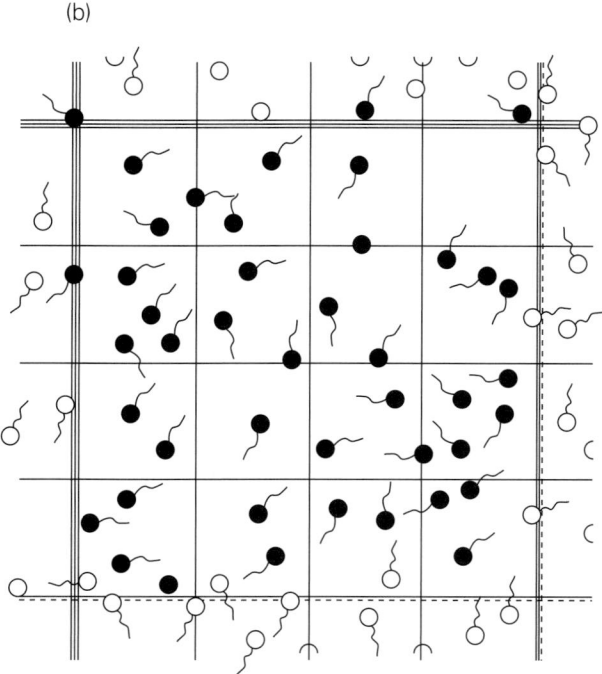

Figure 1 The central part of the Neubauer chamber. (a) The five squares marked by letters A to E in which sperm are counted; (b) one of the 0.2 mm × 0.2 mm of the above squares, bounded by double lines and subdivided into 16 small squares in which sperm are counted. Black head sperm are counted, white head sperm are ignored

distributed; whether or not each sample represents the entire specimen; and how skillful the technician is in making the dilution and applying the drops in the chamber correctly.

Being reusable and a low cost chamber, the hemocytometer is still prevalent in laboratories and clinical offices. In able hands it is considered a reliable tool for sperm counting despite the tedious and time-consuming procedure and good training that its use entails. Hence, most doctors refrain from using it. The multiple steps of preparation, including dilution, cause cumulative errors that are hard to avoid. Freund and Carol[5] showed deviations of 15–20% among counts by different technicians, while Jequier and Ukombe[6] reported on a coefficient of variations of up to 44% among counts from the same specimen. The need to wait 20–30 min for sperm precipitation before counting is started limits its use in busy laboratories and sperm banks.

The Makler Counting Chamber (Sefi Medical Instruments, Haifa, Israel) appeared about 20 years ago[7] and was developed to overcome the main drawbacks of the hemocytometer, eliminating the need for semen dilution and the long wait for sperm precipitation before counting can be initiated. This shallowest available chamber is only 10 μm deep, one-tenth the depth of the hemocytometer. It allows even distribution and clear vision of sperm in one focal plane. Hence, a sperm count from undiluted semen can be performed very rapidly.

The chamber is made of two discs of flat glass, one fixed in the center of the lower part and the other, surrounded by a metal ring, constituting the cover (Figure 2). Imprinted in its center is a 1 mm² grid, subdivided into 100 squares, each 0.1 mm × 0.1 mm (Figure 3). The two discs are held apart by four pins covered by a 10 μm-thick layer of highly resistant quartz (Figure 4). When the cover is placed firmly upon these pins and observed against a reflected fluorescent light, color fringes (Newton's rings) are seen at the four contact points. This confirms that a 10 μm-deep space has been formed and no particles above this size can intervene between the two surface areas.

A drop of about 3–5 μl from the well-mixed undiluted specimen is deposited on the center of the lower glass and covered immediately. Using an × 20 objective and × 10 eyepiece, evaluation of sperm concentration and of motility start instantly. Sperm are counted in a row of ten squares while motility is estimated by scanning several view areas. This is repeated in two or three more rows and the average is taken. Each row of ten squares confines one millionth of a ml (1.0 mm long × 0.1 mm wide × 0.01 mm deep = 10⁻³ mm³ = 10⁻⁶ ml). Hence, the number of sperm in a row of ten

Figure 2 The Makler Counting Chamber. A drop from the undiluted semen is applied on the center of the main body before placing the cover glass

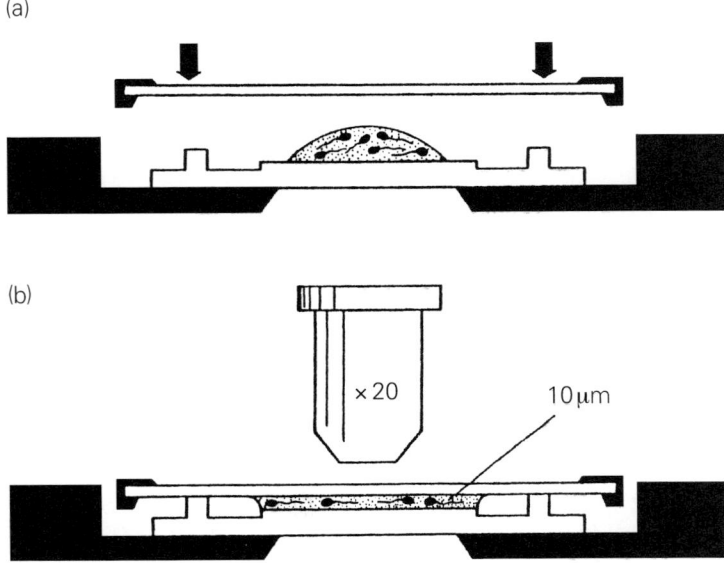

Figure 3 Cross-section of the Makler chamber. (a) A drop of semen placed upon the center of the lower part. (b) The drop becomes 10 μm thick by the cover glass, placed upon the four slightly elevated pins. Sperm cells are spread in one focal plane where they can move freely

squares indicates their concentration in 10^6/ml. Aggressive sperm that cannot be counted are immobilized by splitting the original specimen and placing one test tube in a cup of hot water at 50–60°C for about 10–15 min. Adding formaldehyde or other spermicides should be avoided, as they cause sperm clumping. In cases of

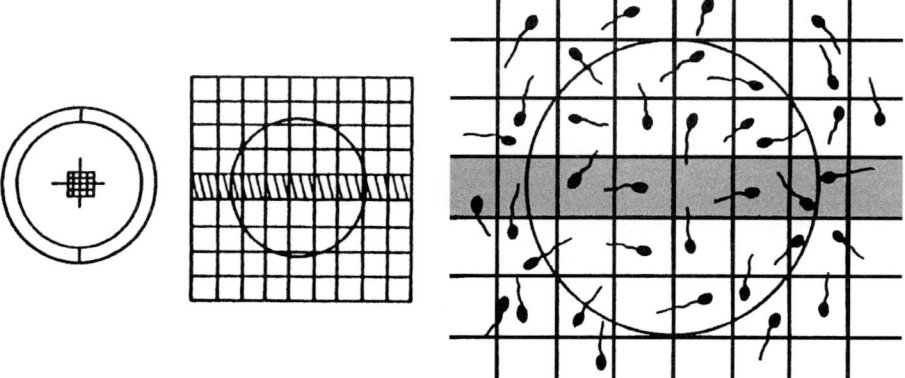

Figure 4 The 1 mm² grid in the center of the cover of the Makler chamber, subdivided into 100 squares, 0.1 mm × 0.1 mm each. The number of sperm in a line of ten squares indicates their concentration in millions/ml

extremely low counts, sperm are counted in the entire 100 squares of the grid and five zeroes are added to obtain concentration in millions/ml.

The main advantage of the Makler Chamber is its ability to perform rapid sperm counts from undiluted specimen. Sperm motility can also be evaluated while the moving sperm are examined under standard conditions (see 'Evaluation of sperm motility', below). By using a phase-contrast microscope, preliminary morphology of unstained and undistorted sperm in their natural wet environment can also be assessed. Being reusable hundreds of times makes the chamber quite economical, too.

The main disadvantage of the Makler Chamber is the need to give the optical parts a thorough cleaning and to check the correct placement of the cover glass before each count, while looking for color fringes (Newton's rings) is obligatory to avoid inaccuracy. If these steps, which users may not be aware of, are not adhered to, errors may occur.

The microcell is another type of a shallow counting chamber[8] that does not need semen dilution (Fertility Technologies, Natick, MA). The shape of an ordinary slide, the cell is disposable, being designed for a single use. Most contain a fixed, unchangeable space of 20 μm between two slides. For sperm count, a special grid must be placed in the microscope eyepiece. The final results are obtained from the number of sperm counted in a certain area of this grid and the cell depth.

Due to the fixed predetermined depth there is no need to check for its thickness after the sample is deposited. As with the 10 μm chamber, semen dilution is not required and the analysis can be performed immediately and rapidly. The main disadvantage is the need to place a grid in the eyepiece and to remove it when counting is finished. Since the area covered by this grid is specific for each microscope, it needs to be calibrated and a factor must be used to calculate the final result. With not many users aware of this requirement, results may vary by up to ± 20% of the actual value.

The 20 μm thickness of the microcell is beyond the depth of field of the × 20 objective so that many sperm are seen quite blurred. Also, as the two parts of the microcell are non-separable, very viscous samples cannot be inserted through the thin gap between the tightly fixed slides.

Other means of sperm concentration assessment

The Coulter Counter for blood cell analysis[9] was once considered for semen analysis (Coulter Electronics, Hialeah, FL). However, results were very frequently inaccurate due to the Coulter Counter's inability to differentiate between sperm and debris.

CASA systems which are mainly used for sperm motility evaluation (see below) also provide readouts of sperm concentration.

Some non-microscopic means for determining sperm concentration have also been suggested. Color-induced chemical reactions with sperm DNA have been measured fluorometrically or colorimetrically and calibrated into a scale used to test unknown specimens[10]. However, their practical use is still very limited.

Normal values

For more than 50 years the studies of MacLeod and Gold[11–13] were the benchmark in andrology regarding the three main parameters of human sperm. Accordingly, 20 millions/ml was accepted as the lower normal value for fertile males. However, this value indicated the probability of a natural, non-assisted pregnancy and has never been considered the strict cutoff between fertility and infertility. More recent studies[14,15] have pointed to values as low as 10 millions/ml, and Burris and co-workers[16] reported on five out of 31 pregnancies (16%) when sperm counts revealed $< 1 \times 10^6$/ml. Daily practice shows that there is no lower limit, except a total lack of live sperm, where pregnancy could not happen.

For practical purposes the following terms are being used:

Normospermia: above 20 million/ml.

Oligospermia: less than 20 million/ml.

Polyzoospermia: high level of sperm concentration of 250 million/ml or above. The significance of polyzoospermia as a cause of infertility is still a matter for debate.

Azoospermia: total absence of sperm in the ejaculate. This finding must be treated with great caution. Due to the feasibility to treat severe oligospermia by ICSI, the examiner must scan several samples in an attempt to discover even a single spermatozoon. When such scrutiny fails, it is suggested that the specimen be spun for 10–20 min at 300 *g* and a thick drop from the sediment examined.

Evaluation of sperm motility

Sperm motility refers to two main parameters. The first is the ratio between motile and non-motile sperm, i.e. the percentage of motility. The second is the movement quality of motile sperm, their speed and track patterns.

Until the last two decades, sperm motility was assessed subjectively, unlike sperm concentration, which is determined by precise means. Usually, for clinical purposes, a drop from a well-mixed specimen is examined on a simple slide covered by a cover slip. Under microscopic × 200 power, several view areas are scanned and the fraction of motile sperm to the nearest 10% is estimated. The movement quality is graded + 1 to + 4 as follows:

(1) + 1 – sluggish movement with minimal displacement;

(2) + 2 – moderate forward progression mainly of linear pattern;

(3) + 3 – good forward progression mainly of linear pattern;

(4) + 4 – vigorous progression, very high velocity with linear or jerky pattern.

The main problem with this method is the lack of standard conditions for the examined sperm. Free sperm movement depends on the pressure applied on the cover slip. The higher the pressure, the greater the friction that impairs sperm free movement[17]. On the other hand, low pressure and a thick drop induce overestimation of the percentage of moving sperm. Due to the appearance of blurred sperm in the thick sample, many moving sperm are overlooked. This becomes a serious problem when one needs to compare different specimens or to follow samples from the same specimen periodically.

To examine sperm motility under standard conditions to the analyzed drop, the use of the 10 μm-deep Makler Chamber is suggested. The drop always spreads uniformly into that thickness, no matter how powerful the pressure applied on the cover. Sperm swim in a free, frictionless space and are clearly seen in one focal plane. Basically, under microscopic observation, sperm *seem* to move in two dimensions. In fact, sperm swim in screw-like three-dimensional movements[18], occupying a space of at least 50 μm. Nevertheless, even in shallower chambers, where their tail and body touch the walls, sperm do not exhibit any sign of slowing down or of reduced efficacy of movement. Velocity is affected only when the space becomes less than 10 μm[17].

Objective determination of sperm motility

Many methods for sperm motility determination have been developed during the last three decades, among them being the non-visual approach in which a certain effect by the moving sperm is detected either photoelectrically[19,20] or by laser Doppler systems[21]. Most of these systems are not in practical use to date, however.

Among visual methods are those based on cinematography[22], still camera time-lapse photography[23] and multiple exposure photography (MEP)[24]. MEP was the first technique to provide full details on sperm parameters such as sperm concentration, percent of motility, average sperm velocity in μm/s and frequency distribution of sperm velocities. It was the incentive for the development of CASAs that eliminated the tedious phase of taking photographs. The first commercial computer, the 'CELL SOFT automated semen analyzer' (Cryo Resources, NY), was developed in 1985 and studied immediately in two centers[25,26]. Later on, other CASAs appeared on the market, the most well known being Hamilton Thorne HTM (Hamilton-Thorne Research, Beverly, MA), the CellTrak System (CTS Motion Analysis Corp., Santa Rosa, CA) and the Stromberg-Mika system (Mika Medical, Bad Feilnbach, Germany). The CASAs provided a descriptive analysis of sperm movement characteristics and tracks, defined as follows[27] (Figure 5):

(1) VSL (straight linear velocity): the time-average velocity of the sperm head along a straight line from the first to the last position;

(2) VCL (curvilinear velocity): the time-average velocity of the sperm head along its actual trajectory;

(3) VAP (average path velocity): the time-average velocity of the sperm head along its average velocity;

(4) LIN (linearity): linearity of the curvilinear velocity expressed as the ratio VSL/VCL;

(5) STR (straightness): straightness of the average path expressed as the ratio VSL/VAP;

(6) ALH (amplitude of lateral head displacement): amplitude of variations of the actual sperm-head trajectory about its average trajectory.

The data provided by CASAs seem now somewhat exaggerated. To date, most of the computer analyzers have not been of much value to the clinician, the resulting information not helping the consulting doctor to predict the male partner's fertilization potential. Their main value lies in the realms of pure basic science and in research on the physiology of sperm movements.

Normal values

Based on MacLeod's subjective assessment[12], 60% motile sperm was accepted as the lower normal rate for fertile men. However, objective motility determination by the MEP method showed that examiners' judgment tended to overestimate moving sperm by about 15% because human eyes are trapped by the moving sperm, ignoring many of the non-motile sperm[28]. Hence, it seems that MacLeod's evaluations may have also been somewhat overestimated. To date, 40–45% motile sperm is widely accepted as the low borderline for fertile men.

Two terms are commonly used to describe abnormal sperm motility:

Asthenospermia: low motility grade and percentage of less than 40% motile sperm.

Necrospermia: total lack of motile sperm.

The latter term should be used only after its validation by the sperm vitality test (see below).

Sperm longevity (survival) test

Quite often one can find fresh sperm with good initial motility that drops abruptly. To quantify this phenomenon, it is necessary to follow the change of sperm motility with time. The 'sperm longevity' or 'survival' test is performed either on sperm in the semen or on

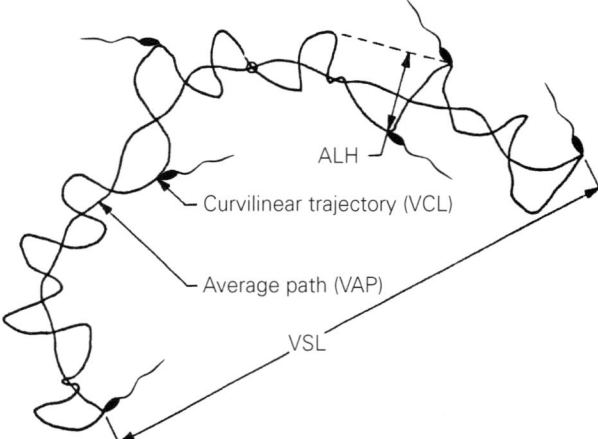

Figure 5 Various sperm movement patterns analyzed by computer-assisted semen analyzers as described and interpreted in the text. Adapted with permission from Davis RO, Niswander PW, Katz DF. New measures of sperm motion. I. Adaptive smoothing and harmonic analysis. *J Androl* 1992;13:139–52

sperm washed in artificial media. A useful index of longevity is provided by comparing the percentage of motile sperm 1 h and 4 h after ejaculation[29]. If sperm motility at 1 h is 60% and after 4 h only 20%, then the index of longevity is 20:60 or 33% survival rate.

Sperm vitality test

Before the era of ICSI, there was little interest in exploring whether non-motile sperm are just motionless or really dead. Non-motile sperm were unable to fertilize eggs even when brought in contact during standard IVF. There was no way to activate non-motile sperm. Adding xanthines, like caffeine, enhanced only sluggish moving sperm; totally immotile sperm remained unexcited.

Since the introduction of ICSI, it has become possible to achieve fertilization from live immotile sperm such as in Kartagener's syndrome, where the fault is in the sperm tail axoneme. A simple way to identify live immotile sperm is by the supravital test[30]. A drop of semen is released onto a slide, mixed with another drop containing 0.5% eosin-Y and secured with a cover slip. After 2 min, by microscopic observation, dead sperm appear red-stained and live sperm remain transparent, since eosin can penetrate only dead cell membranes. It has been suggested that a drop of 10% nigrosin be added to the eosin–semen mixture and a dry smear made. The dark background that surrounds the dry sperm enables an easy differentiation between transparent and red sperm. However, it is important not to increase the concentration of eosin-Y, or the stain itself would kill a certain part of the live sperm and change the live–dead ratio.

The hypo-osmotic swelling test

In some aspects this is related to the live–dead test and it has been suggested that they be combined[31]. The hypo-osmotic swelling test (HOST) examines intact membrane integrity in live sperm, which is indispensable for fertilization. It was introduced by Jeyendran and co-workers[32], and consists of the microscopic scanning of sperm 30 min after they have been placed in a hypo-osmotic medium. The 150 osmol/kg medium is prepared by mixing 0.735 g of NaCl and 1.351 g of fructose in 100 ml of distilled water. This hypo-osmotic medium causes a typical swelling of intact sperm at the midpiece and the tip of the tail. It may also produce a curling of the tail. In normal specimens one can find more than 50% swollen, live sperm, with intact membrane. In some centers this procedure is used to pick up live immotile sperm for ICSI[33].

Assessment of sperm morphology

This parameter is the most difficult to assess. It is the only one that cannot be evaluated by absolute standards despite the great progress of CASAs. All attempts to program these machines to read the entire sperm configuration, including the head, midpiece and tail, have failed. At the moment, sperm morphology assessment still depends on the subjective judgment of the examiner. The main difficulty is how to differentiate correctly between normal and abnormal forms. Amid the two extremes – entirely normal looking sperm and totally abnormal forms – there is a wide gray zone in which sperm may be qualified as either normal or abnormal by different examiners. These variations may lead to contradictory results of sperm morphology from the same semen.

Routinely, morphology is assessed from a dry smeared, Papanicolaou-stained preparation, examined under a $\times 1000$ power light microscope and by oil immersion. Unless the specimen is severely oligospermic, about 200 sperm should be explored. A sperm is considered normal if it has the following configurations: an oval head 4–5 μm long and 2.5–3.5 μm wide; a midpiece about 3 μm long, directly extending from the head's rear pole and free of cytoplasm droplets; and a delicate single tail 40–50 μm long, projecting from the midpiece. A sperm is considered abnormal *if it does not look normal*. Abnormal forms display a wide variety of shapes and are classified according to head, midpiece and tail defects as follows:

Head defects: Giant head – macrocephalosperm; tiny head – microcephalosperm; round head – globocephalosperm (acrosomeless); elongated head; pear-shaped head; amorphous head; double head.

Midpiece defects: Angular (broken) midpiece; thin midpiece; cytoplasm droplet.

Tail defects: Double tail; short tail; coiled tail; fuzzy-shaped tail.

Morphology evaluation based on verbal description is not reliable unless the examiner was trained according to one of the illustrated color atlases[34–36], or a study published by Freund[37], which was based on a comprehensive co-operative study. The manual contained hundreds of black and white photomicrographs of stained sperm, the photomicrographs being circulated among experts at centers all over the world. Sperm photos were judged independently, each being scored between 0 and 100 according to the percentage of experts that considered it normal.

Normal values

For several decades, 60% normal forms, as suggested by MacLeod[13], was considered the low normal value for fertile males. It is still not entirely clear that sperm shape reflects fertility potential. With regard to an individual spermatozoon, there is no evidence that if it seems normal by visual appraisal it means that it can also fertilize and vice versa: a spermatozoon that looks abnormal is not necessarily unable to fertilize.

The Kruger group[38] from South Africa developed a new method for performing sperm morphology evaluations according to 'strict criteria' that can be studied with a special computer program. The method is based on the concept that the presence of even the slightest defect disqualifies a sperm from being normal. Such minimal defects are determined from the head dimensions and shape, as well as from any minor abnormality of the midpiece or tail, cytoplasm droplet, etc. This method brought about a remarkable change of attitude towards the minimum requirement of normal sperm for fertilization. The minimum value of normal forms by strict criteria, as found by Enginsu[39] for successful IVF treatment[39], was 14%.

David and co-workers[40] proposed a system called the 'multiple anomaly index' (MAI) or 'teratozoospermia index' (TZI) which takes into account the number of defects in each sperm. In the standard mode, each sperm is considered abnormal, no matter how few or how many defects it exhibits. The TZI is calculated from the total number of defects, divided by the number of abnormal forms analyzed. Hence, the index falls between 1.00,

where each abnormal sperm shows a single defect only, and 3.00, where each abnormal sperm exhibits defects of all parts, head, midpiece and tail. As for example in a case where 320 sperm were explored, the number of head defects was 295, midpiece defects 22 and tail defects 45, giving a total of 362 defects, from which a TZI figure of 1.13 was calculated (362 divided by 320). It was suggested that < 1.50 on the TZI is a good predictor for pregnancy, 1.60 is borderline and > 1.80 is poor.

The current consensus by the World Health Organization (WHO) concerning the lower value of normal forms for *in vivo* pregnancy[41] is 30%. Many andrologists believe that 14% is too low, the filtering effect of the cervical mucus effectively preventing the penetration of abnormal sperm so that, in the case of 14% normal forms, very few sperm would cross the cervical canal and the chance to reach the egg would be slim. An excessively abnormal form is named *teratospermia*.

When a specimen shows a combination of oligospermia, asthenospermia and teratospermia, the term OTA syndrome is frequently used.

Assessment of sperm morphology by CASA

Until recently, the objective evaluation of sperm morphology by the CASA system did not meet in full the WHO requirements or Kruger's criteria. Most CASA systems are programed to assess sperm morphology of the dimensions and shape of the head only, being unable to identify reliably the midpiece and tail defects. Correct evaluation is dependent on a thorough staining that can be achieved only with great care[42]. The Cellsoft Morphologizer II (Cryo Resources, NY) did not show any advantage over the manual method[43]. Recently, however, it was reported that the latest versions of the HTM 10 IVOS and the CellForm-Human system (Motion Analysis Corp., Santa Ana, CA) have met the criteria, in the first instance of Kruger, and in the second of the WHO[44].

Other findings during microscopic semen analysis

The presence of round cells in the semen indicates either white blood cells (WBCs) or germinal cells. In the wet sample or stained smear, WBCs and germinal cells may look the same and it can be difficult to differentiate between the two. However, it is important to do so since leukocytes are involved with genital tract infection, which requires treatment with antibiotics. Wolf and co-workers[45] showed that the most reliable method of identifying leukospermia was peroxidase staining, the peroxidase-producing WBCs appearing brown and the germinal cells being stained pink. A count of one million WBC per ml or above is considered an abnormal finding.

EXTENDED SEMEN ANALYSIS

This includes biochemical and microbiological evaluations, as well as sperm function tests, none of which are performed routinely.

Biochemical evaluation

The seminal fluid contains numerous detectable biochemical substances of great interest to biologists and physiologists, but only a few are of much clinical importance and very rarely need assessment. Most biological tests are tedious and require special facilities, not always available in an ordinary laboratory.

Fructose

This is the main source of sperm energy, produced primarily in the seminal vesicles under the stimulation of male sex hormones. Therefore it has been used in the past as an indicator of the normal function of Leydig cells. Normally its range is 140–280 mg/100 ml. The main purpose of fructose assessment is to discover and locate congenital obstructive azoospermia. A total absence of fructose in the semen should arouse strong suspicions that the occlusion is at the common ejaculatory duct (high obstruction), which blocks both the seminal vesicles and the vas deferens.

Other substances

Several substances in the seminal fluid can be used as markers. The more important are zinc (2.4 mol per ejaculate), sorbitol, inositol, citric acid (52 μmol per ejaculate) and acid phosphatase – all are good indicators of prostate function. Carnitine, glycerylphosphorylcholine (GPC) and α-glucosidase (above 20 μmol per ejaculate) are mainly secreted by the epididymis and low levels indicate a dysfunction of this organ. Creatine phosphatase is concerned with sperm fertilizability[46].

Microbiological evaluation

The discovery of WBCs in the semen is an indication to refer the patient for bacterial culture. Even when the seminal fluid is collected by aseptic means, contamination can hardly be avoided. Bacterial cultures can also be grown from expressed prostatic secretion obtained by prostatic massage.

The relationship between genital tract infection and male infertility has already been well established[47]. Whether male infertility is related directly to bacteria, chlamydia or ureaplasma urealyticum, or is due to the formation of antisperm antibodies, is still under debate. Sperm have been incubated *in vitro* for several hours with various pathogenic bacteria up to 10^6/ml without showing any deterioration in motility or vitality[48]. It seems that the harmful effect is due to the presence of leukocytes in the seminal fluid, which creates excessive oxygen free radicals (see 'Reactive oxygen species', below).

Sperm function tests

Microscopic sperm analysis does not ensure that a tested specimen can really fertilize an egg. It is assumed that the specimen's capability to do this can be confirmed only by additional special tests. The function tests, then, are designed to investigate whether trials for conception are justified or whether infertility is simply inexplicable. In cases of men with OTA syndrome, the tests set out to explore if there is any chance that their sperm will eventually impregnate their partners. Finally, the function tests indicate at what stage a couple should choose the ultimate solution – ICSI treatment.

The sperm–cervical mucus penetration test

This is one of the oldest function tests and was originally intended to check the properties of the cervical mucus and its hostility to moving sperm. Later on it was recognized as a real sperm function test and used for checking the penetrability of sperm into the cervical mucus. It was shown that sperm penetration into mucus of optimal quality correlated well with pregnancy rate[49], as well as with sperm penetration into the zona pellucida (ZP)[50]. Therefore, it was decided to perform the test with mucus from a female donor[51].

In the post-coital form of the test (Huhner test), cervical mucus is aspirated from the cervical canal several hours (from two to eight) after intercourse, and spread upon a slide. In a positive test, 10–20 moving sperm in every view area are observed by microscope at × 200 power. In the *in vitro* slide test, named after Miller and Kurzrok, cervical mucus and a drop from a fresh specimen are laid side by side upon a slide. A cover slip is placed gently over the two, so that the drops only just touch each other and do not mix. In a few minutes, sperm migrate into the mucus through several 'phalanxes' formed between the two media by the surface tension along their borders.

In the Kremer test[52] the cervical mucus is aspirated into a glass capillary and dipped into a small reservoir containing a portion of the tested semen. The distance the sperm migrate along the tubing, or their number at a certain location after 30 min, is measured and recorded. In addition, according to Kremer, the appearance of sperm with shaking movements in the mucus indicates an antisperm antibody reaction. Assessment of sperm migration into a ready-to-use capillary containing preserved bovine cervical mucus can be performed in a similar way as the Kremer test.

Testing the sperm fertilizability

In most mammals, in order to acquire the ability to fertilize the oocyte, sperm must undergo a series of structural and functional changes known as 'capacitation'[53]. What exactly induces capacitation is still not clear but it is known to include the following phases, in this order: sperm hyperactivation, zona binding, acrosome reaction, sperm–egg fusion, and membrane penetration to form a pronucleus.

Hyperactivation of sperm movements

This important phenomenon has been the subject of scrutiny for three decades now. Hyperactivated movements have been described by such terms as: extremely vigorous, high amplitude, whiplash frenzied, twisted, jerky, exaggerated lateral head movements, back and forth excited movements, decreased linear progression, etc. It was Burkman and co-workers[54] who showed the link between hyperactivated sperm, acrosome reaction, hemizona binding and IVF success.

Hyperactivation of sperm movements can easily be identified by direct microscopic observation of a wet preparation in all kinds of chambers. Its peak occurs after spontaneous capacitation by incubation in artificial media at 37°C, and 5% CO_2, for 2–3 h. Burkman[55] used the HTM computer to assess hyperactivation according to sperm movement characteristics, but it was possible only to evaluate the percentage of hyperactive sperm and not also to quantify the hyperactive intensity of the entire population. Recently, however, Mazzili and co-workers[56] have developed a computer-aided method, incorporating the MEP technique, which quantifies sperm hyperactivation intensity. Each sperm is assessed individually according to the sequential deviations of its head's long axis every twelfth of a second. The spermis scored by summing up the angles of these deviations during one second, and the score then used to calculate the mean hyperactivity of the entire population and to follow the individual sperm's changes with time.

The zona-binding test

The hyperactivated sperm that has thrust itself between the cumulus cells and reached the ZP must bind to the surface area of the zona in order to penetrate it. The sperm can achieve penetration only if it undergoes the acrosome reaction (see below). The ability to bind to the ZP surface is considered an important sperm function and one that can also be tested. Being a species-specific phenomenon, this *in vitro* laboratory test must be performed only with zonae from non-living preserved human eggs (hemizona-binding assay, HZA)[57]. Half of the dissected hemizonae are incubated with tested sperm and the other half with sperm of males with proven fertility. A reading is made after several hours' incubation, looking for the ratio between the bound sperm in both groups. A ratio of 0.36 was chosen arbitrarily as the normal threshold. Failure to bind does not necessarily mean that the fault is specific to the sperm, it being probable that faulty receptors (ZP3) of the zona itself are responsible. The fact that the sperm is able to bind to the ZP surface is not evidence that it can penetrate the entire zona thickness with its own propulsive force. For this purpose, a specific bioassay, showing the capability of sperm for zona penetration, has been developed[58].

The acrosome reaction

In this process, the outer layer of the acrosome becomes perforated and dissolved so that proteolytic enzymes, including hyalorunidase and acrosin, pour out from what remains of its body[59]. It is yet not clear where, when and what triggers this process. It is assumed that it occurs either by contact with the cumulus cells, the ZP or the follicular fluid, or else even spontaneously. The proteolytic enzymes enable the sperm to bind to the outer surface of the ZP, to penetrate this investment and finally to fuse with the egg membrane. The acrosome reaction can be identified by various means: electron microscopy, special staining technique (triple staining)[60] or assay of acrosomal enzymes. According to Schill[61], 10–35% of the sperm population should reveal this reaction.

A special method, the acrosome reaction ionophore challenge test (ARIC test), was described by Cummins and co-workers[62]. It is based on exciting these cells to undergo an artificial acrosome reaction. It occurs after incubating collected motile sperm for 30 min, in a solution containing the cation ionophore A23187 and dimethyl sulfoxide. The former replicates the stimulus that naturally induces this reaction, probably via the zonal glycoprotein ZP3. Smears stained with conjugated *Pisum sativum* and examined under ultraviolet light are checked for fluorescent shining heads. A fully stained (shining) cap indicates intact sperm; partially stained caps are partially reacted; and non-stained sperm indicate completely reacted sperm. Both the triple staining and the ARIC test are complex and tedious methods and not suitable for daily use in an ordinary clinical laboratory.

The zona-free hamster egg penetration test (SPA)

This is the most famous sperm-function test and was once considered a very promising means to check sperm fertility potential and predict the chance for conception in unexplained infertility. Its main use was in cases of low-quality semen, to make final decisions for IVF treatment or donor insemination. It was also employed as a tool to judge the efficacy of various sperm preparation processes. Though sperm–egg penetration is a species-specific phenomenon, penetration of hamster egg by human sperm became possible when the ZP was removed by trypsin[63]. Later on, fresh ova were substituted by transportable freeze-dried hamster eggs.

In practice, fresh collected eggs are divided into two groups of 25. Ova from one group are incubated for 3 h with the tested motile sperm, and the control group with sperm from fertile donors. Thereafter, the ova are microscopically checked for the number of pronuclei formed by the penetrating sperm and the number of the two groups are compared. Despite its prevalence, the drawback of this test was the lack of standard protocols of performance and the dispute as to the cutoff between positive and negative results. Though considered abnormal when the percentage of penetrated ova was below 14%[64], pregnancies were reported with results of zero penetration.

Reactive oxygen species

In essence, reactive oxygen species (ROS) assay is not considered a pure sperm function test. However, it was found that the presence of excessive oxygen free radicals and superperoxides in the seminal fluid interfered with sperm functions[65]. This disturbance is involved with lipid peroxidation by leukocyte contamination and abnormal sperm in the seminal fluid, where the generated hydrogen peroxide is not entirely destroyed by sperm antioxidant enzymes (superoxide dismutase, glutathione peroxidase and catalase). As a result, the sperm head membrane is damaged and its ability to fuse with the egg investment is impaired.

This complicated assay was described in full by Sukcharoen and co-workers[66]. It is performed by using a luminometer to assess the ROS generated by sperm and leukocytes incubated with luminol and stimulated by FLMP (formyl-methionyl-leucyl-phenylalanine). It is important to recognize that the preparation of sperm by centrifugation can cause damage to the cells in the seminal fluid by which ROS are released, leaving the sperm exposed to superoxidation.

CONCLUSIONS

The breakthrough by ICSI evoked a remarkable change in attitude towards the male partner. Until the era of this novel treatment, andrologists steadily tried to conceive new tests or to elaborate the existing ones to solve problems that still awaited clarification. The appearance of ICSI raised the question about the future need of extended sperm analysis as well as sperm function tests. Obviously, what was applicable for *in vivo* conception or IVF became irrelevant for ICSI. Sperm that can fertilize eggs by ICSI do not need to show their capability to penetrate cervical mucus, to undergo hyperactivation, to bind to the ZP, to undergo the acrosome reaction and to penetrate zona-free hamster eggs. It seems that a sperm that cannot achieve these assignments is still able to impregnate and fertilize the female partner after being injected directly into her egg. It is most probable that successful ICSI can be achieved even with sperm that do not qualify as normal by Kruger's strict criteria or exhibit the poorest movement characteristics by CASA systems. In this case, are advanced sperm analysis and function tests still justified?

In fact, many centers and clinics have totally abandoned the performance of these tests. Hardly any new semen analysis or sperm function tests have been conceived since the appearance of ICSI. Zona drilling, sub-zona sperm injection, intrafallopian sperm injection, as well as most of the tubal reconstructive microsurgery and many of the existing extended semen analyses are prone to disappear. The current trend, then, is to perform just the indispensable routine semen analysis as a screening, before deciding when to refer patients to the most promising ICSI treatment. It is therefore essential that the basic semen analysis be performed very thoroughly.

References

1. Palermo G, Joris H, Devroey P, Van Steirteghem AC. Pregnancies after intracytoplasmic injection of single spermatozoon into an oocyte. *Lancet* 1992;340:17–18
2. Lilja H, Abrahamson PA, Lundwall A. Semenoglobin, the predominant protein in human semen. *J Biol Chem* 1989;264: 1894–900
3. Tauber PF, Propping D, Schumacher GFB, Zaneveld LJD. Biochemical aspects of the coagulation and liquefaction of human semen. *J Androl* 1980;1:281–8
4. Makler A, David R, Blumenfeld Z, Better OS. Factors affecting sperm motility. VII. Sperm motility as affected by changes of pH and osmolarity of semen and urine specimens. *Fertil Steril* 1981;36:507–15
5. Freund M, Carol B. Factors affecting hemocytometer counts of sperm concentration in human semen. *J Reprod Fertil* 1964;8:149–55
6. Jequier AM, Ukombe EB. Errors inherent in the performance of a routine semen analysis. *Br J Urol* 1983;55:434–6
7. Makler A. The improved 10 mic chamber for rapid sperm count and motility evaluation. *Fertil Steril* 1980;33: 337–8
8. Ginsburg KA, Armant DR. The influence of chamber characteristics on the reliability of sperm concentration and movement measurements obtained by manual and videomicrography analysis. *Fertil Steril* 1990;53:882–7
9. Sundquist T, Fjallbrant B, Magnusson KE. Computer-aided counting with the Coulter Counter of low numbers of spermatozoa in human semen. *Int J Androl* 1981;4:18–24
10. Paz (Frenkel) G, Homonai ZT, Korenblum H, Kraicer PF. Comparison of colorimetric and fluorometric methods for estimation of sperm concentration in human ejaculates and epididymal sperm. *Int J Androl* 1978;1:570–5
11. MacLeod J, Gold RZ. The male factor in fertility and infertility: II. Spermatozoa counts in 1000 men of known fertility and in 1000 cases of infertile marriage. *J Urol* 1951;66:436–54
12. MacLeod J, Gold RZ. The male factor in fertility and infertility: III. An analysis of mobile activity in the spermatozoa of 1000 men of known fertility and in 1000 men of infertile marriage. *Fertil Steril* 1951;2:187–97
13. MacLeod J, Gold RZ. The male factor in fertility and infertility: IV. Sperm morphology in fertile and infertile marriage. *Fertil Steril* 1951;2:294–414
14. Zuckerman Z, Rodriguez-Rigau LJ, Smith KD, Steinberger E. Frequency distribution of sperm counts in fertile and infertile males. *Fertil Steril* 1977;28:1310–13
15. Bostflow E, Serup J, Rebbe H. Relation between sperm count and semen volume and pregnancies obtained during a twenty-year follow-up period. *Int J Androl* 1982;5: 267–75

16. Burris AS, Clark RV, Vantman D, *et al*. A low sperm concentration does not preclude fertility in men with isolated hypogonadotropic hypogonadism after gonadotropin therapy. *Fertil Steril* 1988;50:343–7

17. Makler A. The thickness of the microscopically examined sample and its relationship to motility estimation. *Int J Androl* 1978;1:213–17

18. Ishijama S, Mohri H. Beating patterns of mammalian spermatozoa. In Gegnon C, ed. *Controls of Sperm Motility; Biological and Clinical Aspects*. Boca Raton, FL: CRC Press, 1990:29–42

19. Bae-Sagie D, Mayevsky A, Bartoov B. A new optical technique for evaluating collective motility of ram and bull ejaculated spermatozoa. *Int J Androl* 1980;3:198–209

20. Bartoov B, Ben-Barak J, Mayevsky A, *et al*. A sperm motility index: a new parameter for human sperm evaluation. *Fertil Steril* 1991;56:108–12

21. Jouannet P, Volonchine B, Deguent P, *et al*. Light scattering determination of various characteristic parameters of spermatozoa in sperm of human semen. *Andrologia* 1977;9:36–49

22. Burjos MH, Tovar ES. Sperm motility in the rat epididymis. *Fertil Steril* 1974;25:985–91

23. Milligan MP, Harris SJ, Dennis KJ. The effect of temperature on the velocity of human spermatozoa as measured by time-lapse photography. *Fertil Steril* 1978;30:592–4

24. Makler A. Use of the elaborated MEP method in routine sperm motility analysis and for research purposes. *Fertil Steril* 1980;33:160–8

25. Knuth UA, Yeung C, Nieschlag E. Computerized semen analysis: objective measurement of semen characteristics is biased by subjective parameter setting. *Fertil Steril* 1987;48: 118–24

26. Mathur S, Carlton C, Ziegler J, *et al*. A computerized sperm motion analysis. *Fertil Steril* 1986;46:484–8

27. Davis RO, Niswander PW, Katz DF. New measures of sperm motion. I. Adaptive smoothing and harmonic analysis. *J Androl* 1992;13:139–52

28. Makler A. New multiple exposure photography method for objective sperm motility determination. *Fertil Steril* 1978;30: 192–4

29. Makler A. Index of longevity – a definition of a new index in assessment of sperm motility. *Int J Androl* 1979;2:21–5

30. Blom E. A one minute live dead sperm stain by means of eosin–nigrosin. *Fertil Steril* 1950;1:176–8

31. Chan PJ, Tredway DR, Corselli I, *et al*. Combined supravital staining and hypo-osmotic swelling. *Human Reprod* 1991; 6:1115–18

32. Jeyendran RS, Van der Ven HH, Perez-Paleaz M, *et al*. Development of an assay to assess the functional integrity of the human sperm membrane and its relationship to the other semen characteristics. *J Reprod Fertil* 1984;70:219–28

33. Jequier AM. *Male Infertility – A Guide for the Clinician*. Oxford, UK: Blackwell Science, 2000:356–7

34. Ludwig G, Frick J. *Spermatology Atlas and Manual*. Berlin: Springer-Verlag, 1990:43–120

35. Kalla NR, Shah BS, Verma HK. *Practical Spermatology*. Ludhiana, India: USG Publishers & Distributors, 1998: 119–40

36. *WHO Laboratory Manual for Examination of Human Semen and Sperm-Cervical Mucus Interaction*, 4th edn. Cambridge: Cambridge University Press, 1999:21–7

37. Freund M. Standards for rating of human sperm morphology – a cooperative study. *Int J Fertil* 1966;11:97–118

38. Menkveld R, Stander FSH, Kotze TJW, *et al*. The evaluation of morphological characteristics of human spermatozoa according to strict citeria. *Human Reprod* 1990; 5:586–92

39. Enginsu ME, Dumoulin JCM, Pieters MH, *et al*. Predictive value of morphologically normal sperm concentration in the medium for *in vitro* fertilization. *Int J Androl* 1993;16: 113–20

40. David G, Bisson JP, Czyglik F, *et al*. Anomalies morphologique de spermatozoide humain. *J Gynecol Obstet Biol Reprod* 1975;4(Suppl. 1):17–36

41. Mortimer D, Menkveld D. Sperm morphology assessment – historical perspective and current opinion. *J Androl* 2001; 22:192–205

42. Davis RO, Gravance CG. Standardization of specimen preparation, staining sampling methods improves automated sperm-head morphology analysis. *Fertil Steril* 1993;59: 412–17

43. Wang C, Leung A, Tsoi WL, *et al*. Computer-assisted assessment of human sperm morphology comparison with visual assessment. *Fertil Steril* 1991;55:983–8

44. Davis RO, Thal DM, Bain DE, *et al*. Accuracy and precision of the human automated sperm morphology instrument. *Fertil Steril*. 1992;58:763–9

45. Wolf H, Panhans H, Zebhauser M, Meuer M. Comparison of three methods to detect white blood cells in semen: leukocyte esterase dipstick test, granulocytic elastase enzyme immunoassay, and peroxidase cytochemistry. *Fertil Steril* 1992;58:1260–2

46. Huszar G, Vigue L, Corrales M. Sperm creatine phosphatase activity in fertile and infertile oligospermic men. *J Androl* 1990;33:40–6

47. Wilkins SS, Toth A. Relationship between genital tract infection, sperm antibodies in seminal fluid and infertility. *Fertil Steril* 1983;40:805–41

48. Makler A, Urbach Y, Lefler E, Merzbach D. Factors affecting sperm motility. VI. Sperm viability under influence of bacterial growth in human ejaculates. *Fertil Steril* 1981;35: 666–73

49. Hull MGR, Savage PE, Bromham DR. The prognostic value of the postcoital test: a prospective based on time specific conception rates. *Br J Obstet Gynaecol* 1982;89:299–305

50. Schats R, Aitken RJ, Templeton AA, *et al*. The role of cervical mucus–sperm interaction in infertility of unknown etiology. *Br J Obstet Gynaecol* 1984;91:371–6

51. Aitken RJ, Warner PE, Reid R. Factors influence the success of sperm–cervical mucus interaction in patients exhibiting unexplained infertility. *J Androl* 1986;7:3–10

52. Kremer J, Jager S. Sperm–cervical mucus interaction in particular in the presence of antisperm antibodies. *Hum Reprod* 1988;3:69–73

53. Yanagimach R. Mammalian fertilization. In Knobil E, Neil J, eds. *The Physiology of Reproduction*. New York: Raven Press Ltd, 1988:135–85

54. Burkman LJ, Johnson D, Fulgham D, *et al*. Temporal relationship between zona binding, hyperactivated motion (HA) and acrosome reaction (AR) in sperm from fertile men. Presented at the *Serono Symposium on Gamete Physiology*, Newport Beach, November 6, 1988

55. Burkman LJ. Discrimination between non hyperactivated and classical hyperactivated motility patterns in human spermatozoa using computerized analysis. *Fertil Steril* 1991;55:363–71

56. Mazzili F, Rossi T, Delfino M, *et al*. A new objective method for scoring sperm hyperactivation based on head axis angle deviation. *Int J Androl* 2001;24:89–96

57. Burkman LJ, Coddington CC, Franken DA, *et al*. The hemizona assay (HZA): development of a diagnostic test for the binding of human spermatozoa pellucida to predict fertilization potential. *Fertil Steril* 1988;49:688–97

58. Liu DY, Baker HWG. A new test for the assessment of sperm zona-pellucida penetration: relationship with results of other sperm tests and fertilization *in vitro*. *Hum Reprod* 1994;9:489–96

59. Suarez SS, Wolf DP, Meizel S. Induction of acrosome reaction in human spermatozoa by fraction of follicular fluid. *Gamete Res* 1986;14:107–21

60. Talbot P, Chacon RS. Triple stain technique for evaluating normal acrosome. Reaction of human sperm. *J Exp Zool* 1981; 215:201–8

61. Schill B. Some distribution of acrosomal development and function in human spermatozoa. *Hum Reprod* 1991;6: 969–78

62. Cummins JM, Pember S, Jequier AM, *et al*. A test for a human sperm acrosome reaction following ionophore challenge: relationship to fertility and other seminal parameters. *J Androl* 1991;12:98–103

63. Yanagimachi R, Yanagimachi H, Rogers BJ. The use of zona-free animal ova as a test system for the assessment of the fertilizing capacity of human spermatozoa. *Biol Reprod* 1976;15:471–6

64. Rogers BJ. Sperm penetration assay: its usefulness reevaluated. *Fertil Steril* 1985;43:821–40

65. Iwasaki A, Gagnon C. Formation of reactive oxygen species in spermatozoa in infertile patients. *Fertil Steril* 1992;57: 409–16

66. Sukcharoen N, Keith J, Irvine DS, Aitken RJ. Predicting the fertilizing potential of human sperm suspensions *in vitro*: importance of sperm morphology and leukocyte contamination. *Fertil Steril* 1995;63:1293–300

Assessment techniques and clinical significance of antisperm antibodies

8

Andrea Lenzi, Francesco Lombardo, Loredana Gandini and Franco Dondero

ANTISPERM ANTIBODY ASSESSMENT TECHNIQUES

The greatest challenge in modern reproductive immunology is to establish a universally accepted protocol of antisperm antibody (ASA) detection assays. Controversy continues to surround such assays because the antigen or antigenic systems specifically relating to immune infertility have still not been identified, probably due to the heterogeneity of ejaculate sperm populations. Individual populations may have different degrees of fertilizing ability and maturity and different surface antigen expression.

As the purification of a molecular antigen known to be involved in immune infertility has not yet been achieved, antisperm immunity must be studied using the sperm cell as the antigen. However, the use of cellular rather than extracted antigen may lead to inter-assay variability in the antigen component employed. This problem can be addressed by testing a biological sample using various methods in parallel and repeating each test using different donors. A previous standardization trial of this type showed that the same sample tested with different donor sperm may lead to a difference in titer of up to one or two dilutions[1].

Possible instability of antibody titer and percent binding are additional issues which could prevent the successful interpretation of follow-up results during therapy. However, we have only encountered a problem with such titer variations in low titer or percent binding immunization. Where high titer immunization was encountered, no variation of one dilution (Table 1) was observed.

Few methods proposed in the last 20 years have been found to be robust. For this reason, standardization of ASA detection has been the goal of many recent serum exchange workshops. One such was held by the Italian Society of Endocrinology, using the Laboratory of Seminology and Reproductive Immunology Unit at 'La Sapienza' University of Rome as the reference laboratory. Samples of negative and titrated positive blood sera from the reference laboratory's collection were utilized. These had previously been tested in 1998 during an International Workshop of the Sperm Antibody Testing Committee of the American Society of Immunology of Reproduction. Sera were coded, divided into aliquots and lyophilized (one aliquot was immediately re-tested to evaluate any lyophilization-induced modifications). The gelatine agglutination test (GAT) and tray agglutination test (TAT) were used for ASA detection. To simplify analysis, only samples giving the same ASA titer with both methods were selected for further work. Three laboratories used the methods proposed while a fourth used only the indirect immunobead test (i-IBT) and another used a commercially available enzyme-linked immunosorbent assay (ELISA). As reported in Table 2, laboratories using GAT and TAT obtained unanimity in positive and negative results obtained and good ASA titer correlation. The indirect immunobead test (i-IBT) method also gave unanimity of positive and negative results and excellent correlation between percentage of binding activity and ASA titer. However, the laboratory using the ELISA assay obtained a very poor correlation, with four false negatives and two false positives.

As stated previously, no method for ASA detection is universally accepted. A useful protocol is the employment of at least one macroscopic (GAT) and one microscopic (TAT) test or the i-IBT assay to detect circulating

Table 1 Antibody titers and binding percentage in five patients followed up without any therapy for 6 months using d-IBT, GAT and TAT detection assays

Patients		d-IBT IgG		d-IBT IgA		GAT bs titer	TAT bs titer
		T	M	T	M		
1	a	64	20	70	8	512	512
	b	65	24	66	20	1024	512
2	a	87	10	80	10	128	128
	b	73	27	77	21	128	128
3	a	—	100	42	50	256	128
	b	4	96	30	55	256	256
4	a	—	100	11	85	128	128
	b	17	82	18	73	128	128
5	a	42	57	42	50	256	256
	b	39	56	44	48	128	128

(a) first results; (b) results after 6 months. bs, blood serum; d-IBT, direct immunobead test; GAT, gelatine agglutination test; Ig, immunoglobulin; TAT, tray agglutination test; T, tail; M, mixed

antibodies, even though this may be the subject of some criticism. One or more of the 'direct' tests, such as direct (d-)IBT or SpermMar (CGA-Fisenze-Italia), should also be utilized in the study of male infertility.

As previously reported, none of the commercially available direct and indirect ELISA kits detects antibodies related to patient infertility. ELISA and radioimmunoassay (RIA) tests for ASA detection customized in-house give better results[2–4] only when prepared and used in highly specialized laboratories. Even then, the presence of whole sperm antigens in the test tube[4], an indispensable condition for the detection of ASA-related infertility, can only be reasonably guaranteed by the use of whole spermatozoa (fixed or in suspension).

Immunofluorescence assays coupled with cyto-fluorimetric systems may be useful as a future tool. However, problems with standardization and cost-to-benefit ratios may inhibit their use for routine ASA screening[5].

CORRELATION BETWEEN ASSESSMENT TECHNIQUES

Although around 10% of male infertile patients have ASA, the correlation of laboratory results to the clinical situation is still not fully understood, mainly because the application of general immunologic rules to the field of fertility is fraught with problems. In fact, as stated above, great difficulties in the interpretation of laboratory results are created by this 'autovaccination' against sperm antigens, which may differ from one individual to another and even in different spermatogenetic cycles in the same man. However, as with all immunologic reactions, the degree of 'protection' against the antigen, expressed as the level of immunization (i.e., antibody titer), is relative. The role of ASA in the determination of infertility has in fact been questioned in the literature in response to some cases of proven fertility in ASA-positive men. Problems of execution, comparison and correlation of data obtained by different ASA detection methods only add to such doubts. For all these reasons our group carried out a study to correlate the results of ASA detection assays.

In 1985 we conducted a study which examined the correlation between direct immunoglobin G mixed antiglobulin reaction (d-IgG MAR) test results and semen analysis in 395 male partners of infertile couples[6]. Selection was only based on the compatibility of seminal characteristics (sperm concentration $\geq 10 \times 10^6$/ml, sperm motility $\geq 10\%$) with test execution. Spontaneous sperm autoagglutination showed a significant correlation

Table 2 Results of various antisperm antibody detection assays presented at the serum exchange workshop held by the Italian Society of Endocrinology in 1994

Samples	Reference Lab. GAT	Reference Lab. TAT	Lab. F i-IBT	Lab Y. ELISA	Lab. N GAT	Lab. N TAT	Lab. R GAT	Lab. R TAT	Lab. S GAT	Lab. S TAT
1	Neg	Neg	Neg	1:64	Neg	Neg	Neg	Neg	Neg	Neg
2	Neg	Neg	Neg	1:264	Neg	Neg	Neg	Neg	Neg	Neg
3	1:128	1:128	80%	1:128	1:64	1:128	1:128	1:128	1:256	1:256
4	1:64	1:64	43%	Neg	1:64	1:256	1:64	1:64	1:256	1:256
5	Neg	Neg	Neg	Neg	Neg	Neg	Neg	Neg	Neg	Neg
6	1:256	1:236	70%	Neg	1:64	1:256	1:256	1:256	1:1024	1:1024
7	1:512	1:512	79%	Neg	1:128	1:1024	1:512	1:512	1:1024	1:1024
8	Neg	Neg	Neg	Neg	Neg	Neg	Neg	Neg	Neg	Neg
9	Neg	Neg	Neg	Neg	Neg	Neg	Neg	Neg	Neg	Neg
10	1:1024	1:1024	88%	Neg	1:64	1:1024	1:1024	1:1024	1:2048	1:2048

ELISA, enzyme-linked immunosorbent assay; i-IBT, indirect immunobead test; GAT, gelatine agglutination test; TAT, tray agglutination test; Neg, negative

with a positive MAR test result. In 1989, we carried out a wide-scale comparative study on the application of GAT and TAT to blood serum (bs) and seminal plasma (sp) of male patients examined in our laboratory[7]. The 623 subjects were divided into the following patient groups: (1) fertile; (2) randomly selected with a diagnosis of infertility; (3) previous diagnosis of ASA-induced infertility without any other andrologic pathology; (4) vasectomized subjects studied 3–12 months after surgery; (5) unilateral testicular hypotrophy following torsion of the spermatic chord; (6) previous unilateral orchi-epididymitis; (7) unilateral idiopathic varicocele; (8) inflammation of the accessory glands; (9) previous unilateral cryptorchidism; or (10) previous unilateral testicular injury. The most significant results can be summarized as follows: an antisperm immune reaction was found in all groups. This was highest in vasectomized subjects (45.8%) and those suffering from accessory gland inflammation (30.6%). Separate elaboration of GAT and TAT results confirmed that for a full clinical picture, use of both tests is necessary as it is impossible to establish if use of one technique alone will detect positivity and if one method is elective for each single pathology and biological fluid. Blood serum showed a higher percent positivity than seminal plasma. No pathology or group of pathologies showed an

exclusive biological fluid of positivity; TAT titration of positive samples gave higher titers in blood serum, both overall and in the various groups. Subjects with infertility due to antibodies (without any other andrologic pathologies) showed the highest titers, followed by post-vasectomy subjects and patients with genital tract inflammation. The highest correlation between the titration endpoint in blood serum and seminal plasma for each method was found in subjects with ASA-induced infertility. A strict and significant correlation between antibody titers in both media was found in subjects positive to both tests in the same biological fluid. H-H was found to be the most frequent TAT agglutination type in both the patient series and each group. Finally, no correlation was found between pathology type and agglutination type.

Also in 1989 we studied the correlation between three direct methods for the detection of ASA bound to the sperm surface[8]: IgG MAR test, IgG SpermMar test and immunobead test using only the IgG class (d-IgGIBT). The three assays were carried out by different technicians in blind experimental protocols. We found that the IgG IBT, the IgG MAR and the IgG SpermMar tests were strictly correlated both in percent of positivity and in reaction degree (percent motile spermatozoa bound respectively to beads or to red blood cells). On study of the localization of the immune reaction markers (polyacrylamide beads, latex particles or red blood cells) we found a prevalence of mixed and tail positivity with the IgG MAR and IgG SpermMar assays and of tail positivity with the IgG IBT assay. We believe that this is a result of differences in the test systems of the three assays rather than a real difference in the antigen–antibody system detected by each method. In fact, the MAR and the SpermMar assays use the same system, which is based on the reaction between the immunoglobulin on the surface of sperm, red blood cells or latex particles, and anti-immunoglobulin. In contrast, IBT employs a direct reaction between the immunoglobulin on the sperm surface and the anti-immunoglobulin bound to the beads.

In 1991 we published the results of a correlation study of direct IBT (IgG, IgA and IgM), direct IgG Mar and two indirect methods (GAT and TAT)[9]. A total of 605 infertile men, divided into two groups, was studied. 'Group 1' consisted of 428 subjects who underwent d-IBT, GAT and TAT assays of both blood serum and seminal plasma. 'Group 2' comprised 177 subjects who underwent d-IBT and d-IgG MAR assays. A notable correlation was found between d-IBT anti-IgG and -IgA results and GAT and TAT results in both serum and seminal plasma. Good correlation was also found for GAT and TAT positivity titer and percent binding to d-IBT, but only for GAT bs and TAT bs titers higher than 1:64 and for GAT sp and TAT sp titers higher than 1:32. Below these values, few samples were positive by d-IBT. This may be due to the different affinities of ASA to antigen epitopes expressed on sperm. Where there is a high affinity, ASA in the seminal plasma might be almost completely bound to the sperm surface, thus reducing the antibody titer detectable by the indirect techniques. In the case of low affinity, the opposite situation would occur. The Group 1 study confirmed the high correlation of d-IBT and d-IgG Mar assays demonstrated in 1989[8].

In 1995 we concluded this 10-year cycle of clinical, biological and statistical correlation studies by trying to establish if biological models can explain antibody test results and if a predictive expectation threshold can be established to explain the apparent discrepancies between direct and indirect test results[10]. Six hundred and sixty-seven patients were screened with the d-IBT

assay for ASA bound to the sperm surface and the GAT and TAT assays were performed in sera; 134 (20.1%) patients were found to have a clinically significant positivity by d-IBT (binding percentage ≥ 20%). At least one of the indirect tests (GAT and TAT) gave a clinically positive result (i.e., ASA titer ≥ 1:32). Positive d-IBT results (i.e., ≥ 20%) were found in both immunoglobulin classes for 110 of these, while 10 patients showed d-IBT positivity in only one immunoglobulin class. Further analysis showed that discrepancies between direct and indirect test results are associated with differences in the immunoglobulin class of immunization or with the site of the epitopes involved. In particular, with respect to the GAT/TAT results, two distinct possibilities arise with a titer lower than 1:32: (1) either both d-IBT anti-IgG and anti-IgA have a binding percentage less than 20%; or (2) only one class is clinically positive at a medium or low level. With a GAT/TAT titer of 1:32 or 1:64, d-IBT can give a high anti-IgG binding percentage with a low level of anti-IgA (or vice versa) or both classes have a medium level of positivity. With a GAT/TAT titer higher than 1:128, d-IBT consistently shows a higher binding percentage to both anti-IgG and anti-IgA. From the viewpoint of d-IBT results, if both classes are lower than 20%, we are unlikely to find a high GAT/TAT titer. If d-IBT demonstrates a big difference between anti-IgG and anti-IgA binding percentages, the blood serum titer by GAT/TAT is likely to be medium or low. Binding site data indicate that a wide-ranging immunization involving all possible sperm surface antigens is related to higher levels of systemic immunization. In fact, mixed positivity is related to higher GAT and TAT titers, and a simultaneously higher d-IBT percentage to both the classes. In contrast, patients with only one d-IBT class positivity show a prevalence of tail and head localizations, confirming a partial response of the immune system to the triggering sperm antigens.

From all this we can draw the following conclusions:

(1) ASA presence does not necessarily correspond to specific seminal characteristics.

(2) ASA may be found in all andrologic pathologies, and in both blood serum and seminal plasma, though with different degrees of immunization. The

importance of the antibobdy titer is thus underlined also for this field of immunology.

(3) Direct test results (IBT, MAR test, SpermMar test) show strict correlation. Any differences observed in the site of positivity are mainly due to differences in the test systems.

(4) The results of indirect tests (GAT and TAT) show strict correlation for percent positivity, titer and biological fluid.

(5) The study of correlations and particularly discrepancies between direct and indirect test results leads to some interesting observations:

(a) The observed inverse correlation between d-IBT data and indirect test results in seminal plasma is partly justified by the immunologic affinity between ASA in seminal plasma and the antigens expressed by the sperm populations in the ejaculate at the moment of testing;

(b) Cases of discrepancy between d-IBT and indirect test (GAT and TAT) results in blood serum can be explained by the immunoglobulin classes and antigen epitopes involved in such immunization.

(6) Finally, our results demonstrate that either local and systemic ASA production shows strict correlation or ASA passes from blood to the genital compartment. For this reason, where d-IBT and other direct tests cannot be performed, i.e., in patients with very poor or no sperm motility, the immunologic situation of the sperm cell can in any case be extrapolated from data on ASA circulation.

CLINICAL SIGNIFICANCE OF ANTISPERM ANTIBODIES

Rumke and Hellinga, in 1959, were the first to demonstrate that autoimmunity to sperm is found in a significant number of infertile men[11], and suggested that ASA can interfere with the fertilizing ability of the spermatozoa.

The association of ASA and low sperm motility has been noted[12] and the improvement in motility by use of continuous, long-term, low-dose steroid therapy has been demonstrated[13]. However, even though ASA presence is often associated with hypomotility or changes in motility, several cases have revealed normal motility under microscopic examination, a result confirmed by computerized evaluation (computer-assisted sperm analysis, CASA) for all kinetic parameters measured by the system[14].

ASA can affect penetration of the cervical mucus[15–17]. In fact, gametes may be made more vulnerable to phagocytosis in the female reproductive tract by ASA-coated sperm[18]. Sperm-bound IgA antibodies are associated with poor cervical mucus penetration[19–22]. Wang and co-workers found that ASA directed against the head of the sperm cell were of prime importance[21], while Clarke suggested that ASA directed against the tail are involved in this phenomenon[23]. Barratt and co-workers reported a negative correlation between IgG on the sperm tail and forward motility and cervical mucus penetration[17].

A high percentage of spermatozoa with vibratory motility is presumably caused by the cross-linking of motile, antibody-coated spermatozoa with the cervical mucous gel via the Fc part of the immunoglobulins[24,25]. A relationship between sperm-bound autoantibodies of the IgA immunoglobulin class and a poor sperm–cervical mucus contact test has been found in various studies[20], as has a relationship between sperm antibodies and poor post-coital test results[19]. Clarke[16] demonstrated with the capillary test that IgA immunoglobulin class sperm-bound antibodies were associated with poor *in vitro* cervical mucus penetration[16]. Although antibodies at the sperm tail endpiece do not interfere with *in vitro* cervical mucus penetration, autoantibodies at the mainpiece play a vital role in obstruction of *in vitro* cervical mucus penetration.

This strict correlation was not confirmed by our study using the slide test for study of sperm–cervical mucus interaction[26]. In fact, it is extremely difficult to identify clear differences (which are related to poor results of the sperm–mucus interaction test) between the immunoglobulin classes or between the binding sites above a cut-off of 60% of ASA detected by direct IBT. Our results confirm our previous observations on the main role of the intensity of immunization on other immunologic parameters (prevalent binding site and prevalent immunoglobulin class)[27].

Immunologic pathologies are often treated with immunosuppressive steroid treatments. Unfortunately, no double-blind, randomized study has so far been conducted and hence the efficacy of this therapy has not been validated. Additionally, as it is not easy to select homogeneous patient groups, success rates are variable and inconsistent and improved semen quality is seen more often than reduced levels of autoimmune reactions. Corticosteroids at various dosages and with various modes of administration are the therapeutic protocols most commonly used in humans. Although the results reported are interesting given the percentage of pregnancies obtained, their real efficacy on autoimmune pathology is not proven, bearing in mind that these drugs are used more for their capacity to reduce inflammation than for their ability to suppress the immune response.

Due to the improvement and dissemination of *in vitro* fertilization techniques, it is now possible to study directly the effect of antibody-bound sperm at the level of *in vitro* gamete interaction.

CONCLUSIONS

ASA detection is especially useful in this age of assisted reproductive technologies (ART). Not only can ASA have a negative impact on *in vivo* spermatozoa fertilizing ability, but it may also have a major role in the failure of various insemination/fertilization techniques[25,28]. It should be pointed out, however, that other authors have reached different conclusions[29–31]. The presence of ASA in the serum of women, and the related ASA presence in follicular fluid, imposes the need for a particularly careful washing of oocytes. In fact, ASA bound to the acrosomal region of sperm of the male partner of couples undergoing ART could interfere with the mechanism of sperm–oocyte interaction under *in vitro* fertilization conditions[32]. In both cases therefore particular attention is needed, and *ad hoc in vitro* capacitation protocols have been used to obtain fertilization[32].

In conclusion, the detection of ASA is essential for all laboratories dedicated to human reproduction. Such detection can be performed reliably using the modern techniques now available.

References

1. Shulman S. Infertility as caused by sperm antibodies. *Gynecol Obstet Invest* 1986;22:113–18
2. Witkin SS, Zelikovsky G, Good RA, *et al*. Demonstration of 11S IgA antibody to spermatozoa in human seminal fluid. *Clin Exp Immunol* 1981;44:368–74
3. Clark JS, Clark DA, Hendry WF. Antisperm antibodies detected by ZER ELISA kit are not those detected by TAT. *Am J Reprod Immunol* 1987;13:76–7
4. Gandini L, Lenzi A, Lombardo F, *et al*. Radioimmunobinding test for antisperm antibody detection: analysis and critical revision of various methodological steps. *Andrologia* 1991;23:61–8
5. Hjort T, Hansen KB. Immunofluorescent studies on human spermatozoa. I. The detection of different spermatozoal antibodies and their occurrence in normal and infertile women. *Clin Exp Immunol* 1971;8:9–23
6. Cerasaro M, Valenti M, Massacesi A, *et al*. Correlation between the direct IgG MAR test (mixed antiglobulin reaction test) and seminal analysis in men from infertile couples. *Fertil Steril* 1985;44:390–5
7. Dondero F, Lenzi A. Comparative study on the application of gelatin and tray agglutination test in blood serum and seminal plasma of male patients. *Biol Immunol Reprod* 1989;15:21–7
8. Lenzi A, Gandini L, Lombardo F, *et al*. Correlation between tests for the detection of antibodies bound to the sperm surface. *Acta Eur Fertil* 1989;20:141–5
9. Dondero F, Lenzi A, Gandini L, *et al*. A comparison of the direct immunobead test and other tests for sperm antibodies detection. *J Endocrinol Invest* 1991;14:443–9
10. Gandini L, Lenzi A, Culasso F, *et al*. Study of antisperm antibodies bound to the sperm cell surface and their relationship to circulating ASA. *Am J Reprod Immunol* 1995;34:375–80
11. Rumke PH, Hellinga G. Autoantibodies against spermatozoa in sterile men. *Am J Clin Pathol* 1959;32:357–63
12. Mathur S, Williamson HO, Baker ME, *et al*. Sperm motility on postcoital testing correlates with male autoimmunity to sperm. *Fertil Steril* 1984;41:81–7
13. Hendry WF, Stedronska J, Hughes L, *et al*. Steroid treatment of male subfertility caused by antisperm antibodies. *Lancet* 1978;ii:498–500
14. Lombardo F, Gandini L, Anticoli L, *et al*. Can computer analyzed sperm motility be normal in seminal samples with high percentage of antisperm antibody bound to sperm surface? *EOS* 1992;2:115
15. Matson PL, Junk SM, Spittle JW, Yovich JL. Effects of antisperm antibodies in seminal plasma upon sperm function. *Int J Androl* 1988;11:101–6

16. Clarke GN. Immunoglobulin class and regional specificity of antispermatozoal autoantibodies blocking cervical mucus penetration by human spermatozoa. *Am J Reprod Immunol Microbiol* 1988;16:135–42

17. Barratt CLR, Dunphy BC, McLeod I, Cooke ID. The poor prognostic value of low to moderate levels of sperm surface-bound antibodies. *Hum Reprod* 1992;7:95–8

18. London SN, Haney AF, Weinberg JB. Macrophages and infertility: enhancement of human macrophage-mediated sperm killing by antisperm antibodies. *Fertil Steril* 1985; 43:274–8

19. Kremer J, Jager S. Characteristics of anti-spermatozoal antibodies responsible for the shaking phenomenon with special regard to immunoglobulin class and antigen-reactive sites. *Int J Androl* 1980;3:143–52

20. Jager S, Kremer J, Kuiken J, van Slochteren-Draaisma T. Immunoglobulin class of antispermatozoal antibodies from infertile men and inhibition of *in vitro* sperm penetration into cervical mucus. *Int J Androl* 1980;3:1–9

21. Wang C, Barker HWG, Jennings G, *et al*. Interaction between human cervical and sperm surface antibodies. *Fertil Steril* 1985;44:484–8

22. Parslow JM, Poulton TA, Besser GM, Hendry WF. The clinical relevance of classes of immunoglobulins on spermatozoa from infertile and vasovasostomized males. *Fertil Steril* 1985;43:621–7

23. Clarke GN. Sperm antibodies and human fertilization. *Am J Reprod Immunol* 1988;17:65–71

24. Jager S, Kremer J, Kuiken J, *et al*. The significance of the Fc part of ASA for the shaking phenomenon in the sperm–cervical mucus contact test. *Fertil Steril* 1981;36:792–7

25. Clarke GN, Lopata A, McBain JC, *et al*. Effect of sperm antibodies in males on human *in vitro* fertilization (IVF). *Am J Reprod Immunol Microbiol* 1985;8:62–6

26. Lenzi A, Gandini L, Lombardo F, *et al*. Preliminary data on sperm–mucus interaction and sperm-bound antibodies regarding percent positivity, Ig class and site of the reaction. In Spera G, Gnessi L, eds. *Unexplained Infertility: Basic and Clinical Aspects*. New York: Raven Press 1985:265–71

27. Lenzi A, Gandini L, Claroni F, *et al*. Immunological usefulness of semen manipulation for artificial insemination homologous (AIH) in subjects with antisperm antibodies bound to sperm surface. *Andrologia* 1988;20:314–18

28. Dondero F, Lenzi A, Gandini L, Lombardo F. Antisperm antibodies and assisted reproduction. In Colpi G, Pozza D, eds. *Diagnosing Male Infertility* Basel: Karger Press, 1992: 178–85

29. Mandelbaum SL, Diamond SP, DeCherney AH. Relationship of antisperm antibodies to oocyte fertilization in *in vitro* fertilization-embryo transfer. *Fertil Steril* 1987; 47:644–52

30. Acosta A, Oehninger S, Morshedi M, *et al*. Assisted reproduction in the diagnosis and treatment of the male factor. *Obstet Gynecol Surv* 1989;44:1–9

31. Van der Merwe JP, Kruger TF, Windt EM, *et al*. Treatment of male sperm autoimmunity by using the gamete intrafallopian transfer procedure with washed sperm. *Fertil Steril* 1990; 53:682–9

32. Lenzi A, Gandini L, Lombardo F, *et al. In vitro* sperm capacitation to treat antisperm antibodies bound to the sperm surface. *Am J Reprod Immunol* 1992;28:51–5

Fine needle aspiration cytology of the testis

9

Carlo Foresta, Andrea Bettella, Marco Rossato and Davide Spolaore

INTRODUCTION

The clinical analysis of male reproductive functions in the assessment of male infertility consists of physical examination, one or more seminal evaluations and determination of plasma levels of follicle stimulating hormone (FSH), luteinizing hormone (LH) and testosterone. The knowledge of testicular structure is important in the investigation of male infertility and so surgical biopsy can help to assess testicular damage and to express a prognostic evaluation[1–3]. However, this method is not useful for routine analysis of spermatogenic activity because of its invasiveness, especially when performed on both sides.

Several methods of aspiration biopsy of the testis have been proposed in the past as less invasive ways to obtain material for histologic or cytologic evaluation[4,5], but they did not receive enough clinical acceptance because they were considered too traumatic. Later, fine needle aspiration (FNA) of the testis, pioneered by Obrant and Persson[6,7], was proposed as a low-invasive technique but for many years this technique was not included in routine diagnostic practice, probably for two reasons. First, the cytologic analysis of germinal cells was proven to be difficult because isolated germ cells, as seen in cytologic smears, showed different features from those observed in histologic sections[8]. Second, the cytologic specimens gave no information about tubular diameter, thickness of the tubular basement membrane or status of the interstitial tissue[9].

The evaluation of imprint smears prepared at the time of surgical biopsy allowed a description of the cytologic features of normal seminiferous epithelium[8,9]. In these studies cytologic analysis was compared with the respective histologic findings, observing high statistical reproducibility and complete agreement with histologic pictures.

Recently, in light of these cytologic descriptions, we have proposed fine needle aspiration cytology (FNAC) of the testis as a parameter in the assessment of the testicular status in infertile subjects[10–12]. In fact, this method can be considered as a low-invasive diagnostic tool in male infertility evaluation and can provide cytologic material for a qualitative and quantitative analysis of spermatogenesis in these patients.

Spermatogenesis is a long and complex process, in which a highly ordered process of cell division and maturation produces spermatozoa. Three major steps in the spermatogenic process have been identified: (i) the proliferation of the spermatogonia population by mitosis; (ii) meiosis, in which there is a reduction of chromosome number from diploid to haploid (this phase involves the division of primary spermatocytes to form secondary spermatocytes, then these germinal cells divide again to form round spermatids); (iii) spermiogenesis, a complex spermatid transformation in spermatozoa. Three main types of spermatogonia can be identified at the base of the seminiferous epithelium[13,14]: the A dark (Ad) spermatogonia that are thought to be the reserve stem cells and in normal circumstances do not have any proliferating activity[15]; these cells produce the A pale (Ap) spermatogonia that divide and differentiate into type B spermatogonia (B). By mitotic division these cells produce the primary spermatocytes. During the first meiotic phase diploid primary spermatocytes undergo a division producing the secondary spermatocytes; these cells contain a haploid chromosomal set in duplicate form. In the second meiotic division secondary spermatocytes divide to form haploid spermatids. The latter lead to the maturation into spermatozoa by the process of spermiogenesis, without further cell division[13,16–18]. This process involves the nuclear condensation, the development of the acrosome and the formation of the flagellum; in the end the spermatid sheds a large part of its cytoplasm, which is phagocytosed by Sertoli cells.

FNAC allows the identification of all germinal cells in their different maturation steps as well as that of Sertoli cells; furthermore, it permits the characterization of specific tubular damage, similarly to histologic preparations, such as Sertoli cell-only syndrome, hypospermatogenesis or maturation arrest. Therefore, FNAC of the testis represents a minimally invasive and reliable indicator of the status of the seminiferous epithelium, allowing the identification of different groups of infertile patients.

FINE NEEDLE ASPIRATION CYTOLOGY: THE METHOD

Bilateral FNAC is performed on both testes under sterile conditions. Informed consent has to be obtained from each patient. The testis, secured by an assistant, is percutaneously punctured with a 23-G (0.6 mm) butterfly needle attached to a 20 ml syringe. The site of the puncture corresponds to the central area of the testis, on the side opposite to the epididymis. Direct pressure applied by hand provides immediate hemostasis. Patients are then allowed to rest for 10 min. The retrieved material (30–50 μl) is placed on two or more microscopic slides for each testis, allowed to air-dry for 24 h, stained with May–Grünwald and Giemsa. After being washed with distilled water, the smears are air-dried for a few minutes and then examined under a light Orthoplan microscope (Wild-Leitz, Wetzlar, Germany) at magnifications of × 125, × 400 and × 1250. At least 200 spermatogenic and Sertoli cells are counted per smear. Tubular cells are recognized based on their staining and morphologic aspects, such as cell diameter, chromatin pattern and cytoplasm size. According to the histologic description, cytologic analysis allows identification of the following spermatogenic cell types: spermatogonia, primary and secondary spermatocytes, early and late spermatids (corresponding to the Sa–Sb and Sc–Sd steps of spermatogenesis, respectively) and spermatozoa. The

relative number of each cell type is expressed as a percentage. Sertoli cells are expressed as the Sertoli index (SEI; the number of Sertoli cells/100 spermatogenic cells), which has been found to be a reliable index of the tubular germ cell potential and represents an attempt to simplify the interpretation of cytologic results. Since the number of Sertoli cells is constant per unit of tubular length in the adult male, an increase in the SEI indicates tubular germ cell depopulation (hypospermatogenesis). The proportion of spermatozoa is expressed as the spermatic index (SI; the number of spermatozoa/100 spermatogenic cells) reflecting the final maturation step and the spermiogenic process.

After FNAC the patients are investigated for evidence of development of scrotal hematomas, inflammatory signs and hydrocele. A small group of patients was studied for possible complications because of percutaneous puncture, by testicular ultrasonography a week later and by sperm autoantibody analysis 3 months later. No complications were reported as a result of percutaneous puncture of the testis.

QUALITATIVE ANALYSIS

Tubular cells were recognized based on their staining and morphologic aspects such as cell diameter, chromatin pattern and cytoplasm size.

(i) *Spermatogonia*: Three main types of spermatogonia can be identified in human seminiferous epithelium, classified as A dark, A pale and B spermatogonia[13]. The nuclei possess a diameter of 15 to 18 μm and a fine or finely threaded chromatin pattern. Dark and pale spermatogonia can be easily distinguished according to their chromatin density as seen in histologic sections[14]: (a) dark spermatogonia exhibit deeply stained, dark-violet chromatin with a denser and deeper-stained seminular area, located peripherically along the nuclear membrane; and (b) pale spermatogonia have pale-stained chromatin, with a deeper-stained seminular peripheral area and a washed-out central area. The cytoplasm of both cell types is scanty or medium-sized and less basophilic than that of more mature cells. Type B spermatogonia are characterized by a nucleus with several dark clumps of chromatin distributed around the nuclear membrane which appears more delineated; usually a singular nucleolus is present within the central area of the nucleus. Spermatogonia usually have a single, round or oval nucleus, but cells undergoing mitosis or showing multinucleation are common.

(ii) *Primary spermatocytes*:   These cells are easily identified because their nuclear structure shows a round shape (15 to 19 µm diameter), a deeply stained threaded chromatin and an eccentric nucleolus. Chromatin patterns differ according to the stages of meiotic division, so primary spermatocytes are rarely seen at rest: throughout the prophase, chromatin threads become increasingly thicker, longer and most frequently parallel to each other. Metaphase and anaphase can be clearly recognized and through anaphase two new secondary spermatocytes are formed, each containing a haploid number of chromosomes. Binucleated primary spermatocytes undergoing meiosis, connected by a cytoplasmic bridge, are common. The cytoplasm, similar to spermatogonia, appears scanty but more basophilic, especially in the periphery of the cell.

(iii) *Secondary spermatocytes*:   These cells are seldom observed in cytologic smears, owing to their short life span. Their nuclear mass is approximately half of that of primary spermatocytes and can be recognized by a chromatin pattern that appears to be uniformly and finely granulated. The nucleus is round, centrally placed, without visible nucleoli and is smaller than primary spermatocytes but larger than spermatids, so that these cells are not always easily distinguished from round spermatids. The cytoplasm is slight and basophilic. Binucleation is common because each cell completes the second maturation division to produce two spermatids.

(iv) *Spermatids*:   Similar to the histologic classification proposed by Clermont[13], the distinction between Sa, Sb, Sc and Sd spermatids seems to correspond to specific morphologic and staining features of these cells in cytologic specimens[16–18]. To simplify their identification we have labeled spermatids as early and late, which corresponds to the Sa–Sb and Sc–Sd steps of spermiogenesis, respectively. Generally, early spermatids are twice as small as secondary spermatocytes (9 to 10 µm diameter), because of their haploid set of chromosomes, and have a round and slightly triangular nucleus, often eccentrically located, showing a homogeneous, finely granular chromatin without nucleoli. Late spermatids show a smaller, elongated shape and a uniform, darker chromatin. Because the nuclear shrinkage occurs quickly the cytoplasm of these cells appears abundant, sometimes vacuolized and gathered on the same side of the growing tail. Cytoplasmic fragments can separate from cells as typical basophilic residual bodies. The complete development of the acrosomal region, clearly visible as a fainter-stained cap, the more compact chromatin pattern and the discard of the residual cytoplasm[13] seem to be effective criteria to distinguish between late spermatids and spermatozoa. Multinucleated cells, originating from clumped spermatids, can often be observed. They may alternatively show two, three or four nuclei if one or more spermatozoa have already been released.

(v) *Spermatozoa*:   Towards the end of spermiogenesis, germ cells are released from the seminiferous epithelium as spermatozoa, whose morphologic characteristics (head, neck and tail) are well defined. They sometimes show a small cytoplasmic residue on the neck region.

(vi) *Sertoli cells*:   These cells are easily recognizable since they show a round, or less frequently, a kidney-shaped nucleus, with a regular finely granular chromatin. It always possesses a single and conspicuous, eccentrically located nucleolus. The cytoplasm appears to be very large, pale or slightly basophilic, with a triangular or elongated shape and well-defined borders, and often contains tiny vacuoles of various sizes. In some cases the cytoplasm may include spermatids or mature spermatozoa whose tails protrude in the direction of the seminiferous lumen. When seen in groups, indistinct cell borders can be observed in Sertoli cells.

QUANTITATIVE ANALYSIS

Forty normozoospermic infertile subjects (16 autoimmune, 24 idiopathic) have been considered as controls

Table 1 Cell types as observed in smears of normozoospermic subjects and correlation coefficients of first and second differential counts*

Type of cells	Left testis	Right testis	Mean	Correlation between counts[†]
Spermatogonia	2.3 ± 2.2	2.6 ± 2.2	2.5 ± 2.2	0.723
Primary spermatocytes	7.0 ± 4.1	4.9 ± 3.1	6.0 ± 3.6	0.871
Secondary spermatocytes	3.9 ± 2.1	3.9 ± 2.5	3.9 ± 2.3	0.748
Early spermatids	15.6 ± 6.3	14.9 ± 5.3	15.2 ± 5.8	0.884
Late spermatids	37.2 ± 13.4	38.5 ± 12.5	37.9 ± 12.9	0.907
Spermatozoa (spermatic index)	34.3 ± 13.9	35.3 ± 12.9	34.8 ± 13.3	0.923
Sertoli cells (Sertoli index)	30.9 ± 18.6	29.9 ± 10.9	30.4 ± 11.6	0.941

*Values are mean ± standard deviation of percentages of various cell types; [†]averge correlation is 0.856

for testicular cytologic analysis[10] (Table 1). Studies on the reproducibility of the cytologic analysis have been performed and have reported a very low coefficient of variation between differential counts[10].

The quantitative analysis of spermatogenesis performed in normozoospermic subjects showed a constant pattern characterized by increasing percentages of cells from spermatogonia to spermatids, according to the progressive maturation of germ cells throughout spermatogenesis. The Sertoli index of the normozoospermic control group was 30.4 ± 11.6%. No differences were observed between the percentages of germ and Sertoli cells seen in aspirates obtained from the right and left testicles[10].

Early and late spermatids, if considered together, clearly appear to be the most commonly encountered population, coinciding with the maturative progression of the germ cells during spermatogenesis. No difference was detected between the percentages of primary and secondary spermatocytes and between those and late spermatids and spermatozoa. The former finding may be explained with the short lifespan of secondary spermatocytes (24 h) compared to that of primary spermatocytes (2 weeks or more); the latter observation seems to be related to the absence of any cell division during spermiogenesis and to a quick sperm release in the seminiferous lumen after maturation.

Sertoli cells, whose quantification has been expressed as the Sertoli index, represent approximately one-third of the spermatogenic cells as a whole. Because the number of Sertoli cells in post-pubertal age is constant per unit of tubular length[19,20], an increase of the Sertoli index may indicate a reduction of the germ cells in the tubules and hence hypospermatogenesis.

Comparing qualitative and quantitative cytologic analyses with those obtained from normozoospermic subjects we have identified five different appearances in infertile patients:

(1) *Sertoli cell-only syndrome*: Characterized by a complete lack of germ cells, this appearance is characterized by the presence of Sertoli cells only (Figure 1A). When bilateral, this condition is always associated with azoospermia and low testicular volume. These patients exhibit higher FSH plasma levels and lower inhibin B concentration compared to normozoospermic subjects[10,12,21].

(2) *Hypospermatogenesis*: This condition is characterized by a great number of Sertoli cells, higher proportions of immature germ cells and a reduction of spermatids and spermatozoa (Figure 1B). These findings are in agreement with a significantly higher Sertoli index. This cytologic condition is often bilateral and associated with higher FSH plasma levels whereas inhibin B concentration and mean testicular volume are lower than in normozoospermic subjects[10–12,21,22]. Hypospermatogenesis is always associated with oligozoospermia or, if severe, with azoospermia.

(3) *Spermatogonial or primary spermatocyte arrest*: This maturation block is characterized by higher percentages of spermatogonia and/or primary spermatocytes, with a reduction of percentages of the other germinal cells (Figure 1C and D). If complete this condition is associated with azoospermia, while a partial arrest at this first maturation step can be compatible with oligozoospermia. These patients

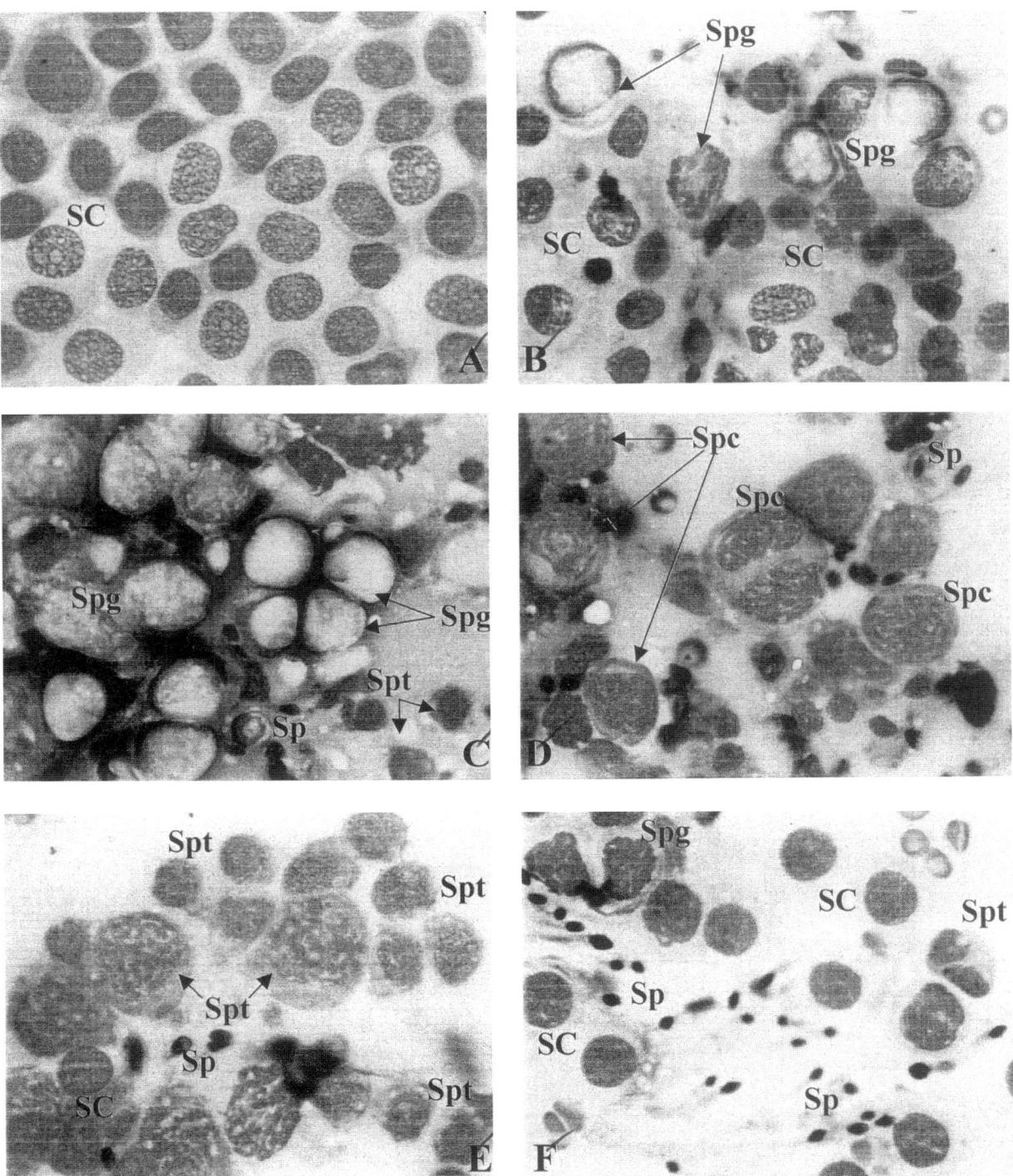

Figure 1 Representative appearance of cytologic findings. (A) A cytologic picture of Sertoli cells-only syndrome, characterized by complete lack of germ cells; (B) hypospermatogenesis, a great number of Sertoli cells (high Sertoli index) with a prevalence of spermatogonia and spermatocytes and a reduction of mature spermatozoa; (C) a cytologic aspect of spermatogonial arrest, characterized by higher percentages of spermatogonia with reduction of other germinal cells; (D) spermatocytes arrest; (E) a cytologic picture of a case of spermatidic arrest, in which a greater number of round spermatids are seen with a reduction of mature spermatozoa; (F) a normal germ line aspect: many spermatozoa are present (high spermatic index) owing to intratubular stasis (obstructive form). May–Grünwald and Giemsa staining, × 1250 magnification. Spg, spermatogonia; Spc, spermatocytes; Spt, spermatids; Sp, spermatozoa; SC, Sertoli cell

show higher FSH plasma levels and lower inhibin B concentration and mean testicular volume with respect to normozoospermic subjects (although the latter did not reach statistical significance)[10–12,21,22].

(4) *Spermatidic arrest*: This late maturation arrest is characterized by a significant increase in the proportions of round and elongated spermatids (Sa–Sb and Sc–Sd) with reduction or total absence of spermatozoa (Figure 1E). In these patients FSH and inhibin B plasma levels and the mean gonadal volume are not different from control subjects[10–12,21,22].

(5) *Normal germ line*: In this condition the proportions of the various spermatogenic cells are similar to those seen in normozoospermic men (Figure 1F). In azoospermic patients this picture is associated with an increased percentage of mature sperm (high SI) indicating an intratubular stasis because of obstruction of the efferent ducts. Occasionally, we have observed this cytologic picture in oligozoospermic subjects, however, a further semen analysis 3 and 6 months later demonstrated a spontaneous improvement of seminal parameters; therefore, we hypothesize that some noxia (i.e., drugs, systemic acute illness or others) could have resulted in an acute damage of spermatogenic cells and previously the cytologic analysis may have 'photographed' the tubular improvement before semen count had been restored. No difference in FSH and inhibin B plasma levels and mean testicular volume are observed in this group with respect to control subjects[10–12,21].

CONCLUSIONS

Testicular fine needle aspiration is a useful diagnostic tool in the assessment of infertile patients. The cytologic findings enable us to identify different groups of patients based upon the percentages of tubular cells.

The characterization of each cell type appeared to be easy: no problem was encountered in recognizing spermatogonia and primary spermatocytes, even when observed during division or when observing spermatids in the different stages of development. A careful quantification procedure of the various cell types was then undertaken, revealing high reproducibility. The percentage of germ cells observed in smears of normozoospermic

subjects showed increasing values from spermatogonia to spermatozoa, in agreement with the maturation progression of the spermatogenic process. The ratio of Sertoli to spermatogenic cells has been termed the Sertoli index (SEI) and constitutes an attempt to simplify the understanding of the cytologic results: because the number of Sertoli cells in adults is constant per unit of tubular length, an increase in the Sertoli index may reveal the existence of tubular germ cell depopulation and thus hypospermatogenesis.

The clinical application of this cytologic description and quantification of testicular germ cells enabled us to identify, in all infertile patients examined, specific cytologic pictures that are related to the nature of the testicular damage. Several pathologic conditions altering the spermatogenic process, such as varicocele, cryptorchidism, orchitis or testicular trauma, determine different alterations of the seminiferous epithelium (quantitative reduction of germ cells and/or maturational disturbances of spermatogenesis) producing a reduction in sperm production[10–12]. In a large percentage of cases the pathogenesis of testicular damage remains unknown (idiopathic) but also in these cases the alteration of seminiferous tubules may be quite different[11,22].

Azoospermia is found in approximately 15–20% of men evaluated for infertility[23]. The advances in assisted fertilization techniques, using microfertilization or microinjection of oocytes, can also permit the treatment of infertility in azoospermic subjects, not only when mature spermatozoa are present in the epididymis, as in segmental obstruction of the efferent ducts, but also in other conditions characterized by the presence of mature spermatozoa and even late spermatids in the testicular parenchyma[24–29]. In this regard evaluation of the spermatogenic activity in azoospermic subjects by cytologic analysis could be of primary diagnostic and prognostic value, and it may be helpful for determining the therapeutic treatment. In a cytologic picture in which only Sertoli cells are found, corresponding to a histologic picture known as Sertoli cell-only syndrome, high FSH plasma levels and low inhibin B concentrations and testicular volumes agree with this diagnosis, but a similar hormonal pattern and a reduced gonadal volume may be present in other situations characterized by azoospermia, as in patients of groups 2 and 3. In subjects of group 2, an important quantitative reduction of the spermatogenic line is observed, as revealed by an increase of the Sertoli index. This cytologic relief represents the histologic picture of hypospermatogenesis associated with maturation disturbances. In group 3 the cytologic analysis demonstrates the presence of an important alteration in the maturation process resulting in strongly increased percentages of the more immature forms, the total absence of mature spermatozoa and an important reduction of spermatids. These findings indicate the presence of an important maturation arrest during the first phases of spermatogenesis. The normality of the Sertoli index may be the result of an enhanced amount of cells at spermatogonial and spermatocytic levels. In the subjects of this group the testicular volume is moderately reduced but FSH plasma levels are elevated. Therefore, in azoospermic men showing higher FSH plasma levels and reduced inhibin B concentrations and testicular volume, cytologic analysis allows us to distinguish three different conditions: (i) Sertoli cell-only syndrome, which is an irreversible form of infertility; (ii) severe germ cell depopulation (hypospermatogenesis); and (iii) spermatogonial/spermatocytic arrest that, if mature sperm is present, may be hypothetically treatable with assisted fertilization techniques. In these conditions the retrieval of spermatozoa is very difficult since few areas of spermatogenesis may be present and their localization is impossible. Some authors have suggested multiple biopsies or testicular aspiration to retrieve spermatozoa in men with non-obstructive azoospermia, but the results in terms of sperm retrieval are very poor[30,31]. Recently, we have proposed the evaluation of intratesticular blood vessel distribution by color power Doppler ultrasound before performing any method of intratesticular spermatozoa retrieval for intracytoplasmic sperm injection (ICSI); in fact the presence of blood vessels may indicate the possible presence of residual spermatogenic areas and so increase the possibility of finding mature spermatozoa by FNAC in these regions[32].

The cytologic analysis of azoospermic subjects of groups 4 and 5 does not show germ cell depopulation, but exhibits two different situations: in group 4, higher percentages of spermatids and lack of spermatozoa suggest a maturation arrest at the spermatidic level. In this group, FSH and inhibin B plasma levels are in the normal range, and so is the testicular volume. In group 5

we observe no maturation disturbance, but an increased percentage of mature cells (high SI). This finding is indicative of intratubular stasis because of obstruction of the efferent ducts: the vesiculodeferentiography in some of these subjects confirmed the obstruction. Since hormonal and clinical parameters are similar in these two groups of subjects, the cytologic analysis is the only parameter that permits discrimination between azoospermia caused by spermatidic arrest and azoospermia caused by obstruction of the seminal tract.

All these findings demonstrate that in azoospermic subjects the cytologic evaluation of the testes represents a minimally invasive and reliable parameter of the status of the seminiferous epithelium and enables the identification of different classes of azoospermic subjects. No differences in fertilization, cleavage or implantation rates have been found after ICSI performed with sperm retrieved by FNAC and open biopsy in azoospermic men with normal spermatogenesis[28]. FNAC is relatively simple and less time-consuming and expensive compared to microsurgical epididymal sperm aspiration or testicular open biopsy. Thus, this method may be considered an attractive alternative to surgical procedures in patients with these clinical characteristics.

As for azoospermia, the bilateral cytologic evaluation of the testis in oligozoospermic patients enables the identification of different classes of infertile subjects. In fact, oligozoospermia represents the endpoint of different alterations of the seminiferous epithelium such as tubular germ cell depopulation or maturation disturbances at different levels of spermatogenesis. The knowledge of the tubular status is useful in establishing the kind of tubular damage and to display a rationale for pharmacologic treatment of oligozoospermic subjects. At present the treatment of oligozoospermia represents a still unresolved problem, probably because any treatment has to consider each specific tubular alteration responsible for the reduction of sperm production. FSH indirectly influences the spermatogenic process and Sertoli cells are believed to be the target cells for this gonadotropin[33] but attempts to treat oligozoospermic subjects with FSH are still controversial[33-35]. Clinical parameters, such as testicular volume or FSH and inhibin B plasma levels, are important in the management of this group of patients; however these findings do not permit us to distinguish between specific testicular tube alterations.

In idiopathic oligozoospermic patients with normal FSH plasma levels, FSH treatment increased the spermatogonial population in all cases, but only when oligozoospermia was caused by hypospermatogenesis without maturation disturbances was the increase in the spermatogonial population coupled to an activation of spermatogenic and spermiogenic processes producing an increase of ejaculated sperm concentration[36,37]. In the presence of hypospermatogenesis associated with disturbances of spermatid differentiation, the enhancement of the spermatogonial population by FSH was associated with an important increase in the percentage of spermatocytes and spermatids without modifications of ejaculated sperm concentration. Recently, we have confirmed the effectiveness of FSH treatment in oligozoospermic subjects with germ cell depopulation without maturation disturbances, independently from the cause determining the testicular damage[22,38]. The presence of high FSH plasma levels seems to be a clear index of the insensitivity of Sertoli cells to a further stimulation by FSH[22,38], but the presence of normal FSH and inhibin B plasma levels is not always predictive of success, because they do not distinguish subjects affected by isolated hypospermatogenesis (who will respond to FSH therapy) from those affected by hypospermatogenesis associated with maturation disturbances (who will not respond to FSH therapy). So, only knowledge of the specific tubular alterations allows us to distinguish between patients responsive to FSH treatment in terms of an increase of sperm production.

Testicular bilateral FNAC appears to be an important diagnostic tool in the assessment of the testicular tubular status of some groups of oligozoospermic patients, valuable for conditions such as unilateral cryptorchidism and varicocele. In ex-cryptorchid subjects cytologic evaluation permits us to identify not only testicular damage in the orchidopexied testis but also in the controlateral descended testis, suggesting that this condition is the endpoint of different pathologic situations, including testicular intrinsic (congenital alterations) and extrinsic causes (anatomic alterations)[39]. Furthermore, FNAC permits us to suggest that the sperm concentration in these patients is related to the tubular status of the normally descended testis. In addition, testicular FNAC gives important prognostic parameters in the selection of infertile patients affected by varicocele. In fact, only subjects with maturation disturbances at the spermatid level present an increase in sperm count and other seminal parameters after sclerotherapy[40].

In conclusion, testicular cytologic analysis allows the identification of different clinical groups of infertile subjects. This classification may be very important in determining the prognosis and the therapy of these patients. In azoospermia, FNAC permits the choice between surgical treatment and the use of assisted fertilization techniques by retrieval of epididymal or intratesticular spermatozoa or spermatids. In oligozoospermic subjects testicular cytologic analysis can help to display a rationale for pharmacologic or surgical treatment and to identify testicular damage in certain infertile groups.

References

1. Johnsen SG. Testicular biopsy score count. A method for registration of spermatogenesis in human testes: normal values and results in 335 hypogonadal males. *Hormones* 1970;1:2–9
2. Meinhard E, McRae CU, Chisholm GD. Testicular biopsy in evaluation of male infertility. *Br Med J* 1977;3:577–82
3. Pesce CM. The testicular biopsy in the evaluation of male infertility. *Semin Diagn Pathol* 1987;4:264–74
4. Huhner M. Aspiration of the testicles in the diagnosis and prognosis of sterility. *J Urol* 1928;19:31–41
5. Nseyo UO, Englander LS, Huben RP, *et al.* Aspiration biopsy of the testis: another method for histologic examination. *Fertil Steril* 1984;42:281–4
6. Obrant KO, Persson PS. Zytologische Untersuchung des Hodens durch Aspirationsbiopsie zur Beurteilung der Fertilität. *Urol Int* 1965;20:176–89
7. Persson PS, Ahren C, Obrant KO. Aspiration biopsy smear of testis in azoospermia. *Scand J Urol Nephrol* 1971;5:22–6
8. Schenck U, Schill WB. Cytology of the human seminiferous epithelium. *Acta Cytol* 1988;32:697–706
9. Papic Z, Katona G, Skrabalo Z. The cytologic identification and quantification of testicular cell subtypes: reproducibility and relation to histologic findings in the diagnosis of male infertility. *Acta Cytol* 1988;32:689–96
10. Foresta C, Varotto A, Scandellari C. Assessment of testicular cytology by fine needle aspiration as a diagnostic parameter in evaluation of the azoospermic subjects. *Fertil Steril* 1992; 57:858–65
11. Foresta C, Varotto A. Assessment of testicular cytology by fine needle aspiration as a diagnostic parameter in evaluation of the oligozoospermic subjects. *Fertil Steril* 1992;58:1028–33

12. Foresta C, Ferlin A, Bettella A, *et al*. Diagnostic and clinical features in azoospermia. *Clin Endocrinol* 1995;43:537–43

13. Clermont Y. The cycle of the seminiferous epithelium in man. *Am J Anat* 1963;112:35–51

14. Clermont Y. Spermatogenesis in man. A study of the spermatogonial population. *Fertil Steril* 1966;17:705–21

15. Schatt S, Weinbauer GF. Immunohistochemical localization of proliferating cell nuclear antigen as a tool to study cell proliferation in rodent and primate testes. *Int J Androl* 1994;17:214–22

16. Heller CG, Clermont Y. Kinetics of the germinal epithelium in man. *Rec Prog Horm Res* 1964;20:545–71

17. de Krester DM. Ultrastructural features of human spermiogenesis. *Z Zellforsch* 1969;98:477–505

18. Holstein AF. Ultrastructural observations on the differentiation of spermatids in man. *Andrologia* 1976;8:157–65

19. Rowley MJ, Heller CG. Quantification of cells of the seminiferous epithelium of the human testis in employing the Sertoli cells as a constant. *Z Zellforsch* 1971;115:461–72

20. Berndtson WE, Igboeli G, Parker WG. The number of Sertoli cells in mature Holstein bulls and their relationship to quantitative aspects of spermatogenesis. *Biol Reprod* 1987;37:60–7

21. Foresta C, Bettella A, Petraglia F, *et al*. Inhibin B levels in azoospermic subjects with cytologically characterized testicular pathology. *Clin Endocrinol* 1999;50:695–701

22. Foresta C, Bettella A, Rossato M, *et al*. Inhibin B plasma concentrations in oligozoospermic subjects before and after therapy with follicle stimulating hormone. *Hum Reprod* 1999;14:101–7

23. Stanwell-Smith RE, Hendry WF. The prognosis of male subfertility: a survey of 1025 men referred to a fertility clinic. *Br J Urol* 1984;56:422–8

24. Temple-Smith PD, Southwick GJ, Yates CA, *et al*. Human pregnancy by *in vitro* fertilization (IVF) using sperm aspirated from the epididymis. *J In Vitro Fert Embryo Transf* 1985;2:119–22

25. Jow WW, Steckel J, Schlegel PN, *et al*. Motile sperm in human testis biopsy specimens. *J Androl* 1993;14:194–8

26. Tournaye H, Devroey P, Liu J, *et al*. Microsurgical epididymal sperm aspiration and intracytoplasmic sperm injection: a new effective approach to infertility as a result of congenital bilateral absence of the vas deferens. *Fertil Steril* 1994;61:1045–51

27. Ogura A, Matsuda J, Yanagimachi R. Birth of normal young after electrofusion of mouse oocytes with round spermatids. *Proc Natl Acad Sci USA* 1994;91:7460–2

28. Craft I, Tsirigotis M, Bennett V, *et al*. Percutaneous epididymal sperm aspiration and intracytoplasmic sperm injection in the management of infertility due to obstructive azoospermia. *Fertil Steril* 1995;63:1038–42

29. Tournaye H, Clasen K, Aytoz A, *et al*. Fine needle aspiration versus open biopsy for testicular sperm recovery: a controlled study in azoospermic patients with normal spermatogenesis. *Hum Reprod* 1998;13:901–4

30. Friedler S, Raziel A, Strassburger D, *et al*. Testicular sperm retrieval by percutaneous fine needle sperm aspiration compared with testicular sperm extraction by open biopsy in men with non-obstructive azoospermia. *Hum Reprod* 1997;12:1488–93

31. Turek PJ, Cha I, Ljung BM, Ohl D. Systematic fine-needle aspiration of the testis: correlation to biopsy and results of organ 'mapping' for mature sperm in azoospermic men. *Urology* 1997;49:743–8

32. Foresta C, Garolla A, Bettella A, *et al*. Doppler ultrasound of the testis in azoospermic subjects as a parameter of testicular function. *Hum Reprod* 1998;13:3090–3

33. Bartoov B, Eltes F, Lenenfeld E, *et al*. Sperm quality of subfertile males before and after treatment with human follicle-stimulating hormone. *Fertil Steril* 1994;61:727–34

34. Acosta AA, Oehninger S, Ertunc H, *et al*. Possible role of pure human follicle-stimulating hormone in the treatment of severe male-factor infertility by assisted reproduction: preliminary report. *Fertil Steril* 1991;55:1150–6

35. Acosta AA, Khalifa E, Oehninger S. Pure human follicle stimulating hormone has a role in the treatment of severe male infertility by assisted reproduction: Norfolk's total experience. *Hum Reprod* 1992;7:1067–72

36. Foresta C, Bettella A, Ferlin A, *et al*. Evidence for a stimulatory role of follicle-stimulating hormone on the spermatogonial population in adult males. *Fertil Steril* 1998;69:636–42

37. Foresta C, Bettella A, Merico M, *et al*. Recombinant human FSH in the treatment of male infertility. *Fertil Steril* 2002;77:238–44

38. Foresta C, Bettella A, Merico M, *et al*. FSH in the treatment of oligozoospermia. *Mol Cell Endocrinol* 2000;161:89–97

39. Foresta C, Ferlin A, Garolla A, *et al*. Functional and cytologic features of the controlateral testis in cryptorchidism. *Fertil Steril* 1996;66:624–9

40. Bettella A, Merico M, Spolaore D, *et al*. Assessment of testicular cytology by fine needle aspiration as a diagnostic parameter in the evaluation of patients affected by varicocele. *Arch Ital Urol Androl* 2001;73:3–13

Sperm testing by hyaluronic acid binding: andrologic laboratory assessment and sperm selection for ICSI

10

Gabor Huszar, Attila Jakab, Cilar Celik-Ozenci, Denny Sakkas, Tamas Kovacs and Lynne Vigue

INTRODUCTION

This is an overview of the ongoing work towards the clinical utilization of the hyaluronic acid (HA) binding assay of human sperm. First we will discuss the work of the Huszar laboratory on the objective biochemical markers of sperm function and fertility in the past 15 years. Subsequently, we will review the current data supporting the utility of HA binding for sperm testing and sperm selection for intracytoplasmic spermatozoa injection (ICSI). The key points of this chapter are as follows:

(1) Cytoplasmic retention as evidence of sperm immaturity, and the two-wave expression pattern of the testis-specific HspA2 chaperone protein during meiosis and late spermiogenesis;

(2) Cellular maturation and plasma membrane remodeling, and their contributions to the fertilization function of human sperm;

(3) The relationship between sperm immaturity and increased frequencies of chromosomal aneuploidy;

(4) Semen analysis and the assessment of sperm maturity by HA binding in a double chamber device;

(5) The selection of mature individual sperm with low levels of chromosomal aneuploidy and high DNA integrity.

CYTOPLASMIC RETENTION AND OTHER BIOCHEMICAL MARKERS OF SPERM CELLULAR MATURATION

The primary interest of our laboratory has been the development of objective biochemical markers of human sperm maturity and function which would predict male fertility, independently from the traditional semen criteria of sperm concentration and motility. In measurements of sperm creatine-N-phosphotransferase or creatine kinase (CK), we found significantly higher sperm CK activities in men with diminished fertility[1,2]. We addressed the reasons underlying the sperm CK activity differences by labeling the enzymatic active site of sperm CK with [^{14}C]FDNB followed by autoradiography. In another approach, we visualized the CK in individual sperm with CK-immunocytochemistry[3]. The CK immunostaining patterns indicated that the high sperm CK activity was a direct consequence of increased cytoplasmic protein and CK concentrations in the spermatozoon (Figure 1). The combination of increased CK and protein concentrations, coupled with the diminished fertility, suggested to us that we had identified a sperm developmental defect in the last phase of spermiogenesis when the cytoplasm (unnecessary for the mature sperm) normally is extruded and left in the adluminal area as 'residual bodies'[4].

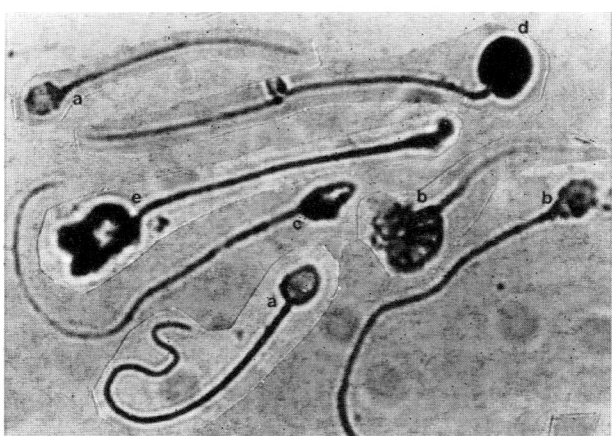

Figure 1 Montage of CK immunostained sperm with different pattern at × 1000. a, normal sperm with lightly stained or clear heads; b, sperm with different degrees of CK stippling; c,d, sperm with different degrees of solid CK-staining; and 3, amorp-sperm. Reproduced with permission from Huszar G, Vigue L. Incomplete development of human spermatozoa is associated with increased creatin phosphokinase concentraction and abnormal head morphology. *Mol Reprod Dev* 1993;34:292–8

Upon electrophoretic analysis of human sperm extracts, in addition to the CK-B isoform, we found another ATP-containing protein, which was proportional to the incidence of mature sperm characterized by low CK activity and no cytoplasmic retention in the semen samples[5]. We have recently identified this developmentally regulated protein as the 70 kDa testis-expressed chaperone protein, which in man is called HspA2[6]. The close inverse correlation between the proportions of sperm with cytoplasmic retention and low expression of HspA2 and those sperm with lack of cytoplasmic retention and increased expression of HspA2 indicated that cytoplasmic extrusion and the commencement of the HspA2 synthesis are related, developmentally regulated, spermiogenetic events. In three independent studies the correlation between HspA2 levels and CK activity was $r = -0.69, -0.71$ and -0.76 ($p < 0.001$; $n = 159, 134$ and 119)[5,7,8].

HspA2 which, due to its electrophoretic properties and ATP content, we initially assumed to be an unusual form of sperm-specific CK-M isoform (several properties have also indicated that it was not a conventional CK-M[5]), proved to be a most useful objective biochemical marker. We have shown that mature and immature sperm are different with respect to HspA2

ratio, as expressed by the concentrations of sperm CK and HspA2 (% HspA2/(HspA2 + CK-B)), morphologic and morphometric attributes, zona pellucida-binding properties and fertility[8,9]. Furthermore, we have established that in spermiogenesis, simultaneously with cytoplasmic extrusion and the commencement of HspA2 synthesis, the sperm plasma membrane also undergoes a maturation-related remodeling. This remodeling step facilitates the formation of the sites and receptors for zona binding and for hyaluronic acid binding in mature sperm[10]. We have also shown, along with another laboratory, that immature sperm have increased rates of lipid peroxidation[11,12]. Finally, we established that all sperm maturational events, related to the decline of CK activity and increase in HspA2 expression, are completed by the time the sperm enter the caput epididymis[13].

CELLULAR MATURATION AND DEVELOPMENT OF THE FERTILIZATION FUNCTION

The predictive value of CK activity, representing cytoplasmic retention, was tested in couples where the oligospermic male partner had been treated with intra-uterine insemination. Irrespective of whether pregnancy had been achieved, the sperm concentration and motility parameters were identical in all of the men, but those whose female partners had conceived were found to have a four-times lower level of sperm CK activity ($p < 0.001$). Also, a logistic regression analysis indicated that sperm CK activity, but not sperm concentrations, contributed significantly to the predictive power[14].

The value of the HspA2 ratio in the assessment of male fertility was tested in two blind studies of couples undergoing *in vitro* fertilization (IVF). In the first, based only on their sperm HspA2 ratios, we classified 84 husbands from two different IVF centers (without any information on their semen parameters or reproductive histories) into 'high likelihood' (> 10% HspA2 ratio) and 'low likelihood' (< 10% HspA2 ratio) fertility groups. All pregnancies occurred in the 'high likelihood' group, none in the 'low likelihood' group. In the 'high likelihood' group, if at least one oocyte was fertilized, indicating the lack of oocyte defects in the wife, the predictive rate of HspA2 ratio for pregnancy was a very high 30.4% per cycle. An additional important utility of the HspA2 ratio became apparent: nine of the 22 'low likelihood' men were normospermic but had diminished fertility. Thus, the HspA2 ratio provided, for the first time, a diagnostic tool for unexplained male infertility (infertile men with normal semen)[8]. We have also demonstrated morphometric differences between mature and diminished maturity spermatozoa[15].

More recently, we re-examined the utility of CK-M ratios in predicting IVF failure in 119 couples treated at Yale. Similar to the 1992 study[8], none of the partners of the 25 men with < 10% CK-M ratios achieved pregnancy whether the men had low or high sperm concentrations[16]. The value of sperm CK studies has also been confirmed by other laboratories[17–19].

To identify the steps of the fertilization process, at which the low HspA2 immature sperm are deficient, we explored human sperm–oocyte binding. With the study of sperm–hemizona complexes, we established that only the clear-headed (low CK, in figure 1a), mature sperm were able to bind to the zona[9] (Figure 2). Sperm with retained cytoplasm were deficient in the oocyte-binding site. In a further study, we confirmed that a remodeling of the plasma membrane occurs in human sperm, simultaneously with cytoplasmic extrusion, during spermiogenetic maturation. This was demonstrated by the close correlation ($r = 0.8$) between CK concentration or the HspA2 ratio and the density of the sperm plasma membrane-specific enzyme, $\beta 1,2$-galactosyltransferase, in sperm fractions of various maturities[10]. Such remodeling apparently facilitates the formation of the zona pellucida- and HA-binding sites[9,20]. This finding explains two major characteristics of sperm with diminished maturity: cytoplasmic retention and deficiency in zona pellucida binding.

In general, chaperone proteins facilitate the assembly and intracellular transport of proteins. Indeed, the expression of HspA2 is simultaneous with major sperm protein movements, underlying cytoplasmic extrusion and remodeling of the human sperm plasma membrane. This in turn facilitates the development of the zona pellucida-binding site. We believe that retention of the cytoplasm, and the lack of zona-binding sites in immature sperm, are likely to be related to the diminished expression of HspA2, and also to diminished DNA integrity, as a consequence of the impaired delivery

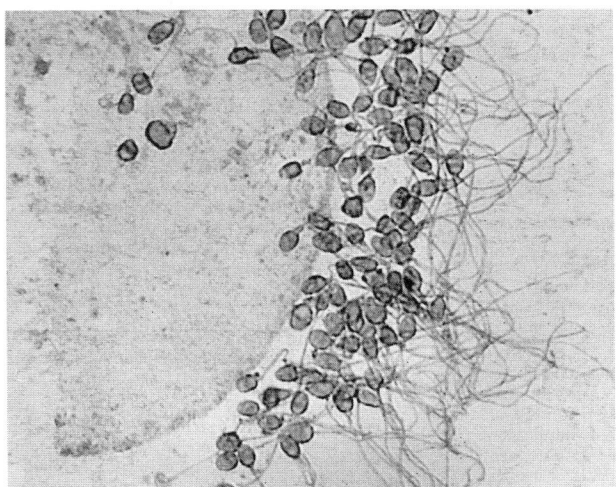

Figure 2 CK-B isoform immunostained sperm–hemizona complexes. Note the lack of dark CK-stained immature sperm with cytoplasmic retention among the bound sperm

of DNA repair enzymes during and following meiosis. In order to confirm our finding regarding the expression of HspA2 during terminal spermiogenesis, we also examined the expression pattern of HspA2 in human testicular tissue (Figure 3). Varying low levels of immunostaining were evident in spermatocytes and spermatids, reflecting the presence of HspA2 in the synaptonemal complexes, but the staining was particularly striking in the cytoplasm of elongating spermatids and mature sperm about to be released from the adluminal compartment[5,6].

From the perspective of male infertility, it is important that synthesis of the Hsp70-2 family of proteins is developmentally regulated and that Hsp70-2 appears during meiotic prophase as a component of the synaptonemal complexes. The testis-expressed Hsp70-2 protein has been identified in the mouse, and it is expressed in pachytene spermatocytes during the meiotic phase of spermatogenesis and in spermatids and mature sperm[21]. The apparent functions of Hsp70-2 in mice are maintaining the synaptonemal complexes and assisting chromosome crossing-over during meiosis and spermatocyte development. Accordingly, the targeted disruption of the *hsp70-2* gene causes arrested sperm

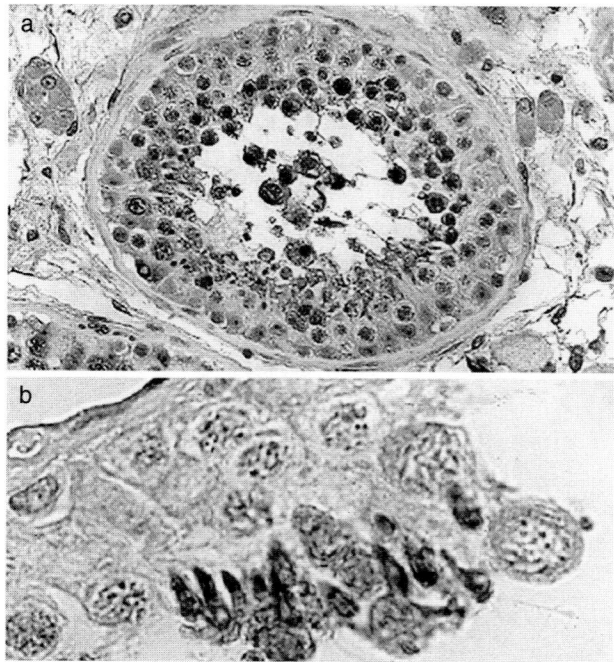

Figure 3 Human testicular biopsy tissues immunostained with human HspA2 antiserum. Panel a and b in the composite represent 200× and 1000× magnifications to illustrate the tubular structure, and the staining pattern of the adluminal area. HspA2 expression begins in meiotic spermatocytes, but is predominant during terminal spermiogenesis in the elongated spermatids and sperm

maturation and azoospermia. These events could be related to faulty meiotic recombination in spermatocytes, disruption of the meiotic cell cycle regulatory machinery or perhaps to a more direct disruption of the apoptotic machinery in spermatocytes or even in spermatids or ejaculated immature sperm. Regarding human sperm, our laboratory was the first to demonstrate the expression pattern of the HspA2 protein in human testis and sperm and to correlate the expression level of HspA2 to sperm function[6]. Because we had already identified maturational differences in cytoplasmic content, plasma membrane remodeling, DNA integrity and aneuploidy rates, we considered the question of whether the plasma membrane structure differences and features specific for mature sperm could facilitate the selection of mature sperm for ICSI.

THE RELATIONSHIP BETWEEN SPERM IMMATURITY, DNA INTEGRITY AND CHROMOSOMAL ANEUPLOIDIES

Because HspA2 is a component of the synaptonemal complex in rodents, assuming that this is also the case in man, we hypothesized that the frequency of chromosomal aneuploidies will be higher in immature versus mature sperm[22]. We examined this proposition in sperm arising from semen and from 80% Percoll pellets (enhanced in mature sperm) of the same ejaculate in 10 oligozoospermic men. Immature sperm with retained cytoplasm, which signifies spermiogenetic arrest, were identified by immunocytochemistry. We have evaluated with fluorescence *in situ* hybridization (FISH) approximately 7000 sperm nuclei in each of the 20 fractions (142 086 sperm in all), using centromeric probes for the X, Y and 17 chromosomes. The proportions of immature sperm (as detected by cytoplasmic retention) were $45.4 \pm 3.4\%$ versus $26.6 \pm 2.2\%$ in the two groups (medians: 48.2% versus 25%, $p < 0.001$, $n = 300$ sperm evaluated per fraction, 6000 sperm in all). There was also a concomitant decline in total disomy, total diploidy and total aneuploidy frequencies in the 80% Percoll versus semen fractions (0.17 versus 0.54%, 0.14 versus 0.26% and 0.31 versus 0.81%, respectively; $p < 0.001$ in all comparisons). The mean decline of aneuploidies was 2.7-fold. Regarding our hypothesis that aneuploidies are related to sperm immaturity, there was a close correlation between the incidence of immature sperm and disomies ($r = 0.7$, $p < 0.001$), indicating that disomies originate primarily in immature sperm. Thus, the idea that the common factor underlying sperm immaturity and aneuploidies is the diminished expression of the HspA2 appears to be valid[22].

A DOUBLE CHAMBER DEVICE FOR SEMEN ANALYSIS AND FOR A SPERM HA-BINDING ASSAY

Our current ideas on sperm maturation in men are summarized in Figure 4. Looking for the underlying reason for diminished zona binding by immature sperm, we have established that in spermiogenesis, simultaneously with the cytoplasmic extrusion and the

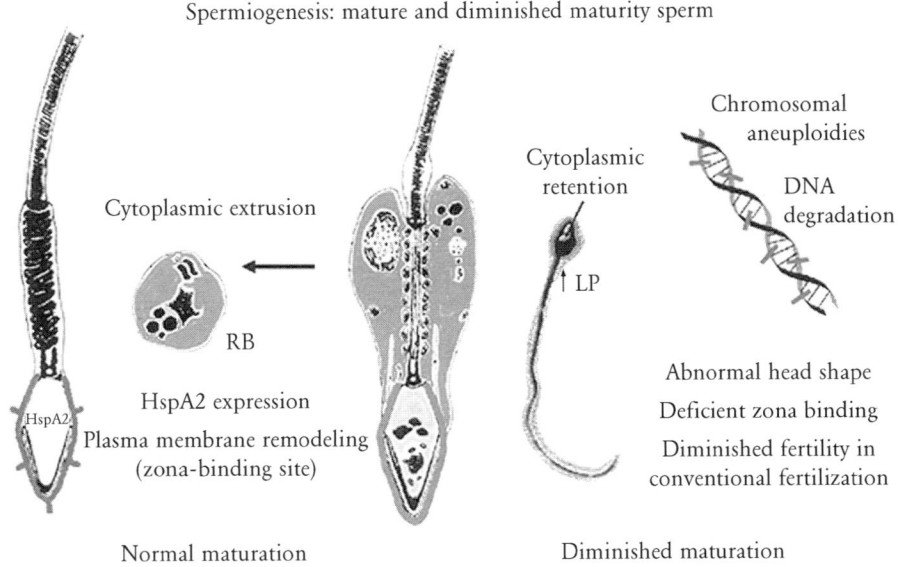

Spermiogenesis: mature and diminished maturity sperm

Normal maturation

Diminished maturation

Figure 4 A model of normal and diminished maturation of human sperm. In *normal* sperm maturation HspA2 is expressed in the synaptonemal complex of spermatocytes, supporting meiosis. HspA2 is also likely to be involved in the processes of late spermiogenesis, such as cytoplasmic extrusion (represented by the loss of residual body, RB), plasma membrane remodeling and the formation of the zona pellucida-binding site. *Diminished maturity* sperm lacks HspA2 expression, which causes meiotic defects and a higher rate of retention of CK and other cytoplasmic enzymes, increased levels of lipid peroxidation (LP) and consequent DNA fragmentation, abnormal sperm morphology and deficiency in zona-binding and HA-binding sites. HA, hyaluronic acid

commencement of HspA2 synthesis, the sperm plasma membrane also undergoes a maturation-related remodeling that promotes the formation of the zona-binding and HA-binding sites. Thus, in immature sperm with cytoplasmic retention there is a low density of zona-binding sites and also of HA receptors[6,21].

Based on the above concepts, we examined three key points:

(1) The ability of sperm to bind permanently to solid-state HA;

(2) The diagnostic utility of sperm binding to HA, in a double chamber device in which the A side provides the measures of sperm concentration and motility (thus motile sperm concentration), and the B side is coated with HA in order to test the proportion of mature sperm exhibiting HA binding;

(3) The potential correlation between sperm CK activity or sperm HspA2 expression (proven clinical utility in predicting diminished fertility), and the proportion of sperm bound to HA.

The data indicate that:

(1) *Sperm binds to HA* There were three sperm populations: (a) sperm permanently bound to HA; (b) sperm that exhibited no binding; and (c) a small proportion of sperm (< 5%) that was initially bound to HA, was released shortly and rebound again. We interpreted these three patterns as mature sperm with a high density of HA receptors, immature sperm with deficient maturity and plasma membrane remodeling and sperm of intermediate maturity with a low density of HA receptors (Figure 5).

(2) *Correlation of binding with the HspA2 ratio and CK activity* When we studied the relationship between sperm binding and CK activity or between sperm binding and the HspA2 ratio ($n = 56$ men), we found a strong correlation with both CK activity ($r = -0.80$, $p < 0.001$) and the HspA2 ratio ($r = 0.54$, $p < 0.001$), which are the well-characterized sperm maturity markers (Figures 6 and 7). The fact that the correlation was closer with CK activity, as compared to the HspA2 ratio, is an expected effect because the

A side   B side

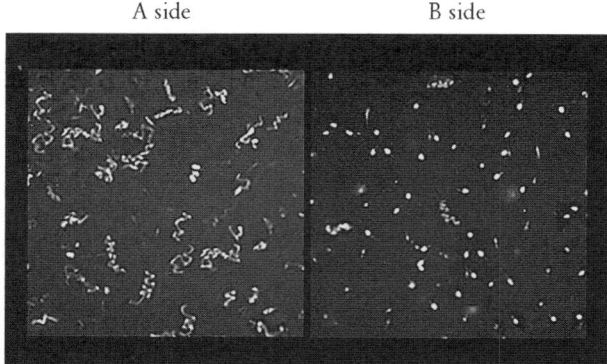

Figure 5 Sperm movement patterns in the double chamber device. A side, conventional semen analysis chamber with motile sperm. B side, mature sperm are bound, and diminished motility sperm remain motile in the HA-coated chamber. Sperm are stained with cyber green DNA stain (Molecular Probes, Eugene, OR) that permeates viable sperm. HA, hyaluronic acid

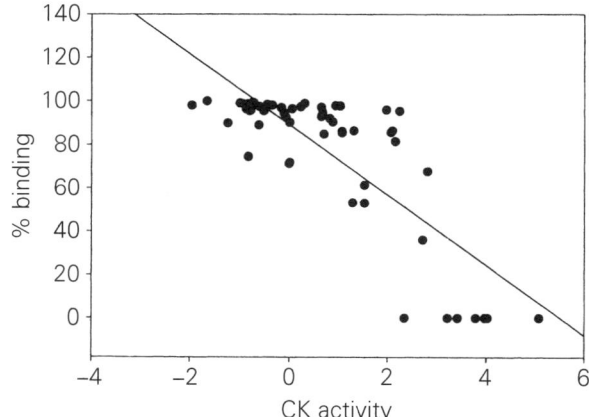

Figure 6 Correlation between CK activity and the percentage of HA binding of sperm ($n = 56$ men, $r = -0.80$, $p < 0.001$)

membrane remodeling and the formation of the HA receptor sites occur simultaneously with cytoplasmic extrusion, whereas HspA2 expression occurs in two waves (in meiotic spermatocytes as a part of the synaptonemal complex and in terminal spermatogenesis when the cytoplasm is extruded), and thus is not directly related to membrane remodeling.

(3) *Diagnostic utility of sperm binding to HA* Finally, we evaluated the CK activities, HspA2 ratios and the percentage of sperm binding to the HA-coated slides of 56 men. With respect to binding, we classified the sperm populations as follows: > 90% ($n = 32$) were excellent, between 60 and 90% ($n = 14$) were intermediate and < 60% ($n = 10$) were diminished binders. A summary of the CK activity and HspA2 ratio data of these sperm categories is presented in Table 1. In line with our previous findings with respect to CK and HspA2, these were largely independent from sperm concentration. Among men within the < 20 million sperm/ml concentration range ($n = 18$ of 56 men), we identified three excellent, seven moderate and eight diminished HA binders.

SELECTION OF SPERM WITH LOW ANEUPLOIDY FREQUENCIES FOR ICSI

Previously we found that mature, but not immature, spermatozoa showed increased velocity and retention of long-term motility in response to HA[20]. We suggested that this effect was receptor-mediated. Based on the association between sperm maturation and plasma membrane remodeling, we made the hypothesis that the

Table 1 Summary of the HA binding, CK activity and HspA2 ratio data of 56 men

	n	*Concentration* (10^6/ml)	*Motility* (%)	*CK activity* (IU/10^8 sperm)	*HspA2 ratio*	*Binding* (%)
Excellent binders to HA (> 90%) range (90.3–99.8)	32	72 ± 9.6 (11.5–220)	61 ± 2.7	1.5 ± 0.4	48 ± 3.7	97 ± 0.5
Moderate binders to HA (60–90%) range (61.3–90.0)	14	16 ± 1.9 (4–30)	39 ± 5.1	4.3 ± 1.2	29 ± 5.5	80 ± 2.4
Diminished binders to HA (< 60%) range (00.0–53.1)	10	4 ± 1.5 (0.2–13)	21 ± 4.1	40 ± 14.6	14 ± 1.8	14 ± 7.4
Total	56	37 ± 3.1	48 ± 3.0	9 ± 3.2	37 ± 3.1	78 ± 4.3

presence of the HA receptor in mature but not immature sperm, and a respective device with an HA-coated surface, facilitated the selection of single mature sperm with a high DNA integrity and a low frequency of chromosomal aneuploidies for ICSI.

As established, there is a relationship between the proportion of immature sperm with cytoplasmic retention and the frequency of chromosomal aneuploidies in human sperm[22]. This relationship is based on the dual role of the HspA2 chaperone, which supports meiosis as a component of the synaptonemal complex, and facilitates plasma membrane remodeling as well as the formation of the zona pellucida- and HA-binding sites during spermiogenesis[6] (Figure 3). The increased rate of chromosomal aberrations and other potential consequences of using immature sperm for ICSI is of major concern.

In ongoing studies, we have tested the efficiency of HA sperm selection with respect to elimination of sperm with chromosomal aneuploidies and diploidies. Washed sperm of eight moderately oligospermic men (OS, sperm concentration ± standard error of the mean (SEM): $20.6 \pm 1.7 \times 10^6$/ml; motility: $54.1 \pm 2.5\%$) and 80% isolate gradient sperm pellets from seven normospermic IVF patients (ISL80, sperm concentration ± SEM: $118 \pm 21.4 \times 10^6$/ml; motility: $59.1 \pm 4.9\%$) were evaluated. Sperm suspended in human tubal fluid medium (HTF) were placed over HA spots bonded to Petri dishes (Biocoat Co., PA). After incubation for 15 min, the HA-attached sperm were collected using an ICSI micropipette. Aliquots of the sperm suspension and HA-bound sperm were examined after FISH, using centromeric probes for the X, Y and 17 chromosomes. Data were analysed using the chi-square test.

In each man we evaluated the initial sperm suspension (mean: 3000 sperm, 45 000 sperm in the 15 men), and all HA-bound sperm collected in the eight OS men (mean: 753; range: 224–1142) and seven ISL80 men (mean: 644; range 224–1128). In the OS group, the proportion of disomies showed a mean 6.9-fold reduction, and the sex chromosome disomies alone a 3.6-fold reduction. Diploidies in the HA-selected samples were 6.4-fold lower compared to the initial semen sample. With respect to the ISL80 group, the disomy rates declined in the HA-bound fractions (Table 2). The decrease for sex chromosomes was approximately four-fold, even though the ISL80 samples were 80% isolate pellets of normospermic men (thus 'ideal sperm'). Diploid sperm decreased six-fold in both groups ($p < 0.001$).

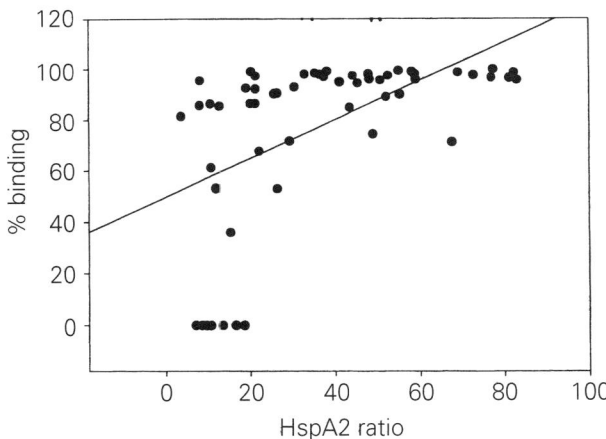

Figure 7 Correlation between HspA2 and the percentage of HA binding of sperm (n = 56 men, r = 0.54, $p < 0.001$)

Table 2 Disomy and diploidy frequencies in sperm arising from semen and after hyaluronic acid selection

	Group OS			*Group ISL80*		
	Disomy			*Disomy*		
	Sex	*17*	*Diploidy*	*Sex*	*17*	*Diploidy*
Initial (%)	0.35	0.23	0.81	0.25	0.12	0.58
HA-bound (%)	0.09	0.04	0.13	0.06	0.07	0.10
Reduction	4.0 ×	5.3 ×	6.1 ×	4.1 ×	1.7 ×	6.1 ×
p Value (chi^2)	< 0.001	< 0.001	< 0.001	< 0.001	ns	< 0.001

We can conclude that HA selection eliminated sperm with disomy and diploidy. The four-fold decline of sex chromosome disomies is consistent with the increase of chromosomal aberrations in ICSI children. In spite of the sample differences, the aneuploidy and diploidy rates in the HA-bound fraction declined to a narrow low 0.04–0.13% range, which is comparable to normal fertile men. Thus, HA sperm selection provides a new, safe and efficient solution for the selection of mature sperm for ICSI.

CONCLUSIONS

We identified the expression of a 70 kDa chaperone protein, HspA2 (formerly CK-M), in mature human sperm. We have established the central role of HspA2 as a measure of sperm cellular maturity, DNA integrity and aneuploidy frequency and function – including fertilizing potential. The presence of HspA2 in the synaptonemal complex is likely to provide the link between the lack of HspA2 expression and the increased frequency of aneuploidies in diminished maturity sperm. The spermiogenetic events of cytoplasmic extrusion and the remodeling of the plasma membrane, which facilitates the formation of zona pellucida-binding sites in human sperm, are related. Finally, the presence of the HA receptor on the plasma membrane of sperm that completed cellular maturation, coupled with the HA-coated surface method, will facilitate the selection of single mature sperm for ICSI. Thus, we can conclude that:

(1) Mature sperm permanently bind to HA-coated surfaces.

(2) There is a close correlation between the ability of sperm to bind to HA, the fertility markers of sperm CK activity and the HspA2 ratio. CK activity has a closer correlation because cytoplasmic extrusion and plasma membrane remodeling are directly related spermiogenetic events (Figure 3). In other studies we have shown that in HA-bound fraction sperm with chromosomal aneuploidy and diploidy are reduced by approximately four- to six-fold compared to semen sperm factions.

(3) The new HA chamber provides a platform for both conventional semen analysis and HA-binding mediated sperm function tests.

Further studies of the relationship between HA binding and hemizona binding of semen samples are in progress.

ACKNOWLEDGEMENTS

The authors wish to thank Sevil Cayli for the kind assistance in preparing the illustrations.

References

1. Huszar G, Corrales M, Vigue L. Correlation between sperm creatine phosphokinase activity and sperm concentrations in normospermic and oligospermic men. *Gamete Res* 1988; 19:67–75

2. Huszar G, Vigue L, Corrales M. Sperm creatine phosphokinase activity as a measure of sperm quality in normospermic, variablespermic, and oligospermic men. *Biol Reprod* 1988;38:1061–6

3. Huszar G, Vigue L. Incomplete development of human spermatozoa is associated with increased creatine phosphokinase concentration and abnormal head morphology. *Mol Reprod Dev* 1993;34:292–8

4. Clermont Y. The cycle of the seminiferous epithelium in man. *Am J Anat* 1963;112:35–51

5. Huszar G, Vigue L. Spermatogenesis-related change in the synthesis of the creatine kinase B-type and M-type isoforms in human spermatozoa. *Mol Reprod Dev* 1990;25:258–62

6. Huszar G, Stone K, Dix D, Vigue L. Putative creatine kinase M-isoform in human sperm is identified as the 70-kilodalton heat shock protein HspA2. *Biol Reprod* 2000;63:925–32

7. Lalwani S, Sayme N, Vigue L, *et al*. Biochemical markers of early and late spermatogenesis: relationship between the lactate dehydrogenase-X and creatine kinase-M isoform concentrations in human spermatozoa. *Mol Reprod Dev* 1996;43:495–502

8. Huszar G, Vigue L, Morshedi M. Sperm creatine phosphokinase M-isoform ratios and fertilizing potential of men: a blinded study of 84 couples treated with *in vitro* fertilization. *Fertil Steril* 1992;57:882–8

9. Huszar G, Vigue L, Oehninger S. Creatine kinase immuno-cytochemistry of human sperm–hemizona complexes: selective binding of sperm with mature creatine kinase-staining pattern. *Fertil Steril* 1994;61:136–42

10. Huszar G, Sbracia M, Vigue L, *et al.* Sperm plasma membrane remodeling during spermiogenetic maturation in men: relationship among plasma membrane beta 1,4-galactosyltransferase, cytoplasmic creatine phosphokinase, and creatine phosphokinase isoform ratios. *Biol Reprod* 1997;56:1020–4

11. Huszar G, Vigue L. Correlation between the rate of lipid peroxidation and cellular maturity as measured by creatine kinase activity in human spermatozoa. *J Androl* 1994;15:71–7

12. Aitken J, Krausz C, Buckingham D. Relationships between biochemical markers for residual sperm cytoplasm, reactive oxygen species generation, and the presence of leukocytes and precursor germ cells in human sperm suspensions. *Mol Reprod Dev* 1994;39:268–79

13. Huszar G, Patrizio P, Vigue L, *et al.* Cytoplasmic extrusion and the switch from creatine kinase B to M isoform are completed by the commencement of epididymal transport in human and stallion spermatozoa. *J Androl* 1998;19:11–20

14. Huszar G, Vigue L, Corrales M. Sperm creatine kinase activity in fertile and infertile oligospermic men. *J Androl* 1990;11:40–6

15. Gergely A, Kovanci E, Senturk L, *et al.* Morphometric assessment of mature and diminished-maturity human spermatozoa: sperm regions that reflect differences in maturity. *Hum Reprod* 1999;14:2007–14

16. Ergur AR, Dokras A, Giraldo JR, *et al.* Sperm maturity and treatment choice of IVF or ICSI: diminished sperm HspA2 chaperone levels predict IVF failure. *Fertil Steril* 2002;77:910–18

17. Gomez E, Buckingham DW, Brindle J, *et al.* Development of an image analysis system to monitor the retention of residual cytoplasm by human spermatozoa: correlation with biochemical markers of the cytoplasmic space, oxidative stress, and sperm function. *J Androl* 1996;17:276–87

18. Orlando C, Krausz C, Forti G, Casano R. Simultaneous measurement of sperm LDH, LDH-X, CPK activities and ATP content in normospermic and oligozoospermic men. *Int J Androl* 1994;17:13–18

19. Sidhu RS, Sharma RK, Agarwal A. Relationship between creatine kinase activity and semen characteristics in subfertile men. *Int J Fertil* 1998;43:192–7

20. Sbracia M, Grasso J, Sayme N, *et al.* Hyaluronic acid substantially increases the retention of motility in cryopreserved/thawed human spermatozoa. *Hum Reprod* 1997;12:1949–54

21. Dix DJ, Allen JW, Collins BW, *et al.* Targeted gene disruption of Hsp70-2 results in failed meiosis, germ cell apoptosis, and male infertility. *Proc Natl Acad Sci USA* 1996;93:3264–8

22. Kovanci E, Kovacs T, Moretti E, *et al.* FISH assessment of aneuploidy frequencies in mature and immature human spermatozoa classified by the absence or presence of cytoplasmic retention. *Hum Reprod* 2001;16:1209–17

Technologies for human
in vitro fertilization

IVF in humans: technologies for oocyte retrieval, *in vitro* insemination and embryo transfer

11

Vishvanath Karande and Norbert Gleicher

INTRODUCTION

Over the past 25 years, a great deal of progress has been made in the field of human *in vitro* fertilization (IVF) and, as a result, new technologies have been developed that are now performed routinely at many assisted reproduction centers worldwide. Since the birth of the world's first IVF baby, Louise Brown, in 1978, thousands of babies conceived through IVF have been delivered each year. Indeed, at that time Patrick Steptoe was a visionary when he said, 'This is the first time we have solved all the problems at once. We are at the end of the beginning – not the beginning of the end.' Remarkable advances have since been made in every step involved in IVF and this is reflected in the increase in pregnancy rates reported world-wide.

In this chapter, we will discuss oocyte retrieval, *in vitro* insemination, and embryo transfer (ET). Very little has changed in the oocyte retrieval technique in the past 10 years, so the procedure will be discussed only briefly. There are several chapters in this book that address laboratory issues, therefore we will only briefly review *in vitro* insemination. In the past few years, the importance of transferring embryos properly and the potential impact of this simple step on success rates are being realized. The ET technique, especially with ultrasound guidance, will therefore constitute the bulk of this chapter.

OOCYTE RETRIEVAL

The recovery of oocytes was originally done via laparoscopy[1]. With the advent of ultrasound guidance, initially transabdominally and now transvaginally, the procedure has become less invasive and relatively simple. Gregory Pincus first reported on the culture of human oocytes obtained via laparotomy more than 25 years ago[2]. Patrick Steptoe was the first to use the laparoscope to visualize the female pelvis and see the ovarian follicles with the potential of aspirating follicles from them[3]. In fact, the first IVF baby was conceived with oocytes obtained via laparoscopy in a natural cycle[4]. For the next few years, oocyte retrievals were performed exclusively under laparoscopic guidance until the advent of ultrasound guidance, which relegated laparoscopic techniques to being only of historical significance[5].

The use of a transabdominal probe with needle aspiration of oocytes through the vagina was first reported in 1983 by Gleicher and co-workers at our Center[6]. This was followed by Dellenbach and co-workers who reported a large series using this technique[7]. This method is seldom used now due to the advent of vaginal ultrasound probes. The only indication for this technique is probably in patients who have ovaries high up in the pelvis that are not visualized vaginally. Parsons and co-workers reported on transabdominal–perurethral ultrasound-directed oocyte retrieval. They performed 242 cases with no serious complications[8].

In 1985, Wikland and co-workers reported on the use of a vaginal transducer for oocyte retrieval[1]. They also described the method of guiding the aspirating needle alongside the transducer to achieve very accurate puncture of follicles. This is now universally the preferred technique for oocyte recovery as the ovaries are close to the probe and are easily visualized.

Technique

At the Center for Human Reproduction, transvaginal ultrasound-guided oocyte retrieval is performed as an office-based procedure under general anesthesia. The procedures are scheduled 34 h after the administration of human chorionic gonadotropin (hCG) with the patients being instructed to fast overnight. The patient empties her bladder prior to the procedure to avoid catheterization. Aseptic precautions are followed throughout the procedure. The patient is placed in a dorsal lithotomy position and fully draped. A speculum is inserted in the vagina which is irrigated thoroughly with normal saline. Antibacterial solutions are not used because of their potential toxicity to the oocytes. A trial transfer is often performed at this time and the data are recorded[9]. Most patients prefer light general anesthesia without intubation. If analgesia alone is requested, the patient is also given a paracervical block. A prophylactic antibiotic is routinely administered intravenously (IV) pre-operatively. We are currently using a single IV dose of cefazolin (Ancef 1 g, Baxter Healthcare Corporation, Deerfield, IL). Patients allergic to penicillin are given a single IV dose of clindamycin (Cleocin Phosphate 600 mg, Pharmacia & Upjohn Company, Kalamazoo, MI).

We have over the years used several ultrasound machines and have no special preferences. The vaginal transducers usually have a frequency of 5 or 7.5 MHz although there are machines with variable frequency transducers. The transducer is covered with a sterile polythene sleeve, with a small amount of coupling gel to cover the end of the probe. The needle guide is attached to the ultrasound probe, which is inserted in the vagina. The pelvis is initially scanned to orient oneself with the position of the uterus, the uterine lining, the ovaries and the number of follicles. A Cook single lumen needle (Cook Ob/Gyn, Spencer, IN) is inserted through the needle guide and the ultrasound probe is manipulated until the follicles are in close proximity to the needle guide lines. These are single-use needles with sharp points and have an echogenic tip which is easily seen on ultrasound. The needle is then gently inserted into a follicle and its contents aspirated. The probe is then manipulated to bring the next follicle near the needle line and the procedure repeated. A concerted effort is made to aspirate as many follicles as possible with a single puncture in the ovary. This also minimizes the number of punctures in the vaginal wall. The needle is withdrawn and flushed with media prior to reinsertion. The follicular fluid passes into plastic tubes which have been placed in a 'hot block' (Cook Ob/Gyn, Spencer, IN) to maintain their temperature at 37°C. The full tubes are passed to the embryologist in the adjacent laboratory, and the oocytes are identified. All follicles, including the smaller ones, are systematically aspirated. Finally, the cul-de-sac is aspirated of any visible fluid. The transducer is then replaced by a speculum and the vagina inspected for any bleeding. If there is bleeding in the vagina, it is controlled by applying pressure with a sponge stick. Rarely, an absorbable stitch is required.

In patients with few follicles, we will sometimes flush a follicle with media in an attempt to recover oocytes. We have seldom resorted to using double-lumen needles for oocyte retrievals. With the use of needles with sharp tips, mobile ovaries that tend to 'run away' have not been much of a problem. Care must be taken to avoid puncturing the large pelvic vessels whose lumen, when viewed in cross-section, may be mistaken by a novice for a follicle.

IN VITRO INSEMINATION

An embryologist initially examines the follicular aspirates under a stereo dissecting microscope with a transmitted illumination base and heated stage[10]. The aspirates are then poured into two or three Petri dishes. The thin layer of fluid is quickly and carefully scanned for the presence of oocytes. Initial scanning is at low-power magnification ($\times 6$–12). Subsequent oocyte identification can be done using higher magnification ($\times 25$–50). The oocytes usually appear within varying quantities of cumulus cells and, if very mature, may be pale and difficult to visualize (immature oocytes are dark and also difficult to see). Granulosa cells are clearer and more 'fluffy', present in amorphous, often iridescent clumps. Blood clots, especially from the collection needle, should be carefully dissected with 23-gauge needles to check for the presence of cumulus cells.

When an oocyte/cumulus complex is found, its stage of maturity is assessed by noting the volume, density, and condition of the surrounding coronal and cumulus cells. If the egg can be seen, the presence of a single polar body indicates that it has reached the stage of metaphase II[11].

Certain quality control mechanisms are essential to ensure proper handling of gametes. The embryologist should be familiar with each treatment plan and make sure that all appropriate consents have been signed. Semen assessment is critical and the laboratory personnel should be aware if the sample needs any special preparation or precautions. Laboratory case notes, media, culture vessels and tubes for sperm preparation are prepared during the afternoon prior to each case, with clear labeling throughout. Tissue culture dishes or plates are equilibrated in the CO_2 incubator overnight. Before beginning each retrieval, the embryologist must ensure that the heating blocks, stages, and trays are warmed to 37°C. Also, the collection test tubes and Petri dishes for scanning aspirates must be pre-warmed.

The choice of culture systems is a matter of individual preference, but the two systems commonly used are either microdroplets under oil or four-well plates. Oocytes are routinely inseminated with a concentration of 100 000 normal motile sperm per ml. If the sperm characteristics are sub-normal, the insemination concentration may be increased accordingly. Preovulatory oocytes are inseminated after 3–4 h of preincubation *in vitro*. Traditionally, inseminated oocytes are incubated overnight in the presence of a prepared sperm sample. However, sperm binding to the zona pellucida normally takes place within 1–3 h of insemination, and the oocytes can be washed free of excess sperm after 3 h of incubation[11].

Complete failure of fertilization is a disastrous, albeit rare, occurrence in our practice. This is so because we have a very liberal policy regarding the use of intracytoplasmic sperm injection (ICSI). Reinsemination of the oocytes or the performance of 'rescue ICSI' has been attempted by several groups with very limited success.

The evaluation of oocytes, scoring of fertilization on day 1, assessing embryo quality, and selection for transfer are beyond the scope of this chapter. For a detailed description of sperm preparation see Chapter 12.

EMBRYO TRANSFER

Historically, the ET technique is a topic that is less studied than other aspects of assisted reproduction. Over recent years, this has changed and it is now generally accepted that paying a lot of attention to individual variations in transfer technique can have a positive impact on success rates with IVF[12,13]. The old technique of performing transfers with the patient in a knee–chest position and mandatory bed rest for 4–12 h post ET is no longer used. Until recently there have been few randomized studies with large numbers evaluating variations in transfer technique. Here, we will review some of the details of ET technique that impact upon outcomes. The role of ultrasound guidance will be discussed in detail, and recommendations regarding the best transfer technique will be made.

What factors are important for successful embryo transfer?

Kovacs reported on the results of a survey of the program directors of each IVF unit in Australia and New Zealand[14]. The questionnaire asked them their attitude with respect to 12 factors, which constitute the ET matrix. They were requested to rate each step on a scale of 1–10, where 1 was irrelevant and 10 was very important. Table 1 represents the cumulative experience of the 50 clinicians with more than 500 years of 'hands on' IVF practice. The most highly ranked factor was the removal of hydrosalpinges prior to starting a cycle. Number two on the list was the absence of bleeding/blood on the catheter. This is probably a strong indicator as to whether or not the transfer was traumatic. The use of ultrasound was almost at the bottom of the list. This was considered to be technically difficult when done transvaginally. Surprisingly, the use of abdominal scanning was criticized as the bladder needed to be filled and the catheter tip was not easily visualized on the scan. ET techniques have changed significantly since Kovacs' data were published and the use of ultrasound guidance is becoming routine. The impact of hydrosalpinx on success rates is beyond the scope of this chapter. We routinely remove hydrosalpinges that are visualized on ultrasound before proceeding with an IVF cycle[15,16]. Before continuing with the discussion on the use of

Table 1 The relative importance of various factors constituting embryo transfer as rated by the total, mean score and standard deviation (SD). The maximum possible score for each variable was 500

Priority		Mean score	SD	Total score
1	Removal of hydrosalpinges before treatment	6.8	2.8	340
2	Absence of bleeding/blood on catheter	6.6	2.5	330
3	Type of catheter used	6.1	2.7	255*
4	Not touching the fundus	5.8	3.2	292
5	Avoiding the use of a tenaculum	5.7	2.9	283
6	Removal of all mucus from cervix	5.2	3.2	258
7	Ultrasound details of cavity before treatment	4.3	2.8	216
8	Leaving catheter in place for at least 1 min	4.2	3.1	211
9	30 min rest after transfer	3.8	2.8	192
10	Dummy transfer before treatment	3.1	3.1	157
11	Ultrasonic monitoring of transfer	2.6	2.2	125
12	Antiprostaglandins to prevent contractions	1.9	1.5	93

*Only 42 clinicians responded to this question. Reproduced with permission of Oxford University Press/Human Reproduction, from Kovacs GT. What factors are important for successful embryo transfer after *in-vitro* fertilization? *Hum Reprod* 1999;14:590–2, ©European Society of Human Reproduction and Embryology

ultrasound guidance, let us consider some of the other factors that affect ET success rates.

Absence of bleeding/blood on catheter

The presence of blood inside the catheter tip after a transfer is associated with lower pregnancy rates[17,18]; the blood is probably an indicator of trauma during transcervical passage of the catheter. But a difficult transfer alone does not necessarily result in lower pregnancy rates. Tur-Kaspa and co-workers[19] reported on a large series of patients where they categorized transfers as 'easy' or 'difficult' as assessed by the physician performing the transfer. They also evaluated the impact of cervical dilatation and the performance of multiple transfers due to retained embryos. The surprising finding was the similarity in pregnancy rates in patients with 'easy' (23.3%), 'difficult' (23.6%), cervical dilatation (23.8%), or multiple transfers (29.6%) (Table 2). Groutz and co-workers[20] on the other hand, reported on 41 women who required cervical dilatation at the time of oocyte retrieval and only one achieved an intrauterine pregnancy. Details on the presence or absence of blood at the catheter tip were not provided.

Type of catheter used

At our Center there was a definite increase in pregnancy rates when we switched to the Edwards–Wallace (Bourn) ET catheter (Smiths Industries Medical Systems (SIMS), Lancing, UK)[13]. Over the past few years, there has been a worldwide trend toward the use of 'soft' catheters[21]. The possible advantage of this type of catheter is its ability to follow the direction of the cervix and enter the cavity. Rigid catheters tend to penetrate the endometrial surface, and the catheter tip becomes plugged with mucus or endometrial tissue and causes bleeding. Wisanto and co-workers[22] studied three types of catheters in 400 patients retrospectively. The pregnancy rate with the Frydman catheter (32.3%) was better than the one achieved with the Edwards–Wallace catheter (19.2%), which did better than the TDT catheter (19.4%).

For difficult cervical insertions, we use the Wallace stylet. This consists of an outer sheath with a firm, coated rod in its lumen that provides stability to negotiate highly convoluted cervixes. In rare instances, we use the Frydman catheter which has a wireguide that is used to negotiate the cervix. Details of the transfer technique are discussed later.

Table 2 Pregnancy rates and outcomes for the different categories of embryo transfer

Pregnancy outcome	Easy	Difficult	Cervical dilatation	Multiple	Total
Ongoing or delivered	118/171 (69)	11/17 (64.6)	3/5 (60)	5/8 (62.5)	137/201 (68.2)
Ectopic	0	2/17 (11.8)	0	0	2/201 (1)
Missed abortion	26/171 (15.2)	2/17 (11.8)	2/5 (40)	3/8 (37.5)	33/201 (16.4)
Chemical	27/171 (15.8)	2/17 (11.8)	0	0	29/201 (14.4)
Total	171/734 (23.3)	17/72 (23.6)	5/21 (23.8)	8/27 (29.6)	201/854 (23.5)

Data expressed as *n* (%). Reproduced with permission of Oxford University Press/Human Reproduction, from Tur-Kaspa I, Yuval Y, Bider D, *et al*. Difficult or repeated sequential embryo transfers do not adversely affect *in-vitro* fertilization pregnancy rates or outcome. *Hum Reprod* 1998;13:2452–5, ©European Society of Human Reproduction and Embryology

Not touching the fundus

The general consensus seems to be that the best area to deposit the embryos is 1 cm beyond the internal os and 1 cm below the fundus. This is usually 5–7 cm from the external os (which is the first or second mark on the outer sheath of the Edwards–Wallace catheter). Waterstone and co-workers[23] reported that the pregnancy rate between two clinicians differed based on the technique used. One clinician advanced the catheter until resistance was felt and then withdrew 5 mm before injection. His success rate (24%) was lower than that of another clinician (46%) who routinely deposited embryos 5 cm beyond the cervical os. The pregnancy rate of the first physician increased to 46% when the low transfer technique was used. Rosenlund and co-workers[24] on the other hand could not correlate success rates with the site of deposition in a study using ultrasound measurement.

The deposition of embryos close to the fundus may increase the risk of ectopic pregnancy[25]. The site of embryo placement is also important as the embryos seem to implant in the area where the air bubble was seen immediately after transfer[26].

Avoiding the use of a tenaculum

The use of a tenaculum or some kind of a 'claw' instrument to hold the cervix and straighten out the canal to facilitate transfer should be avoided; holding the cervix with a tenaculum stimulates uterine contractions. Lesny and co-workers[27] recently confirmed this and they performed mock ETs on 20 patients in the mid-luteal phase of the cycle. The patients were assessed with a transvaginal scan for 2 min to obtain baseline junctional zone activity, and the images were recorded on videotape. The cervix was then grasped with Littlewood's tissue forceps and the uterine position was corrected. The instrument was released and the recording continued for a further 2 min. The images were digitized and converted to five times normal speed to allow analysis of junctional zone contractions. The data clearly showed an increase in the total number of contractions and the number of cervico-fundal, opposing, and random contractions (Table 3). The mechanism for this increase in contractions remains elusive. Fanchin and co-workers[28] have shown that uterine contractions at the time of ET negatively impact upon success rates. The same group assessed uterine contractility during the luteal phase of ovarian stimulation[29]. They noted that a slight, yet significant, decrease in uterine contraction frequency, observed from day of hCG (4.4 ± 0.2 contractions/min) to hCG + 4 (3.5 ± 0.2 contractions/min), was followed by a more pronounced, additional decrease between hCG + 4 and hCG + 7 (1.5 ± 0.2 contractions/min; $p < 0.001$). They concluded that it is possible that such a uterine relaxation assists blastocyst implantation.

Removal of all mucus from cervix

Poindexter and co-workers[30] showed the presence of embryos in the cervical mucus as well as the vagina and the speculum after transfer. The removal of cervical mucus seems to be important. Cervical mucus plugging the catheter tip may result in an increase in retained embryos, damage to embryos and improper placement. Cervical mucus adherent to the embryos may result in embryo expulsion after transfer ('sling-shot' effect).

Table 3 Total number of junctional zone (JZ) contractions in a group of 20 patients before and after the application of a tenaculum to the cervix

Pattern of JZ contractions	Before application	After application	p value*
Cervico-fundal	4	34	0.005
Fundo-cervical	0	9	0.067
Random	24	45	0.001
Opposing	35	58	0.007
Total	63	146	0.0003

*Wilcoxan matched pairs signed-rank test. Reproduced with permission of Oxford University Press/Human Reproduction, from Lesny P, Killick SR, Robinson J, *et al.* Functional zone contractions and embryo transfer: is it safe to use a tenaculum? *Hum Reprod* 1999;14:2367–70, ©European Society of Human Reproduction and Embryology

Mansour and co-workers[31] performed dummy transfers and showed that prior aspiration of cervical mucus led to methylene blue in the cervix on 23% of the transfers. Without aspiration, 57% of patients demonstrated methylene blue in the cervix.

McNamee and co-workers[32] recommend a vigorous cervical lavage with culture medium to remove all cervical mucus. A 10 ml syringe containing culture medium with the outer Wallace sheath is inserted 2 cm into the cervical canal. This can sometimes result in medium entering the uterine cavity although this does not seem to effect pregnancy rates. In addition to removing cervical mucus, cervical lavage may also reduce the amount of bacteria present in the cervical canal. The presence of bacteria at the catheter tip has been shown to negatively impact upon pregnancy rates[33,34]. A recent multicenter randomized study prospectively evaluated 253 patients[35]. Surprisingly, the group with cervical irrigation showed a lower pregnancy rate (45 vs. 57%, $p = 0.051$).

At our Center we carefully remove cervical mucus using plastic Q-tips. In patients with copious cervical mucus, we use a 10 ml syringe to aspirate the mucus. We do not irrigate the cervix.

Ultrasound details of the cavity before treatment

Evaluation of the uterine cavity by hysterosonography, hysterosalpingography or hysteroscopy is routinely performed by most centers. This is done to rule out polyps, fibroids, adhesions, and other intrauterine abnormalities. We routinely evaluate the cavity with one of these techniques. In experienced hands, all three techniques are effective.

Leaving the catheter in place for at least a minute

Martinez and co-workers[36] investigated the influence that the time interval before withdrawal of the catheter after ultrasound-guided ET might have on pregnancy rates with IVF. They prospectively randomized 100 women about to undergo transfer of at least two optimal embryos into two groups. Fifty-one women had the catheter slowly withdrawn immediately after embryo deposit. In 49 women, there was a 30 s delay before catheter withdrawal. The pregnancy rates for transfer in the two groups were 60.8 and 69.4% respectively, with no significant differences. The results indicate that either the waiting interval was insufficient to detect differences, or the retention time before withdrawing the catheter is not a factor that influences pregnancy rate.

Thirty minute rest after transfer

In the early days of IVF, it was routine to keep the patient in bed rest often in a Trendelenberg position for 4 h post transfer. The rest period seems to have been reduced by most programs to 30–60 min. Sharif and co-workers[37] showed no decrease in pregnancy rates with no bed rest after transfer. This was also confirmed in another study by Botta and Grudzinkas[38]. With the use of transabdominal ultrasound guidance, we routinely require the patients to have a full bladder prior to ET and allow patients to rest for a few minutes if they so desire. We have noticed no decrease in pregnancy rates in patients who get up immediately after transfer to void.

Dummy transfer before treatment

Mansour and co-workers demonstrated, in a large randomized study, the benefit of a dummy transfer[9]. Dummy transfers reduced the incidence of difficult transfers from 29.8% to nil, and the pregnancy rates increased significantly from 13.1 to 22.8%. This can be carried out in a prior cycle or at the time of retrieval. The use of ultrasound guidance necessitates a full bladder and this sometimes changes the direction of the cervical canal. If the dummy transfer is carried out with the bladder empty, one should be aware that the cervical direction might change. This is not a major problem as the catheter tip can usually be easily followed into the uterine cavity. A full bladder itself can sometimes make a difficult transfer easy by straightening out the cervix. Transmyometrial transfers have been recommended in patients where the cervix is completely inaccessible[39].

Ultrasound monitoring of transfer

Ultrasound guidance for ET was initially described by Strickler and co-workers[40] and then by Leong and co-workers[41] more than 15 years ago. Patients underwent transvaginal ET, which was monitored by transabdominal ultrasound. This necessitated the need for a full bladder at the time of transfer for proper visualization of the catheter tip. Initial publications showed marginal improvements in success rates[42,43]. However, more recently, several publications have shown statistically significant improvement in pregnancy rates with ultrasound guidance[21,44,45]. This is confirmed by the fact that an increasing number of programs in the United States now routinely use ultrasound guidance.

Kan and co-workers[42] randomly assigned 187 patients to ultrasound-guided or clinical touch transfers. There was a non-significant trend towards higher pregnancy rates with ultrasound guidance in the whole group (37.8 versus 28.9%) and in subsets of older women (38.1 versus 20.4%). Kan and co-workers suggest that ultrasound guidance should be used in clinically difficult transfers and in older women as it seemed to improve the pregnancy rate over clinical touch transfers (Table 4). Hurley and co-workers[43] showed some improvement in patients with single ETs. Lindheim and co-workers[44] showed improved pregnancy rates with ultrasound guidance in donor oocyte cycles. In patients with easy transfers, ultrasound guidance improved implantation (28.8 versus 18.4%) and pregnancy rates (63.1 versus 36.1%). Coroleu and co-workers[45] conducted a large prospective randomized study and compared 182 patients who had ultrasound-guided ET with 180 patients who had clinical touch ET. There were no significant differences between the two groups in terms of

Table 4 Ultrasound versus clinical touch embryo transfer: outcome

	'Ultrasound' group	'Clinical' group
Pregnancy rate	37/98 (37.8)	28/97 (28.9)
Implantation rate	53/260 (20.4)	41/253 (16.2)
Pregnancy rate		
One embryo transferred	1/5 (20)	2/12 (16.7)
Two embryos transferred	5/24 (20.8)	4/14 (28.6)
Three embryos transferred	31/69 (44.9)	22/71 (31)
Pregnancy rate		
< 37 years old	21/56 (37.5)	17/43 (39.5)
≥ 37 years old	16/42 (38.1)	11/54 (20.4)
Ease of transfer		
'Easy' and 'difficult'	13/26 (50)	6/22 (27.3)
'Difficult' only	6/11 (54.5)	1/10 (10)

Data expressed as *n* (%). Values in parentheses are percentages. There were no significant differences between the groups. Reproduced with permission of Oxford University Press/Human Reproduction, from Kan AK, Abdalla HI, Gafar AH, *et al*. Embryo transfer: ultrasound-guided versus clinical touch. *Hum Reprod* 1999;14:1259–61, ©European Society of Human Reproduction and Embryology

age, cause of infertility, or characteristics of the IVF cycle. The pregnancy rate, however, was significantly higher among the ultrasound-guided group (50%) than the clinical touch group (33.7%). Furthermore, there was a significant improvement in the implantation rate: 25.3% in the ultrasound guidance group compared with 18.1% in the clinical touch group. Wood and co-workers[21] also noted a statistically significant improvement in pregnancy rates when they switched to soft catheters and used ultrasound guidance. This study, however, was a retrospective analysis of data and lacked randomization.

Sub-endometrial transfers are associated with lower pregnancy rates than when the embryos are transferred into the cavity itself[46]. With ultrasound visualization of the tip it is possible to avoid this from occurring. In cases with sub-endometrial transfer, the transfer-associated air bubble remains beneath the endometrium following withdrawal of the transfer catheter. These authors used a Jansen–Anderson K-JITS 2000 catheter set and analyzed 121 ETs. The transfers were performed using tactile assessment. Once the catheter was in a position the operator believed to be in the endometrial cavity, the speculum was removed and a transvaginal ultrasound probe was inserted to assess the position of the catheter. They concluded that tactile assessment of ET catheter placement is unreliable. In 17.4% of patients, the outer guiding catheter inadvertently abutted the fundal endometrium. The outer guiding cannula indented the endometrium in 24.8% and the transfer catheter embedded the endometrium in 33.1% of patients. Unavoidable sub-endometrial transfers occurred in 22.3% of transfers.

In a subsequent publication, Woolcott and Stanger[47] tracked the movement of embryo-associated air bubbles on standing after transfer. Ninety-three patients undergoing 101 consecutive ETs were evaluated. After undergoing ultrasound-guided ET, the patients underwent a second scan in a standing position immediately after transfer, allowing the movement of the embryo-associated air to be assessed. No movement occurred in 94.1% (95/101) of transfers, movement of < 1 cm occurred in 4% (4/101) of transfers and movement of 1–5 cm occurred in 2% (2/101) transfers. In none of the patients did the embryo-associated air move out of the uterine cavity either into the cervix or the intramural portion of the fallopian tube. The authors concluded that standing shortly after ET does not play a significant role in the final position of embryo-associated air and is unlikely to be a factor in determining the position of embryos transferred to the uterine cavity during ET.

A recent development is the availability of a coaxial catheter system with an echo-dense tip for ultrasound-guided ET[48]. The Cook Echo-Tip® catheter (Cook Ob/Gyn, Spencer, IN) is a modification of the universally used soft-tip Wallace catheter (SIMS Portex, Hythe, UK) and has an echogenic band at its tip. This is supposed to make visualization of the tip easier and facilitate ET. Since the tip is echo-dense, it is easily seen in obese patients as well as in patients where the bladder is not optimally full. Letterie and co-workers[48] performed 20 transfers with a clinical pregnancy rate of 45% per cycle. They noted that the outer sheath of the system was well visualized during passage through the cervix and into the lower uterine segment due to the thickness of the catheter. With the echo-dense tip, immediate recognition of the tip of the inner sheath was achieved in all patients. With small movements of the ultrasonographic transducer in the transverse plane, the echo-dense catheter tip could be easily tracked during passage through the entire uterine cavity into the fundal region during the first pass. The ability to identify the tip of the inner catheter by the movements of the transducer minimized the amount of to-and-fro motion necessary to identify the catheter tip. This is unlike conventional catheters where one has to move the tip often for proper identification. The authors commented that this would maximize the ability to transfer embryos atraumatically, which is extremely important in embryos that have undergone micromanipulation.

We prospectively compared the Cook Echo-Tip catheter with the Edwards–Wallace catheter in a randomized study[49]. In order to eliminate any variation in success rates due to 'physician factor', a single physician performed all the ETs[13]. All ETs were performed on day 3 after retrieval with the patient in a dorsal lithotomy position. A similar technique was used in all patients. Transabdominal ultrasound guidance took place using a 3.5 MHz linear array probe (RT-3200 Advantage I, General Electric, Milwaukee, WI) with the patients having a full bladder. The cervical mucus was carefully cleaned prior to transfer using plastic Q-tips. In patients with copious mucus, a 10 ml syringe was used to aspirate the mucus. Initially, ET was attempted with the loaded intact catheter connected to a tuberculin syringe. If the first pass failed to negotiate the cervix with ease, the catheter was gently withdrawn. A Wallace stylet (SIMS, Lancing, UK) was then inserted under ultrasound guidance until its tip reached the base of the endometrial stripe. The obturator was then removed and replaced by the floppy inner catheter, which was loaded with the embryos. In either case, the embryos were gently expelled into the uterine cavity with the catheter tip positioned within 1–2 cm of the fundus. Care was taken to avoid white-knuckling the fingers while expelling the embryos. We were careful to avoid touching the uterine fundus. The catheter was then slowly withdrawn over 30 s with a rotating motion. A tenaculum was not used in any of the cases. Care was also taken to avoid a sub-endometrial transfer and the embryos were placed 1–2 cm from the fundus by monitoring the 'embryo bubble' that is formed by the fluid and/or the air surrounding the embryos. The catheters were closely inspected under a stereomicroscope to confirm that all embryos had been expelled and for the presence of blood at the tip. Patients were encouraged to void and leave within a few minutes of the transfer.

As expected, the catheter tip was always seen with the Echo-Tip, but not clearly visualized in five (9.4%) patients in the Wallace catheter group. This was more evident in patients who were obese, had a partly filled bladder, or a retroverted uterus. The stylet was used in 14 (26%) patients in the Wallace group and 16 (34%) with the Echo-Tip catheter. The embryologists initially observed that the echogenic band of the Cook Echo-Tip catheter interfered with the loading of embryos but over a period of time adjusted to it. Pregnancy and implantation rates, however, were similar in both groups (Table 5).

A recent publication surveyed ET practice in the UK[50]. The factor that received the highest rating was the need for a standardized protocol for all unit staff regarding ET policy. The second critical factor was the presence of blood on the embryo catheter at the end of the ET process. Not touching the uterine fundus, the type of catheter used and avoiding the use of a tenaculum rounded off the top five. Ultrasound monitoring of ET was listed number 14 after removal of all mucus from the cervix, removal of hydrosalpinges before treatment, catheter rotation, ultrasound details of the cavity before treatment, leaving the catheter in place for 30 s, a full bladder for ET, dummy transfer before actual ET, and dummy transfer before the treatment cycle. The least important factor was prolonged bed rest following ET. This publication would be more interesting if the authors had compared ET techniques between programs with high and low pregnancy rates.

SUMMARY

The ultimate goal of a successful ET is to deposit embryos into the uterine cavity without trauma. This apparently simple task is fraught with subtle variations that can significantly affect outcomes. Data from large programs have shown wide variation in success rates based upon the individual physician doing the transfer[13,51]. Ultrasound guidance during ET is a valuable tool that can improve success rates. Direct observation of the catheter tip can be useful particularly in patients with variations of normal anatomy, for example a long cervix, or an acutely anteverted or retroverted uterus. This will minimize the possibility of transferring embryos in a blind passage in the cervix, detect when the catheter tip is folding back on itself, and make it possible to avoid sub-endometrial transfer of the embryos[46]. However, ultrasound-guided ET does require a certain degree of eye–hand co-ordination and like any other new technique has a learning curve[51]. At our Center, we have been routinely using this technique and our current pregnancy rates are the highest they have been since the inception of the IVF program more than 15 years ago.

Table 5 Comparison of the Cook Echo-Tip® and the Wallace catheter for ultrasound-guided embryo transfer

	Cook Echo-Tip		Wallace	
	Pregnancy rate	*Implantation rate*	*Pregnancy rate*	*Implantation rate*
Female age (years)				
< 30	6/11 (55)	9/23 (39)	8/11 (73)	12/27 (44)
30–34	12/23 (52)	17/46 (37)	7/18 (39)	11/43 (25)
35–39	5/8 (63)	6/24 (25)	8/16 (50)	12/38 (32)
≥ 40	0/5 (0)	0/12 (0)	5/8 (63)	6/26 (23)
Number of embryos				
1	1/5 (20)	1/5 (20)	0/4 (0)	0/4 (0)
2	17/31 (55)	23/62 (37)	13/25 (52)	20/50 (40)
3	2/6 (33)	4/18 (22)	10/16 (62)	15/48 (31)
4	3/5 (60)	4/20 (20)	5/8 (63)	6/32 (19)
Total	23/47 (49)	32/105 (31)	28/53 (53)	41/134 (31)

Data expressed as *n* (%)

References

1. Wikland M, Enk L, Hamberger L. Transvesical and transvaginal approaches for the aspiration of follicles by use of ultrasound. *Ann N Y Acad Sci* 1985;442:182–94
2. Edwards R, Steptoe P. *A Matter of Life. The Story of a Medical Breakthrough*, 1st edn. London: Hutchinson, 1980;40
3. Steptoe PC. Laparoscopy and ovulation. *Lancet* 1968;2:913
4. Steptoe PC, Edwards RG. Birth after reimplantation of a human embryo. *Lancet* 1978;2:366
5. Brinsden PR. Oocyte recovery and embryo transfer techniques for *in vitro* fertilization. In Brinsden PR, ed. *A Textbook of In Vitro Fertilization and Assisted Reproduction*. New York: Parthenon Publishing, 1999;171–84
6. Gleicher N, Friberg J, Fullan N, *et al*. Egg retrieval for *in vitro* fertilization by sonographically controlled culdocentesis. *Lancet* 1983;2:508–9
7. Dellenbach P, Nisand I, Moreau L, *et al*. Transvaginal sonographically controlled ovarian follicle puncture for egg retrieval. *Fertil Steril* 1985;44:656–62
8. Parsons J, Riddle A, Booker M, *et al*. Oocyte retrieval for *in vitro* fertilization by ultrasonically guided needle aspiration via the urethra. *Lancet* 1985;1:1076–7
9. Mansour R, Aboulghar M, Serour G. Dummy embryo transfer: a technique that minimizes the problems of embryo transfer and improves the pregnancy rate in human *in vitro* fertilization. *Fertil Steril* 1990;54:678–81
10. Elder KT. Laboratory techniques: oocyte collection and embryo culture. In Brinsden PR, ed. *A Textbook of In Vitro Fertilization and Assisted Reproduction*. New York: Parthenon Publishing, 1999;185–201
11. Elder K, Dale B. *In Vitro Fertilization*, 2nd edn. Cambridge: Cambridge University Press, 2000;152–91
12. Naaktgeboren N, Broers FC, Hijnsbroek I, *et al*. Hard to believe, hardly discussed, nevertheless very important for the IVF/ICSI results: embryo transfer technique can double or halve the pregnancy rate. *Hum Reprod* 1997;12(Suppl.):S149
13. Karande V, Morris R, Chapman C, *et al*. Impact of the 'physician factor' on pregnancy rates in a large assisted reproductive technology program: do too many cooks spoil the broth? *Fertil Steril* 1999;71:1001–9
14. Kovacs GT. What factors are important for successful embryo transfer after *in vitro* fertilization? *Hum Reprod* 1999;14:590–2
15. Strandall A, Lindhard A, Waldenstrom U, *et al*. Hydrosalpinx and IVF outcome: a prospective, randomized multicentre trial in Scandinavia on salpingectomy prior to IVF. *Hum Reprod* 1999;14:2762–9
16. Strandell A, Lindhard A. Salpingectomy prior to IVF can be recommended to a well-defined subgroup of patients. *Hum Reprod* 2000;15:2072–4
17. Goudas VT, Hammitt DG, Damario MA, *et al*. Blood on the embryo transfer catheter is associated with decreased rates of embryo implantation and clinical pregnancy with the use of *in vitro* fertilization-embryo transfer. *Fertil Steril* 1998;70:878–82
18. Englert Y, Puissant F, Camus M, *et al*. Clinical study on embryo transfer after human *in vitro* fertilization. *J In Vitro Fertil Embryo Trans* 1986;3:243–6
19. Tur-Kaspa I, Yuval Y, Bider D, *et al*. Difficult or repeated sequential embryo transfers do not adversely affect *in vitro* fertilization pregnancy rates or outcome. *Hum Reprod* 1998;13:2452–5
20. Groutz A, Lessing JB, Wolf Y, *et al*. Cervical dilation during ovum pick-up in patients with cervical stenosis: effect on pregnancy outcome in an *in vitro* fertilization embryo transfer program. *Fertil Steril* 1997;67:909–11
21. Wood EG, Batzer FR, Go KJ, *et al*. Ultrasound-guided soft catheter embryo transfers will improve pregnancy rates in *in vitro* fertilization. *Hum Reprod* 1999;14:107–12
22. Wisanto A, Janssens R, Deschacht J, *et al*. Performance of different embryo transfer catheters in a human *in vitro* fertilization program. *Fertil Steril* 1989;52:79–84
23. Waterstone J, Curson R, Parsons J. Embryo transfer to low uterine cavity. *Lancet* 1991;337:1413
24. Rosenlund B, Sjoblom P, Hilensjo T. Pregnancy outcome related to site of embryo deposition in the uterus. *J Assist Reprod Genet* 1996;13:511–13
25. Yovich JL, Turner SR, Murphy AJ. Embryo transfer technique as a cause of ectopic pregnancies in *in vitro* fertilization. *Fertil Steril* 1985;44:318–21
26. Baba K, Ishihara O, Hayashi N, *et al*. Where does the embryo implant after embryo transfer in humans? *Fertil Steril* 2000;73:123–5
27. Lesny P, Killick SR, Robinson J, *et al*. Junctional zone contractions and embryo transfer: is it safe to use a tenaculum? *Hum Reprod* 1999;14:2367–70
28. Fanchin R, Righini FO, Taylor S, *et al*. Uterine contractions at the time of embryo transfer alter pregnancy rates after *in vitro* fertilization. *Hum Reprod* 1998;13:1968–74
29. Fanchin R, Ayoubi J-M, Righini C, *et al*. Uterine contractility decreases at the time of blastocyst transfers. *Hum Reprod* 2001;16:1115–19
30. Poindexter A, Thompson D, Gibbons W, *et al*. Residual embryos in failed embryo transfer. *Fertil Steril* 1986;46:262–7
31. Mansour RT, Aboulghar MA, Serour GI, *et al*. Dummy embryo transfer using methylene blue dye. *Hum Reprod* 1994;9:1257–9
32. McNamee P, Huang T, Carwile A. Significant increase in pregnancy rates achieved by vigorous irrigation of endocervical mucus prior to embryo transfer with a Wallace catheter in an IVF-ET program. *Fertil Steril* 1998;70:S228
33. Egbase P, Al-Sharhan M, Al-Othman S, *et al*. Incidence of microbial growth from the tip of the embryo transfer catheter after embryo transfer in relation to clinical pregnancy rate following *in vitro* fertilization and embryo transfer. *Hum Reprod* 1996;11:1687–9
34. Fanchin R, Harmas A, Benaoudia F, *et al*. Microbial flora of the cervix assessed at the time of embryo transfer adversely affects *in vitro* fertilization outcome. *Fertil Steril* 1998;70:866–70
35. Glass KB, Green CA, Fluker MR, *et al*. Multicenter randomized controlled trial of cervical irrigation at the time of embryo transfer. *Fertil Steril* 2000;74:S31

36. Martinez F, Coroleu B, Parriego M, *et al*. Ultrasound-guided embryo transfer: immediate withdrawal of the catheter versus a 30 second wait. *Hum Reprod* 2001;16:871–4

37. Sharif K, Afnan M, Lashen H, *et al*. Is bed rest following embryo transfer necessary? *Fertil Steril* 1998;69:478–81

38. Botta G, Grudzinskas G. Is a prolonged bed rest following embryo transfer useful? *Hum Reprod* 1997;12:2489–92

39. Kato O, Takatsuka R, Asch R. Transvaginal-transmyometrial embryo transfer: the Towako method; experience of 104 cases. *Fertil Steril* 1993;59:51–3

40. Strickler RC, Christianson C, Crane JP, *et al*. Ultrasound guidance for human embryo transfer. *Fertil Steril* 1985; 43:54–61

41. Leong M, Leung C, Tucker M, *et al*. Ultrasound-assisted embryo transfer. *J In Vitro Fertil Embryo Transf* 1986;3: 383–5

42. Kan AKS, Abdalla HI, Gafar AH, *et al*. Embryo transfer: ultrasound-guided versus clinical touch. *Hum Reprod* 1999; 14:1259–61

43. Hurley V, Osborn J, Leoni M, *et al*. Ultrasound-guided embryo transfer: a controlled trial. *Fertil Steril* 1991;55: 559–62

44. Lindheim SR, Cohen MA, Sauer MV. Ultrasound guided embryo transfer significantly improves pregnancy rates in women undergoing oocyte donation. *Int J Gynecol Obstet* 1999;66:281–4

45. Coroleu B, Carreras O, Veiga A, *et al*. Embryo transfer under ultrasound guidance improves pregnancy rates after *in vitro* fertilization. *Hum Reprod* 2000;15:616–20

46. Woolcott R, Stanger J. Potentially important variables identified by transvaginal ultrasound-guided embryo transfer. *Hum Reprod* 1997;12:963–6

47. Woolcott R, Stanger J. Ultrasound tracking of the movement of embryo-associated air bubbles on standing after transfer. *Hum Reprod* 1998;13:2107–9

48. Letterie GS, Marshall L, Angle M. A new coaxial system with an echodense tip for ultrasonographically guided embryo transfer. *Fertil Steril* 1999;72:266–8

49. Karande VC, Hazlett D, Gleicher N. A prospective randomized comparison of the Wallace catheter and the Cook Echo-Tip catheter during ultrasound-guided embryo transfer. *Fertil Steril* 2002; in press

50. Salha OH, Lamb VK, Balen AH. A postal survey of embryo transfer practice in the UK. *Hum Reprod* 2001;16:686–90

51. Papageorgiou TC, Hearns-Stokes RM, Leondires MP, *et al*. Training of providers in embryo transfers: what is the minimum number of transfers required for proficiency? *Hum Reprod* 2001;16:1415–19

In vitro processing of human sperm for assisted reproduction   12

Marco Rossato, Carlo Galeazzi and Carlo Foresta

INTRODUCTION

For many years, different assisted reproductive technologies (ART) have been developed to resolve couples' infertility and over the years these techniques have been modified with the evolution of basic and clinical research in reproductive biology together with technical developments. As a result of the difficulty or impossibility of treating some male and female forms of infertility, the therapy procedures in ART involve sperm processing techniques that range from intrauterine insemination (IUI), through gamete intrafallopian transfer (GIFT) and *in vitro* fertilization and embryo transfer (IVF-ET), to intracytoplasmic sperm injection (ICSI).

Although the presence of viable, motile and morphologically normal sperm does not mean that they are fully competent for oocyte fertilization, one of the most important steps in ART is the selection of such highly viable, motile and morphologically normal spermatozoa to further use. Furthermore, whichever ART is to be utilized, an important purpose of all known sperm selection techniques is to separate sperm from the seminal plasma since it contains a number of so-called 'decapacitating' factors that prevent the sperm undergoing capacitation, a complex and still undefined series of processes essential for sperm to acquire the full competence required for oocyte penetration.

Each known sperm separation method has precise indications and there is a need for the individualization of each technique with the semen sample to be processed. In this chapter we review the main preparation techniques utilized for sperm selection. In the last part of the chapter we will introduce some particular aspects of sperm preparation methods which deal with the possible increase of the fertilizing potential of sperm.

SPERM WASHING

The simplest method of sperm preparation is a simple washing procedure, in which semen is diluted with an appropriate culture medium and then centrifuged through two or three cycles (300–500 g for 5–10 min) with the final sperm pellet being resuspended in the culture medium at the desired final concentration after discarding the supernatant. Culture media utilized for sperm preparation techniques can be human tubal fluid, HAM's F-10, Earle's medium, BWW, Dulbecco's medium, M-199 or any other sterile culture medium that is commercially available. The use of added proteins (i.e., albumin) is not considered of primary importance, and also with regard to its potential hazardous effects[1–3].

This technique successfully separates sperm from the seminal plasma but the repeated centrifugation cycles result in the collection of unselected sperm and other cell types, including leukocytes that have been found to have deleterious effects on sperm thus impairing their fertilizing potential[4,5]. For these reasons, sperm washing techniques should be avoided for all laboratory applications that require the selection of viable and functionally active sperm.

SPERM MIGRATION

Sperm migration from semen – swim-up technique

This original sperm selection technique has been used for many decades to select sperm. It relies on the ability of motile sperm to migrate from a semen layer placed beneath a given aliquot of culture medium to the medium itself after an incubation of as long as 30–60 min. This technique allows the selection of a highly motile sperm population that is devoid of any other cell type, of immotile and dead sperm and of potentially dangerous leukocytes[6,7]. This technique often gives low results when the processed semen samples are from severely oligozoospermic or asthenozoospermic subjects.

As a modification of the above described technique, some authors have suggested using a preparation of sodium hyaluronate at a concentration of 1 mg/ml in culture medium as the migration medium to make use of the positive effects on sperm function described for hyaluronate[8–10]. This technique utilizing hyaluronate (SpermSelect, Mizarra Med. Uppsala, Sweden) has been investigated and suggested as a sperm selection method during ART techniques such as IUI[9,10].

Sperm migration from washed sperm pellet – swim-up from pellet

In contrast to the previous method, it has been suggested that sperm should be allowed to migrate after simple washing from the pellet to the culture medium layered over the sperm pellet[11]. Although the swim-up from the pellet could be useful to select motile sperm from some semen samples with low sperm concentrations, it should be avoided for the same reasons described above for sperm washing techniques.

Sperm migration – swim-down

The opposing view of the swim-up procedure has led some authors to suggest the introduction of a laboratory procedure for sperm selection based upon the ability of sperm to swim from a semen layer into an underlying layer of medium providing that the latter's viscosity has been increased[12]. Other authors have reported a modification of this method, describing better results in terms of motile sperm concentration isolated using this modified swim-down technique based upon an initial centrifugation step to concentrate sperm with a simple washing procedure before the swim-down procedure.

ADHERENCE METHODS

Glass wool filtration

This technique of sperm preparation involves the filtration of a sperm suspension (with or without a preliminary washing step) through a package of glass wool fibers[13,14]. This method is not suitable for severely oligozoospermic samples since it leads to poor yields. Furthermore, it has been demonstrated that glass wool fibers may damage the sperm plasma membrane[15] and the use of glass wool fibers (SpermPrep columns, SpermPrep, Lexington, KY) has not been reported to have optimal sperm recovery[14]. Thus, for these reasons, this method has now been abandoned.

DENSITY GRADIENTS

Albumin gradients

Many years ago the use of gradients of albumin solution (bovine or human) to isolate motile sperm from the remaining semen components was suggested[16]. Albumin gradient solution with culture medium was utilized to obtain layers containing albumin concentrations rising from 7.5% to 17%. Semen is washed with a single-step centrifugation in culture medium and then layered at the top of the albumin gradient. After incubation at 37°C for 30 min the sperm layer is removed. Albumin gradients are then further incubated for 1 h and then the 17% albumin layer (supposed to contain motile sperm) is collected, washed with culture medium and then the final pellet is resuspended with the culture medium.

Over the years this method has been replaced by the less time-consuming Percoll gradient technique as described below and the isolation method using albumin gradients has now been abandoned[17].

Percoll gradients

The main innovation in sperm preparation techniques was the introduction of modified colloidal silica particles in the late 1970s when Pertoft and co-workers proposed the use of colloidal silica particles coated with polyvinylpyrrolidone[18]. The results obtained by the application of the Percoll gradients to sperm preparations led to the commercialization of the particles. This technique comprises the production of a single Percoll concentration (continuous Percoll separation) or of different Percoll gradients (90%, 70%, 50%; discontinuous Percoll separation) through which the sperm sample is processed to select a population of highly motile sperm. To this aim the Percoll stock solution is diluted 9:1 with 10× culture medium to produce a 100% Percoll solution that is isotonic to 1× culture medium. This can be used as it is (continuous Percoll separation) or diluted with the culture medium to obtain the desired density gradients (discontinuous Percoll separation). Typically, 1 ml layers of 90%, 70% and 50% Percoll solution are stratified in a vial and the ejaculate is layered over the lower density gradient and then the sample is centrifuged at 300 g for 20 min. After centrifugation, the sperm pellet and the higher Percoll gradient are collected and washed with culture medium, resuspending the final sperm pellet in culture medium at the desired concentration.

A variance of this technique, the so-called 'mini-Percoll', has been proposed, using small aliquots (typically 0.3 ml) of each Percoll gradient as described above. The 'mini-Percoll' technique has been used for sperm processing in case of severely oligozoospermic semen samples[19,20].

The main intrinsic chemico-physical properties of Percoll have made this compound ideal for use in sperm selection; in fact it does not alter the osmotic pressure of each gradient and its very low viscosity does not interfere with sperm-processing procedures. However, although this method has been succesfully utilized for many years for sperm isolation, Percoll has been withdrawn from clinical use indications by its producing company, Pharmacia Biotech in October 1996.

After the withdrawal of Percoll for clinical use, other colloidal silica preparations have been proposed for clinical use and are commercially available: Pure Sperm (Nicadon Int AB, Goteborg, Sweden), Isolate (Irvine Scientific, Santa Ana, CA) and Perwash (Conception Technologies, La Jolla, CA). These are used as for Percoll and are ready-to-use preparations. Since their introduction, different studies have documented the usefulness of these colloidal preparation for sperm isolation in different ART, reporting similar efficacy in terms of their sperm selection success[21,22].

Other substances have been, and are being, used for sperm isolation with density gradients: Nycodenz, using the iodinated cyclic hydrocarbon iodexol, has been shown to be able to select sperm for different ART[23] and has also been commercialized (OptiPrep, Nycomed Pharma, Oslo, Norway). One adverse effect of this preparation, and of its analog iodixonol (Accudenz, Nycomed Pharma, Oslo, Norway), is that it has a high osmotic force and thus care has to be given to the preparation of sperm incubation medium during all the preparation steps to maintain the optimal conditions that support sperm fertilizing ability[24,25].

Other products have been introduced on the market: Enhance Cell Isolation Product (Conception Technologies, San Diego, CA) and Isocare (In Vitro Care, San Diego, CA). The results obtained using these preparations have to be compared with those of the other well-tested agents used for sperm selection to evaluate and demonstrate their usefulness.

One of the most important steps during sperm isolation through density gradients is the centrifugation force to be used: as originally reported for Percoll gradients, an optimal centrifugation occurs at 300 g for 20 min. This centrifugation protocol is also utilized for the other colloidal silica preparations introduced in the market for sperm isolation procedures. No beneficial effects have been reported for different centrifugation protocols, by varying speed and/or time of centrifugation, providing that the centrifugation speed does not exceed 800 g[26].

SELECTIVE SPERM ISOLATION FOR INTRACYTOPLASMIC SPERM INJECTION

ICSI has revolutionized ART and has became the treatment of choice for severe male infertility. In this respect, the success rate of this technique has been proved to be higher when motile (and thus viable) sperm are used[27], while the success rate when utilizing immotile sperm has been reported to be variable[28–30]. Since sperm viability has a major role in the outcome of ICSI[31], the use of live sperm for this procedure is recommended. It is obvious that motile sperm should be used when available, but this is not always possible in some forms of complete absence of motile sperm, as observed in patients affected by Kartagener's and immotile cilia syndrome[32,33]. In these patients the identification of live sperm is of primary importance to give rise to viable embryos.

To this aim different methods have been suggested for the identification and selection of live sperm from semen samples with total asthenozoospermia. The main technique utilized takes advantage of the ability of live sperm to swell when exposed to a hypo-osmotic medium, as suggested previously by Jeyendran and co-workers[34]. On exposure of sperm to hypo-osmotic medium an influx of water within the cytoplasm results in sperm swelling and curling of the tail[35]. This allows the differentiation between swollen (and therefore viable) and not swollen (and therefore dead) immotile sperm without influencing sperm function. Exposure to hypo-osmotic medium may even be beneficial for sperm, inducing their activation and fertilizing ability[36,37]. This technique consists of the exposure of sperm to a hypo-osmotic cell culture medium prepared by mixing the medium with sterile distilled water to obtain a final osmolarity close to 150 mOsm[38] and then transferring the swollen (live) sperm into a drop of polyvinyl-pyrrolidone before sperm injection into the oocyte cytoplasm. With this technique pregnancies and the birth of normal babies has been reported[33,39]. The availability of this simple but functional technique of immotile viable sperm selection is much more important when considering the selection of sperm extracted from the epididymis or testis and then cryopreserved before ICSI[35]. The identification of swollen sperm from frozen–thawed samples could improve fertilization and pregnacy rates in these cases.

SPERM RECOVERY FROM RETROGRADE EJACULATION

Retrograde ejaculation involves the passage of sperm into the urinary bladder during ejaculation as a result of a defective bladder neck (underlying reasons are neurologic such as diabetes complications or the use of α-adrenergic blockers). Although in these patients sperm can be retrieved from the epididymis or testis, thus avoiding complex procedures of urine alkalinization, here we describe a procedure for sperm retrieval from the urine of these patients. The main difficulty is the neutralization of the acid pH and high osmolarity of urine, which alter sperm function[40,41]. This method involves the administration, on the day of the sperm retrieval procedure, of a solution of 6–10% sodium bicarbonate in 250 ml of water. After 2–3 h the patient has to provide a urine sample, without completely emptying the bladder. If the urine pH is 7.5–8.2 and its osmolarity is between 300 and 500 mOsm, the patient has to masturbate and ejaculate into a vial containing a culture medium buffered at pH 7.6–8.2 to check the presence of sperm; he then has to collect urine within another vial containing the same buffered medium. This urine sample is divided into different aliquots that are then centrifuged at 300 *g* for 10 min.

If the urine pH and osmolarity are not within the recommended range, the amount of sodium bicarbonate administered can be modified to adjust the urine to the preferred values.

SPERM PREPARATION IN IMMUNOLOGIC INFERTILITY

The treatment of choice in cases of male immunologic infertility has been a matter of debate for many years during which different pharmacologic therapies have been suggested to reduce the immune response in these patients[42]. These pharmacologic treatments have not been shown to be of benefit and have potential hazardous side-effects for these patients. A number of laboratory techniques have been proposed for antisperm-antibody-free sperm isolation or antisperm antibody removal but these have been of little success, although different data have been reported[43–46]. Some studies have attempted to remove sperm-bound antisperm antibodies by processing with washing centrifugations, Percoll gradients, or enzymatic treatment, all with poor results[47–49]. Current knowledge seems to suggest that the treatment of choice in cases of immunologic infertility is ICSI[42,50].

ENHANCEMENT OF SPERM FERTILIZING ABILITY

The main purpose of each sperm isolation technique, beyond the separation of non-sperm cells, is that of obtaining highly motile sperm for use in different ART. This depends on the sperm selection technique utilized but more especially on the basal seminal characteristics of the sample utilized for the selection. It is more difficult to obtain high-quality sperm samples from poor-quality semen samples. To this aim in previous years a number of authors have attempted to enhance sperm fertilizing ability by incubation with agents known to increase their motility. Focus has been concentrated on the methylxanthine derivatives, such as theophylline caffeine and above all pentoxyphylline, which has been demonstrated to increase sperm motility *in vitro*[51–55]. Methylxanthine derivatives have been shown to raise sperm motility by reducing cAMP degradation after the inhibition of cAMP phosphodiesterase, thus leading to an intracellular increase of cAMP, levels of which have been shown to correlate positively with sperm motility[51–55]. Other authors have demonstrated that *in vitro* sperm treatment with pentoxyphylline increased the acrosome reaction and fertilizing ability in humans[56–58]. However, some authors have reported detrimental effects of these agents, in particular of pentoxyphylline on oocytes and embryos[59], thus suggesting that caution should be exercised in the use of these drugs to activate sperm and in the sperm–oocyte interaction in humans, at least until their full effects have been elucidated. At the moment the precise role of these substances in ART for the treatment of male-factor infertility is yet to be determined.

Other authors have suggested the use of follicular fluid for sperm treatment before ART. This technique has been shown to increase sperm fertilizing ability and pregnancy rates[60] by incubating sperm for 15 min in the presence of human follicular fluid or control medium after a simple washing procedure. The rationale for this technique lies in the presence of progesterone (and other substances) within the follicular fluid. Progesterone has been previously shown to be an inducer of the acrosome reaction in human sperm[61–64] and is putatively responsible for increased fertilizing potential of sperm after incubation with follicular fluid.

More recently, a novel technique of sperm activation using a non-toxic agent has been suggested[65] based on the use of adenosine-triphosphate (ATP). Extracellular ATP has been demonstrated to be a potent and rapid activator of human sperm fertilizing ability[66,67]. Sperm preincubation with extracellular ATP at a concentration of 2.5 mmol/l for 60 min has been proven to result in higher fertilization and implantation rates when utilized during *in-vitro* fertilization–embryo transfer (IVF-ET) techniques with respect to sperm incubation with control medium in cases of male-factor infertility[65]. These results suggest that incubation with extracellular ATP can be a non-toxic and physiologic method for the treatment of sperm during ART procedures in order to increase their fertilizing potential.

CONCLUSIONS

Over the years various methods have been introduced for the preparation of normal and pathologic ejaculates for use in ART techniques. As briefly described in the present chapter, and as currently known from the literature, the most appropriate method for sperm selection during ART has to be considered as that recovering a high number of viable, motile and morphologically normal spermatozoa without altering sperm fertilizing ability and DNA integrity. In this respect all techniques utilizing sperm centrifugation through density gradients have been demonstrated to be safer by many different and independent studies (providing that Percoll is not used for sperm preparation for clinical applications in humans as declared by Percoll's manufacturer, Pharmacia Biotech). The swim-up technique has also been demonstrated to be effective in the selection of viable and motile sperm (depending on the starting seminal characteristics). The role of other laboratory methods for sperm selection is still to be demonstrated. With this in mind it could be of help to present some data reporting the efficacy of different sperm selection techniques in a group of infertile subjects affected by oligo-asthenozoospermia. Collective data regarding seminal characteristics, plasma membrane functional aspects (as evaluated with the hypo-osmotic swelling test) and chromatin ability to decondense after exposure to decondensing agents of sperm obtained utilizing different sperm isolation techniques are presented in Table 1. It can be observed that the simple washing procedure does not enhance the percentage of sperm with good characteristics. Nevertheless, this technique can be utilized when the seminal sample shows very severe oligozoospermia with only a few sperm detectable in the seminal sample, always bearing in mind the considerations discussed in the section regarding this

Table 1 Standard and functional seminal characteristics before and after different sperm isolation techniques from semen samples of 25 infertile subjects affected by oligo-asthenozoospermia

Method	Sperm no. (×10⁶/ml)	Sperm no. (×10⁶/ejaculate)	Sperm isolated (×10⁶)	% Motility (grade a + b)	% Normal morphology	% Viability	% HSO	% Decondensed sperm
Washing	—	—	26.4 ± 5.8	27.6 ± 13.3	44.2 ± 9.1	40.1 ± 18.6	52.3 ± 13.3	76.0 ± 4.2
Swim-up	—	—	8.5 ± 4.2	78.6 ± 12.5**	78.6 ± 5.1*	72.1 ± 13.3**	80.2 ± 9.6**	77.4 ± 5.1
Swim-up from pellet	—	—	7.6 ± 5.5	70.2 ± 23.3**	77.6 ± 6.4*	78.8 ± 12.1**	83.4 ± 10.1**	78.8 ± 6.3
Glass filtration	—	—	18.5 ± 6.6	57.4 ± 18.2*	73.5 ± 7.7	66.2 ± 8.6*	78.3 ± 9.2**	80.7 ± 5.0
Density gradient (Isolate)	—	—	16.2 ± 5.4	68.2 ± 13.4**	76.2 ± 6.1*	79.9 ± 6.2*	82.5 ± 8.8**	81.6 ± 6.3
Basal control	14.4 ± 4.8	31.7 ± 8.6	—	37.1 ± 10.3	45.0 ± 12.3	58.3 ± 12.2	55.1 ± 11.2	76.4 ± 5.3

*$p < 0.01$ versus basal control; **$p < 0.001$ versus basal control; HSO, hypo-osmotic swelling

procedures on the potential damaging effects of this technique. In case of moderate oligo-asthenozoospermia, the better sperm isolation methods seem to be swim-up and centrifugation through density gradients. These techniques allow the isolation of a sufficient number of sperm with high motility and normal morphologic characteristics together with optimal plasma membrane functional properties as shown by the percentages obtained with the hypo-osmotic swelling test. As can be observed from the data presented in the table, whichever technique is utilized for sperm selection they are all, apart from the washing procedure, able to recover sperm with optimal biological characteristics, making the sperm suitable for use in different ART techniques. However, the use of motile and morphologically normal sperm is not synonymous with good fertilizing ability since the ability of sperm to fertilize the egg depends not only on these characteristics but also on other important and complex processes that standard laboratory procedures do not explore.

References

1. Wuthrich B, Stern A, Johansson SG. Severe anaphylactic reaction to bovine serum albumin at the first attempt of artificial insemination. *Allergy* 1995;50:179–83
2. de Blay F, Tomb R, Vouillot C, *et al.* Urticaria and angioedema during insemination with fluid containing bovine serum albumin. *Contact Dermatitis* 1993;28:119
3. Sonenthal KR, McKnight T, Shaughnessy MA, *et al.* Anaphylaxis during intrauterine insemination secondary to bovine serum albumin. *Fertil Steril* 1991;56:1188–91
4. Aitken RJ, Clarkson JS. Cellular basis of defective sperm function and its association with the genesis of reactive oxygen species by human spermatozoa. *J Reprod Fertil* 1987;81:459–69
5. Chan SYW, Tucker MJ. Differential sperm performance as judged by the zone-free hamster egg penetration test relative to differing sperm penetration techniques. *Hum Reprod* 1992;7:255–60
6. Lopata A, Patullo MJ, Chang A, *et al.* A method for collecting motile spermatozoa from human semen. *Fertil Steril* 1976;27:677–84
7. Russell LD, Rogers BJ. Improvement in the quality and fertilization potential of a human sperm population using the rise technique. *J Androl* 1987;8:25–33
8. Wikland M, Wik O, Steen Y, *et al.* A self-migration method for preparation of sperm for *in-vitro* fertilization. *Hum Reprod* 1987;2:191–5
9. Sbracia M, Grasso J, Sayme N, *et al.* Hyaluronic acid substantially increases the retention of motility in cryopreserved/thawed human spermatozoa. *Hum Reprod* 1997;12:1949–54
10. Huszar G, Willetts M, Corrales M. Hyaluronic acid (SpermSelect) improves retention of sperm motility and velocity in normospermic and oligospermic specimens. *Fertil Steril* 1990;54:1127–34
11. Chan PJ, Tredway DR, Su BC, *et al.* A double method sperm wash for artificial insemination. *Arch Androl* 1992;29:43–8
12. Gonzales GF, Pella RE. Swim-down: a rapid and easy method to select motile spermatozoa. *Arch Androl* 1993;30:29–34

13. Paulson JD, Polakoski KL. A glass wool column procedure for removing extraneous material from the human ejaculate. *Fertil Steril* 1977;28:178–81

14. Smith S, Hosid S, Scott L. Use of postseparation sperm parameters to determine the method of choice for sperm preparation for assisted reproductive technology. *Fertil Steril* 1995;63:591–7

15. Sherman JK, Paulson JD, Liu KC. Effect of glass wool filtration on ultrastructure of human spermatozoa. *Fertil Steril* 1981;36:643–7

16. Dmowski WP, Gaynor L, Lawrence M, *et al*. Artificial insemination homologous with oligospermic semen separated on albumin columns. *Fertil Steril* 1979;31:58–62

17. Berger T, Marrs RP, Moyer DL. Comparison of techniques for selection of motile spermatozoa. *Fertil Steril* 1985;43:268–73

18. Pertoft H, Laurent TC, Laas T, *et al*. Density gradients prepared from colloidal silica particles coated by polyvinyl-pyrrolidone (Percoll). *Anal Biochem* 1978;88:271–82

19. Ord T, Patrizio P, Marello E, *et al*. Mini-Percoll: a new method of semen preparation for IVF in severe male factor infertility. *Hum Reprod* 1990;5:987–9

20. Mortimer D. Sperm recovery techniques to maximize fertilizing capacity. *Reprod Fertil Dev* 1994;6:25–31

21. Sharma RK, Seifarth K, Garlak D, *et al*. Comparison of three sperm preparation media. *Int J Fertil Womens Med* 1999;44:163–7

22. Claassens OE, Menkveld R, Harrison KL. Evaluation of three substitutes for Percoll in sperm isolation by density gradient centrifugation. *Hum Reprod* 1998;13:3139–43

23. Gellert-Mortimer ST, Clarke GN, Baker HW, *et al*. Evaluation of Nycodenz and Percoll density gradients for the selection of motile human spermatozoa. *Fertil Steril* 1988;49:335–41

24. Sbracia M, Sayme N, Grasso J, *et al*. Sperm function and choice of preparation media: comparison of Percoll and Accudenz discontinuous density gradients. *J Androl* 1996;17:61–7

25. Smith TT, Byers M, Kaftani D, *et al*. The use of iodixanol as a density gradient material for separating human sperm from semen. *Arch Androl* 1997;38:223–30

26. Mortimer D. Sperm preparation methods. *J Androl* 2000;21:357–66

27. El-Nour AM, Al Mayman HA, Jaroudi KA, *et al*. Effects of the hypo-osmotic swelling test on the outcome of intra-cytoplasmic sperm injection for patients with only nonmotile spermatozoa available for injection: a prospective randomized trial. *Fertil Steril* 2001;75:480–4

28. Kahraman S, Tasdemir M, Tasdemir I, *et al*. Pregnancies achieved with testicular and ejaculated spermatozoa in combination with intracytoplasmic sperm injection in men with totally or initially immotile spermatozoa in the ejaculate. *Hum Reprod* 1996;11:1343–6

29. Vandervorst M, Tournaye H, Camus M, *et al*. Patients with absolutely immotile spermatozoa and intracytoplasmic sperm injection. *Hum Reprod* 1997;12:2429–33

30. Shulman A, Feldman B, Madgar I, *et al*. In-vitro fertilization treatment for severe male factor: the fertilization potential of immotile spermatozoa obtained by testicular extraction. *Hum Reprod* 1999;14:749–52

31. Poe-Zeigler R, Nehchiri F, Hamacher P, *et al*. Effects of sperm viability on fertilization and embryo cleavage following intracytoplasmic sperm injection. *J Assist Reprod Genet* 1997;14:277–81

32. von Zumbusch A, Fiedler K, Mayerhofer A, *et al*. Birth of healthy children after intracytoplasmic sperm injection in two couples with male Kartagener's syndrome. *Fertil Steril* 1998;70:643–6

33. Cayan S, Conaghan J, Schriock ED, *et al*. Birth after intracytoplasmic sperm injection with use of testicular sperm from men with Kartagener/immotile cilia syndrome. *Fertil Steril* 2001;76:612–14

34. Jeyendran RS, Van der Ven HH, Perez-Pelaez M, *et al*. Development of an assay to assess the functional integrity of the human sperm membrane and its relationship to other semen characteristics. *J Reprod Fertil* 1984;70:219–28

35. Ma S, Nigro MK, Rowe T, *et al*. Selection of viable sperm from frozen–thawed immotile spermatozoa based on the phenomenon of sperm tail curling in men who underwent testicular biopsy and epididymal sperm aspiration. *Fertil Steril* 2000;74:172–3

36. Rossato M, Di Virgilio F, Foresta C. Involvement of osmo-sensitive calcium influx in human sperm activation. *Mol Hum Reprod* 1996;2:903–9

37. Foresta C, Rossato M. Calcium influx pathways in human spermatozoa. *Mol Hum Reprod* 1997;3:1–4

38. Tsai YL, Liu J, Garcia JE, *et al*. Establishment of an optimal hypo-osmotic swelling test by examining single spermatozoa in four different hypo-osmotic solutions. *Hum Reprod* 1997;12:1111–13

39. Kay VJ, Irvine DS. Successful *in-vitro* fertilization pregnancy with spermatozoa from a patient with Kartagener's syndrome: case report. *Hum Reprod* 2000;15:135–8

40. Makler A, David R, Blumenfeld Z, Better OS. Factors affecting sperm motility. VII. Sperm viability as affected by change of pH and osmolarity of semen and urine specimens. *Fertil Steril* 1981;36:507–11

41. Braude PR, Ross LD, Bolton VN, *et al*. Retrograde ejaculation: a systematic approach to non-invasive recovery of spermatozoa from post-ejaculatory urine for artificial insemination. *Br J Obstet Gynaecol* 1987;94:76–83

42. Lombardo F, Gandini L, Dondero F, *et al*. Antisperm immunity in natural and assisted reproduction. *Hum Reprod Update* 2001;7:450–6

43. Foresta C, Varotto A, Caretto A. Immunomagnetic method to select human sperm without sperm surface-bound auto-antibodies in male autoimmune infertility. *Arch Androl* 1990;24:221–5

44. Zavos PM, Correa JR, Zarmakoupis-Zavos PN. Antisperm antibody treatment mode: levels of antisperm antibodies after incubation with TEST-yolk buffer and filtration using the SpermPrep II method. *Fertil Steril* 1998;69:517–21

45. Ryan M, Drudy L, Cottell E, *et al*. Preparation of anti-body free spermatozoa by *in vitro* immunodepletion using immunobeads. *Andrologia* 1994;26:247–50

46. Grundy CE, Robinson J, Gordon AG, *et al*. Selection of an antibody-free population of spermatozoa from semen samples of men suffering from immunological infertility. *Hum Reprod* 1991;6:593–6

47. Adeghe AJ. Effect of washing on sperm surface auto-antibodies. *Br J Urol* 1987;60:360–3
48. Bronson RA, Cooper GW, Rosenfeld DL, *et al*. The effect of an IgA1 protease on immunoglobulins bound to the sperm surface and sperm cervical mucus penetrating ability. *Fertil Steril* 1987;47:985–91
49. Haas GG Jr, D'Cruz OJ, Denum BM. Effect of repeated washing on sperm-bound immunoglobulin G. *J Androl* 1988;9:190–6
50. Bates CA. Antisperm antibodies and male subfertility. *Br J Urol* 1997;80:691–7
51. Nassar A, Morshedi M, Mahony M, *et al*. Pentoxifylline stimulates various sperm motion parameters and cervical mucus penetrability in patients with asthenozoospermia. *Andrologia* 1999;31:9–15
52. Fisch JD, Behr B, Conti M. Enhancement of motility and acrosome reaction in human spermatozoa: differential activation by type-specific phosphodiesterase inhibitors. *Hum Reprod* 1998;13:1248–54
53. Calogero AE, Fishel S, Hall J, *et al*. Correlation between intracellular cAMP content, kinematic parameters and hyperactivation of human spermatozoa after incubation with pentoxifylline. *Hum Reprod* 1998;13:911–15
54. Fleming S, Green S, Hall J, *et al*. Sperm function and its manipulation for microassisted fertilization. *Baillières Clin Obstet Gynaecol* 1994;8:43–64
55. Jiang CS, Kilfeather SA, Pearson RM, *et al*. The stimulatory effects of caffeine, theophylline, lysine-theophylline and 3-isobutyl-1-methylxanthine on human sperm motility. *Br J Clin Pharmacol* 1984;18:258–62
56. Tesarik J, Mendoza C. Sperm treatment with pentoxifylline improves the fertilizing ability in patients with acrosome reaction insufficiency. *Fertil Steril* 1993;60:141–8
57. Tesarik J, Mendoza C, Carreras A. Effects of phosphodiesterase inhibitors caffeine and pentoxifylline on spontaneous and stimulus-induced acrosome reactions in human sperm. *Fertil Steril* 1992;58:1185–90
58. Yovich JM, Edirisinghe WR, Cummins JM, *et al*. Influence of pentoxifylline in severe male factor infertility. *Fertil Steril* 1990;53:715–22
59. Scott L, Smith S. Human sperm motility-enhancing agents have detrimental effects on mouse oocytes and embryos. *Fertil Steril* 1995;63:166–75
60. Blumenfeld Z, Nahhas F. Pretreatment of sperm with human follicular fluid for borderline male infertility. *Fertil Steril* 1989;51:863–8
61. Blackmore PF, Beebe SJ, Danforth DR, *et al*. Progesterone and 17 alpha-hydroxyprogesterone. Novel stimulators of calcium influx in human sperm. *J Biol Chem* 1990;265:1376–80
62. Osman RA, Andria ML, Jones AD, *et al*. Steroid induced exocytosis: the human sperm acrosome reaction. *Biochem Biophys Res Commun* 1989;160:828–33
63. Foresta C, Rossato M, Di Virgilio F. Ion fluxes through the progesterone-activated channel of the sperm plasma membrane. *Biochem J* 1993;294:279–83
64. Foresta C, Rossato M, Mioni R, *et al*. Progesterone induces capacitation in human spermatozoa. *Andrologia* 1992;24:33–5
65. Rossato M, La Sala GB, Balasini M, *et al*. Sperm treatment with extracellular ATP increases fertilization rates in *in-vitro* fertilization for male factor infertility. *Hum Reprod* 1999;14:694–7
66. Foresta C, Rossato M, Chiozzi P, *et al*. Mechanism of human sperm activation by extracellular ATP. *Am J Physiol* 1996;270:C1709–14
67. Foresta C, Rossato M, Di Virgilio F. Extracellular ATP is a trigger for the acrosome reaction in human spermatozoa. *J Biol Chem* 1992;267:19443–7

Culture media for the human embryo

13

David K. Gardner and Michelle Lane

INTRODUCTION

In order to culture any cell successfully it is important to understand the nutrient requirements and physiology of the cell in question. The mammalian preimplantation embryo therefore represents an intriguing challenge as its nutrient requirements and physiology change as development proceeds. Similarly the environment in which the embryo finds itself *in vivo* is a highly dynamic one providing the embryo with a changing gradient of pyruvate, lactate, glucose, amino acids and oxygen.

It is the aim of this chapter to outline changes in the metabolism, physiology and requirements of the preimplantation embryo. It is proposed that media which are successful in supporting the human embryo not only cater to the nutrient requirements of the embryo at a particular stage of development, but also minimize intracellular stress, thereby facilitating normal cellular function.

DYNAMICS OF EMBRYO PHYSIOLOGY AND MATERNAL ENVIRONMENT

In the human, fertilization occurs in the ampullary–isthmic region of the fallopian tube and the subsequent embryo is then transported along the tube into the uterus with the help of the tubal epithelial secretions, cilia and peristaltic contractions of the tubular musculature. The first 3 to 4 days of embryonic development from syngamy to the morula stage occurs in the fallopian tube[1]. For the majority of this time the cells of the embryo (blastomeres) remain unattached and are held in close proximity by the zona pellucida. The blastomeres of early embryos have a physiology somewhat analogous to that of single-celled organisms. It is only after the

embryo undergoes compaction with the generation of the first transporting epithelium that the cells of the newly formed morula take on a more somatic cell-like physiology[2,3]. This means that prior to compaction the embryo is potentially more susceptible to its environment. The significance of compaction with regards to embryo physiology cannot be overstated. Differences in embryo physiology before and after compaction are therefore listed in Table 1. Around the morula stage the embryo descends into the uterus, undergoing the energy-dependent process of blastocoele formation and the loss of the zona pellucida, by hatching via hydrostatic pressure and by the action of trophectoderm and uterine proteases. The zona-free blastocyst subsequently attaches and implants in the endometrium of the uterus.

In parallel to changes in the embryo's physiology and metabolism are changes in the levels of nutrients

Table 1 Differences in embryo physiology before and after compaction

Before compaction	After compaction
Low biosynthetic activity	High biosynthetic activity
Low QO_2	High QO_2
Pyruvate and lactate preferred nutrients	Glucose preferred nutrient
Non-essential amino acids beneficial	Both non-essential and essential amino acids beneficial
Maternal genome	Embryonic genome
Individual cells	Transporting epithelium
One cell type	Two distinct cell types: ICM and trophectoderm

ICM, inner cell mass; QO_2, respiratory quotient

available to the embryo (Table 2). In the oviduct, pyruvate and lactate levels are at their highest at the time when the embryo is present, and glucose at its lowest[4]. In contrast, in the uterus, pyruvate and lactate levels are at their lowest and glucose at its highest at the time when the embryo is present. Such changes in nutrient availability mirror the changes in nutrient preference of the human embryo (see below).

Following on from this, it is perhaps not too surprising that the conditions most suited to the development of the zygote do not support adequate blastocyst differentiation. On the contrary, the very conditions that facilitate the development and differentiation of a viable blastocyst in culture are not well suited to the development of the zygote. This means that should one wish to culture human embryos past the 8-cell stage, i.e., post compaction, then more than a single medium formulation is required.

METABOLISM DURING PREIMPLANTATION DEVELOPMENT

Concomitant with changes in physiology, the mammalian embryo undergoes changes in nutrient requirements and utilization during the preimplantation period. The human embryo, like that of several other mammalian species, has an initial preference for pyruvate over glucose as a nutrient[5]. However, as development proceeds the human embryo exhibits an increasing capacity to utilize glucose and post compaction glucose is the

main carbohydrate utilized[5,6]. The reason for the change in carbohydrate utilization can be explained in terms of the changing physiology of the embryo as it develops. The zygote exhibits relatively low levels of both energy production and biosynthesis prior to embryonic genome activation and expression. As a direct result of this there is a high ATP/ADP ratio within the embryo[7], which in turn allosterically inhibits the flux of glucose through the glycolytic pathway. As the embryo becomes increasingly energetically and transcriptionally active, protein synthesis increases, and as the blastocoele is formed through the action of the basolateral ATPases, there is an increasing demand for energy (ATP). Consequently, the ATP/ADP ratio in the later-stage embryo falls and an increased glycolytic flux and glucose uptake occurs.

Alternative explanations proposed for the observed switch in carbohydrate utilization are the lack of glucose carriers on the plasma membrane of the early embryo and/or the synthesis of sufficient enzymes for glucose metabolism by the later stages[8]. However, it is unlikely that these are the underlying mechanisms involved in the switch, as a carrier for glucose appears to be present at all stages of development and there is more than sufficient total enzyme activity present to accommodate glucose metabolism through glycolysis[9]. What is probably more important to the embryo than total enzyme activity is their allosteric regulation and the appearance of new isoforms of enzymes as development proceeds. The significance of different enzyme isoforms and their relative changes in abundance in terms of embryo physiology/ metabolism has not been fully determined, although it is evident that the isoforms of some enzymes do change during the pre- and peri-implantation period[10]. Different isoforms vary in their response to specific regulators of enzyme function and could therefore help to explain the observed metabolic responses of different stages of development to their environment in culture.

Significantly, it has been established in the mouse that the different cell types of the blastocyst have different metabolic profiles[11]. It was determined that whereas the trophectoderm converted around 50% of the glucose consumed to lactate, the inner cell mass (ICM) was totally glycolytic in nature. In the mouse[12] and human[13] the ICM constitutes around 35% of the cells of the blastocyst.

Table 2 Differences between the oviduct and uterine environments

Component	Oviduct	Uterus
Glucose (mmol/l)	0.5[4]	3.15[4]
Pyruvate (mmol/l)	0.32[4]	0.10[4]
Lactate (mmol/l)	10.5[4]	5.2[4]
Oxygen (%)	8[120]	1.5[120]
Carbon dioxide (%)	12[129]	10[130]
pH	7.5[129]	7.1[131]
Glycine (mmol/l)	2.77[132]	19.33[27]
Alanine (mmol/l)	0.5[132]	1.24[27]
Serine (mmol/l)	0.32[132]	0.80[27]

HOMEOSTASIS DURING PREIMPLANTATION DEVELOPMENT

Prior to implantation the embryo is essentially a free-living organism. As such the embryo possesses several means of regulating its internal environment[3,14,15]. The control of intracellular pH (pHi) is used here as an example of not only how the embryo regulates its internal environment, but also how the environment in which the embryo finds itself impacts its intracellular environment.

The resting pHi of embryos from mammalian species appears to be around 7.2 (mice[16]; hamster[17,18]; cow[19]; human[20]). Studies on all mammalian cells examined with the exception of the erythrocyte have demonstrated that protons are not in equilibrium across the cell membrane[21,22]. In other words, external pH (pHo) does not equal pHi. This fact has also been demonstrated in mammalian embryos including the human, where it is observed that the pHo of the external medium does not equate to pHi. In fact in the mammalian embryo, pHo within the range of 7.0–7.4 does not alter pHi. However, this information that protons do not exist in equilibrium with the mammalian embryo is frequently overlooked. It is a common misconception in the field of embryology and assisted reproductive technologies (ART) that the culture media must be at a pH of 7.4 to permit normal embryo development. However, it is apparent that medium within the range of 7.0–7.4 will not alter pHi and therefore these levels of pHo will not alter development. However, modifying pHo outside this range does effect pHi and results in decreased rates of development[17,23–25]. Therefore, it would appear advisable to adjust culture media to a pHo more similar to the physiologic pHi of 7.2 in order to minimize the amount of energy spent by the embryo to maintain its internal environment.

Specific components of the culture medium have been shown to have significant effects on pHi regulation of the embryo. Two examples are lactic and amino acids in the medium, which have been shown to alter pHi. Being permeant lactate can readily enter the embryo and because it is a weak acid it can subsequently reduce pHi when present in the culture medium at concentrations > 1 mmol/l[25,26]. Both the L and D isoforms of lactate affect pHi equally, though only the L isoform is biologically active. Amino acids are key regulators of pHi. A proportion of some non-essential amino acids such as taurine and glycine exist as zwitterions at physiologic pH. Zwitterions are able to move readily across the cell membrane and are negatively charged so that they are able to bind protons and therefore buffer pHi. Therefore, a high concentration of these types of amino acids are able to buffer against an increase in protons in the cell and prevent an increase in pHi. The amino acids that can buffer protons, taurine, glycine and glutamine, are the amino acids present at high concentrations in the female reproductive tract[27]. Therefore, it is likely that *in vivo* these amino acids are able to regulate pHi. The capacity of amino acids in the culture medium to buffer pHi has been demonstrated in mouse pronucleate-stage embryos. Incubation of embryos in the presence of the weak acid 5,5-dimethyl-2,4-oxazolidinedione (DMO) results in a significant decrease in pHi. Addition of non-essential amino acids (which contain glycine) to the culture medium containing DMO reduces the resultant acidification and therefore increases the ability of the embryos to buffer protons and maintain pHi[28].

Furthermore, the addition of amino acids to the culture medium prevents the efflux of endogenous amino acids[29] and therefore assists in maintaining the intrinsic buffering capacity of the cytoplasm. Additionally, these amino acids are able to prevent against toxicity of Na^+ ionic stress by acting as osmolytes. It is therefore advisable to include amino acids in any medium for gametes or embryos.

In order to maintain pHi at a constant and steady level, the cell must be able to regulate and neutralize intracellular acid such as that derived from metabolic by-products. Cells have two mechanisms to regulate pHi; short-term and long-term regulation. Short-term regulation of changes in pHi are achieved by the physico-chemical buffering of the cytoplasm as well as buffering by the organelles themselves. For long-term stability and regulation of pHi it is necessary to either extrude the protons or buffer them by the import of bicarbonate. This is achieved by specific transport systems in the cell membrane. However, these transporters can take several minutes to restore pHi to the physiologic level while the intrinsic buffering capacity of the cytoplasm can buffer pHi in a matter of seconds. Long-term regulation is achieved through specific transporter systems such as the Na^+/H^+ antiporter and HCO_3^-/Cl^- exchanger.

Unlike embryos, mammalian oocytes do not appear to posses any transport mechanisms for regulating pHi. Studies on both mouse and hamster oocytes have failed to detect either Na^+/H^+ antiporter[18] or HCO_3^-/Cl^- exchanger[16,30] activity, in spite of the fact that the mRNA for both transporters are present[16,31]. Activity of these two pHi regulatory systems was not detected until several hours after egg activation[16,18]. For both the Na^+/H^+ antiporter and the HCO_3^-/Cl^- exchanger, activity was not associated with protein synthesis or cytoskeletal movement. Rather the calcium oscillations initiated by activation of the oocyte that continue for around 5–6 h after fertilization[32–34] appear to be involved in the activation of the transporters. These calcium oscillations signal second messenger systems that result in activation of existing protein. For the Na^+/H^+ antiporter, activity is activated by a calcium-dependent protein kinase C pathway[18]. Therefore, the mammalian oocyte and early embryo do not have any mechanism for the regulation of pHi prior to around 6–10 h after fertilization.

However, *in vivo* the mammalian oocyte and early embryo are not without buffering of pHi as they are surrounded by cumulus cells that persist for several hours after fertilization. Cumulus cells are surrounded by a matrix of glycoproteins and glycosaminoglycans, such as hyaluronate, which provide a gel-like protective layer around the oocyte itself[35,36]. It is proposed that *in vivo* the oocyte and early embryo are protected by the surrounding cumulus and are not required to regulate pHi themselves until after the transport systems are initiated at around 8 h after egg activation. Therefore the cumulus may play a significant protective role for the oocyte and embryo that is lost when embryos are denuded and placed in culture. This will have a significant impact on procedures in the *in vitro* fertilization (IVF) laboratory such as intracytoplasmic sperm injection (ICSI) that routinely involve denuding oocytes. It is proposed that under such circumstances, amino acids should be present in the surrounding medium to facilitate the buffering of pHi.

Having considered embryo metabolism and homeostasis, it is important to focus on the role of specific medium components in order to determine what should be included in embryo culture media and when. The following sections will focus on the roles of carbohydrates and amino acids.

ROLE OF PYRUVATE AND LACTATE IN PREIMPLANTATION DEVELOPMENT

Pyruvate enters the embryo both passively and by means of a facilitated carrier[37,38] and is the preferred nutrient of the cleavage-stage embryo of several species including the human[5]. Although lactate is readily taken up, and can be metabolized to some degree, it cannot support the first cleavage division in the mouse[39]. Once inside the embryo pyruvate and lactate are interconverted by the enzyme lactate dehydrogenase (LDH) through the following reaction:

$$\text{Pyruvate} + \text{NADH} + \text{H}^+ \underset{}{\overset{\text{LDH}}{\rightleftharpoons}} \text{Lactate} + \text{NAD}^+$$

A primary function of the interconversion of pyruvate and lactate in cells is to regenerate NAD^+ for subsequent use in glycolysis when under anaerobic conditions and this is therefore of greatest significance at the blastocyst stage. Cytosolic regeneration of NAD^+ is required as the cytoplasmic and mitochondrial pools of NADH are not shared. Rather, the reducing power between these two distinct cellular compartments is transferred through a specific system such as the malate–aspartate shuttle. Recent studies have revealed that this shuttle is involved in the metabolism of carboxylic acids and certain amino acids in the mouse zygote and cleavage-stage embryo. Furthermore, a reduction of activity of this shuttle at the blastocyst stage may be responsible for aberrant levels of lactate production by blastocysts developed *in vitro*[9,40].

It has been shown that the mouse zygote and blastocyst differ in their ability to metabolize pyruvate and lactate and that such differences can only be accounted for by a change in the intracellular NAD:NADH ratio, which in turn is affected by the ratio of pyruvate:lactate[41]. Therefore by changing the ratio of certain medium components one can inadvertently change the ratio of important intracellular regulators. For example, changing the concentration of lactate in the culture medium can have a significant effect on mouse embryo viability and this effect is stage-specific[42]. Such studies bring into question the potential pitfalls of using one of these carboxylic acids in the absence of the other and emphasize the significance that gradients of carboxylic acids have on embryo physiology and development.

As discussed, being weak acids both pyruvate and lactate can reduce the pHi of the embryo when they are present in culture media at high concentrations[25,26]. This is particularly pertinent for lactate, which can be present in some culture media at over 20 mmol/l. Lactate routinely comes in the form of D- and L-isomers, both of which can decrease pHi[25]. It is therefore important to use only the biologically active form, the L-isomer, in order to reduce effects on pHi.

Finally, it has been shown that pyruvate is a powerful antioxidant, being able to reduce intracellular levels of hydrogen peroxide in the embryo[43,44]. The presence of pyruvate in embryo culture medium therefore confers a significant degree of protection against oxidative stress as well as serving as a vital energy source.

ROLE OF GLUCOSE IN PREIMPLANTATION DEVELOPMENT

There is widespread belief that glucose is toxic to the human embryo in culture. This originated from culture studies on the hamster by Bavister and co-workers, who demonstrated that in the presence of phosphate 5.5 mmol/l glucose could induce developmental arrest[45,46]. However, such arrest has subsequently been attributed to phosphate alone[47]. Indeed in the absence of phosphate, glucose has been shown to be beneficial to subsequent fetal development in the hamster[47], so rather than being detrimental, glucose confers a benefit to the developing hamster embryo in culture. The addition of physiologic levels of glucose (0.5 mmol/l) to embryo culture medium has recently been endorsed by Bavister[48].

In vivo glucose is present in both oviduct and uterine fluids[4]. Furthermore, both oocytes and embryos have a specific carrier for this hexose[37,49–51]. Glucose is typically considered as an energy source. However, glucose is also used by the embryo throughout the preimplantation period for the synthesis of triacylglycerols and phospholipids and to provide precursors for complex sugars of mucopolysaccharides and glycoproteins. Metabolism of glucose through the pentose phosphate pathway (PPP) generates ribose moieties required for nucleic acid synthesis and the NADPH required for the biosynthesis of lipids and other complex molecules[52–54]. The production of nucleic acids is one of the key biosynthetic roles of glucose. The NADPH generated by glucose utilization is also required for the reduction of intracellular glutathione, an important antioxidant for the embryo[55].

During invasive implantation, such as that in the human and rodents, there is little vasculature in the vicinity of the implantation site for several hours[56,57]. During this initial period of implantation anaerobic glycolysis will be the only available means of energy production for the blastocyst. It is plausible that during this time the blastocyst uses its endogenous glycogen to provide the glucose required. Consistent with this hypothesis there is a loss of viability if the mouse embryo is cultured without glucose from 8-cell stage to blastocyst[58]. This indicates that the embryo is forced to use its endogenous glycogen store during preimplantation development in order to generate free glucose for subsequent metabolism, thereby compromising the embryo at implantation. A similar result has been subsequently reported for the hamster, where 1-cell embryos were cultured to the blastocyst stage in the presence or absence of 0.5 mmol/l glucose. Hamster embryos cultured for the entire preimplantation period in the absence of glucose had a significantly reduced viability compared to those embryos exposed to 0.5 mmol/l glucose[47].

It has been demonstrated that the concentration of glucose in the culture medium affects its rate of consumption by the embryo[59]. Therefore increasing the concentration of glucose in the medium can result in increased glucose uptake and utilization. The genes for the glucose transporter are transcribed in the human oocyte and embryo[49–51], and kinetic studies have indicated the presence of the glucose transporter in the mouse embryo throughout development[37]. The maximal activities of several key enzymes required for glucose metabolism have been determined in the mouse and human[60–63]. In all cases the activities of the three rate-limiting enzymes of glucose metabolism (hexokinase, phosphofructokinase and pyruvate kinase) have been determined to be higher than the amount of glucose utilized during the preimplantation period. In all probability glucose utilization during the preimplantation period is not regulated by transport across the plasma membrane nor by the absence of sufficient enzyme activity. Rather substrate availability (concentration) and the specific regulation of enzyme activity appear to control glucose utilization by the preimplantation embryo[64]. Significantly amino acids affect how the embryo utilizes glucose (see below)[65].

ROLE OF AMINO ACIDS IN PREIMPLANTATION DEVELOPMENT

Human embryos can develop in culture for 2 to 3 days in the absence of amino acids and give rise to babies. As a result of this the significance of amino acids during preimplantation embryo development has been overlooked for many years. This is regrettable, as amino acids are amongst the most important regulators of the mammalian preimplantation embryo. Indeed, up to just 10 years ago amino acids were noticeably absent from media formulated for mammalian preimplantation embryos, this in spite of the fact that amino acids are abundant in the fluids of the female reproductive tract[27,66,67]. Noticeably, some amino acids such as glycine and taurine are present in millimolar amounts. Furthermore, the oocyte and embryo maintain an endogenous pool of amino acids[68] and possess specific transport systems to take up amino acids from their surroundings[69].

It was shown as early as 1966[70] that amino acids were required for the attachment and outgrowth of mouse blastocysts in culture. Gwatkin and Haidri[71] subsequently demonstrated that glutamine, isoleucine, methionine and phenylalanine promoted nuclear maturation of the hamster oocyte. Juetten and Bavister[72] went on to determine the effects of this group of amino acids on hamster embryo development. Following studies on the rat[73], the mouse[42,74–76] and sheep[77] showed that amino acids were not only beneficial during the culture of various stages of development, but also significantly increased the resultant viability of embryos.

Further research has revealed several roles for amino acids during the preimplantation period including: biosynthetic precursors[78], sources of energy[79], regulators of energy metabolism[65], osmolytes[80], buffers of pHi[25], antioxidants[81] and chelators[82].

Significantly, a biphasic requirement for amino acids during the preimplantation period has now been established[12,83]. The zygote and cleavage-stage embryo benefit from the inclusion of Eagle's non-essential amino acids and glutamine. Of interest, this group of amino acids bears a striking homology to those present at high levels in the female reproductive tract. Although the term non-essential and essential amino acids as defined by Eagle[84] have little to do with the requirements of the mammalian embryo, they serve as convenient groups in which to place amino acids. Figure 1 shows the effects of different amino acid groups on the development of mouse zygotes. What is evident from such work is that the addition of 20 amino acids to media for the zygote and cleavage stages does not have the same beneficial effect as the addition of just seven amino acids (non-essential amino acids and glutamine). In contrast to the beneficial effects of the non-essential amino acids, Eagle's essential amino acids confer little or no benefit to the cleavage-stage embryo. One reason for this is that essential amino acids do not fulfil the same cellular niches for the embryo as the non-essential amino acids,

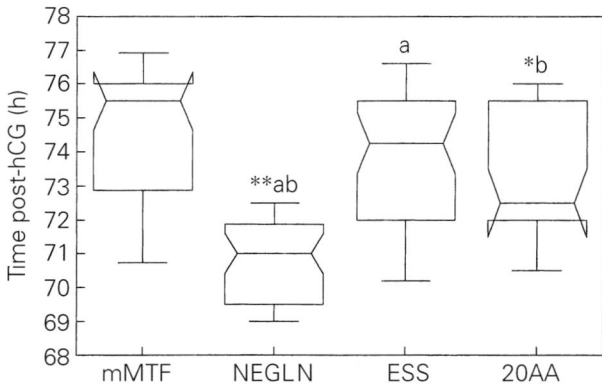

Figure 1 Effect of amino acids on cleavage rates from the 4- to 8-cell stage; *n* = at least 30 embryos per treatment group. Notches represent the interquartile range, therefore including 50% of the data. Whiskers represent 5% and 95% quartiles. Line across the box represents the mean. Non-essential amino acids and glutamine/modified mouse tubal fluid (NEGLN/mMTF) is supplemented with non-essential amino acids and glutamine; essential amino acids (ESS)/mMTF is supplemented with essential amino acids without glutamine; 20AA-mMTF is supplemented with all 20 amino acids. *Significantly different to mMTF ($p < 0.05$); **significantly different to mMTF ($p < 0.01$); [a]like pairs significantly different ($p < 0.01$); [b]like pairs significantly different ($p < 0.05$). Culture with non-essential amino acids stimulated cleavage rates to the 8-cell stage. Addition of essential amino acids. hCG, human chorionic gonadotropin. Reproduced with permission from Lane M, Gardner DK. Nonessential amino acids and glutamine decrease the time of the first three cleavage divisions and increase compaction of mouse zygotes *in vitro*. *J Assist Reprod Genet* 1997;14:398–403

e.g., it is non-essential amino acids such as glycine, glutamate, serine and proline that act as intracellular pH buffers and osmolytes. One reason why the presence of 20 amino acids is not as effective as the non-essential group on their own has been attributed to competition at the carrier level by the essential amino acids[85], which are typically present at higher levels than the non-essential group. It would therefore be concluded that reducing the concentration of the essential amino acids should increase development of the cleavage-stage embryo, and indeed this appears to be the case[83,86]. However, optimal development of the zygote and cleavage stages may well be optimized in their absence.

After the 8-cell stage the mammalian embryo benefits from the presence of a more complex array of amino acids[83], with Eagle's essential amino acids being found to stimulate the development of the ICM[12]. Significantly, equivalent rates of implantation to *in vivo* developed blastocysts were obtained when mouse zygotes were cultured with non-essential amino acids up to the 8-cell stage, followed by culture to the blastocyst in the presence of 20 amino acids[12]. Eagle's non-essential amino acids could perhaps be best classified as facilitators of blastomere function. The concentration of amino acids in Eagle's medium are listed in Table 3.

Devreker and co-workers[87] have recently shown that human embryo development in culture is enhanced when embryos are first exposed to non-essential amino acids followed by a more complex array of amino acids, such as those found in the media G2 and Ham's F-10 (Blastocyst medium).

MACROMOLECULES

Macromolecules should be considered as an integral component of culture media, rather than simply as a medium supplement, as they have a profound effect on development. Most embryo culture media are routinely supplemented with serum albumin. The use of serum should be avoided due to the detrimental effects on the embryos such as perturbed energy metabolism[77], ultrastructural damage to the mitochondria[88–90] and abnormal accumulation of lipids[89,90]. Embryos cultured with serum also have a reduced tolerance to cryopreservation[90]. Additionally, in animal models the use of serum in the culture media has been associated with

Table 3 Concentration of amino acids in Eagle's medium

Non-essential amino acids	Concentration (mmol/l)	Essential amino acids*	Concentration (mmol/l)
Alanine	0.1	Arginine	0.6
Asparagine	0.1	Cystine	0.1
Aspartate	0.1	Histidine	0.2
Glutamate	0.1	Isoleucine	0.4
Glycine	0.1	Leucine	0.4
Proline	0.1	Lysine	0.4
Serine	0.1	Methionine	0.1
		Phenylalanine	0.2
		Threonine	0.4
		Tryptophan	0.05
		Tyrosine	0.2
		Valine	0.4

*Glutamine, an essential amino acid as defined by Eagle[84], is typically used at a concentration of 0.5–1.0 mmol/l. The effectiveness of essential amino acids in promoting mouse and cow embryo development has been shown to increase by halving the concentration listed by Eagle[83,86]

increases in pregnancy and neonatal complications[91]. However, with the development of more physiologic culture media containing amino acids the use of serum is no longer advocated. Instead serum can be replaced with serum albumin in the media resulting in high rates of blastocyst development and subsequent pregnancy and implantation rates. Recently recombinant human albumin has become available and has been shown to be as effective as blood-derived albumin in supporting embryo development[92,93]. Significantly, embryos cultured in the presence of recombinant albumin exhibit an increased tolerance to cryopreservation[94]. Furthermore, and of greatest significance, the use of recombinant albumin not only eliminates the variation that exists between lots of protein derived from blood[95,96], but it also eliminates the potential of disease transmission that exists with the use of any blood-derived product.

In addition to albumin, the glycosaminoglycan hyaluronan is an important macromolecule for embryo culture media. Hyaluronan is a high molecular mass polysaccharide and can be obtained endotoxin- and prion-free from a yeast fermentation procedure. In the mouse hyaluronan levels in the uterus increase at the time of implantation[97]. It has been shown that not only can hyaluronan replace albumin in a mouse and bovine embryo culture systems, but that its use for embryo transfer results in a significant increase in embryo implantation[98]. Furthermore, similar to results with recombinant albumin, the presence of hyaluronan in the culture medium increases the cryosurvivability of blastocysts[98,99]. It has been demonstrated that recombinant albumin and hyaluronan confer a synergistic benefit to the embryo[93,94,98]. Therefore, hyaluronan should also be considered to be an integral component of embryo culture media.

LESSONS LEARNED FROM MOTHER AND EMBRYO TO DEVELOP MORE SUITABLE CULTURE CONDITIONS

Historically the culture media used in clinical ART have fallen into two categories, simple and complex (as described in detail by Gardner and Lane[100]). Simple media, an example of which is HTF, are those based on the culture media of the 1960s, formulated to support the embryos of inbred strains of mice. Complex media, examples of which include Ham's F-10 (blastocyst medium) and minimal essential medium (MEM), are those developed to support somatic cells. The reality is that the fertilized oocyte and cleavage-stage embryo benefit from a medium that is neither 'simple' nor 'complex' but rather one somewhere in between. In contrast, as discussed above, post compaction, the embryo benefits from a more complex-type medium, though not necessarily one specifically designed with the requirements of somatic cells in mind.

Culture media for the zygote and cleavage-stage embryo

The composition of three commonly used human embryo culture media, HTF, P1 and G1, are listed in Table 4. At first glance such media appear very different, although they have been used very effectively in the culture of viable human embryos from day 1 to day 3. Unfortunately, there are no prospective, randomized trials comparing the efficacy of these media for cleavage-stage embryos. The issue of potential glucose toxicity in medium HTF has been addressed in two ways. In medium P1 glucose, along with phosphate, has been

removed. In medium G1, glucose is present at the level found in the human fallopian tube, and phosphate is present. The presence of non-essential amino acids and ethylenediaminetetraacetic acid (EDTA) in medium G1 enables the embryo to control the utilization of the glucose present. In contrast to medium G1, medium P1 contains a single amino acid, taurine, at the concentration found to stimulate the hamster embryo in culture[101]. Taurine has been shown to have a beneficial effect on the mouse and human embryo[102,103], although its beneficial effects are not evident when glutamine is also present in the culture medium[104]. This example further highlights the plasticity of the embryo in being able to use different components of the media for the same function.

Culture media for the blastocyst stage

Acceptable implantation and pregnancy rates have been reported with the transfer of cleavage-stage human embryos to the uterus on either day 2 or 3 of development[105,106] (Table 5). However, the human

Table 4 Composition of culture media for the cleavage-stage human embryo

Component (mmol/l)	HTF[133]	P1[134]	G1[111]	G1.2[112]
NaCl	101.6	101.6	85.16	90.08
KCl	4.69	4.69	5.5	5.5
KH$_2$PO$_4$	0.37	—	—	—
NaH$_2$PO$_4$.2H$_2$O	—	—	0.5	0.25
CaCl$_2$.2H$_2$O	2.04	2.04	1.8	1.8
MgSO$_4$.7H$_2$O	0.20	0.20	1.0	1.0
NaHCO$_3$	25.0	25.0	25.0	25.0
Na pyruvate	0.33	0.33	0.32	0.32
Na lactate	21.4	21.4	10.5	10.5
Glucose	2.78	—	0.50	0.50
Glutamine	—	—	1.0	0.5
Taurine	—	0.05	0.1	0.1
Non-essential amino acids	—	—	All	All
Citrate	—	0.5	—	—
EDTA	—	—	0.01	0.01
Human serum albumin/SSS	5 g/l	6 g/l	5 g/l	5 g/l

Non-essential amino acids (Ala, Asn, Asp, Glu, Gly, Pro, Ser) are present at 0.1 mmol/l as specified by Eagle[84]. EDTA, ethylenediaminetetraacetic acid; SSS, serum substitute supplement

Table 5 Embryo implantation rates for uterine transfer following selection on different days of development

Day of transfer	Implantation rate (%)	Reference
1	28	Scott and Smith, 1998[105]
3	48	Gerris *et al.*, 1999[106]
5	70	Gardner *et al.*, 2000[116]

embryo does not reside in the uterus until around day 4 of development[1]. It has been demonstrated in animal models that the transfer of the embryo before compaction to the uterus of a recipient female is not as consistent with subsequent fetal development as the transfer of blastocysts. This has therefore raised the issue of whether human embryos should be transferred at the blastocyst stage.

It has been shown that the human embryo can develop to the blastocyst stage in a variety of culture conditions. However, the resultant viability of such blastocysts is very different. This fact has contributed significantly to the confusion regarding the role of blastocyst transfer in human ART. The reason that embryos can develop in culture under a wide variety of conditions is due to their plasticity, this being the ability of the embryos to compensate for the lack of a preferred nutrient or regulator by the use of an alternative. This represents an obvious developmental advantage to the embryo as it develops *in vivo* and is exposed to changing gradients of nutrients as it progresses though the lumen of the female reproductive tract. However, if the embryo is forced to adapt too much it loses viability. Therefore, although human embryos can develop in a wide variety of culture conditions, the developmental potential of the resultant blastocysts varies enormously. For example, Bolton and co-workers[107] found that it was possible to obtain 40% blastocyst development using Earle's salts supplemented with pyruvate and 10% maternal serum. However, the resultant implantation and pregnancy rate were only 7%. In other words, although blastocysts were obtained, their developmental capacity and viability was significantly below that of embryos transferred on day 3 in the same program. The reason for this is that the culture medium employed was not sufficiently complete to support appropriate blastocyst differentiation. However, with the development of sequential media systems

it is possible to culture embryos to the blastocyst stage while maintaining high levels of viability (Figure 2).

While different culture conditions have been used in sequence to support mammalian embryos[108], the concept of sequential media[109], that are designed specifically with the changing needs of the embryo in mind, has proved extremely successful. In our laboratory the design of such media has focused on the dynamics of embryo physiology and metabolism, and the reduction of intracellular stress, as discussed earlier in this chapter. Two media were therefore formulated for the pre- and post-compacted embryo, G1 and G2[110,111]. These media were subsequently modified[112] (G1.2 and G2.2), and the formulations are shown in Tables 4 and 6.

A gradient of carbohydrates exists from G1 to G2, based upon that which exists in the human female reproductive tract. Similarly, the embryo is exposed to an increasingly complex array of amino acids as development proceeds. EDTA is present in medium G1, as it has been shown to stimulate the cleavage-stage embryos of both mice[58,74,113] and cows[114]. It has been determined that EDTA is an inhibitor of glycolytic kinases and helps prevent aberrant levels of glycolysis[114,115]. However, the inclusion of EDTA in medium for embryos from the 8-cell stage onwards (medium G2) is not advisable, as the embryo becomes increasingly glycolytic in nature, especially the ICM[11]. Therefore, continued exposure to EDTA negatively impacts blastocyst development and impairs ICM formation[114].

There exist several potential advantages of blastocyst culture and transfer, including: (1) embryo selection, the ability to identify those embryos with limited developmental potential; (2) synchronization of embryonic stage with the female tract thereby reducing cellular stress on the embryo; (3) minimizing exposure of the embryo to a hyperstimulated uterine environment; (4) reduction in uterine contractions thereby reducing the chance of the embryo being expelled; (5) ability to undertake cleavage-stage embryo biopsy without the need for cryopreservation when the biopsied blastomere has to be sent to a different locale for analysis; (6) assessment of true embryo viability by assessing the embryo post genome activation.

Table 7 details the results to date using blastocyst transfer in our clinic. Embryo development in culture is determined by the genetics of the parents. It is therefore difficult to establish baseline data on human blastocyst development because humans represent the most genetically diverse group of mammals. This is further compounded by the fact that, by default, patients having an ART cycle are subfertile, with differing etiologies and typically of advanced reproductive age. However, with the establishment of oocyte donation as a treatment for some patients there exists a population of oocytes from fertile young women, which are theoretically more competent and have fewer chromosomal abnormalities. After 211 blastocyst transfers following oocyte donation, an implantation rate (fetal heart/blastocyst transferred) of

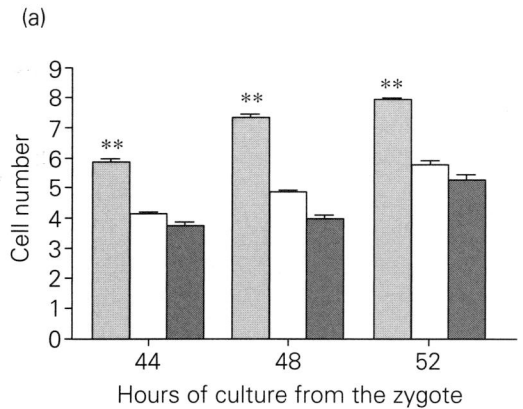

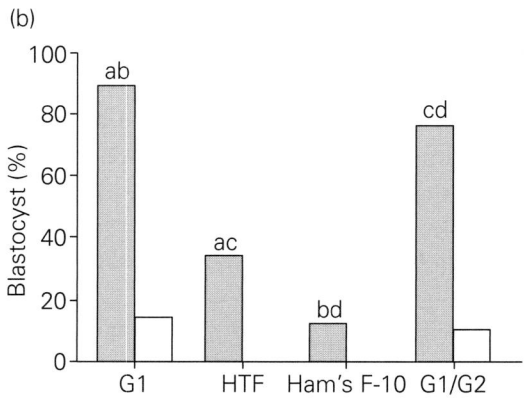

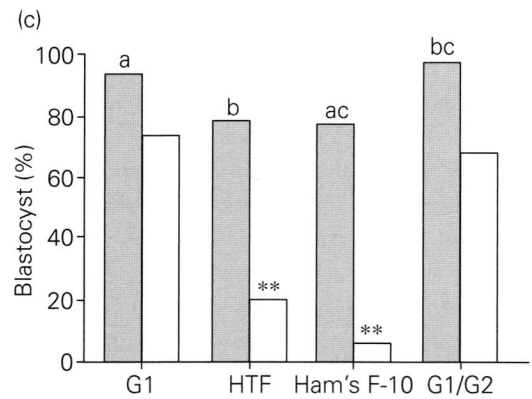

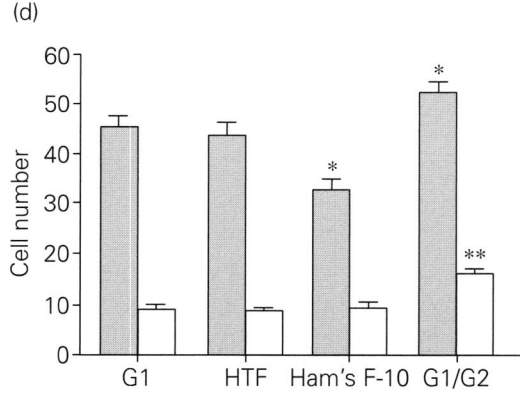

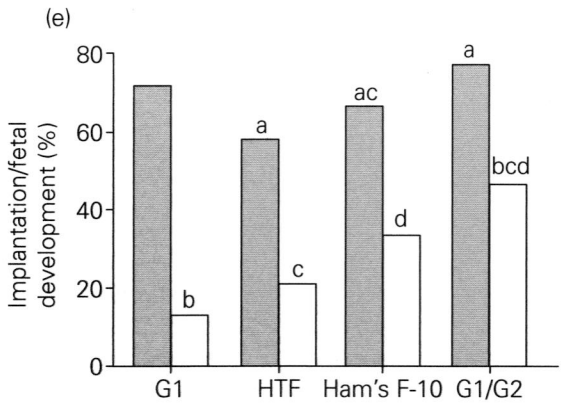

61% was attained, resulting in a clinical pregnancy rate of 80% with a mean of just 2.1 blastocysts transferred. Such high implantation rates are equivalent to those obtained when high scoring blastocysts are transferred to patients using their own oocytes (70%)[116] and to those obtained following the transfer of *in vivo* developed blastocysts (60%)[117]. Therefore, for oocyte donors blastocyst culture is an effective means of increasing implantation rates and reducing the number of embryos that need to be transferred. This latter fact is most poignant given the advanced age of the recipient. This group of patients also serves to demonstrate the efficacy or otherwise of the culture system used. Figure 3 shows the morphology of human blastocysts on the morning of day 5 (4 days of culture from the pronucleate stage).

THE CULTURE SYSTEM IS GREATER THAN THE MEDIA ALONE

The culture media used for the development of embryos is only one part of the culture system[100]. Other parts of the culture system such as gas phase and incubation volume:embryo ratio have a significant impact on subsequent embryo development. For example, the CO_2 concentration can have a profound impact on the media as it forms part of the buffering system of the media HCO_3^-/CO_2. Therefore, changes in the level of CO_2 cause alterations in the pH of the media. For example, most media for embryo culture contain a bicarbonate concentration of 25 mmol/l. Changing the CO_2 concentration from 5% to 6% results in the media pH changing from 7.4–7.45 to 7.25–7.3. Therefore, as the pHi of the mammalian embryo appears to be around 7.2 less stress

Figure 2 (opposite) Effect of sequential culture media on the development of F1 (C57BL/6 × CBA/Ca) mouse zygotes *in vitro*. Zygotes were collected at 20 h post-hCG. All media were supplemented with BSA (2 mg/ml). All embryos were transferred to fresh medium after 48 h of culture, with the exception of embryos in medium G1, where the embryos were transferred to either medium G1 or G2. To compensate for this, twice the number of embryos was originally cultured in medium G1, although only a designated 50% of these embryos were used in the statistical analysis of the 44 to 52 h data set. (a) Embryo cell number after 44, 48 and 52 h of culture. Values are mean ± SEM. n = 200 embryos/medium. Media: G1 (light gray bar); HTF (open bar); Ham's F-10 (dark gray bar). Significantly different from other media; **$p < 0.01$. (b) Embryo development after 72 h of culture. n = 150 embryos/medium. G1/G2; embryos cultured for 48 h in medium G1 and then transferred to medium G2. Blastocyst (gray bar), hatching blastocysts (as a percentage of total blastocysts; open bar). Like pairs are significantly different: a, c, d, $p < 0.05$; b, $p < 0.01$. (c) Embryo development after 92 h of culture. n = 150 embryos/ medium. G1/G2; embryos cultured for 48 h in medium G1 and then transferred to medium G2. Blastocyst (gray bar), hatching blastocysts (as a percentage of total blastocysts; open bar). Like pairs are significantly different: a, b, c, $p < 0.05$. Significantly different from medium G1 and G1/G2; **$p < 0.01$. (d) Cell allocation in the blastocyst after 92 h of culture. n = 150 embryos/medium. G1/G2; embryos cultured for 48 h in medium G1 and then transferred to medium G2. Trophectoderm (solid bars), ICM (open bars). Significantly different from other media; *$p < 0.05$; **$p < 0.01$. (e) Viability of cultured blastocysts. n = at least 60 blastocysts transferred per treatment. G1/G2; embryos cultured for 48 h in medium G1 and then transferred to medium G2. Implantation (gray bar), fetal development per implantation (open bar). Like pairs are significantly different: a, d, $p < 0.05$; b, c, $p < 0.01$. When mouse embryos were cultured in medium G1 for the entire preimplantation period to the blastocyst, although the embryos form healthy looking blastocysts, most implantations were lost, i.e., they did not have a sufficient ICM to form a viable fetus. The lack of adequate ICM development stems from both the lack of sufficient glucose and the presence of EDTA (both affecting glycolysis) and the omission of essential amino acids. In contrast, those mouse embryos that were switched to medium G2 after 48 h of culture formed blastocysts at the same rate and of equivalent morphologies as those in medium G1 for the entire culture period. However, for those blastocysts developed in medium G1 and switched to medium G2, very few implantations were lost due to the development of a large ICM, thereby maintaining a very high pregnancy rate. hCG, human chorionic gonadotropin; BSA, bovine serum albumin; SEM, standard error of the mean; ICM, inner cell mass; EDTA, ethylenediaminetetraacetic acid. Reproduced with permission of Oxford University Press/Human Reproduction, from Gardner DK, Lane M. Culture of viable human blastocysts in defined sequential serum-free media. *Hum Reprod* 1998;13(Suppl.3):148–59, © European Society of Human Reproduction and Embryology

Table 6 Composition of G2.2 and blastocyst medium

Component (mmol/l)	G2.2	Blastocyst medium*
Alanine	0.1	0.1
Alanyl–glutamine	1.0	—
Arginine	0.6	1.0
Asparagine	0.1	0.1
Aspartate	0.1	0.1
Biotin	—	0.0001
Calcium chloride	1.8	0.3
Calcium lactate	—	1.0
Choline chloride	0.0072	0.005
Copper sulfate	—	0.00001
Cysteine	—	0.2
Cystine	0.1	—
Folic acid	0.0023	0.003
Glucose	3.15	6.1
Glutamate	0.1	0.1
Glutamine	—	1.0
Glutathione	—	1.0
Glycine	0.1	0.1
Histidine	0.2	0.1
i-Inositol	0.01	0.003
Iron sulfate	—	0.003
Isoleucine	0.4	0.02
Leucine	0.4	0.1
Lysine	0.4	0.1
Magnesium sulfate	1.0	1.1
Methionine	0.1	0.03
Nicotinamide	0.0082	0.005
Pantothenate	0.0042	0.003
Phenylalanine	0.2	0.03
Potassium bicarbonate	—	5.0
Potassium chloride	5.5	3.8
Potassium phosphate	—	0.6
Proline	0.1	0.1
Pyridoxine	0.0049	0.001
Riboflavin	0.00027	0.001
Serine	0.1	0.1
Sodium bicarbonate	25.0	20.0
Sodium chloride	90.08	116.6
Sodium lactate	5.87	—
Sodium phosphate	0.25	1.1
Sodium pyruvate	0.10	1.0
Thiamine	0.003	0.003
Thioctic acid	—	0.001
Threonine	0.4	0.03
Thymidine	—	0.003
Tryptophan	0.5	0.003
Tyrosine	0.2	0.01
Valine	0.4	0.03
Vitamin B-12	—	0.001
Zinc sulfate	—	0.0001

*Blastocyst medium is a modification of Ham's F-10 containing glutathione

Table 7 Summary of blastocyst transfer data at the Colorado Center for Reproductive Medicine

	IVF patients*	Oocyte donors
No. of patients	401	211
No. of patients having embryo transfer	395	211
Mean age (± SEM) in years	33.4 ± 0.2	40.6 ± 0.3 recipients
Age range	20–43	27–50
FSH (mean ± SEM) (IU/l)	6.7 ± 0.1	6.1 ± 0.1 donors
Patients with ICSI (%)	40.4	39.3
No. of pronucleate embryos (mean ± SEM)	14.5 ± 0.3	15.2 ± 0.4
Blastocyst development on day 5 (%)	44.1	51.7
Blastocyst development on day 6 (%)	8.1	8.3
Total blastocyst development (%)	52.2	60.0
No. of embryos transferred (mean ± SEM)	2.2 ± 0.03	2.1 ± 0.03
Patients with embryo freezing (%)[†]	75.3	85.3
Mean no. of blastocysts frozen (mean ± SEM)	4.3 ± 0.2	5.6 ± 0.3
Implantation rate (fetal sac) (%)[‡]	50.1	62.1
Implantation rate (fetal heart) (%)[‡]	46.4	60.8
Clinical pregnancy rate (%)[§]	68.6	79.6

*IVF patients had at least ten follicles; [†]Only blastocysts scoring 3BB or higher by the afternoon of day 6 were cryopreserved; [‡]Implantation rates are expressed as fetal sac or heart/blastocyst transferred. The calculations included every patient who had an embryo transfer and not just those who subsequently became pregnant; [§]Includes six patients in the blastocyst culture group who did not have an embryo transferred on day 5 due to embryonic arrest at the cleavage stages. Clinical pregnancy was determined by the presence of a fetal heartbeat. IVF, *in vitro* fertilization; SEM, standard error of the mean; FSH, follicle stimulating hormone; ICSI, intracytoplasmic sperm injection. Reproduced from Gardner DK, Lane M, Schoolcraft WB. Physiology and culture of the human blastocyst. *J Reprod Immunol* 2002;55: 85–100, with permission from Elsevier Science

will be placed on the embryo by using a CO_2 concentration (6%) that results in an pH_o level of around 7.2. Similarly, the oxygen concentration used has a profound effect on embryo metabolism and development. Interestingly, the human and F1 mouse embryo can develop in atmospheric oxygen (20%). Consequently this has led to some confusion regarding the optimal concentration for

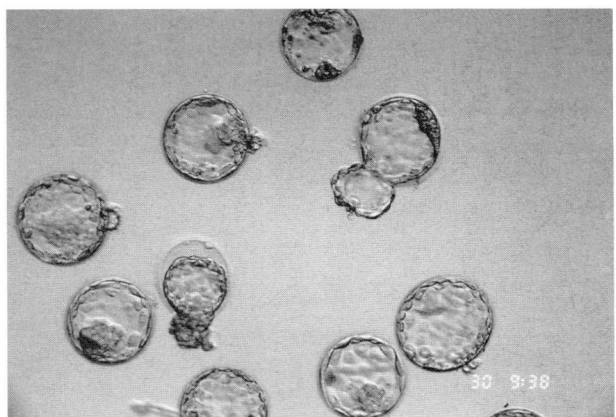

Figure 3 Photomicrograph of human blastocysts. Embryos were cultured for 4 days from the pronucleate stage. For the first 48 h embryos were cultured in groups of four in 50 μl drops of medium G1.2 in an atmosphere of 6% CO_2, 5% N_2 and 89% N_2. The second 48 h of development were in medium G2.2 under the same conditions as G1.2. Blastocysts possess distinct inner cell masses and several have initiated hatching from the zona pellucida

embryo culture. The concentration of oxygen in the lumen of the rabbit oviduct is reported to be 2–6%[118,119] whereas the oxygen concentration in the oviduct of hamster, rabbit and rhesus monkey is 8%[120]. Furthermore, the oxygen concentration in the uterus is significantly lower than in the oviduct, ranging from 5% in the hamster and rabbit to 1.5% in the rhesus monkey[120,121]. Several studies on different species have demonstrated that culture at a reduced oxygen concentration (5–7%) results in enhanced embryo development *in vitro*[100], and that exposure to reduced oxygen *in vitro* leads to higher fetal development[122]. It is our experience that human embryos cultured in a low oxygen environment (5%) produce blastocysts with significantly more cells than those embryos cultured in a high oxygen environment (20%)[123]. Considering the physiology of the reproductive tract and the beneficial effects of using a reduced oxygen concentration as determined in controlled studies, it is advisable to culture embryos at low oxygen concentrations, around 5%.

The culture of mammalian embryos in reduced volumes of medium and/or in groups significantly increases blastocyst development[124–127], as well as increasing blastocyst cell number[126]. Furthermore

culturing embryos in reduced volumes increases subsequent viability after transfer[126]. It has been proposed that the benefit of growing embryos in small volumes and/or in groups is due to the production of specific embryo-derived autocrine/paracrine factor(s) that stimulate development. The culture of embryos in large volumes will result in a dilution of the factor so that it becomes ineffectual[110]. This phenomenon is not confined to the mouse, in which several embryos reside in the female tract at one time, but has also been reported for the sheep and cow, which like the human are monovular[77,128]. It has been shown in both the mouse and cow that increasing the embryo:incubation volume ratio specifically stimulates the development of the ICM. This explains the increased viability of embryos cultured in reduced volumes in groups (Figure 4)[128].

CONCLUSIONS

The human embryo undergoes dramatic changes in its physiology and metabolism as it develops from the zygote to the blastocyst stage. Concomitantly the developing embryo is exposed to a continually changing environment as it passes through the maternal reproductive tract. Meeting the changing demands of the embryo is one aspect of being able to successfully maintain embryo development in culture. Minimizing intracellular stress and maintaining the appropriate activities of energy-generating pathways is fundamental to successful development in culture.

Sequential media have been designed with the above concepts in mind and have been shown to support the development of viable human blastocysts in culture, culminating in high implantation rates equivalent to the *in vivo* developed blastocyst[117]. Common themes among the various sequential media now available for clinical use include low levels of glucose and the inclusion of one or more non-essential amino acids in media designed for the cleavage stages. In contrast, media designed to support blastocyst development and differentiation include higher glucose levels and a wider array of amino acids. It would appear prudent to include macromolecules, such as recombinant albumin together with hyaluronan, in such media due to their synergistic benefit *in vitro* and their ability to increase embryo cryotolerance.

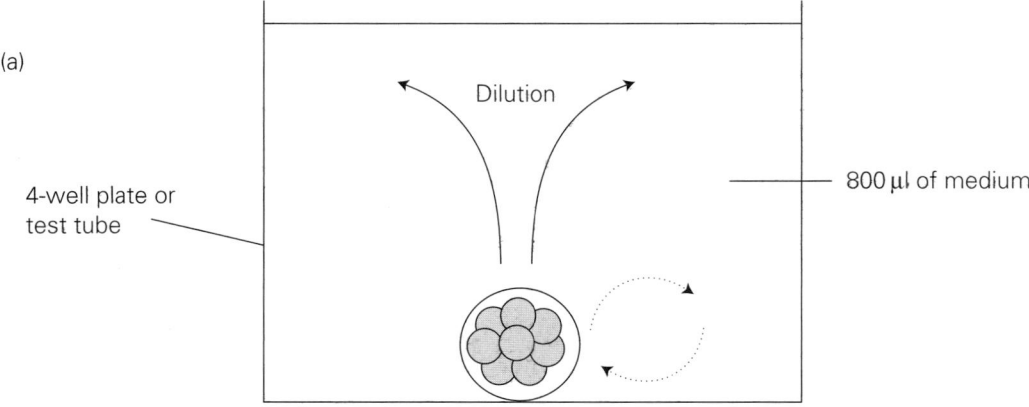

(a)

4-well plate or test tube

Dilution

800 µl of medium

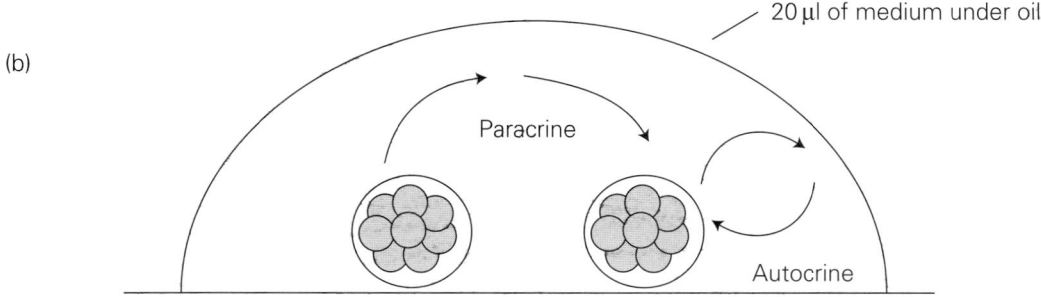

(b)

20 µl of medium under oil

Paracrine

Autocrine

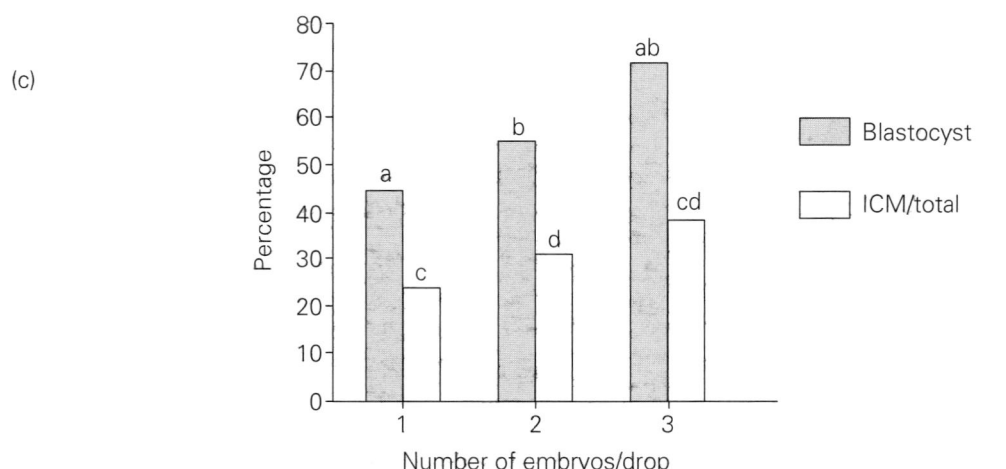

(c)

Figure 4 (opposite) Effect of incubation volume and embryo grouping on embryo development and differentiation. (a) A single embryo cultured in a 4-well plate or test tube. Any factor produced by the embryo will become ineffectual due to dilution. (b) Culture of embryos in reduced volumes and/or in groups increases the effective concentration of embryo-derived factors, facilitating their action in either a paracrine or autocrine manner. (c) Effect of embryo grouping on bovine blastocyst development and differentiation. Bovine embryos were cultured either individually or in groups of two or four in 50 μl drops of medium[128]. Like pairs of letters are significantly different; $p < 0.05$. ICM, inner cell mass. Reproduced with permission from Gardner DK. Improving embryo culture and enhancing pregnancy rate. In Shoham Z, Howles C, Jacobs J, eds. *Female Infertility Therapy: Current Practice*. London, Martin Dunitz, 1998:283–99

Finally, it is important to consider media as a part of the culture system. Other factors in the laboratory, such as a gaseous environment and the incubation:embryo volume ratio, can have a significant impact on how media perform.

References

1. Croxatto HB, Diaz S, Fuentealba B, *et al*. Studies on the duration of egg transport in the human oviduct. I. The time interval between ovulation and egg recovery from the uterus in normal women. *Fertil Steril* 1972;23:447–58
2. Gardner DK. Changes in requirements and utilization of nutrients during mammalian preimplantation embryo development and their significance in embryo culture. *Theriogenology* 1998;49:83–102
3. Lane M. Mechanisms for managing cellular and homeostatic stress *in vitro*. *Theriogenology* 2001;55:225–36
4. Gardner DK, Lane M, Calderon I, Leeton J. Environment of the preimplantation human embryo *in vivo*: metabolite analysis of oviduct and uterine fluids and metabolism of cumulus cells. *Fertil Steril* 1996;65:349–53
5. Hardy K, Hooper MA, Handyside AH, *et al*. Non-invasive measurement of glucose and pyruvate uptake by individual human oocytes and preimplantation embryos. *Hum Reprod* 1989;4:188–91
6. Gardner DK, Lane M, Stevens J, Schoolcraft WB. Non-invasive assessment of human embryo nutrient consumption as a measure of developmental potential. *Fertil Steril* 2001;76:1175–80
7. Leese HJ, Biggers JD, Mroz EA, Lechene C. Nucleotides in a single mammalian ovum or preimplantation embryo. *Anal Biochem* 1984;140:443–8
8. Scott LA. Oocyte and embryo culture. In Keel BA, May JV, De Jonge CJ, eds. *Handbook of the Assisted Reproduction Laboratory*. Boca Raton: CRC Press, 2000:197–219
9. Gardner DK, Pool TB, Lane M. Embryo nutrition and energy metabolism and its relationship to embryo growth differentiation and viability. *Semin Reprod Med* 2000;18:205–18
10. Auerbach S, Brinster RL. Lactate dehydrogenase isozymes in the early mouse embryo. *Exp Cell Res* 1967;46:89–92
11. Hewitson LC, Leese HJ. Energy metabolism of the trophectoderm and inner cell mass of the mouse blastocyst. *J Exp Zool* 1993;267:337–43
12. Lane M, Gardner DK. Differential regulation of mouse embryo development and viability by amino acids. *J Reprod Fertil* 1997;109:153–64
13. Hardy K, Handyside AH, Winston RM. The human blastocyst: cell number, death and allocation during late preimplantation development *in vitro*. *Development* 1989;107:597–604
14. Lane M, Gardner DK. Regulation of ionic homeostasis by mammalian embryos. *Semin Reprod Med* 2000;18:195–204
15. Lane M, Gardner DK. Blastomere homeostasis. In Gardner DK, Lane M, eds. *ART and the Human Blastocyst*. New York: Springer-Verlag, 2001:69–90
16. Phillips KP, Baltz JM. Intracellular pH regulation by HCO_3^-/Cl^- exchange is activated during early mouse zygote development. *Dev Biol* 1999;208:392–405
17. Lane M, Baltz JM, Bavister BD. Regulation of intracellular pH in hamster preimplantation embryos by the sodium hydrogen (Na^+/H^+) antiporter. *Biol Reprod* 1998;59:1483–90
18. Lane M, Baltz JM, Bavister BD. Na^+/H^+ antiporter activity in hamster embryos is activated during fertilization. *Dev Biol* 1999;208:244–52
19. Lane M, Bavister BD. Regulation of intracellular pH in bovine oocytes and cleavage stage embryos. *Mol Reprod Dev* 1999;54:396–401
20. Phillips KP, Leveille MC, Claman P, Baltz JM. Intracellular pH regulation in human preimplantation embryos. *Hum Reprod* 2000;15:896–904
21. Boron WF, Roos A. Comparison of microelectrode, DMO, and methylamine methods for measuring intracellular pH. *Am J Physiol* 1976;231:799–809
22. Roos A, Boron WF. Intracellular pH. *Physiol Rev* 1981;61:296–434
23. Leclerc C, Becker D, Buehr M, Warner A. Low intracellular pH is involved in the early embryonic death of DDK mouse eggs fertilized by alien sperm. *Dev Dyn* 1994;200:257–67

24. Zhao Y, Chauvet PJ, Alper SL, Baltz JM. Expression and function of bicarbonate/chloride exchangers in the preimplantation mouse embryo. *J Biol Chem* 1995;270: 24428–34

25. Edwards LJ, Williams DA, Gardner DK. Intracellular pH of the preimplantation mouse embryo: effects of extracellular pH and weak acids. *Mol Reprod Dev* 1998;50: 434–42

26. Gibb CA, Poronnik P, Day ML, Cook DI. Control of cytosolic pH in two-cell mouse embryos: roles of H(+)-lactate cotransport and Na⁺/H⁺ exchange. *Am J Physiol* 1997;273:C404–19

27. Miller JG, Schultz GA. Amino acid content of preimplantation rabbit embryos and fluids of the reproductive tract. *Biol Reprod* 1987;36:125–9

28. Edwards LJ, Williams DA, Gardner DK. Intracellular pH of the mouse preimplantation embryo: amino acids act as buffers of intracellular pH. *Hum Reprod* 1998;13:3441–8

29. Kolajova M, Baltz JM. Volume-regulated anion and organic osmolyte channels in mouse zygotes. *Biol Reprod* 1999;60:964–72

30. Lane M, Baltz JM, Bavister BD. Bicarbonate/chloride exchange regulates intracellular pH of embryos but not oocytes of the hamster. *Biol Reprod* 1999;61:452–7

31. Barr KJ, Garrill A, Jones DH, Orlowski J, Kidder GM. Contributions of Na⁺/H⁺ exchanger isoforms to preimplantation development of the mouse. *Mol Reprod Dev* 1998;50:146–53

32. Miyazaki S, Igusa Y. Fertilization potential in golden hamster eggs consists of recurring hyperpolarizations. *Nature* 1981;290:702–4

33. Jones KT, Carroll J, Merriman JA, et al. Repetitive sperm-induced Ca²⁺ transients in mouse oocytes are cell cycle dependent. *Development* 1995;121:3259–66

34. Bos-Mikich A, Whittingham DG, Jones KT. Meiotic and mitotic Ca²⁺ oscillations affect cell composition in resulting blastocysts. *Dev Biol* 1997;182:172–9

35. Salustri A, Yanagishita M, Underhill CB, et al. Localization and synthesis of hyaluronic acid in the cumulus cells and mural granulosa cells of the preovulatory follicle. *Dev Biol* 1992;151:541–51

36. Laurent C, Hellstrom S, Engstrom-Laurent A, et al. Localization and quantity of hyaluronan in urogenital organs of male and female rats. *Cell Tissue Res* 1995; 279:241–8

37. Gardner DK, Leese HJ. The role of glucose and pyruvate transport in regulating nutrient utilization by preimplantation mouse embryos. *Development* 1988;104:423–9

38. Leese HJ, Barton AM. Production of pyruvate by isolated mouse cumulus cells. *J Exp Zool* 1985;234:231–6

39. Biggers JD, Whittingham DG, Donahue RP. The pattern of energy metabolism in the mouse oocyte and zygote. *Proc Natl Acad Sci USA* 1967;58:560–7

40. Lane M, Gardner DK. Regulation of substrate utilization in mouse embryos by the malate aspartate shuttle. *Biol Reprod* 2000;62(Suppl. 1):371

41. Lane M, Gardner DK. Lactate regulates pyruvate uptake and metabolism in the preimplantation mouse embryo. *Biol Reprod* 2000;62:16–22

42. Gardner DK, Sakkas D. Mouse embryo cleavage, metabolism and viability: role of medium composition. *Hum Reprod* 1993;8:288–95

43. Kouridakis K, Gardner DK. Pyruvate in embryo culture media acts as antioxidant. *Proc Fertil Soc Aus* 1995;14:29

44. O'Fallon JV, Wright RWJ. Pyruvate revisited: a non-metabolic role for pyruvate in preimplantation embryo development. *Theriogenology* 1995;43:288

45. Schini SA, Bavister BD. Two-cell block to development of cultured hamster embryos is caused by phosphate and glucose. *Biol Reprod* 1988;39:1183–92

46. Seshagiri PB, Bavister BD. Glucose inhibits development of hamster 8-cell embryos *in vitro*. *Biol Reprod* 1989;40: 599–606

47. Ludwig TE, Lane M, Bavister BD. Differential effect of hexoses on hamster embryo development in culture. *Biol Reprod* 2001;64:1366–74

48. Bavister BD. Glucose and culture of human embryos. *Fertil Steril* 1999;72:233–4

49. Hogan A, Heyner S, Charron MJ, et al. Glucose transporter gene expression in early mouse embryos. *Development* 1991;113:363–72

50. Aghayan M, Rao LV, Smith RM, et al. Developmental expression and cellular localization of glucose transporter molecules during mouse preimplantation development. *Development* 1992;115:305–12

51. Dan-Goor M, Sasson S, Davarashvili A, Almagor M. Expression of glucose transporter and glucose uptake in human oocytes and preimplantation embryos. *Hum Reprod* 1997;12:2508–10

52. Hume DA, Weidemann MJ. Role and regulation of glucose metabolism in proliferating cells. *J Natl Cancer Inst* 1979;62:3–8

53. Reitzer LJ, Wice BM, Kennell D. The pentose cycle: control and essential function in HeLa cell nucleic acid synthesis. *J Biol Chem* 1980;255:5616–26

54. Morgan MJ, Faik P. Carbohydrate metabolism in cultured animal cells. *Biosci Rep* 1981;1:669–86

55. Rieger D. Relationship between energy metabolism and development of the early embryo. *Theriogenology* 1992;37: 75–93

56. Rogers PW, Murphy CR, Gannon BJ. Absence of capillaries in the endometrium surrounding the implanting rat blastocyst. *Micron* 1982;13:373–4

57. Rogers PW, Murphy CR, Gannon BJ. Changes in the spatial organization of the uterine vasculature during implantation in the rat. *J Reprod Fertil* 1982;65: 211–14

58. Gardner DK, Lane M. Alleviation of the '2-cell block' and development to the blastocyst of CF1 mouse embryos: role of amino acids, EDTA and physical parameters. *Hum Reprod* 1996;11:2703–12

59. Vella P, Lane M, Gardner DK. Induction of glycolysis in the day-3 mouse embryo by glucose. *Biol Reprod* 1997;57 (Suppl. 1):26

60. Brinster RL. Hexokinase activity in the preimplantation mouse embryo. *Enzymologia* 1968;34:304–8

61. Martin KL, Hardy K, Winston RM, Leese HJ. Activity of enzymes of energy metabolism in single human preimplantation embryos. *J Reprod Fertil* 1993;99:259–66

62. Chi MM, Manchester JK, Yang VC, *et al*. Contrast in levels of metabolic enzymes in human and mouse ova. *Biol Reprod* 1988;39:295–307

63. Brinster RL. Phosphofructokinase activity in the pre-implantation mouse embryo. *Wilhelm Roux Arch Entwicklungsmech Org* 1971;166:300–2

64. Biggers JD, Gardner DK, Leese HJ. Control of carbohydrate metabolism in preimplantation mammalian embryos. In Rosenblum IY, Heyner S, eds. *Growth Factors in Mammalian Development*. Boca Raton: CRC Press, 1989: 19–32

65. Gardner DK, Lane M. The 2-cell block in CF1 mouse embryos is associated with an increase in glycolysis and a decrease in tricarboxylic acid (TCA) cycle activity: alleviation of the 2-cell block is assocated with the restoration of *in vivo* metabolic pathway activities. *Biol Reprod* 1993;48 (Suppl. 1):152

66. Casslen BG. Free amino acids in human uterine fluid. Possible role of high taurine concentration. *J Reprod Med* 1987;32:181–4

67. Moses DF, Matkovic M, Cabrera-Fisher E, Martinez AG. Amino acid contents of sheep oviductal and uterine fluids. *Theriogenology* 1997;47:336

68. Schultz GA, Kaye PL, McKay DJ, Johnson MH. Endogenous amino acid pool sizes in mouse eggs and preimplantation embryos. *J Reprod Fertil* 1981;61: 387–93

69. Van Winkle LJ. Amino acid transport in developing animal oocytes and early conceptuses. *Biochim Biophys Acta* 1988; 947:173–208

70. Gwatkin RB. Amino acid requirements for attachment and outgrowth of the mouse blastocyst *in vitro*. *Cell Comp Physiol* 1966;68:335–44

71. Gwatkin RB, Haidri AA. Requirements for the maturation of hamster oocytes *in vitro*. *Exp Cell Res* 1973;76:1–7

72. Juetten J, Bavister BD. The effects of amino acids, cumulus cells, and bovine serum albumin on *in vitro* fertilization and first cleavage of hamster eggs. *J Exp Zool* 1983;227: 487–90

73. Zhang X, Armstrong DT. Presence of amino acids and insulin in a chemically defined medium improves development of 8-cell rat embryos *in vitro* and subsequent implantation *in vivo*. *Biol Reprod* 1990;42:662–8

74. Mehta TS, Kiessling AA. Development potential of mouse embryos conceived *in vitro* and cultured in ethylenediaminetetraacetic acid with or without amino acids or serum. *Biol Reprod* 1990;43:600–6

75. Gardner DK, Lane M. Amino acids and ammonium regulate mouse embryo development in culture. *Biol Reprod* 1993;48:377–85

76. Lane M, Gardner DK. Increase in postimplantation development of cultured mouse embryos by amino acids and induction of fetal retardation and exencephaly by ammonium ions. *J Reprod Fertil* 1994;102:305–12

77. Gardner DK, Lane M, Spitzer A, Batt PA. Enhanced rates of cleavage and development for sheep zygotes cultured to the blastocyst stage *in vitro* in the absence of serum and somatic cells: amino acids, vitamins, and culturing embryos in groups stimulate development. *Biol Reprod* 1994;50: 390–400

78. Crosby IM, Gandolfi F, Moor RM. Control of protein synthesis during early cleavage of sheep embryos. *J Reprod Fertil* 1988;82:769–75

79. Rieger D, Loskutoff NM, Betteridge KJ. Developmentally related changes in the uptake and metabolism of glucose, glutamine and pyruvate by cattle embryos produced *in vitro*. *Reprod Fertil Dev* 1992;4:547–57

80. Van Winkle LJ, Haghighat N, Campione AL. Glycine protects preimplantation mouse conceptuses from a detrimental effect on development of the inorganic ions in oviductal fluid. *J Exp Zool* 1990;253:215–19

81. Liu Z, Foote RH. Development of bovine embryos in KSOM with added superoxide dismutase and taurine and with five and twenty percent O_2. *Biol Reprod* 1995;53: 786–90

82. Lindenbaum A. A survey of naturally occurring chelating ligands. *Adv Exp Med Biol* 1973;40:67–77

83. Steeves TE, Gardner DK. Temporal and differential effects of amino acids on bovine embryo development in culture. *Biol Reprod* 1999;61:731–40

84. Eagle H. Amino acid metabolism in mammalian cell cultures. *Science* 1959;130:432–7

85. Van Winkle LJ. Amino acid transport regulation and early embryo development. *Biol Reprod* 2001;64:1–12

86. Lane M, Hooper K, Gardner DK. Effect of essential amino acids on mouse embryo viability and ammonium production. *J Assist Reprod Genet* 2001;18:519–25

87. Devreker F, Hardy K, Van den Bergh M, *et al*. Amino acids promote human blastocyst development *in vitro*. *Hum Reprod* 2001;16:749–56

88. Dorland M, Gardner DK, Trounson A. Serum in synthetic oviduct fluid causes mitochondrial degeneration in ovine embryos. *J Reprod Fertil Abstr Ser* 1994;13:70

89. Thompson JG, Gardner DK, Pugh PA, *et al*. Lamb birth weight is affected by culture system utilized during *in vitro* pre-elongation development of ovine embryos. *Biol Reprod* 1995;53:1385–91

90. Abe H, Yamashita S, Satoh T, Hoshi H. Accumulation of cytoplasmic lipid droplets in bovine embryos and cryotolerance of embryos developed in different culture systems using serum-free or serum-containing media. *Mol Reprod Dev* 2002;61:57–66

91. Hasler JF. *In-vitro* production of cattle embryos: problems with pregnancies and parturition. *Hum Reprod.* 2000;15 (Suppl. 5):47–58

92. Gardner DK, Lane M. Recombinant human serum albumin and hyaluronan can replace blood-derived albumin in embryo culture media. *Fertil Steril* 2000;74(Suppl. 3):O86

93. Hooper K, Lane M, Gardner DK. Toward defined physiological embryo culture media: replacement of BSA with recombinant albumin. *Biol Reprod* 2000;62(Suppl 1):249

94. Gardner DK, Maybach JM, Lane M. Hyaluronan and rHSA increase blastocyst cryosurvival. *Proc 17th World Congress on Fertility and Sterility* 2001:226

95. Batt PA, Gardner DK, Cameron AW. Oxygen concentration and protein source affect the development of preimplantation goat embryos in vitro. *Reprod Fertil Dev* 1991;3:601–7

96. McKiernan SH, Bavister BD. Different lots of bovine serum albumin inhibit or stimulate *in vitro* development

of hamster embryos. *In Vitro Cell Dev Biol* 1992;28A: 154–6

97. Zorn TM, Pinhal MA, Nader HB, *et al.* Biosynthesis of glycosaminoglycans in the endometrium during the initial stages of pregnancy of the mouse. *Cell Mol Biol* 1995;41: 97–106

98. Gardner DK, Rodriegez-Martinez H, Lane M. Fetal development after transfer is increased by replacing protein with the glycosaminoglycan hyaluronan for mouse embryo culture and transfer. *Hum Reprod* 1999;14:2575–80

99. Stojkovic M, Kolle S, Peinl S, *et al.* Effects of high concentrations of hyaluronan in culture medium on development and survival rates of fresh and frozen–thawed bovine embryos produced *in vitro. Reproduction* 2002; 124:141–53

100. Gardner DK, Lane M. Embryo culture systems. In Gardner DK, Trounson AO, eds. *Handbook of In Vitro Fertilization.* Boca Raton: CRC Press, 1993:85–114

101. Bavister BD, McKiernan SH. Regulation of hamster embryo development *in vitro* by amino acids. In Bavister BD, ed. *Preimplantation Embryo Development.* New York: Springer-Verlag, 1993:57–72

102. Dumoulin JC, Evers JL, Bras M, *et al.* Positive effect of taurine on preimplantation development of mouse embryos *in vitro. J Reprod Fertil* 1992;94:373–80

103. Dumoulin JC, van Wissen LC, Menheere PP, *et al.* Taurine acts as an osmolyte in human and mouse oocytes and embryos. *Biol Reprod* 1997;56:739–44

104. Devreker F, Hardy K. Effects of glutamine and taurine on preimplantation development and cleavage of mouse embryos *in vitro. Biol Reprod* 1997;57:921–8

105. Scott LA, Smith S. The successful use of pronuclear embryo transfers the day following oocyte retrieval. *Hum Reprod* 1998;13:1003–13

106. Gerris J, De Neubourg D, Mangelschots K, *et al.* Prevention of twin pregnancy after *in-vitro* fertilization or intracytoplasmic sperm injection based on strict embryo criteria: a prospective randomized clinical trial. *Hum Reprod* 1999; 14:2581–7

107. Bolton VN, Wren ME, Parsons JH. Pregnancies after *in vitro* fertilization and transfer of human blastocysts. *Fertil Steril* 1991;55:830–2

108. Bavister BD. Stage-specific culture media and reactions of embryos to them. In Jansen R, Mortimer D, eds. *Towards Reproductive Certainty Fertility and Genetics Beyond 1999.* New York: Parthenon, 1999:367–77

109. Gardner DK, Leese HJ. Concentrations of nutrients in mouse oviduct fluid and their effects on embryo development and metabolism *in vitro. J Reprod Fertil* 1990;88: 361–8

110. Gardner DK. Mammalian embryo culture in the absence of serum or somatic cell support. *Cell Biol Int* 1994;18: 1163–79

111. Barnes FL, Crombie A, Gardner DK, *et al.* Blastocyst development and birth after *in-vitro* maturation of human primary oocytes, intracytoplasmic sperm injection and assisted hatching. *Hum Reprod* 1995;10:3243–7

112. Gardner DK, Schoolcraft WB, Wagley L, *et al.* A prospective randomized trial of blastocyst culture and transfer in *in-vitro* fertilization. *Hum Reprod* 1998;13: 3434–40

113. Abramczuk J, Solter D, Koprowski H. The beneficial effect of EDTA on development of mouse one-cell embryos in chemically defined medium. *Dev Biol* 1977;61:378–83

114. Gardner DK, Lane MW, Lane M. EDTA stimulates cleavage stage bovine embryo development in culture but inhibits blastocyst development and differentiation. *Mol Reprod Dev* 2000;57:256–61

115. Lane M, Gardner DK. Inhibiting 3-phosphoglycerate kinase by EDTA stimulates the development of the cleavage stage mouse embryo. *Mol Reprod Dev* 2001;60: 233–40

116. Gardner DK, Lane M, Stevens J, *et al.* Blastocyst score affects implantation and pregnancy outcome: towards a single blastocyst transfer. *Fertil Steril* 2000;73:1155–8

117. Buster JE, Bustillo M, Rodi IA, *et al.* Biologic and morphologic development of donated human ova recovered by nonsurgical uterine lavage. *Am J Obstet Gynecol* 1985;153:211–17

118. Mastroianni LJ, Jones R. Oxygen tension in the rabbit fallopian tube. *J Reprod Fertil* 1965;9:99

119. Ross RN, Graves CN. O_2 levels in the female rabbit reproductive tract. *J Anim Sci* 1974;39:994

120. Fischer B, Bavister BD. Oxygen tension in the oviduct and uterus of rhesus monkeys, hamsters and rabbits. *J Reprod Fertil* 1993;99:673–9

121. Maas DH, Storey BT, Mastroianni LJ. Oxygen tension in the oviduct of the rhesus monkey (*Macaca mulatta*). *Fertil Steril* 1976;27:1312–17

122. Harlow GM, Quinn P. Foetal and placental growth in the mouse after pre-implantation development *in vitro* under oxygen concentrations of 5 and 20%. *Aust J Biol Sci* 1979; 32:363–9

123. Gardner DK, Lane M, Johnson J, *et al.* Reduced oxygen tension increases blastocyst development, differentiation, and viability. *Fertil Steril* 1999;72:S30–1

124. Wiley LM, Yamami S, Van Muyden D. Effect of potassium concentration, type of protein supplement, and embryo density on mouse preimplantation development *in vitro. Fertil Steril* 1986;45:111–19

125. Paria BC, Dey SK. Preimplantation embryo development *in vitro*: cooperative interactions among embryos and role of growth factors. *Proc Natl Acad Sci USA* 1990;87: 4756–60

126. Lane M, Gardner DK. Effect of incubation volume and embryo density on the development and viability of mouse embryos *in vitro. Hum Reprod* 1992;7:558–62

127. Salahuddin S, Ookutsu S, Goto K, *et al.* Effects of embryo density and co-culture of unfertilized oocytes on embryonic development of *in-vitro* fertilized mouse embryos. *Hum Reprod* 1995;10:2382–5

128. Ahern TJ, Gardner DK. Culturing bovine embryos in groups stimulates blastocyst development and cell allocation to the inner cell mass. *Theriogenology* 1998;49: 194

129. Maas DH, Storey BT, Mastroianni LJ. Hydrogen ion and carbon dioxide content of the oviductal fluid of the rhesus monkey (*Macaca mulatta*). *Fertil Steril* 1977;28:981–5

130. Garris DR. Uterine blood flow, pH, and pCO_2 during nidation in the guinea pig: ovarian regulation. *Endocrinology* 1984;114:1219–24

131. Dale B, Menezo Y, Cohen J, *et al*. Intracellular pH regulation in the human oocyte. *Hum Reprod* 1998;13:964–70

132. Iritani A, Nishikawa Y, Gomes WR, VanDemark NL. Secretion rates and chemical composition of oviduct and uterine fluids in rabbits. *J Anim Sci* 1971;33:829–35

133. Quinn P, Kerin JF, Warnes GM. Improved pregnancy rate in human *in vitro* fertilization with the use of a medium based on the composition of human tubal fluid. *Fertil Steril* 1985;44:493–8

134. Pool TB, Atiee SH, Martin JE. Oocyte and embryo culture: basic concepts and recent advances. *Infertil Reprod Med Clin N Am* 1998;9:181–203

135. Gardner DK, Lane M, Schoolcraft WB. Physiology and culture of the human blastocyst. *J Reprod Immunol* 2002;55:85–100

136. Lane M, Gardner DK. Nonessential amino acids and glutamine decrease the time of the first three cleavage divisions and increase compaction of mouse zygotes *in vitro*. *J Assist Reprod Genet* 1997;14:398–403

137. Gardner DK, Lane M. Culture of viable human blastocysts in defined sequential serum-free media. *Hum Reprod* 1998;13(Suppl. 3):148–59

138. Gardner DK. Improving embryo culture and enhancing pregnancy rate. In Shoham Z, Howles C, Jacobs H, eds. *Female Infertility Therapy: Current Practice*. London, Martin Dunitz, 1998:283–99

Quality assessment of human embryos: state of the art and future perspectives

14

Lewis C. Krey

INTRODUCTION

At the beginning of an *in vitro* fertilization (IVF) cycle the embryologist is presented with cohorts of male and female gametes and is then expected to perform three primary tasks. The first task is to inseminate the eggs successfully. The second task is to culture the resultant embryos in a manner that maintains the integrity of those embryos that possess the 'biologic potential' to initiate implantation and pregnancy. The third and final task is to identify the embryos that most likely possess that 'biologic potential' and select them for transfer to the uterus.

During the past 15–20 years significant procedural advances have been made that allow embryologists to complete the first two tasks more successfully on a day-to-day basis. The development of intracytoplasmic sperm injection (ICSI) for severe male-factor patients has dramatically increased the pregnancy rate for couples with severe male-factor infertility[1]. Recent improvements in embryo culture have also had a significant impact. In particular sequential serum-free media systems have been developed that address the changes in metabolism that occur during early embryonic development. As a result embryos can be cultured to the blastocyst stage prior to transfer, resulting in higher pregnancy and implantation rates[2,3]. In contrast to these advances, no innovative new approach to the third task – embryo selection – has been adopted into daily usage by embryologists in IVF clinics world wide. As was the case 20 years ago, embryo selection during routine IVF cycles is based on microscopic evaluation that focuses on counts of blastomere numbers and/or the presence of morphologic markers that can be linked to pregnancy outcome or the absence of degenerative anomalies. This review will examine whether other approaches can be used to select embryos for transfer. The discussion of each approach will focus on the biologic rationale underlying the selection process, logistic considerations involved in performing the procedures and, finally, their future clinical potential.

EMBRYO SELECTION BY METABOLISM

An embryo's ability to develop, implant and initiate a pregnancy is linked to its metabolic rate, and the machinery that maintains this rate is inherited from the egg. This relationship is clearly demonstrated in studies in which intracellular markers of metabolism have been measured. For example, measurements of cellular ATP levels within cohorts of murine and human eggs have revealed a strong relationship between ATP levels in the test eggs and the ability of other eggs in the cohort to implant[4]. Unfortunately, measurements like these are of little practical value for present considerations since the embryos must be lysed in order to measure the level of intracellular substrate. In contrast, embryo metabolism can also be assessed by measuring the usage of extracellular molecules that the embryo must take up because they are necessary to supply its energy needs. Such measurements have been performed for glucose and for oxygen.

Retrospective studies with cattle and murine blastocysts have demonstrated a positive correlation between glucose consumption and fetal developmental potential[5,6]. More importantly for present considerations, when murine blastocysts were prospectively selected for transfer based on a non-invasive assessment of glucose uptake and lactate production, there was a significant increase in implantation rate[7]. Recently, Gardner and co-workers[8] reported similar research findings using donated cryopreserved human embryos. These authors reported that pyruvate and glucose uptakes during a 4 h period on day 4 of culture are significantly higher in embryos that progress to blastocyst stage. Moreover, glucose uptake is highest in embryos that progress to the highest grade (4AA) on day 5. Because there was a large variation in glucose consumption among 13 identically scored blastocysts that developed from a single patient's cryopreserved embryos, the authors suggest that glucose consumption may be an appropriate method for screening individual blastocysts for transfer. However, additional observations that glucose consumption by individual blastocysts increases from day 4 to day 6 and that glucose consumption by non-cryopreserved embryos is higher than that of cryopreserved embryos indicate that strict criteria must be imposed in the conduct and selection of embryos for evaluation before one can attribute any physiologic significance to an absolute value for glucose consumption.

In another elegant series of studies, Trimarchi and co-workers have described the development of a self-referencing electrophysiologic technique to measure the levels of dissolved oxygen in the medium surrounding individual murine embryos[9–11]. In this non-invasive procedure an electrode placed in close proximity to the embryo measures the movement of oxygen between the embryo and medium; gradients of oxygen depletion can be measured that extend greater than 50 μm from the embryo into the medium[11]. Further studies revealed that the amount of oxygen consumed by mitochondrial oxidative phosphorylation increases as an embryo progresses from cleavage stage to the blastocyst stage (30% to 60–70% of the oxygen consumed, respectively). In contrast, non-oxidative phosphorylation mechanisms consume a relatively constant amount of oxygen throughout this period of development. As a result of these patterns, blastocysts were found to deplete the surrounding medium of twice the level of oxygen than did cleavage-stage embryos[11]. In additional studies these authors adapted the self-referencing electrode to look at other physiologic gradients, e.g., Ca^{2+} [9,10], as well as K^+ efflux following chemically induced apoptosis[12]. More recent work on pancreatic β-cells indicates that this approach also has the potential to quantify real-time glucose utilization by embryos[13].

All of the above approaches have provided important information about the physiology and metabolism of early embryos. In particular the self-referencing electrode appears to offer considerable promise as a tool to characterize many sequelae to normal developmental or artificially imposed changes in embryonic physiology in a non-invasive fashion. However, whether any of these procedures can be adapted to perform routine embryo selection on a day-to-day basis in a clinical IVF program is open to question, although the approach by Gardner and co-workers offers the greatest hope because they have demonstrated that analysis can be completed prior to the day of selection and that within-stage differences in glucose uptake are measurable and physiologically relevant.

Although the foregoing studies provide optimism about alternative approaches to embryo selection, the transition of each procedure from research laboratory to clinical embryology laboratory may be difficult. Each procedure utilizes sophisticated and, unfortunately, expensive equipment, and this equipment requires in turn personnel specially trained to operate it in an accurate and reliable manner. In this era of increasing governmental oversight into clinical reproductive medicine and IVF in particular, the use of such equipment will likely be subject to regulatory control. In the USA, where even glass micropipettes and transfer catheters have to be approved for use by the Federal Drug Administration, it is likely that both the self-referencing electrode and the fluorescence microscopic set-up used by Gardner and co-workers will have to be approved specifically for embryo selection purposes. Government licensing of laboratories may also become an issue with laboratory competency evaluated on several levels. Unfortunately, these concerns will be repeated throughout this review since they also apply to the embryo selection procedures described in the next two sections.

EMBRYO SELECTION BY SECRETION PRODUCTS

Quantification of a molecular marker that is present in and secreted by normal embryos during early development, and that can be linked to pregnancy outcome, is another potential approach for selection of embryos for uterine transfer. As with the glucose and oxygen monitoring studies above, this approach does not entail sampling the embryo directly but rather monitoring the spent culture media for the presence or absence of a secretagogue. Recent studies have suggested that leptin may be such a molecule.

Leptin is a 16 K peptide (146 amino acids) whose primary physiologic role is in the regulation of body mass[14]. However, leptin has also been reported to play a role in reproduction[15–17]. In mammals, leptin has been identified in follicular fluid, in granulosa cells and in oocytes in which its intracellular distribution is polarized[18]. Although leptin mRNA has been identified in cumulus cells, it cannot be quantified in oocytes or embryos[19]. In the oocyte, leptin distribution is co-polarized with that of the transcriptional activator STAT3, suggesting that it may be involved in controlling gene transcription during early embryonic development[18]. The polarity to leptin distribution persists through numerous embryonic divisions; eventually leptin-positive blastomeres can be identified by immunostaining in the outer trophectoderm of a blastocyst, but not in its inner cell mass[18].

The intracellular actions of leptin are mediated by cell membrane leptin receptors and these receptors have been isolated in mammalian eggs and embryos[18]. The presence of receptors provides further support that this peptide may be involved in the paracrine or autocrine regulation of gene transcription in oocytes and embryos. Leptin receptors and their mRNA have also been identified in human endometrial tissue biopsies[20,21]. Co-culture studies have noted that endometrial leptin secretion is influenced by an embryo and this relationship depends on whether the embryo is arrested or has developed into a blastocyst capable of implantation following transfer[20].

Gonzalez and co-workers recently reported that leptin is secreted by human embryos during culture to day 5[20]. More importantly for present considerations, these investigators noted that leptin concentrations in spent media collected on day 5 (+120 h post-retrieval) from individually cultured blastocysts correlate in a positive fashion with pregnancy outcome, with higher levels present in hatching blastocysts than in arrested embryos[20]. However, further studies are necessary to verify the existence of this relationship and to define the range of variation among blastocysts.

Several steps must be taken before embryonic leptin secretion can be used as a prospective laboratory screen to predict pregnancy outcome. In the Gonzalez and co-workers study, secretion was monitored from day 4 to day 5 in culture and the leptin results determined by enzyme-linked immunosorbent assay (ELISA) after embryo transfer had been performed[20]. Current leptin immunoassays recommend overnight incubation and it seems unlikely that they can be successfully performed between the time of media harvesting on the morning of day 5 and embryo transfer that afternoon, especially since ELISAs often present a striking loss of sensitivity and reliability when their incubation periods are shortened. Unfortunately, embryonic leptin secretion was not reported for earlier days of culture. It would be important to know if leptin secretion is a blastocyst-specific phenomenon or whether it occurs earlier in development. Observations that the outer blastomeres of morulae are leptin-positive suggest that day 3 to day 4 culture may also provide useful results[18]. However, because many embryos that progress to morula stage actually arrest at that stage (in the New York University School of Medicine laboratory 30–40% of morulae arrest was noted), there may be little value to screening at the day 3 to 4 transition. In addition, differences in leptin secretion have only been reported between blastocysts and arrested embryos[20]. The sensitivity and reliability of the assay have to be verified to distinguish between embryo-to-embryo differences across the many stages of blastocyst development. Embryo cohorts can also vary from patient-to-patient and this will also require assessment and verification. Clearly, future work in these areas is needed to determine whether leptin, or any other secretory marker, can be used to select embryos for transfer.

The incorporation of leptin screening into a clinical IVF setting would also have financial, labor and regulatory consequences. The leptin ELISA kits are expensive (US $110–450/kit for 100 tubes), and the cost per sample rises further when only a few samples are assayed on given days. The plate reader can also be an expensive piece of equipment. In large IVF clinics leptin assays would require separate space as well as technicians trained in ELISA techniques. In the view of many government regulatory agencies the use of such a kit for diagnostic purposes would likely require additional laboratory licensure and the imposition of a range of regulatory requirements for staff hiring, equipment testing and day-to-day monitoring procedures to ensure that routine testing for leptin was conducted in a precise and reliable manner.

EMBRYO SELECTION BY GENETICS

Aneuploidy (i.e., an abnormal number of chromosomes) is a well-documented genetic anomaly in human eggs and embryos and is thought to be the primary factor underlying the decline in fertility experienced during advancing maternal age. The incidence of aneuploidy in embryos that develop from bi-pronucleated zygotes in IVF or IVF + ICSI cycles has been shown to correlate with embryo development and morphology as well as with increasing maternal age[22,23]. In these studies the highest rates were observed in embryos that had arrested or developed at a slow rate with accompanying dysmorphologies; moreover, the rate of aneuploidy increased significantly with increasing maternal age not only in these dysfunctional embryos but also in morphologically and developmentally normal cleavage-stage embryos. Many aneuploidies arise in oocytes as a result of inappropriate chromosome segregation during the first meiotic division, although inappropriate segregation of chromatids during the second meiotic division has also been observed[24–26]. Post-fertilization mitotic divisions can also be abnormal resulting in monosomic–trisomic cell pairs[27]. These mitotic aneuploidies and the embryonic mosaicism they create also appear to be a product of poor-quality oocytes. Because research to date has employed only a limited number of fluorescently labeled probes that are specific for only their respective chromosomes, it is possible that this genetic problem occurs at an even greater frequency than has been described.

In 1993 Munne and co-workers[28] first demonstrated that fluorescence *in situ* hybridization (FISH) can assess ploidy in individual blastomeres within the time frame of an IVF cycle. Because FISH can quantify chromosome number in interphase nuclei, there is no need to prepare a metaphase spread. As a result, every biopsied blastomere is suitable for analysis. To date several investigators have used FISH to obtain a genetic diagnosis for preimplantation embryos (PGD) prior to transfer in an IVF cycle. To accomplish this, cleavage-stage embryos are biopsied on day 3 and the blastomeres are tested for ploidy by simultaneous hybridization with as many as nine different chromosome-specific, labeled probes[29]. Only 'normal' embryos which generate diploid cells are selected for transfer and in some cycles, embryo transfer may be cancelled because there are no 'normal' embryos present.

What is important for present considerations are the findings that, by using this genetic screening approach to select embryos, one can significantly increase the pregnancy rate per transfer and reduce the rate of miscarriage once pregnancy ensues[30,31]. The biologic bases for the pregnancy per transfer and miscarriage findings are understandable, because one is excluding the aneuploid embryos destined not to implant or to implant only to result in miscarriage. However, such results do not necessarily translate into an increased pregnancy rate per retrieval, the benchmark parameter for laboratory success. The only explanation for an increase in pregnancy rate per retrieval in the PGD patients would be if embryos were transferred to the uterus that would not be selected normally on the bases of morphologic and developmental criteria. It is important to mention that the PGD studies cited above deal with patients approaching 40 years of age or older[30,31]. At least at NYU-SM, the majority of patients of this age have fewer quality embryos for transfer than are allowable by our guidelines. Since screening for aneuploidy only reduces the number of suitable embryos, the end result after PGD is the transfer of embryos that would be transferred regardless or no transfer at all. Clearly, neither scenario would be expected to increase the pregnancy rate per retrieval. However, some older patients do generate large numbers (> 10) of eggs and embryos with quality morphology for transfer, and in the IVF cycles of these patients, embryo selection by PGD might result in an increased pregnancy rate. However, although large embryo numbers are more common in young women (i.e., less than 37 years of age), the chances for a similar increase in pregnancy rate is considerably smaller because the rate of aneuploidy in this population is relatively low. Clearly, before an IVF program commits to PGD for embryo selection, it must seriously consider the personal and cycle characteristics of the patient population to be screened. Not all patients will benefit from such a selection process and in some pregnancy success may even be compromised by unintentional trauma associated with the biopsy procedure. Polar-body biopsy represents a reasonable alternative to minimize this concern but such a procedure can only be used to monitor for aneuploidies that originate in the oocyte[26].

Since embryo transfer was performed on day 3 in the foregoing PGD studies, one must wonder if similar increases in pregnancy and decreases in miscarriage rates would have occurred if the day of embryo selection was simply extended to day 5. Although mosaicism has been observed in blastocysts[32], it does not appear to happen as frequently as has been observed in embryos at cleavage stage. Such findings suggest that a biologic screening process takes place during the cleavage–morula–blastocyst transitions that clears out many genetically compromised embryos. As a result, many investigators have reported that the implantation rate of blastocysts is higher than that of cleavage embryos. During the years 2000–2001 at NYU-SM, the clinical pregnancy (+ fetal heartbeat) and implantation rates for patients 41 years of age or older were twice as high for day 5 transfer of blastocysts than for day 3 transfer of cleavage-stage embryos.

As is the case with approaches in the preceding sections, the incorporation of FISH staining into the daily IVF procedures would be a major undertaking. Experienced staffing is needed for blastomere or polar-body biopsy and fixation as well as for the FISH procedure. FISH is not simple, and even in the most experienced laboratories, a 'no result' outcome can occur due to sample loss at different procedural steps or to chromosome overlap at fixation resulting in hybridization patterns that are uninterpretable or misleading[33]. A separate facility for fixation, hybridization and fluorescence microscopy is advised, and the cost of labeled probes can be dramatic if one selects a large number of chromosomes for screening. Government regulatory issues also may arise and require

laboratory licensing to perform a genetic diagnosis, especially if there is a financial charge for this diagnostic procedure.

In contrast to the previous two selection approaches discussed above, the biopsy step necessary for this screening for aneuploidy may be viewed as an invasive procedure, especially if a blastomere(s) is removed; whether there are long-term consequences of non-intentional trauma during this step on embryo viability remains to be determined. In addition, the approaches described in the preceding section can also be used to select embryos from all IVF patients, regardless of age. Considering the relationship between patient age and aneuploidy frequency, such an argument cannot be defended for this genetic screening procedure. As a result any IVF clinic that is considering adding this screening procedure to its program should seriously consider the number of their patients who would actually benefit from such a selection procedure and whether this number justifies the financial outlay and other considerations that PGD entails.

EMBRYO SELECTION BY MORPHOLOGY

Although the previous sections suggest that new embryo selection procedures are possible, it still appears that, at this point in time, the easiest and most reliable approach to select embryos for transfer utilizes morphologic and developmental criteria. However, this approach has also evolved with time as new relevant morphologic markers have been identified that correlate with pregnancy outcome. The development of sequential media and culture to day 5 has also been particularly significant because it permits the evaluation of embryos that have successfully completed the transition from maternal to embryonic genomic control.

Many morphologic markers have been described as positive (blastomere 'pitting') or negative (degree of fragmentation, presence of multinucleated blastomeres) predictors for pregnancy success. One such marker, the spatial organization of nucleoli in the male and female pronuclei after fertilization, has been reported by Scott and co-workers to be an excellent prognostic indicator for developmental progression to the blastocyst stage and for implantation success[34]. This morphologic marker has proved to be of important practical value to IVF programs in countries in which government regulations limit the numbers of pronuclear zygotes that can be cultured to embryo status for transfer during an IVF cycle. A recent multicenter study conducted in Germany, where such regulations exist, reports that using these criteria, one can identify zygotes that have a $\geq 40\%$ chance for pregnancy when cultured and then transferred as cleavage-stage embryos[35].

The selection of blastocysts for transfer based on their morphologic and developmental criteria has also improved pregnancy rates. In general these criteria are based on the stage of blastocyst development through the hatching stage as well as the morphology of the inner cell mass and trophectoderm[3]. Using such a grading system at NYU-SM, we have observed that pregnancy and implantation rates vary depending on the development and morphology of the lead, or highest rated, and secondary blastocysts on day 5. Embryo cohorts led by morulae have the lowest rates (< 30 and ~ 15%, respectively) and expanded or hatching blastocysts the highest (reaching ~ 65% per transfer and $\geq 50\%$, respectively). This range of pregnancy rates was consistent for patients up to age 41; however, the implantation rate at each blastocyst level did appear to decline by 10–20% as patient age increased. In general these rates virtually doubled those obtained from concurrent IVF cycles with day 3 transfers and such an increase was even observed in patients 41 years of age or older. In a retrospective analysis of IVF cycles involving blastocyst transfer, we also found that in approximately 30% of the cycles the transferred blastocysts would not have been among the cohort of embryos selected for transfer if the latter had been conducted on day 3[36]. Further evidence that a biology-based selection of embryos occurs between day 3 and day 5 is suggested by the observation that approximately $\geq 30\%$ of the embryos that reach morula status on day 4 arrest at that stage. This selection may be genetically based. Support for such a conclusion is provided by Evsikov and Verlinsky[32] who found aneuploid mosaicism in only 10.5% of the blastocysts examined with an average of five aneuploid cells in each affected embryo. Such a frequency of mosaicism is considerably lower than the 30% rate reported by Munne and co-workers for cleavage-stage embryos[27].

Embryo culture procedures can vary dramatically among IVF programs. Some laboratories culture many embryos together in batches. Such a system eases the embryologists' workload at several different levels; however, it is difficult, if not impossible, to track the development of individual embryos cultured in this manner. Other laboratories culture embryos individually, and although this entails more work, the end results can be used to describe the developmental and morphologic path followed most frequently by embryos that implant and initiate a pregnancy. At NYU-SM we have charted embryo development and morphology in this manner and find it useful when selecting embryos for transfer. For example, if there are multiple blastocysts of similar stage and quality on day 5, we will incorporate information about their status on days 2 and 3 into the decision-making process. These data can also be used to establish embryonic criteria on day 3, e.g., number of embryos, quality and number of blastomeres, that is needed to continue culture to day 5. In compiling this database, some surprising results emerged; for example, ≥ 10-cell embryos on day 3 often fail to progress to blastocyst stage. As a result we do not consider them optimal candidates for transfer, even on day 3. Because a variety of local factors such as culture systems and scheduled monitoring times can have an impact on the picture of embryonic development that emerges, each IVF laboratory should chart their own morphology and developmental database on which to base their own decisions.

CONCLUSIONS

Embryos can be selected for transfer during an IVF cycle using many different approaches. Some approaches measure an embryo's physiologic parameters and are biologically feasible, but the methodologies involved require further testing before they can be transferred to the IVF laboratory, and once there, they may be subject to regulatory oversight. However, such approaches have proved valuable in research studies and have generated important data about the physiology of early embryonic development. Although the selection of embryos using genetic criteria, such as PGD for aneuploidy using FISH, has been shown to improve the pregnancy rate and reduce the miscarriage rate in several IVF programs, this approach may actually be of real value for only a relatively small portion of a patient population and may best be left to a few centers skilled in the technique. Finally, the current 'gold standard' – selecting embryos on the basis of division rate and the presence/absence of morphologic markers – is still undergoing improvement as more relevant morphologic markers and longer culture periods provide additional opportunities to obtain reliable information on which to base this critical decision.

References

1. Van Steirteghem A, DeVos A, Staessen C, *et al*. Is ICSI the ultimate ART procedure ? In Kempers RD, Cohen J, Haney AF, Younger JB, eds. *Fertility and Reproductive Medicine*. Amsterdam: Elsevier, 1998:27–38

2. Gardner DK, Lane M. Culture of viable blastocysts in defined sequential serum-free media. *Hum Reprod* 1998; 13(Suppl. 3):148–59

3. Gardner DK, Schoolcraft WB. *In vitro* culture of human blastocysts. In Jansen R, Mortimer D, eds. *Towards Reproductive Certainty: Fertility and Genetics Beyond 1999*. Carnforth: Parthenon Press, 1999:378–88

4. Van Blerkom J, Davis PW, Lee J. ATP content of human oocytes and developmental potential and outcome after *in-vitro* fertilization and embryo transfer. *Hum Reprod* 1995; 10:415–24

5. Renard JP, Philippon A, Menezo Y. *In vitro* glucose uptake of glucose by bovine blastocysts. *J Reprod Fertil* 1980;58:161–4

6. Gardner DK, Leese HJ. Assessment of embryo viability prior to transfer by the non-invasive measurement of glucose uptake. *J Exp Zool* 1987;242:102–5

7. Lane M, Gardner DK. Selection of viable mouse blastocysts prior to transfer using a metabolic criterion. *Hum Reprod* 1996;11:1975–8

8. Gardner DK, Lane M, Stevens J, *et al*. Noninvasive assessment of human embryo nutrient consumption as a measure of developmental potential. *Fertil Steril* 2001;76:1175–80

9. Smith PJS, Hammar K, Porterfield DM, *et al*. A self-referencing, non-invasive, ion-selective electrode for single cell detection of trans-plasma membrane calcium flux. *Microsc Res Tech* 1999;46:398–417

10. Trimarchi JR, Liu L, Porterfield DM, *et al*. A non-invasive method for measuring preimplantation embryo physiology. *Zygote* 2000;8:15–24

11. Trimarchi JR, Liu L, Porterfield DM, *et al*. Oxidative phosphorylation-dependent and independent oxygen consumption by individual preimplantation mouse embryos. *Biol Reprod* 2000;62:1866–74

12. Trimarchi JR, Liu L, Smith PJ, *et al*. Noninvasive measurement of potassium efflux as an early indicator of cell death in mouse embryos. *Biol Reprod* 2000;63:851–7

13. Jung SK, Trimarchi JR, Sanger RH, *et al*. Development and application of a self-referencing glucose microsensor for the measurement of glucose consumption by pancreatic beta-cells. *Anal Chem* 2001;73:3759–67

14. Minocci A, Savia G, Lucatoni R, *et al*. Leptin plasma concentrations are dependent on body fat distribution in obese patients. *Int J Obes Relat Metab Disord* 2000;9:1139–44

15. Mantzoros CS, Cramer DW, Liberman RF, *et al*. Predictive value of serum and follicular fluid leptin concentrations during assisted reproductive cycles in normal women and in women with polycystic ovarian syndrome. *Hum Reprod* 2000;15:539–44

16. Imani B, Eijkemans MJC, DeJong FH, *et al*. Free androgen index and leptin are the most prominent endocrine predictors of ovarian response during clomiphene citrate induction of ovulation in normogonadotropic oligoamenorrheic infertility. *J Clin Endocrinol Metab* 2000;85:676–82

17. Karlsson C, Lindell K, Svensson E, *et al*. Expression of functional leptin receptors in the human ovary. *J Clin Endocrinol Metab* 1997;82:4144–8

18. Antczak M, Van Blerkom JV. Oocyte influences on early development: the regulatory proteins leptin and STAT3 are polarized in mouse and human oocytes and differentially distributed within the cells of the preimplantation stage embryo. *Mol Hum Reprod* 1997;2:1067–86

19. Cioffi J, Van Blerkom JV, Antczak M, *et al*. The expression of leptin and its receptors in preovulatory human follicles. *Mol Hum Reprod* 1997;3:467–72

20. Gonzalez, RR, Caballero-Campo P, Jasper M, *et al*. Leptin and leptin receptor are expressed in human endometrium and endometrial leptin secretion is regulated by the human blastocyst. *J Clin Endocrinol Metab* 2000;85:4883–8

21. Kitawaki J, Koshiba H, Ishihara H, *et al*. Expression of leptin receptor in human endometrium and fluctuation during the menstrual cycle. *J Clin Endocrinol Metab* 2000; 85:1946–50

22. Munne S, Alikani M, Tomkin G, *et al*. Embryo morphology, developmental rates, and maternal age are correlated with chromosome abnormalities. *Fertil Steril* 1995;64:382–91

23. Munne S, Marquez, C, Reing A, *et al*. Chromosome abnormalities in embryos obtained after conventional *in vitro* fertilization and intracytoplasmic sperm injection. *Fertil Steril* 1998;69:904–8

24. Angell R. Predivision in human oocytes at meiosis 1: a mechanism for trisomy formation in man. *Hum Genet* 1991: 86:383–7

25. Mahmood R, Brierly CH, Faed MJW, *et al*. Mechanisms of maternal aneuploidy: FISH analysis of oocytes and polar bodies in patients undergoing assisted conception. *Hum Genet* 2000;106:620–6

26. Verlinsky Y, Kuliev A. Polar body biopsy. In Gardner DK, Weissman A, Howles CM, Shoham Z, eds. *Textbook of Assisted Reproductive Techniques. Laboratory and Clinical Perspectives*. London: Martin Dunitz Ltd, 2001:333–40

27. Munne S, Cohen J. Chromosome abnormalities in human embryos. *Human Reprod Update* 1998;4:842–55

28. Munne S, Lee A, Rosenwaks Z, *et al*. Diagnosis of major chromosome aneuploidies in human preimplantation embryos. *Hum Reprod* 1993;8:2185–91

29. Bahce M, Escudero T, Sandalinas M, *et al*. Improvements of preimplantation diagnosis of aneuploidy by using microwave hybridization, cell recycling and monocolor labeling of probes. *Mol Human Reprod* 2000;9: 849–54

30. Gianaroli L, Magli C, Ferraretti AP, *et al*. Preimplantation diagnosis for aneuploidies in patients undergoing *in vitro* fertilization with a poor prognosis: identification of the categories for which it should be proposed. *Fertil Steril* 1999;72:837–44

31. Munne S, Magli C, Cohen J, *et al*. Positive outcome after preimplantation diagnosis of aneuploidy in human embryos. *Hum Reprod* 1999;14:2191–9

32. Evsikov S, Verlinsky Y. Mosaicism in the inner cell mass of human blastocysts. *Hum Reprod* 1998;11:3151–5

33. Munne S, Sandalinas M, Cohen J. Chromosome abnormalities in human embryos. In Gardner DK, Weissman A, Howles CM, Shoham Z, eds. *Textbook of Assisted Reproductive Techniques. Laboratory and Clinical Perspectives*. London: Martin Dunitz Ltd, 2001:297–318

34. Scott L, Alvero R, Leondires M, *et al*. The morphology of human pronuclear embryos is positively related to blastocyst development and implantation. *Hum Reprod* 2000;15: 2394–403

35. Montag M, van der Ven H. Evaluation of pronuclear morphology as the only selection criteria for further embryo culture and transfer: results of a prospective multicenter study. *Human Reprod* 2001;16:2384–9

36. Song E, McCaffrey C, Krey LC, *et al*. Blastocyst transfer: does early embryonic evaluation and/or ultimate embryo selection affect prognosis? *Fertil Steril* 2001;76(Suppl. 3S):s87

Cryobiology of human reproduction

Technology for the cryopreservation of human embryos and gametes   15

Eleonora Porcu, Patrizia M. Ciotti, Raffaella Fabbri, Giuseppe Damiano, Angela Scarano and Stefano Venturoli

INTRODUCTION

The cryopreservation technique offers several advantages in an *in vitro* fertilization (IVF) program. The current use of ovulation induction regimens yields a high number of oocytes available for fertilization. In order to reduce the number of multiple births, most IVF clinics today transfer maximally three embryos. Following the first report of a pregnancy as a result of cryopreservation[1], supernumerary embryos are now routinely frozen, stored and used in subsequent cycles. Severe ovarian hyperstimulation syndrome (OHSS) during ovulation induction can be a serious complication. Since it has been observed that pregnancy will aggravate OHSS and also prolong its duration, freezing of all embryos and delaying transfer by one or more cycles seems preferable. Cryopreservation will also allow patients to have more than one child after IVF without additional ovarian stimulation and oocyte retrieval. Unfortunately, embryo storage has several moral, legal and religious implications involving both patients and clinicians. In some countries the application of embryo cryopreservation has been strictly limited and even prohibited in some places.

Human oocyte cryopreservation, in contrast, could be an alternative solution to these problems and may have several other clinical applications. In fact, this is the only method to preserve the reproductive capacity for women at risk of losing their fertility because of premature ovarian failure, pelvic diseases, surgery or antineoplastic treatments. Oocyte storage has faced technical difficulties as documented by the low number of births achieved after oocyte cryopreservation in the past.

FREEZING BIOLOGY

In the cryopreservation procedure the main steps are: (1) preliminary exposure to the cryoprotectants, substances that reduce cell damage caused by ice crystals; (2) progressive temperature reduction down to −196°C; (3) storage; (4) subsequent thawing after a variable length of time; and (5) diluting and washing off the cryoprotectants to restore a physiologic microenvironment and allow subsequent development.

The two most critical moments affecting cell survival are the phase of initial cooling and the return to physiologic conditions. When reducing the temperature from −5 to −15°C, ice nucleation is first induced in the extracellular medium by a process called 'seeding'. When temperature is gradually decreased, ice formation increases and the solutes concentrate in the extracellular medium. This determines an osmotic gradient: water moves out from the cytoplasm and the cell shrinks because of dehydration. If freezing is performed slowly, water diffusion out of the cell does not allow the formation of big ice crystals inside the cytoplasm.

A detailed mathematical model has been developed to calculate the rate of change in cell volume, as a function of the permeability, surface area and temperature[2]. For those cells with a low surface/volume ratio, e.g., female gametes, it is necessary to have a low cooling rate to allow the diffusion of a sufficient amount of water out of the cell. For each cell type, increasing the freezing rate decreases survival[3]. On the other hand, the survival rate decreases if thawing velocity is too low because ice crystals in the cytosol have enough time to enlarge. Two events occurring during thawing can reduce cell survival: recrystallization and osmotic shock. Recrystallization means that water shifts back into the cell and solidifies around the small ice crystals already formed, increasing their size. The likelihood of recrystallization depends both on cooling and thawing rates. It is possible to avoid this event by careful dehydration and rapid thawing of the cell. Osmotic shock may occur if, after thawing, the cryoprotectant penetrating the cell during cooling cannot diffuse out quickly enough to prevent the influx of water and swelling of the cell. A careful review of the literature provides conflicting information about the most suitable and less harmful methods to preserve cell integrity.

FREEZING TECHNOLOGY

Several freezing and thawing methods for oocytes, zygotes and pre-embryos are available: (1) slow/rapid; (2) ultra-rapid; and (3) vitrification. Cryoprotectants to prevent and reduce cellular damage of the biological specimens are: glycerol, dimethyl sulfoxide (DMSO), 1,2-propanediol (PROH), polyvinylpyrrolidone (PVP) and glucose.

A slow-freezing/rapid-thawing protocol with PROH and sucrose as cryoprotectants is by far the most commonly used. It results in rates of embryo survival that vary from 70 to 80% and pregnancy rates from 15 to 20%. The freezing and thawing solutions are prepared in phosphate buffered saline (PBS) sterilized with 0.22p filters and kept at a temperature of +4°C. They are used at room temperature within 48 h (freezing solutions: 0.2 mol/l sucrose, 0.1 mol/l PROH; thawing solutions: 0.2 mol/l sucrose, 0.5–1.0 mol/l PROH). For feasibility, freezing and thawing are carried out on plates with four tanks which facilitate the passage in the various tanks containing solutions with different concentrations of cryoprotectant. In every tank, a final volume of 0.5 ml facilitates the moving and the loading of the specimens in the straws.

Protein supplementation to the freezing and thawing solutions protects and facilitates specimens during moving and loading. Commonly, the patient's anti-complementary and sterilized serum, synthetic human serum or human albumin are used. The concentrations can vary between 6% and 20%. The freezing devices used are ampoules and straws. In the case of oocytes and pre-embryos, one or more specimens are loaded per straw. With a syringe and an adaptor the straw is loaded: the liquid containing the specimen is isolated with air zones, to keep it away from the extremities of the straw and reduce the risk of escape. It is also important to place all the specimens in the central part of the straws to keep them in the same position in the freezing chamber to allow them to crystallize simultaneously. Different straw sizes are commercially available (0.15–0.5 ml). The straws can be sealed with a welder or with a colored cork. Identification can be carried out by placing a sort of flag on the straw, or through the cork used to close it. On either, the details are written with permanent ink. Three or four straws are placed inside every colored goblet. On the latter, a small band is placed to write the name of the patient and the number of embryos and to facilitate the identification of the goblet in order to minimize exposure to room temperature. The number and quality of the cryo-preserved samples, the color of the goblet, the number of the canister or of the bank in which they are placed, a copy of the graphic of the temperature curve and all patient data (full name, age, address, telephone number, technique used, number of embryos transferred on natural cycle and other details) must be registered on files. An accurate storage method is crucial for the success of the program.

In order to carry out the slow-freezing methods, it is necessary to have a programable freezing unit consisting of (1) an active pump that pressurizes the liquid nitrogen; (2) an active pump for the automatic filling of the containers with liquid nitrogen; (3) cryogenic containers (tanks or containers for specimen storage); and (4) the possibility of a weekly supply of liquid nitrogen. To maximize survival after storage, the freezing procedure and reproducibility have to be controlled. Computer-assisted freezing units reduce the phenomenon of latent heat with the consequent formation of ice crystals inside the cytoplasm. This phenomenon occurs in all biological materials at temperatures between −7 and −12°C and results in a temperature increase of about 20°C when the transition from the semicolloidal state (from −1°C to −7 or −8°C) to the solid state occurs. Through programed freezing, the liquid nitrogen freezing unit creates a drop in temperature inside the biological structures which is opposing the increase in heat, preventing ice crystal formation.

The device consists of an electronic switchboard and of a freezing chamber containing a heating element and a spout through which expanding liquid nitrogen flows when the valve placed at the base of the same chamber is opened. The cycles of opening and closing the valve and the heating cycles are regulated by the electronic switchboard according to the programed freezing curve. The switchboard uses two PRT (platinum resistance to temperature) probes, one placed at the base of the freezing chamber and the other one inserted in the biological specimen, which continuously monitor the temperatures of both. Alternating and/or overlaying the cycles determines the programed temperature decrease in order to reach the cryogenic temperatures (from −130 to −180°C), suitable for specimen storage inside the appropriate cryobiological containers.

The freezing chambers can work both horizontally and vertically. They can be loaded with both straws and ampoules. Seeding (the induction of a nucleus of crystallization, which occurs normally at −7/−8°C) can be either automatic or manual. The microprocessor control modulus determines flexibility, accuracy and reproducibility of the system. The system offers protection in case of blackout through an internal energy supply. During short blackouts (less than one minute), the system works normally without any effects on specimen survival. For longer blackouts, the system stops until electricity comes back and restarts from that point onwards. It also supplies a graphic for the control of the program, of specimens and of chamber temperatures, with the date and the time. At least a weekly supply of liquid nitrogen has to be guaranteed. The possibly automatic filling and refilling of the containers has to be carried out at need during periods of activity and about every 10 days during periods of inactivity. During periods of activity, about 10 liters of liquid nitrogen are used, while a sealed container in an air-conditioned room consumes about 300 ml of liquid nitrogen per day.

Liquid nitrogen is a very dangerous substance and has to be handled with care. In the liquid form, it can cause severe burns and, in the gaseous form, serious breathing problems. For safety and health reasons, it is absolutely necessary to place the equipment and the cryogenic containers in a separate room, with controlled temperature and a ventilation system in cases of emergency. The workers have to wear adequate gloves and eye protection.

HUMAN EMBRYO FREEZING

Human embryos can be successfully cryopreserved by protocols using PROH, DMSO or glycerol. PROH seems appropriate for zygotes or early-cleaved embryos, DMSO for cleaved embryo stages and glycerol for blastocysts.

The protocol using PROH (1.5 mol/l) and sucrose (0.1 mol/l) is quite successful when embryos are slow cooled (−0.3°C/min) to −30°C and is especially suitable for pronucleated and two- or three-day-old embryos. Optimal survival rates are achieved only when sucrose is combined with PROH. PROH alone reduces survival rates to only 10% of totally intact embryos as compared to 44% with the PROH–sucrose protocol.

DMSO was the first cryoprotective agent used in humans, mostly in cleaved embryo stages. It still remains difficult to determine which method produces the best results. Glycerol is preferably used for the advanced stages of preimplantation development. New cryoprotectants and molecules such as polymers or antifreeze proteins have barely been attempted in humans.

The ultra-rapid freezing method is used for human embryo freezing in IVF programs only in exceptional cases. The same is true for vitrification procedures, with poor results for multicellular embryos. The slow-freezing/fast-thawing technique using PROH and sucrose as cryoprotectants seems the best to achieve an acceptable embryo survival rate. However, it is important to keep in mind that the freezing/thawing procedure can damage blastomeres. A choice must be made and only embryos with two or more cells and low fragmentation should be cryopreserved.

Appropriate embryo selection and correct freezing and thawing procedures allow obtaining an acceptable embryo survival and pregnancy rate. With regard to the duration of embryo storage, it is noteworthy that a pregnancy in a 44-year-old woman following the transfer of embryos which were kept frozen for 7.5 years was recently reported[4]. The efficiency of a freezing program is evaluated on the morphologic integrity of the embryo at thawing, and on its ability to further cleave *in vitro* and *in vivo*. Embryos can be considered 'surviving' when they keep at least half of their initial blastomeres intact after thawing and dilution of the cryoprotectants ('survival index' = 50%). On the other hand the survival rate is also expressed as the percentage of 'surviving' embryos among all frozen/thawed embryos and this is usually at least 65%. Surviving zygotes appear intact after thawing, with a clear cytoplasm and no zona pellucida breaches, and are able to cleave *in vitro* during the next 24 h of culture. Blastocyst survival is more difficult to evaluate.

High survival rates as well as high pregnancy rates are documented with frozen–thawed embryos[5–8]. A day 2/four-cell or day 3/eight-cell embryo is ideal for freezing. The optimal thawed embryo to transfer is a totally preserved embryo with 100% of intact blastomeres. On average thawed multicellular embryos survive in 70% of cases and about 90% of patients have a transfer. The mean birth rate is 10–15% all over the world.

The transfer of electively cryopreserved embryos in patients at risk of developing severe OHSS allows to obtain pregnancy and delivery rates indistinguishable from those obtained with fresh embryos. The transfer of cryopreserved embryos can be performed during a natural cycle without any pharmacologic support, during an artificial cycle employing gonadotropin-releasing hormone (GnRH) analog desensitization and micronized estradiol or during a semi-artificial cycle with micronized estradiol only[8]. A progesterone supplementation is given in both pharmacologic protocols.

HUMAN OOCYTE FREEZING

The main factors that seem to influence the outcome of cryopreservation regard the oocyte and the technique. Oocyte size is a critical parameter in the freezing process and it may influence the likelihood of intracellular ice formation and the overall survival rate. Male gametes in human species are 180 times smaller than female ones due to a lower amount of cytosol, and their survival rate is considerably higher. Moreover, mouse oocytes whose size is smaller than those of humans and rabbits show a higher survival rate[9].

Cryopreserving aged oocytes results in a considerable decrease in fertilization rate and an increase in abnormal fertilization and polyploidy[10]. An alternative approach to mature gamete storage is represented by the cryopreservation of oocytes at prophase I, when meiosis is arrested at diplotene, and chromosomes are within the membrane-bound nucleus. Following the first disappointing results, Toth and co-workers[11] reported interesting outcomes: prophase I oocytes, collected from

thin slices of ovarian tissue, are able to survive cryopreservation and mature to metaphase II after thawing.

The efficacy of cryoprotectants depends on the temperature at which these compounds are added to the freezing medium. Pickering and co-workers[12] demonstrated that human oocytes exposed to DMSO at 37°C lose their potential to fertilize. The authors suggest that the addition of DMSO to the medium must be performed at a temperature lower than 10°C, to avoid impairment of fertilization rate. The optimal concentration of the cryoprotectant depends on cell and species type[13].

In 1988 Sathananthan and co-workers[14] demonstrated that the duration of exposure to DMSO affects the damage being done to the meiotic spindle: while after 10–20 min at 1.5 mol/l the spindle maintains its normal morphology, 60 min are sufficient to induce severe alterations, not reversible in the majority of oocytes. Van der Elst and co-workers[15] reported that exposure of the oocyte to PROH (1.5 mol/l at 0°C) for a short time (12 min) is harmless. The removal from the cytoplasm of the permeating cryoprotectant with a stepwise dilution is an important step in the freezing process[12]. In fact, if after thawing oocytes are placed directly in a medium without the cryoprotectant, they could swell and burst because of the osmotic effect exerted by permeating agents. At this step, the use of non-permeating molecules, such as sucrose, rises the osmotic pressure of the external medium, opposing the inflow of water to the cytosol and preventing oocyte burst.

Freezing/thawing rate affects the diffusion of the water through the cell membrane. The optimal rate of thawing depends on the freezing procedure because this step conditions the amount of ice inside the oocyte. Generally, if slow cooling is arrested at temperatures of −30°C to −40°C, some water remains in the cell; thus, thawing must be rapid to avoid the development of large ice crystals in the cytoplasm. Instead, if dehydration continues down to −80°C, thawing must be slower to guarantee a suitable rehydration: this allows the cell to restore its initial volume gradually. Thawing is actually the limiting step of the whole process.

Several protocols for oocyte cryopreservation have been used, based on different rates of freezing and thawing. Oocyte storage is often performed with a slow freeze–rapid thaw procedure. Chen[16] achieved the first worldwide pregnancy with this protocol. Siebzehnruebl and co-workers[17] adopted the same strategy. Although uncommon and rarely reported in the literature, the slow-freezing/slow-thawing method permitted to obtain the second reported pregnancy with a frozen oocyte[18]. The authors used 1.5 mol/l of DMSO as cryoprotectant and oocyte thawing was performed at room temperature.

Ultra-rapid freezing avoids the formation of ice crystals and induces a glassy, amorphous medium using high concentrations of cryoprotectants. Trounson[19] first applied this strategy to human oocyte cryopreservation by direct immersion of ova in liquid nitrogen (ultra-rapid freezing). Rapid thawing was performed at 37°C in a water bath. Nine of 18 mature human ova thus treated survived to thawing, but all of them degenerated in culture.

Vitrification is a process in which a highly concentrated solution of cryoprotectants solidifies during freezing without the formation of ice crystals, in a super-cooled, highly viscose fluid. A high cooling rate (nearly −1500°C/min) and high concentrations of cryoprotectants such as DMSO, acetamide, propylene glycol and polyethylene glycol are required for vitrification. The theoretic principles of vitrification were outlined by Rall and Fahy[20] in 1985.

However, results are discordant and the toxicity of the cryoprotectants is confirmed by experimental studies, which have reported acceptable survival and fertilization rates but low cleavage rates[3,19]. The cleavage block may be related to the irreversible damage induced in the cytoskeleton by a combination of the cooling and the vitrification solution. The intact oocyte appears shiny, without cell disruption. The damaged oocyte presents zona fractures, dark, contracted and pyknotic cytoplasm. Several cell structures can be damaged during the whole process of freezing/thawing and by the cryoprotectants. These damages can result in morphologic and functional abnormalities.

Concern has been expressed about the possibility of inducing disarray among the microtubules of the meiotic spindle, since the oocytes cryopreserved at metaphase II present the 23 dichromatidic chromosomes strictly bound to the microtubules. This structure is extremely sensitive to temperature. However Gook and co-workers[10,21] reported normal karyotypes and the absence of stray chromosomes in cryopreserved oocytes. The authors suggest that the meiotic spindle of human oocytes seems to be more resistant compared to murine gametes, and they do not confirm in humans the results that were previously obtained in mice.

Another possible effect of cryopreservation is the untimely resumption of meiosis, with parthenogenetic activation (PA). Gook and co-workers[22] demonstrated that sole exposure to PROH does not induce parthenogenesis, while freezing may cause it in 27–29% of the cases. Another subcellular structure usually involved in cryopreservation-induced damage is the cytoskeleton, whose structures (microtubules, actin microfilaments and intermediate filaments) are extremely sensitive to low temperatures.

Hunter and co-workers[23] suggest that the arrest of development in embryos derived from cryopreserved oocytes is related to subtle perturbations in cyto-architectonics: i.e., cytoskeleton abnormalities may prevent the breakdown of the pronuclear membrane and the joining of maternal and paternal chromosomes; alternatively, they can prevent correct zygote cleavage or induce the formation of several endocytotic vacuoles in the cytoplasm after pronuclei formation. During fertilization the exocytosis of the cortical granules (zonareaction) situated in the periphery of the oocyte prevents the entry of more than one spermatozoon (polyspermy block). A study of human and murine oocytes performed by electron microscopy detected reduction and morphologic abnormalities of the granules after thawing[24].

On the other hand in 1993 Gook and co-workers[21] documented plenty of cortical granules in all the cryopreserved oocytes, postulating that the freezing process does not affect the release of these structures. They also demonstrated in a later study that the high percentages of abnormal fertilizations are related to the protracted *in vitro* culturing of the oocytes rather than to the cooling procedure[10]. *In vitro* culture and aging of the oocytes could lead to damage of the cortical granule. Several researchers pointed out the risk of damages to the zona pellucida[25,26] in oocyte cryopreservation. Zona damage is believed to be caused either by the formation of fracture planes in the ice or by the ice crystals able to trap or pierce the cell. The addition of polymers such as dextran and PVP reduces the size of the ice crystals and exerts cryoprotective properties[25].

A survival rate of cryopreserved oocytes varies considerably from less than 25% to more than 80%. Chen[27] reports one of the best survival rates in the literature (76%). The author froze only mature (MII) oocytes of very good quality. A lower survival rate (25%) is reported by Al-Hasani and co-workers[9], who, however, used only excess oocytes of IVF patients, which were usually of poor quality and immature. The author compares DMSO and PROH, reaching a survival rate of 28% with the former (40/144) and 32% with the latter (12/38). Low survival rates were also reported by Kazem and co-workers[26] (34.4%) and by Tucker and co-workers[28]. Both teams used PROH. Gook and co-workers[22,29] have reported a survival rate varying from 48 to 95%. They used PROH and obtained better results with denuded oocytes (69% versus 48%). In addition, the authors found that aged oocytes survived better than fresh oocytes but had a low fertilization rate.

A considerable improvement of the mean survival rate (70–80%) has recently been achieved in our group by increasing the sucrose concentration and cryoprotectant exposure[30]. The fertilization rate of cryopreserved oocytes with IVF is extremely variable ranging from 13 to 71%[26,27]. However, in most studies, the variability is between 30 and 55%, being on average lower than the fertilization rate obtained with fresh oocytes. Abnormal fertilization, usually polyploidy, ranges from 5–10.8% to 15.3%[9,22,26]. A reduced fertilization rate and a sometimes increased abnormal fertilization rate have been linked to possible damage of the zona pellucida and of the cortical granules which would prevent a correct interaction with spermatozoa.

Intracytoplasmic sperm injection (ICSI) has recently been proposed as a solution to those problems. With this technique, Gook and co-workers[29] obtained 50% of normal fertilization together with 21% of abnormal fertilization. Compared to traditional IVF, embryos obtained with ICSI showed a better embryo development. Similar experiences of fertilization rates (43.2% of normal fertilization[26], and 65% fertilization rate and three pregnancies[28], all ending in abortion) have been documented[26,28]. In our experience, ICSI-induced damage of thawed oocytes was 7%, by far lower than that reported by others (32%[26] and 26.3%[29]). Our normal fertilization rate is 64.3%, similar to that obtained by Tucker and co-workers[28]. Our abnormal fertilization rate (7.2%) resembles that found in IVF and ICSI of fresh oocytes. Most embryos are of good quality and cleave regularly. In 1997 we reported the birth of a healthy female from ICSI of frozen oocytes[31]. Subsequently our team published pregnancies from frozen oocytes and epididymal[32], testicular[33] and frozen sperm[34]. In more than 200 oocyte-thawing cycles, the pregnancy rate of our oocyte cryopreservation program approaches that obtained from frozen embryos. Oocyte storage makes assisted reproductive technologies (ART) more flexible and devoid of ethical problems.

Freezing oocytes in patients at risk of developing severe OHSS is a valid alternative to storing a large number of embryos often abandoned at a later stage. Storing eggs in oncologic patients undergoing chemotherapy or radiotherapy is probably the only present, reliable choice to save fertility.

References

1. Trounson A, Mohr L. Human pregnancy following cryopreservation, thawing and transfer of an eight-cell embryo. *Nature* 1983;305:707–9
2. Mazur P. Limits to life at low temperatures and at reduced water contents and water activities. *Orig Life* 1980;10:137–59
3. Friedler S, Giudice L, Lamb E. Cryopreservation of embryos and ova. *Fertil Steril* 1988;49:743–64
4. Ben-Ozer S, Vermesh M. Full term delivery following cryopreservation of human embryos for 7.5 years. *Hum Reprod* 1999;14:1650–2
5. Macas E, Imthurn B, Borsos M, *et al.* Impairment of the developmental potential of frozen zygotes obtained after intracytoplasmic sperm injection. *Fertil Steril* 1998;69:630–5
6. Hoover L, Baker A, Check JH, *et al.* Clinical outcome of cryopreserved pronuclear stage embryos resulting from intracytoplasmic sperm injection. *Fertil Steril* 1997;67:621–4
7. Ferraretti AP, Gianaroli L, Magli C, *et al.* Elective cryopreservation of all pronucleate embryos in women at risk of ovarian hyperstimulation syndrome: efficiency and safety. *Hum Reprod* 1999;14:1457–60
8. Simon A, Hurwitz A, Pharhat M, *et al.* A flexible protocol for artificial preparation of the endometrium without prior gonadotropin-releasing hormone agonist suppression in

women with functioning ovaries undergoing frozen–thawed embryo transfer cycles. *Fertil Steril* 1999;71:609–13

9. Al Hasani S, Diedrich K, van der Ven, *et al*. Cryopreservation of human oocytes. *Hum Reprod* 1987;2:695–700

10. Gook D, Osborn S, Bourne H, Johnston W. Fertilization of human oocytes following cryopreservation; normal karyotypes and absence of stray chromosomes. *Hum Reprod* 1994;9: 684–91

11. Toth TL, Lazendorf SE, Sandow BA, *et al*. Cryopreservation of human prophase I oocytes collected from unstimulated follicles. *Fertil Steril* 1994;61:1077–82

12. Pickering S, Braude P, Johnson M. Cryoprotection of human oocyte: inappropriate exposure to DMSO reduces fertilization rates. *Hum Reprod* 1991;6:142–3

13. Vincent C, Pruliere G, Pajot-Augy E, *et al*. Effects of cryoprotectants on actin filaments during cryopreservation of one-cell rabbit embryos. *Cryobiology* 1990;27:9–23

14. Sathananthan AH, Trounson A, Freeman L, Brady T. The effects of cooling human oocytes. *Hum Reprod* 1988;8:968–77

15. Van Der Elst J, Van den Abbeel E, Nerinckx S, Van Steirteghem A. Parthenogenetic activation pattern and microtubular organization of the mouse oocyte after exposure to 1,2-propanediol. *Cryobiology* 1992;29:549–62

16. Chen C. Pregnancy after human oocyte cryopreservation. *Lancet* 1986;i:884–6

17. Siebzehnruebl ER, Todorow S, Van Uem J, *et al*. Cryopreservation of human and rabbit oocytes and one-cell embryos: a comparison of DMSO and propanediol. *Hum Reprod* 1989;4:312–17

18. Van Uem JFHM, Siebzehnrubl ER, Schun B, *et al*. Birth after cryopreservation of unfertilized oocytes. *Lancet* 1987;i: 752–3

19. Trounson A. Freezing human eggs and embryos. *Fertil Steril* 1986;46:1–12

20. Rall WF, Fahy GM. Ice-free cryopreservation of mouse embryos at −196°C by vitrification. *Nature* 1985;313:573–5

21. Gook D, Osborn S, Johnston W. Cryopreservation of mouse and human oocytes using 1,2 propanediol and the configuration of the meiotic spindle. *Hum Reprod* 1993;8:1101–9

22. Gook D, Osborn S, Johnston W. Parthenogenetic activation of human oocytes following cryopreservation using 1,2-propanediol. *Hum Reprod* 1995;10:654–8

23. Hunter JE, Bernard A, Fuller B, *et al*. Fertilization and development of the human oocyte following exposure to cryoprotectants, low temperatures and cryopreservation: a comparison of two techniques. *Hum Reprod* 1991;6: 1460–5

24. Al Hasani S, Diedrich K. Oocyte storage. In Grudzinskas JG, Yovich JL, eds. *Gametes – The Oocyte*. Cambridge, UK: Cambridge University Press, 1995:376–95

25. Dumoulin JCM, Janssen JMB, Pieters HEC, *et al*. The protective effects of polymers in the cryopreservation of human and mouse zonae pellucidae and embryos. *Fertil Steril* 1994;62:793–8

26. Kazem R, Thompson LA, Srikantharajah A, *et al*. Cryopreservation of human oocytes and fertilization by two techniques: *in-vitro* fertilization and intracytoplasmic sperm injection. *Hum Reprod* 1995;10:2650–4

27. Chen C. Pregnancies after human oocyte cryopreservation. *Ann N Y Acad Sci* 1987;541:541–9

28. Tucker M, Wright G, Morton P, *et al*. Preliminary experience with human oocyte cryopreservation using 1,2 propanediol and sucrose. *Hum Reprod* 1996;11:1513–15

29. Gook D, Schiewe MC, Osborn S, *et al*. Intracytoplasmic sperm injection and embryo development of human oocytes cryopreserved using 1,2-propanediol. *Hum Reprod* 1995;10: 2637–41

30. Fabbri R, Porcu E, Marsella T, *et al*. Human oocyte cryopreservation: new perspectives regarding oocyte survival. *Hum Reprod* 2001;16:411–16

31. Porcu E, Fabbri R, Seracchioli R, *et al*. Birth of a healthy female after intracytoplasmic sperm injection of cryopreserved human oocytes. *Fertil Steril* 1997;4:724–6

32. Porcu E, Fabbri R, Petracchi S, *et al*. Ongoing pregnancy after intracytoplasmic injecton of testicular spermatozoa into cryopreserved human oocytes. *Am J Obstet Gynecol* 1999;180: 1044–5

33. Porcu E, Fabbri R, Petracchi S, *et al*. Ongoing pregnancy after ICSI of epididymal spermatozoa into cryopreserved human oocytes. *J Assisted Reprod Genetics* 1999;16:283–5

34. Porcu E, Fabbri R, Damiano G, *et al*. Clinical experience and applications of oocyte cryopreservation. *Mol Cell Endocrinol* 2000;169:33–7

Cryopreservation of testicular tissue 16

Outi Hovatta

WHY CRYOPRESERVATION OF TESTICULAR TISSUE?

In humans, cryopreservation of testicular tissue is needed for the preservation of fertility in prepubertal boys who undergo chemotherapy or radiotherapy and often lose their sperm production[1]. In postpubertal boys and men, cryopreservation of semen is also clinically well established and widely used. There are, however, situations after puberty when no ejaculated spermatozoa can be obtained. Cryopreservation of testicular tissue or spermatozoa extracted from testicular tissue is indicated in these cases[2].

Cryopreservation of testicular tissue or spermatozoa is always indicated when a testicular biopsy is carried out as a result of azoospermia[3]. Repeated testicular biopsies can be avoided, because intracytoplasmic sperm injection (ICSI) is feasible even in cases where only a few spermatozoa are available.

In animal science, large amounts of new information can be obtained by trans-species transplantation of cryopreserved testicular cells and tissues[4]. These techniques might offer opportunities for creating transgenic animals[5]. Cryopreserving testicular sperm has already helped in achieving posthumous offspring of rare animals[6].

TESTICULAR TISSUE OF ADULT MEN

For azoospermic men, excellent pregnancy results can be obtained when their testicular sperm, either fresh or cryopreserved, is used for ICSI[3,7–9].

Testicular tissue from azoospermic men has been cryopreserved, using glycerol as a cryoprotectant, as cell suspension[7,10–14] or as pieces of tissue[3,15–19]. Healthy children have been born as a result of ICSI carried out using spermatozoa after both freezing methods.

Although the main indication of testicular tissue freezing in adult men is obstructive or non-obstructive azoospermia, there are specific situations in which cryopreservation of testicular tissue or sperm is a method for fertility preservation. Men with testicular cancer[20,21] or Klinefelter's syndrome[22–24] would particularly benefit from it.

Using a relatively easy needle-biopsy technique gives enough tissue for histology, ICSI and cryopreservation at the same time[3,25]. Although it is a relatively non-invasive technique and does not appear to cause permanent damage to the testis[26], repeated biopsies should be avoided as much as possible. At least after open testicular biopsy, testicular function may be impaired[27].

PREPUBERTAL BOYS

For prepubertal boys who do not yet have sperm production and who are going to lose the spermatogenetic cells during chemotherapy or radiotherapy, cryopreservation of testicular tissue is the only option for fertility preservation[2,28–30].

Survival rates after chemotherapy and radiotherapy have improved and there is now a relatively large number of young boys and adolescents who would benefit from cryopreservation of testicular tissue. Approximately 70% of these boys are cured of the disease, but many have permanent gonadal damage, depending on the type of treatment[1,30–34]. The procarbazine-containing regimens in particular tend to abolish spermatogenesis, but permanent damage is also seen after preparation for stem cell transplantation[1]. It is extremely important for the quality of life of these survivors to maintain their fertility and normal sexual life.

At the start of treatment it is not known how much chemotherapy will be needed. In leukemia especially, it is not known at the time when treatment starts which patients will subsequently require stem cell transplantation. This is why cryopreservation of spermatozoa is recommended before any chemotherapy. Similarly, cryopreservation of testicular tissue might be indicated for all prepubertal boys receiving chemotherapy. This is important in all types of malignancies.

Ethical aspects are extremely important when tissue from young boys is being frozen[35]. Cryopreservation of testicular cells is applied in clinical practice, but there is no guarantee of restoring fertility, and that has to be explained to the family and to the adolescent patients. A young child cannot give his consent for the procedure and the procedure has to be carried out after parental consent.

Testicular biopsy is an invasive procedure for a young boy, and there have to be good reasons for it. The biopsy can be carried out under general anesthesia at the same time as insertion of the central line for chemotherapy. There is a risk of bleeding during the taking of the biopsy, especially in hematologic disorders.

As regards boys in early puberty, cryopreservation of testicular tissue might already be a beneficial option. Spermatogenesis is already occurring to some extent in the testes of boys at very early stages of pubertal development[36–38]. These boys also understand the procedure and give their consent. They often have mature or almost mature spermatids in the testicular tissue and ICSI using these is already feasible.

CLINICAL USE OF CRYOPRESERVED TESTICULAR TISSUE

If there are spermatozoa or spermatids in the tissue, these can, of course, simply be used for ICSI. Early spermatids can be matured *in vitro* before ICSI[39].

Testicular tissue from prepubertal boys is still a challenge. In animal models, reimplantation of cryopreserved testicular cells into the testes of the donor has already been successful. Brinster and co-workers succeeded in initiating spermatogenesis in infertile mice by transplanting donor testicular cells into the testes of mice previously made infertile by an intraperitoneal injection of busulphan[4,40]. Donor testicular cells taken from prepubertal and adult mice and frozen and stored in liquid nitrogen were also capable of generating spermatogenesis in recipient seminiferous tubules, although no offspring were reported. Xenotransplantation in other species also proved successful[41,42]. Testicular cells from hamsters and mice have been used in establishing spermatogenesis in recipient animals, also after cryopreservation[43–45]. Despite the occurrence of spermatogenesis, no live offspring have been obtained in these experiments.

Non-frozen testicular cells of rat, bovine, monkey and human testes have been successfully injected through the rete testis under ultrasound control[46], showing that the injection is technically feasible. Brook and co-workers succeeded in filling about 50% of the tubules by injecting blue dye into human testes which had been removed because of prostatic cancer[47].

In Manchester, testicular tissue from boys suffering from Hodgkin's disease or non-Hodgkin's lymphoma has been cryopreserved, and five of them have received cell suspension back to their testes[34].

In the era of ICSI, also cryopreserved pieces of testicular tissue might also be reimplanted, and spermatozoa for ICSI could subsequently be aspirated from this tissue. This technique might also make it possible to restore the function of Leydig cells, which is also not completely normal after the heavy treatments[31,48].

In hematologic malignancies, there is a risk of replacing malignant cells together with the cryopreserved cell suspension or tissue[34,49,50]. In other malignancies the risk is lower. In the future, it will be important to identify the presence of even small numbers of malignant cells in the suspensions before reimplantation.

Spermatogenesis *in vitro*, starting from spermatogonia, would be an excellent option for boys with hematologic malignancies which carry a high risk of being transmitted back. Despite several promising attempts in experimental animals, spermatogenesis *in vitro* starting from premeiotic cells has not been successful[51–54].

Theoretically, xenotransplantation to mature spermatozoa from premeiotic cells could be possible. Cryopreserved human testicular tissue might be transplanted to immunodeficient mice or rats, and the resulting spermatids or spermatozoa could be used for ICSI in cases of malignancies with a high risk of relapse.

However, xenotransplantation bears risks, such as infections. An animal model might be used to test whether the cryopreserved tissue contains malignant cells.

FREEZING TECHNIQUES

When testicular tissue from infertile men is cryopreserved for possible ICSI, the purpose is to cryopreserve mature spermatids and spermatozoa. Freezing spermatozoa, which are very small cells, either in semen or as washed preparations, by using glycerol as a cryoprotectant has been a widely used technique from the early 1950s[55].

In non-obstructive azoospermia, the numbers of testicular spermatozoa which can be obtained are often very small. Techniques to cryopreserve these few spermatozoa have been developed[56–58] using storage in cell-free zona pellucida.

The optimal freezing techniques for pieces of testicular tissue, and for testicular cell suspensions or isolated testicular spermatozoa, are probably different.

Glycerol does not penetrate tissues especially well, but is likely to work for spermatids and spermatozoa in suspensions of testicular cells. Crabbe and co-workers compared the motility of spermatozoa after cryopreservation of cell suspensions prepared from human testicular biopsies, to that of spermatozoa cryopreserved within a piece of testicular tissue when glycerol was used as a cryoprotectant[59]. The spermatozoa frozen as a testicular cell suspension had better motility after thawing. Good morphology was, however, maintained within testicular tissue after freezing using glycerol[16]. In most of the published studies, glycerol has been used.

We have also applied a protocol similar to that used in the freezing of human ovarian tissue[60] with propanediol–sucrose as a cryoprotectant in a serum-containing medium (human serum), for testicular tissue. It appeared to give good tissue morphology after thawing.

Whole tissue can be frozen within a few minutes after a diagnostic biopsy, while another part can be used in a wet preparation to see if there are spermatozoa in the biopsy specimen. On the other hand, a cell suspension prepared from the biopsy specimen, containing also the spermatozoa, is relatively easy to freeze using programs similar to those used in cryopreservation of semen.

Freezing whole tissue would be optimally carried out using cryoprotectants such as propanediol, dimethyl sulfoxide (DMSO) or ethylene glycol, which penetrate the tissue more efficiently than glycerol that is usually used in the cryopreservation of semen.

Tissue-freezing programs are needed when tissue from prepubertal boys is frozen, because the purpose is to cryopreserve spermatogonia, which have different properties from those of spermatids or spermatozoa, and the cryoprotectant has to penetrate the whole tissue.

With regard to suspensions of testicular cells, Avarbock and colleagues used DMSO and rapid freezing of mouse testicular cell suspension, and they obtained 30% cryosurvival[61]. The cells were cooled rapidly to −70°C, maintained at that temperature for 12 h, and then transferred to liquid nitrogen. They did not carry out comparisons between different methods. Starting the process at 0°C might have improved the results with DMSO, as this cryoprotectant is less toxic at lower temperatures. Ogawa and colleagues cryopreserved hamster testicular cells using the same method[43]. They showed 43% viability after thawing, as revealed by the trypan blue exclusion test. These cells were integrated to the seminiferous tubules of recipient mice similarly to non-frozen cells, but no mature normal sperm could be seen.

Brook and co-workers reported 87% survival in mouse and 66% in human testicular cell suspensions after enzymatic dispersal of testicular tissue[47]. They used a trypan blue exclusion test as the marker of survival. The mean viability of the cells after slow programed cooling using four different cryoprotectants (glycerol, DMSO, 1,2-propanediol and ethylene glycol, all 1.5 mol/l, in Leibovitz medium containing 4% fetal calf serum) was 52–58%.

We have used programed freezing using propanediol and sucrose as the cryoprotectant. It proved to give the best survival rate for testicular cell suspension, when compared with glycerol using slow or rapid freezing, or rapid freezing using DMSO (unpublished results). Of the mouse testicular cells, 65% and of the human cells, 60%, survived. The same propandiol–sucrose freezing method was also successful for testicular tissue. The technique is the same as that we have used in the cryopreservation of human ovarian tissue[15]. Also when glycerol was used, the programed freezing proved better than the rapid technique.

CONCLUSIONS

Testicular spermatozoa or spermatids can be cryopreserved for ICSI whenever a testicular biopsy is taken. In cases of azoospermia it is important to carry out a diagnostic needle biopsy before the ovarian stimulation of the female partner is started. If spermatozoa are found, they can be frozen to avoid repeated biopsies. There are specific freezing techniques for very small numbers of spermatozoa. Testicular tissue can be cryopreserved from prepubertal boys undergoing chemotherapy or radio-therapy. Transplantation back to the testes can be carried out if the risk of transmitting back the original malignancy is considered low. Especially in hematologic malignancies, *in vitro* maturation of spermatozoa from the tissue would be an option, but this technique is not yet feasible. Testicular spermatozoa or cell suspensions can be cryopreserved using techniques similar to those used routinely in semen cryopreservation. In the cryopreservation of testicular tissue, slow programed freezing is probably better.

References

1. Howell SJ, Shalet SM. Testicular function following chemotherapy. *Hum Reprod Update* 2001;7:363–9
2. Hovatta O. Cryopreservation of testicular cells in cancer patients. *Hum Reprod Update* 2001;7:378–83
3. Tuuri T, Moilanen J, Kaukoranta S, *et al*. Testicular biopsy gun needle biopsy in collecting spermatozoa for intra-cytoplasmic injection, cryopreservation and histology. *Hum Reprod* 1999;14:1274–8
4. Brinster RL, Avarbock MR. Germline transmission of donor haplotype following spermatogonial transplantation. *Proc Natl Acad Sci USA* 1994;91:11303–7
5. Nagano M, Brinster CJ, Orwig KE, *et al*. Transgenic mice produced by retroviral transduction of male germ-line stem cells. *Proc Natl Acad Sci USA* 2001;98:13090–5
6. Zomborsky Z, Zubor T, Toth J, Horn P. Sperm collection from shot red deer stags (*Cervus elaphus*) and utilisation of sperm frozen and subsequently thawed. *Acta Vet Hung* 1999; 47:263–70
7. Gianaroli L, Magli MC, Selman HA, *et al*. Diagnostic testicular biopsy and cryopreservation of testicular tissue as an alternative to repeated surgical openings in the treatment of azoospermic men. *Hum Reprod* 1999;14:1034–8
8. Prins GS, Dolgina R, Studney P, *et al*. Quality of cryopreserved testicular sperm in patients with obstructive and non-obstructive azoospermia. *J Urol* 1999;161:1504–8
9. Palermo GD, Schlegel PN, Haripashad JJ, *et al*. Fertilization and pregnancy outcome with intracytoplasmic sperm injection for azoospermic men. *Hum Reprod* 1999;14:741–8
10. Gil-Salom M, Romero J, Minguez Y, *et al*. Pregnancies after intracytoplasmic sperm injection with cryopreserved testicular spermatozoa. *Hum Reprod* 1996;11:1309–13
11. Podsiadly BT, Woolcot RJ, Stanger JD, Stevenson K. Pregnancy resulting from intracytoplasmic injection of cryopreserved spermatozoa recovered from testicular biopsy. *Hum Reprod* 1996;11:1306–8
12. Friedler S, Strassburger D, Raziel A, *et al*. Intracytoplasmic injection of fresh and cryopreserved testicular spermatozoa in patients with nonobstructive azoospermia – a comparative study. *Fertil Steril* 1997;68:892–7
13. Khalifeh FA, Sarraf M, Dabit ST. Full-term delivery following intracytoplasmic sperm injection with spermatozoa extracted from frozen–thawed testicular tissue. *Hum Reprod* 1997;12: 87–8
14. Ben-Yosef D, Yogev L, Hauser R, *et al*. Testicular sperm retrieval and cryopreservation prior to initiating ovarian stimulation as the first line approach in patients with non-obstructive azoospermia. *Hum Reprod* 1999;14:1794–801
15. Hovatta O, Foudila T, Siegberg R, *et al*. Pregnancy resulting from intracytoplasmic injection of spermatozoa from a frozen–thawed testicular biopsy specimen. *Hum Reprod* 1996; 11:2472–3
16. Salzbrunn A, Benson DM, Holstein AF, Schulze W. A new concept for the extraction of testicular spermatozoa as a tool for assisted fertilization (ICSI). *Hum Reprod* 1996;11:752–5
17. Allan JJ, Cotman AS. A new method for freezing testicular biopsy sperm: three pregnancies with sperm extracted from cryopreserved sections of seminiferous tubule. *Fertil Steril* 1997;68:741–4
18. Oates RD, Mulhall J, Burgess C, *et al*. Fertilization and pregnancy using intentionally cryopreserved testicular tissue as the sperm source for intracytoplasmic injection in 10 men with non-obstructive azoospermia. *Hum Reprod* 1997;12: 734–9
19. Perraguin-Jayot S, Audebert A, Emperaire JC, Parneix I. Ongoing pregnancies after intracytoplasmic injection using cryopreserved testicular spermatozoa. *Hum Reprod* 1997;12: 2706–9
20. Baniel J, Sella A. Sperm extraction at orchiectomy for testis cancer. *Fertil Steril* 2001;75:260–2
21. Kohn FM, Schroeder-Printzen I, Weidner W, *et al*. Testicular sperm extraction in a patient with metachronous bilateral testicular cancer. *Hum Reprod* 2001;16:2343–6
22. Damani MN, Mittal R, Oates RD. Testicular tissue extraction in a young male with 47,XXY Klinefelter's syndrome: potential strategy for preservation of fertility. *Fertil Steril* 2001;76:1054–6
23. Friedler S, Raziel A, Strassburger M, *et al*. Outcome of ICSI using fresh and cryopreserved–thawed testicular spermatozoa

in patients with non-mosaic Klinefelter's syndrome. *Hum Reprod* 2001;16:2616–20

24. Rosenlund B, Hreinsson JH, Hovatta O. Birth of a healthy male after frozen-thawed blastocyst transfer following intracytoplasmic injection of frozen–thawed testicular spermatozoa from a man with non–mosaic Klinefelter's syndrome. *J Assist Reprod Genet* 2002;19:149–51

25. Rosenlund B, Kvist U, Ploen L, *et al*. Percutaneous cutting needle biopsies for histological assessment and sperm retrieval in men with azoospermia. *Hum Reprod* 2001;16:2154–9

26. Westlander G, Rosenlund B, Soderlund B, *et al*. Sperm retrieval, fertilization, and pregnancy outcome in repeated testicular sperm aspiration. *J Assist Reprod Genet* 2001;18:171–7

27. Schlegel PN, Su L-M. Physiological consequences of testicular sperm extraction. *Hum Reprod* 1997;12:1688–92

28. Picton H, Kim SS, Gosden RG. Cryopreservation of gonadal tissue and cells. *Br Med Bull* 2000;56:603–15

29. Hovatta O. Cryopreservation of testicular cells. *Mol Cell Endocrincol* 2000;169:113–15

30. Gosden RG. Trade-offs in cancer and reproduction. *Hum Reprod Update* 2001;7:360–2

31. Siimes MA, Dunkel L, Rautonen J. Risk factors for endocrine testicular dysfunction in adolescent and adult males who have survived malignancies in childhood. *J Cancer* 1992;5:28–31

32. Meirow D, Schenker JG. Cancer and male infertility. *Hum Reprod* 1995;10:2017–22

33. Howell SJ, Shalet SM. Gonadal damage from chemotherapy and radiotherapy. *Endocr Metab Clin N Am* 1998; 27:927–43

34. Radford JA, Shalet SM, Lieberman BA. Fertility after treatment for cancer *Br Med J* 1999;319:935–6

35. Bahadur G, Ralph D. Gonadal tissue cryopreservation in boys with paediatric cancers. *Hum Reprod* 1999;14:11–17

36. Muller J, Skakkebaek N. Quantification of germ cells and seminiferous tubules by stereological examination of testicles from 50 boys who suffered sudden death. *Int J Androl* 1983; 6:143–56

37. Janczewski Z, Bablok L. Semen characteristics in prepubertal boys. *Arch Androl* 1985;15:199–205

38. Nielsen CT, Skakkebaek N, Richardson DW, *et al*. Onset of release of spermatozoa (spermarche) in boys in relation to age, testicular growth, pubic hair and height. *J Clin Endocrinol Metab* 1986;62:532–5

39. Tesarik J, Mendoza C, Greco E. *In vitro* maturation of immature human male germ cells. *Mol Cell Endocrinol* 2000; 166:45–50

40. Brinster RL, Zimmermann JW. Spermatogenesis following male germ cell-transplantation *Proc Natl Acad Sci USA* 1994; 91:11298–302

41. Jiang F-X, Short RV. Male germ cell transplantation in rats: apparent synchronisation of spermatogenesis between host and donor seminiferous epithelia. *Int J Androl* 1995;18:326–30

42. Dobrinski I, Avarbock MR, Brinster RL. Transplantation of germ cells from rabbits and dogs into mouse testes. *Biol Reprod* 1999;61:1331–9

43. Ogawa T, Dobrinski I, Avarbock MR, Brinster RL. Xenogeneic spermatogenesis following transplantation of hamster germ cells to mouse testis. *Biol Reprod* 1999;60:515–21

44. Ogawa T, Dobrinski I, Avarbock MR, Brinster RL. Transplantation of male germ line stem cells restores fertility in infertile mice. *Nat Med* 2000;61:29–34

45. Russel LD, Griswold MD. Spermatogonial transplantation – an update for the millennium. *Mol Cell Endocrinol* 2000; 161:117–20

46. Schlatt S, Rosiepen G, Weinbauer GF, *et al*. Germ cell transfer into rat, bovine, monkey and human testes. *Hum Reprod* 1999;14:144–50

47. Brook PF, Radford JA, Shalet SM, *et al*. Isolation of germ cells from human testicular tissue for low temperature storage and autotransplantation. *Fertil Steril* 2001;75:269–74

48. Howell SJ, Radford JA, Ryder WD, Shalet SM. Testicular function after cytotoxic chemotherapy: evidence of Leydig cell insufficiency. *J Clin Oncol* 1999;17:1493–8

49. Söder O. Male germ cell transplantation. *Eur J Endocrinol* 1997;136:41–2

50. Jahnukainen K, Hou M, Pettersen C, *et al*. Transmission of leukemia of fresh and cryopreserved spermatogenic cells from leukemic donors to syngenic recipients in a rat model. *Cancer Res* 2001;61:706–10

51. Miura T, Yamauchi K, Takahashi H, Nagahama Y. Hormonal induction of all stages of spermatogenesis *in vitro* in male Japanese eel (*Anguilla japonica*). *Proc Natl Acad Sci USA* 1991;88:5774–8

52. Boitani C, Politi MG, Menna T. Spermatogonial cell proliferation in organ culture of immature rats. *Biol Reprod* 1993;48:761–7

53. Abe SI, Ji ZS. Initiation and stimulation of spermatogenesis *in vitro* by mammalian follicle-stimulating hormone in the japanese newt, *Cynops pyrrhogaster*. *Int J Dev Biol* 1994;38:201–8

54. Zhou B, Hutson JM. Human chorionic gonadotropin (hCG) fails to stimulate gonocyte differentiation in newborn mouse testes in organ culture. *J Urol* 1995;153:501–5

55. Royere D, Barthelemy C, Hamamah S, *et al*. Cryopreservation of spermatozoa. A review. *Hum Reprod Update* 1996;2:553–9

56. Cohen J, Garrisi GJ, Congedo-Ferrara TA, *et al*. Cryopreservation of single human spermatozoa. *Hum Reprod* 1997;12:994–1001

57. Liu J, Zheng XZ, Baramki TA, *et al*. Cryopreservation of small number of fresh human spermatozoa and testicular spermatozoa cultured *in vitro* for 3 days in an empty zona pellucida. *J Androl* 2000;21:409–11

58. Borini A, Sereni E, Bonu, Flamigni C. Freezing a few spermatozoa retrieved by TESA. *Mol Cell Endocrinol* 2000;169: 27–32

59. Crabbe E, Verheyn G, Tournaye H, Van Steirteghem A. Freezing of testicular tissue as a minced suspension preserves sperm quality better than whole-biopsy freezing when glycerol is used as cryoprotectant. *Int J Androl* 1999; 22:43–8

60. Hovatta O, Silye R, Krausz T, *et al*. Cryopreservation of human ovarian tissue using dimethylsulphoxide and propanediol–sucrose as cryoprotectants. *Hum Reprod* 1996; 11:1268–72

61. Avarbock MR, Brinster JB, Brinster RL. Reconstitution of spermatogenesis from frozen spermatogonial stem cells. *Nat Med* 1996;2:693–6

Cryopreservation of ovarian tissue   17

Raffaella Fabbri, Graziella Bracone, Cosetta Iannascoli and Stefano Venturoli

INTRODUCTION

Ovarian tissue cryopreservation is a new method of medically assisted reproduction[1] in which one of the essential goals is the preservation of ovarian function, and hence the fertility, of young women who will be undergoing anticancer treatment. There are several potential options for preserving fertility[2] in these patients: embryo cryopreservation, oocyte cryopreservation, chemoprotection and transposition of gonads and ovarian tissue cryopreservation.

The cryopreservation of embryos is a well-established technique with a relatively good efficiency. However, it has limited application in cancer patients because it requires *in vitro* fertilization (IVF), with its lengthy protocol. Before IVF, ovulation induction is required; it takes about 2 months, which could present an unacceptable delay[3] in cancer treatment. In addition, if ovarian stimulation is carried out, malignant growth of estrogen-sensitive tumors such as carcinoma of the breast may, at least in theory, be enhanced. IVF is not suitable for prepubertal girls and often unacceptable for single women who do not wish to use donor spermatozoa. In addition the storage of human frozen embryos raises many ethical and legal issues. In particular, if cancer treatment fails and the mother dies, 'orphan embryos' are created, the disposal of which raises moral and ethical dilemmas. In view of these problems a preferable alternative may be the storage of oocytes prior to insemination.

Until now cryopreservation of oocytes has had a very limited success. Mature human oocytes are extremely sensitive to temperature change and have little capacity for repairing cellular damage[2]. Cryoprotectant exposure and/or ice crystal formation during the freeze/thaw procedure can lead to depolarization of the meiotic spindle, disruption of chromatid separation during meiosis and potential induction of aneuploidy. Furthermore, zona

hardening can occur as a result of the premature release of cortical granules from ooplasm.

However, human oocyte cryopreservation represents an attractive option to the range of infertility treatments available at present. Recently Fabbri's group obtained a higher survival rate cryopreserving oocytes in the presence of a doubled sucrose concentration in the freezing solution (60%)[4]. The survival rate was even higher when the sucrose concentration was tripled (82%). In addition, a longer exposure time (from 10.5 to 15 min) to cryoprotectants, before lowering the temperature, significantly increased the oocyte survival rate (70%). Unfortunately ovulation induction is required also for oocyte cryopreservation, encountering the same problems as mentioned for embryo cryopreservation.

Surgical transposition can be used[5] to place at least one ovary as far out of the main radiation field as possible to limit radiation damage to the ovaries. However, because of scattered radiation, a substantial loss of ovarian function might still occur.

Ovarian tissue cryopreservation could be a practical alternative to oocyte storage and avoids the ethical dilemmas of embryo storage. Contrary to the difficulties with oocytes[5], primordial follicles that are found in the ovarian cortex are less sensitive to cryopreservation damage. Therefore, it is possible to freeze and store ovarian tissue from cancer patients, prior to therapy, to restore fertility and replant the cryopreserved thawed tissue once the patients are cured. It is noteworthy that retrieval of ovarian tissue involves only a simple laparoscopic operation, without the side-effects of hormonal stimulation for IVF. Freezing of ovarian tissue has several potential advantages over oocyte storage. The patients do not need to undergo the superovulation protocol required for the collection of mature oocytes (MII). Furthermore, in the ovarian tissue many thousands of primordial follicles with a high number of oocytes are present. These follicles are undifferentiated, contain a low number of cytoplasmic organelles, metabolic activity is quiescent, they are small and are lacking in the zona pellucida; so they should be more tolerant to freeze–thaw procedures than MII oocytes.

The first problem for cryopreserving ovarian tissue is to minimize injury from freezing. When isolated cells or tissues are frozen[2], a number of biological and chemical changes take place. Cooling retards the degradation process and preserves the structure of cells. However, it can destroy their function irreversibly when it goes below freezing temperature without cryoprotective agents. Although cells frozen and stored at the temperature of liquid nitrogen (−196°C) can maintain their viability for many years, the rate at which they are frozen and thawed is crucial for survival. The main causes of freezing injury are intracellular ice crystal formation and salt deposits.

Although cryoprotectants are imperative for successful freezing[2] many cryoprotectants can be cytotoxic. This cytotoxic property is enhanced by prolonged exposure times at temperatures above 0°C. Over recent years[3] significant advances have been made in the field of cryobiology, most importantly the introduction of a controlled-rate freezing apparatus and the development of more efficient cryoprotective agents. An effective cryoprotectant depends on a number of key properties[2]: high water solubility to depress the freezing point, high permeability to minimize the osmotic gradient and low toxicity. Although many compounds have some of these properties, dimethyl sulfoxide (DMSO), 1,2-propanediol (PROH), ethylene glycol (EG) and glycerol are most commonly used for the freezing of living cells. In 1996, Newton[6] investigated a few methods for storing human ovarian tissue at liquid nitrogen temperature by comparing four cryoprotectants (DMSO, PROH, EG, glycerol). The viability of primordial follicles was assessed after grafting of frozen–thawed tissue slices into immunologically tolerant animals. There were no significant differences between the results with EG, DMSO and PROH, despite a large number of observations, but survival in glycerol was poor. Glycerol, although chemically less toxic[2], may not be as effective for ovarian tissue cryopreservation because of slower penetration and higher osmotic stress.

WHO MAY BENEFIT FROM OVARIAN TISSUE CRYOPRESERVATION?

The cryopreservation of ovarian tissue could be applied to the patients (adult and children) who need to undergo anticancer treatments (Table 1). Two of the principal treatments against cancer[7] are radiation therapy and chemotherapy. Here, ovarian tissue cryopreservation (OTCP) might allow follicular reserves to be protected before the patient undergoes such treatments. The problem is that it is difficult to assess the extent of follicular destruction for a given patient, as individuals vary in their susceptibility to radiation and chemotherapy. It is estimated that ovarian radiation of 9 Gy renders humans infertile, but pregnancies have been reported after significant irradiation exposure. Similarly, although a total dose of 20 g of cyclophosphamide is considered to make patients infertile, exceptions have also been reported. The main chemotherapeutic agents[5] that induce ovarian damage are alkylating agents such as cyclophosphamide, melphalan and chlorambucil; other drugs such as *cis*-platinum and vinca alkaloids have also been implicated. Hence the patients affected with systemic diseases (Hodgkin's disease and non-Hodgkin's lymphomas), extrapelvic diseases (breast cancer,

Table 1 Indications for ovarian tissue cryopreservation

Extrapelvic diseases	Bone cancer (osteosarcoma-Ewing sarcoma), follicular thyroid carcinoma, kidney cancers, breast cancer, melanoma, neuroblastoma, bowel malignancy
Pelvic diseases (non-gynecologic malignancy)	Pelvic sarcoma, sarcoblastoma, rhabdomyosarcoma, sacral tumors, rectosigmoid tumors
Gynecologic malignancy	Early cervical carcinoma, early vaginal carcinoma, early vulvar carcinoma, endometrial adenocarcinoma, selected cases of ovarian carcinoma (STIA), exophytic ovarian borderline tumors (stage I)
Systemic diseases	Hodgkin's disease, non-Hodgkin's lymphoma
Chromosomal abnormalities	Deletions, inversion and complete or mosaic Turner's syndrome
Inflammatory diseases	Crohn's disease, lupus erythematosus, rheumatoid arthritis, periarteritis nodus

follicular thyroid carcinoma, hepatocellular carcinoma in a cirrhotic liver, renal carcinoma) and pelvic diseases (colon carcinoma, invasive cervix carcinoma, vaginal lymphoma, endometrial adenocarcinoma, unilateral or bilateral borderline ovarian exophytic tumors which partially involve the ovarian parenchyma) can cryopreserve their ovarian tissue before chemo- or radiotherapy. Also patients with some non-malignant conditions[7] (thalassemias, systemic lupus erythematosus, nephritic syndromes, periarteritis nodulous) may need treatment consisting of long-term use of alkylating agents. When the total cyclophosphamide dose is expected to exceed 10 g, it is reasonable to consider OTCP for protecting fertility.

Certain chromosomal abnormalities[7], most notably on chromosome X (such as deletions, inversion and complete or mosaic Turner syndrome), cause premature menopause. Congenital galactosemias and certain autoimmune and infectious diseases also cause abnormally rapid exhaustion of follicles. In some of these situations it is possible to predict premature menopause and, in theory, the ovarian tissue from these patients could be extracted for use in a later pregnancy.

For OTCP, the patient's age is a determining factor. It is likely that if ovarian tissue is extracted after 35 years of age, the follicular population is already greatly reduced. In contrast, child ovary contains many primordial follicles and extraction would enable a great number of follicles to be stockpiled. On the other hand oocyte or embryo cryopreservation is unavailable for children.

OVARIAN TISSUE CRYOPRESERVATION

Background

The first studies on ovarian tissue cryopreservation[8] were carried out in rodents during the 1950s. Ovarian follicles were found to survive freeze–thawing, albeit in low numbers, and if tissue was transplanted into ovariectomized recipients cyclical function and fertility were restored.

In 1990, Carrol[9] froze and subsequently thawed isolated primary mouse follicles in the presence of DMSO. Similar proportions of freshly collected and

frozen–thawed primary follicles undergo folliculogenesis in the absence of other ovarian tissue. Some of the mature oocytes recovered from these follicles were fertilized *in vitro* and, after transfer to pseudopregnant recipients at the two-cell stage, developed into live young. Subsequently, the viability of oocytes derived from frozen–thawed primary follicles was assessed.

In 1994, Gosden[10] tested the feasibility of storing sheep ovaries, that are comparable to those of humans in size, composition and follicle density. Cortical slices were prepared from the right ovaries of six lambs and either grafted directly to the ovarian pedicles of origin or cooled slowly to liquid nitrogen temperatures in medium containing DMSO. Three weeks later, the contralateral ovary was removed and replaced with frozen–thawed slices from the same animal. Two of the animals mated during their second estrus cycle 3–4 months later and the remainder had at least one ovulatory cycle. The pregnancies reached full-term development, one lamb being derived from an ovulation in a fresh graft and the other from a frozen–thawed graft. The results showed that cortical grafts returned to the orthotopic site either directly or after frozen storage restored the estrous cycle and fertility.

The cryopreservation of ovarian tissue was also applied with success on marmoset monkeys[11]. Pieces of marmoset ovary were frozen by slow cooling in DMSO. The follicles in fresh and frozen tissue were counted and examined for morphologic appearance in stained serial sections. The proportion of normal follicles was similar in fresh and frozen tissue examined immediately after thawing. Fresh and frozen tissue was transplanted underneath the kidney capsules of estrogenic, ovariectomized, immunodeficient mice. The establishment of grafts was similar and estrogenic activity was observed in the recipients 20 and 16 days after transplantation of fresh and frozen grafts, respectively.

The encouraging results reported by animal studies justified an extension of the work to human tissue. Many studies have been carried out to analyze the best way of cryopreserving ovarian tissue. Pieces of human cortical tissue[12] were frozen using one or two different cryoprotectants, either DMSO or a combination of PROH and sucrose. After cryopreservation the ovarian pieces were thawed and studied histologically. Specimens observed before and after cryopreservation with either protectant showed no signs of tissue necrosis. Follicles at similar developmental stages were found before and after freezing. Oocytes also had the same appearance after freezing and thawing. These results suggest that cryopreservation of human ovarian tissue is feasible.

In 1997, Oktay[13] reported the first successful isolation of human primordial follicles. Follicles that were isolated by a combination of a gentle enzymatic technique and manual dissection appeared to be viable when assessed by fluorescent viability stains and electron microscopy.

Thin slices of human ovarian cortex[14] were evaluated following cryopreservation in PROH/sucrose under various conditions. To examine the effect of cooling rates, the cortical slices after dehydration were loaded into vials and frozen using: (1) slow rate; (2) intermediate rate (achieved by suspending the vial in the liquid nitrogen vapor for about 12 h followed by submerging it in liquid nitrogen); and (3) rapid rate (achieved by plunging the vial directly into liquid nitrogen). The intermediate and rapid rates of cooling both resulted in more damage to the oocyte and granulosa than the slow rate. Slices were prepared prior to cryopreservation using two dehydration regimens: (1) a two-step method and (2) a one-step method.

In this study, single-step dehydration in PROH/sucrose for 90 min and slow cooling/rapid thawing resulted in the highest proportion of intact human primordial and primary follicles.

Recent studies have found good follicular survival[15] in frozen–thawed ovarian tissue; to optimize the process, an effective cryopreservation method needs to be developed. The diffusion of four cryoprotective agents (DMSO, EG, PROH and glycerol) into human tissue at both 4°C and 37°C was investigated. Also studied was the effect of adding different concentrations of the non-penetrating cryoprotective agent sucrose to the freezing media, using the release of lactate dehydrogenase as a measure of its protective effect. At 4°C PROH and glycerol penetrated the tissue significantly slower than either EG or DMSO. At the higher temperature of 37°C all four cryoprotectants penetrated at a faster rate, however concern about enhanced toxicity prevents the use of these conditions in practice. Thus, the results suggest that the best method of preparing tissue for freezing is exposure to 1.5 mol/l solutions of EG or DMSO at 4°C for 30 min. It was decided to use the temperature of 4°C in order to minimize toxicity.

Clinical applications

Theoretically ovarian tissue, after thawing, can be used in three different ways: (1) as an autograft; (2) for '*in vitro*' maturation of the follicles; and (3) as a xenograft.

Autograft

The autograft consists of restituting[1] the ovarian tissue to the patient. The ovarian tissue could be grafted into its normal site, to the ovarian pedicle (orthotopic site), which would allow the possibility of pregnancy without further medical assistance. Alternatively the ovarian tissue can also be grafted into a site other than its normal position (heterotopic site), for example omentum bursa, anterior surface of the uterus, deltoid muscle, kidney capsule, musculus rectus abdominal sheath or under the skin of the forearm. The heterotopic transplantation needs recourse to IVF to obtain pregnancies because it resumes only the endocrine function. Although autografting into an orthotopic or heterotopic site[2] seems most promising, the clinical application for cancer patients is problematic because of the potential risk of transmission of microscopic metastatic disease. Hence, it is necessary to develop screening methods to detect minimal residual disease (MRD) in ovarian tissue to eliminate the risk of cancer cell transmission with transplantation. Currently, molecular genetic techniques such as nested polymerase chain reaction (PCR), flow cytometry, fluorescence *in situ* hybridization (FISH) and cytogenetics have been applied to detect MRD before autologous peripheral stem cell or bone marrow transplant.

The first case of orthotopic transplant[16] of thawed ovarian tissue in the human was reported by Oktay in 1999. The patient, 29 years old, had a previous right salpingo-oophorectomy and left cystectomy (with wedge resection at the age of 17), to remove dermoid cysts. Later, the patient developed exercise-induced hypothalamic amenorrhea. After the salpingo-oophorectomy, ovarian cortical strips were cryopreserved, using a slow freezing protocol and 1.5 mol/l PROH as a cryoprotectant. Six months later the patient was prepared for ovarian transplantation. Sixty of the 72 vials were thawed, yielding 80 pieces; then the tissues were sutured to form three separate strings. By laparoscopy, the grafts were sutured into a pocket which was created by a dissection in the pelvic side wall, posterior to the broad ligament.

Grafts were visualized immediately after the surgery and the blood flow to the grafts could be distinguished 3 weeks later by Doppler ultrasound examination. Fifteen weeks after the procedure, the patient was stimulated with human menopausal gonadotropin (hMG). After 11 days of stimulation, a dominant follicle emerged in the ovarian graft and hormone replacement was discontinued. Serial ultrasounds demonstrated continual follicle growth, but the hMG dose had to be gradually increased to sustain this. After 24 days of stimulation, the average follicle diameter reached 17 mm. Ovulation was confirmed by sonographic demonstration of a collapsed follicle: the patient had spontaneous menses 16 days after the first hCG injection and every 25 to 28 days thereafter.

Ten months after the transplant the patient was monitored again. Levels of follicle stimulating hormone (FSH), luteinizing hormone (LH) and estradiol on the second day of menstruation indicated normal ovarian reserve. Ultrasound monitoring suggested that a dominant follicle of 5–6 mm in diameter had developed in the graft. Unfortunately, because the patient declined to discontinue hormone replacement and further stimulation, the team could not confirm the presence of a functional follicle by estradiol and progesterone measurements.

The first case of heterotopic transplant[17] of thawed ovarian tissue in the human was carried out by Oktay in 1999. He transplanted ovarian tissue heterotopically under the forearm skin in a 35-year-old women affected with stage IIIB squamous cell cervical carcinoma. The patient consented to transplantation prior to pelvic radiotherapy to preserve her ovarian function. Both ovaries were removed laparoscopically and their cortices were prepared in 16 strips. A 1 cm vertical incision was made over the brachioradialis muscle, 5 cm below the antecubital fossa. Ovarian strips were wedged subcutaneously, using a suture pull-through technique. The patient was started on 1 mg of micronized estradiol on postoperative day 2. Approximately 10 weeks after the

transplant, she reported the presence of a painless bulge at the site of the transplant. Ultrasound examination showed a 15 mm dominant follicle and four other antral follicles, measuring 5 to 7 mm, at which point estrogen replacement was discontinued. Repeated hormonal analyses showed estrogen production, as well as normalization of FSH and LH concentrations. Serial ultrasound examinations showed continual development of new antral follicles as large as 15.5 mm. During 18-month follow-up, a dominant follicle developed each month. Percutaneous oocyte aspirations yielded a mature oocyte.

A second case of heterotopic transplant was reported[17] in 2000. Ovarian tissue was grafted to the forearm of a 37-year-old patient with recurrent benign ovarian serous cysts. The transplantation technique was similar to that in the first patient. After 5 months, the patient felt a lump growing at the transplant site. One month later, an ultrasound scan showed a 7.5 mm follicle in the forearm, which grew to 9 mm in 2 days. Hormone replacement was discontinued. During the subsequent month the patient reported spontaneous menstruation. On day 13 of that cycle, a 9 mm follicle was noted by ultrasound and her hormone measurements indicated a mid-cycle surge. The patient menstruated spontaneously 2 weeks later, and every 25 to 28 days thereafter, but the ovarian graft was still functional 10 months after the transplantation.

In conclusion, ovarian transplantation to the forearm resumes endocrine function. With optimization of ovarian stimulation and percutaneous oocyte retrieval techniques, this procedure may also restore fertility in the near future when used in conjunction with IVF.

Recently, Radford[18] described the success of orthotopic reimplantation of ovarian cortical strips after thawing in a woman treated with cytotoxic chemotherapy for malignant disease. Measurement of concentrations of sex steroids and gonadotropins confirmed ovarian failure, and hormone replacement therapy (HRT) was started. After 5 months she asked to be considered for reimplantation of ovarian cortical strips to see whether this procedure could restore ovarian function and eliminate the need for HRT. HRT was therefore stopped and hormonal measurement returned to an ovarian-failure profile. Two ovarian cortical strips were thawed and reimplanted, one into the left ovary and another at the site of the right ovary.

Seven months after reimplantation of ovarian cortical strips, the patient reported frequent hot flushes. Subsequently, these symptoms disappeared and a further month later a serum estradiol concentration of 217 pmol/l was detected: a month later the patient menstruated. Nine months after the reimplantation, however, the FSH and LH concentrations had returned to those seen with ovarian failure.

The delay in onset of ovarian function after reimplantation is probably due to hypoxia, which causes loss of all but the primordial follicles. These structures are thought to be the most resistant to damage and in humans require between 6 and 12 months to grow and differentiate to a stage consistent with ovulation.

In vitro maturation

In vitro growth and maturation (IVM) of human primordial follicles[2], followed by IVF, is an attractive and desirable option, but it is technically challenging because of the length of the growing phase and lack of knowledge about the optimal conditions for growth and maturation of human oocytes (see chapter 19).

Several studies were carried out on maturation of isolated follicles from fresh and cryopreserved animal ovarian tissue. *In vitro* culture of human fresh preantral follicles has been reported by Roy and Treacy[19] in 1993. The follicles were enzymatically isolated and cultured *in vitro* for up to 120 h in a serum-free medium with insulin–transferin–selenium (ITS), with or without FSH. Culture medium was changed every 24 h. The results showed that FSH improved the health of granulosa cells, the growth and the steroidogenic function of the follicles.

To develop a procedure for isolating small human follicles[20] and to determine their growth requirements, preantral and early antral follicles were isolated manually and cultured for a few weeks. It was shown that preantral follicles can grow in culture to the early antral stages and that small antral follicles are capable of enlarging in culture. Human FSH was essential for antrum formation and growth. Supplementation of human LH in addition to human FSH had a beneficial effect on growth, enhancing antral formation and supporting the development of small preantral follicles. Only 10–20% of the early antral follicles cultured contained an oocyte. It was presumed that the oocytes in such follicles are lost because of early atresia *in vivo*. Attempts to isolate human follicles should be applied to follicles at earlier stages because at these stages in current and previous studies, no oocyte loss was observed in human follicles before the antral stage. It is, however, technically difficult to isolate preantral, primordial and primary human follicles from the surrounding connective tissue. The enzymatic method of isolation using collagenase and deoxyribonuclease requires cooling and is lengthy, reducing the viability of the follicles. Enzymatic isolation of primordial follicles has been successful in the pig and it is possible that this technique could be modified and applied to the human ovary.

In vitro maturation of human ovarian tissue[21] from fresh and frozen–thawed tissue was carried out in 1997. These slices were cultured for 4–21 days in either α-minimum essential medium (α-MEM) or Earle's balanced salt solution (EBSS) with added pyruvate. Both media were supplemented with 10% human serum, insulin, gonadotropins and antibiotics. The results indicate that it is possible to culture human primary and primordial follicles.

An ovarian tissue slice culture system[22] was used to examine the effects of media composition, FSH and serum substitution on the development of small human follicles *in vitro*. Human ovarian cortex biopsies were cut into small pieces and cultured for 5, 10 or 15 days. Control (non-cultured) and cultured tissue was fixed, serially sectioned and stained. Comparison of the ability of α-MEM, Waymouth's or EBSS culture media to support follicle growth demonstrated significantly increased initiation and growth of follicles in α-MEM during the first 10 days of culture. The supplementation of α-MEM with FSH significantly reduced levels of atresia and increased the mean diameter of healthy follicles. Follicles present in tissue cultured for 10 days with human serum albumin and ITS were significantly larger, more developed and showed less atresia than those cultured with serum alone. Human unilaminar follicles cultured in collagen gels[23] showed an increase in the number of granulosa cell layers and in the oocyte size in 40% and 38.7% of the follicles from fresh and frozen–thawed tissue, respectively, during a 24 h culture period.

Recently it was reported[24] that the success of *in vitro* growth of human follicles is likely to be achieved through the use of a multistage culture strategy. The first step was to initiate and to maintain the primordial follicle growth *in vivo*: from primordial to secondary follicle stages. The follicles were then isolated and matured *in vitro* to the antral follicle stage; the final step was to mature the oocytes within their cumulus cells.

The ability to completely grow[2] and mature human immature oocytes *in vitro*, however, will not be available until the development of an optimal culture system, which depends on the acquisition of a full understanding of the signal and control mechanism of follicle growth.

Xenograft

Tranplantation of frozen–thawed ovarian tissue[2] into an animal host with subsequent maturation and collection of oocytes can offer considerable advantages to cancer survivors. With this technique, the possibility of cancer transmission and relapse can be eliminated because cancer cells cannot penetrate the zona pellucida and some technical difficulties of *in vitro* growth and maturation of primordial follicles can be bypassed.

It has been demonstrated not only that human ovarian tissue survives in severe-combined immuno-deficiency (SCID) mice but that the follicles can grow to antral-secretory stages. Therefore, the richly vascularized subcapsular region of the kidney has been a favored site for xenografts in rodents. Theoretically, *in vivo* animal culture systems for the growth and maturation of human primordial follicles should be easier to manage than *in vitro* culture systems. It is also of concern that animal pathogens can be transmitted to human tissue with xenografting.

Hence, this method is not yet perfected[1] and it additionally raises ethical concerns that would need to be settled before envisioning practical application.

Among these three methods, only autografting appears to be currently realizable.

Our experience

In 1998, our research group started the cryopreservation of ovarian tissue obtained from oncology patients. The ovarian cortex, retrieved by laparoscopy from the first three patients with ages ranging from 30 to 36 years, was cryopreserved using 1.5 mol/l DMSO as a cryoprotectant[25] and a slow freezing/rapid thawing procedure. The tissue was cut into small specimens (1–2 mm thick) and put into precooled cryovials containing the cryopreservation solution. The vials were placed in a rolling system for 30 min at 4°C to allow the cryoprotectants to enter the tissue. A slow freezing procedure was carried out using a programmable freezer (Planer Cryo 10 1.7 Series III). The specimens were rapidly thawed in air at room temperature for 2 min and then immersed in a water bath at 37°C until the ice had thawed. The morphologic analysis, carried out in fresh and frozen–thawed specimens, showed that DMSO treatment provided a fairly good preservation even if more interstitial edema was observed in the cryopreserved samples compared to the fresh specimens. This edema is probably due to a disruption of the plasmatic membrane of the stromal cells during the cryopreservation procedure. However, even if many follicles showed nucleate oocytes there were others that showed anucleate oocytes with packaged chromatin as a sign of nuclear degeneration.

In a following study we compared the effects of DMSO versus PROH as cryoprotectants[26]. The samples were put into precooled cryovials containing 1.5 mol/l PROH to which was added 0.2 mol/l sucrose or 1.5 mol/l DMSO. The vials were cryopreserved by a slow freezing/rapid thawing procedure. After a cryopreservation period of about 9 months a small piece of tissue from each patient was thawed and fixed in formalin to evaluate the influence of the cryopreservation method on follicle survival.

The results showed that PROH treatment provided a better tissue preservation compared to DMSO (Figure 1). Particularly, many viable oocytes and granulosa cells (GCs) were found in the ovarian tissue frozen in PROH. The mean follicle numbers observed in three serial histologic sections, with a surface area of 10 mm^2 each, were 5.22 ± 3.36 in the fresh tissue and 4.27 ± 3.62 and 4.88 ± 3.23, respectively, in PROH and DMSO cryopreserved tissues. The differences were not statistically significant. The mean follicle percentage in frozen PROH and DMSO ovarian tissue was 93% and 81%, respectively, when compared to 100% in fresh tissue. However the percentage of follicles observed in each

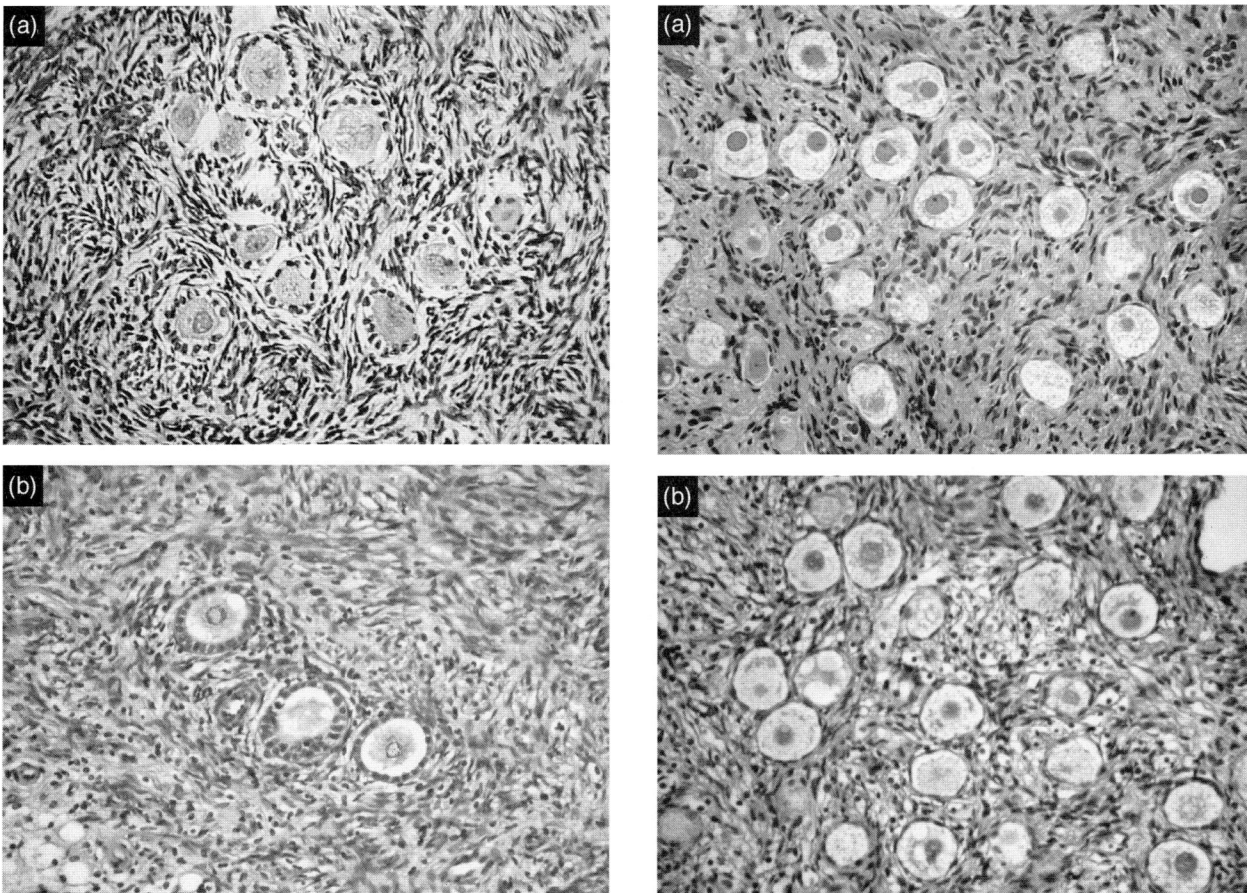

Figure 1 Morphologic appearance of human ovarian tissue cryopreserved in dimethyl sulfoxide (DMSO) or 1,2-propanediol (PROH). (b) PROH treatment provides better tissue preservation compared to DMSO. Original magnification 20 ×

Figure 2 Morphologic appearance of child ovarian tissue before (a) and after (b) cryopreservation using 1,2-propanediol and sucrose as cryoprotectants. Child ovary contains many primordial follicles. Original magnification 20 ×

stage of development in the thawed tissues was not statistically different from the fresh tissue.

Considering these results, we later examined the PROH cryopreservation effects on ovarian tissue by morphologic and immunohistochemical analyses[27]. Ovarian tisssue was obtained from 10 oncology patients, aged from 21 to 39 years, affected with different malignant diseases (breast cancer, colon cancer, kidney cancer, Hodgkin's disease, non-Hodgkin's lymphoma, vaginal lymphoma, endometrial adenocarcinoma and ovarian borderline tumor).

Morphologic assessment of follicles and stromal cells, before and after freezing–thawing, suggested that human ovarian tissue survived well after PROH cryo-

preservation (Figure 2). The follicle distribution and density are similar in fresh and frozen–thawed tissue, indicating that there is no decrease in follicle population after cryopreservation.

Using immunohistochemical analysis, we investigated, in ovarian tissue, the presevation of antigenicity using anti-estrogen receptor (ER) and anti-progesterone receptor (PR) antibodies, cell proliferation by anti-Ki67 and cellular anti-apoptotic activity by anti-Bcl2 protein. For estrogen and progesterone receptors the follicles showed a high percentage of negative staining for both ER and PR. The stromal cells were predominantly negatively stained for ER and diffusely stained for PR without showing significant differences between the fresh and frozen–thawed samples, indicating that the antigenicity is maintained in the thawed tissue also. Furthermore, we found a positive staining for Bcl2 and Ki67 in both granulosa and stromal cells without significant differences between fresh and frozen–thawed samples. These data indicate that ovarian tissue could resume the normal cellular cycle.

We are now evaluating the morphologic ultrastructural aspect of ovarian tissue to confirm our results in terms of good ovarian tissue cryopreservation.

Ethical and legal considerations

As has happened with other new reproductive technologies, the clinical application of ovarian cryopreservation[2] and transplantation contains potential legal and ethical issues. This is the beginning of a new technology that has numerous unresolved ethical and legal issues, including clinical indications, safety, age limits, time limits for storage and tissue custody.

Defining clinical indications and resolving safety issues will be an ongoing effort, along with improvement of the technology.

The patients undergoing ovarian tissue cryopreservation should be provided with counseling during which the experimental nature of the technique is explained. The time limits of storage for a good ovarian tissue preservation are unknown. These patients are likely to be storing material for many years, thus it is important that they realize that the chances of a successful outcome are uncertain, particularly when the materials stored appertain to pediatric patients who cryopreserve their tissue for a long time. It is important to ensure that the patient consents to an upper limit of time and also to the disposal of tissue in the circumstance of death or mental incapacity.

Ethical and moral dilemmas suggest that the procedure should only be offered to patients with sound medical justification who want to preserve fertility with no other option.

References

1. Aubard Y, Piver P. Ovarian tissue cryopreservation and gynecologic oncology: a review. *Eur J Obstet Gynecol Biol Reprod* 2001;97:5–14.
2. Kim S, Battaglia DE, Soules MR. The future of human ovarian cryopreservation and tranplantation: fertility and beyond. *Fertil Steril* 2001;75:1049–55
3. Newton H. The cryopreservation of ovarian tissue as a strategy for preserving the fertility of cancer patients. *Hum Reprod* 1998;4:237–47
4. Fabbri R, Porcu E, Marsella T, et al. Human oocyte cryopreservation: new perspectives regarding oocyte survival. *Hum Reprod* 2001;16:411–16
5. Abir R, Fisch B, Raz A. Preservation of fertility in women undergoing chemotherapy: current approach and future prospects. *J Assist Reprod Genet* 1998;15:469–77
6. Newton H, Aubard Y, Rutherford A, et al. Low temperature storage and grafting of human ovarian tissue. *Hum Reprod* 1996;7:1487–91
7. Aubard Y, Poirot C, Piver P, et al. Are there indications for ovarian tissue cryopreservation? *Fertil Steril* 2001;76:414–15
8. Newton H, Illingworth P. *In-vitro* growth of murine pre-antral follicles after isolation from cryopreserved ovarian tissue. *Hum Reprod* 2001;16:423–9
9. Carrol J, Whittingham DG, Wood MJ, et al. Extra ovarian production of mature viable mouse oocytes from frozen primary follicles. *J Reprod Fertil* 1990;90,321–7
10. Gosden RG, Baird DT, Wade JC, et al. Restoration of fertility to oophorectomised sheep by ovarian autografts stored at −197°C. *Hum Reprod* 1994;9:597–603
11. Candy CJ, Wood MJ, Whittingham DG. Follicular development in cryopreserved marmoset ovarian tissue after transplantation. *Hum Reprod* 1995;10:2334–8
12. Hovatta O, Silye R, Krausz T, et al. Cryopreservation of human ovarian tissue using dimethylsulphoxide and propanediol–sucrose as cryoprotectant. *Hum Reprod* 1996;11:1268–72
13. Oktay K, Nugent D, Newton H, et al. Isolation and characterization of primordial follicles from fresh and cryopreserved human ovarian tissue. *Fertil Steril* 1997;67:481–6
14. Gook DA, Edgar DH, Stern C. Effect of cooling rate and dehydration regimen on the histological appearance of human ovarian cortex following cryopreservation in 1,2-propanediol. *Hum Reprod* 1999;14:2061–8
15. Newton H, Fisher J, Arnold JRP. Permeation of human ovarian tissue with cryoprotective agents in preparation for cryopreservation. *Hum Reprod* 1998;13:376–80
16. Oktay K, Aydin BA, Karlikaya G. A technique for laparoscopic transplantation of frozen-banked ovarian tissue. *Fertil Steril* 2001;75:1212–16
17. Oktay K, Economos K, Kan M, et al. Endocrine function and oocyte retrieval after autologous transplantation of ovarian cortical strips to the forearm. *J Am Med Assoc Prelim Commun* 2001;286:1490–3

18. Radford JA, Lieberman BA, Smith ARB, *et al.* Orthotopic reimplantation of cryopreserved ovarian cortical strips after high-dose chemotherapy for Hodgkin's lymphoma. *Lancet* 2001;357:1172–5

19. Roy SK, Treacy BJ. Isolation and long-term culture of human preantral follicles. *Fertil Steril* 1993;59:783–90

20. Abir R, Franks S, Mobberley MA, *et al.* Mechanical isolation and *in-vitro* growth of preantral and small antral human follicles. *Fertil Steril* 1997;68:682–8

21. Hovatta O, Silye R, Abir R, *et al.* Extracellular matrix improves survival of both stored and fresh human primordial and primary ovarian follicles in long-term culture. *Hum Reprod* 1997;12:1032–6

22. Wright CS, Hovatta O, Margara R, *et al.* Effects of follicle-stimulating hormone and serum substitution on the *in-vitro* growth of human ovarian follicles. *Hum Reprod* 1999;14:1555–62

23. Abir R, Roizman P, Fisch B, *et al.* Pilot study of isolated early human follicles cultured in collagen gels for 24 hours. *Hum Reprod* 1999;14:1299–301

24. Picton HM, Gosden RG. *In vitro* growth of primordial follicles from frozen-banked ovarian tissue. *Mol Cell Endocrinol* 2000;166:27–35

25. Fabbri R, Marsella T, Diano C, *et al.* Human ovarian tissue banking using DMSO as cryoprotectant. Presented at the *Mammalian Follicle and Oocyte Culture* meeting, Kilpisjärvi, Finland, June 1999

26. Fabbri R, Marsella T, Diano C. Human ovarian tissue cryopreservation. Presented at the *11th World Congress on In Vitro Fertilization and Human Reproductive Genetics*, Australia, May 1999, 234

27. Fabbri R, Tibaldi E, Iannascoli C, *et al.* Morphological evaluation of fresh and frozen/thawed human ovarian tissue in oncology patients. Presented at the *1st SIERR National Congress*, Rome, Italy, December 2000

In vitro maturation of human gametes

In vitro maturation of human spermatozoa: an update

18

Jan Tesarik, Carmen Mendoza and Ermanno Greco

INTRODUCTION

Several studies using *in vitro* culture of mammalian testicular tissue and reporting different degrees of spermatogenic cell maturation during culture have been published over the past 30 years (reviewed in reference 1). Interest in these techniques was recently revived when it became evident that immature spermatozoa and spermatogenic cells can fertilize oocytes when introduced into them by appropriate micromanipulation methods, and that *in vitro* culture can enhance fertilization under these conditions. *In vitro* maturation was first proposed to enhance the fertilizing ability of human round spermatids[2,3], but later it also proved to be useful for other indications. Detailed reviews of the indications for which sperm *in vitro* maturation is used, the methods and the clinical outcomes have been published recently[1,4]. This chapter focuses on new aspects of the sperm *in vitro* maturation technique and provides updates on the outcomes obtained over the past year.

APPLICATIONS OF THE *IN VITRO* MATURATION TECHNIQUE

Despite the fact that *in vitro* maturation of human germ cells was originally developed to improve assisted reproduction outcomes with spermatids[2,3], the unexpected observation was made that germ cells from some patients with maturation arrest at the primary spermatocyte stage can overcome the developmental block and differentiate into developmentally competent spermatids in merely 2 days of culture[5,6]. The first birth of a healthy child fathered by a man with maturation arrest at the primary spermatocyte stage was achieved with the use of this technique[5]. The unexpected rapidity of this process as compared to *in vivo* conditions was widely debated and

was explained by the presence of supraphysiologic concentrations of follicle stimulating hormone (FSH) and testosterone in the culture medium[1,4]. We recently confirmed, on a larger group of cases, the original observation that only patients with maturation arrest at the pachytene stage of primary spermatocyte development can benefit from *in vitro* maturation, as opposed to patients whose germ cells arrest at earlier stages of spermatocyte maturation or at the spermatogonium stage[6].

Recently we have shown that testicular sperm *in vitro* maturation reduces the frequency of spermatozoa carrying apoptotic DNA damage as compared to fresh samples[7], and healthy twins were born after *in vitro* maturation of spermatozoa was used for intracytoplasmic sperm injection (ICSI) in a case with a high degree of sperm apoptosis and previous repeated failures of ICSI with fresh spermatozoa[8]. The beneficial effect of sperm *in vitro* culture on the outcome of assisted reproduction in cases of substantial sperm apoptosis has now been confirmed in a larger series of patients[9].

Taken together, the original application of sperm *in vitro* maturation, which was limited to patients with post-meiotic maturation arrest at the spermatid stage, can now be extended to certain cases of meiotic arrest and to those with complete spermatogenesis but a high degree of sperm apoptosis (Table 1). Of course the prognosis is different for each of these possible indications (see below).

IN VITRO CULTURE CONDITIONS

The technique of *in vitro* maturation of human spermatozoa, in its currently most efficient version, has been

Table 1 Pathological conditions in which *in vitro* maturation of testicular germ cells can be useful to enhance germ cell reproductive capacity

(1) Post-meiotic maturation arrest
    (a) Arrest at the round spermatid stage
    (b) Arrest at the elongating or elongated spermatid stage

(2) Meiotic maturation arrest
    (a) Arrest at the primary spermatocyte stage – only when the pachytene stage of the first meiotic division is present
    (b) Arrest at the secondary spermatocyte stage?

(3) Hypospermatogenesis with massive apoptosis of testicular spermatozoa

Table 2 Laboratory conditions for *in vitro* culture of testicular germ cells

Sample preparation	Mechanical disintegration with glass slides
Temperature	30–32°C
Media	Standard IVF media if CO_2 incubator
	HEPES-buffered media if water bath in air
Media supplements	FSH, 25–500 IU/l
	Testosterone, 1–10 μmol/l

FSH, follicle stimulating hormone

developed by using samples of testicular tissue from men with obstructive azoospermia and apparently normal spermatogenesis[2,3]. The basic features of this technique are summarized in Table 2. The unusual acceleration of meiotic and post-meiotic differentiation events was shown to be dependent on the presence of high concentrations of FSH (> 25 IU/l) and testosterone (1 μmol/l) in culture medium. Yet, the concentration of both hormones used in those studies was relatively close to the physiologic levels[2,3].

When the technique was applied to testicular tissue samples from men suffering from primary testicular failure, however, the *in vitro* concentrations of FSH and testosterone which were sufficient for samples from men with normal spermatogenesis failed to produce any detectable effect in many cases. This led to attempts to overcome the poor *in vitro* responsiveness of testicular tissue by increasing the concentration of both hormones. In a recent controlled study we have found that an improvement of *in vitro* differentiation of germ cells from men with primary testicular failure can be obtained by increasing the concentration of FSH up to 500 IU/l and that of testosterone up to 10 μmol/l[10].

The effect of FSH on the *in vitro* differentiation of germ cells is supposed to be mediated by FSH receptors located on Sertoli cells via as yet undetermined intermediates. This FSH effect appears to be mediated by the classical FSH receptor, coupled to a G-protein- and cyclic adenosine monophosphate (cAMP)-mediated pathway, because FSH in culture medium could be replaced with pentoxifylline, an adenylate cyclase inhibitor increasing the concentration of cAMP in treated cells[11]. In contrast, insulin did not exert any effect on *in vitro* germ cell survival and differentiation, even when added at high concentrations known also to act at the insulin-like growth factor receptor[11].

The need for the pharmacologic concentrations of FSH to promote *in vitro* differentiation of germ cells from men with primary testicular failure may be explained by the compensatory increase in serum FSH concentrations in most of these cases. In the constant presence of high serum FSH concentrations, FSH receptors on Sertoli cells may be desensitized, but may still resume responsiveness when FSH concentration is increased to highly supraphysiologic levels. This hypothesis, however, remains to be confirmed in an appropriate experimental system.

The mechanism by which the *in vitro* differentiation of germ cells from men with primary testicular failure is improved by pharmacologic concentrations of testosterone[10] remains to be determined.

Post-meiotic differentiation can also occur *in vitro* in the absence of added hormones[12–14]. However, meiotic differentiation apparently does not occur under these conditions; the presence of an underlying Vero cell monolayer is needed for optimal germ cell survival and differentiation, and the spermiogenetic events apparently occur more slowly as compared to the rates in FSH- and testosterone-enriched culture media.

UPDATE ON CLINICAL OUTCOMES

The clinical efficacy of sperm *in vitro* maturation for individual indications needs to be appreciated in the context of the efficiency of assisted reproduction using freshly obtained germ cells. Despite the relatively acceptable success rates with elongated spermatid injection (ELSI), the results with round spermatid injection (ROSI) before the introduction of *in vitro* sperm maturation were disappointing (Table 3). No assisted reproduction technique was applicable to cases of maturation arrest at the primary spermatocyte stage. Elevated proportions of spermatozoa with apoptosis-related DNA damage were known to be associated with poor assisted reproduction outcomes[15–18].

Even though the number of analyzable cases of assisted reproduction with *in vitro* matured spermatozoa is still relatively low, the data available strongly suggest that the inclusion of the *in vitro* maturation step improves clinical outcome of spermatid conception and provides, for the first time, a potentially successful treatment option to men with certain types of meiotic maturation arrest (Table 4). Moreover, *in vitro* culture facilitates the selection of non-apoptotic spermatozoa for assisted reproduction treatment[7] and improves assisted reproduction outcomes in cases of substantial germ cell apoptosis[8,9]. In fact, the latter indication offers the best prognosis as to the success of assisted reproduction with *in vitro* matured spermatozoa as compared to the other clinical indications (Table 4).

Table 3 Success rates of human ROSI and ELSI before the introduction of *in vitro* germ cell maturation

Technique	Fertilization rate	Pregnancy rate	Birth rate
ROSI	599/1919 (31.2%)	13/257 (5.1%)	5/257 (1.9%)
ELSI	395/667 (59.2%)	36/98 (36.7%)	24/98 (24.5%)

ELSI, elongated spermatid injection; ROSI, round spermatid injection. Data derived from references 31–35 and unpublished data

Table 4 Assisted reproduction outcomes with *in vitro* cultured testicular germ cells in different clinical indications

Indication	Cases treated	Pregnancies	Babies born	Abnormalities
Meiotic arrest	23	2	3	None
Post-meiotic arrest	25	5	6	None
Substantial sperm apoptosis	21	11	12	None

Data derived from references 5–9 and unpublished data

OVERCOMING APOPTOSIS – A COMMON DENOMINATOR OF SPERM *IN VITRO* MATURATION SUCCESS

Programed cell death (apoptosis) accompanies normal spermatogenic events by eliminating damaged or developmentally incompetent germ cells and adjusting the proportion between the germ and Sertoli cells in the spermatogenic epithelium[19]. In pathologic situations, apoptosis appears to be the final event following various testicular and systemic pathologies[20]. Substantial apoptosis of germ cells is associated with maturation arrest[21,22]. However, high frequencies of mature spermatozoa showing different types of apoptosis-related cell damage were also observed in men with complete, though mostly disturbed, spermatogenesis[23–26], and this condition was shown to be associated with lower sperm concentration[27], production of reactive oxygen species[26] and impaired fertilizing ability[15,18].

As in other cell types, apoptosis is likely to be induced in germ cells when a cell's checking mechanism detects an anomaly incompatible with further development into a functionally competent gamete. However, many gamete abnormalities which are incompatible with normal *in vivo* fertilization can be overcome, under *in vitro* conditions, by micromanipulation-assisted fertilization. With this perspective in mind, overcoming apoptosis of certain types of germ cells may enable micromanipulation-assisted fertilization and normal embryonic development.

The time between apoptosis induction and execution is usually long, but the length of the *in vivo* spermatogenic process leaves enough time for apoptosis to prevent immature germ cells from achieving final stages of sperm maturation. On the other hand, the unusual acceleration of spermatogenic events during *in vitro* maturation allows those germ cells that respond to the given culture conditions to achieve final maturational stages before the apoptosis-activated autodestruction pathways can eliminate the cell components required for normal fertilization and post-fertilization development (Figure 1).

It has to be noted in this context that most of the time spent on *in vivo* spermatogenesis is likely to be dedicated to internal proofreading mechanisms controlling the quality of individual developmental events at predetermined developmetal checkpoints. One such checkpoint is known to be situated at the metaphase-to-anaphase transition of the first meiotic division. If this checkpoint is suppressed, e.g., when mouse primary spermatocytes are incubated *in vitro* in the presence of okadaic acid, a protein phosphatase inhibitor, the metaphase-to-anaphase transition is markedly accelerated, probably by increasing the prevalence of the phosphorylated, enzymatically active form of the p34[cdc2] serine–threonine protein kinase, the catalytic subunit of metaphase-promoting factor (MPF)[28]. By analogy, the unusual speed of spermatogenic events observed in

human explanted and *in vitro* cultured testicular tissue makes it questionable whether all naturally acting developmental checkpoints are operational in these particular conditions. In the absence of these checkpoints, minor developmental abnormalities occurring at the primary spermatocyte and round spermatid stages might be tolerated, and the switches of developmental programs to the apoptotic pathway, normally resulting from such abnormalities, might be avoided (Figure 1). In the presence of added stimulatory factors, which are potentially deficient *in vivo*, and in the absence of

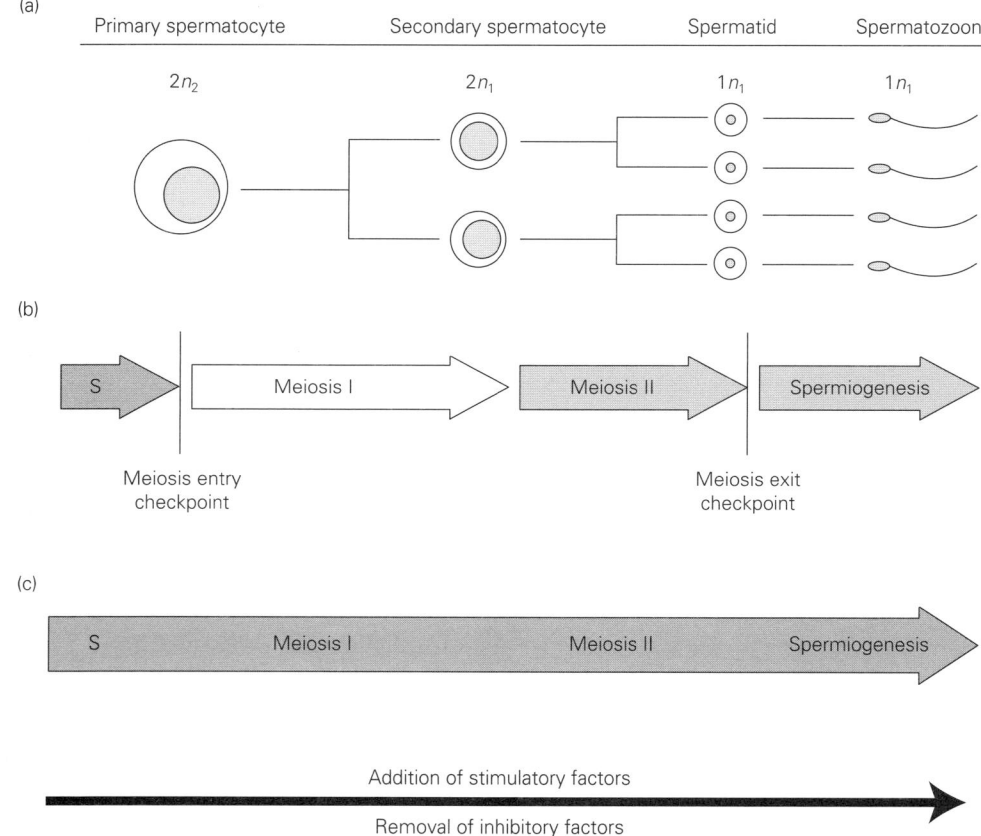

Figure 1 Schematic representation of the possible reasons for the observed differences in the characteristics of human spermatogenesis occurring *in vivo* and *in vitro*. Only the final phase of the spermatogenic process is included. (a) The meiotic and post-meiotic phase of spermatogenesis beginning with the last S-phase in primary spermatocytes (2*n*, diploid), including the first and the second meiotic divisions leading to the formation of secondary spermatocytes (2*n*, haploid) and round spermatids (1*n*, haploid), respectively, and ending by post-meiotic sperm differentiation during spermiogenesis. (b) *In vivo* spermatogenesis showing two major developmental checkpoints located at meiosis entry and meiosis exit, respectively. In pathologic situations, germ cell development is preferentially blocked at these checkpoints, and the developmental program is switched to apoptosis. (c) *In vitro* spermatogenesis during which the checkpoints are eliminated. This leads to a dramatic acceleration of certain developmental processes, accompanied by a lack of co-ordination between them. This condition also offers an opportunity to damaged cells to escape apoptosis. All these phenomena have been observed during *in vitro* culture of human testicular cells

inhibitory factors, which are potentially abundant *in vivo*, the development of such rescued germ cells may continue up to a stage at which they can be used for micromanipulation-assisted fertilization.

FUTURE PROSPECTS

Despite the recent progress with *in vitro* maturation of human spermatozoa and with the use of *in vitro* matured spermatozoa for assisted reproduction, there are still many cases which cannot be helped in this way because the germ cells do not mature to spermatids or spermatozoa during *in vitro* culture or there are no germ cells in the testis at all. In the former condition, further refinement of the *in vitro* culture techniques may reduce the proportion of treatment-resistant cases. In particular, it might be helpful to develop culture systems in which germ cells are capable of surviving and continuing their differentiation during prolonged periods of time, similar to those required for normal *in vivo* spermatogenesis. It has been shown recently that isolated round spermatids can survive and differentiate *in vitro* into late elongated spermatids when cultured for 7 days in microdrops of Vero cell-conditioned medium[14]. The evaluation of different culture and co-culture systems, as well as different hormone and growth factor enrichments for culture media used for *in vitro* maturation of human spermatozoa, is currently under way.

For men with a total lack of the germ line in the testis, alternative strategies, based on haploidization of somatic cells to be used as gametes, are currently being investigated. The combined use of a gamete and a somatic cell in assisted reproduction was originally suggested as a means for reconstruction of new oocytes consisting of an artificially haploidized somatic nucleus and an enucleated oocyte from a donor; these oocytes need to be fertilized by a spermatozoon to form an embryo[29]. Recent data have shown that somatic cells can also be used as substitutes for the male gamete[30]. The technique may thus become the only possible treatment conserving the genetic parenthood of men with a total lack of germ line in the testis.

References

1. Tesarik J, Mendoza C, Greco E. *In-vitro* maturation of immature human male germ cells. *Mol Cell Endocrinol* 2000;166:45–50

2. Tesarik J, Greco E, Rienzi L, *et al*. Differentiation of spermatogenic cells during *in-vitro* culture of testicular biopsy samples from patients with obstructive azoospermia: effect of recombinant follicle stimulating hormone. *Hum Reprod* 1998;13:2772–81

3. Tesarik J, Guido M, Mendoza C, Greco E. Human spermatogenesis *in vitro*: respective effects of follicle-stimulating hormone and testosterone on meiosis, spermiogenesis, and Sertoli cell apoptosis. *J Clin Endocrinol Metab* 1998;83:4467–73

4. Tesarik J, Mendoza C, Greco E. Immature germ cell conception – *in vitro* germ cell manipulation. *Baillières Clin Endocrinol Metab* 2000;14:437–52

5. Tesarik J, Bahceci M, Özcan C, *et al*. Restoration of fertility by *in-vitro* spermatogenesis. *Lancet* 1999;353:555–6

6. Tesarik J, Bahceci M, Özcan C, *et al*. *In-vitro* spermatogenesis. *Lancet* 1999;353:1708

7. Tesarik J, Mendoza C, Greco E. *In vitro* culture facilitates the selection of healthy spermatids for assisted reproduction. *Fertil Steril* 1999;72:809–13

8. Tesarik J, Cruz-Navarro N, Moreno E, *et al*. Birth of healthy twins after fertilization with *in vitro* cultured spermatids from a patient with massive *in vivo* apoptosis of postmeiotic germ cells. *Fertil Steril* 2000;74:1044–6

9. Tesarik J, Greco E, Mendoza C. Assisted reproduction with *in-vitro*-cultured testicular spermatozoa in cases of severe germ cell apoptosis: a pilot study. *Hum Reprod* 2001;16:2640–5

10. Tesarik J, Nagy P, Abdelmassih R, *et al*. Pharmacological concentrations of follicle-stimulating hormone and testosterone improve the efficacy of *in-vitro* germ cell differentiation in men with maturation arrest. *Fertil Steril* 2002;77:245–51

11. Tesarik J, Mendoza C, Greco E. The effect of FSH on male germ cell survival and differentiation *in vitro* is mimicked by pentoxifylline but not insulin. *Mol Hum Reprod* 2000;6:877–81

12. Aslam I, Fishel S. Short-term *in-vitro* culture and cryopreservation of spermatogenic cells used for human *in-vitro* conception. *Hum Reprod* 1998;13:634–8

13. Cremades N, Bernabeu R, Barros A, Sousa, M. *In-vitro* maturation of round spermatids using co-culture on Vero cells. *Hum Reprod* 1999;14:1287–93

14. Cremades N, Sousa M, Bernabeu R, Barros A. Developmental potential of elongating and elongated spermatids obtained after *in-vitro* maturation of isolated round spermatids. *Hum Reprod* 2001;16:1938–44

15. Sun JG, Jurisicova A, Casper RF. Detection of deoxyribonucleic acid fragmentation in human sperm: correlation with fertilization *in vitro. Biol Reprod* 1997;56:602–7

16. Aitken RJ, Gordon E, Harkiss D, *et al.* Relative impact of oxidative stress on the functional competence and genomic integrity of human spermatozoa. *Biol Reprod* 1998;59:1037–46

17. Host E, Lindenberg S, Smidt-Jensen, S. DNA strand breaks in human spermatozoa: correlation with fertilization *in vitro* in oligozoospermic men and in men with unexplained infertility. *Acta Obstet Gynecol Scand* 2000;79:189–93

18. Host E, Lindenberg S, Smidt-Jensen, S. The role of DNA strand breaks in human spermatozoa used for IVF and ICSI. *Acta Obstet Gynecol Scand* 2000;79:559–63

19. Kierszenbaum AL. Apoptosis during spermatogenesis: the thrill of being alive. *Mol Reprod Dev* 2001;58:1–3

20. Gandini L, Lombardo F, Paoli D, *et al.* Study of apoptotic DNA fragmentation in human spermatozoa. *Hum Reprod* 2000;15:830–9

21. Tesarik J, Greco E, Cohen-Bacrie P, Mendoza C. Germ cell apoptosis in men with complete and incomplete spermiogenesis failure. *Mol Hum Reprod* 1998;4:757–62

22. Jurisicova A, Lopes S, Meriano J, *et al.* DNA damage in round spermatids of mice with targeted disruption of the *Pp1cγ* gene and in testicular biopsies of patients with non-obstructive azoospermia. *Mol Hum Reprod* 1999;5:323–30

23. Gorczyca W, Traganos F, Jesionowska H, Darzynkiewicz Z. Presence of DNA strand breaks and increased sensitivity of DNA *in situ* to denaturation in abnormal human sperm cells: analogy to apoptosis in somatic cells. *Exp Cell Res* 1993;207:202–5

24. Aravindan GR, Bjordahl J, Jost LK, Evenson DP. Susceptibility of human sperm to *in situ* DNA denaturation is strongly correlated with DNA strand breaks identified by single-cell electrophoresis. *Exp Cell Res* 1997;236,231–7

25. Sakkas D, Mariethoz E, St John JC. Abnormal sperm parameters in humans are indicative of an abortive apoptotic mechanism linked to the Fas-mediated pathway. *Exp Cell Res* 1999;15:350–5

26. Barroso G, Morshedi M, Oehninger S. Analysis of DNA fragmentation, plasma membrane translocation of phosphatidylserine and oxidative stress in human spermatozoa. *Hum Reprod* 2000;15:1338–44

27. Oosterhuis GJ, Mulder AB, Kalsbeek-Batenburg E, *et al.* Measuring apoptosis in human spermatozoa: a biological assay for semen quality? *Fertil Steril* 2000;74:245–50

28. Wiltshire T, Park C, Caldwell KA, Handel MA. Induced premature G2/M-phase transition in pachytene spermatocytes includes events unique to meiosis. *Dev Biol* 1995;169:557–67

29. Tesarik J, Nagy ZP, Sousa M, *et al.* Fertilizable oocytes reconstructed from patient's somatic cell nuclei and donor ooplasts. *Reprod Biomed Online* 2001;2:160–4

30. Lacham-Kaplan O, Daniels R, Trounson, A. Fertilization of mouse oocytes using somatic cells as male germ cells. *Reprod Biomed Online,* 2001;2:203–9

31. Amer M, Soliman E, El-Sadek M, Mendoza C, Tesarik J. Is complete spermiogenesis failure a good indication for spermatid conception? *Lancet* 1997;350,116

32. Sousa M, Barros A, Takahashi K, *et al.* Clinical efficacy of spermatid conception: analysis using a new spermatid classification scheme. *Hum Reprod* 1999;14:1279–86

33. Al-Hasani S, Ludwig M, Palermo I, *et al.* Intracytoplasmic injection of round and elongated spermatids from azoospermic patients: results and review. *Hum Reprod* 1999;14(Suppl. 1):97–107

34. Ghazzawi IM, Alhasani S, Taher M, Souso S. Reproductive capacity of round spermatids compared with mature spermatozoa in a population of azoospermic men. *Hum Reprod* 1999;14:736–40

35. Gianaroli L, Selman HA, Magli MC, *et al.* Birth of a healthy infant after conception with round spermatids isolated from cryopreserved testicular tissue. *Fertil Steril* 1999;72:539–41

In vitro maturation of human follicles and oocytes

19

Geraldine M. Hartshorne

INTRODUCTION

The development and growth of oocytes and their follicles are inextricably linked. Follicles are formed around the mid-trimester of human gestation, when somatic cells encircle the nascent oocytes and remain closely associated with them for as long as the oocyte/follicle unit remains viable. Many crucial hurdles must be overcome if a follicle is to grow and survive the stringent selection processes to produce a mature fertile oocyte. This chapter will provide an overview of the main stages of follicle and oocyte development, and document our progress towards supporting some key stages *in vitro*. The goal of complete *in vitro* development of human female gametes is many years away, but would have potential applications in both research and clinical settings. These would include the management of fertility and subfertility as well as related problems such as aging and menopause. Clinical developments in these areas are currently restricted by the limited supply of human oocytes for study, a situation which could be alleviated by improved *in vitro* methods for the prodution of viable research oocytes.

BASIC PHYSIOLOGY OF FOLLICLE FORMATION AND GROWTH

Follicle formation

Oocyte and follicle formation in the fetal gonad start when primordial germ cells arrive after migration from the yolk sac[1]. Their path requires interaction with certain integrins to ensure migration, and survival factors such as stem cell factor (SCF, also known as Kit ligand) support mitosis along the way as well as appropriate migration[2]. In humans, these migrant germ cells arrive at the gonad around the fourth week post-fertilization[3]. There, they continue to divide by mitosis, greatly increasing in numbers, and become oogonia, the precursors of the oocytes. These also divide, though at a slower rate than previously, and form clusters known as nests. The nests generally comprise oogonia or oocytes which are probably sister cells having cytoplasmic continuity via bridges and developing in synchrony[4,5]. Oogonia enter meiosis towards the end of the first trimester or early in the second[6]. While cells in the nests tend to undergo meiosis in concert, across the whole ovary, the process is asynchronous, and various stages of meiosis are present concurrently until around birth. Oocytes do not complete meiosis prenatally, but those destined to form the ovarian reserve become arrested at the end of meiotic prophase I, termed primary oocytes, arrested in the diplotene stage.

Germ cell size increases during meiotic prophase I[3] but the oocyte at this stage is still small (approximately 25 μm diameter), little larger than the ovarian somatic cells, has no zona pellucida (ZP) and resides in its 'nest' with other oocytes[3,7]. Specialized somatic cells of the gonad surround the individual oocytes with very fine processes to form the earliest primordial follicles from shortly before 20 weeks of gestation[3]. A peak of between 2 and 7 million oocytes is present in a human around this time[8]; however, by birth, around 70% of these will have been lost to atresia, principally via apoptosis[7,9–11]. Oogonia and oocytes start to undergo atresia from the fifth month of gestation[3], and some of the surviving follicles begin to grow almost immediately, resulting in a variety of follicle stages present in the infant ovary. Only a minority remain in their primordial state, forming the 'ovarian reserve'.

A proportion of these primordial follicles initiates growth each day throughout life until the menopause, but only those follicles which start to grow within the woman's reproductive years will be capable of contributing to future generations. It is uncertain whether the population of primordial germ cells surviving beyond puberty is in some way selected or different to those suffering early demise; however, as a small minority of the original pool, this possibility is important. It would be particularly relevant if oocytes from fetal or prepubertal sources were used in future for the production of viable oocytes with the intention of producing offspring. In animals, the use of oocytes from immature individuals would be especially useful in reducing the inter-generation time for valuable stock or rare and endangered breeds. From a practical perspective, the fetal or juvenile ovary is the age at which oocytes are most plentiful, and so it is the richest source of immature oocytes for study. However, our knowledge of these early stages of follicle formation and the factors controlling the ovarian reserve and the initiation of follicle growth is rudimentary, especially in humans.

Follicle growth

The early stages of growth, as follicles exit the primordial pool, occur very slowly and are difficult to recognize histologically[12]. Primordial follicles comprise a single layer of squamous granulosa cells surrounding an oocyte which has not initiated its major growth phase. Signs of early follicle activation include growth of the oocyte and its nucleus beyond a threshold size and a preponderance of cuboidal cells among the granulosa cell layer (such as shown in the follicle in Figure 1), although the occurrence of occasional cuboidal cells or even a primary follicle appearance may not be a firm indication of follicle growth initiation[13]. Molecular markers for the onset of growth are under investigation, such as proliferating cell nuclear antigen (PCNA), a marker of cell division[14]. However, the controlling factors remain uncertain and it is not known whether the oocyte, the granulosa cells or some external signal, e.g., growth differentiation factor-9 (GDF-9)[15], represent the initiating factor. Moreover, there may be differences in the way that follicles are recruited to grow between fetal and adult ovaries, because a significantly higher proportion

of active follicles is evident in fetal ovaries[14]. Follicles appear programed to grow, perhaps through progressive release from local inhibitory feedback mechanisms from other follicles[16] or in response to some positive signal such as SCF or transforming growth factor α (TGFα)[14,17]. The possibility of a 'production line' or time-dependent process remains[18,19].

Follicle growth entails the rapid mitosis of granulosa cells which results in a follicle of increasing size. During follicle growth, the granulosa cells remain closely apposed and surrounded by a basement membrane, forming a solid spherical structure with the oocyte in the center. Initially, follicle and oocyte growth occur concurrently and these are tightly aligned processes[20]. The ZP is secreted by the oocyte during its growth period, which occurs mostly in the preantral follicle stages[21]. In humans, the oocyte is mostly fully grown by the time the follicle becomes antral, usually at around 200–1000 μm, but extended processes of the granulosa cells passing through the ZP remain in contact with the oocyte's surface via gap junctions and have a critical role in oocyte–cumulus communication[22]. Once the oocyte approaches full size, its growth slows while the follicle continues to enlarge by granulosa cell division, and the granulosa cells also begin to produce follicular fluid which accumulates to form pools in the intercellular spaces. The oocyte remains central until eventually displaced towards one side as the follicular fluid-filled antrum expands; however, it is surrounded by several layers of specialized granulosa cells which then become

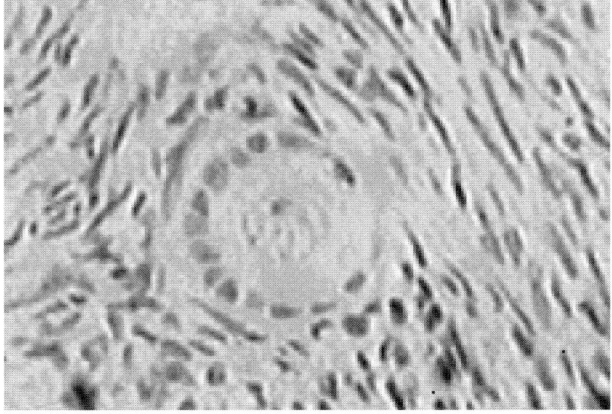

Figure 1 Section through a human follicle in transition between primordial and primary stages, present in an ovarian cortical biopsy from a volunteer patient aged 30 years undergoing sterilization. Courtesy of Dr. P. Starr

known as the cumulus oophorus, the inner layer of which is known as the corona.

Specific receptors present in the granulosa cell layers mediate follicle development at different stages. The primordial follicle is non-responsive to gonadotropins[23,24], and the factors controlling its growth are not clear, although a variety of growth factors and other molecules have been implicated[25]. Once follicle stimulating hormone (FSH) sensitivity is achieved, around the secondary to preantral stage in humans, the growing follicle becomes responsive to the waves of pituitary gonadotropins associated with the menstrual cycle which control the later stages of its growth, and the follicle selection process[26].

The enlarged follicle is also the major site of sex steroid production, producing estrogens and progestagens which are active systemically for e.g., secondary sexual characteristics, brain differentiation and signaling to the uterus in preparation for implantation. Luteinizing hormone (LH) and FSH cooperate in the control of follicular steroid production. While the follicle's basic endocrinology is controlled by pituitary feedback mechanisms of systemic hormones including steroids, inhibins etc., a multiplicity of local factors also contribute. These modulate the sensitivity of granulosa cells to stimuli, cytokine levels in the follicular fluid, etc., to the point where no two follicles in the ovary have identical internal microenvironments, even through they may have been exposed to similar systemic influences.

Ovulation in young women generally occurs monthly, unless pregnancy, breastfeeding or pathologic anovulation intervenes. Hence, the vast majority of oocytes in the ovarian reserve are never ovulated and their follicles are lost through atresia. Before the initiation of follicle growth, it appears that atresia can arise in the oocytes of primordial follicles resulting in follicle degeneration. After a follicle has begun to grow, only a very small proportion of degenerate cells among the granulosa cell population is observed in healthy follicles; however, whether cell death in granulosa cells is a cause or an effect of follicle demise is not clear. The control of follicle cell death is being studied extensively because of its potential to enable the control of ovarian function in both animals and humans. Knockout of *bax*, a part of the pro-apoptotic pathway, results in larger numbers of surviving follicles, indicating selection to be mediated by

an apoptotic mechanism, but the nature of the control is uncertain as *bax* is found in both oocytes and granulosa cells[27]. Overexpression of *bcl-2* in mouse ovarian somatic cells but not oocytes also results in increased follicle survival, increased litter sizes and possibly germ cell tumorigenesis, indicating a major role of the somatic cells in the normal balance of follicular atresia[28]. Atretic oocytes in growing follicles are only found together with evidence of follicle atresia[11]; however, the apoptotic pathways in oocytes and granulosa cells appear to be separate and distinct, only the latter being caspase 3-dependent[29].

The growth of a secondary follicle to preovulatory size in the human is believed to take around 85 days, but the extended period of follicle growth initiation which precedes this may be much longer[12]. Throughout all these stages of follicle and oocyte growth, with the exception of atresia, the oocyte remains in the same stage of meiosis, namely diplotene of meiotic prophase I, and is still termed a primary oocyte, despite increasing its volume by about 400-fold and changing its nature and potential markedly.

OOCYTE MATURATION

Oocytes become competent to mature, i.e., progress beyond diplotene of meiotic prophase I to reach metaphase II, when they have achieved a certain threshold size which is sufficient to overcome the size-related checkpoint to oocyte maturation[30]. Human oocytes considered fully grown in the histologic series of follicle growth discussed above[20] had an average diameter of 80 μm; however, immature oocytes collected from unstimulated ovaries average around 107 μm diameter (Cavilla *et al.*, unpublished) and the diameter of a fully grown mature human oocyte collected for *in vitro* fertilization (IVF) procedures is normally around 111–114 μm[31]. This difference could be explained either by shrinkage during histologic processing, and/or further growth of oocytes at a later stage of follicle growth. Moreover, Lintern Moore's series[20] was performed in infant ovaries and an age effect could be present. No similarly detailed study has been presented for adult humans.

Maturational competence, first to metaphase I, then to metaphase II, competence for fertilization of the mature oocyte and embryo development to various stages are attained sequentially in humans, as is well described in mammalian species[32–34]. Human oocytes of increasing diameter and those arising from increasingly large follicles are more likely to have attained competence in each activity[31,35–38], suggesting an ongoing process of oocyte development throughout all stages of follicle growth. For this reason, it may be difficult to find a single set of conditions which support the whole sequence of oocyte development *in vitro*. The influence of follicle cells is key to oocyte development and therefore attempts should focus on supplying environments which support the somatic components optimally as this will reflect positively on oocyte development. Even a short period of removal from the follicle environment can impact upon oocyte development and drastically reduce developmental competence.

The competence of the oocyte to progress through meiosis can be dissociated from its ability to decondense a sperm, form pronuclei, cleave, form a blastocyst and form a viable conceptus[30]. Such important aspects of the oocyte's development are collectively known as 'cytoplasmic maturation'. This imprecise term covers, for example, the appropriate processing of messenger RNAs previously accumulated during oocyte growth, the formation and peripheral arrangement of cortical granules, the development of calcium signaling mechanisms for oocyte activation, the ability for embryo cleavage and mitosis and the ability to unpackage the fertilizing sperm DNA and enable embryonic genome activation[39–41]. While incompetent oocytes can be forced by molecular manipulation to resume meiosis[42], their cytoplasmic immaturity remains evident in subsequent aspects of development.

Few features of oocytes or follicles may be used to indicate competence. The multitude of parallel redundant pathways which have evolved to support the fundamental processes of oocyte and follicle development limit the predictive value of individual markers and the complexity and variety of follicular components is astonishing. One possible marker for a potentially competent follicle is the extent of its local blood supply, relating to the oxygenation, pH and temperature of the follicle and its oocyte[43,44]. These factors are known to affect oocyte survival and normal function[45–47]. Sufficient oxygenation is essential to enable the maturing oocyte to produce a normal spindle, as shown in cultured

mouse oocytes[48]. The ability to identify readily an incompetent oocyte or follicle with a simple minimally invasive test would be a welcome development.

Fully grown oocytes mature during the final 37 h of follicle development, corresponding *in vivo* to the time between the LH surge (or other ovulatory stimulus) and ovulation. The LH surge from the pituitary gland acts upon receptors in the granulosa cells of the preovulatory follicle, triggering a cascade of irreversible events which culminate in ovulation of the follicle and complete (meiotic and cytoplasmic) maturation of the oocyte. LH receptors are most concentrated in the outer layers of the granulosa compartment, closest to the perifollicular blood supply. The cumulus cells surrounding the oocytes have relatively few LH receptors, and inappropriate exposure of mouse oocyte–cumulus complexes to FSH may result in inappropriate maturation associated with precocious and abnormal LH receptor expression[49]. Sensitivity of the follicle to LH requires prior exposure to FSH and the isoforms of FSH may have specific roles in preparing the follicle and oocyte for later development[50]. It is not yet understood how the follicle and its granulosa cells normally transmute the signal from the outer LH receptors to the cumulus cells and the oocyte itself to cause the changes elicited by the LH surge, but both release from inhibitors and positive signals promoting meiotic progression are implicated. The granulosa cells are connected by gap junctions and in some respects act as a syncitium, despite major variations in the characteristics of granulosa cells in different parts of the follicle[51,52]. Local factors within the follicle, which is known to be extremely rich in numerous bioactive growth factors, cytokines and macromolecules, fine-tune the follicular response at the cellular level[53]. Attempts have been made to identify some of the key components of oocyte maturation by comparing the compositions of mature and immature oocytes and their follicles[38,54]. This approach, combined with single-cell technology, may in future allow more detailed understanding of the molecular mechanisms involved in competence attainment by maturing human oocytes[55].

Oocyte maturation occurs reliably *in vivo* and only occasionally do meiotically immature oocytes persist after an appropriate ovarian stimulation regimen. The competence of the embryos resulting from IVF of *in vivo* matured oocytes is difficult to assess, but as around 50% of them become blastocysts *in vitro* and perhaps 25% lead to implantation and birth, competence seems to be nearly equivalent to *in vivo* embryo development[56]. A different scenario is encountered when the final stages of oocyte maturation occur *in vitro*.

IN VITRO MATURATION

In clinical IVF, fertile, fully grown oocytes are collected from follicles when ovulation is imminent. These oocytes are at a 'mature' stage, which in humans is metaphase II of meiosis. Such oocytes are ready to become fertilized by a competent sperm; however, their lifespan is short as they remain capable of fertilization and further development for only around 24 h.

The need to collect oocytes at a precise time just before ovulation is inconvenient, and in modern practice relies upon hormone stimulation to gain several oocytes concurrently. Early work in humans[57] showed that, in common with various other mammals, immature human oocytes will undergo the final stages of meiotic maturation (from prophase I to metaphase II) spontaneously *in vitro*, with timing similar to that occurring *in vivo*, suggesting a 'programed' mechanism. However, in common with several other species, such as cattle, rabbits, rats and hamsters, human oocytes obtained in this way appeared to lack developmental competence[58]. This has been explained in terms of the dissociation of meiotic maturation and cytoplasmic maturation mentioned above; however, the conditions required to assure complete cytoplasmic maturation *in vitro* have not yet been optimized.

During the development of human IVF, *in vitro* maturation (IVM) was abandoned in favor of carefully timed aspirations of mature oocytes[59]. Since then, various attempts at IVM in humans have produced only modest success; however, over the past 10 years, considerable effort has been focused upon revisiting IVM and a summary of recent work in this area is presented in Tables 1 and 2. Some more encouraging data are now accumulating, although unreliability has resulted in this method not gaining wide acceptance, and limited developmental competence continues to be a problem[38,60].

Recent successful clinical protocols of IVM have generally collected immature oocytes from large antral follicles of patients during the mid-late follicular phase of the menstrual cycle or after a withdrawal bleed in amenorrheic patients (Table 1). Some authors now routinely give human chorionic gonadotropin (hCG) before oocyte collection, although they still collect immature oocytes, possibly because of follicular immaturity to respond to the ovulatory stimulus. Uterine preparation for implantation is usually achieved with exogenous hormone replacement therapy from the day of oocyte collection, although occasional success with cryopreservation of embryos for a postponed transfer has been reported (Tables 1 and 2). The culture medium applied for oocyte maturation *in vitro* is usually 'complex' including amino acids and vitamins, and often includes hormone supplementation with gonadotropins and estrogen with variable protein supplementation. Oocyte maturation is most successful when the immature oocyte at collection is closely surrounded by a dense cluster of unexpanded cumulus cells, which remain in the culture until maturation has occurred. An example of such an oocyte is shown in Figure 2. The endogenous factors which mediate effects within the cumulus are numerous; however, while the oocyte may respond to a few factors directly, its responses are usually orchestrated by the cumulus cells. Identification of receptors expressed at the oocyte surface is difficult and oocytes are notoriously 'sticky' for non-specific antibody binding. Moreover, the gap junctional communication with cumulus cells allows interaction via non-oolemma receptors and is even more difficult to study.

The presence of closely attached cumulus cells generally indicates that the originating follicle is likely to be non-atretic; however, naked or poorly surrounded oocytes from atretic or abnormal follicles can also mature *in vitro* and produce successful pregnancies[61] with the inclusion of somatic cells with naked oocytes potentially beneficial[62]. For example, the peripheral follicles of patients with polycystic ovaries may yield oocytes suitable for IVM[63]. This apparent anomaly is not yet fully explained, but might result in part from the initiation of spontaneous oocyte maturation when close communication with the follicle is lost. Oocytes in atretic follicles sometimes appear to initiate maturation[64] and this may help to start the process which is then continued *in vitro*. It is possible that the deliberate induction of incipient atresia by FSH withdrawal in the follicular phase may be beneficial to IVM[65]. If the maturation process is begun

Table 1 Comparison of published studies with *in vitro* maturation of non-frozen human oocytes applied for for clinical treatment and resulting in embryo transfer

References	Patients/cycles	Drugs in vivo	Oocyte collection	Culture conditions	Maturation	Fertilization	Cleavage	Pregnancy
Veeck et al., 1983[100]	44 assorted; IVF	hMG/hCG	74 immature	22–35 h; Ham's F10	—	IVF	—	Two (immature eggs)
Cha et al., 1991[101]	Gynecologic operations		270 eggs from 23 removed ovaries	32–48 h; Ham's F10 + 50% FF or 20% FCS	FF (55.8%); cord (35.9%)	FF (81%); cord (31.6%)	1 × 5 ET to recipient with POF	One set of triplets
Trounson et al., 1994[63]	19 PCO and 23 ovulatory, with many 2–10 mm follicles by u/s	E2 and P4 from day 2 post-OCR	2.8–15.3/patient	21–54 h; Eagle's MEM; Earle's salts; gln; 10% FCS, hMG 75 mIU/ml; hCG 500 mIU/ml; E2 1 µg/ml; granulosa cells	65–81%	34%	56% to ≥ 8c	One
Russell et al., 1997[102]	14 assorted lacking dominant follicles (≥ 1.2 cm) on day 5–7	E2 from day 5–7; P4 from 1–2 day post-OCR	Day 3 n = 83; day 5–7 n = 78	52 h; Eagle's or TCM199; FSH or hMG 75 mIU/ml; hCG 500 mIU/ml; E2 1 µg/ml; 3% SSS	34/83 (39.7%) 48/78 (61.5%)	25/34 (75.7%) 36/48 (75%) ICSI	16/25 (64%) 33/36 (92%)	One
Cha et al., 1998[61]	72 PCOS		n = 832	TCM199; 20% FCS; 10 U/ml PMSG; 10 U/ml hCG	499/832 (60%)	364 (73%)	90%	16/64 (25%/transfer)
Thornton et al., 1998[103]	123 cycles spontaneous ovulatory, < 40 years, no male factor; cancel if LH rise pre-hCG	hCG midcycle 36 h pre-OCR	111 immature eggs from 59 of 123 OCRs; 55 GV, 56 MI	24 h–5 days; medium not described; 25 GVs medium only, 30 GVs 50% FF, 56 MIs medium only	8/25 (32%) 9/30 (30%) 25/56 (45%)	5/8 (62%) 7/9 (78%) 20/25 (80%)		Two from immature oocytes only
Chian et al., 1999[104]	20/25 PCO aged < 41	P4 withdrawal; hCG 36 h to OCR	Day 10–14	24–48 h; TC 199; 20% patient serum, pyr 25 mmol/l; FSH + LH; 75 mIU/ml	209/249 (84%)	ICSI		10 (40%) (two miscarried)
Mikkelsen et al., 1999[105]	32	± rFSH, day 3 until lead follicle = 10 mm; OCR 24 h later	n = 115	36–48 h; TCM199; pyr 0.3 mol/l; E2 1 µg/ml; rFSH 75 mIU/ml; hCG 500 mIU/ml; 10% patient serum	89/115 (77%)	ICSI	65	6/32 (19%/cycle)
Jaroudi et al., 1999[106]	18/21 assorted etiology	GnRHa + hMG, risk of OHSS	n = 171	44 h; HTF; + 10% SSS; hMG 75 mIU/ml; hCG 500 mIU/ml	121 (70.7%)	71/121 (58.7%) ICSI	53/71 (75%) 44 transferred	2/17 (12%/transfer) 2/21 (9%/cycle)
Chian et al., 2000[66]	17/24 PCOS	± 10 000 IU hCG	day 10–14	24–48 h; TCM199; 20% FCS; 75 mIU/ml FSH+LH	86/102 + hCG 56/81 – hCG	ICSI		5 (38.5%) 3 (27.3%)

Study	Patients	Protocol	n	Medium / maturation time	Maturation	Fertilization (IVF/ICSI)	Frozen / cleaved	Pregnancy / miscarriage
Suikkari et al., 2000[60]	6 regular cycles; 6 irregular/PCO	rFSH from LH + 11 or P4 + 9; withdraw for 2–5 days when lead follicle = 10 mm	n = 127 non-atretic oocytes; 67 from regular, 69 from irregular	44 h; TCM199; 10% FCS; rFSH 75 mIU/ml; rLH 500 mIU/ml; 0.29 mmol/l pyr	43/67 (64%) / 47/60 (78%)	31/43 (72%) / 27/47 (57%) ICSI	All frozen 2 PN or cleaved	1/15 miscarriage
Smith et al., 2000[107]	48/55 regular cycles, no PCO; 29 cycles 28 h; 26 cycles 36 h	E2 from OCR, P4 from day 2 post-OCR	172 (107 used) 136 (84 used)	28 h or 36 h; TCM 199; pyr 0.3 mmol/l; E2 1 µg/ml; rFSH 75 mIU/ml; hCG 500 mIU/ml; 10% patient serum	78/107 (73%) / 65/84 (77%)	56/78 (72%) / 51/65 (78%) ICSI	50/56 (89%) / 49/51 (96%)	4/21 (19%/transfer) / 4/20 (20%/transfer)
Cha et al., 2000[108]	64/94 PCOS	10–13 day post-menses or post withdrawal bleed	n = 1280 (1139 used)	48 h; TCM199 + Earle's salts, pyr 0.29 mmol/l; 10 IU/ml PMSG; 10 IU/ml hCG; 20% FCS	708/1139 (62.2%)	481/708 (68%) ICSI	266/302 (88.1%) 4.9 ± 2.5 ET	23/85 (27%/transfer) (17 deliveries, 3 twin, 5 miscarriages, 1 ectopic)
Mikkelsen et al., 2000[109]	75/87 normal unstimulated; male and/or tubal; 18–37 y; normal endocrinologically, ≤ three attempts	Lead follicle = 10 mm; endo ≥ 5 mm, E2 from OCR, P4 from day 2 post-OCR	n = 532 (388 used)	28–36 h; TCM199 or Medicult IVM medium; pyr 0.3 mmol/l; E2 1 mg/ml; rFSH 75 mIU/ml; hCG 500 mIU/ml; 10% patient serum or 0.5% hSA	234/388 (60%)	180/234 (77%) ICSI	156/180 (87%) (125 transfers ≤ 2 ET)	11/63 (13%/cycle, 17%/transfer); 9 deliveries; 2 miscarried
Mikkelsen et al., 2001[110]	100/132 male factor and/or tubal, regular cycles, ≤ 3 attempts, no PCOS	E2 from OCR, P4 from day 2 post-OCR; OCR when follicle = 10 mm, endo ≥ 5 mm	n = 714 (509 used)	28–36 h; TCM199; pyr 0.3 mmol/l; E2 1 µg/ml; rFSH 75 mIU/ml; LH 500 mIU/ml; 10% patient serum	306/509 (60%)	223/306 (73%) ICSI	193/223 (87%)	15/83 (18%/transfer, 11%/cycle); 3 miscarried
Mikkelsen et al., 2001[111]	A: 53; B: 28; regular cycles, no endocrine anomalies	As above; A without and B with 100% increase in E2 relative to day 3	n = 308	28–30 h; TCM199; pyr 0.3 mmol/l; E2 1 mg/ml; rFSH 75 mIU/ml; hCG 500 mIU/ml; 10% patient serum	A: 107/184 (58%); B: 56/124 (45%)	3/107 (68%); 35/56 (62.5%) ICSI	56/73 27/35	9/53 (17%/cycle); 0/28 (0%)
Yoon et al., 2001[112]	51/63 natural cycles	OCR day 7–13 when lead follicle = 10 mm; hCG 48 h *after* OCR; progynova from OCR; P4 from day 2 post-OCR	n = 568 (506 used)	24–56 h; YS medium 70% FF	206/506 at 24 h 362/506 at 48 h 376/506 at 56 h	273/376 (72.6%) IVF or ICSI	243/273 (89%); mean ET = 3.6 embryos	9/51 (17.6%/transfer) 3 miscarried

IVF, *in vitro* fertilization; hMG, human menopausal gonadotropin; hCG, human chorionic gonadotropin; FF, follicular fluid; FCS, fetal calf serum; ET, embryo transfer; POF, premature ovarian failure; PCO, polycystic ovaries; u/s, ultrasound; E2, estrogen; P4, progesterone; OCR, oocyte recovery; MEM, minimal essential medium; gln, glutamine; TCM, tissue culture medium; SSS, synthetic serum substitute; PCOS, polycystic ovary syndrome; PMSG, pregnant mares serum gonadotropin; LH, luteinizing hormone; GV, germinal vesicle; MI, meiosis I; pyr, sodium pyruvate; rFSH, recombinant follicle stimulating hormone; GnRHa, gonadotropin-releasing hormone analog; OHSS, ovarian hyperstimulation syndrome; HTF, human tubal fluid medium; rLH, recombinant luteinizing hormone; PN, pronuclei; endo, endometrium; IVM, *in vitro* maturation; hSA, human serum albumin

Table 2 Case reports with *in vitro* maturation of non-frozen human oocytes applied for clinical treatment and resulting in embryo transfer

References	Patients/cycles	Drugs in vivo	Oocyte collection	Culture conditions	Maturation	Fertilisation	Cleavage	Pregnancy
Paulson et al., 1994[113]	1	Natural cycle; serial u/s; OCR 36 h post-hCG	3 (1 × MII, 1 × MI, 1 × GV)	48 h; 50% FF (Cha et al., 1991[101])	MII and MI not fertilized; 1/1 GV IVM	1/1 IVF	1/1 frozen at 2 PN	1/1
Barnes et al., 1995[114]	1/3	nil	Day 10–12 of spontaneous menstual cycle, 19/20/13 eggs	46 h; TCM199 + 10% FCS + 75 mIU/ml rFSH, 500 mIU/ml hCG, 0.29 mmol/l pyr	8(47%) 11 (55%) 10 (77%)	0 IVF 5 (45%) ICSI 8 (80%) ICSI	0 3 One blast	One term
Nagy et al., 1996[115]	1	GnRHa, hMG, hCG	14 GV oocytes 36 h post-hCG	30 h; cumulus cell coculture	9/14 (64%)	7/8 ICSI	5/7 (71%); four transferred	One delivered
Kodama et al., 1996[116]	One at risk of OHSS			30% FF + granulosa cells	1	1	One cryo-preserved	None
Edirisinghe et al., 1997[117]	One ICSI patient	GnRHa + FSH flare protocol	11 MII 2 MI 3 GV	24 h; HTF + 10% human serum	(6/11 mature MII) 5/5 IVM	4/5 ICSI	Frozen as 2 PN	FET 2 embryos surviving; one delivery
Liu et al., 1997[118]	1	GnRHa + metrodin + hMG, no hCG	5	48 h, B2 + FSH 50 IU/ml + hCG 50 IU/ml	5/5	1/5 2 PN 2/5 3 PN ICSI	Cryopreserved at four-cell stage	One
Jaroudi et al., 1997[119]	1 OHSS	GnRHa/hMG	Transvaginal u/s		9/10	7/9 ICSI		One miscarriage
Hwang et al., 1997[120]	One egg donor; one egg recipient; POF	Syringe aspiration at Cesarian section; HRT in recipient	n = 7	48 h; TCM199; PMSG 10 IU/ml; hCG 10 IU/ml; bSA 20%	2/7	2/2 ICSI	2/2	One
Abdul-Jalil et al., 2001[121]	1 regular cycles with PCO	hCG 36 h before OCR; OCR on day 9; E2 from OCR, P4 from ICSI	n = 12	24 h; TCM199; FSH + LH 75 mIU/ml; pyr 25 mol/l (sic); 20% patient serum	6/12 (50%)	4/6 (67%) ICSI	3 ET	One twin pregnancy
Chian et al., 2001[122]	1 PCOS	P4 for 10 days for withdrawal bleed; hCG 36 h before OCR	OCR day 11 post-bleed; n = 63 (53 used)	24–48 h; TC199; FSH + LH 75 mIU/ml 20% patient own serum	41/53 (77%)	31/41 (76%) ICSI	15 cultured 16 cryopre-served 2 PN	3 ET miscarriage; 3 ET no pregnancies; 3 ET delivered

u/s, ultrasound; OCR, oocyte recovery; hCG, human chorionic gonadotropin; MI, meiosis I; MII, meiosis II; GV, germinal vesicle; FF, follicular fluid; IVM, *in vitro* maturation; IVF, *in vitro* fertilization; PN, pronuclei; TCM, tissue culture medium; FCS, fetal calf serum; rFSH, recombinant follicle stimulating hormone; pyr, sodium pyruvate; ICSI, intracytoplasmic sperm injection; GnRHa, gonadotropin-releasing hormone analog; hMG, human menopausal gonadotropin; OHSS, ovarian hyperstimulation syndrome; FET, frozen embryo transfer; bSA, bovine serum albumin; POF, premature ovarian failure; HRT, hormone replacement therapy; PMSG, pregnant mares serum gonadotropin; PCO, polycystic ovaries; E2, estrogen; P4, progesterone; FSH, follicle stimulating hormone; LH, luteinizing hormone; PCOS, polycystic ovary syndrome; ET, embryo transfer

in vivo with a dose of hCG, but continued *in vitro*, results appear to be improved[66]. Similarly, oocytes derived from follicles previously exposed to a standard stimulation regimen, yet remaining immature at collection, can be effectively matured *in vitro*.

Oocytes which have matured *in vitro* are normally fertilized by intracytoplasmic sperm injection (ICSI) rather than standard IVF for several reasons: the peripheral organization of cortical granules may be incomplete *in vitro*, raising the risk of polyspermy and changing the

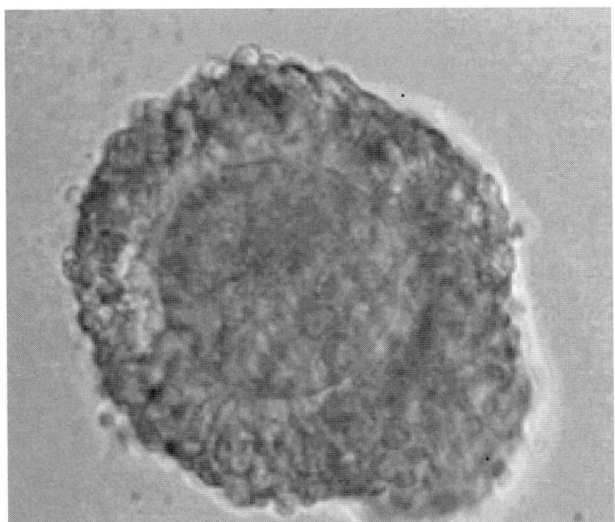

Figure 2 Human immature oocyte surrounded by dense cumulus, retrieved from a volunteer patient with polycystic ovaries. Courtesy of Miss J. Cavilla

cortical environment; the function of cumulus cells in preparing the oocyte and sperm for fertilization may be compromised by suboptimal *in vitro* conditions; and the ZP may harden prematurely *in vitro*. ICSI can result in quite reliable fertilization rates; however, there is an elevated incidence of abnormal fertilization in *in vitro* matured oocytes, even after ICSI, which may result from post maturity[67] or problems in extrusion of the second polar body. The germinal vesicle, initially in the center of the oocyte, moves to the periphery before breaking down as the oocyte resumes meiosis from prophase I. The first polar body is formed at telophase I and its dimensions relate to the peripheral positioning of the spindle which mediates separation of the chromosomes, and the cytoskeleton which holds it in the peripheral position. Irregular polar body formation, particularly an enlarged first polar body, may occur in *in vitro* matured oocytes, possibly because of disturbances in this peripheral positioning mechanism[68]. Moreover, an elevated incidence of arrest of fertilized oocytes prior to cleavage is a common finding (see Table 1).

A consensus has not yet been reached on ideal conditions for IVM of human oocytes. Most workers in clinical IVM add FSH and LH or hCG to the cultures in an attempt to mimic the surge conditions *in vitro*. These are considered relatively 'safe' options, as the patient has already been exposed to them. While dose–response data

in humans are lacking, and the role of FSH is complex[50], in rhesus monkeys, addition of gonadotropins *in vitro* proved a significant advantage for fertilization and embryo development[69]. In macaques, FSH combined with LH for IVM promoted faster post-fertilization development[70]. Less well characterized substances whose effects, though positive, are less well understood have not yet been used in clinical trials, e.g., epidermal growth factor (EGF)[71] and meiosis activating sterol (MAS)[72].

We are at an early stage in the application of IVM for human pregnancies and it is most important that the data available be reported in full and children resulting from this new technique be carefully monitored. This is particularly important for IVM for two reasons. Firstly, some animal species in which IVM is applied regularly have presented abnormalities in offspring as well as reduced developmental competence[73], although the relationship of such problems to the IVM process rather than some other *in vitro* manipulation is not established[74]. Second, the molecular events occurring in oocyte maturation are not fully understood. The possible disturbance of some crucial aspect may therefore have unpredictable effects upon development.

Concerns such as these become greater the further back in follicle and oocyte development the maturation *in vitro* is attempted, because more of the events relating to oocyte development are unknown in the earlier stages. The control of oocyte growth is a crucial area which remains almost completely unstudied in humans[75] and the methods used as well as the results obtained are likely to differ for IVM of fully grown, growing and primordial oocytes. The one animal (mouse) produced via culture of a primordial follicle using a sequence of culture systems[76] has done little to alleviate such concerns, being of dubious health and morbidly obese.

FOLLICLE GROWTH *IN VITRO*

Since the process of removing a fully grown oocyte from its follicle, even for the short period preceding ovulation, has major effects upon the efficiency of oocyte maturation and subsequent embryonic competence, the possibility of culturing oocytes together with their follicles or follicle-derived cells has been considered for many years[77]. For large follicles, any culture period must necessarily be short, otherwise diffusion-limited processes will rapidly fail[78]. However, there are few applications for removing a fully grown follicle from the ovary, except for research purposes.

For animals in which the maximum follicle size is relatively small, follicles may sometimes be cultured to full size *in vitro*. Outstanding success has been obtained with this approach in mice, allowing individual isolated primary follicles to reach full size (~ 500 μm) *in vitro*, with oocytes gaining meiotic competence and maturing *in vitro* in response to an ovulatory signal[79]. Developmental potential of the oocytes produced has been demonstrated by the production of live young from two separate follicle culture systems, one which maintains the spherical structure of the follicle, and one which allows the follicle wall to collapse[80,81]. Each of these types of culture has advantages, the spherical system maintaining the normal follicular architecture, but being limited by diffusion if follicles require to grow beyond a certain size, and lacking clear visualisation of oocytes (see Figure 3a). The flat culture system allows follicles to develop and retain many of their normal features without limitations of diffusion through many cell layers; however, the microenvironment of the follicular fluid-filled antrum is lost and the basement membrane normally breaks down (see Figure 3b). The basement membrane in spherical cultures may be supported by the addition of ascorbate, an antioxidant[82]. The availability of oxygen has recently been shown to be critical for the normal formation of the meiotic spindle in follicles cultured to provide mature oocytes[48]. These observations highlight the differing requirements of follicles at different stages of development. Collective culture of mouse oocyte–somatic cell complexes in a 'non-follicular' form, where oocytes from primary follicles develop on a bed of granulosa cells, is also successful in this species[83].

For species with larger maximum follicle sizes, above about 500 μm, which could not readily be grown to full size *in vitro*, a non-intact follicle culture system will be required, at least for the later stages of follicle growth. The culture of human follicles will fall into this category and an example of such a method has recently been published for porcine follicles, resulting in blastocyst development from 13% of *in vitro* matured follicles, which may offer hope for the application of such techniques for our own species[84]. However, for the time being, the most immediate challenge for human follicle growth *in vitro* is achieving normal follicles of a substantial size from *in vitro* development of primordial follicles. Primordial follicles constitute the major population in the ovary at all stages, and are the only type reliably surviving cryopreservation in ovarian cortical biopsies, such as those stored for women facing sterilizing therapies. They are therefore among the most accessible starting material in humans and of practical importance in establishing the uses of such material in clinical applications.

Primordial follicles of mammals are small enough for individual culture *in vitro*; however, for species with extensive ovarian stroma, such as humans, the isolation

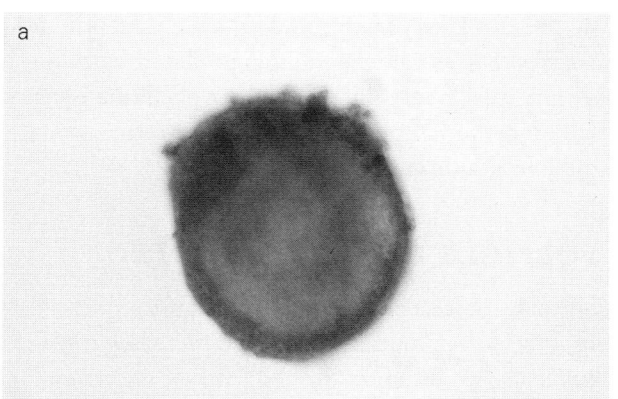
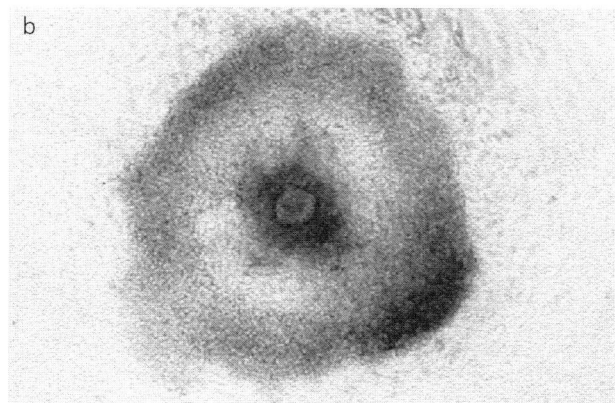

Figure 3 (a) Mouse follicle cultured in an intact system. The oocyte is visible, surrounded by a partially formed antrum with the granulosa layers translucent and the follicular wall darker around the external surface. (b) Mouse follicle cultured in adherent culture system, with a clearly visible oocyte and antrum-like organization of follicles. The basement membrane is breached allowing medium to become continuous with the follicle interior. Figure 3b reproduced with permission of Oxford University Press/Human Reproduction, from Mitchell L, Kennedy CR, Hartshorne GM. Effects of varying gonadotrophin dose and timing on antrum formation and ovulation efficiency of mouse follicles *in vitro*. *Hum Reprod* 2002;17:1181–8, ©European Society of Human Reproduction and Embryology

of follicles has proven difficult[85–87], and the results obtained culturing follicles within tissue pieces or slices have been more successful[88]. Hovatta[89] reported development of primordial and primary follicles to secondary and occasional antral follicles in cortical slices cultured with extracellular matrix, which appears to be the most promising system reported to date. Primordial follicles and early developmental stages have limited metabolic requirements which may enable their survival within a dense tissue even without additional measures to ensure oxygenation and nutrient supply. However, in this approach, follicle culture is collective and the potential for interfollicular communication and signaling affecting growth *in vitro* cannot be controlled at present.

Apparently viable follicles resulting from culture of human primordial follicles have been found after 3 weeks of culture[90]. However, the normality of follicles after even a short period of growth *in vitro* is uncertain, and various physical alterations in granulosa cell appearance have been noted[91] as well as premature oocyte extrusion, a recognized sign of follicle demise *in vitro*[88].

In the majority of species studied so far, many primordial follicles appear to initiate growth rapidly, within a day or two, once they are placed into an *in vitro* environment[92,93]. It is unclear whether this is because of loss of an inhibitory signal, or because the apparent ovarian 'reserve' (i.e., the number of follicles present in the *in vitro* system) is greatly reduced, potentially resulting in accelerated follicle recruitment[94]. Even when ovarian tissue is transplanted into immunodeficient mice, an accelerated rate of growth initiation occurs[95]. Only one method of culture so far has been successful in restoring the normal rate of growth initiation, and this involves transplantation into the allantois of a chick egg[93]. The reasons why this approach works are not yet certain but possibly circulatory linkage of the transplant to the chick gonad results in more appropriate systemic feedback loops, or an apparent increase in the ovarian reserve.

Serum is normally included in the culture media, and to date, the most successful systems for human follicles have been undefined[96]. This is also true of the successful mouse follicle culture systems upon which the human technology is loosely based[50,82,97]. However, rat, bovine and baboon follicle culture systems which lack serum are being developed and used to assess the roles of growth factors on follicle growth *in vitro*[92,98,99]. Such experiments will eventually result in a greater understanding of the controlling factors for follicle growth.

FUTURE DIRECTIONS

Oocytes are precious cells, nurtured over a period of years by specialized somatic cells to produce, occasionally, a ripe mature oocyte at its full potential, invested with the unique ability to be fertilized and form new life. Even with the advent of ICSI and nuclear transfer in humans, the secrets of the ooplasm remain unknown and the mature oocyte remains the key ingredient for fertility.

At present, neither the rate of follicle growth initiation nor the rate of follicle growth can be modified *in vivo* to influence the rate of production of oocytes. These factors control a woman's fertility and reproductive lifespan. Gonadotropic stimulation results in avoidance of atresia in follicles which would otherwise degenerate, but cannot promote new follicle growth. If the rate of loss of follicles from the ovarian reserve could be slowed until such time as fertility was required, the potential benefits might include not only greater control of reproductive choices, but perhaps avoidance of some of the age-related problems with fertility, such as the increasing risk of fetal anomalies and pregnancy loss due to aneuploid oocytes. If follicle growth could be hastened, fertility on demand could become accessible, and the prospects for growth of human oocytes by follicle culture would improve, as the time required for prolonged culture would be shortened.

An ability to grow follicles and oocytes to maturity outside the body would result in greater flexibility of reproductive technology, easing the processes of oocyte donation and ovarian stimulation, and avoiding the risks of ovarian hyperstimulation. The risk of accidental reimplantation of cancerous cells in women having cortical tissue transplants could be avoided by follicle culture *in vitro*.

In humans, these developments remain distant, and will raise moral and ethical dilemmas for individuals, resulting children and society. Yet in animals, the advantages would be more immediate and perhaps more generally acceptable, including the potential to expand threatened populations and dwindling biodiversity more rapidly and efficiently than natural breeding programs alone can support, as well as to optimize livestock breeding for commercial benefit or pest control.

Most of all, in future, application of the methods described here and the resultant increased availability of oocytes for research would enable a massive increase in the volume and quality of research, which currently suffers from a serious shortage of material for study. This would in turn produce novel developments as well as an urgently needed better understanding of human oocytes as the key cells without which reproduction is currently impossible.

References

1. Witschi E. Migration of the germ cells of human embryos from the yolk sac to the primitive gonadal folds. *Contrib Embryol Carnegie Inst Washington* 1948;32:67–80
2. McLaren A. Germ and somatic cell lineages in the developing gonad. *Mol Cell Endocrinol* 2000;163:3–9
3. Motta PM, Makabe S, Nottola SA. The ultrastructure of human reproduction I. The natural history of the female germ cell: origin, migration and differentiation inside the developing ovary. *Hum Reprod Update* 1997;3:281–95
4. Gondos B. Germ cell differentiation and intercellular bridges. In: Van Blerkom J, Motta PM, eds. *Ultrastructure of Reproduction, Gametogenesis, Fertilization and Embryogenesis*. The Hague: Martinus Nijhof Publishers 1984: 31–45
5. Pepling ME, Spradling AC. Female mouse germ cells from synchronously dividing cysts. *Development* 1998;125: 3323–8
6. Manatoya T, Potter EL. Oocytes in prophase of meiosis from squash preparations of human fetal ovaries. *Fertil Steril* 1963;14:378–92
7. Pepling ME, Spradling AC. Mouse ovarian germ cell cysts undergo programmed breakdown to form primordial follicles. *Dev Biol* 2001;234:339–51
8. Block E. Quantitative morphological investigations of the follicular system in women: variations at different ages. *Acta Anat* 1952;14:108–23
9. Forabosco A, Sforza C, DePol A, et al. Morphometric study of the human neonatal ovary. *Anat Rec* 1991;231: 201–8
10. Morita Y, Perez GI, Paris F, et al. Oocyte apoptosis suppressed by disruption of the acid sphingomyelinase gene or by sphingosine-1-phosphate therapy. *Nat Med* 2000;6: 1109–14
11. Reynaud K, Driancourt MA. Oocyte attrition. *Mol Cell Endo* 2000;163:101–8
12. Gougeon A. Dynamics of follicular growth in the human: a model from preliminary results. *Hum Reprod* 1986;1:81–7
13. Gougeon A, LeFevre B, Testart J. Morphometric characteristics of the population of the small ovarian follicles in cynomolgus monkey: influence of a gonadotrophin releasing hormone agonist and gonadotrophins. *J Reprod Fertil* 1992;95:567–75
14. Gougeon A, Busso D. Morphologic and functional determinants of primordial and primary follicles in the monkey ovary. *Mol Cell Endocrinol* 2000;163:33–41
15. Aaltonen J, Laitinen MP, Vuojolainen K, et al. Human growth differentiation factor 9 (GDF-9) and its novel homologue GDF-9B are expressed in oocytes during early folliculogenesis. *J Clin Embryol Metab* 1999;84: 2744–50
16. Baker SJ, Spears N. The role of intra-ovarian interactions in the regulation of follicle dominance. *Hum Reprod Update* 1999;5:153–65
17. Parrott JA, Skinner MK. Kit-ligand/stem cell factor induces primordial follicle development and initiates folliculogenesis. *Endocrinology* 1999;140:4262–71
18. Henderson SA, Edwards RG. Chiasma frequency and maternal age in mammals. *Nature* 1968;218:22–8
19. Polani PE, Crolla JA. A test of the production line hypothesis of mammalian oogenesis. *Hum Genet* 1991;88:64–70
20. Lintern-Moore S, Peters H, Moore GPM, Faber M. Follicular development in the infant human ovary. *J Reprod Fertil* 1974;39:53–64
21. Wassarman PM. Zona pellucida glycoproteins. *Ann Rev Biochem* 1988;57:415–42
22. Albertini DF, Combelles CM, Benecchi E, Carabatsos MJ. Cellular basis for paracrine regulation of ovarian follicle development. *Reproduction* 2001;121: 647–53
23. Oktay K, Newton H, Mullan J, Gosden RG. Development of human primordial follicles to antral stages in SCID/hpg mice stimulated with follicle stimulating hormone. *Hum Reprod* 1998;13:1133–8
24. Matthews CH, Borgato S, Beck-Peccoz Adams M, et al. Primary amenorrhoea and infertility due to a mutation in the β subunit of follicle stimulating hormone. *Nat Genet* 1992;5:83–6
25. Van den Hurk R, Abir R, Telfer EE, Bevers MM. Primate and bovine immature oocytes and follicles as sources of fertilizable oocytes. *Hum Reprod Update* 2000; 4:457–74
26. Zeleznik AH, Hillier SG. The role of gonadotropins in the selection of the preovulatory follicle. *Clin Obstet Gynecol* 1984;27:927–40
27. Perez GI, Robles R, Knudson CM, et al. Prolongation of ovarian lifespan into advanced chronological age by Bax-deficiency. *Nat Genet* 1999;21:200–3
28. Hsu SY, Lai RH, Finegold M, Hsueh AJ. Targeted overexpression of Bcl-2 in ovaries of transgenic mice leads to decreased follicle apoptosis, enhanced folliculogenesis and increased germ cell tumorigenesis. *Endocrinology* 1996; 137:4837–43

29. Matikainen T, Perez GI, Zheng TS, *et al*. Caspase-3 gene knockout defines cell lineage specificity for programmed cell death signaling in the ovary. *Endocrinology* 2001; 142:2468–80

30. Fulka J Jr, First NL, Moor RM. Nuclear and cytoplasmic determinants involved in the regulation of mammalian oocyte maturation. *Mol Hum Reprod* 1998;4:41–9

31. Wolf JP, Bulwa S, Rodrigues D, Jouannet P. Human oocytes cytometry and fertilisation rate after subzonal insemination. *Zygote* 1995;3:101–9

32. Pavlok A, Lucas-Hahn A, Niemann H. Fertilization and developmental competence of bovine oocytes derived from different categories of antral follicles. *Mol Reprod Dev* 1992; 31:63–7

33. Eppig JJ, Schultz RM, O'Brien M, Chesnel F. Relationship between the developmental programs controlling nuclear and cytoplasmic maturation of mouse oocytes. *Dev Biol* 1994;164:1–9

34. Gilchrist RB, Nayudu PL, Hodges JK. Maturation, fertilization and development of marmoset monkey oocytes *in vitro*. *Biol Reprod* 1997;56:238–46

35. Durinzi KL, Saniga EM, Lanzendorf SE. The relationship between size and maturation *in vitro* in the unstimulated human oocytes. *Fertil Steril* 1995;63:404–6

36. Tsuji K, Sowa M, Nakano R. Relationship between human oocyte maturation and different follicular sizes. *Biol Reprod* 1985;32:413–17

37. Scott RT, Hofmann GE, Muasher SJ, *et al*. Correlation of follicular diameter with oocyte recovery and maturity at the time of transvaginal follicular aspiration. *J In Vitro Fertil Embryo Transf* 1989;6:73–5

38. Trounson A, Andieresz C, Jones G. Maturation of human oocytes *in vitro* and their developmental competence. *Reproduction* 2001;121:51–75

39. Van Blerkom J, Davis PW, Merriam J. The developmental ability of human oocytes penetrated at the germinal vesicle stage after insemination *in vitro*. *Hum Reprod* 1994;9: 697–708

40. Herbert M, Gillespie JI, Mudoch AP. Development of calcium signalling mechanisms during maturation of human oocytes. *Mol Hum Reprod* 1997;3:965–73

41. Gandolfi TA, Gandolfi F. The maternal legacy to the embryo: cytoplasmic components and their effects on early development. *Theriogenology* 2001;5:1255–76

42. DeVantery Arrighi C, Campana A, Schorderet-Slatkine S. A role of MEK-MAPK pathway in okadaic acid-induced meiotic resumption of incompetent growing mouse oocytes. *Biol Reprod* 2000;63:658–65

43. Chui DKC, Pugh ND, Walker SM, *et al*. Follicular vascularity – the predictive value of transvaginal power Doppler ultrasonography in an *in vitro* fertilization programme: a preliminary study. *Hum Reprod* 1997;12: 191–6

44. Van Blerkom J, Antczak M, Schrader R. The developmental potential of the human oocyte is related to the dissolved oxygen content of follicular fluid: association with vascular endothelial growth factor levels and perifollicular blood flow characteristics. *Hum Reprod* 1997; 12:1047–55

45. Van Blerkom J, Davis PW, Lee J. ATP content of human oocytes and developmental potential and outcome after

in vitro fertilisation and embryo transfer. *Hum Reprod* 1995;10:415–24

46. Tarin JJ, Vendrell FJ, Ten J, *et al*. The oxidising agent tertiary butyl hydroperoxide induces disturbances in spindle organisation, meiosis and aneuploidy in mouse oocytes. *Mol Hum Reprod* 1996;2: 895–901

47. Malamitsi-Puchner A, Sarandakou A, Baka SG, *et al*. Concentrations of angiogenic factors in follicular fluid and oocyte–cumulus complex culture medium from women undergoing *in vitro* fertilisation: association with oocyte maturity and fertilisation. *Fertil Steril* 2001;76: 98–101

48. Hu Y, Betzendahl I, Cortvrindt R, *et al*. Effects of low O_2 and ageing on spindles and chromosomes in mouse oocytes from preantral follicle culture. *Hum Reprod* 2001;16: 737–48

49. Eppig JJ, Hosoe M, O'Brien MJ, *et al*. Conditions that affect acquisition of developmental competence by mouse oocytes *in vitro*: FSH, insulin, glucose and ascorbic acid. *Mol Cell Endocrinol* 2000;163:109–16

50. Vitt UA, Nayudu PL, Rose UM, Kloosterboeeer HJ. Embryonic development after follicle culture is influenced by follicle stimulating hormone isoelectric point range. *Biol Reprod* 2001;65:1542–7

51. Gersak K, Tomazevic T. Subpopulations of human granulosa-luteal cells in natural and stimulated *in vitro* fertilization-embryo transfer cycles. *Fertil Steril* 1996;65: 608–13

52. Rodgers RJ, Irving-Rodgers HF, van Wezel IL, *et al*. Dynamics of the membrana granulosa during expansion of the ovarian follicular antrum. *Mol Cell Endocrinol* 2001; 171:41–8

53. Hillier SG. Gonadotropic control of ovarian follicular growth and development. *Mol Cell Endocrinol* 2001;179: 39–46

54. Rzucidlo SJ, Gibbons J, Stice SL. Comparison by restriction fragment differential display RT-PCR of gene expression pattern in bovine oocytes matured in the presence of absence of fetal calf serum. *Mol Reprod Dev* 2001;59:90–6

55. Monk M, Holding C, Goto T. Isolation of novel developmental genes from human germ cell, oocyte and embryo cDNA by differential display. *Reprod Fertil Dev* 2001;13:51–7

56. Formigli L, Roccio C, Belotti G, *et al*. Non-surgical flushing of the uterus for pre-embryo recovery: possible clinical applications. *Hum Reprod* 1990;5:329–35

57. Edwards RG. Maturation *in vitro* of mouse, sheep, cow, pig, rhesus monkey and human ovarian oocytes. *Nature* 1965;58:349–51

58. Armstrong DT, Zhang X, Vanderhyden BC, Khamsi F. Hormonal actions during oocyte maturation influence fertilisation and early embryonic development. *Ann N Y Acad Sci* 1991;626:137–58

59. Steptoe PC, Edwards RG. Birth after the reimplantation of a human embryo. *Lancet* 1978;ii:366

60. Suikkari AM, Tulppala M, Tuuri T, *et al*. Luteal phase start of low dose FSH priming of follicles results in an efficient recovery, maturation and fertilisation of immature human oocytes. *Hum Reprod* 2000;15:747–51

61. Cha KY, Chian RC. Maturation *in vitro* of immature human oocytes for clinical use. *Hum Reprod Update* 1998; 4:103–20

62. Janssenswillen C, Nagy ZP, Van Steirteghem A. maturation of human cumulus-free germinal vesicle stage oocytes. *Hum Reprod* 1995;10:375–8

63. Trounson A, Wood C, Kausche A. *In vitro* maturation and the developmental competence of oocytes recovered from untreated polycystic ovarian patients. *Fertil Steril* 1994; 62:353–62

64. Gougeon A, Testart J. Germinal vesicle breakdown in oocytes of human atretic follicles during the menstrual cycle. *J Reprod Fertil* 1986;78:389–401

65. LeFevre B, Gougeon A, Testart J. *In vitro* oocyte maturation: some questions concerning the initiation and prevention of this process in humans. *Hum Reprod* 1987; 2:495–7

66. Chian RC, Buckett WM, Tulandi T, Tan SL. Prospective randomised study of human gonadotrophin priming before immature oocyte retrieval from unstimulated women with polycystic ovary syndrome. *Hum Reprod* 2000;15:165–70

67. Goud P, Goud A, Van Oostveldt P, *et al*. Fertilization abnormalities and pronucleus size asynchrony after intracytoplasmic sperm injection are related to oocyte postmaturity. *Fertil Steril* 1999;72: 245–52

68. Battaglia DE, Klein NA, Soules MR. Changes in centrosomal domains during meiotic maturation in the human oocyte. *Mol Hum Reprod* 1996;2:845–51

69. Morgan PM, Warikoo PK, Bavister BD. *In vitro* maturation of ovarian oocytes from unstimulated rhesus monkeys: assessment of cytoplasmic maturity by embryonic development after *in vitro* fertilisation. *Biol Reprod* 1991;45:89–93

70. Weston AM, Zelinski-Wooten MB, Hutchison JS, *et al*. Developmental potential of embryos produced by *in vitro* fertilisation from gonadotrophin releasing hormone agonist-treated macaques stimulated with recombinant human follicle stimulating hormone alone or in combination with luteinising hormone. *Hum Reprod* 1996;11:608–13

71. Smitz J, Cortvrindt R, Hu Y. Epidermal growth factor combined with recombinant human chorionic gonadotrophin improves meiotic progression in mouse follicle-enclosed oocyte culture. *Hum Reprod* 1998;13:664–9

72. Cavilla JL, Kennedy CR, Baltsen M, *et al*. The effects of meiosis activating sterol on *in vitro* maturation and fertilisation of human oocytes from stimulated and unstimulated ovaries. *Hum Reprod* 2001;16:547–55

73. Holm P, Walker SK, Seamark RF. Embryo viability, duration of gestation and birth weight in sheep after transfer of *in vitro* matured and *in vitro* fertilised zygotes cultured *in vitro* or *in vivo*. *J Reprod Fertil* 1996;107: 175–81

74. Sinclair KD, Young LE, Wilmut I, McEvoy TG. *In utero* overgrowth in ruminants following embryo culture; lessons from mice and a warning to men. *Hum Reprod* 2000; 15(Suppl. 5):68–86

75. Picton H, Briggs D, Gosden R. The molecular basis of oocyte growth and development. *Mol Cell Endocrinol* 1998; 145: 27–37

76. Eppig JJ, O'Brien MJ. Development *in vitro* of mouse oocytes from primordial follicles. *Biol Reprod* 1996;54: 197–207

77. Moor RM, Trounson AO. Hormonal and follicular factors affecting maturation of sheep oocytes *in vitro* and their subsequent developmental capacity. *J Reprod Fertil* 1977; 49:101–9

78. Gosden RG, Byatt-Smitt JG. Oxygen concentration gradient across the ovarian follicular epithelium: model, prediction and implications. *Hum Reprod* 1986;1:65–8

79. Rose UM, Hanssen RGJM, Kloosterboer HJ. Development and characterisation of an *in vitro* ovulation model using mouse ovarian follicles. *Biol Reprod* 1999;61:503–11

80. Spears N, Boland NI, Murray AA, Gosden RG. Mouse oocytes derived from *in vitro* growth primary follicles are fertile. *Hum Reprod* 1994;9:527–32

81. Cortvrindt R, Smitz J, Van Steirteghem AC. *In vitro* maturation, fertilisation and embryo development of immature oocytes from early preantral follicles from prepubertal mice in a simplified culture system. *Hum Reprod* 1996;11: 2656–66

82. Murray AA, Molinek MD, Baker SJ, *et al*. Role of ascorbic acid in promoting follicles integrity and survival in intact mouse ovarian follicles *in vitro*. *Reproduction* 2001;121: 89–96

83. Eppig JJ. Mouse oocyte maturation, fertilisation and preimplantation development *in vitro*. In Richter JD, ed. *Comparative Methods Approach to the Study of Oocytes. A*. Oxford:Oxford University Press, 1999:3–9

84. Wu J, Emery BR, Carrell DT. *In vitro* growth, maturation, fertilization and embryonic development of oocytes from porcine preantral follicles. *Biol Reprod* 2001;64: 375–81

85. Roy SK, Treacy BJ. Isolation and long term culture of human preantral follicles. *Fertil Steril* 1993;59:783–90

86. Oktay K, Nugent D, Newton H, *et al*. Isolation and characterisation of primordial follicles from fresh and cryopreserved human ovarian tissue. *Fertil Steril* 1997;67:481–6

87. Abir R, Fisch B, Nitke S, *et al*. Morphological study of fully and partially isolated early human follicles. *Fertil Steril* 2001;75:141–6

88. Hovatta O, Wright C, Kransz T, *et al*. Human primordial primary and secondary ovarian follicles in long term culture: effect of partial isolation. *Hum Reprod* 1999;14: 2519–24

89. Hovatta O. Cryopreservation and culture of human primordial and primary ovarian follicles. *Mol Cell Endocrinol* 2000;169:95–7

90. Hovatta O, Silye R, Abir R, *et al*. Extracellular matrix improves survival of both stored and fresh human primordial and primary ovarian follicles in long term culture. *Hum Reprod* 1997;12:1032–6

91. Abir R, Roizman P, Fisch B, *et al*. Pilot study of isolated early human follicles cultured in collagen gels for 24 hours. *Hum Reprod* 1999;14:1299–301

92. Fortune JE, Kito S, Wandji SA, Srsen V. Activation of bovine and baboon primordial follicles *in vitro*. *Theriogenology* 1998;49:441–9

93. Fortune JE, Cushman RA, Wahl CM, Kito S. The primordial to primary follicle transition. *Mol Cell Endocrinol* 2000;163:53–60

94. Faddy MJ. Follicle dynamics during ovarian ageing. *Mol Cell Endocrinol* 2000;163:43–8

95. Oktay K, Newton H, Gosden RG. Transplantation of cryopreserved human ovarian tissue results in follicle growth initiation in SCID mice. *Fertil Steril* 2000;73: 599–603

96. Wright CS, Hovatta O, Margara R, *et al*. Effects of follicle stimulating hormone and serum substitution on the *in vitro* growth of human ovarian follicles. *Hum Reprod* 1999;14: 1555–62

97. Mitchell LM, Kennedy CR, Hartshorne GM. Effects of varying gonadotrophin dose and timing on antrum formation and ovulation efficiency of mouse follicles *in vitro*. *Hum Reprod* 2002;17:1181–8

98. Wandji SA, Srsen V, Nathanielsz PW, *et al*. Initiation of growth of baboon primordial follicles *in vitro*. *Hum Reprod* 1997;12: 1993–2001

99. McGee EA, Smith R, Spears N, *et al*. Mullerian inhibitory substance induces growth of rat preantral ovarian follicles. *Biol Reprod* 2001;64:293–8

100. Veeck LL, Worston JW Jr, Witmyer J, *et al*. Maturation and fertilisation of morphologically immature human oocytes in a program of *in vitro* fertilisation. *Fertil Steril* 1983;39:594–602

101. Cha KY, Koo JJ, Ko JJ, *et al*. Pregnancy after *in vitro* fertilisation of human follicular oocytes collected from non-stimulated cycles, their culture *in vitro* and their transfer in a donor oocyte program. *Fertil Steril* 1991;55:109–13

102. Russell JB, Knezevich KM, Fabian KF, Dickson JA. Unstimulated immature oocyte retrieval: early versus midfollicular endometrial priming. *Fertil Steril* 1997;67: 616–20

103. Thornton MH, Francis MM, Paulson RJ. Immature oocyte retrieval: lessons from unstimulated IVF cycles. *Fertil Steril* 1998;70:647–50

104. Chian RC, Gulecki B, Buckett WM, Tan SL. Priming with hCG before retrieval of immature oocytes in women with infertility due to polycystic ovarian syndrome. *N Engl J Med* 1999;341:1624–6

105. Mikkelsen AL, Smith SD, Lindenberg S. *In-vitro* maturation of human oocytes from regularly menstruating women may be successful without follicle stimulating hormone priming. *Hum Reprod* 1999;14:1847–51

106. Jaroudi KA, Hollanders JMG, Elnour AM, *et al*. Embryo development and pregnancies from *in vitro* matured and fertilised human oocytes. *Hum Reprod* 1999;14:1749–51

107. Smith SD, Mikkelsen AL, Lindenberg S. Development of human oocytes matured *in vitro* for 28 or 36 hours. *Fertil Steril* 2000;73:541–4

108. Cha KY, Han SY, Chung HM, *et al*. Pregnancies and deliveries after *in vitro* maturation culture followed by *in vitro* fertilization and embryo transfer without stimulation in women with polycystic ovary syndrome. *Fertil Steril* 2000; 73:978–83

109. Mikkelsen AL, Smith S, Lindenberg S. Impact of oestradiol and inhibin A concentrations of pregnancy rate in *in vitro* oocyte maturation. *Hum Reprod* 2000;15: 1685–90

110. Mikkelsen AL, Andersson AM, Skakkebaek NE, Linderberg S. Basal concentrations of oestradiol may predict the outcome of *in vitro* maturation in regularly menstruating women. *Hum Reprod* 2001;16:862–7

111. Mikkelsen AL, Lindenberg S. Influence of the dominant follicle on *in vitro* maturation of human oocytes: a prospective non-randomised study. *Reprod Biomed Online* 2001;3:199–204

112. Yoon HG, Yoon SH, Son WY, *et al*. Pregnancies resulting from *in vitro* matured oocytes collected from women with regular menstrual cycle. *J Assist Reprod Genet* 2001;18: 325–29

113. Paulson RJ, Sauer MV, Francis MM, *et al*. Factors affecting pregnancy success of human *in vitro* fertilisation in unstimulated cycles. *Hum Reprod* 1994;9:1571–5

114. Barnes FL, Crombie A, Gardner DK, *et al*. Case Report Blastocyst development and birth after *in vitro* maturation of human primary oocytes, intracytoplasmic sperm injection and assisted hatching. *Hum Reprod* 1995;10: 3243–7

115. Nagy ZP, Janssenwillen C, Liu J, *et al*. Pregnancy and birth after intracytoplasmic sperm injection of *in vitro* matured germinal vesicle stage oocytes: case report. *Fertil Steril* 1996;65:1047–50

116. Kodama H, Fukuda J, Karube H, *et al*. *In vitro* fertilization of *in vitro* matured oocytes obtained from the follicles without hCG exposure for prevention of severe ovarian hyperstimulation syndrome: a case report. *J Obstet Gynaecol Res* 1996;22:61–5

117. Edirisinghe WR, Junk SM, Matson RP, Yovich JL. Case report: Birth from cyropreserved embryos following *in-vitro* maturation of oocytes and intracytoplasmic sperm injection. *Hum Reprod* 1997;12:1056–8

118. Liu J, Katz E, Garcia JE, *et al*. Successful *in vitro* maturation of human oocytes not exposed to human chorionic gonadotrophin during ovulation induction, resulting in pregnancy. *Fertil Steril* 1997;67:566–8

119. Jaroudi KA, Hollanders JMG, Sieck UV, *et al*. Pregnancy after transfer of embryos which were generated from *in vitro* matured oocytes. *Hum Reprod* 1997;12:857–9

120. Hwang JL, Lin YH, Tsai YL. Pregnancy after immature oocyte donation and intracytoplasmic sperm injection. *Fertil Steril* 1997;68:1139–40

121. Abdul-Jalil AK, Child TJ, Phillips S, *et al*. Ongoing twin pregnancy after ICSI of PESA-retrieved spermatozoa into *in-vitro* matured oocytes: case report. *Hum Reprod* 2001; 16:1424–6

122. Chian RC, Gulekli B, Buckett WM, Tan SL. Pregnancy and delivery after cryopreservation of zygotes produced by *in vitro* matured oocytes. *Hum Reprod* 2001;16: 1700–2

Genetics of human reproduction

Genetic factors of female infertility 20

Valeriy N. Zaporozhan and Ruslan V. Sobolev

INTRODUCTION

The intensive development of cytogenetic and molecular diagnostic techniques in the study of the human genome has made it possible to accumulate information on the chromosomal and genetic abnormalities that lead to the disorders in sex differentiation and in female reproductive function. Genetic abnormalities are detected in a minor percentage of married couples, as the sole cause of infertility. Discussing the problems associated with multifactorial infertility, experts are convinced that reproductive system disorders of hypothalamo-hypophyseal origin are accompanied by changes in secondary sexual characteristics and infertility conditioned by genetic pathology.

The molecular genetic investigation of female patients having reproductive dysfunctions acquired special significance due to the development of wide-spectrum techniques correcting such pathology. The same techniques are used for predicting the treatment outcomes. It is impossible to discover a chromosomal abnormality in a patient with reproductive dysfunction without performing cytogenetic and molecular tests. Thus, the chromosomal abnormalities in phenotypically healthy individuals occur either in a balanced condition (translocation, inversion) or in an occult 'minimal' mosaicism, or in the presence of an additional marker chromosome.

All this information is necessary for the doctor to have a differentiated approach to patients attending their clinic with the aim of restoring fertility and achieving pregnancy, especially in those who also need courses in assisted reproductive technologies (ART). Along with the standard examination and laboratory tests, molecular genetic investigations are necessary because of a high risk of patients giving birth to children with inherited reproductive system pathology, a fact that should be considered in particular as a side-effect of the performed therapeutic techniques.

The reproductive dysfunction arises when there are numeric and structural aberrations in sex chromosomes, in inherited forms with faulty differentiation of gonads in females and in endocrine dysfunctions on different levels of the hypothalamic–pituitary–ovarian axis resulting from the changes in the structure and function of genes.

This chapter describes the main genetic abnormalities responsible for female reproductive dysfunction and details the latest research performed in this field. It also considers the genetic defects determining structural and hormonal disorders and the resultant dysfunctions and the most frequently encountered clinical syndromes observed in such cases, including spontaneous miscarriages, polycystic ovary syndrome and premature ovarian failure. The present state of molecular diagnostics considers such conditions to be the result of single or multiple gene defects.

Abnormal pubescence or abnormal reproductive function usually arises in the context of cases of hypothalamic secretion or pituitary reception dysfunction, but can also be a result of changes in gonadotropin production or reception by the gonads.

Abnormalities in steroid synthesis also play a significant role. Phenotypic manifestations of different endocrine anomalies depend on the level of penetrance of this defect, on the degree of its manifestation and on the essence of the mutation that has occurred. As a result we observe variations of clinical manifestation starting from insignificant ovarian dysfunction to well-manifested forms of hypogonadism, virilization syndrome, female pseudohermaphroditism and absolute infertility.

The chapter concludes with a discussion on the need for highly qualified counseling for parents about to undergo genetic testing, and, especially, for parents who have already been tested and been diagnosed as having some chromosomal abnormality or gene defect.

NUMERIC CHROMOSOMAL ABERRATIONS

Turner's syndrome

The phenotype of a female of low stature with the signs of ovarian dysgenesis, valgus leg deformation, neck lymphatic swelling and other secondary stigmata of dysembryogenesis was first described in the 1930s by Ullrich and Turner. Later this condition was named 'Turner's syndrome'. After introducing the cytogenetic test technique the karyotype 45,XO was determined in women of such phenotype. This anomaly occurs in 1 per 2000 live births and this proves the existence of a certain percentage of X chromosome anomalies in an embryo and also indicates the fact that such cases do not develop in a later gestational term. Most adult 45,XO patients have fibrous cord instead of normal gonads, the cords being 2.0–3.0 cm in length and about 0.5 cm in width. Histologic examination has revealed that these cords present overlying fibers of dense, fibrous tissue, morphologically resembling the normal ovarian stroma. Germinal cells have been found in embryos with 45,XO karyotype, but their absence in adults has been proven[1]. The pathology of the germinal cell death has been studied and described by Ohno[2]. It does not stem from a deficiency in germinal cell formation, but from increased follicular atresia[2].

The manifestation of phenotypic characteristics in patients varies widely and depends on the degree of the impairment of the structural chromosome, and on the presence of a complete or mosaic form of the syndrome. In the latter form, different tissues and organs of the individual contain either normal or aneuploid cells. In such cases, two-cell clones are present, one containing a normal karyotype, the other an abnormal one. Not all the tissues and organs of such an individual contain a corresponding percentage of normal and aneuploid cells which reflect the degree of mosaicism. Some patients have a normal menarche and satisfactory ovarian function and in exceptional cases there were pregnancies registered. Such a weak phenotypic manifestation is mainly a characteristic of insignificant anomalies in the X chromosome structure. These structural anomalies

may be deletions, duplications, inversions, translocations or ring chromosomes which are associated with chromosomal breakage and a considerably unbalanced content of genes in the X chromosome. In some cases the phenotype and mental abilities of the patient are not affected when the pathologic chromosome is inactivated[3]. However, if there is a ring X chromosome and/or an X autosomal translocation the frequency of mental retardation and congenital malformations rises considerably. Such anomalous phenotypes can be explained by the absence of inactivation or partial inactivation of the pathologic chromosome at the expense of the normal ones.

On the other hand, clinical evidence of the absence of one X chromosome confirms the necessity for two homologous intact alleles to be present for normal functioning of the genes in the active chromosome.

It is necessary to note the presence of aberrations that are detected in one cell. Single-cell aberrations, according to some authors, reflect chromosomal instability or cultivational artefacts[4]. Other authors confirm an elevated amount of single-cell aberrations in patients with recurrent miscarriages[5]. Among the single-cell aberrations not only aneuplodies are determined in the sex chromosomes, but also inversions, translocations and deletions. One cannot be sure as yet that a direct dependence exists between the presence of single-cell aberrations and reproductive dysfunction; however, some published data would appear to confirm such a supposition[4,6]. It can therefore be said that single-cell aberrations at least *influence* the reproductive function and possibly decrease fertility. One cannot neglect such aberrations as they can serve as a sign of weak mosaicism, which is impossible to diagnose given the techniques that are currently available.

Gonadal dysgenesis in females with 46,XY karyotype

Women with 46,XY karyotype have gonadal dysgenesis which is considered to be the result of mutations in genes controlling ovarian development. Such malformation of the ovaries leads to an individual with perverted genitalia; thus, karyotypically such an individual is a male, while phenotypically the person is female. Such patients suffer primary amenorrhea and have a high risk of tumor development in the germinal gonads.

The genetic basis of this anomaly is probably a mutation in *sry*, *dax-1* and *sox9* genes. The *sry* mutation comprises about 20% of all the cases with gonadal dysgenesis and 46,XY karyotype[7]. The *sry* gene is mapped to the Yp11.3 locus and it encodes a DNA-binding protein. The non-differentiated cells of gonads develop into Leydig cells under the influence of the *sry* gene or into granulosa cells in the case of the gene's absence. These cells give rise to three different cell lines. In male fetuses they form spermatogonia, Leydig cells and pericanalicular cells, while in female fetuses they form oogonia, theca cells and stromal cells. The *sry* gene is activated only after the non-differentiated gonads have been formed and, thus, this gene is the main factor in ovarian development at the early stage of embryogenesis. The *sox9* gene is close in essence to *sry*, and prevents testes development, but does not affect the ovaries.

STRUCTURAL CHROMOSOMAL ABERRATIONS

Structural disorders of the X chromosome's short arm

X chromosome short arm deletion carriers (46,Xdel(Xp)) may have different phenotypes. The degree of phenotype change depends on how the Xp presence is manifested. The most common breakage point in terminal deletions is Xp11. In the case of 46,Xdel(Xp) (p11) only the proximal part of the X chromosome short arm is affected. A chromosome with such a deletion becomes either acrocentric or telocentric.

The chromosomes with more distal deletions, 46,Xdel(Xp21, 22.1 and 22.3), have also been described. The X autosomal translocations lead to the interstitial deletion of the X chromosome's short arm.

Approximately half of the patients with the X chromosome short arm deletion 46,Xdel(Xp)(p11) have primary amenorrhea and gonad dysgenesis. The other half develop mammary glands and menses; however, they do not menstruate regularly[8,9]. It was found that about 50% of patients with deletions of the X chromosome short arm suffer primary amenorrhea, and 45% have secondary amenorrhea[10]. At the same time the ovarian function is retained to a considerable

degree compared to those who have Turner's syndrome (45,XO).

Females with a more distal deletion of the X chromosome's short arm (46,Xdel(Xp21 and 22.1)) have a more frequent menstrual cycle, but suffer from infertility and secondary amenorrhea. Thus, the terminal locus of the X chromosome's short arm plays a by no means unimportant role in the development of the ovaries[8–15].

Most females with such a deletion are of short stature and this allows us to suggest that the gene(s) determining the height of the patient are in that particular locus of the X chromosome's short arm. It is very important clinically to keep in mind that a patient of short stature and with a normal ovarian–menstrual cycle may be a carrier of such a deletion[16,17]. Both mothers and daughters can be the carriers of similar deletions of the X chromosome's short arm, combined not only with the X autosomal translocations, but also with the terminal deletions. Family cases of the short arm distal deletions were described for the first time in 1977. Among the 10 cases described later, two were observed in both mother and daughter. Six of the 10 described were *de novo* mutations[14]. There are family cases with both of the deletions of the short arm in the Xp11 locus and one in the Xp22–12 locus. The combinations of such deletions have been described several times[18–20]. The phenotypic manifestations vary even among relatives.

Isochromosome X

Patients with isochromosome 46,Xins(Xq) always have gonadal cords and primary amenorrhea (Figure 1). In rare cases these patients menstruate[10,21]. Most of them are of very short stature, even shorter than those who

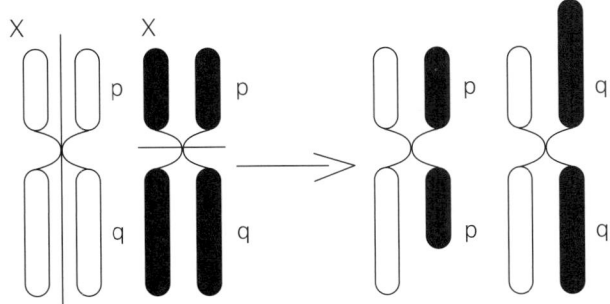

Figure 1   Formation of iso X chromosomes

suffer from Turner's syndrome. Other phenotypic characteristics are analogous to those of Turner's syndrome[21]. Somatic anomalies are of the same frequency, patients having 46,Xins(Xq) corresponding with those having 45,XO.

A nearly complete absence of gonadal development is observed in patients with 46,Xins(Xq), which is the opposite of those having 46,Xdel(Xp)(p11), who partially develop mammary glands and menstruate. The contrast becomes more evident if these patients are compared with the carriers of more distally located short arm deletions. The phenotypic difference could be explained by the gonadal development determinants being positioned in several different loci on the X chromosome's short arm, when the gonadal development determinant is absent in one locus ins(Xq) but present in del(Xp)(p11). Thus, the duplication of the long arm of the X chromosome, which is an isochromosome, cannot compensate for the absence of genetic material on the X chromosome's short arm. This is also true for the determinants located on the short and long arms of the X chromosome.

Alternatively, not all of the isochromosome loci are fully activated. If complete compensation of the chromosome's inactivation in females with 46,XX exists, patients with X chromosome short and long arm deletions will have gonadal dysgenesis as a simple manifestation of aneuploidy of sex chromosomes. More than that, incomplete compensation in inactivation of one of the chromosomes is confirmed by the anomalies in polysomy syndromes of sex chromosomes (47,XXY, 48,XXXY, 49,XXXXY). It seems doubtful that the duplication of the X chromosome's long arm (ins(Xq)) is capable of resulting in an anomaly, taking into consideration that patients with 47,XXX karyotype are phenotypically healthy.

Structural defects of the X chromosome's long arm

The deletions of the X chromosome's long arm have been described in detail in the literature[8–13] and differ in terms of their content. If the localization of the breakage point leads to terminal deletion and is formed from Xq13, then the derived chromosome will resemble chromosome 17 or 18; the breakage point in the Xq21 locus leads to the formation of a chromosome resembling

chromosome 16. Practically all deletions taking place at Xq13 are associated with primary amenorrhea, underdeveloped mammary glands and absolute ovarian dysfunction[9,10]. That is why it can be stated with certainty that the Xq13 locus is vitally important for supporting the normal functioning of the ovaries.

The locus responsible for proper ovarian function can be situated more proximally at Xq21, but not more distally. The evidence for this is the fact that patients with X chromosome long arm deletions del(Xq)(q21–q24) do menstruate. Menstruating females with del(Xq)(q21) probably have a preserved part of the chromosome with the gene which supports the ovarian function, while females having del(Xq)(q13 or q21) and suffering primary amenorrhea will have lost this locus. Analysis of X autosomal translocations leads to results similar to those found with terminal deletions. In the case of seven X autosomal translocations involving Xq21–22, it was shown that the breakage points leading to ovarian dysfunction were located on the long arm of the X chromosome[22], leading to the conclusion that eight different genes can determine X chromosome condition. These genes are located on the Xq21 locus.

In more distal deletions of the X chromosome's long arm a premature ovarian failure is observed[10,11,13,19,23]. Although at present the limits of short regions are not yet clear it is convenient conventionally to divide terminal deletions which come from the loci Xq13–21, Xq22–25, Xq26–28[1]. It may be that the distal part of the long arm of the X chromosome is less important in providing ovarian function than the proximal part, which is clearly important in the provision of adequate ovarian function.

In addition to the cases of terminal deletions arising at different sites, two cases of interstitial deletions have been described[23,24]. These deletions may complicate correct diagnosis if molecular techniques are not used. Familial cases of distal deletions of the X chromosome's long arm are rarely encountered. Some of the familial cases of Xq deletions have been derived from autosomal translocations on the X chromosome's long arm. Also, familial cases of terminal or interstitial deletions are possible[24]. The familial cases of terminal and interstitial deletions of the X chromosome long arm have been described at different loci from Xq25 to Xq27. The most frequent breakage point of these is Xq27.

The distal deletions of the X chromosome's long arm have a less manifested negative character in terms of phenotypic expression and stature than the proximal ones. Such patients are not phenotypically alike to patients with Turner's syndrome. One of the studied genes pertaining to that part of the arm is *diaph2*, which produces a protein responsible for determining cellular polarity, correct cytokinesis and reorganization of actin cytoskeleton. In one study, a patient with a familial form of Xq21 autosomal translocation was found to have a structural disorder of the last intron of the *diaph2* gene[2]. On the other hand, in another study only one patient out of four cases of Xq21–Xqter deletions exhibited a disorder in gene structure[25]. The difficulty of defining clearly all the functions of the *diaph2* gene lies in the fact that it is expressed in many tissues of the body.

FRAXA syndrome

The syndrome of a fragile X chromosome (FRAXA) is an X-linked disease which is inherited in a dominant fashion. The molecular basis for this condition is the expansion of a trinucleotide repeat (CGG) at the beginning of a non-translated region (5′ UTR) of the *fmri* gene located at Xq27.3[26,27]. In the average population the length of CGG repeats is extremely polymorphic, ranging from six to 54 repeats. It is usually stably inherited by the progeny[28]. Carriers have the length of this trinucleotide repeated within the range of 60 to 200 repeats. The condition is termed a premutation since the premutated alleles may undergo expansion when passed from carrier to progeny. The expansion with over 200 repeats is termed a complete mutation and results in hypermethylation of the CGG trinucleotide in the upstream direction of the *fmr* gene and, consequently, reading its transcriptional inactivation[28,29]. The fragile X chromosome premutation carriers do not show clinical symptoms, although there is evidence that the premutated alleles' presence is linked with premature ovarian failure. Indeed, a number of studies have shown a relative frequency of premature ovary failure syndrome among FRAXA carriers compared to the average in the overall population[30–32]. One of the latest studies has shown that up to 16% of females who are carriers of premutated FRAXA have their menopauses before the age of 40[33].

Structural aberrations in autosomes

The balanced autosomal reciprocal translocations which are quite rare in the fertile female population belong to the chromosomal aberrations that are more frequently encountered in female reproductive dysfunction[1,3,34] (Figure 2). They present a mutual exchange between chromosomes and their distal parts. As they are balanced types, the chromosomal material is retained. If the cells that have complete or partial translocation undergo meiosis with gamete formation, part of them will be monosomic and part trisomic. Thus, at fertilization both trisomic and monosomic zygotes will be formed.

Robertsonian translocations

These are also characteristic for the group of patients with reproductive dysfunction[4,35]. They present one of the variants of interchromosomal exchange when the connection of long arms of acrocentric chromosomes takes place. In this case chromosomes with one or two centromeres can be formed. Centric fusion is one of the most widely spread types of chromosome restructuring in man. Participation of all acrocentric chromosomes in the formation of such translocations points to the fact that the very structure of such types of chromosome serves as a basis for the occurrence of such an aberration (Figure 3). One of the clinical examples of a Robertsonian translocation is a centric fusion of arms of chromosome 21 with one of the chromosomes from group D or G. Such translocation can arise in any of the parents. The peculiarity stems from the fact that this karyotype has 45 chromosomes, but the translocation is a balanced one in terms of genetic information; thus, the

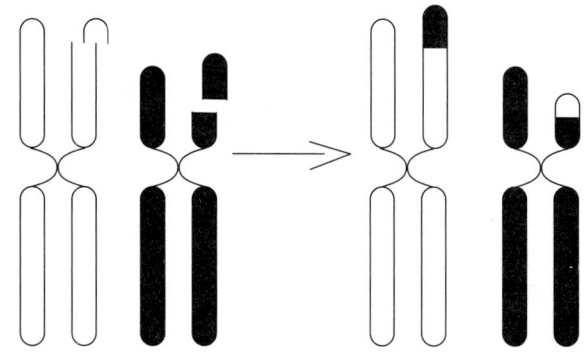

Figure 2   Formation of a reciprocal translocation (balanced type)

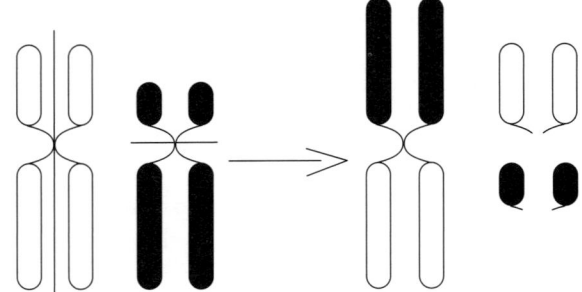

Figure 3   Formation of a Robertsonian translocation

breakage of the short arms in acrocentric chromosomes is not manifested phenotypically. Most of the balanced translocation carriers are phenotypically healthy, but careful analysis shows that insignificant anomalies exist and in some cases mental and physical retardation are more apparent compared with normal karyotype carriers.

There are two leading factors which condition the decreased fertility of the couple when one of the partners is a Robertsonian translocation carrier. The first factor is aneuploid segregation of the parental chromosome having Robertsonian translocation, and the other is a post-zygotic factor resulting in uncontrolled distribution of chromosomes at the early stages of splitting[36]. The parental zygotes with such translocations happen to be the carriers of unbalanced translocations and, correspondingly, are only a little bit viable or else not viable at all. Such married couples doubtless need ART to realize their reproductive function, and these should be accompanied by preimplantation genetic analysis methods[37].

Pericentric inversions

The next aberration type in this examination of patients with infertility is inversion[3,4,35,38]. Pericentric inversions arise as a result of a breakage of two arms and a 180° turn of part of the chromosome, which contains a centromere. Paracentric inversion arises as a result of the 180° turn of a part of one arm, which builds in after the interstitial deletion. In diagnosis paracentric inversions of various sizes are found, but they do not manifest themselves phenotypically and have no influence on reproductive function.

Pericentric inversions manifest themselves phenotypically in different ways, ranging from mental retardation to a complete absence of any anomaly. This type of inversion is seen in mothers who had their children delivered with developmental defects, and/or whose family members also had such defects or in females with habitual miscarriages. The pericentric inversions are also seen in chromosome 9. Practically, this type of inversion does not reduce the number of chromosomes in meiosis and does not lead to prenatal death in heterozygotic embryos. It is a most common structural, balanced chromosome aberration, and is looked upon as a paraphysiological variant of a normal karyotype. However, a lot of data have been amassed[38–40] on the carriers of this inversion, who have different physiologic anomalies, including reproductive function. In particular, habitual miscarriages and subfertility are referred to[38,40].

The frequency of cases of the pericentric inversion on chromosome 9 (in v9phqh) in infertile patients is of special interest from the viewpoint of the heterochromatization phenomenon. The peculiarity of this phenomenon is in the fact that gene activity in the euchromatic region close to heterochromatin is repressed. In the case of a partial inversion of the heterochromatin region, the genes in the loci of the chromosome breakage lose their habitual orientation and change their functional activity.

The gene responsible for the synthesis of prostaglandin precursors is localized in the terminal region of the heterochromatin on chromosome 9. Thus, it can be supposed that if this gene turns out to be at the locus of the inversion its functional activity or the activity of other neighboring sequences will be changed, and this fact might not be unimportant for the reproductive process. Females having partial pericentric inversion of chromosome 9 in their case histories present spontaneous miscarriages. When analyzing the abortuses from such mothers, the percentage of female fetus carriers of the inversion is considerably higher than the male one. Long clinical observation is witness to the fact that the progeny of parents with such an inversion are confronted with infertility more often than their parents[38].

Our own observations allow us to consider the presence of the fragile chromosome 4 as a probable sign of reproductive dysfunction of 46,XX;FRA 4(q2.7–2.8). Up to the present time there is no indication in the literature concerning the existence of a link between this aberration and a pathologic phenotype. The connection between the fragile part in 4(q2.7–2.8) (Figure 4) and infertility still remains unknown, but the discovered anomaly needs further study.

Special interest is aroused by the high frequency of anomalies in the structure of pericentromere regions of heterochromatin on the long and short arms of chromosome 1: 1ph–, 1ph+, 1qh–, 1qh+. This frequency differs to a great extent when compared with the normal population frequency[39,41]. It possibly explains the unfavorable reproductive prognosis in this particular group of females. Molecular cytogenetic studies of the extreme variant of chromosome 1 have been carried out using the 'classic' satellite and a satellite DNA (DNA probes pUC 1.77 and R1-12, respectively). It was established that an enlargement of the heterochromatin region is associated with the number of copies of the 'classic' satellite DNA; moreover, these two different types of repeated DNA vary independently of each other.

According to our studies performed at the Odessa State Medical University in the Ukraine, the changes in the acrocentric chromosomes which are on the border of individual polymorphism and pathology are encoun-tered more frequently in females with reproductive dysfunction than in the average general population. The main morphologic peculiarity of these changes is either an increase or decrease in satellites or satellite threads on chromosomes 13, 14, 15, 18, 21 and 22 (Figure 5). It should also be noted that all the enumerated variants are not single in their manifestation, but are usually combined with each other. There is no clear dependence between the combinations of these structural anomalies and the reproductive dysfunction.

The high frequency of acrocentric chromosome associations has also been noticed in patients with reproductive failure. If the associations of acrocentric chromosomes are moderately high in number they are located without hindering other chromosomes, but if they are many in number, they occur with a secondary strangulation of chromosome 9 (Figure 6), and less frequently with the secondary band of chromosome 1.

Such instability of the genome and the presence of a certain quantity of deviations are witness to the instability of the genome in general and this may promote non-segregation of chromosomes in cell division. As a result one cannot deny the possibility that damage to sex cells occurs and that this, in turn, decreases the possibility of fertilization or even makes fertilization impossible.

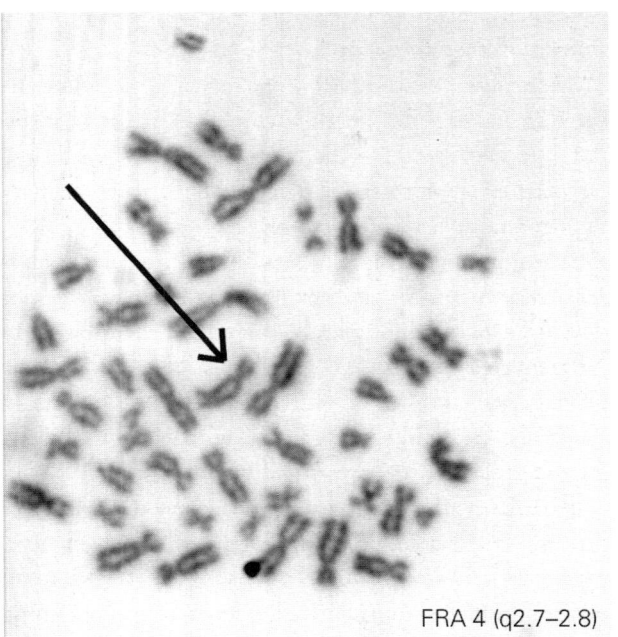

FRA 4 (q2.7–2.8)

Figure 4 Fragile span of the chromosome A locus q2.7–2.8

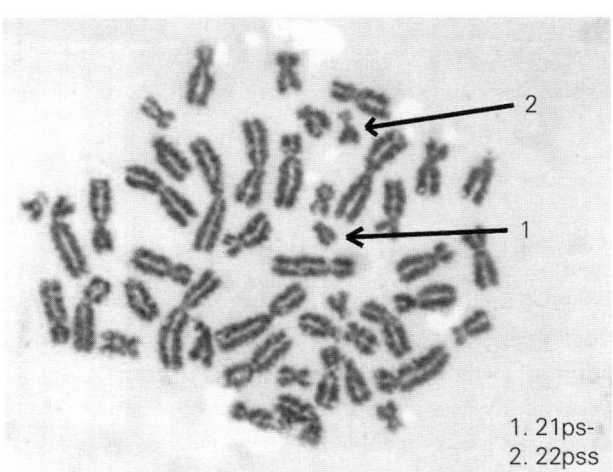

1. 21ps-
2. 22pss

Figure 5 Chromosome 21 is missing satellite material on the short arm (1); chromosome 22 has doubled satelllite material on the short arm (2; possible translocation of the satellite)

Dysfunction of autosomal genes

XX dysgenesis of gonads (Mendelian)

Gonadal dysgenesis, similar in histologic structure to that experienced by patients with sex chromosome aneuploidy (Turner's syndrome, 45,XO), can be observed in patients with a normal karyotype (46,XX). Mosaicism is excluded in such patients after performing a complex examination, although it is impossible to exclude mosaicism at the embryo stage. Patients are diagnosed as having 46,XX gonad dysgenesis, of which there are many forms, but the dysgenesis prototype is not associated with somatic anomalies and is evidently inherited in an autosomal-recessive fashion. Females with this pathology are usually of normal stature[42], with no signs of Turner's syndrome whatsoever.

A clear dependence is observed in families with such pathology in consanguineous marriages, and it is confirmed by the inheritance type following an autosomal-dominant pattern. Segregational analysis revealed that two-thirds of 46,XX gonad dysgenesis cases are the result of genetic changes, while one-third are a consequence of an infectious process, infarction or an infiltrative autoimmune process[1]. Various manifestations of this syndrome are seen at clinical examination of such patients, and these are of particular interest. For instance, one patient may have connective tissue cords in the projection of the supposed gonads, while her relative would have primary amenorrhea and a manifested hypoplasia of the ovaries (having a considerable number of oocytes)[43–46]. If the mutant gene responsible for the 46,XX gonad dysgenesis has different degrees of expression it may also be the gene which determines the premature ovarian failure syndrome.

Syndromal 46,XX gonadal dysgenesis

There exists a whole group of syndromes in females with a normal karyotype who have gonadal dysgenesis. Among them is Perrault syndrome, which comprises gonadal dysgenesis and neurosensory deafness[42,47–50]. Other syndromes include 46,XX gonadal dysgenesis and cerebellar ataxia[51]; 46,XX gonadal dysgenesis, microcephaly and arachnodactyly[1]; 46,XX gonad dysgenesis and epibulbar dermoid[52,53]; and 46,XX gonad dysgenesis, low stature and metabolic acidosis[54]. These syndromes are inherited in an autosomal-recessive fashion, as they are often encountered in close relatives. However, we cannot exclude a chromosomal deletion or some non-Mendelian mechanism of inheritance altogether in such cases.

Blepharphimosis–ptosis is always accompanied by ovarian dysfunction[55,56]. After performing DNA tests, the gene responsible for blepharophimosis–ptosis was localized on the long arm of chromosome 3 (3q21–24). Still another syndrome associated with the germ cell dysfunction is linked with ataxia and telangiectasia. Ataxia telangiectasia is an autosomal-recessive disease characterized by unstable gait, progressing neuromotor degeneration and dilation of blood vessels of the conjunctiva and facial skin, as well as by cellular and humoral immune response deficiency. The gene responsible for the development of this syndrome is located on the long arm of chromosome 11 (11q22–23) and belongs to the family of protein kinases, participating in different metabolic processes, including regulation of the cell cycle.

Inherited diseases of the autosomal-dominant type include Cockayne's syndrome[56], which is characterized by dwarfism, microcephaly, mental retardation, retinitis pigmentosa, photosensibilization, early senility and high sensitivity to ultraviolet rays. These patients also have ovarian dysfunction in conjunction with atrophy and fibrosis of the ovaries[57].

Martsolf's syndrome[58] phenotypically manifests itself by low stature, microbrachiocephaly, cataract and facial anomalies due to the relative prognathism related to maxillar hypoplasia. These patients have primary hypogonadism[59,60].

Chromosomal instability, immunodeficiency, hypersensitivity to ionizing radiation and a tendency to tumor development phenotypically manifest themselves in Nijmengen's syndrome[61], with patients showing a primary insufficiency of ovarian function[62,63]. Werner's syndrome is typified by low stature, early senility, sclerodermia and severe ovarian hypofunction[64].

Patients with Rothmund–Thompson's syndrome[65] present with telangiectasia, erythema, uneven skin pigmentation, low stature, cataracts, short and small upper and lower limbs and a tendency to osteosarcoma development. Such patients have primary hypogonadism and sexual maturation delay in combination with a manifested ovarian hypofunction[66].

Carbohydrate–glycoprotein deficiency, type I, syndrome (phosphomannomutase deficiency) is characterized by various neurologic manifestations, such as ataxia, hypotension–hyporeflexion, paralysis and joint contractures, in addition to hypogonadism[67].

Bloom's syndrome[68–70] is characterized by dolichocephalism, low stature, sensitive facial erythema and chromosomal instability (a high rate of exchange between sister chromosomes). Patients have a tendency to experience tumor development and to manifest ovarian function insufficiency.

46,XX agonadism syndrome

This condition may occur in persons having the 46,XX karyotype[71–73], although it is more common in those having the 46,XY karyotype. The external genitalia present in the form of an underdeveloped penis, covered slightly by the labia, and resembling unfinished labioscrotal fold fusion[1]. But for some rudimentary structures one can barely detect the internal genitalia[74].

GENE DYSFUNCTION AFFECTING THE HYPOTHALAMIC–HYPOPHYSEAL–OVARIAN AXIS

The first hormone on this axis, which regulates the reproductive function, is the gonadotropin-releasing hormone (GnRH). Its deficiency is associated with the development of hypogonadotropic hypogonadism that is quite often described in males. This is an X-linked disease inherited in a recessive fashion. Kallmann's syndrome[75] is a classic description of the condition, which is accompanied by olfactory bulb malformation that leads to anosmia. The product of the *kal* gene is responsible for the migration of GnRH producing neurons from the olfactory placode into the brain.

Females who have the neuron migration disorder exhibit insufficient development of genitalia, or even a complete absence, and also amenorrhea. The degree of the GnRH decrease in the serum of such patients varies but a prescription of pulsating GnRH doses corrects the condition.

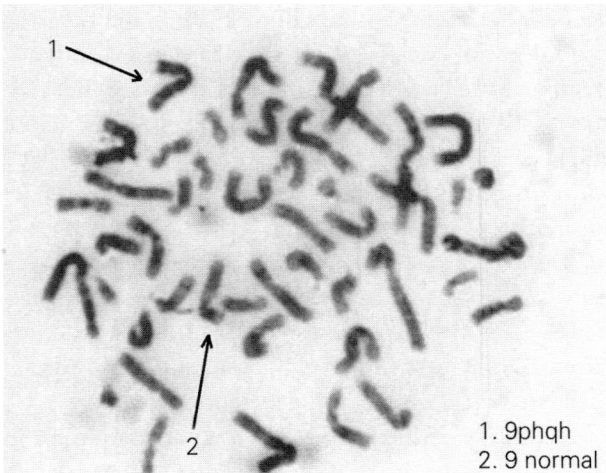

1. 9phqh
2. 9 normal

Figure 6 Enlargemement of the heterochromatic region of the long and short arms of chromosome 9 (1); normal structure of chromosome 9 (2)

The normal GnRH function is mediated through its receptor

The product of the gene determining the formation and functioning of GnRH receptors is the G-protein-coupled receptor, which contains six transmembrane regions. The mutations in the first[76] and the third intracellular regions are known, as well as those in the transmembrane domain[77], which are found in females with hypogonadotropic hypogonadism.

It is accepted that the mutated GnRH receptor is encountered in 7% of familial cases of hypogonadotropic hypogonadism. Such patients suffer primary infertility and amenorrhea. The level of estradiol is within the normal range with a sufficiently high level of gonadotropins in blood serum[76].

The gonadotropins – follicle stimulating hormone and luteinizing hormone

Follicle stimulating hormone (FSH) and luteinizing hormone (LH) are heterodimers with α- and β-subunits. The α-subunit is identical in both FSH and LH; the β-subunit defines the specificity of the given hormonal action. The activity of the hormone is manifested when the subunits unite to form a dimer.

The gene encoding the synthesis of the β-subunit of FSH is located on chromosome 11 and consists of three exons. The exon 3 deletion (which breaks the sequence of 61–86 codons by introducing an immature stop codon) in the homozygotic female carrier is manifested by primary amenorrhea and infertility[78,79]. This is linked to the fact that the mutant gene product is a β-subunit of FSH, which is unable to bind with the α-subunit and form an active hormone. Females with isolated FSH insufficiency at birth have normal anatomic structure of the gonads, internal and external genitalia. In pubescence primary amenorrhea and the postmenopausal LH level are observed. In females who are heterozygotic carriers of this mutation an irregular menstrual cycle and infertility are noted.

Deficiency in FSH evoked by polymorphism in codon 58, exon 3, was found alongside the 2 bp deletion in codon 61 and a missense mutation in codon 51[80]. In one case the patient presented with primary amenorrhea and secondary sex features – with pubarche without telarche, and mammary gland development according to Tanner's classification (stage 1).

The gene encoding the FSH receptor is located on chromosome 2. The mutation of the FSH receptor in the gene is a missense mutation in the encoding region of the extracellular ligand-binding domain of this receptor[46]. This mutation is an exchange of cytosine for thymine in codon 566, leading to a substitution of alanine for valine in position 189. The expression of the mutated gene in transfected cells leads to a decrease in binding of ligands and transduction signals. The homozygote female carriers of this mutation have hypergonadotropic ovarian dysgenesis. The abnormal development of the follicular apparatus of the ovaries and a severe form of hypogonadism characterize this pathology. The specific feature in such patients is a high level of FSH and LH and a low level of estrogen, which is clinically manifested by disorders in the maturation of sexual genitalia. Histologic tests have confirmed the presence of primordial follicles in patients having the mutant receptor[81]. At the same time these patients have primary amenorrhea and infertility. An inactivating mutation has been discovered which leads to the substitution of thymine for cytosine, thus promoting the substitution of phenylalanine for serine in codon 591. This has been observed in nine of 13 females who had stromal tumors of the ovaries[82]. The activating mutation of the FSH receptor gene has not yet been found. It can be supposed that these patients will develop ovarian cysts, signs of

spontaneous hyperstimulation of the ovaries and an early menopause as the result of follicle pool depletion at an early stage.

The gene encoding the LH α-subunit is located on chromosome 6, while the gene encoding the β-subunit is on chromosome 19. Only one mutation is known in the gene of the LH β-subunit, which leads to the substitution of glutamine for arginine in position 54 of the polypeptide chain. The mutant hormone loses its ability to bind with the receptors of the target cell. Glutamine in position 54 is an obligatory element for the β-subunits of glycoprotein hormones (FSH, LH, thyroid stimulating hormone (TSH), human chorionic gonadotropin (hCH)) and ensures their biological activity. The loss in LH bioactivity leads to dysfunction in steroidogenesis. In heterozygotic female carriers this mutation is not manifested as it is inherited in an autosomal-recessive way. Manifested polymorphism is characteristic of the LH β-subunit gene[83,84]. Two more frequently encountered allele variants of the gene have been found and described in females with the immunologic LH anomalous variant, and two point mutations have been found in exon 3, leading to the substitution of tryptophan for alanine in position 8 as well as to the substitution of isoleucine for threonine in position 15[85]. Variants of the LH β-subunit polymorphism result in an increase of its bioactivity *in vitro* with a simultaneous decrease of its half life *in vivo*[86]. Data analysis of phenotype features of carriers of allelic variant LH β-subunits has revealed that part of them exhibit features of reproductive dysfunction. The main disturbances are menstrual cycle disorders, infertility and habitual miscarriages. A new mutation has been found in codon 1502, exon 3, of the LH β-subunit, leading to a substitution of serine for glycine in position 102[87]. This mutation was found in females with a normal menstrual cycle, infertility, a normal or elevated LH level in serum and a low estradiol level at the early follicular phase.

Not withstanding the clinical observations, it still remains unclear if the LH β-subunit polymorphism leads to a stable pathology in the reproductive system and to infertility in the carriers. This is connected to the high frequency of these variants among the population, who otherwise function normally in terms of reproduction. At the same time there are clinical data[88] testifying that the general activity of polymorphic LH variants with amino acid substitution is much lower than the norm. Added to which is the consideration of a pleiotropic influence of hormonal metabolism in females on the phenotypic manifestation of the LH β-subunit polymorphism[88].

The highest quantity of mutations has been described for the LH receptor gene. These mutations are of two types: those which activate the receptor and those which inactivate it. Mutations in genes resulting in LH receptor activation in females have not yet been described, the literature citing clinical cases of such mutations in males only. In females with an inactivating gene mutation of the LH receptor, normal morphologic structure of the ovaries is observed with the presence of a satisfactory amount of antral follicles, and high FSH and LH levels in blood serum in the context of a very low level of estradiol. Females who are homozygote carriers of mutant alleles of the LH receptor gene have a slightly changed phenotype, but they do have primary amenorrhea and lack fertility. DNA testing of the gene encoding the LH receptor has revealed a mutation in the sixth transmembrane domain leading to the substitution of alanine for proline at position 539[89]. One mutation has been described where codon 554 led to a stop-codon appearance in a patient with amenorrhea and normal secondary sex features[90]. In this case, the estrogen level was within the normal range at the early follicular phase and this was sufficient for normal development in puberty. But at the same time, enlargement of the ovaries and multiple ovarian cysts were revealed[91].

In general, the gonadotropin receptors present a separate group of G-protein-coupled transmembrane receptors. Among them is the receptor of TSH. The mutations lead either to activation or suppression of the TSH receptor function, not only causing conditions in which the function of the directly dependent organs is impaired (for instance, hyperfunction of the thyroid gland[92] or rare congenital forms of hypothyreosis[93,94]) but also affecting reproductive function. Among mutations activating G-protein is McCune– Albright syndrome, which is characterized by a polysteotic fibrotic dysplasia with 'milk and coffee' spots[95]. Sex steroid hypersecretion, early sexual maturation of both males and females, and the hyperfunction of other glands of endocrine secretion (thyroid, parathyroid and pituitary) are all characteristic features of such patients.

Heterogenic mutations leading to the loss of gene function encoding the α- and β-subunits of the stimulating G-protein lead to a generalized hypofunction of the G-protein-coupled receptors at the inherited Albright osteodystrophy[96].

GENETIC FACTORS AFFECTING SEX STEROID SYNTHESIS

The synthesis of sex steroid hormones is a cascade of biochemical reactions resulting in the transformation of cholesterol into steroid hormones of adrenal glands and ovaries. If only one link in the chain fails it will lead to changes in quantitative and qualitative characteristics of the produced hormones. Each link in the synthesis of sex steroids is under the influence of specific enzymes, which belong to the groups of cytochromes P450. One of them is the side-chain cleavage (scc) enzyme responsible for the so-called slow phase of chronic steroid synthesis. When stimulated by the tropic hormones the steroid acute regulatory protein (StAR) is involved in the process. StAR is responsible for transport of cholesterol to the inner membrane of the mitochondria and for its subsequent quick conversion into pregnenolone. The *star* gene and its protein are expressed on theca cells, granulosa cells of mature follicles and in corpus luteum[97], which indicates their functional role in the physiology of the ovaries.

Congenital lipoid hyperplasia of adrenal glands is an autosomal-recessive and potentially lethal disease, which is characterized by a rapid decrease in biosynthesis of steroid adrenal hormones and sex glands.

The complex of the P450-scc enzyme is encoded by the *cyp11a* gene and is responsible for the first link in the synthesis of steroids, i.e., of turning cholesterol into pregnenolone. The condition is characterized by a P450-scc enzyme deficiency and occurs very rarely, but again is considered potentially lethal.

The enzyme 3β-hydroxysteroid dehydrogenase (3β-HSD) turns the five-membered carbohydrate complex into a four-membered one. It is necessary for the normal synthesis of steroids of adrenal glands and sex glands. Two genes found to encode 3β-HSD synthesis are located on chromosome 1 in the locus p11–13. The first gene is expressed in the tissues of the skin, kidneys and placenta, the second is expressed only in gonads and adrenal glands. Low activity of the former can affect pregnancy due to the lowering of the progesterone-producing function of the placenta. Low activity or complete absence of the latter leads to dysfunction or absence of sex hormone and adrenal steroid synthesis.

The cytochrome P450c17 enzyme, which is encoded by the *cyp17* gene, is necessary for the conversion of progestin to androgens. Matrix RNA of P450c17 and the protein itself are found exclusively in theca cells in females. The absence of this enzyme in the granulosa cells confirms the hypothesis that these cells are not capable of the *de novo* synthesis of estradiol from cholesterol.

17α-Hydroxylase deficiency is a rare autosomal-recessive disease primarily diagnosed as congenital adrenal hyperplasia because of decreased cortisol production resulting in subsequent stimulation of the adrenal growth by elevated serum adrenocorticotropic hormone (ACTH) concentrations[98,99]. At this status, over-production of deoxycorticosterone causes a certain degree of hypertension. In females, initial differentiation of the genital tract is normal, but sexual development during puberty may be disrupted owing to the inability to produce sex steroids. At a later age, this syndrome is characterized by hypogonadism. Up to now, many cases of P450c17 deficiency and more than 14 different mutations in the *cyp17* gene have been described. It is obvious that some of the mutations result in only partially reduced enzyme activity. Some estrogens are even produced in partial forms of *cyp17* deficiency, resulting in normal female sexual development and, sometimes, regular menstrual cycles.

17β-Hydroxysteroid dehydrogenase is responsible for the conversion of androstenedione to testosterone and estrone to estradiol. Autosomal-recessive mutations in the 17β-HSD 3 gene have been described in several cases[100,101].

Cytochrome P450 aromatase enzyme is encoded by the *cyp19* gene. It catalyzes three hydroxylation steps converting C19 androgens to estrogens. This enzyme is found in ovarian granulosa cells and in testis Sertoli and Leydig cells as well as in various extragonadal tissues, such as adipose tissue, liver, muscle, brain and placenta. Granulosa cells are therefore capable of converting theca cell-derived androgens to estrogens[102]. In cases of aromatase enzyme gene mutation patients present with sexual infantilism, primary amenorrhea, elvated serum gonadotropin levels, distince delay in bone age, large ovarian cysts and tall stature[103]. Direct sequencing of the *cyp19* aromatase gene indicated two mutations with single-base changes at position 1303, cytosine being substituted by thymine, and cysteine being encoded instead of arginine, while at position 1310 guanine was substituted by adenine, which encodes tyrosine instead of cysteine in exon[104]. These are missense mutations and mutant proteins lose their function. An additional mutation in exon 9 has been detected in the *cyp19* gene[105,106]. The patient presented with non-adrenal female pseudohermaphroditism at birth and showed progressive virilization and absent pubertal development. Multiple large cystic masses were observed in the ovaries of this patient in the context of elevated endogenous gonadotropin secretion[105,107].

5α-Reductase enzymes are encoded by two genes: *srd5a* on chromosome 5 and *srd5a2* located on chromosome 2[108]. These isoenzymes are present in various organs, including gonads, brain and skin, and convert testosterone to the more potent 5α-dihydrotestosterone (DHT). 5α-Reductase deficiency is an autosomal-recessive condition. Affected females present with decreased body hair but with normal ovarian function and with their child-bearing status unhindered, implying that 5α-reductase deficiency plays no role in female sexual differentiation and function[109–111].

The cytochrome enzymes P450c21 and P450c11 show adrenal-specific expression. The 21-hydroxylase enzyme (P450c21) stimulates the conversion of the four-carbohydrate cycle steroids progesterone and 17OH-progesterone to 11-deoxycorticosterone and 11-deoxycortisol. The structural gene (*cyp21* or *cyp21b*) encoding the P450c21 enzyme and a pseudogene (*cyp21p* or *cyp21a*) is closely linked to the histocompatibility complex – on the short arm (p21.3 region) of chromosome 6.

The cytochrome P450c21 enzyme is encoded by the *cyp21* gene, as described above. The majority of mutations in the *cyp21* gene are deletions or transfers of deleterious sequences from the adjacent pseudogene to the active gene (referred to as gene conversion[112,113]). Consequently, a small or large region of *cyp21p* is transferred to the defective *cyp21* gene, a unique mechanism in genetic disorders. Mutations correlate with clinical disease symptoms and the severity of endocrine abnormalities[114]. In particular, the adenine to cytosine substitution in exon 7 is associated with the late-onset form of the disease and the deletion in exon 3 with the salt-losing form of adrenogenital syndrome. Congenital adrenal hyperplasia is a 21-hydroxylase deficiency. Described mutations in *cyp21* include deletions in intron 2 and in all but exons 5 and 9, an insertion in exon 7, a nonsense mutation in codon 318 of exon 8 and eight missense mutations. Affected females are diagnosed more frequently because of the prominent signs of virilization which can include complete sex reversal (male phenotype). Late-onset or 'non-classical' adrenal hyperplasia is observed in several groups of patients and is caused by mutations that only partially reduce 21-hydroxylase enzyme activity and that were detected in exons 1, 7, 8 and 10 of the *cyp21* gene[115–117]. The mutations give rise to symptoms of hyperandrogenemia only after puberty; these patients may be diagnosed as having polycystic ovary syndrome (PCOS)[107].

The cytochrome P450c11 enzyme is encoded by the *cyp11b1* gene. Mutations causing 11β-hydroxylase deficiency are responsible for approximately 5% of cases of classic adrenal hyperplasia. This defect results in decreased cortisol secretion, elevated serum levels of ACTH, upstream steroids and hyperandrogenism. More than 13 different mutations have been identified[118,119]. Mild forms of 11β-hydroxylase deficiency have also been implemented in hyperandrogenism in PCOS[120].

Any information on mutations of the human estrogen receptor is absent in the available literature, but a second subtype of the estrogen receptor – referred to as the estrogen receptor β (ERβ) – was discovered recently[121]. The gene has been mapped to the long arm of chromosome 14, in contrast to the *erα* gene on chromosome 6. This discovery may also explain tissue-specific expression of estrogenic/antiestrogenic effects of certain compounds such as tamoxifen and raloxifene. ERβ expression has been found predominantly in the human ovary (stroma and granulosa cells)[122], suggesting that many of the effects of estrogens on the female reproductive system may be mediated by this receptor.

All androgens act through the X-linked androgen receptor, the N-terminal domain of which contains a polyglutamine tract encoded by a highly polymorphic CAG trinucleotide repeat sequence in exon 1[123].

Recently, variations in this CAG microsatellite sequence, while remaining within the normal polymorphic range of 11–38 CAG repeats, have been inversely correlated with receptor activity. Thus, short sequences are associated with high intrinsic activity of the androgen receptors and increased severity and earlier age of onset of the androgen-regulated tumors, whereas long CAG sequences are associated with low activity of androgen receptors. The association between short CAG repeat length and the subset of anovulatory patients with low serum androgens has been proven, which suggests that the mechanism of the polycystic ovaries in these patients could be due to the increased intrinsic androgenic activity, associated with the androgen receptors' alleles[124]. Affected females exhibit minimal clinical symptoms, and the familial form of this disease is transmitted in an X-linked fashion through fertile heterozygous females. In some cases such patients present with complete androgen insensitivity and primary amenorrhea, with moderate breast development, along with absent pubic and axillary hair. External genitalia are female, the uterus is absent, the karyotype is 46,XY and endocrine investigation has revealed elevated serum LH and androgen levels[125]. Mutations causing such effect involve codons in exon 3.

CONCLUSIONS

In the past decade an increasing number of genetic factors have been implicated in the impairment of female reproductive function. Some of these factors have been well characterized and described by the identification of genes and their corresponding products. Others have been localized only to a specific region or regions on the chromosomes, while still others that are recognized as being of genetic origin remain completely unidentified. New technologies have made it possible to detect single-gene defects at the level of sex chromosomes or the synthesis and action of all hormones involved in the hypothalamic–hypophyseal–ovarian axis.

Understanding the numerous genetic factors that influence reproductive function, the essential need for genetic counseling becomes obvious. Genetic counseling itself is a communication process focused on helping families to understand the nature of the genetic or congenital disorder which one or both partners may have, its transmission and the options for management of the condition – possibly assisting them in choice of option. Traditionally, genetic counseling was aimed at parents fearing their children might have inherited genetic or chromosomal pathology, or to parents whose offspring were already showing signs of hereditary or congenital disease. It is not always necessary to send patients to the clinical geneticist because there are certain cases when the gynecologist must be able to provide them with necessary information. Basic steps for genetic counseling are: building up a trusting relationship between counselor and couple, providing basic information on the reason the couple are being subjected to such counseling, and confirming the diagnosis irrespective of how obvious the diagnosis may be. Then one needs to take a complete family history, and look for possible incidences of the same pathology in close or distant relatives, for consanguinity, spontaneous abortions, ethnicity, etc. If any supplementary test, analysis or procedure is necessary, the couple must be informed about purposes, risks and possible outcomes. Counseling should never be directive in tone or attempt to push the family to certain decisions.

The complexity of the genetic causes of infertility or subfertility is obvious and requires a multidisciplinary approach, a successful collaboration between gynecologist, andrologist, urologist, clinical geneticist, and maybe other specialists involved in ART. This will help couples achieve a broad understanding of the risks for themselves and their children, whether already born or still to be conceived.

References

1. Simpson JL, Rajkovic A. Ovarian differentiation and gonadal failure. *Am J Med Genet* 1999;89:186–200
2. Ohno S. *Sex Chromosomes and Sex-linked Genes*. Berlin: Springer-Verlag, 1967
3. Hens L, Bonduelle M, Liebaers I, *et al*. Chromosome aberrations in 500 couples referred for *in vitro* fertilization or related fertility treatment. *Hum Reprod* 1998;3:451–7
4. Peshka B, Leygraaf JK, Van der Ven, *et al*. Type and frequency of chromosome aberrations in 781 couples undergoing intracytoplasmic sperm injection. *Hum Reprod* 2001;14:2257–63
5. Higgins MD, Palmer CG. Single cell translocations in couples with multiple spontaneous abortions. *Hum Genet* 1987;75:24–7
6. Lange R. Chromosome studies in *in-vitro* fertilization patients. *Hum Reprod* 1993;8:572–4
7. Goodfellow PN, Lowell-Badge R. SRY and sex determination in mammals. *Annu Rev Genet* 1993;27:71–92
8. Simpson JL. Genetic control of sexual development. In Teoh ES, Ratnam SS, Goh VHH, eds. *Advances in Fertility and Sterility: Releasing Hormones and Genetics and Immunology in Human Reproduction*. Lancaster, UK: Parthenon Publishing, 1987;165–73
9. Simpson JL. Phenotypic–karyotypic correlation of gonadal determinants: current status and relationship to molecular studies. In Sperling K, Vogel F, eds. *Human Genetics* (Proceedings of the Seventh International Congress, Berlin, 1986). Heidelberg, Berlin: Springer-Verlag, 1987:224–32
10. Ogata T, Matsuo N. Turner syndrome and female sex chromosome aberration deduction of the principal factors involved in the development of clinical features. *Hum Genet* 1995;95:607–29
11. Simpson JL. Genetics of oocyte depletion. In Lobo RA, ed. *Perimenopause*. Norwell, MA: Springer, 1997:36–45
12. Simpson JL. Genetics of female infertility. In Filicori M, Flamigni C, eds. *Proceedings of the Conference, Treatment of Infertility: The New Frontiers*. Boca Raton, FL: Communications Media for Education, Inc., 1998:37–52
13. Simpson JL. Genetic programming in ovarian development and oogenesis. In Lobo RA, Marcus R, Kelsev J, eds. *Menopause*. San Diego: Academic Press, 2000
14. James RS, Coppin B, Dalton P, *et al*. A study of females with deletions of the short arm of the X-chromosome. *Hum Genet* 1998;102:507–16
15. Zinn AR, Ouyang B, Ross JL, *et al*. Del (X)(p21.2) in a mother and two daughters with variable ovarian function. *Clin Genet* 1997;52:235–9
16. Fraccaro M, Maraschio P, Pasquali F, Scappaticci S. Women heterozygous for deficiency of the (Xpter X21) region of the X chromosome are fertile. *Hum Genet* 1977;39:283–92
17. Simpson JL, LeBeau MM. Gonadal and statural determinants on the X chromosome and their relationship to *in vitro* studies showing prolonged cell cycles in 45,X, 46,X, del(X)(p11), 46,X,del(X)(q13) and q(22) fibroblasts. *Am J Obstet Gynecol* 1981;141:930–9
18. Zinn AR, Ouyang B, Ross JL, *et al*. Del (X)(p21.2) in a mother and two daughters with variable ovarian function. *Clin Genet* 1997;52:235–9
19. Fitch N, de Saint VJ, Richer CL, *et al*. Premature menopause due to small deletion in long arm of the X chromosome: a report of three cases and a review. *Am J Obstet Gynecol* 1982;142:968–72
20. Schwartz C, Fitch N, Phelan MC, *et al*. Two sisters with a distal deletion at the Xq26/Xq27 interface: DNA studies

indicate that the gene locus for factor IX is present. *Hum Genet* 1987;76:54–7

21. Simpson JL. Gonadal dysgenesis and abnormalities of the human sex chromosomes: current status of the phenotypic–karyotypic correlations. *Birth Defects* 1975;11:23–59

22. Sala C, Arrigo G, Torn G, *et al.* Eleven X chromosome breakpoints associated with premature ovarian failure (POF) map to a 15-Mb YAC contig spanning Xq21. *Genomics* 1997;40:123–31

23. Leppig DA, Disteche CM. Ring X and other structural X chromosome abnormalities: X inactivation and phenotype. *Semin Reprod Med* 2001;19:147–57

24. Tharapel AT, Anderson KP, Simpson JL, *et al.* Deletion (X) (q26 1q28) in a proband and her mother: molecular characterisation and phenotypic–karyotypic deductions. *Am J Hum Genet* 1993;52:463–71

25. Marozzi A, Furlan D, Manfredini E, *et al.* Diaphanous gene disruption in a woman with Xq deletion and secondary hypergonadotropic amenorrhea. *Am J Hum Genet* 1998;64:A114(abstr.812)

26. Verkerk AJ, Pieretti M, Sutcliffe JS, *et al.* Identification of a gene (*FMR-1*) containing a CCG repeat coincident with a breakpoint cluster region exhibiting length variation in fragile X syndrome. *Cell* 1990;65:905–14

27. Warren ST, Ashley CT. Triplet repeat expansion mutations: the example of fragile X chromosome. *Annu Rev Neurosci* 1995;18:77–99

28. Fu YH, Kuhl DP, Pizutti A, *et al.* Variation of the CGG repeat at the fragile X site results in genetic instability: resolution of the Sherman paradox. *Cell* 1991;67:1047–58

29. McConkie-Rossel A, Lachiewicz AM, Spiridigliozzi GA, *et al.* Evidence of methylation of the *FMR-1* locus is responsible for variable phenotypic expression of the fragile X syndrome. *Am J Hum Genet* 1993;53:800–9

30. Cronister A, Schereiner R, Wittenberger M, *et al.* Heterozygous fragile X female: historical, physical, cognitive and cytogentic features. *Am J Med Genet* 1991; 38:269–74

31. Schwartx CE, Dean J, Howard-Peebles PN, *et al.* Obstetrical and gynecological complications in fragile X carriers: a multicenter study. *Am J Med Genet* 1994;51: 400–2

32. Partington MW, Moore DV, Turner GM. Confirmation of early menopause in fragile X carriers. *Am J Med Genet* 1996;64:370–2

33. Allingham-Hawkins DJ, Babul-Hirji R, Chitayat D, *et al.* Fragile X premutation is a significant risk factor for premature ovarian failure: the international collaborative POF in fragile X study – preliminary data. *Am J Med Genet* 1999; 83:322–5

34. Schreurs A, Legius E, Meuleman C, *et al.* Increased frequency of chromosomal abnormalities in female partners of couples undergoing *in vitro* fertilization or intracytoplasmic sperm injection. *Fertil Steril* 2000;74:94–6

35. Gekas J, Thepet T, Turleau C, *et al.* Chromosomal factor of infertility in candidate couples for ICSI: an equal risk of constitutional aberrations in women and men. *Hum Reprod* 2001;16:82–90

36. Conn CM, Harper JC, Winston RM, Delhanty JD. Infertile couples with Robertsonian translocations: pre-

implantation genetic analysis of embryos reveals chaotic cleavage divisions. *Hum Genet* 1998;102:117–23

37. Scriven PN, Flinter FA, Braude PR, Ogilvies CM. Robertsonian translocations – reproductive risks and indications for preimplantation genetic diagnosis. *Hum Reprod* 2001;16:2267–73

38. Uehara S, Akai Y, Takeyama Y, *et al.* Pericentric inversion of chromosome 9 in prenatal diagnostics and infertility. *Exp Med* 1992;166:417–27

39. Hsu LY, Denn PA, Tannenbaum HL, *et al.* Chromosomal polymorphisms of 1, 9, 16 and Y in 4 major ethnic groups: a large prenatal study. *Am J Med Genet* 1987; 26:95–101

40. Teo SH, Tan M, Knight L, *et al.* Pericentric inversion 9 – incidence and clinical significance. *Ann Acad Med Singapore* 1995;24:302–4

41. Vorsanova SG, Beresheva AK, Kazantceva LZ, *et al.* Molecular–cytogenetic diagnostics of the chromosomal anomalies in couples with reproductive function impairment. *Probl Reprod* 1998;22:41–6

42. Simpson JL. Gonadal dysgenesis and sex chromosome abnormalities phenotypic–karyotypic correlations. In Vallet HL, Porter IH, eds. *Genetic Mechanisms of Sexual Development.* New York: Academic Press, 1979;365–405

43. Simpson JL, Christakos AC, Horwith M, Silvernan F. Gonadal dysgenesis associated with apparently chromosomal complements. *Birth Defects Orig Art Ser* 1971;7: 215–28

44. Boczkowski K. Pure gonadal disgenesis and ovarian dysplasia in sisters. *Am J Obstet Gynecol* 1970;106:626–8

45. Aittomaki K. The genetics of XX gonadal dysgenesis. *Am J Hum Genet* 1994;54:844–51

46. Aittomaki K, Luccena JL, Pakarinen P, *et al.* Mutation in the follicle stimulating hormone receptor gene causes hereditary hypergonadotropic ovarian failure. *Cell* 1995; 82:959–68

47. Christacos AS, Simpson JL, Younger JB, Christian CB. Gonadal dysgenesis as an autosomal recessive condition. *Am J Obstet Gynecol* 1969;104:1027–30

48. Pallister PD, Opitz JM. The Perrault syndrome: autosomal recessive ovarian dysgenesis with facultative, non sex-limited sensorineural deafness. *Am J Med Genet* 1979; 4:239–46

49. McCarthy DJ, Opitz JM. Perrault syndrome in sisters. *Am J Med Genet* 1985;22:629–31

50. Nishi Y, Hamamoto K, Kajiyama M, Kawamura I. The Perrault syndrome: clinical report and review. *Am J Med Genet* 1988;31:623–9

51. Skre H, Bassoe HH, Berg K, Frovig AG. Cerebella ataxia and hypergonadotropic hypogonadism in two kindreds: chance concurrence, pleiotropism or linkage? *Clin Genet* 1976;9:234–44

52. Quayle SA, Copeland KC. 46,XX gonadal dysgenesis with epibulbar dermoid. *Am J Med Genet* 1991;40:75–6

53. Pober BR, Zemel S, Hisama FM. 46, XX gonadal dysgenesis, short stature and recurrent metabolic acidosis in two sisters. *Am J Hum Genet* 1998;63:abstr.117

54. Zlotogora J, Sagi M, Cohen T. The blepharophimosis, ptosis, and epicanthus inversus syndrome: delineation of two types. *Am J Hum Genet* 1983;35:1020–7

55. Panidis D, Rousso D, Vavilis D, *et al*. Familial blepharophimosis with ovarian failure. *Hum Reprod* 1983; 9:2034–7

56. Nance MA, Berry SA. Cockayne syndrome: review of 140 cases. *Am J Med Genet* 1992;42:68–84

57. Sugarman GI, Landing BH, Reed WB. Cockayne syndrome: clinical study of two patients and neuropathologic findings in one. *Clin Pediatr (Phila)* 1977;16:225–32

58. Martsolf JT, Hunter AG, Haworth JC. Severe mental retardation, cataracts, short stature, and primary hypogonadism in two brothers. *Am J Med Genet* 1978;1: 291–9

59. Hennekam KC, van de Meeberg AG, van Doorne JM, *et al*. Martsolf syndrome in a brother and sister: clinical features and pattern of inheritance. *Eur J Pediatr* 1988;47: 539–43

60. Harbord MG, Baraitser M, Wilson J. Microcephaly, mental retardation, cataracts, and hypogonadism in sibs: Martsolf's syndrome. *J Med Genet* 1989;26:397–400

61. Weemacs CM, Hustinx TW, Scheres JM, *et al*. A new chromosomal instability disorder: the Nijmegen breakage syndrome. *Acta Pediatr Scand* 1981;70:557–64

62. Chrzanowska KH, Kleijer WL, Krajewska-Walasek M, *et al*. Eleven Polish patients with microcephaly, immunodeficiency, and chromosomal instability: the Nijmegen breakage syndrome. *Am J Med Genet* 1995;57:462–71

63. Conley ME, Spinner NB, Emanuel BS, *et al*. A chromosomal breakage syndrome with profound immunodeficiency. *Blood* 1986;67:1251–6

64. Goto M, Tammoto K, Honuchi Y, Sasazuki T. Family analysis of Werner's syndrome: a survey of 42 Japanese families with a review of the literature. *Clin Genet* 1981;9: 8–15

65. Hall JG, Pallister PD, Clarren SK, *et al*. Congenital hypothalamic hamartoblastoma, hypopituitarism, imperforate anus and postaxial polydactyly: a new syndrome? I. Clinical, causal, and pathogenetic considerations. *Am J Med Genet* 1980;7:47–74

66. Starr DG, McClure JP, Connor JM. Non-dermatological complications and genetic aspects of the Rothmund–Thomson syndrome. *Clin Genet* 1985;27:102–4

67. Kristiansson B, Stibler H, Wide L. Gonadal function and glycoprotein hormones in the carbohydrate-deficient glycoprotein (CDG) syndrome. *Acta Paediatr* 1995;84:655–9

68. German J. Bloom's syndrome. I. Genetical and clinical observations in the first twenty seven patients. *Am J Hum Genet* 1969;21:196–227

69. German J. Bloom syndrome: a Mendelian prototype of somatic mutational disease. *Medicine* 1993;72:393–406

70. German J, Bloom D, Passarge E. Bloom's syndrome. XL Progress report for 1983. *M Clin Genet* 1984;25:166–74

71. Duck SC, Sekkan GS, Wilbois R, *et al*. Pseudohermaphroditism with testes and 46,XX karyotypes. *J Pediatr* 1975;87:58–62

72. Levinson G, Zaratc A, Guzman-Toledano R, *et al*. An XX female with sexual infantilism, absent gonads, and lack of Mullerian ducts. *J Med Genet* 1976;13:68–9

73. Mendonca BB, Barbosa AS, Arnhold IJ, *et al*. Gonadal agenesis in XX and XY sisters: evidence for the involvement of an autosomal gene. *Am J Med Genet* 1994;52:39–43

74. Sarto GE, Opitz JM. The XY gonadal agenesis syndrome. *J Med Genet* 1973;10:288–93

75. Layman LC. Genetics of human hypogonadotropic hypogonadism. *Am J Med Genet* 1999;89:240–8

76. De Roux N, Young J, Misrahi M. A family with hypogonadotropic hypogonadism and mutations in the gonadotropin-releasing hormone receptor. *N Engl J Med* 1997;337:1597–602

77. Layman LC, Cohen DP, Jin M, *et al*. Mutations in gonadotropin-releasing hormone receptor gene cause hypogonadotropic hypogonadism. *Nat Genet* 1998;18:14–15

78. Matthews CH, Borgato S, Beck-Peccoz P. Primary amenorrhoea and infertility due to a mutation in the beta-subunit of follicle-stimulating hormone. *Nat Genet* 1993;5:83–6

79. Yamomoto M. Mutations in the follicle-stimulating hormone receptor genes in patients with gonadal dysfunction. *Nippon Rinsho* 2002;60:272–6

80. Layman LC, Lee EJ, Peak DB, *et al*. Delayed puberty and hypogonadism caused by mutations in the follicle-stimulating hormone beta-subunit gene. *N Engl J Med* 1997;337:607–11

81. Aittomaki K, Herva R, Stenman UH, *et al*. Clinical features of primary ovarian failure caused by a point mutation in the follicle-stimulating hormone receptor gene. *J Clin Endocrinol Metab* 1996;81:3722–6

82. Kotlar TJ, Young RH, Albanese C, *et al*. A mutation in the FSH receptor occurs frequently in human ovarian sex cord tumors. *J Clin Endocrinol Metab* 1997;82:1020–6

83. Okuno A, Komori S, Sakata K, *et al*. Genetic analysis of a variant luteinizing hormone in an infertile woman. *Arch Gynecol Obstet* 2001;265:148–50

84. Crambler DW, Petterson KS, Barbieri RL, Huhtaniemi IT. Reproductive hormones, cancers and conditions in relation to a common genetic variant of luteinizing hormone. *Hum Reprod* 2000;15:2103–7

85. Pettersson K, Ding YQ, Huhtaniemi I. An immunologically anomalous LH variant in a healthy woman. *J Clin Endocrinol Metab* 1992;74:164–71

86. Haavisto AM, Pettersson K, Bergendahl M, *et al*. Occurrence and biological properties of a common genetic variant of luteinizing hormone. *J Clin Endocrinol Metab* 1995;80:1257–63

87. Liao W-X, Roy A, Chan C, *et al*. A new molecular variant of luteinizing hormone associated with female infertility. *Fertil Steril* 1998;69:102–6

88. Kopteva AV, Dzenis IG, Baharev VA. Genetic disturbances of the hypothalama–hypophyseal regulation of the reproductive system. *Probl Reprod* 2000;33:28–33

89. Toledo SP, Brunner HG, Kraaij R, *et al*. An inactivating mutation of the luteinizing hormone receptor causes amenorrhea in a 46,XX female. *J Clin Endocrinol Metab* 1996;81:3850–4

90. Latronico AC, Anasti J, Arnhold IJ, *et al*. Testicular and ovarian resistance to luteinizing hormone caused by inactivating mutations of the luteinizing hormone-receptor gene. *N Engl J Med* 1996;334:507–12

91. Prado Arnhold IJ, Latronico AC, Batista MC, *et al*. Ovarian resistance to luteinizing hormone: a novel cause of amenorrhea and infertility. *Fertil Steril* 1997;67:394–7

92. Parma J, Duprez L, Sande J, *et al*. Somatic mutations in the thyrotropin receptor gene cause hyperfunctioning thyroid adenomas. *Nature* 1997;365:649–51

93. Takamatsu K, Nishikawa M, Horimoto M, Ohsava N. Familial unresponsiveness to thyrotropin by autosomal recessive inheritance. *J Clin Endocrinol Metab* 1993;77: 1569–73

94. Spiegel AM. Inborn errors of signal transduction: mutations in G proteins and G protein-coupled receptors as a cause of disease. *J Inherit Metab Dis* 1997;20:113–21

95. Weinstein LS, Shenker A, Gejman PV, et al. Activating mutations of the stimulatory G protein in the McCune–Albright syndrome. *N Engl J Med* 1991;325:1688–95

96. Miric A, Vechio JD, Levine MA. Heterogeneous mutations in the gene encoding the alpha-subunit of the stimulatory G protein of adenylyl cyclase in the Albright hereditary osteodystrophy. *J Clin Endocrinol Metab* 1993; 76:1560–8

97. Pollack SE, Furth EE, Kallen CB, et al. Localization of the steroidogenic acute regulatory protein in human tissue. *J Clin Endocrinol Metab* 1997;82:4243–51

98. Yanase T, Simpson ER, Waterman MR. 17α-Hydroxylase/17,20-lyase deficiency: from clinical investigation to molecular definition. *Endocr Rev* 1991;12:91–108

99. Biglieri EG. 17α-Hydroxylase deficiency: 1963–1966. *J Clin Endocrinol Metab* 1998;82:48–50

100. Geissler, Davis DL, Wu L, et al. Male pseudo-hermaphroditism caused by mutations of testicular 17β-hydroxysteroid dehydrogenase 3. *Nat Genet* 1994;7:34–9

101. Andersson S, Geissler WM, Wu L, et al. Molecular genetics and pathophysiology of 17β-hydroxysteroid dehydrogenase 3 deficiency. *J Clin Endocrinol Metab* 1996; 81:130–6

102. Dewailly D, Vantyghem-Haudiquet MC, Sainsard C, et al. Clinical and biological phenotypes in late-onset 21-hydroxylase deficiency. *J Clin Endocrinol Metab* 1986;63: 418–23

103. Conte FA, Grumbach MM, Ito Y, et al. A syndrome of female pseudohermaphroditism, hypergonadotropic hypogonadism and multicystic ovaries associated with missense mutations in the gene encoding aromatase (P450arom). *J Clin Endocrinol Metab* 1994;78:1287–92

104. Ito Y, Fisher CR, Conte F, et al. Molecular basis of aromatase deficiency in an adult female with sexual infantilism and polycystic ovaries. *Proc Natl Acad Sci USA* 1993;90:11673–7

105. Morishima A, Grumbach MM, Simpson ER, et al. Aromatase deficiency in male and female siblings caused by a novel mutation and the physiological role of estrogens. *J Clin Endocrinol Metab* 1995;80:3689–98

106. Mullis PE, Yoshimura N, Kuhlmann B, et al. Aromatase deficiency in a female who is compound heterozygote for two new point mutations in the p450arom gene: impact of estrogens on hypergonadotropic hypogonadism. multicystic ovaries, and bone densitometry in childhood. *J Clin Endocrinol Metab* 1997;82:1739–45

107. Bulun SE. Aromatase deficiency in women and men: would you have predicted the phenotypes? *J Clin Endocrinol Metab* 1996;81:867–71

108. Russell DW, Berman DM, Bryant JT, et al. The molecular genetics of steroid 5α-reductase. *Recent Prog Horm Res* 1994;49:275–84

109. Katz MD, Cai L-Q, Zhu Y-S, et al. The biochemical and phenotypic characterization of females homozygous

110. Milewich L, Mendonca BB, Arnhold I, et al. Women with steroid 5α-reductase 2 deficiency have normal concentrations of plasma 5α-dehydroprogesterone during the luteal phase. *J Clin Endocrinol Metab* 1995;80:3136–9

111. Hochberg Z, V'hayen R, Reiss N, et al. Clinical, biochemical and genetic findings in a large pedigree of male and female patients with 5α-reductase 2 deficiency. *J Clin Endocrinol Metab* 1996;81:2821–7

112. White PC. Analysis of mutations causing steroid 21-hydroxylase deficiency. *Endocr Res* 1989;15:239–56

113. Strachan PC. Molecular pathology of congenital adrenal hyperplasia. *Clin Endocrinol* 1990;32:373–93

114. Withchel SF, Bhamidipati DK, Hoffman EP, Cohen JB. Phenotype heterogenity associated with the splicing mutation in congenital adrenal hyperplasia due to 21-hydroxylase deficiency. *J Clin Endocrinol Metab* 1996;81: 4081–8

115. Speiser PVT, Dupont J, Zhu D, et al. Disease expression and molecular genotype in congenital adrenal hyperplasia due to 21-hydroxylase deficiency. *J Clin Invest* 1992;90: 584–95

116. Azziz R, Dewailly D, Owerbach D. Nonclassical adrenal hyperplasia: current concepts. *J Clin Endocrinol Metab* 1994;78:810–15

117. Azziz R, Owerbach D. Molecular abnormalities of the 21-hydroxylase gene in hyperandrogenic women with an exaggerated 17-hydroxyprogesterone response to short-term adrenal stimulation. *Am J Obstet Gynecol* 1995;172: 914–18

118. Geley S, Kapelari K, Johrer K, et al. CYP11B1 mutations causing congenital adrenal hyperplasia due to 11β-hydroxylase deficiency. *J Clin Endocrinol Metab* 1996; 81:2896–901

119. Merke DP, Tajima T, Chhabra A, et al. Novel CYP11B1 mutations in congenital adrenal hyperplasia due to steroid 11β-hydroxylase deficiency. *J Clin Endocrinol Metab* 1998; 83:270–3

120. Franks S, Charani N, McCarthy M. Candidate genes for polycystic ovaries syndrome. *Hum Reprod Update* 2001; 7:405–10

121. Kuiper G, Enmark E, Pelto-Huikko M, et al. Cloning of a novel estrogen receptor expressed in rat prostate and ovary. *Proc Natl Acad Sci USA* 1996;93:5925–30

122. Enmark E, Pelto-Huikko M, Grandien K, et al. Human estrogen receptor β-gene structure, chromosomal localization, and expression pattern. *J Clin Endocrinol Metab* 1997;82:4258–65

123. Hickey T, Chandy A, Norman RJ. The androgen receptor CAG repeat polymorphism and X-chromosome inactivation in Australian Caucasian women with infertility related to polycystic ovary syndrome. *J Clin Endocrinol Metab* 2002;87:161–5

124. Mifsud A, Ramirez S, Yong EL. Androgen receptor gene CAG trinucleotide repeats in anovulatory infertility and polycystic ovaries. *J Clin Endocrinol Metab* 2000;85:3484–8

125. Ko TM, Yang YS, Wu MY, et al. Complete androgen insensitivity syndrome. Molecular characterization in two Chinese women. *J Reprod Med* 1997;42:424–8

Molecular biology of the human Y chromosome

21

Marijo Kent-First

INTRODUCTION

The search for, and characterization of, fertility-related gene(s) on the Y chromosome continues. Contrary to historical perception, the Y chromosome is rich with genes required not only for testis determination, but testis differentiation and, specifically, spermatogenesis and male-specific organ function. Though many azoospermia factor (AZF) restricted deletions are large and encompass multiple subregions, protocols containing larger panels of sequence tagged sites (STS) designed to reflect the gene content of the chromosome allow for the resolution of smaller deletions restricted to subregions of AZFa, AZFb and AZFc. This has led to a more precise characterization of phenotype/genotype (microdeletion) correlations of the most proximal AZF deletion subinterval, namely AZFa. Testing for Y chromosome microdeletions has become a standard for clinical embryology; however, prognosis will undoubtedly improve as individual laboratories standardize their methods to accurately and precisely detect these microdeletions and their extent, with constant referral to the genome databases and the Y chromosome STS maps that are now readily available.

Our resolution of the Y chromosome has dramatically improved during the last half century. Until the last decade, it was commonly believed that this small evolutionary remnant of the X chromosome contained only the gene(s) involved in testis determination. The remaining Y chromosome was thought to contain merely 'junk' – repetitive heterochromatic DNA, specifically GATA or GACA repeats. However, after the discovery of the testis determining factor (TDF), namely sex determining region of the Y chromosome (SRY)[1,2], the debate concerning the Y chromosome's gene content and gene function became refocused to include region(s)

required for male fertility (testis differentiation) and the maintenance of male-specific organs such as the testis and prostate gland.

At least 33 genes are linked to the human Y chromosome and they are subdivided into three groups (Table 1, Figure 1). Group 1 (Table 1a) consists of genes linked to the only regions of the Y chromosome that pair with the X chromosome during male meiosis, namely the pseudoautosomal pairing region(s) (PAR). Genes in this group have homologs on the X chromosome PAR. The PAR regions are located at the telomeres of the short (p) and long (q) arms of the X and the Y chromosomes and are referred to as PAR1 and PAR2, respectively. PAR1, which is 2.6 Mb in length, contains 10 genes and PAR2, which is approximately 400 kb in length, contains two genes. The PARs comprise about 5% of the total Y chromosome DNA. These genes have mainly 'housekeeping' functions that are required in both genders and relate to enzyme metabolism, stature, growth factors and cellular energy. Group 2 (Table 1b) consists of 10 genes linked to the non-recombining region of the Y chromosome (NRY) with similar but non-identical homologs on the X chromosome. Group 2 genes on the Y chromosome are single-copy and most are ubiquitously expressed in a broad range of tissues. The X-linked homologs to Group 2 genes may or may not escape X inactivation in the female and are implicated in a range of diseases associated with female infertility, failures of imprinting and mental retardation[3]. Group 3 (Table 1c) contains 11 Y-linked genes including *sry* which switched the gender-ambiguous gonad towards testis determination during early embryogenesis. The functions of the remaining 10 genes in this group are less well defined but are

Table 1 Organization of Y chromosome-linked genes

(a) Group 1: X–Y identical genes

Gene	Gene name	Comments
Pseudoautosomal region 1 – short arm		
pgpl	Pseudoautosomal GTP-binding protein-like	A conserved GTP-binding protein gene; most distal gene in PAR
shox/phog	Short stature homeobox/pseudoautosomal homeobox containing osteogenis gene	A homeobox-containing gene involved in bone growth and stature – Turner syndrome gene candidate
xe7	xe7	A gene derived from an inactive X chromosome cDNA library, nuclear located product
csf2a	GM-CSF receptor 2, α subunit	A receptor gene for the granulocyte-macrophage colony stimulating factor, a growth and differentiation factor
il3ra	Interleukin 3 receptor α subunit	A cytokine receptor gene sharing homology to *csf2a* gene; *il3* binds to its receptors and promotes hematopoietic cell proliferation
ant3	Adenine nucleotide translocase	A member of the ADP/ATP translocase family involved in cellular energy metabolism
asmtl	*asmt*-like	A gene encoding a putative fusion protein that shares homology with the bacterial maf/orfE at its amino terminal end and to *asmt* at its carboxyl terminus
asmt	Acetylserotonin methyltransferase	Coding for an enzyme involved in the last step of melatonin synthesis
tramp	tramp	A single-exon gene encoding a protein with homology to transposases of the Ac family
mic2	Monoclonal Imperial Cancer fund 2 (order of discovery)	Coding for a surface antigen detected by a monoclonal antibody, 12E7
Pseudoautosomal region 2 – long arm		
il9r	Interleukin 9 receptor	Receptor for a growth factor for T-cells, erythroid and myeloid precursor cells
sybl1	Synaptobrevin-like 1	Coding for a synaptobrevin-like protein

(b) Group 2: X–Y homologous genes

Gene	Gene name	Comments
Non-recombining region (NRY) – short arm		
rps4y	Ribosomal protein S4 Y isoform	Ribosomal protein gene postulated to be involved in Turner syndrome; X homolog is *rps4x*
zfy	Zinc finger Y	Coding for a nuclear transcription factor harboring a DNA-binding domain with 13 zinc fingers; X homolog is *zfx*
prky	Protein kinase Y	*prky* and its X homolog, *prkx*, are members of the cAMP-dependent serine/threonine protein kinase superfamily
amely	Amelogenin Y	Expresses only in developing tooth buds; X homolog is *amelx*; possibly related to tooth size locus

Non-recombining region (NRY) – long arm

dffry	Drosophila fat facets-related Y	Homologous to Drospophila deubiquinating enzyme required for eye development and oogenesis; X homolog is *dffrx*
dby	Dead box Y	Coding for a protein homologous to RNA helicases harboring conserved DEAD (Asp–Glu–Ala–Asp) motifs
uty	Ubiquitous TPR motif Y	Coding for a protein containing the TPR motif, implicated in protein–protein interacton and as a H–Y antigen
tb4y	Thymosin B4 Y isoform	X homolog sequestering actin
smcy	Selected mouse cDNA Y	Human homolog of the mouse *smcy* gene coding for a minor histocompatability H–Y antigen; may serve as a spermatogenic factor and play a role in mitosis and be involved in oncogenesis of the testis tumors and gonadoblastoma
eif1ay	Translation initiation factor 1A	X homolog; an essential translation initiation factor

(c) Group 3: Y-specific genes

Gene	Gene name	Comments
Non-recombining region – short arm		
sry	Sex determining region Y	Evolutionarily conserved gene coding for the testis determining factor (TDF)
tspy	Testis-specific protein Y-encoded	Repeated gene coding for a putative cyclin-B binding protein homologous to that of oncogene *set*; postulated to direct the spermatogonial cells to enter meiosis
pry	PTP-BL-related Y	Coding for a protein homologous to PTP-BL, a putative protein tyrosine phosphatase, also present on the long arm
tty1	Testis transcript 1	Repetitive transcripts without any protein coding sequences, also present on the long arm
tty2	Testis transcript 2	Repetitive transcripts without any protein coding sequences, also present on the long arm
Non-recombining region – long arm		
bpy1	Basic protein Y1	Coding for a 125-residue basic protein, repeated gene
bpy2	Basic protein Y2	Coding for a 106-residue basic protein, repeated gene
cdy	Chromodomain Y	Coding for a protein with chromodomain and putative catalytic domain; may modify DNA/protein during spermatogenesis; repeated gene
xkry	XK-related Y	Coding for a protein homologous to XK, a membrane transporter protein; repeated gene
rbm	RNA binding motif	Repeated gene coding for a RNA binding protein, a candidate for the azoospermia factor (AZF)
daz	Deleted in azoospermia	A gene coding for another RNA binding protein, a candidate for the azoospermia factor (AZF)

Data derived from Lau C, Zhang J. Expression analysis of thirty one Y chromosome genes in human prostate cancer. *Mol Carcin* 2000;27:308–21.

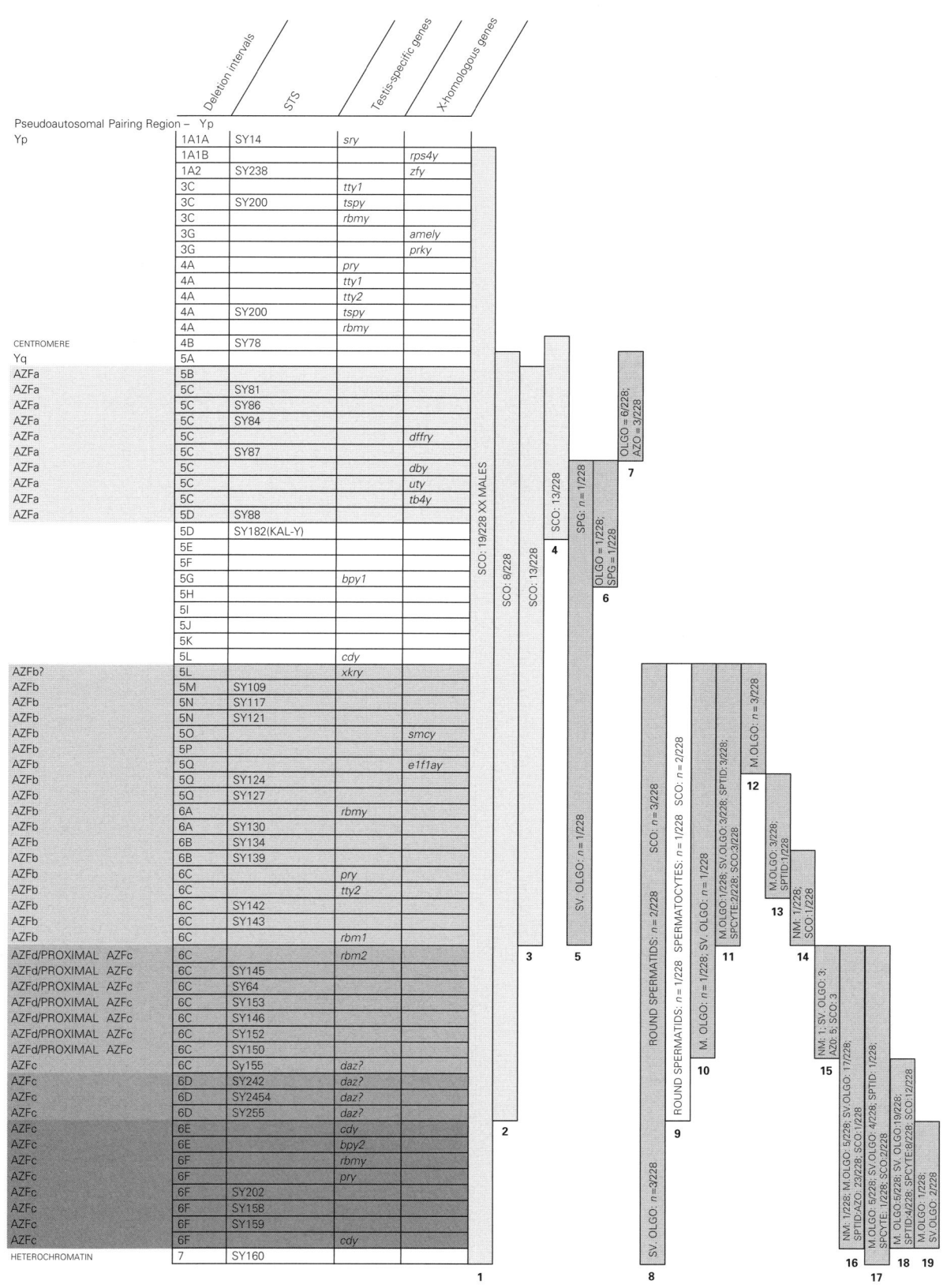

postulated to play roles in male-specific organs including the testis and, specifically, spermatogenesis[4]. Most of the genes within Group 3 have multiple and highly variable copy numbers and are expressed exclusively in the testis.

THE Y CHROMOSOME AND TESTIS DETERMINATION

The clinical relevance of the 'master Y-linked switch' in testis determination is well established. Because *sry* is located in the region immediately adjacent to PAR1, it may become involved in abnormal meiotic recombination events which result in X-bearing sperm cells that harbor translocated *sry*. The 46,XX male syndrome occurs in one in 20 000 males[6]. Up to 20% of azoospermic males presenting with feminized phenotype including gynacomastia, hypogonadism and/or for assisted reproduction are XX males whose condition is due to a translocation of the distal region of Yp to the X chromosome. Though the majority of XX males are *sry*-positive, a small proportion are *sry*-negative and are thought to have mutations in genes upstream in the regulatory pathway to *sry*, while a few have point mutations in *sry*. Testicular phenotype varies depending on the size of the Y to X translocation; however, the majority of XX males have Sertoli cells only (SCO). Conversely, the reciprocal meiotic event can manifest itself with the XY female and the XY sex reversal syndrome[6,7].

Though the XY female is usually diagnosed by karyotyping, the underlying cause of sex reversal is genetic. Occasionally, XY females present for assisted reproduction having received a differential diagnosis of testicular feminization (TFM) earlier in life due to mutations associated with the androgen receptor (AR) gene or gonadal dysgenesis. The former syndrome results in abdominal testes or streaks associated with an absence of Müllerian derivatives, and parenthood can be achieved only by surrogacy or adoption. XY females with an intact AR gene usually arise because either of deletions or of point mutations in *sry* – though occasionally XY females present with *sry* and its protein intact[5]. This implies that other genes may interact with *sry* to induce testis formation. Although frequently the XY female presents with complete gonadal dysgenesis, studies using the horse model of the XY sex reversal syndrome indicate that a range of gonad type from fertile ovary to abdominal testis may occur[6,7]. In individuals more completely sex-reversed, both spontaneous and artificially stimulated ovulation can occur and result in karyotypically normal and sex-reversed offspring[6,7].

THE Y CHROMOSOME AND TESTIS DIFFERENTIATION

In 1976, studies by Tiepolo and Zuffardi of azoospermic men with large deletions encompassing Yq suggested that this part of the chromosome is required for spermatogenesis[8]. With the progress made by the

Figure 1 Relationship between the Y chromosome and infertility-associated macro- and microdeletions occurring in 228 men and derived from studies that provided precise clinical data including, in cases of azoospermia, histologic evaluation of the testis. Data included further analysis of the cases in Figure 4 and 16 additional XX males, also mapped in our laboratory. Testis phenotype is coded as follows: bars 1–4 = sertol cell only (SCO); bars 5–8 and 10–19 = a range in testis phenotype from SCO to oligozoospermia and a few cases of individuals with relatively normal sperm counts but abnormal sperm morphology; bar 5 = azoospermia associated with the presence of germ cells and arrested spermatogenesis. Data regarding the specific phenotypes associated with each deletion type are given within each bar which diagrams the deletion extent. Abbreviations are as follows: SCO, Sertoli cell only; SV. OLGO, severe oligozoospermia defined by sperm counts from 1 to 5 million; SPG, spermatogenic arrest with only spermatogonia present; OLGO, oligospermia; AZO, azoospermia; M. OLGO, mild oligozoospermia associated with sperm counts from 5 million to < 20 million; SPTID, spermatogenic arrest beyond the round spermatid stage; SPCYTE, spermatogenic arrest beyond spermatocyte stage; and NM, severe male factor infertility associated with sperm counts > 20 million and > 95% abnormal sperm morphology. Phenotypes are given as a ratio when space permits with the numerator being the number of individuals with a given phenotype and the denominator being the total number of individuals included in the diagram (*n* = 228). Deletions are mapped relative to deletion intervals, commonly used sequence tagged sites (STS), testis-specific genes and genes with an X homolog. Subregions within AZF are highlighted in different shades of gray. Sizes of the deletion intervals are not necessarily to scale; however, the deletion boundaries are indicated according to Vogt, and to Tilford, and their co-workers[10,72] and to naturally occurring patient deletions. When the boundary extent is approximated, a '?' is given (for example, AZFb?). Centromere and heterochromatin are shown in small caps

Human Genome Organization (HUGO) to map the Y chromosome, the molecular tools, namely STS derived from bacterial artificial chromosomes (BACs) or yeast artificial chromosomes (YACs), were used in studies which provided definition to the region containing AZF[9,10]. Mapping studies using phenotype-selected patients have led to an improved understanding of the AZF region. However when these studies began, only two gene families, namely *RNA binding motif (rbm)*[11] and *deleted in azoospermia (daz)*[12], were linked to AZF. Both genes have multiple and highly variable copy numbers distributed across the Y chromosome; however, the *daz* gene family tend to occur in a cluster in deletion interval 6. Both genes are expressed in the germ cells in the testis and code for RNA binding proteins. The timing of gene expression and gene function differs in that *daz* is required for germ cells to progress into the meiotic phase of spermatogenesis whereas *rbm* is likely to be required for germ cells to complete the meiotic phase of spermatogenesis. Evolutionary studies have shown that *rbm* is conserved on the Y chromosome across all mammalian taxa including marsupials whereas *daz* is conserved only in higher primates. The *daz* gene evolved from the *boule* gene which was first cloned in Drosophila. Expressed in the fly ovary and testis, *boule* plays a role in both male and female gametogenesis. *Dazla* is a *daz* homologous gene which is linked to chromosome 3 in humans and is conserved across mammalian taxa. It is expressed in both the ovary and the testis and is apparently required for oogenesis as well as spermatogenesis. Mapping studies of Yq suggested that naturally occurring deletion intervals exist, and AZF was subdivided into the subintervals AZFa, AZFb and AZFc[13], and the region between AZFb and AZFc which excludes *daz*, namely AZFd[14]. Unfortunately, attempts to correlate specific testis phenotypes with AZF subregions have achieved only limited success. For example, large deletions encompassing multiple AZF regions usually result in SCO or complete spermatogenic arrest associated with azoospermia; however, restricted interstitial deletions limited to any of the three or four AZF subintervals can also result in an equally compromised testis profile. Although less than 10% of all pathology-associated deletions are exclusively restricted to AZFa, the associated testis phenotype is the most severely compromised in that 66.6% of men with AZFa deletions who have no

germ cells, whereas the remaining 33.3% with AZFa-restricted deletions have severe oligozoospermia associated with partial spermatogenic arrest[15,16]. The AZFa region comprises around 1100 kb and has been completely sequenced. Of the genes mapped to the region, Drosophila fat facets related Y (*dffry*), dead box Y (*dby*), ubiquitous TPR motif Y (*uty*) and thymosin B4 Y isoform (*tb4y*) all have X-linked homologs. *Dffry* (*usp9y*) and *dby* play pivotal roles in early spermatogenesis in that 100% of patients with deletion of both *dffry* and *dby* have no germ cells (SCO), whereas if only *dffry* or *dby* is removed, germ cells and limited spermatogenesis associated with severe oligozoospermia result[17]. Testicular phenotypes associated with AZFb, AZFd and AZFc are similar and range from azoospermia associated with SCO or complete spermatogenic arrest to moderate oligozoospermia (Figures 2, 3 and 4a, -b and -c).

During the last decade more that 5000 infertile men have been tested in protocols employing from one to 140 STS[11–14,18–68]. STS utilized in testing protocols are usually selected at random or from several foundation protocols, and may or may not include Y-linked genes with suspected role(s) in spermatogenesis (Tables 2 and 3). Deletion detection rates (DDR) range from 2 to 38% (Figure 2). However, studies which utilize large numbers of STS do not necessarily achieve an elevated or more accurate DDR. DDR from protocols using from 18 to 40 STS, selected based upon naturally occurring patient deletion intervals, tend to be similar and present a more accurate assessment of the extent of an infertile male's deletion (Figures 2 and 4a, -b and -c).

There have been huge advances in mapping the Y chromosome and now we realize that previous estimates of the gene content were very low. Since *sry*, some genes were identified by focused cloning projects. However, huge strides in defining the functional genetic content of the Y chromosome were made by groups that employed exon trapping methods to identify new Y-linked genes expressed in the testis[15,47,69,70] and optical mapping[71]. These methods to characterize the genetic content of NRY have led to the HUGO map[10] and a more extensive version linked to extensive sequence data developed largely by the efforts of the Whitehead Institute, Washington University, and a few other laboratories[72,73]. These maps, which are frequently ignored, are critical when protocols are developed to test for microdeletions

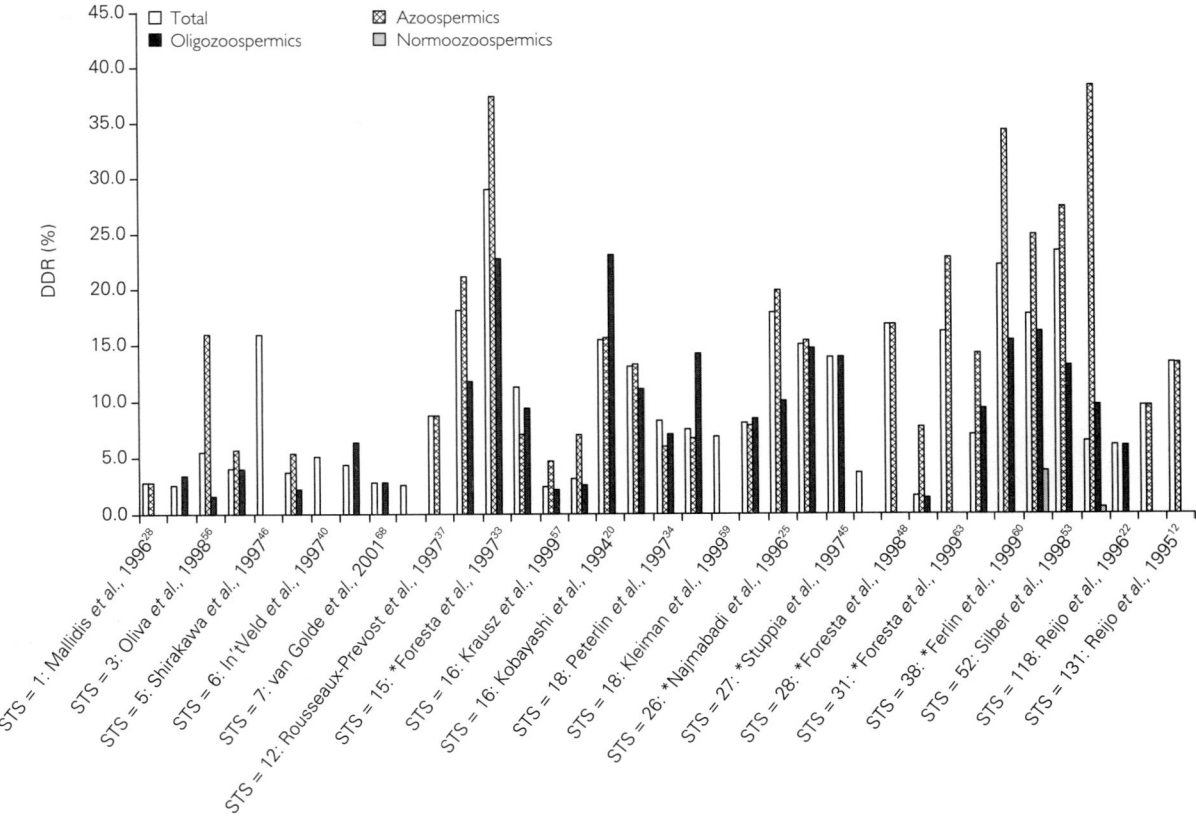

Figure 2 Thirty-seven published protocols using from one to 131 sequence tagged sites (STS) are shown relative to the percent deletion detection rates (% DDR). White bars indicate the total or overall % DDR in infertile men without regard to testicular phenotype, diagonally cross-hatched bars the % DDR in azoospermic men with testicular phenotypes that range from Sertoli cell only (SCO) to spermatogenic arrest with no sperm found in the ejaculate, black bars the % DDR in men with sperm counts from 1 to < 20 million (severe, moderate or mild oligozoospermia) and gray bars the % DDR in men with sperm counts of about 20 million but with > 95% abnormal sperm head morphology. Asterisks (*) indicate studies reporting questionable non-contiguous microdeletions that were not confirmed by Southern blotting experiments

in infertile males. Although the precise function of many of these genes in the human are defined only by studies which correlate testis histology and phenotype in selected patients with microdeletion location and breakpoints defined by STS, animal models have provided clues as to the genes that are likely to be required for cell cycle, spermatogenesis and fertility. The genes linked to Yq AZF occur across the region and some, which are Y-specific, occur in multiple copies and are distributed across the Y chromosome (for example, *rbm*) or cluster in focus deletion intervals (for example, *daz*)[74–76]. When 228 patient deletions are mapped relative to the location of functional genes in AZFa, AZFb and AZFc, the AZF subregions may be further subdivided (Figure 1). As our

understanding of AZF gene function improves, it is likely that the AZF subintervals will be subdivided to an even greater extent based upon gene-specific mutations and even point mutations which can result in aberrant transcription or translation and loss of function[77–80]. The possibility that diagnosis of the integrity of a chromosome can be reduced to the testing of only a few STS which may or may not occur within a gene of functional interest is intriguing but unlikely, and unfortunately has led to inaccurate patient diagnosis. In fact, it is now becoming apparent that our current nomenclature that subdivides AZF may be a misleading oversimplification of a much more complicated group of genes and gene families.

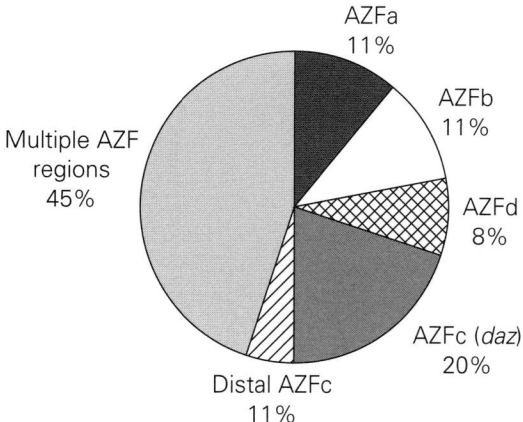

Figure 3 Percent distribution of 212 microdeletions in 112 patients within the AZF and its subintervals as defined in the text and in Table 3. Note that prior to Kent-First[14], when proximal AZFc was referred to as AZFd and deletions distal to *daz* were either ignored or grouped with *daz*, deletions occurring in proximal, middle and distal AZFc would have been depicted in this data set as 33%

GENETIC TRANSMISSION OF Y-LINKED MICRODELETIONS VIA INTRACYTOPLASMIC SPERM INJECTION

Since the template of choice in Y deletion screening protocols is usually DNA derived from blood, the question was asked: Is the mesodermal lineage reflective of the germ cell lineage in an infertile male? In 1996, Reijo and co-workers showed that three infertile men with microdeletions detected in blood-derived DNA also had the same deletions in their sperm[22]. In the same year, in the first genetic screening protocol of a population of intracytoplasmic sperm injection (ICSI)-derived male children and their fathers, Kent-First and co-workers showed the same pathology-associated AZF deletions could be detected in the blood-derived DNA of males, their sperm and in DNA from the blood of their ICSI-derived sons[23,29]. Later studies have confirmed that though men with pathology-associated microdeletions can reproduce with ICSI, sons will inherit the microdeletion from their fathers[61,64,81]. Interestingly, two males in studies of Kent-First and co-workers, who did not have readily detectable microdeletions in blood or pooled sperm, sired sons with readily detectable pathology-associated microdeletions[23,29]. It was suggested that

microdeletions are heritable and can arise by way of two mechanisms: (1) as a *de novo* mutation in the germ line of the fertile father of the infertile patient; and (2) as a *de novo* mitotic error early in the embryogenesis of the infertile patient. In the latter case, the microdeletion occurs as a cryptic mutation in parallel with the wild-type, intact Y chromosome – i.e., a mosaic scenario which will most likely be missed using routine polymerase chain reaction (PCR)[23,29]. Males carrying only microdeleted Y chromosomes will transmit the identical mutation or an enlarged mutation to any sons conceived by assisted reproductive technologies (ART) whereas males who are germ-line mosaics may produce sons who carry the microdeletion or sons with intact Y chromosomes. To test this hypothesis, we amplified individual lymphocytes (mesoderm), hair (ectoderm) and sperm in one of the two patients described above, in six additional infertile males who have produced sons by ICSI and in four fertile control sperm donors. In this study, we confirmed that indeed some men carry deletions in tandem with wild-type cells in varying ratios in lymphocytes, hair and sperm, and can produce normal female and male embryos, and male embryos that inherit the mutation[29]. In this scenario, couples that produce multiple embryos by ART would benefit from preimplantation genetic diagnosis (PGD).

CLINICAL RESEARCH AND DIAGNOSTICS OF TESTIS DIFFERENTIATION AND MALE INFERTILITY

It has become the standard of good practice to test infertile men for Y chromosome integrity in AZF. The DDRs range between 1 and 38% depending on the number and selection of STS employed and the clinical selection criteria of the patients to be tested. Data interpretation is difficult due to wide variations in protocols which contribute to the broad range in DDRs and the reporting of unconfirmed and unlikely non-contiguous deletions[25–27,33,45,48,49,60,62,64]. This problem has prompted two independent multicenter studies. The rationale for both studies was similar – to assess the current diagnostic efficiency of commonly used screening protocols in terms of accuracy (defined by the protocol's ability to detect pathology-associated

(a)

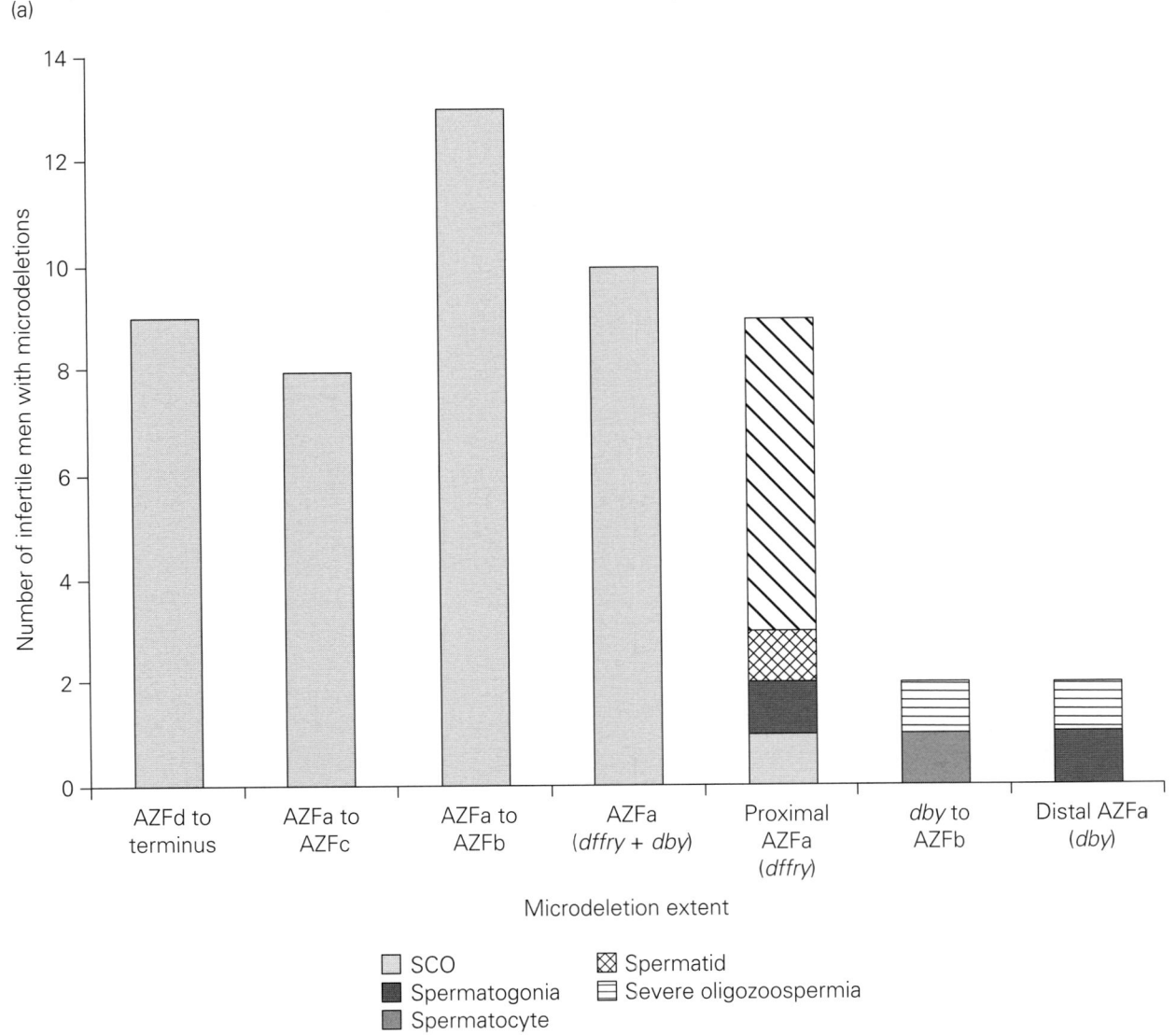

Figure 4 Testicular phenotypes associated with naturally occurring patient deletions: (a) microdeletions involving AZFa; (b) microdeletions involving AZFb; and (c) proximal AZFc (AZFd) and AZFc microdeletions. SCO, Sertoli cell only. Data were derived from studies in which complete testicular profiles were reported

deletions across AZF) and precision (defined by an assay's ability to give reproducible results in replicates in a single experiment and in replicate experiments), and to begin to develop standards in testing for pathology-associated Y chromosome microdeletions. The methods used were different but complementary. The 1999 study by Simoni and co-workers involved 20 laboratories in eight European countries[82]. Two DNA samples with large deletions in AZFb and/or AZFc were sent to the laboratories that had already established screening

protocols in three trials. Surprisingly, the error rate was high with both false-positives and false-negatives reported. Though most laboratories were able to diagnose accurately the deletions in *daz*, several laboratories misdiagnosed microdeletions in AZFb, and the false-positive error rate (defined as misdiagnosing an intact region as deleted) was high (9% across all laboratories).

The second study, co-ordinated by Kent-First and co-workers, involves an ongoing collaboration among

(b)

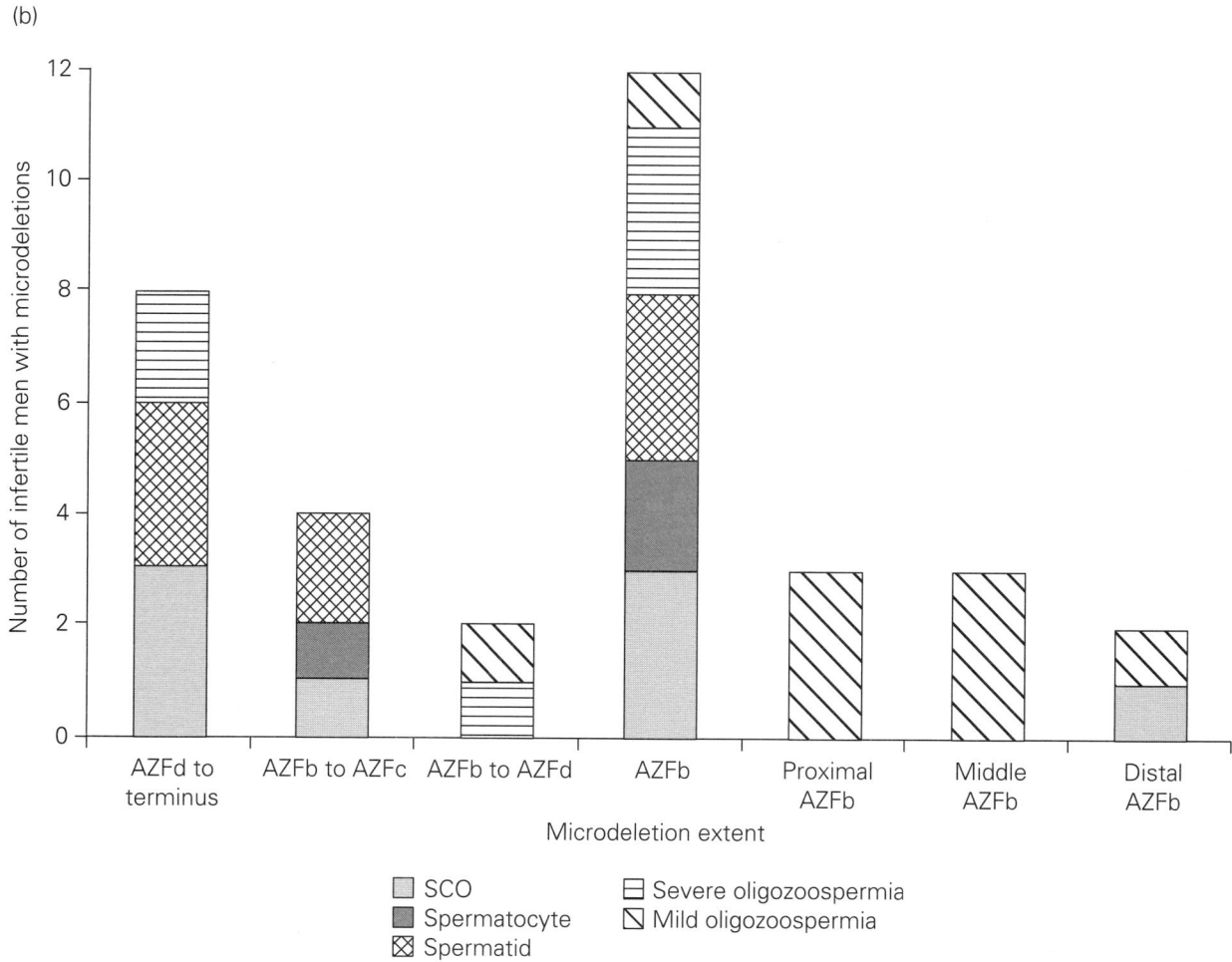

Figure 4 *Continued*

38 laboratories in 19 countries on five continents. The initial focus of the study was aimed at differentiating between informative, non-repetitive or pathology-associated loci and less informative, repetitive or polymorphic loci. A panel of 120 STS was assessed and used to test a large population of 1200 ethnically characterized, fertile males and more than 1500 coded samples previously tested by collaborative laboratories.

Some STS and genes of functional interest are repeated either in clusters or non-specifically across the Y chromosome. When these STS are used in a protocol, a copy occurring in the middle of a large deletion will be deleted; however, the additional copy(ies) which fall outside the deletion will be amplified and thus create what appears to be a non-contiguous deletion. Hence, an ideal panel of markers would not include such loci. Although commonly used STS in *daz* are repeated many times within the *daz* gene cluster, *daz* deletions reported thus far include the entire gene cluster. Although some protocols amplify only one or two regions from the *daz* gene cluster, additional STS from *daz* can serve as a control for misdiagnosing an apparent *daz* deletion. Therefore, as a part of the process of characterizing more than 120 STS, it is important that a substantial focus be placed on discerning repetitive loci (Table 4a and 4b).

Most of the STS used in this study and as part of published screening protocols do not map to known genes, but are merely markers from regions that are likely to contain gene(s) of functional interest, whereas some loci are amplified regions from genes that have been shown to play a role in spermatogenesis (*daz*, *dffry* and others). Some STS serve only as markers for infertility because of their close linkage to a spermatogenesis-required gene, and even regions from

(c)

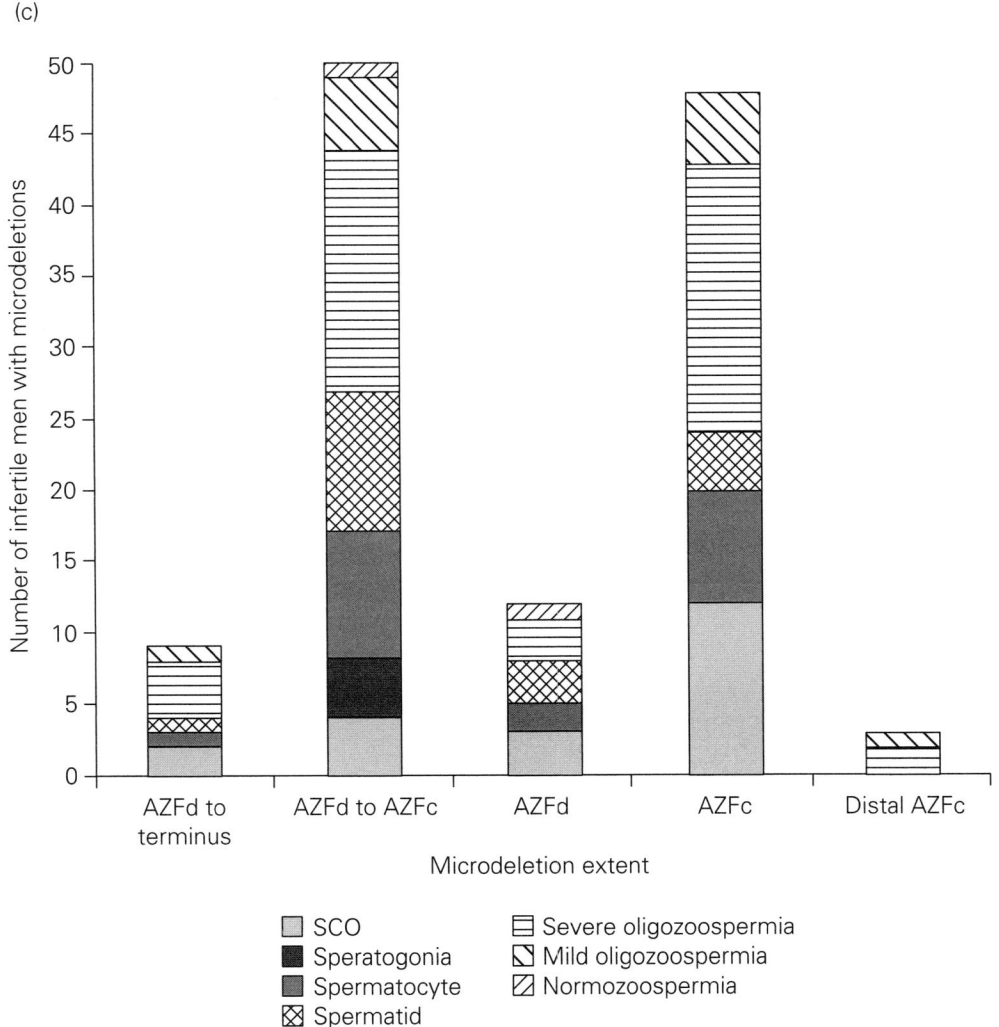

Figure 4 *Continued*

functional genes may exhibit some degree of polymorphism with deletions occurring in both fertile and infertile men. To determine if a statistical correlation with infertility exists when a given marker is deleted in a phenotype-selected sample, it is imperative that the marker is tested in a statistically balanced and large population of control (fertile) males. This testing should be conducted in tandem with a comparably sized population of phenotype-selected individuals and the appropriate statistical tests applied to derive a *p* value. In the initial phase of this study, 1200 proven fertile males were tested for each of the 120 STS. During the course of the study, many of these same STS were tested in additional fertile men in various published protocols in tandem with infertile patients[12–14,19–68]. Thus, at least 60 STS

have been tested in almost 3289 fertile males and 5260 phenotype-selected patients and the probability of correlation of a given deletion with infertility is established. Twelve STS were deleted in one or more normally fertile males and were therefore considered to be polymorphic (Table 4b). Many 'home-brew' protocols based upon several key references have been published to date[12,14,19,23,32]. Unfortunately, there has been a move to use as few markers as possible. In the scramble to create abbreviated panels of STS, some very informative loci have been falsely labeled as being polymorphic and/or repetitive with reference only to unpublished results. For example, it has been suggested that DYS237 (SY153) is polymorphic with reference to unpublished data cited in an editorial[23,29,54]. The former references[23,29] reported a

Table 2   Studies of sequence tagged sites (STS)

Reference	STS (n)	Controls (n)	Screened (n)
STS = 1: Mallidis *et al.*, 1996[28]	1		186
STS = 2: Selva *et al.*, 1997[35]	2	81	81
STS = 3: Oliva *et al.*, 1998[56]	3 (15)	100	186
STS = 3: Landuyt *et al.*, 2000[66]	3 (27)		229
STS = 5: Shirakawa *et al.*, 1997[46]	5	0	25
STS = 5: Simoni *et al.*, 1997[36]	5	86	168
STS = 6: In'tVeld *et al.*, 1997[40]	6	100	58
STS = 7: Kremer *et al.*, 1997[39]	7	100	164
STS = 7: van Golde *et al.*, 2001[68]	7		300
STS = 11: Martinez *et al.*, 1997[43]	11		81
STS = 12: Rousseaux-Prevost *et al.*, 1997[37]	12		23
STS = 13: Stuppia *et al.*, 1996[26,27]	13	21	33
STS = 15: Foresta *et al.*, 1997[33]	15	10	38
STS = 15: Seifer *et al.*, 1999[58]	15	35	53
STS = 16: Krausz *et al.*, 1999[57]	16	10	134
STS = 16: Liow *et al.*, 1998[50]	16	101	202
STS = 16: Kobayashi *et al.*, 1994[20]	16	30	64
STS = 17: Nakahori *et al.*, 1996[30]	17	127	153
STS = 18: Peterlin *et al.*, 1997[34]	18	15	72
STS = 18: Grimaldi *et al.*, 1998[55]	18	19	67
STS = 18: Kleiman *et al.*, 1999[59]	18	32	133
STS = 21: Qureshi *et al.*, 1996[24]	21	80	100
STS = 26: Najmabadi *et al.*, 1996[25]	26	31	60
STS = 27: Stuppia *et al.*, 1998[49]	27	40	126
STS = 27: Stuppia *et al.*, 1997[45]	27	10	50
STS = 28: Vogt 1998[54]	28	200	370
STS = 28: Foresta *et al.*, 1998[48]	28	35	59
STS = 28: Henegariu *et al.*, 1994[19]	29	29	15
STS = 31: Foresta *et al.*, 1999[48]	31	150	240
STS = 31: van der Ven *et al.*, 1997[44]	31	50	204
STS = 35: Brandell *et al.*, 1998[51,52]	35		286
STS = 38: Ferlin *et al.*, 1999[60]	38	100	180
STS = 49: Kent-First *et al.*, 1999[14]	49	920	514
STS = 52: Silber *et al.*, 1998[53]	52	30	81
STS = 85: Pryor *et al.*, 1997[32]	85	200	200
STS =118: Reijo *et al.*, 1996[22]	118	90	35
STS: 118: Mullhall *et al.*, 1997[38]	118	75	83
STS = 131: Reijo *et al.*, 1995[12]	131	50	89
Total		2957	5127

deletion involving SY153 and the flanking region in a man with severe male-factor infertility and his ICSI-derived infant. The cited individual was fertile, but only with ICSI. Similarly, a deletion involving the entire *daz* region has been reported in a multiple-generation family which resulted in severe male-factor infertility in three brothers and their father[67]; several other case reports describe individuals who carry a large deletion and who have sons with ICSI[61,64,65].

Should *daz* be considered polymorphic based upon these reports? Similar to SY153, the *daz* deletions reported in these cases must be considered in the context

Table 3 Sequence tagged sites (STS) utilized in protocols from 22 laboratories according to the current maps of the Y chromosome

Regions	Kent-First[14]	YDDS (Promega)	Simoni[36,81]	Ferlin[60]	Stuppia[45]	Henegariu[19]	Reijo[12]	Van der Ven[44]	Grimaldi[55]	Nakahori[80]	Najmabadi[25]	Qureshi[24]	Foresta[63]	Kremer[39]	Martinez[43]	Peterlin[34]	Shirakawa[46]	Rousseaux-Prevost[37]	Seifer[58]	Oliva[56]	Liow[50]	Krausz[57]
Control	SMCX																					
Yp-TDF	SRY; ZFY	SRY; ZFY	SRY; ZFY	SRY		SRY	SY13, SY14 (SRY); SY238 (ZFY)	SY14 (SRY); SY18	SY14 (SRY)							SY18; SY19; SY70		SY14 (SRY); SY15		SY14		
centromere	SY75; SY78						SY200 (TSPY); SY78	SY69; SY78						SY72		SY78						
AZFa	SY81, SY82	SY81		SY78, SY79, SY81		SY81, SY82		SY81, SY82				SY82	SY78, SY79, SY81			SY81, SY82, SY83						
AZFa		SY86; SY86	SY86; SY86	SY83, SY86		SY86, SY85	SY83, SY85	SY86	SY83, SY86			SY83, SY86, SY85	SY83, SY86	SY86	SY86, SY85			SY85		SY86		
AZFa	SY85, SY84, DFFRY, SY87			SY84, DFFRY, SY87		SY84, SY87	DFFRY	SY84	SY84, SY87			SY84, SY87	SY84, SY87	SY84	SY84					SY84	SY84	SY84
AZFa	SY165, DBY?, UTY?, TB4Y?						DBY, UTY									SY165		GY6				DBY
AZFa				SY88, SY89		SY88	SY88	SY88	SY88			SY88	SY88, SY89									
AZFa	SY182 (KAL-Y)	SY182 (KAL-Y)				SY182 (KAL-Y)	SY90	SY182 (KAL-Y)								SY182 (KAL-Y)						
AZFa	SY97			SY151, SY95, SY98		SY151, SY94, SY95, SY97		SY151, SY94, SY95, SY97					SY151, SY95, SY98, SY100									
AZFb	SY100, SY101, SY102, SY105; SY109; SY113					SY102, SY105; SY109; SY113	CDY; SY102, SY105; SY109	SY102, SY105; SY109	SY112											SY102		

continued

Table 3 Continued

Regions	Kent-First[14]	YDDS (Promega) Simoni[36,81]	Ferlin[60]	Stuppia[45]	Henegariu[19]	Reijo[12]	Van der Ven[44]	Grimaldi[55]	Nakahori[30]	Najimabadi[25]	Qureshi[24]	Foresta[63]	Kremer[39]	Martinez[45]	Peterlin[34]	Shirakawa[46]	Rousseaux-Prevost[37]	Seifer[58]	Oliva[56]	Liow[90]	Krausz[57]
AZFb	XKRY																				XKRY
AZFb	SY116				SY114													SY114			SY114 SY116
AZFb	SY117		SY117		SY117		SY117	SY117				SY117						SY117			
AZFb	SY118																				
AZFb	SY119																SY119				
AZFb	SY121																SY122				
AZFb	SMCY							SMCY	SMCY												
AZFb	E1F1AY																				E1F1AY
AZFb	SY124		SY124									SY124		SY124							
AZFb			SY125		Y6PHc54*		Y6PHc54*					SY125									SY125
AZFb	SY126																SY126				
AZFb	SY127	SY127	SY127		SY127		SY127					SY127	SY127	SY127				SY127			
AZFb	SY128	SY128	SY129	SY129						SY128 SY129		SY129									SY129
AZFb	SY131		SY131	SY131						SY131		SY131	SY131								SY131
AZFb	SY130	SY130	SY130	SY130						SY130							SY130				
AZFb	SY132		SY132	SY132					SY132	SY132					SY132						
AZFb	SY133	SY133	SY134	SY134	SY134				SY134												
AZFb	SY134	SY134	SY134	SY134	SY134		SY134			SY134		SY134		SY134					SY134		SY134
AZFb	SY136		SY164	SY164	SY164			SY164	SY136	SY136 SY164					SY136				SY164		SY136
AZFb			SY138	SY138	SY138				SY137										SY138		SY138
AZFb																					
AZFb								SY55		SY138	SY207										
AZFb	SY139		SY139	SY139	SY139																SY139
AZFb	SY142				SY143	SY142	SY143		SY142	SY142 SY143	SY142	SY142		SY142				SY142	SY143		
AZFb	SY141					SY143	SY143			SY144	SY143								SY143		
AZFb	RBM1		RBM1 RBM2	RBM1	RBM1	RBM1	RBM1		RBM1		RBM1 RBM2	RBM1 RBM2			RBM1 RBM2		RBM1 RBM2	RBM1	RBM1		
AZFd	SY145		SY146	SY153	SY153					SY145 SY146	SY146	SY146							SY146	SY146	
AZFd	SY153		SY153	SY153	SY153		SY153	SY153		SY153	SY153	SY153			SY153			SY153	SY153		
AZFd	SY152		SY152					SY152		SY152	SY152	SY152			SY152		SY152		SY153		SY152
AZFd	SY220		SY220	SY220																	
AZFd	SY221																				
AZFd	SY150		SY150	SY150																	

STS falling within the boundaries of AZFa, AZFb, AZFd and AZFc are indicated in different shades of gray

Table 4 Studies of Y chromosome-linked polymorphic genes

(a)

Repeated genes	Reference	Remarks	Repeated STS	Reference	Remarks
daz	Giacolone, et al., 2000[71]	Clustered repeat	SY272	Reijo et al., 1995[12]	Clustered repeat
rbm	Yen, 1999[78]	Dispersed repeat	SY207	Reijo et al., 1995[12]	Dispersed repeat
dffry	Wong et al., 1999[70]	Dispersed repeat	SY221	Kostiner et al., 1998[77]	Dispersed repeat
pry	Wong et al., 1999[70]	Dispersed repeat	SY78	Kostiner et al., 1998[77]	Dispersed repeat
cdy	Yen, 1999[78]	Dispersed repeat	SY55	Affara et al., 1995[9], Vogt et al., 1997[10]	Dispersed repeat
bpy2	Yen, 1999[78], Kostiner et al., 1998[77]	Clustered repeat	SY75	Affara et al., 1995[9], Vogt et al., 1997[10]	Clustered repeat
tty	Yen, 1999[78]	Dispersed repeat	SY109	Affara et al., 1995[9], Vogt et al., 1997[10]	Dispersed repeat
tspy (SY200)	Yen, 1999[78]	Dispersed repeat	SY112	Affara et al., 1995[9], Vogt et al., 1997[10]	Dispersed repeat
xkry	Lahn and Page, 1998[80], Kostiner, et al., 1998[77]	Dispersed repeat	SY132	Affara et al., 1995[9], Vogt et al., 1997[10]	Dispersed repeat
			SY164	Affara et al., 1995[9], Vogt et al., 1997[10]	Dispersed repeat
			SY138	Affara et al., 1995[9], Vogt et al., 1997[10]	Clustered repeat
			SY150	Reijo et al., 1995[12]	Clustered repeat
			SY144	Yen, 1998[75]	Dispersed repeat
			SY155	Yen, 1998[75], Kent-First et al., 1999[14]	Duplicated in small palindrome in AZFd and AZFc

(b)

Statistically tested/significant level of polymorphism	Reference	Deleted in males fertile only with ICSI	Reference	Deleted in fathers of infertile sons	Reference
SY207	Pryor et al., 1997[32], Kent-First et al., 1999[14]	SY153	Kent-First et al., 1996[23]	SY272/SY207	Pryor et al., 1997[32]
SY272	Pryor et al., 1997[32], Kent-First et al., 1999[14]	SY220–SY158 including CDY1	Saut et al., 2000[67]	Distal AZFb through AZFc	Kobayashi et al., 1994[20]
Amely	Kent-First et al., 1999[14]	SY254, SY255, SY147, SY149	Kamischke et al., 1999[61]	AZFC including dax (SY254, SY255; subfertile	Pryor et al., 1997[32]
SY269	Kent-First et al., 1999[14]	DAZ, BPY2, including SY254, SY255)	Page et al., 1999[64]	SY220–SY158 including CDY1	Saut et al., 2000[67]
DYS287	Hammer et al., 1994[90]	SY153 through SY158	Kleiman et al., 1999b[59]	SY158, SY166, SY167	Stuppia et al., 1996b[27]
DYS7C/50F2C	Casanova et al., 1985[91], Jobling et al., 1996[93]	SY153 through SY158	Jiang et al., 1999[92]		
DYS426	Jobling et al., 1996[93]				
DYF371	Jobling et al., 1996[93]				
DYS425	Jobling et al., 1996[93]				
DYS19	Jobling et al., 1996[93]				
DYS389	Jobling et al., 1998[83]				
DYS390	Jobling et al., 1996[93]				
DYS257	Jobling et al., 1999[94]				
DYF155	Jobling et al., 1998b[83]				
RBM2	Nakahori et al., 1994[21]				

STS, sequence tagged sites; ICSI, intracytoplasmic sperm injection

of the hundreds of azoospermic and severely oligo-zoospermic men who have identical deletions and the 3289 fertile men in which *daz* was intact. Applying Fisher's exact test to these examples, there is a significant probability of correlation of infertility with micro-deletions involving these STS ($p < 0.001$). Similar to the results of Simoni and co-workers[82], the false-positive error rates were elevated (up to 10%); however, additional information could be derived from the blinded collaborative protocol. (1) Pathology-associated deletions were missed most frequently in groups of samples previously tested with abbreviated screening protocols, in terms of STS number and/or selection. (2) False-positive and false-negative error rates were higher in sample sets previously tested with multiplex protocols that did not employ a positive (PCR) control in each multiplex. (3) False-positive and false-negative error rates were higher in sample sets previously tested with individual STS screening protocols which did not employ a positive (PCR) control with each STS or did not repeat the amplifications to confirm results. (4) Error rates were elevated in samples derived from protocols that tested low quality (dilute or highly sheared) DNA. (5) Error rates were elevated in samples screened with home-brew protocols.

Simoni and co-workers[82] concluded by recommending a 'minimal first choice panel' of six STS (two each from AZFa, AZFb and AZFc), with SRY and ZFY/ZFX as positive controls, in an attempt to simplify and standardize testing protocols in Europe. Kent-First and co-workers[23,29] concluded that the choice of STS should be based upon: (1) naturally occurring patient deletion across AZF; (2) location of genes of functional interest across AZF; (3) STS data from the HUGO and the National Center for Biotechnology Information (NCBI) map as well as published deletion maps, avoiding repeated and/or redundant loci; (4) structure and compatible amplification conditions of STS in simplex and multiplex, demonstrated in replicates in experiments performed across multiple days; and (5) inclusion of loci shown to be pathology-associated in published experiments which screen statistically significant numbers of fertile males in tandem with infertile populations and report statistical probabilities.

What data can be extrapolated from the massive numbers of published reports, cumulatively? Many

laboratories experience high error rates. Evidence for this is derived from the two blinded multicenter studies, described above. Further evidence can be found in reports of non-contiguous microdeletions which cannot be explained by map order of STS or by Southern blots, and in reports of unusual deletions involving repeated genes or gene clusters which cannot be confirmed by Southern blot. There is a wide range in DDRs (2–38%) across protocols, which may be attributed to a wide variation in both the number and choice of STS employed (Figure 4). It has also been suggested that microdeletion location and frequency may vary with ethnicity and with Y chromosome haplotype[83,84]. These studies lead us to hypothesize that some genomes may be inherently more unstable than others and thus more predisposed to aberrant sister chromatid exchange and mutation. Finally, there is a lack of consensus concerning the type of patient phenotype to include in a testing protocol. Some protocols suggest that all male-factor infertility cases should be offered testing, whereas others suggest testing only azoospermic men. There is general agreement that pathology-associated micro-deletions are rarely found in normozoospermic infertile males, occurring with the highest frequency in association with severe oligozoospermia and azoospermia (Figure 1, Table 2).

What are the possible solutions to this dilemma? Before standardized protocols are agreed upon one must question whether the purpose of the test is for clinical research or clinical diagnostic applications. Presumably clinical research protocols will be designed to further our knowledge concerning the relevance of Y chromosome microdeletions in assisted reproduction and will be conducted after institutional review. Clinical diagnostic protocols must focus on diagnostic accuracy and precision. Assays such as cytogenetic analysis or tests for single-gene defects such as cystic fibrosis in association with congenital bilateral absence of the vas deferens exemplify good practice, and the assay sensitivity, specificity, precision and accuracy is easily measured according to well-defined standards[3,85,86]. But it is far more difficult to develop genetic tests which meet these standards when the test is indicated for a broad range of phenotypes and does not focus on a single gene or a single phenotype, but rather a large, relatively undefined and poorly understood chromosomal region(s) which contains many functional and non-functional genes and gene families.

The debate concerning the prognostic utility of microdeletion screening protocols is ongoing. However, there is general agreement that male infertility and microdeletions in AZF are correlated and the correlations are statistically valid. There is also a general consensus, supported by published statistical data, that deletions in intervals 5 and/or 6 can result in a range of phenotypes, with azoospermia being the most commonly observed one associated with AZFa region deletions and deletions encompassing multiple deletion intervals (Figure 1). Oligozoospermia is the most frequently detected phenotype associated with deletions in all other AZF regions. For example, men with AZFa-restricted deletions present with the most narrowly defined range of testicular phenotypes, ranging from SCO to the testis with germ cells present and partial spermatogenic arrest associated with severe oligozoospermia (Figure 1). Many screening protocols suggest inclusion of only one or two randomly chosen STS from this region (and others) to detect a deletion and disregard the STS's position relative to the gene candidate(s) for function in the given region. Therefore, the ability to provide meaningful and accurate genetic counseling to the couple with a deletion is limited to only a broad phenotype range when abbreviated testing protocols are used. Focused mapping of the AZFa region provides evidence that if only the proximal half of the region containing *dffry* (*usp9y*) is removed and the distal half containing *dby* remains intact, germ cells will be present even though spermatogenesis is either partially or com-pletely arrested[87] (Figure 1). If both the distal and proximal halves of the region are removed, the testis will be devoid of germ cells. This information is critical when considering the development of standards to be used in a genetic testing protocol.

In light of these data and synteny and gene knockout studies in the mouse, accuracy in genetic testing for deletions in the AZFa region (for example) would be improved by choosing STS which immediately flank or include *dffry* and *dby*[88,89]. For example, SY84 or SY745 immediately flanks *dffry/usp9y* on its proximal border, SY742 or SY87 separates *dffry* and *dby* and SY743, SY88 or even the Kalman-Y locus flanks *dby* on its distal boarder. At the risk of controversy, these data suggest that it is misleading to subdivide AZF into AZFa, AZFb, AZFc and AZFd unless one further subdivides the subregions into proximal, middle and distal subintervals based upon the genes that reside there and naturally occurring patient deletions (Figure 1).

Alternatively, perhaps this opens Pandora's box to an even greater extent! Our ability to assess the integrity of the Y chromosome with improved accuracy and predictability may increase if we step back and look at the chromosome in terms of its functional gene content without regard to AZF subregions. Standards in testing are needed but these standards should evolve over time in an international forum and should be developed in a collaborative fashion by a team consisting of, first, individuals with expertise in testis function with access to testis biopsy material (clinical researchers), and secondly, of individuals with expertise in genomics as it pertains to the Y chromosome.

References

1. Koopman P, Munsterberg A, Capel B, *et al*. Expression of a candidate sex-determining gene during mouse testis differentiation. *Nature (London)* 1990;348:450–2
2. Koopman P, Gubbay J, Vivian N, *et al*. Male development of chromosomally female mice transgenic for *Sry*. *Nature (London)* 1991;351:117–21
3. Kent-First M. The critical and expanding role of genetics in assisted reproduction. *Prenat Diag* 2000;20:536–51
4. Lau C, Zhang J. Expression analysis of thiry one Y chromosome genes in human protstate cancer. *Mol Carcin* 2000;27:308–21
5. Jaffe T, Oates RD. Genetic abnormalities and reproductive failure. *Urol Clin N Am* 1997;21:389–408
6. Kent MG, Shoffner RN, Buoen L, Weber AF. The XY sex reversal syndrome in the domestic horse. *Cytogenet Cell Genet* 1986;42:8–18
7. Kent MG, Shoffner RN, Hunter A, *et al*. XY sex reversal syndrome in the mare: clinical and behavioral studies. *Hum Genet* 1988;79:321–8
8. Tiepolo L, Zuffardi O. Localization of factors controlling spermatogenesis in the nonfluorescent portion of the human Y chromosome long arm. *Hum Genet* 1976;34:119–24

9. Affara N, Bishop C, Brown W, *et al*. Second international workshop of Y chromosome mapping. *Cytogenet Cell Genet* 1995;73(Suppl.):33–76

10. Vogt P, Affara N, Davey P, *et al*. Report of the third international workshop on Y chromosome mapping. *Cytogenet Cell Genet* 1997;79:10–20

11. Ma K, Inglis JD, Sharkey A, *et al*. A Y-chromosome gene family with RNA-binding protein homology: candidates for the azoospermia factor AZF controlling human spermatogenesis. *Cell* 1993;75:1287–95

12. Reijo R, Lee TY, Salo P, *et al*. Diverse spermatogenic defects in humans caused by Y-chromosome deletions encompassing a novel RNA-binding protein gene. *Nat Genet* 1995;10:383–93

13. Vogt PH, Edelmann A, Kirsch S. Human Y-chromosome azoospermia map based on naturally occurring deletions. *Science* 1996;258:52–9

14. Kent-First M, Muallem A, Shultz J, *et al*. Defining regions of the Y-chromosome responsible for male infertility and identification of a fourth AZF region (AZFd) by Y-chromosome microdeletion detection. *Mol Reprod Dev* 1999;53:27–41

15. Sargent C, Boucher C, Kirsch S, *et al*. The critical region of overlap defining the AZFa male infertility interval of proximal Yq contains three transcribed sequences. *J Med Genet* 1999; 36:670–7

16. Kent-First M. The Y chromosome and its role in testis differentiation and spermatogenesis. *Sem Reprod Med* 2000;18: 67–80

17. Blanco P, Shlumukova M, Sargent C, *et al*. Divergent outcomes of intrachromosomal recombination on the human Y chromosome: male infertility and recurrent polymorphism. *J Med Genet* 2000;37:752–8

18. Ma K, Sharkey A, Kirsch S, *et al*. Towards the molecular localization of the AZF locus: mapping of microdeletions in azoospermic men within 14 subintervals of interval 6 of the human Y-chromosome. *Hum Mol Genet* 1992;1:29–33

19. Henegariu O, Hirschmann P, Kilian K, *et al*. Rapid screening of the Y-chromosome in idiopathic sterile men, diagnostic for deletions in AZF: a genetic Y factor expressed during spermatogenesis. *Andrologia* 1994;26:97–106

20. Kobayashi K, Mizuno K, Hida A, *et al*. PCR analysis of the Y-chromosome long arm in azoospermic patients: evidence for a second locus required for spermatogenesis. *Hum Mol Genet* 1994;3:1965–7

21. Nakahori Y, Kobayashi K, Komaki R, *et al*. A locus of the candidate gene family for azoospermia factor (*YRRM2*) is polymorphic with a null allele in Japanese males. *Hum Mol Genet* 1994;3:1709

22. Reijo R, Alagappan RK, Patrizio P, Page DC. Severe oligozoospermia resulting from deletions of azoospermia factor gene on Y chromosome. *Lancet* 1996;347:1290–3

23. Kent-First M, Kol S, Muallem A, *et al*. The incidence and possible relevance of Y-linked microdeletions in babies born after intracytoplasmic sperm injection and their infertile fathers. *Mol Hum Reprod* 1996;2:943–50

24. Qureshi SJ, Ross AR, Ma K, *et al*. Polymerase chain reaction screening for Y chromosome microdeletions: a first step towards the diagnosis of genetically-determined spermatogenic failure in men. *Mol Hum Reprod* 1996;2:775–9

25. Najmabadi H, Huang V, Yen P, *et al*. Substantial prevalence of microdeletions in infertile men with idiopathic azoospermia and oligozoospermia detected using a sequence-tagged site-based mapping strategy. *J Clin Endocrinol Metab* 1996;81:1347–52

26. Stuppia L, Mastroprimiano G, Calabrese G, *et al*. Microdeletions in interval 6 of the Y-chromosome detected by STS–PCR in 6 of 33 patients with idiopathic oligo- or azoospermia. *Cytogenet Cell Genet* 1996;72:155–8

27. Stuppia L, Calabrese G, Franchi P, *et al*. Widening of a Y-chromosome interval-6 deletion transmitted from a father to his infertile son accounts for an oliozoospermia critical region distal to the *RBM1* and *DAZ* genes. *Am J Hum Genet* 1996;59:1393–5

28. Mallidis C, Loreland K, Najmabadi H, Bhasin S. The incidence of the deleted in azoospermia gene in infertile men. *Hum Reprod* 199611:56–7

29. Kent-First MG, Kol S, Muallem A, *et al*. Infertility in intracytoplasmic-sperm-injection-derived sons. *Lancet* 1996; 348:332

30. Nakahori Y, Kuroki Y, Komaki R, *et al*. The Y-chromosome region essential for spermatogenesis. *Horm Res Suppl* 1996; 1:20–3

31. Hargreave TB, Chandley AC, Ross A, *et al*. Y chromosome microdeletions and male subfertility. *Andrologia Suppl* 1996;1:19–21

32. Pryor JL, Kent-First M, Muallem A, *et al*. Prospective analysis of Y chromosome microdeletions in 200 consecutive male infertility patients. *N Engl J Med* 1997;336:534–9

33. Foresta C, Ferlin A, Garolla A, *et al*. Y-chromosome deletions in idiopathic severe testiculopathies. *J Clin Endocrinol Metab* 1997;82:1075–80

34. Peterlin B, Kunej T, Zorn B, *et al*. Sterility associated with Y chromosome abnormalities. In Barratt C, De Jonge C, Mortimer D, Parinaud J, eds. *Genetics of Human Male Infertility*. EDK Press, 1997:66–75

35. Selva J, Kanafani S, Prigent Y, *et al*. Incidence of AZF (azoospermia factor) deletions and familial forms of infertility among patients requiring intracytoplasmic spermatozoa injection (ICSI). *J Assisted Reprod Genet* 1997;14:593–5

36. Simoni M, Gromoll J, Dworniczak B, *et al*. Screening for deletions of the Y-chromosome involving the *DAZ* (deleted in azoospermia) gene in azoospermia and severe oligozoospermia. *Fertil Steril* 1997;67:542–7

37. Rousseaux-Prevost R, Rigot J, Hermand E, *et al*. Molecular mapping of microdeletions of the Y-chromosome in azoospermic men. In Barratt C, De Jonge C, Mortimer D, Parinaud J, eds. *Genetics of Human Male Infertility*. EDK Press, 1997:351–2

38. Mulhall JP, Reijo R, Alagappan R, *et al*. Azoospermic men with deletion of the DAZ gene cluster are capable of completing spermatogenesis: fertilization, normal embryonic development and pregnancy occur when retrieved testicular spermatozoa are used for intracytoplasmic sperm injection. *Hum Reprod* 1997;12:503–8

39. Kremer JA, Tuerlings JH, Meuleman EJ, *et al*. Microdeletions of the Y-chromosome and intracytoplasmic sperm injection: from gene to clinic. *Hum Reprod* 1997;12:687–91

40. In't Veld PA, Halley DJ, Van Hemel JO, *et al.* Genetic counseling before intracytoplasmic sperm injection. *Lancet* 1997;350:490

41. Goncalves J, Medeiros S, Vale F. Screening for microdeletions in oligo-astheno-terato-zoospermic patients. In Barratt C, De Jonge C, Mortimer D, Parinaud J, eds. *Genetics of Human Male Infertility.* EDK Press, 1977:353–5

42. Garolla A, Ferlin C, Foresta C. Y-chromosome microdeletions in idiopathic Sertoli cell-only and severe hypospermatogenesis. In Barratt C, De Jonge C, Mortimer D, Parinaud J, eds. *Genetics of Human Male Infertility.* EDK Press, 1997:358–9

43. Martinez M, Bernabe MJ, Gomez E, *et al.* Preliminary study of somatic chromosomal anomalies and Y-linked microdeletions among infertile men. In Barratt G, De Jonge C, Martine D, Pannaud J, eds. *Genetics of Human Male Infertility.* EDK Press, 1997:360–1

44. Van der Ven K, Montag M, Peschka B, *et al.* Combined cytogentic and Y-chromosome microdeletion screening in males undergoing intracytoplasmic sperm injection. *Mol Hum Reprod* 1997;3:699–704

45. Stuppia L, Gatta V, Mastroprimiano G, *et al.* Clustering of Y chromosome deletions in subinterval E of interval 6 supports the existence of an oligozoospermia critical region outside the *DAZ* gene. *J Med Genet* 1997;34:881–3

46. Shirakawa T, Fujisawa M, Kanzaki M, *et al.* Y chromosome (Yq11) microdeletions in idiopathic azoospermia. *Int J Urol* 1997;4:198–201

47. Brown GM, Furlong RA, Sargent CA, *et al.* Characterization of the coding sequence and fine mapping of the human *DFFRY* gene and comparative expression analysis and mapping to the Sxr^b interval of the mouse Y-chromosome of the *Dffry* gene. *Hum Mol Genet* 1998;7:97–107

48. Foresta C, Ferlin A, Garolla A, *et al.* High frequency of well defined Y-chromosome deletions in idiopathic Sertoli cell-only syndrome. *Hum Reprod* 1998;13:302–7

49. Stuppia L, Gatta V, Calabrese G, *et al.* A quarter of men with idiopathic oligo-azoospermic display chromosomal abnormalities and microdeletions of different types in interval 6 of Yq11. *Hum Genet* 1998;102:566–70

50. Liow S, Ghadessy F, Yong E. Y chromosome microdeletions, in azoospermic or near-azoospermic subjects are located in the AZFc (DAZ) subregion. *Mol Hum Reprod* 1998;4763–8

51. Brandell R, Mielnik, A, Liotta, D, *et al.* AZFb deletions predict the absence of spermatozoa with testicular sperm extraction: preliminary report of a prognostic test. *Hum Reprod* 1988;13:2812–15

52. Brandell R, Mielnik A, Liotta D, *et al.* Effect of partial Y-chromosome deletions on results of treatment for severe male factor infertility. *Program Suppl 16th World Congress on Fertility and Sterility,* San Francisco, October 1988; abstr.0–021

53. Silber S, Alagappan R, Brown A, Page D. Y chromosome deletions in azoospermic and severely oligozoospermic men undergoing intracytoplasmic sperm injection after testicular sperm extraction. *Hum Reprod* 1998;13:3332–7

54. Vogt P. Human chromosome deletions in Yq11, AZF candidate genes and male infertility: history and update. *Mol Hum Reprod* 1998;4:739–44

55. Grimaldi P, Scarponi C, Rossi P, *et al.* Analysis of Yq microdeletions in infertile males by PCR and DNA hybridization techniques. *Mol Hum Reprod* 1998;4:1116–21

56. Oliva R, Margarit E, Ballesca J, *et al.* Prevalence of Y chromosome microdeletions in oligospermic and azoospermic candidates for intracytoplasmic sperm injection. *Fertil Steril* 1998;70:506–10

57. Krausz C, McElreavey K. Y chromosome and male infertility. *Frontiers Biosci* 1999;4:1–15

58. Seifer I, Amat S, Delgado-Viscogliosi P, *et al.* Screening for microdeletions on the long arm of chromosome Y in 53 infertile men. *Int J Androl* 1999;22:148–54

59. Kleiman S, Yogev L, Gamzu R, *et al.* Genetic evaluation of infertile men. *Hum Reprod* 1999;14:33–8

60. Ferlin A, Moro E, Garolla A, Foresta C. Human male infertility and Y chromosome deletions: role of the AZF-candidate genes *DAZ, RBM,* and *DFFRY. Hum Reprod* 1999;14:1710–16

61. Kamischke A, Gromoll J, Simoni M, *et al.* Transmission of a Y chromosomal deletion involving the deleted in azoospermia (*DAZ*) and chromodomain (*CDY1*) genes from father to son through intracytoplasmic sperm injection: case report. *Hum Reprod* 1999;14:2330–2

62. Kim S, Kim K, Paick J. Microdeletions within the azoospermia factor subregions of the Y chromosome in patients with idiopathic azoospermia. *Fertil Steril* 1999;72: 349–53

63. Foresta C, Moro E, Garolla A, *et al.* Y chromosome microdeletions in cryptorchidism and idiopathic infertility. *J Clin Endocrin Metab* 1999;84:3660–5

64. Page DC, Silber S, Brown LG. Men with infertility caused by AZFc deletion can produce sons by intracytoplasmic sperm injection, but are likely to transmit the deletion and infertility. *Hum Reprod* 1999;14:1722–6

65. Chang P, Sauer M, Brown S. Y chromosome microdeletion in a father and his four infertile sons. *Hum Reprod* 1999;14: 2689–94

66. Landuyt L, Lissens W, Stouffs K, *et al.* Validation of a simple Yq deletion screening program in an ICSI candidate population. *Mol Hum Reprod* 2000;6:291–7

67. Saut N, Terriou P, Navarro A, *et al.* The human Y chromosome genes *BPY2, CDY1,* and *DAZ* are not essential for sustained fertility. *Mol Hum Reprod* 2000;6: 789–93

68. Van Golde R, Wetzels A, Graaf R, *et al.* Decreased fertilization rate and embryo quality after ICSI in oligozoospermic men with microdeletions in the azoospermia factor c region of the Y chromosome. *Hum Reprod* 2001;16:289–92

69. Lahn B, Page D. Functional coherance of the human Y chromosome. *Science* 1997;278:675–80

70. Wong J, Blanco P, Affara N. An exon map of the AZFc male infertility region of the human Y chromosome. *Mam Genome* 1999;10:57–61

71. Giacolone J, Delobette S, Gibaja V, *et al.* Optical mapping of BAC clones from the human Y chromosome DAZ locus. *Genome Res* 2000;10:1421–9

72. Tilford CA, Kuroda-Kawaguchi T, Skaletsky H, *et al.* A physical map of the human Y chromosome. *Nature (London)* 2001;409:943–5

73. Kuroda-Kawaguchi T, Skaletsky H, Brown L, *et al*. The AZFc region of the Y chromosome features massive palindrome and uniform recurrent deletions in infertile men. *Nat Genet* 2001; 30:279–86

74. Yen P, Chai N, Salido E. The human DAZ genes, a putative male infertility factor on the Y chromosome, are highly polymorphic in the DAZ repeat region. *Mam Genome* 1997;8:756–9

75. Yen P. A long-range restriction map of deletion interval 6 of the human Y chromosome: a region frequently deleted in azoospermic males. *Genomics* 1998;54:5–12

76. Chai N, Salido E, Yen P. Multiple functional copies of the RBM gene family, a spermatogenesis candidate of the human Y chromosome. *Genomics* 1997;45:355–61

77. Kostiner D, Turek P, Reijo R. Male infertility: analysis of the markers and genes on the human Y chromosome. *Hum Reprod* 1998;13:3032–8

78. Yen P. Advances in Y chromosome mapping. *Curr Opin Obstet Gynecol* 1999;11:275–81

79. Sun C, Skaletsky H, Birren B, *et al*. An azoospermic man with a de novo point mutation in the Y-chromosomal gene *USP9Y*. *Nat Genet* 1999;23:429–32

80. Lahn B, Page D. A human sex-chromosomal gene family expressed in male germ cells and encoding variable charged proteins. *Hum Mol Genet* 2000;9:311–19

81. Cram D, Ma K, deKretser D, *et al*. Transmission of Yq deletions in men with spermogenic disorders through the use of intracytoplasmic sperm injection. In *Proceedings of the 11th World Congress on In Vitro Fertilization and Human Reproductive Genetics*, 1999;59

82. Simoni M, Bakker E, Eurlings M, *et al*. Laboratory guidelines for molecular diagnosis of Y-chromosomal microdeletions. *Int J Androl* 1999;22:292–9

83. Jobling M, Williams G, Schiebel K, *et al*. A selective difference between human Y-chromosomal DNA haplotypes. *Curr Biol* 1998;8:1391–4

84. Kuroki Y, Iwaamoto T, Lee J, *et al*. Spermatogenic ability is different among males in different Y chromosome lineage. *Hum Genet* 1999;44:289–92

85. Patrizio P, Asche R. The relationship between congenital bilateral absence of the vas deferens (CAVD), cystic fibrosis (CF) mutations and epididymal sperm. *Assist Reprod Rev* 1994;4:95–100

86. Oates R, Amos J. The genetic basis of congenital bilateral absence of the vas deferens and cystic fibrosis. *J Androl* 1994; 15:1–8

87. Blagosklonova O, Fellmann F, Clavequin M, *et al*. AZFa deletions in Sertoli cell-only syndrome: a retrospective study. *Mol Hum Reprod* 2000;6:795–9

88. Slee R, Grimes B, Speed R, *et al*. A human *DAZ* transgene confers rescue of the mouse null phenotype. *Proc Natl Acad Sci USA* 1999; 96:8040–5

89. Vogel T, Speed R, Teague P, Cooke H. Mice with Y chromosome deletion and reduced *Rbm* genes on a heterozygous *Dazl1* null background mimic a human azoopermic factor phenotype. *Hum Reprod* 1999;14:3023–9

90. Hammer MF. A recent insertion of an alu element on the Y chromosome is a useful marker for human population studies. *Mol Biol Evol* 1994;11:749–61

91. Casanova M, Leroy P, Boucekkine C, *et al*. A human Y-linked DNA polymorphism and its potential for estimating genetic and evolutionary distance. *Science* 1985;230:1403–6

92. Jiang M, Lien Y, Chen S, *et al*. Transmission of de novo mutations of the *deleted in azoospermia* genes from a severely oligozoospermic male to a son via intracytoplasmic sperm injection. *Fertil Steril* 1999;71:1029–32

93. Jobling MA, Samara V, Panya A, *et al*. Recurrent duplication and deletion polymorphisms on the long arm of the Y chromosome in normal males. *Hum Mol Genet* 1996;5:1767–75

94. Jobling MA, Heyer E, Dieltjes P, Knijff P. Y-chromosome-specific microsatellite mutation rates re-examined using a minisatellite, MSY1. *Hum Mol Genet* 1999;8:2117–20

Preimplantation genetics in human embryology 22

Luca Gianaroli, Maria Cristina Magli, Alessandro Di Gregorio and Anna Pia Ferraretti

INTRODUCTION

The potential transmission of genetic disorders has been a major problem for many couples when contemplating pregnancy. In the last decades, careful evaluation of the family history, age of the mother, karyotype and screening for some of the most common genetic diseases, and the implementation of prenatal diagnosis, have greatly decreased the incidence of affected pregnancies.

Preimplantation genetic diagnosis (PGD) has been designed as an alternative methodology which, in combination with *in vitro* fertilization (IVF) techniques, permits the disorder to be screened before the corresponding embryo is transferred to the uterus of the mother. The original idea arose in the early 1960s when sexing of rabbit embryos at the blastocyst stage was attempted[1]. Twelve years later, the first pregnancies were obtained from the clinical application of PGD to disorders in couples at risk of having offspring with X-linked diseases and cystic fibrosis[2].

The application of PGD is especially important in reproductive medicine due to the frequent association between genetic factors and infertility. Basically, PGD can be considered in two different situations: it can be recommended to couples at risk of having children with genetic diseases, such as cystic fibrosis or thalassemias, as they are healthy carriers of a mutation in the corresponding gene; or, it can be proposed for the diagnosis of chromosomal disorders, both numerical (i.e., aneuploidies) and structural (i.e., inversions or translocations), in couples exposed to the risk of generating aneuploid embryos. This last approach has been used as a tool to improve the IVF clinical outcome by selecting euploid embryos for transfer, based on the hypothesis that aneuploidies are strongly associated with spontaneous

abortions and implantation failures. Although the number of chromosomes analyzed per single cell is restricted, an increased implantation rate and a concomitant reduction in the incidence of spontaneous abortions are the immediate consequences, due to the highest viability of chromosomally normal embryos[4,5]. These results clearly demonstrate that the procedure of embryo biopsy, the first step of PGD, although invasive, does not affect the embryo's potential for development.

EMBRYO BIOPSY

The procedure entails the removal of one blastomere from day 3 regularly cleaving embryos and is generally performed at 62–64 h post-insemination. This time constraint is due to the fact that cell–cell interactions and junctions start to take place toward embryo compaction. Each embryo is biopsied individually in HEPES-buffered medium covered by equilibrated oil; a breach is opened in the zona pellucida using acidic Tyrode's solution (pH 2.35) loaded in a 12 μm inner diameter pipette. As shown in Figure 1, the blastomere selected for biopsy is set at the three o'clock position; the acidic solution is blown toward the zona pellucida at a point corresponding to the blastomere. When the breach is opened (20 μm diameter approximately), the excess of acidic solution in the surroundings is aspirated with the same acidic solution pipette. If fragments are present in the perivitelline space, they are gently removed. Finally, the blastomere is aspirated with a polished glass needle (35–40 μm inner diameter); extreme attention is paid not to damage either the blastomere itself or the

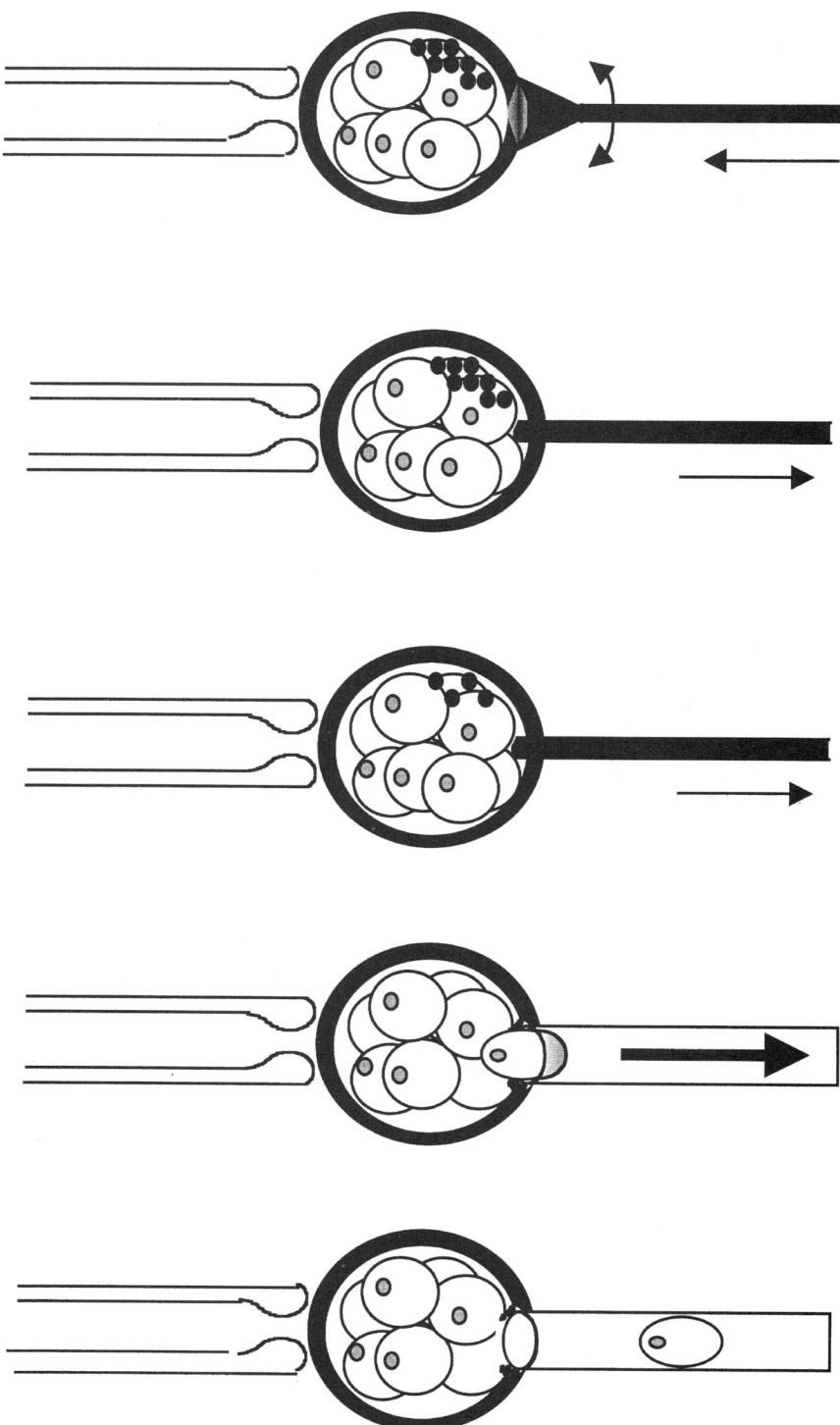

Figure 1   Removal of a blastomere from an eight-cell stage embryo. The opening in the zona pellucida, made with an acidic solution at pH 2.35, permits the aspiration of the blastomere as well as of fragments in the perivitelline space. The presence of the nucleus is a fundamental criterion for the selection of the blastomere to be biopsied

surrounding cells. After biopsy, the embryo is washed, put in fresh medium and incubated until the transfer.

PGD FOR GENETIC DISORDERS

When the biopsy procedure is completed, the removed cell is transferred to a microcentrifuge tube containing lysis buffer to extract DNA. The polymerase chain reaction (PCR) is a technique that amplifies specific DNA sequences to more than a billion times in a few hours by doubling in each cycle the number of DNA copies. Following extraction, the DNA undergoes PCR specifically to amplify the locus where the mutation under investigation maps. Special precautions have to be followed to avoid the possibility of exogenous DNA contamination, whose main sources are of maternal (cumulus cell) or paternal (sperm cell) origin. Operators can also contribute exogenous DNA. This implies that all disposables, medium and oil involved in the culture and micromanipulation of PGD cases are prepared from unopened, sterile packages; the use of sterile gloves and of a specifically designated room with equipment, disposables and reagents to PCR for PGD cases are necessary requirements. All reagents are checked the day before for DNA contamination by running a control PCR to detect the possible presence of human DNA. In addition, embryo culture medium is included during the reaction to verify the presence of external DNA. Similarly, polymorphic markers are considered as an additional method of verifying endogenous DNA amplification.

Following amplification, PCR products are analyzed by the different methods available: heteroduplex formation, restriction digestion and sequencing[3,6,7]. The interpretation of PCR results can be distorted by the possible occurrence of allelic preferential amplification or allele drop out (ADO). The most efficient strategy adopted to make ADO recognizable entails the definition of polymorphic markers which are strongly linked to the gene under study[8]. These markers are mainly represented by short tandem repeats (STRs; short sequences where a 2–5 base pairs unit occurs repeatedly throughout the human genome) that are mostly allocated in intergenic regions; they are amplified simultaneously with the gene to which they are strongly linked in a multiplex PCR. If two or more informative markers are available for the mutation under study, the great majority of ADOs are detected, leading to an accuracy of approximately 97%[9,10].

The single-gene disorders for which PGD has been performed worldwide are reported in Table 1. This list is expanding rapidly according to the incoming information on the association between gene sequencing and corresponding disorders in cases of mutations. As shown in Table 2, 20 PGD cycles from 13 patients have been performed in our institute for the detection of cystic fibrosis, thalassemia, hemophilia and retinoblastoma. Embryo transfer was performed on day 5 in 17 cycles, yielding five clinical pregnancies: two on term and three regularly ongoing. All patients were recommended, in case of pregnancy, to undergo prenatal diagnosis in order to confirm the diagnosis on preimplantation embryos; confirmation was achieved in all the patients. The current data suggest that PGD could be considered not only as a technique aimed at establishing a healthy pregnancy, but also as a general approach toward the prevention of genetic disorders.

PGD FOR ANEUPLOIDY

The numerical analysis of interphase chromosomes is based upon the use of fluorescence-labeled DNA probes in the fluorescence *in situ* hybridization (FISH) technique. The biopsied blastomere needs to be prepared accordingly: lysis is achieved in hypotonic solution and the nucleus is fixed on a glass slide with methanol:acetic acid followed by dehydration in increasing ethanol series (Figure 2). The fluorescent probes are then added and left to hybridize at 37°C after DNA denaturation (Figure 3). After washing off the unbound probe, the diagnosis is made at the fluorescent microscope.

Working on single cells imposes strict limits on the number of chromosomes that can be screened at PGD. However, use of multiple probes permits the simultaneous analysis of different chromosomes and can be followed by successive rounds of the FISH technique using other probes. In this way, diagnosis of up to nine chromosomes is currently possible, including those whose aneuploidies are more frequently detected in spontaneous abortions and on-term pregnancies: X, Y, 13, 15, 16, 18, 21 and 22[11].

Table 1 Gene disorders transmittable to the offspring, which can be analyzed by genetic diagnosis after oocyte and embryo biopsy

Achondroplasia	Central core disease
Agammaglobulinemia	Gaucher's disease
Sickle-cell anemia	Huntington's disease
Fanconi's anemia	Alport's disease
Spinal/bulbar muscular atrophy	Tay–Sachs' disease
Alpha$_1$-antitrypsin deficiency	MELAS
Long chain hydroxyacyl CoA dehydrogenase deficiency	X-linked myotubular myopathy
Ornithine transcarbamilase deficiency	Neurofibromatosis I and II
Deficiency of the mitochondrial trifunctional protein	Multiple endocrine neoplasia type II
Multiple epiphyseal dysplasia	Osteogenesis imperfecta I and IV
Myotonic dystrophy	Familial adenomatous polyposis coli
Becker's muscular dystrophy	Rhetinitis pigmentosa
Duchenne's muscular dystrophy	Rhesus (Rh D)
Hemophilia A and B	Tuberous sclerosis
Epidermolysis bullosa	Crouzon's syndrome
Heart defect exclusion	DiGeorge's syndrome
Familial polyposis–Gardner	Hunter's syndrome MPS II
Phenylketonuria	Lesch–Nyhan's syndrome
Cystic fibrosis	Marfan's syndrome
X-linked hydrocephalus	Digital orofacial syndrome type 1
Incontinentia pigmenti	Stickler's syndrome
Hyperinsulinemic hypoglycemia PHH1	Fragile X syndrome
Early onset Alzheimer's disease	Wiskott–Aldrich syndrome
Charcot–Marie–Tooth's disease 1 and 2A	Thalassemia

Table 2 Preimplantation genetic diagnosis for single-gene disorders (September 1996 to December 2001)

Results	n (%)
Patients	13
Cycles	20
Analyzed embryos	109
Normal embryos	34 (31)
Healthy carrier embryos	49 (45)
Embryos affected with investigated pathology	22 (20)
Embryos with no result	4 (4)
Transferred cycles	17 (80)
Clinical pregnancies	5*
per patient (%)	29
per transfer (%)	38
on-term	2
ongoing	3
Implantation rate (%)	21.7

*All preimplantation genetic diagnosis results have been verified by conventional prenatal diagnosis and confirmation was obtained in every case

Embryo transfer is generally performed on day 4 with the aim of: (1) selecting embryos according to both their chromosomal condition and morphology; (2) having enough time to expand the number of chromosomes diagnosed by using the FISH technique two or three times; and (3) decreasing the risk of damaging the biopsied embryo during the transfer due to more advanced compaction[12].

All pregnant patients are recommended to undergo prenatal diagnosis, taking into consideration the point that PGD patients are at high risk of developing aneuploid embryos; in addition, the diagnosis relies upon a single cell and this makes the error rate not negligible[13].

Overall results

Between September 1996 and December 2001, PGD for aneuploidy was performed on 634 conception cycles in our institute. The assay was informative for 3262

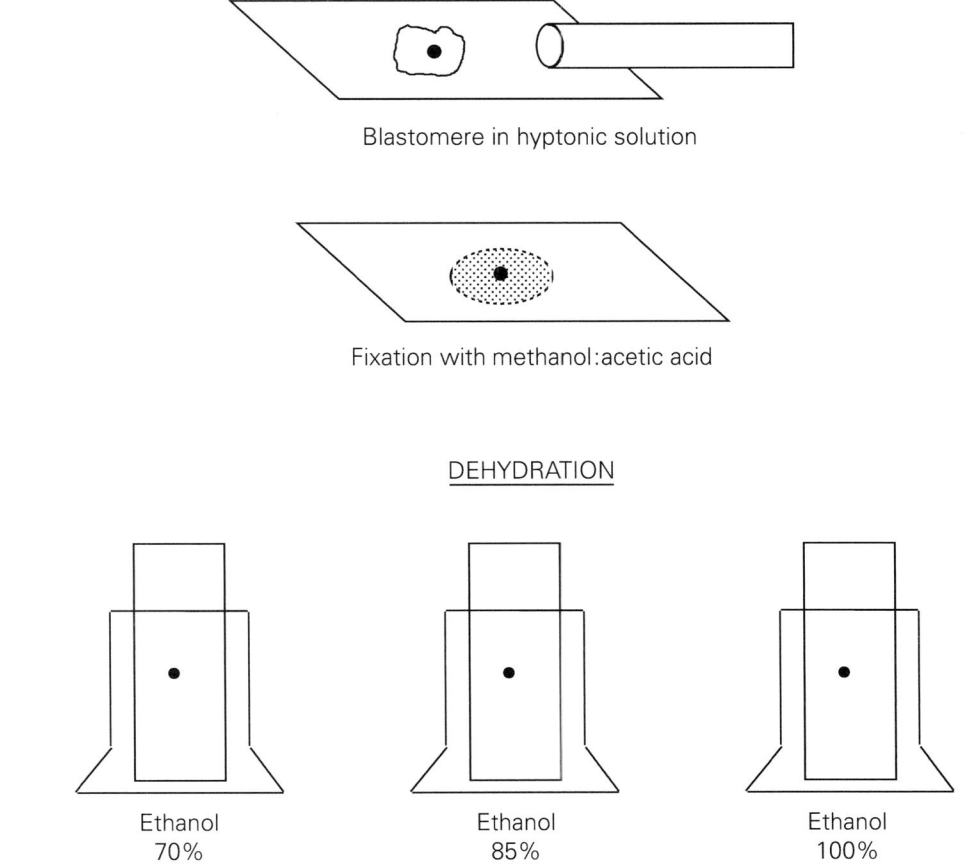

Blastomere in hyptonic solution

Fixation with methanol:acetic acid

DEHYDRATION

Ethanol
70%

Ethanol
85%

Ethanol
100%

Figure 2 The biopsied cell is lysed in hypotonic solution and the nucleus fixed on a glass slide using a methanol:acetic acid solution. Then, it is dehydrated in increasing ethanol series

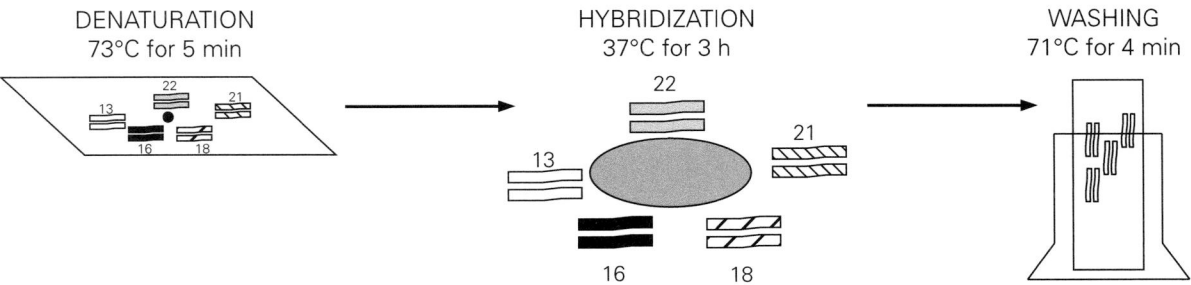

DENATURATION
73°C for 5 min

HYBRIDIZATION
37°C for 3 h

WASHING
71°C for 4 min

Figure 3 The hybridizing solution is put on the nucleus; denaturation follows and hybridization takes place for 3 h at 37°C. A washing step is then performed to eliminate the excess of unbound probe

embryos (99.1% out of 3290): 1042 exhibited a normal chromosomal complement (32%) and an average of 1.8 ± 0.7 euploid embryos were transferred to 418 cycles (66%), yielding 128 clinical pregnancies (31%) and an implantation rate of 22.3% (Table 3). Analysis of the data obtained permitted definition of the categories of patients where a high incidence of aneuploid embryos could be a major cause of reproductive failure.

Indications

Advanced maternal age was the first indication for PGD of aneuploidy based on the observation that a lower reproductive performance is strictly related to an increase in female age. This is probably due to the physiologic decline in the functional and structural quality of the oocyte with aneuploidy increasing proportionally with age[14]. In light of these findings, it was postulated that the selection for transfer of chromosomally normal embryos could reverse the age effect.

PGD was also proposed to other categories of younger patients with a poor prognosis for pregnancy with the aim of verifying whether these conditions could be attributed to a high proportion of chromosomally abnormal embryos[15]. The following indications were considered as inclusion criteria: (1) multiple failures in previous cycles despite the transfer of a suitable number of morphologically normal embryos; (2) an altered karyotype in one or both partners due to balanced translocations or inversions, or the presence of different cell lines of sex chromosomes (also defined as gonosomal mosaicism); and (3) an extremely severe male-factor condition, azoospermia, associated with at least one IVF failure. Other indications to PGD for aneuploidy are still under investigation in the effort of finding the possible cause of reproductive failure and hopefully improving the chances of full-term pregnancy in categories of patients who otherwise have a poor reproductive performance.

Table 4 depicts the results divided for indication to PGD: (1) maternal age ≥ 36 years; (2) ≥ 3 previous IVF failures; (3) an altered karyotype; and (4) MESA–TESE with ≥ 1 previous IVF failure.

Table 3 Preimplantation genetic diagnosis for aneuploidy: FISH and clinical results

Results	n (%)
Cycles	634
Age (mean ± SD)	36.8 ± 4.5
Previous IVF cycles (mean ± SD)	2.3 ± 2.1
FISH analyzed embryos	3290
FISH diagnosed embryos	3262 (99)
FISH normal	1042 (32)
FISH abnormal	2220 (68)
Transferred embryos (mean ± SD)	774 (1.8 ± 0.7)
Transferred cycles	418 (66)
Clinical pregnancies	128 (31)
on-term	92
abortions	19*
Implantation rate (%)	22.3

*Three ectopic pregnancies; one abortion after amniocentesis (the fetal karyotype was normal). FISH, fluorescence *in situ* hybridization; SD, standard deviation; IVF, *in vitro* fertilization

Table 4 Preimplantation genetic diagnosis for aneuploidy: results divided according to indications

Results	≥ 36 years n (%)	≥ 3 IVF cycles n (%)	Altered karyotype n (%)	MESA–TESE n (%)
Cycles	327	81	113	47
Age (mean ± SD)	39.8 ± 2.6	32.3 ± 2.4	34.9 ± 4.1	34.5 ± 4.7
Previous IVF cycles (mean ± SD)	2.6 ± 2.1	4.0 ± 1.4	1.0 ± 1.3	2.3 ± 1.9
FISH analyzed embryos	1717	435	589	212
FISH diagnosed embryos	1702	429	585	211
FISH normal	498 (29)	174 (41)	181 (31)	59 (28)
FISH abnormal	1204 (71)	255 (59)	404 (69)	152 (72)
Transferred embryos (mean ± SD)	364 (1.8 ± 0.7)	129 (2.1 ± 0.8)	147 (1.9 ± 0.8)	44 (1.8 ± 0.7)
Transferred cycles	202 (62)	62 (76)	76 (67)	25 (53)
Clinical pregnancies	62 (31)	20 (32)	31 (41)	6 (24)
on-term	43	16	23	5
abortions	9*	1†	6‡	0
Implantation rate (%)	21.7	21.7	29.2	18.1

*Two ectopic pregnancies; †one ectopic pregnancy; ‡one abortion after amniocentesis (fetal karyotype was normal). IVF, *in vitro* fertilization; MESA–TESE, microsurgical epididymal sperm aspiration–testicular sperm extraction; SD, standard deviation; FISH, fluorescence *in situ* hybridization

Maternal age

A total of 327 cycles underwent PGD because of a maternal age $\geq$ 36 years. In all, 1702 embryos were FISH diagnosed, resulting in 498 chromosomally normal embryos (29%), whereas the remaining 1204 (71%) carried chromosomal abnormalities which were not compatible with either implantation or a healthy pregnancy. Embryo transfer was cancelled in 125 (38% of the started PGD cycles) where no euploid embryos were diagnosed; an average of 1.8 $\pm$ 0.7 embryos were transferred to the remaining 202 cycles, yielding 62 clinical pregnancies (31%) with an implantation rate of 21.7%. Spontaneous abortion occurred in seven pregnancies (11%).

A proportional increase along with maternal age was observed in the incidence of chromosomal abnormalities when the results were arbitrarily divided into four classes of age: 36–37 years, 38–39 years, 40–42 years and $\geq$ 43 years. Accordingly, the pregnancy and implantation rates did not differ among the first three classes of age after PGD whereas a decrease was observed at older ages (Figure 4). These data suggest that the transfer of FISH selected embryos promotes the overcoming of the age factor by giving patients aged between 36 and 42 years the same reproductive performance as younger women[4]. At older ages other factors such as a compromised uterine receptivity or the involvement of other chromosomes in meiotic or mitotic errors contribute to very low chances of pregnancy.

Preliminary data about the frequency of the different chromosome aneuploidies in relation to maternal age have shown that not all the chromosomes studied vary proportionally with age. As reported in a previous study, aneuploidies of chromosomes 15, 21 and 22 increase proportionally with age, chromosomes 1 and 17 are lower in the youngest category, whereas chromosomes 13, 14, 16, 18, X and Y show similar variations irrespective of age[16]. These data suggest that segregation errors could occur at different rates for each chromosome. Extension of this study to other chromosomes, as well as to younger women, could result in identifying those chromosomes, if any, whose numerical variations have a deleterious effect on embryo viability and implantation.

Repeated IVF failures

Young patients who failed to achieve a pregnancy after $\geq$ three IVF cycles have chromosomal abnormalities in more than half of their *in vitro* generated embryos (Table 4). Embryo transfer was performed in 62 out of the 81 PGD started cycles (76%) with 20 clinical

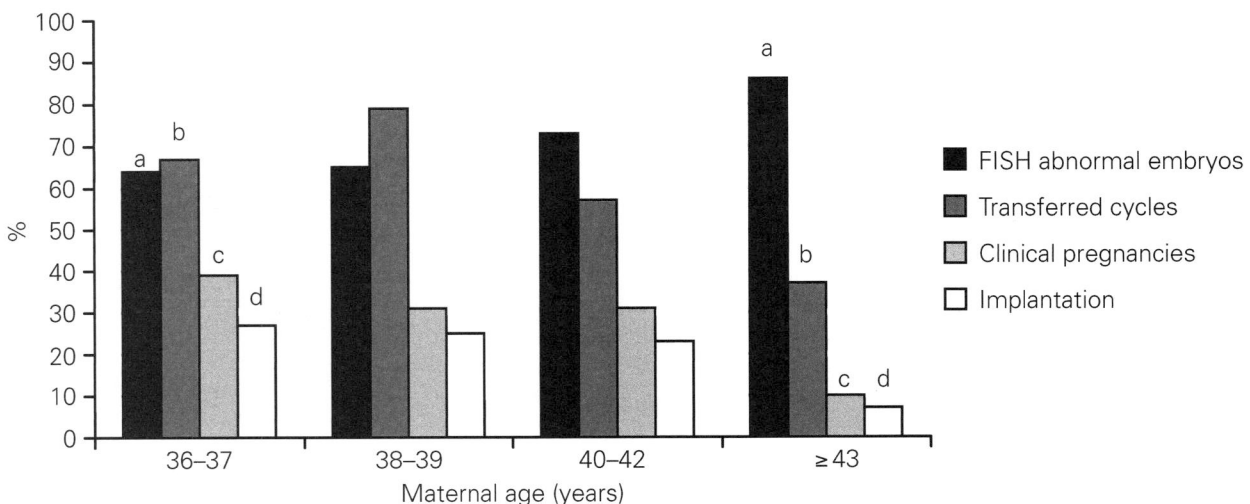

Figure 4 Preimplantation genetic diagnosis: results divided according to maternal age. a, $p < 0.001$; b, $p < 0.005$; c, d, $p < 0.05$. FISH, fluorescence *in situ* hybridization

pregnancies (32%) and an implantation rate of 21.7%. The analysis of the detected abnormalities revealed that mosaicism, haploidy and polyploidy occurred with the highest frequency in this category of PGD patients. Interestingly, the proportion of polyploid embryos was significantly higher (11%) than in other patients' categories undergoing PGD and was independent of the insemination technique (intracytoplasmic sperm injection or conventional IVF). Provided that embryos developed only from regularly fertilized zygotes were selected for PGD, failed cytokinesis after DNA synthesis could be the cause of this condition. This finding and the concomitant high frequency of mosaicism and haploidy suggest a dysfunction in the processes or structures entering cell division.

Altered karyotype

Patients with an altered karyotype due to structural abnormalities or gonosomal mosaicism are recommended to undergo PGD for aneuploidy. In the case of balanced translocations, a high proportion of chromosomes can segregate at meiosis, yielding a significant imbalance that generates monosomy or trisomy in the resulting embryo (Figure 5). A high reproductive risk characterizes carriers of gonosomal mosaicism if the

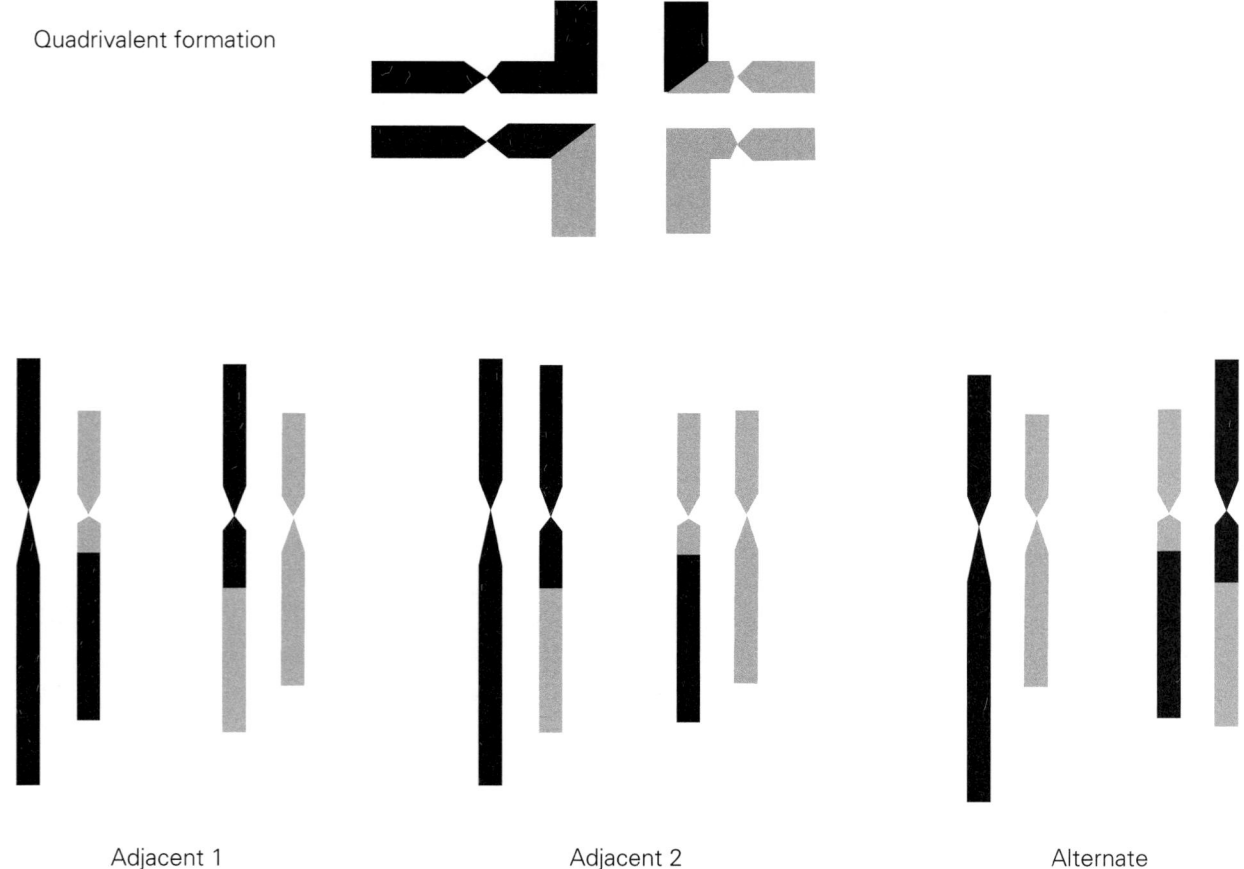

Figure 5 Reciprocal translocations: quadrivalent formation and segregation at meiosis. Six of the most commonly resulting gametes are represented: adjacent 1 and adjacent 2 segregations yield unbalanced gametes, whereas alternate segregation generates gametes with a normal or a balanced chromosomal analysis

Table 5 Preimplantation genetic diagnosis: results in patients with an altered karyotype due to balanced translocations and gonosomal mosaicism

Results	Balanced translocations n (%)	Gonosomal mosaicism n (%)
Cycles	51	54
Age (mean ± SD)	35.3 ± 4.5	34.6 ± 3.6
Generated embryos	304	361
FISH analyzed embryos	259	298
FISH diagnosed embryos	258	295
FISH normal	69 (27)	109 (37)
FISH abnormal	189 (73)	186 (63)
Transferred embryos (mean ± SD)	1.9 ± 0.8	2.2 ± 0.8
Transferred cycles	28 (55)	39 (72)
Clinical pregnancies	11 (39)	19 (49)
on-term	8	16
abortions	3*	2
Implantation rate (%)	27.4	33.7

*One abortion after amniocentesis (fetal karyotype was normal). SD, standard deviation; FISH, fluorescence *in situ* hybridization

Table 6 Preimplantation genetic diagnosis: results in patients with repeated abortions

Results	n (%)
Cycles	580
Age (mean ± SD)	37.6 ± 4.4
FISH analyzed embryos	437
FISH diagnosed embryos	435 (99)
FISH normal	125 (29)
FISH abnormal	310 (71)
Transferred embryos (mean ± SD)	96 (1.8 ± 0.7)
Transferred cycles	54 (67)
Clinical pregnancies	20 (37)
abortions	3
Implantation rate (%)	26.0

SD, standard deviation; FISH, fluorescence *in situ* hybridization

gonosomal aneuploidy was detected; however, monosomy and trisomy appeared as the most common defects, suggesting a predisposition in these patients to generate aneuploid cell lines[4].

Microsurgical epididymal sperm aspiration–testicular sperm extraction

Table 4 shows the data of 47 PGD cycles performed on azoospermic patients who had failed at least one conception cycle. Although a very high percentage of chromosomally abnormal embryos (72%) was detected, the transfer of PGD selected embryos did not exceed a 24% clinical pregnancy rate with an implantation rate of 18.1%. In a recently published study by Gianaroli and co-workers[17], a higher incidence of monosomies and trisomies (45%) was detected in embryos derived from microsurgical epididymal sperm aspiration–testicular sperm extraction (MESA–TESE) compared to those derived from normospermic patients (27%); in addition, the rate of gonosomal aneuploidy increased proportionally with the severity of the male-factor condition, with a frequency of 7.9% in embryos derived from MESA–TESE patients. Therefore, a significant involvement of the male gamete in the etiology of these alterations can be considered as probable. In view of these results, the analysis by PGD of the embryos generated by MESA and TESE should perhaps be taken into consideration in future in the case of general prevention in reproductive medicine.

mosaic cell lines detected in their peripheral blood are also present in the gonads. In both situations, the incidence of chromosomal abnormalities in the analyzed embryos was greater than 60%, with monosomies and trisomies representing the most frequent defects.

Table 5 shows the results obtained by PGD applied to carriers of translocations (51 cycles) and gonosomal mosaicism (54 cycles). Seventy-three percent of the diagnosed embryos in patients with translocations were chromosomally abnormal and embryo transfer was achieved in 28 cycles (55%), resulting in 11 clinical pregnancies (39%) and an implantation rate of 27.4%. These results demonstrate that, due to the high incidence of chromosomal abnormalities, the number of embryos available for chromosomal analysis is crucial for the chances of success in these patients.

The reproductive risk related to gonosomal mosaicism cannot be established in advance and strictly depends on the possible involvement of the gonads in the production of aneuploid cell lines. As shown in Table 5, 63% of the analyzed embryos were chromosomally abnormal. The clinical outcome was very favorable with 19 clinical pregnancies in 39 cycles transferred (49% pregnancy rate). No increase in the incidence of

Other indications

Other indications to PGD for aneuploidy are currently under investigation. Among them, women with a poor response to hormonal stimulation have been included in the effort of verifying whether the low implantation rate in these patients depends on a chromosomal imbalance. The results obtained demonstrated percentages of chromosomally abnormal embryos as high as 68%, implying that more than two-thirds of the few embryos available in these patients are not compatible with implantation[18]. Moreover, the prevalence of monosomies and trisomies exposes these patients, in the case of implantation, to an increased risk of spontaneous abortion and trisomic pregnancies.

Repeated cases of spontaneous abortions have been associated with chromosomal abnormalities. As shown in Table 6, the data derived from PGD on a series of 80 cycles demonstrated a notable percentage of chromosomally abnormal embryos (71%). The transfer of euploid embryos was possible in 54 patients with 20 clinical pregnancies (37%) and an implantation rate of 26.0%; only three ended in spontaneous abortion, indicating that aneuploidy can represent the cause of repeated miscarriages. These results are encouraging, but additional data and a comparative study are required to evaluate the proposed validity of PGD for aneuploidy in this category of patients.

CONCLUSIONS

The results obtained after 10 years of PGD in human IVF lead to the conclusion that this method is a reliable procedure in preventing birth of affected children. In addition, aneuploidy screening is effective in determining a high take-home baby rate when applied to categories of patients where aneuploidy is the main cause of reproductive failure: women of an advanced reproductive age and patients with an altered karyotype due to translocations or gonosomal mosaicism[4]. In these cases, the transfer of euploid embryos provides patients with an extremely poor prognosis of full term pregnancy a reasonable hope of undertaking an assisted conception cycle with the same chances as young, normally karyotyped couples.

Future developments are the subject of a continuous effort aimed at improving the performance of PGD and expanding its applications. New information coming from the knowledge of individual genes and their association with the corresponding product provides the basis toward a wider and wider application of PGD for single-gene disorders.

The new impulse toward the potential use of this technique derives from its applications in special conditions. Indeed the possibility has been proposed of selecting embryos which are less exposed to the onset of disorders later in life. This approach has been adopted for carriers of the p53 suppressor gene and similar mutations are under evaluation. It is also feasible to select embryos according to their HLA haplotype with the aim of matching a sister or brother requiring HLA-compatibility for bone marrow transplantation. These examples give an idea of the potential applications of PGD which go greatly beyond the screening of gene mutations or aneuploidy, toward a general prevention of genetic disorders. Some ethical implications certainly need to be discussed, keeping in mind that the major goal is to provide the best of our own scientific knowledge and technical possibilities for the birth of healthy children.

References

1. Edwards RG, Gardner RL. Sexing of live rabbit blastocysts. *Nature (London)* 1967;214:576–7
2. Handyside AH, Kontogianni EH, Hardy K, *et al*. Pregnancies from biopsied human preimplantation embryos sexed by Y-specific DNA amplification. *Nature (London)* 1990;344:768–70
3. Handyside AH, John G, Lesko MS, *et al*. Birth of a normal girl after *in vitro* fertilisation and preimplantation diagnostic testing for cystic fibrosis. *New Engl J Med* 1992;327:905–10
4. Gianaroli L, Magli MC, Ferraretti AP, *et al*. Preimplantation diagnosis for aneuploidies in patients undergoing *in vitro*

fertilization with a poor prognosis: identification of the categories for which it should be proposed. *Fertil Steril* 1999;72:837–44

5. Munné S, Magli MC, Cohen J, *et al.* Positive outcome after preimplantation diagnosis of aneuploidy in human embryos. *Hum Reprod* 1999;14:2191–9

6. Findlay I, Quirke P. Fluorescent polymerase chain reaction: Part I. A new method allowing genetic diagnosis and DNA fingerprinting of single cells. *Hum Reprod Update* 1996;2: 137–52

7. Kuliev A, Rechitsky S, Verlinsky O, *et al.* Preimplantation diagnosis for thalassemias. *J Assist Reprod Genet* 1998;15: 253–7

8. International Working Group on Preimplantation Genetics. Preimplantation genetic diagnosis – experience of three thousand clinical cycles. *Reprod BioMed Online* 2001;3: 49–53

9. Rechitsky S, Strom C, Verlinsky Y. Accuracy of pre-implantation diagnosis of single-gene disorders by polar body analysis of oocytes. *J Assist Reprod Genet* 1999;16: 192–8

10. Verlinsky Y, Kuliev A. *An Atlas of Preimplantation Genetic Diagnosis.* London: Parthenon Publishing, 2000

11. Munné S, Magli MC, Bahçe M, *et al.* Preimplantation diagnosis of the aneuploidies most commonly found in spontaneous abortions and live births: XY 13, 14, 15, 16, 18, 21, 22. *Prenat Diagn* 1998;18:1459–66

12. Gianaroli L, Magli MC, Munné S, *et al.* Advantages of day 4 embryo transfer in patients undergoing preimplantation genetic diagnosis of aneuploidy. *J Assist Reprod Genet* 1996; 16:170–5

13. Gianaroli L, Magli MC, Ferraretti AP. The *in vivo* and *in vitro* efficiency and efficacy of PGD for aneuploidy. *Mol Cell Endocrinol* 2001;183:13–18

14. Munné S, Alikani M, Tomkin G, *et al.* Embryo morphology, developmental rates, and maternal age are correlated with chromosomal abnormalities. *Fertil Steril* 1995;64:382–91

15. Gianaroli L, Magli MC, Munné S, *et al.* Will preimplantation genetic diagnosis assist patients with a poor prognosis to achieve pregnancy? *Hum Reprod* 1997;12:1762–7

16. Gianaroli L, Magli MC, Ferraretti AP. The role of pre-implantation diagnosis for aneuploidies. *Reprod BioMed Online* 2002;4(Suppl. 3):31–6

17. Gianaroli L, Magli MC, Ferraretti AP, *et al.* Preimplantation diagnosis after assisted reproduction techniques for geneti-cally determined male infertility. *J Endocrinol Invest* 2000;23:711–16

18. Gianaroli L, Magli MC, Ferraretti AP, *et al.* Gonadal activity and chromosomal constitution of *in vitro* generated embryos. *Mol Cell Endocrinol* 2000;161:111–16

Micromanipulation techniques in human IVF

Micromanipulation techniques in human infertility: PZD, SUZI, ICSI, MESA, PESA, FNA and TESE

23

Filippo Ubaldi and Laura Rienzi

INTRODUCTION

Over the last two decades, *in vitro* fertilization (IVF) has been used to provide a therapy for couples with tubal, idiopathic or male-factor infertility. Men with reduced semen quality have a very poor prognosis for conception *in vivo*, and also with IVF the incidence of fertilization is reduced compared to that obtained in couples with tubal infertility[1]. Moreover, an absence of fertilization may occur in about one third of cycles. In these cases, the zona pellucida, a glycoprotein matrix surrounding mammalian oocytes and which performs extremely important functions, represents an insuperable obstacle. During fertilization, the zona pellucida is a mechanical and selective barrier that only capacitated spermatozoa are able to cross[2,3]. Once gamete fusion has occurred, it serves as a confined shelter in which the embryo is protected from microorganisms, viruses and immune cells, and where it can proceed up to morula compaction without blastomere loss. In patients with severe defects in semen quality and sperm function, the complex fertilization processes cannot occur and zona micro-assisted fertilization has the potential to provide a mode of therapy for their infertility.

After almost four decades of manipulation of animal gametes, the first clinical application of micro-insemination techniques to obtain fertilization *in vitro* was reported in 1985[4]. The insertion of a single sperm into the perivitelline space with subsequent fertilization was reported by Laws-King and co-workers[5]. A year later several reports were presented suggesting different approaches to micro-assisted fertilization. Lanzerdorf

and co-workers[6] injected sperms directly into the ooplasm. Tyrode solution was used for 'drilling' a hole in the zona pellucida (zona drilling, ZD) of the human oocytes before transferring them to a suspension of sperm[7]. Cohen and co-workers[8] in 1989 reported on a mechanical technique (partial zona dissection, PZD) which causes rupture of the zona pellucida; this is achieved by piercing opposite ends of the zona pellucida with a microneedle followed by rubbing the latter against the holding pipette until it tears. In the same year Bongso and co-workers[9] reported a pregnancy after transfer of multiple sperms under the zona pellucida (subzonal sperm insemination, SUZI). Despite disappointing clinical results (1 to 5% pregnancy rates) it became the technique of choice for the treatment of severe male-factor infertility until the introduction of intracytoplasmic sperm injection (ICSI) which involves injection of a single spermatozoon through the zona pellucida directly into the oocyte[10]. It became immediately obvious that ICSI resulted in higher fertilization and cleavage rates than did other assisted fertilization procedures such as partial zona dissection and subzonal insemination[11,12] that have therefore been almost abandoned.

Potentially there is a large patient population who may benefit from gamete micromanipulation techniques and in particular from ICSI. The main group of patients is represented by the male-factor population with quantitative and qualitative sperm anomalies. Men with severe oligoasthenoteratozoospermia (less than 5 million spermatozoa and/or less than 500 000 motile spermatozoa), with acrosomeless spermatozoa (globozoospermia)[13–15], with positive antisperm antibody tests[16] and with complete asthenozoospermia, are candidates for gamete micromanipulation techniques. With regard to the latter, although pregnancies and births have been reported with totally immotile spermatozoa[17,18] the fertilization rate is very low. Microinjection of completely immotile spermatozoa is an adverse prognostic factor[19] and the use of hypo-osmotic tests[20] or the use of testicular spermatozoa[21] should be considered. Male infertility due to obstructive azoospermia (congenital bilateral absence of the vas deferens or failed vasostomy or vasoepididymostomy), in which it is possible to recover motile sperms after microsurgical[22] or percutaneous[23] epididymal sperm aspiration (PESA) procedures or after testicular sperm extraction (TESE)[24], percutaneous fine needle aspiration (FNA)[25] or microsurgical testicular sperm extraction[26,27], can be successfully treated by gametes micromanipulation techniques. Moreover, ICSI can be successfully used in patients with non-obstructive azoospermia in which some slightly motile or immotile sperms are obtained after surgical or microsurgical testicular sperm extraction procedures[28,29]. Another group of patients are men with apparently normal or borderline semen who are unable to fertilize *in vitro*. It has been shown that a couple who fails to achieve fertilization following a single IVF treatment has a < 25% chance of fertilization when conventional IVF is attempted again[30]. These patients are difficult to diagnose because generally they have normal semen analyses, no history of urologic disease and negative antisperm antibody tests. Couples with unexplained infertility may achieve significantly higher fertilization rates using ICSI, compared to conventional IVF, with sibling oocytes. In this group of patients complete fertilization failure would have occurred after IVF in one out of four couples if ICSI had not been performed in some sibling oocytes[31,32]. Micromanipulation techniques are also successfully used in the non-male factor infertility population. Recently, it has been suggested that in those cases where very few oocytes are available the ICSI procedure guarantees better results than conventional IVF in non-male factor infertility couples[33]. ICSI is also used to fertilize cryo-thawed mature oocytes[34,35] since the cryopreservation technique can affect the zona

pellucida deleteriously, hindering normal sperm attachment and penetration. *In vitro* maturation of immature oocytes was recently applied clinically and it was found that ICSI was the technique of choice to achieve fertilization of these oocytes[36–39]. The fertilization of oocytes prior to preimplantation genetic diagnosis is another absolute indication to the use of ICSI, since the DNA of spermatozoa stuck to the zona pellucida may contaminate the biopsy[40]. Moreover, high fertilization and pregnancy rates can be achieved with ICSI in patients with disordered zona pellucida-induced acrosome reaction[41].

It has even been suggested that ICSI should completely replace conventional IVF in the future, as pregnancy rates could be improved if normal spermatozoa are injected. However, because of the novelty of the technique, it is currently recommended that, until more data become available, with an apparently normal sperm sample ICSI should not be used in a first cycle. When there is reason to suspect poor fertilization, ICSI can be used in combination with conventional IVF in a split cycle[40].

ZONA DRILLING, PARTIAL ZONA DISSECTION AND SUBZONAL SPERM INSEMINATION

After a few reports of oocyte fertilization with the insertion of spermatozoa in the perivitelline space[5] or directly into the ooplasm[6], the first widely used micromanipulation techniques applied on human oocytes consisted of creating a hole in the zona pellucida to facilitate the spermatozoon to gain access to the oocyte membrane. The zona pellucida was opened by creating a hole chemically (by ZD) applying acidified culture medium to a localized area of the oocyte investment[7]. Because acid medium can damage oocytes or activate them parthenogenetically if it gains access to the oolemma, the oocytes were first exposed to hyperosmotic sucrose solution to shrink them and to increase the perivitelline space thereby facilitating the procedure. This additional treatment could, on the other hand, damage the oocyte ooplasm and should for this reason be avoided. The partial digestion of the zona was also obtained with trypsin or pronase solutions[42]. Alternatively, the hole was created by using a sharpened needle that was passed through and out of the zona pellucida and used to rub the zona against the holding pipette (PZD)[8,43]. It has also been suggested that the hole in the zona pellucida could be created by photo-ablation[44]. Oocytes treated with ZD or PZD were then incubated in culture medium containing a suspension of selected spermatozoa.

In 1988, Bongso and co-workers[9] reported a pregnancy after transfer of multiple sperms under the zona pellucida (SUZI) into the oocyte perivitelline space and several other reports on the application of this technique were published subsequently[45-47]. In this case the zona pellucida was not opened, but was functionally and physically by-passed. To perform the SUZI procedure a micropipette containing one or more spermatozoa was pushed through the zona pellucida into the perivitelline space of the oocyte where the spermatozoa were released. The number of inserted spermatozoa ranged from one to ten. The amount of fluid inserted was kept to a minimum and the oocytes were then incubated in sperm-free medium.

The PZD and SUZI techniques have in common that spermatozoa, whose chances of penetrating the zona pellucida would have otherwise been remote, are given the opportunity to bind to the oolemma. The fertilization and clinical pregnancy rates obtained were nevertheless very low (approximately 20% and 5%, respectively). The limiting factors of these techniques are the requirement for many functional spermatozoa with good progressive motility and the presence of an intact acrosome that does not allow the exposure of the sperm binding sites. Furthermore, these techniques together with a poor fertilization rate resulted in a higher incidence of multiple sperm penetration. For these reasons, immediately after the advent of ICSI which resulted in high fertilization and cleavage rates, ZD, PZD and SUZI were almost abandoned.

INTRACYTOPLASMIC SPERM INJECTION

Although Lanzerdorf and co-workers[6] first reported the insertion of spermatozoa into the ooplasm obtaining oocyte fertilization, Palermo and co-workers[10] reported the first pregnancy with a very effective micro-manipulation technique which consisted of the direct injection of a single spermatozoon into the oocyte cytoplasm. High fertilization and embryo development as well as high clinical pregnancy rates had been achieved since the introduction of the ICSI technique in Brussels[11,12] and in many IVF clinics all over the world. The capability of the injecting spermatozoa to achieve decondensation and male pronucleus formation in the oocyte cytoplasm has offered new possibilities for treating severe male-factor infertility.

TECHNOLOGY AND INSTRUMENTATION

The correct settings and use of the instruments necessary for the micromanipulation procedure are of crucial importance to the results of the technique. The quality of the microneedles, physical properties of the glass, and the inner and outer diameter of the wall are very important to minimize oocyte damage during the manipulation. Two types of micropipette are necessary for the injection procedure: an injection pipette and a holding pipette. The pipettes have to be pulled, beveled, and micro-forged to achieve the desired geometry. The shape of the pipettes is related to their function. The diameter of the injection pipette has to be as small as possible (5–7 μm) to allow sperm head aspiration and minimize the oocyte membrane damage. The zona passage and the piercing of the oolemma can be facilitated by the presence of a spike. The holding pipette has an outer diameter of 60–80 μm and an inner diameter of 10–20 μm. The size of the pipette allows the oocyte to be held by applying just a gentle suction on the zona pellucida. Both needles are bent to an angle of 30–45° in order to allow an angle of injection parallel to the plane of focus. Joris and co-workers[48] described in detail the preparation of the microneedles. This aspect should not be underestimated; the preparation of good quality pipettes requires personal skill and equipment. Alternatively, there are a number of companies supplying good quality micromanipulation pipettes. A second essential tool is the microscope itself. The microscope has to be equipped with a heated stage, correct optics, manipulators and microinjectors. To be safe, the micromanipulation procedure has to be carried out on a heated stage at 37°C. Human oocytes are sensitive to temperature variation and transient cooling of oocytes to room temperature can cause irreversible damage to the meiotic spindle[49]. To perform a proper evaluation of the oocytes and of the procedure itself the micromanipulation has to be carried out on an inverted microscope at × 200 or × 400 magnification equipped with Hoffman modulation contrast. This optics provides a greatly differentiated view. Two manipulators (one for the holding and one for the injection) are necessary to move the pipettes in the x-, y- and z-axes. These manipulators can be electrical or hydraulic and permit coarse and fine movements. To perform aspiration and injection, the holding and the injection pipettes are fitted to a toll holder and are connected to a micrometer type microinjector. The microinjectors are filled with oil. Air bubbles reduce the sensitivity of the system and for this reason should be avoided.

To prevent vibration that could interfere with the injection procedure, the whole set-up should be placed on an anti-vibration table.

OOCYTE TREATMENT FOR INTRACYTOPLASMIC SPERM INJECTION

After oocyte retrieval the cumulus–corona–oocyte complexes (CCOCs) are denudated from cumulus and corona cells to assess nuclear maturity, cytoplasmic morphology and to facilitate handling during the micromanipulation procedure. The denudation process is performed using a combination of enzymatic and mechanical methods. The oocytes are first exposed to culture medium containing hyaluronidase fraction VIII. To enhance enzymatic removal of the cumulus–corona cells, the oocytes are aspirated in and out of a hand-drawn Pasteur pipette with an inner diameter of approximately 135 μm. The diameter of this cumulus-removal pipette is very important. Too small a pipette would strongly compress the oocyte during the aspiration and damage the oocyte cytoskeleton. The oocytes should be exposed to the lowest enzyme concentration and for a very short time. Long exposure of the oocyte to high concentrations of hyaluronidase before the ICSI procedure leads to a high rate of parthenogenetic activation. Van Steirteghem and co-workers[50] were able to decrease the parthenogenetic activation rate from 17% to 3% by lowering the hyaluronidase concentration from 760 IU/ml to 78 IU/ml. To decrease the possible toxic effects from the hyaluronidase on the ICSI outcome it has been suggested that the enzyme concentration to which the oocyte is exposed can be further decreased to 10 IU/ml[51].

Activation of the oocyte might also be caused by the pipetting procedure[47] or mechanical picking of the oocyte[52,53]. By using a pipette with a large diameter (< 1000 μm) for the first step of oocyte denudation it is possible to minimize the mechanical stress and the time of exposure of the oocyte to the denudation process[51]. The timing of the cumulus–corona cell removal has no effect on the ICSI outcome, suggesting that the oocyte-surrounding cells are not necessary for survival, fertilization and embryo development after ICSI and can be removed at any time before the injection[54]. On the other hand we have observed that a preincubation period between oocyte retrieval and ICSI can improve the fertilization rate[55]. This period might be necessary for some oocytes to reach full cytoplasmic maturity, leading to higher activation rates after microinjection.

The nuclear maturity of the denudated oocytes is evaluated before the ICSI procedure. The oocytes are considered to be meiotically mature after extrusion of the first polar body (metaphase II). Very often a non-synchronous oocyte maturation after controlled ovarian hyperstimulation is observed. Approximately 85% of the denudated oocytes are in metaphase II stage (MII) and are immediately suitable for ICSI. The immature collected oocytes can be at metaphase I stage (MI) (about 5%) or germinal vesicle stage (GV) (about 10%). These oocytes could be matured *in vitro* to metaphase II stage prior to becoming available for microinjection. It has been reported that approximately 25–30% of MI oocytes mature in 1–4 h after ovum pick-up[38,56] whereas 50–70% of GV oocytes mature in 30 to 32 h[36]. Attempts have been made to perform ICSI on the *in vitro* matured oocytes from either MI stage[37,38] or GV stage[36,39] resulting in pregnancies and births of healthy babies. Nevertheless, the fertilization ability of these oocytes seems significantly lower compared to the normal fertilization rate obtained with *in vivo* matured oocytes[38]. This observation can be explained by cytoplasmic immaturity of these oocytes despite their nuclear maturation. Moreover, the developmental capacity of the derived embryos has to be evaluated. It has been recently shown[57] that there is a high incidence of genetic abnormalities in the embryos derived from the matured *in vitro* oocytes This observation should be considered before wide clinical application of the injection of matured MI and GV oocytes, and further studies are needed to determine the safety of this technique.

Oocyte morphology can also be evaluated after cumulus–corona cell removal. Dark zona pellucida, large perivitelline space, vacuoles, refractile bodies and an irregular shape can be observed. The correlation between these abnormal oocyte morphotypes, fertilization rates and embryo development is still not clear. Some authors suggested that all oocytes could be fertilized by ICSI independently of their morphologic peculiarities at light microscopy[58,59]. Furthermore, similar embryo quality and viability was observed amongst all the oocyte morphotypes[59] (with the only exception of oocytes displaying vacuoles)[58]. Conversely, other authors[60] reported that although normal fertilization and early embryo development can be achieved in oocytes with abnormal cytoplasmic morphology, the resulting embryos have a lower implantation potential. Moreover, Xia[61] showed that oocyte grading based on the triple factors (first polar body, size of perivitelline space and cytoplasmic inclusions) is correlated with fertilization rates and embryo quality after ICSI. The difference between these studies may be related to the different criteria used to evaluate oocyte morphology. Furthermore, clinical parameters of the patients (age, factor of infertility, total number of oocytes retrieved and estradiol concentration on the day of human chorionic gonadotropin (hCG) administration) seem to be important for oocyte maturity and morphology[61–65].

SEMEN EVALUATION AND PREPARATION FOR INTRACYTOPLASMIC SPERM INJECTION

Semen samples are collected at the time of oocyte retrieval by masturbation after 3–5 days of sexual abstinence and are allowed to liquefy for at least 20 min at 37°C. Evaluation and sperm preparation for ICSI is performed by routine methods. Semen concentration and motility are assessed according to World Health

Organization criteria[66]. Semen morphology is performed according to Kruger's strict criteria[67]. The ICSI procedure requires no specific pretreatment of spermatozoa[68,69]. The sperm sample is first washed by centrifugation to remove the seminal fluid and then layered onto a two- or three-layer discontinuous gradient and centrifuged. This method of preparation permits the selection of predominantly motile sperm cells[70], and increases the percentage of sperm cells with normal morphology[71]. Just before the injection procedure, the concentration of the sperm suspension is, when possible, adjusted to 1×10^6/ml. In the event of a very low sperm count, the semen is simply concentrated by washing twice. It has been recently shown[72] that gradient centrifugation can be omitted in sperm preparation without affecting ICSI results in terms of fertilization and cleavage rates. The selection of morphologically normal motile spermatozoa is in fact performed in a polyvinylpyrrolidone (PVP) droplet at $400 \times$ magnification by the operator just prior to the injection. It has been suggested that because of this selection the outcome of ICSI is not related to any of the three basic sperm parameters (density, motility and morphology), thus the only criterion related to the success of the procedure is the presence of at least one living spermatozoon per oocyte in the sperm suspension[19,73]. This observation has led to the use of ICSI for fresh or frozen–thawed epididymal and testicular sperm[22,28,74].

If only immotile spermatozoa are available in the ejaculate, the fertilization rate after ICSI is considerably compromized[17–19]. Sperm movement is in fact the best sign of viability, allowing the selection of living sperm cells for the injection. Nevertheless, the lack of movement does not necessarily mean cell death. Patients with immotile cilia syndrome frequently have 100% immotile spermatozoa. In this case a viability test (for instance the eosin Y solution test) should be performed in advance to evaluate the chances of choosing a viable spermatozoon during ICSI. In case of the presence of viable spermatozoa in the suspension, these can be selected one by one with the hypo-osmotic swelling test (HOST)[20]. This test is particularly appropriate for these ICSI cases because it does not alter the spermatozoa and the selection can be made directly in the sperm suspension from which the spermatozoon is aspirated into the injection needle[75]. In the cases of total necrozoospermia (total absence of living sperm) testicular sperm extraction has been recommended. By collecting the spermatozoa directly from the testis, acceptable fertilization and clinical pregnancy rates can be obtained[21].

SPERM RETRIEVAL AND PREPARATION IN AZOOSPERMIC PATIENTS

Approximately 2% of infertile couples[76] and about 10% of men seeking fertility evaluation[77] have azoospermia. Azoospermia can be classified as 'obstructive' and 'non-obstructive'. Obstructive azoospermia could be due to vasectomy, congenital absence of the vas deferens, accidental surgical interruption of the vas deferens or epididymis during a hernia or hydrocele operation or epididymal infections. In these cases spermatogenesis is normal and usually spermatozoa are abundant in the epididymal or testicular tubuli permitting a high rate of retrieval success either with microsurgical epididymal sperm aspiration (MESA), PESA, FNA or with surgical or microsurgical TESE. In non-obstructive azoospermia or primary testicular failure, severe defects in spermatogenesis are present and the most common histologic patterns in these patients are hypospermatogenesis maturation arrest and Sertoli cell-only syndrome with or without focal spermatogenesis. In the non-obstructive cases the epididymis is devoid of spermatozoa and sometimes only a few foci with spermatogenesis may be found in the testicles. Failure to find spermatozoa after surgical or microsurgical sperm extraction or after FNA may occur and the spermatozoa failure rate mainly depends on the histologic patterns[78–80].

EPIDIDYMAL SPERM ASPIRATION AND PREPARATION FOR INTRACYTOPLASMIC SPERM INJECTION

In the case of obstructive azoospermia the spermatozoa can be collected directly from the epididymis[22]. The epididymal sperm aspiration can be carried out either by a MESA or PESA procedure without surgical scrotal exploration[23].

The microsurgical procedure can be performed using either general or local anesthesia (spermatic cord infiltration with 0.5% marcaine). After having performed anesthesia, the scrotal contents are extruded through a small incision, the tunica vaginalis is opened and the epididymis is exposed. Using × 10–40 magnification with an operating microscope a tiny incision is made with microscissors in the epididymal serosa to expose first the tubules in the distal head region of the epididymis. After hemostasis, using bipolar coagulation, an epididymal tubule is opened and spermatozoa are aspirated directly from the opening of the epididymal tubule by means of a disposable tip from an intravenous cannula mounted on a 1 ml syringe filled with 0.1 ml of culture medium supplemented with human serum albumin. The epididymal fluid is immediately examined for sperm motility and quality of progression. Typically only a few microliters of epididymal fluid need to be retrieved, because sperm in the epididymal fluid are highly concentrated. If the sperm motility is poor another aspiration is performed proximally to the epididymal caput where spermatozoa with the greatest motility are present. The sperm fractions collected are then treated in the same way as ejaculated semen. In the case that no motile spermatozoa are recovered after several aspirations from the epididymis on either side, a testicular specimen is taken for sperm recovery (see below).

Epididymal sperm aspiration can also be performed percutaneously (PESA) without surgical scrotal exploration[23]. In this case the testis is stabilized and the epididymis is held between the surgeon's thumb and forefinger. A 21-gauge butterfly needle attached to a 20 ml syringe is inserted into the caput of the epididymis and withdrawn gently until fluid can be seen entering the butterfly needle tubing. The procedure is repeated until adequate numbers of sperm are retrieved. This technique offers several advantages: it does not require an operating microscope or expertise in microsurgery, it is repeated easily at low cost and in the case where sperm is not retrieved with PESA, open sperm retrieval or percutaneous testicular sperm aspiration is possible.

When it is possible to retrieve spermatozoa, the microsurgical or percutaneous approach has the advantage of reliable retrieval of large numbers of epididymal spermatozoa for immediate use, as well as for cryopreservation.

TESTICULAR SPERM EXTRACTION/ ASPIRATION AND PREPARATION FOR INTRACYTOPLASMIC SPERM INJECTION

In obstructed azoospermic patients when epididymal sperm aspiration procedures fail, or in non-obstructive azoospermia where the epididymal sperm can never be retrieved because the wall has collapsed, spermatozoa can be retrieved from the testis either by the FNA technique[25] or by surgical or microsurgical TESE[26–29]. Percutaneous puncture and aspiration of the testicular tissue is performed as for PESA using a 21-gauge needle connected to a 20 ml syringe filled with 0.1 ml of culture medium. The technique is similar to the previous one described. With a single puncture the needle is inserted into the testis and suction is applied with the syringe. The needle is then moved gently inside the testis until fluid is seen entering the butterfly needle tubing which is then emptied into a Petri dish and observed under the inverted microscope at × 200–400 magnification to assess the presence of spermatozoa. This procedure is repeated until adequate numbers of sperm are retrieved. In approximately 95% of obstructive azoospermic patients sperm recovery is easily possible with this low cost technique[81]. In cases where sperm are not obtained open sperm retrieval can be performed. While this procedure is very effective in the case of obstructive azoospermia, FNA has a significantly lower yield compared to TESE in non-obstructed azoospermic patients[82].

In cases of non-obstructive azoospermia sperm extraction is performed from the testis by an open excisional technique with multiple small testicular biopsies randomly performed (TESE)[21,24,28,29] or by microsurgical TESE[27]. In both cases a local anesthesia is performed by spermatic cord infiltration with few milliliters of 0.5% marcaine. This produces anesthesia of the testicle, but not of the scrotum. Then, several milliliters of 0.5% marcaine are used to infiltrate the anterior scrotal skin with a 25-gauge needle along a 1–2 cm incision line. For the TESE procedure the testis is positioned with the epididymis and vas deferens directed posteriorly and the scrotal contents are extruded through a small incision in the scrotal skin and in the tunica vaginalis. The albuginea is then opened through a small incision and a small piece of extruding testicular tissue is excised. The testicular specimen is then washed in a Petri dish containing culture medium and fragmented into small pieces. The presence of spermatozoa is assessed under the inverted microscope at × 200–400 magnification. If spermatozoa are observed no further testicular incision is needed. In the case where no spermatozoa are observed repeated sampling is performed in different areas of the testis. One single testicular specimen per testis is taken for histologic analysis. The contents of the Petri dishes are then centrifuged and incubated until the moment of injection. Just before injection the tissue can be further treated to facilitate the identification of spermatozoa. It has been shown that removing erythrocytes from the testicular biopsy specimens is possible to enhance the efficiency of sperm collection in those cycles in which spermatozoa are present[83]. Furthermore, enzymatic digestion of testicular tissue makes it possible to concentrate the specimen facilitating the study of spermatozoa[84]. Nevertheless spermatozoa are not always observed, even after repeated biopsies and different laboratory treatments of the tissue. TESE is not devoid of severe complications (testicular deterioration) caused by either direct interference with the microvascular supply of the seminiferous tubules or increased intratesticular pressure caused by minor amounts of bleeding within the enclosed tunica albuginea or by an even more increased pressure due to the closure of multiple open biopsies with conventional non-microsurgical sutures[85].

Microsurgical TESE is performed using local anesthesia only, as described above. The tunica vaginalis is opened, the testicle exteriorized and the tunica albuginea opened widely on the antimesenteric border to allow extensive visualization of testicular tubules under magnification (× 16–40). After evaluation of tubular dilatation either tiny removal of dilated tubules or large strips of tissue (when no dilated tubules are detectable) can be excised with no damage to blood supply or pressure atrophy. The tunica albuginea is closed with a 9-0 nylon suture after hemostasis using micro-bipolar forceps. In case of mixed Sertoli cell-only syndrome[80], this microsurgical approach allows to distiguish normal tubules with full thickness from the thin Sertoli cell-only tubules avoiding to excise unnecessarily large amounts of tissue. When the azoospermia is due to maturation arrest all tubules are of normal size and the foci of normal spermatogenesis cannot be distinguished with this microsurgical procedure, and larger amounts of testicular tissue are removed. However minimal damage is incurred because blood supply is not interrupted, meticulous hemostasis is performed, the tunica albuginea is closed with a 9-0 nylon suture and consequently there is no increase in intratesticular pressure. Testicular tissue is then prepared as described above.

THE INTRACYTOPLASMIC SPERM INJECTION PROCEDURE

Immediately before the injection procedure 1 µl of the sperm suspension is diluted in a microdroplet containing 4 µl of 10% PVP placed in the centre of a Petri dish while the oocytes are placed in surrounding micro-droplets containing 5 µl of HEPES-buffered culture medium covered by oil. This system enables the gamete to be kept in optimal conditions. The HEPES-buffered culture medium reduces pH variation, while the oil prevents medium evaporation thus maintaining the correct osmolarity. When the sperm suspension is added to the PVP droplet, motile spermatozoa start to migrate into the viscous medium, which decelerates the sperma-tozoa allowing for careful observation and facilitation of aspiration. A single spermatozoon of apparently normal morphology is then selected and immobilized before the injection into the oocyte cytoplasm[86]. To perform immobilization the spermatozoon is first placed transversally to the injection pipette, which is moved over the tail of the spermatozoon. The pipette is then lowered compressing the sperm tail between the pipette and the bottom of the Petri dish (Figure 1). The pipette is then moved over the tail until immobilization is achieved. The sperm immobilization is an important part of the ICSI procedure because it induces permeabilization of the sperm membrane and enhances subsequent nuclear decondensation improving the oocyte fertilization[87]. The immobilized spermatozoon, tail first, is then aspirated into the injection pipette and moved from the PVP droplet into one of the peripheral droplets containing an oocyte. The oocyte is rotated to locate the first polar body at the 6 or 12 o'clock position, held by gentle suction on the holding pipette and the equatorial plane located in focus. The spermatozoon is then ejected slowly, close to the tip and the edge of the pipette, which is then gently pushed deep into the cyto-plasm at the 3 o'clock position. To ensure that the spermatozoon is deposited inside the oocyte cytoplasm and not in the perivitelline space, gentle suction is applied from the injection pipette until the oolemma breaks. The rupture of the oocyte membrane can also be performed by repeated gentle movement of the injection pipette against the oolemma invagination. However, in this case the cytoplasm is aspirated after the rupture. This procedure is necessary for oocyte activation[73]. The cytoplasm organelles and the spermatozoon are then ejected back into the cytoplasm slowly with the smallest amount of medium possible. Thereafter the injection pipette is gently withdrawn and the oocyte released from the holding pipette. The injected oocyte is then rinsed in Hepes-free culture medium and cultured.

Some aspects of the ICSI procedure related to the oocyte require clarification. One of these is the site of sperm deposition during the injection procedure. It was believed that the position of the first polar body predicted the location of the metaphase II meiotic spindle. Therefore, historically, the oocyte was orientated (at 6 or 12 o'clock) in such a way as to reduce the risk of damage to the meiotic spindle[10–12]. The placement of the sperm during ICSI according to the presumed location of the meiotic spindle seems to have an impact on fertilization and embryo quality. Recent

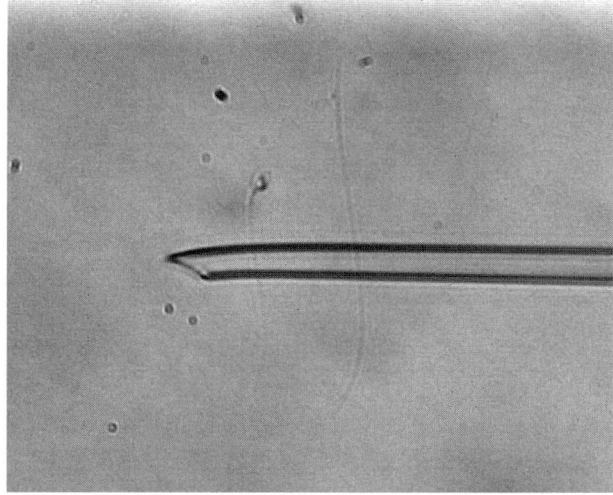

Figure 1 Sperm immobilization technique. The sperm tail is compressed between the injection pipette and the bottom of the Petri dish until complete immobilization is achieved

studies[88–90] have shown that better cleavage can be obtained when the oocyte polar body is positioned at the 6 o'clock position. It appears that development, but not fertilization, is improved by decreasing the distance between the sperm cell and the oocyte meiotic spindle. The study of Stoddart and Fleming[91] is contradictory to this hypothesis. The authors have demonstrated that the orientation of the polar body (6 o'clock versus 12 o'clock) in the oocyte has no significant effect on embryonic developmental potential. The differences in these studies depend on the fact that the first polar body is not an accurate predictor of meiotic spindle location[92,93] since it does not remain attached to the oocyte[94] and can move during the denudation process. Non-invasive localization of the spindle within the cytoplasm can be obtained with the polarized microscope[92,95–97]. The metaphase spindle is in fact composed of microtubules that exhibit positive birefringence when imaged under the polarization microscope. Further studies are needed to clarify which is the best site of sperm deposition in relation to the exact site of spindle location.

Another important aspect is the reaction of the oocyte membrane to the introduction of the injection pipette. Some studies have described different patterns of oolemma rupture during ICSI: (1) sudden rupture, when the membrane is disrupted during the insertion of the pipette; (2) normal breakage, when slight suction to break the membrane is applied; and (3) difficult breakage, when strong aspiration or several penetration attempts are performed[88,98]. The various reactions displayed by the oolemma to the penetrating pipette were predictive of survival and fertilization ability of the injected oocytes. In particular, a sudden rupture of the oocyte membrane was correlated with significantly lower oocyte survival and normal fertilization rates. A higher rate of oocytes displaying three pronuclei was also observed. Table 1 shows the correlation between oolemma characteristics and survival, fertilization and embryo development rates obtained in our center in 1998. Furthermore, a correlation between the modality of stimulation and the oolemma characteristics was observed[98]. The pattern of oolemma breakage can also be influenced by other factors. The deformation of the oocytes and the stress on the oolemma during the

insertion of the injection pipette seem to be related to the quality of the injection pipette itself (size and sharpness), to the method for insertion of the needle into the ooplasm (experience in the use of ICSI is a very important factor) and to the hardness of the zona pellucida.

PVP has been used since the beginning of the clinical application of the microinsemination procedures[6]. As already described, it facilitates manipulation of the spermatozoa and allows good control of the spermatozoa inside the injection pipette. It has been suggested that the injection procedure be performed without using PVP solution[99,100] because it might be harmful to human spermatozoa or affect subsequent fertilization capacity and embryo cleavage[97,101]. The results obtained in this prospective, controlled study seem to be similar to those obtained with the use of the PVP. Nevertheless, experiments on animal models[102] have shown no detrimental effects on either fertilization or development of bovine embryos. Moreover, evidence on the safety of the use of PVP in human ICSI has also come from the obstetric outcomes of pregnancies[103,104]. According to our experience the use of PVP seems important in facilitating sperm immobilization, to have a better control of the oocyte membrane breakage and to reduce the volume of medium injected into the oocyte.

Table 1 Oocyte survival, fertilization and embryo development in relation to oolemma breakage using intracytoplasmic sperm injection

	Types of membrane breakage		
	Sudden	*Normal*	*Difficult*
Injected oocytes	609 (17.2)	2100 (59.3)	833 (23.5)
Oocyte survival rate	539 (91.1)[a]	2033 (96.8)[b]	805 (96.7)[c]
Two pronuclei	391 (64.2)[d]	1509 (71.8)[e]	551 (66.1)[f]
One pronucleus	21 (3.4)	49 (2.3)	28 (3.3)
Three pronuclei	42 (6.8)[g]	49 (2.3)[h]	35 (4.2)[i]
Embryos obtained	343 (92.4)	1401 (92.8)	501 (90.1)

Data expressed as *n* (%). a–b, *p* = 0.001; a–c, *p* = 0.001; d–e, *p* = 0.001; d–f, *p* = not significant; g–h, *p* = 0.001; g–i, *p* = 0.03

REINSEMINATION OF IVF FAILED FERTILIZED OOCYTES BY INTRACYTOPLASMIC SPERM INJECTION

Although the average fertilization rate after standard IVF is about 60–70%, incomplete or partial failure of fertilization may occur. The lower the fertilization rate the less are the chances of obtaining available embryos for the transfer. To improve the fertilization, reinsemination of the unfertilized oocytes has been suggested[105]. Aged, IVF-failed, fertilized oocytes have only a very low chance to fertilize either after standard insemination[106,107] or after a micromanipulation procedure such as partial zona dissection[43] and subzonal insemination[107]. Successful fertilization of unfertilized oocytes has been obtained only with ICSI[108]. ICSI circumvents fertilization failures due to either hardness of the zona pellucida or zona pellucida sperm receptor dysfunction. Furthermore, multiple sperm penetration is avoided by injecting a single spermatozoon. The cleavage rate until day 2 of unfertilized oocytes after ICSI was similar to that obtained with fresh oocytes. Nevertheless, only an extremely low implantation rate was obtained after embryo transfer. It has been suggested[109] that the failure of embryo development derived from unfertilized 1-day-old reinseminated oocytes is due to chromosomal abnormalities. The preliminary data on karyotypes of these embryos have, in fact, shown a high incidence of cytogenetic abnormalities[109,110].

Despite sporadic pregnancies and deliveries of healthy babies that have been reported after ICSI of IVF-failed, fertilized oocytes[111–114], this procedure has almost been abandoned because of the high risk of genetic aberrations of the derived embryos.

PIEZO MICROMANIPULATION FOR INTRACYTOPLASMIC SPERM INJECTION

ICSI of human oocytes can be performed alternatively with a piezoelectric actuator. This technique was first developed using a mouse model[115] and high survival and fertilization rates were achieved compared to those obtained by conventional ICSI, which is characterized by low survival and fertilization rates in mice. The piezo micromanipulator has since been adapted for use in humans[116–118]. Piezo micromanipulation involves a simple and easily made injection pipette of very thin diameter and wall thickness. The action of the piezo actuator enables the pipette to penetrate the oocyte in several steps. First, piezo pulses are applied to allow the penetration of the zona pellucida by the needle. Second, once inside the perivitelline space, the tip of the pipette is gently advanced manually into the oocyte. When the oolemma is extended sufficiently it is broken with one piezo pulse and the sperm is deposited in the ooplasm. No suction of the ooplasm is performed. The results obtained with this technique were comparable to those obtained with the conventional ICSI technique[116] or even improved in terms of survival and fertilization rates[117,118]. The authors suggested that piezo micromanipulation could be used effectively for human oocytes to obtain an improvement in ICSI results. This technique has, in fact, some advantages. First, the needle fabrication is rapid and easy. Second, applying piezo pulses to the tail, sperm immobilization can be easily performed. Third, the deformation of the oocytes during the penetration of the injection needle is reduced. This is the most interesting feature of the technique because in this way oocyte damage is minimized. Finally, because no suction of ooplasm occurs at sperm injection it is presumed that the influence on the cytoskeleton is low. Nevertheless, the technique also presents several disadvantages. During penetration into the zona layer the oolemma can be inadvertently broken by the vibrating tip, causing oocyte degeneration. Because of the small opening of the injection pipette, debris can attach inside the pipette affecting the injection. The use of mercury (to strengthen the tip of the needle) can be harmful to the oocytes and for this reason should be avoided.

LASER-ASSISTED INTRACYTOPLASMIC SPERM INJECTION

As described previously, oocyte survival after ICSI is compromised by a sudden oolemma breakage during penetration of the injection needle into the oocytes. This pattern is often due to a very high resistance of the zona pellucida to the penetrating pipette, resulting in an extensive deformation of the oocyte in the initial phase of injection. When the zona pellucida is finally penetrated, the pipette enters very rapidly inside the oocyte cytoplasm leading to immediate oolemma breakage. The higher degeneration rate observed in these oocytes[88,98] is probably due to the combined effect of the oocyte squeezing and the rapid entry of the injecting pipette that may disrupt the fine oocyte structure. To avoid this situation a modification of the ICSI technique has been introduced in our center[119]. Just before ICSI a laser-drilled hole of approximately 10 μm in diameter is created in the zona pellucida of the oocyte with a 1.48 μm diode laser. The laser device is based on the thermosensitivity of the zona pellucida structure. This system makes it possible to create zona pellucida openings of desired form and size by a relatively simple and rapid manipulation, without any chemical modification of the zona material and in a highly reproducible way[120,121]. The ICSI needle containing the immobilized spermatozoon is then introduced through the hole into the oocyte. The main mechanical barrier represented by the zona pellucida to oocyte penetration is in this way by-passed. Furthermore, only a minimal deformation of the oocyte occurs (Figure 2). The effects on the ooplasm membrane in the proximity of the laser-treated area have been studied in mice[120] and humans[122]. These studies did not reveal any degenerative alterations to this structure.

The results of the laser-assisted ICSI technique have been compared in our laboratory to those obtained with conventional ICSI. For this purpose, patients displaying fragile oocytes (sudden oolemma breakage) were selected. In these cases, half of the oocytes were injected normally (control group) and half were injected with laser-assisted ICSI (Table 2). The oocyte survival rate was statistically increased in the laser-assisted ICSI group whereas no differences were observed in the fertilization of surviving oocytes and embryo development rates between the two groups. This technique also resulted in a reduction in the percentage of abnormal fertilized oocytes (three pronuclei) in this population of patients. Four pregnancies (out of six transfers) have been obtained when only laser-assisted ICSI-derived embryos were transferred. This modification of the ICSI technique appears therefore to be useful for all the patients whose oocytes show inherent fragility.

As described previously oocyte damage can also occur with piezo-assisted ICSI. Nevertheless, for practical reasons, laser-assisted ICSI may be more convenient for all those laboratories that are already equipped with a microsurgical laser device or are considering buying it for another purpose such as assisted hatching or pre-implantation genetic diagnosis.

FERTILIZATION EVENTS DURING INTRACYTOPLASMIC SPERM INJECTION

Meiotic maturation of the oocyte and subsequent activation of the egg by the spermatozoon are two separate events which are absolute prerequisites for normal fertilization. The oocytes are considered to be meiotically mature after a complex process that involves progressive and stage-specific events, including germinal vesicle breakdown, chromosomal condensation, polar body extrusion, metaphase II arrest and cytoplasmic maturation[123]. These changes are associated with changes in the organization of microtubules and microfilaments during the phases of the cell cycle. In particular, the microtubules are involved in chromatin reconstruction during meiotic maturation while the microfilaments are involved in chromosomal movement[124]. As discussed previously, nuclear and cytoplasmic maturation are acquired independently during oocyte maturation[125]. It has been observed that mouse MII oocytes gradually develop the capacity for activation after they have reached MII[126]. After ICSI, the fertilizing spermatozoon performs the following two roles: it contributes paternal DNA and it induces the oocyte to complete the second meiotic division. It has been demonstrated that the spermatozoon releases the factor responsible for triggering oocyte activation[101]. This factor seems to be heat-sensitive, intracellularly active and not species-specific, with an activity that is not identifiable in dead sperm cells[101]. Rupture of the sperm membrane that takes place during the sperm immobilization process seems to be a prerequisite for both the release of sperm-associated oocyte-activating factor and for sperm nucleus decondensation. However, Flaherty and co-workers[53] found that, after ICSI, many unfertilized oocytes had a correctly injected spermatozoon that had undergone partial or complete nuclear

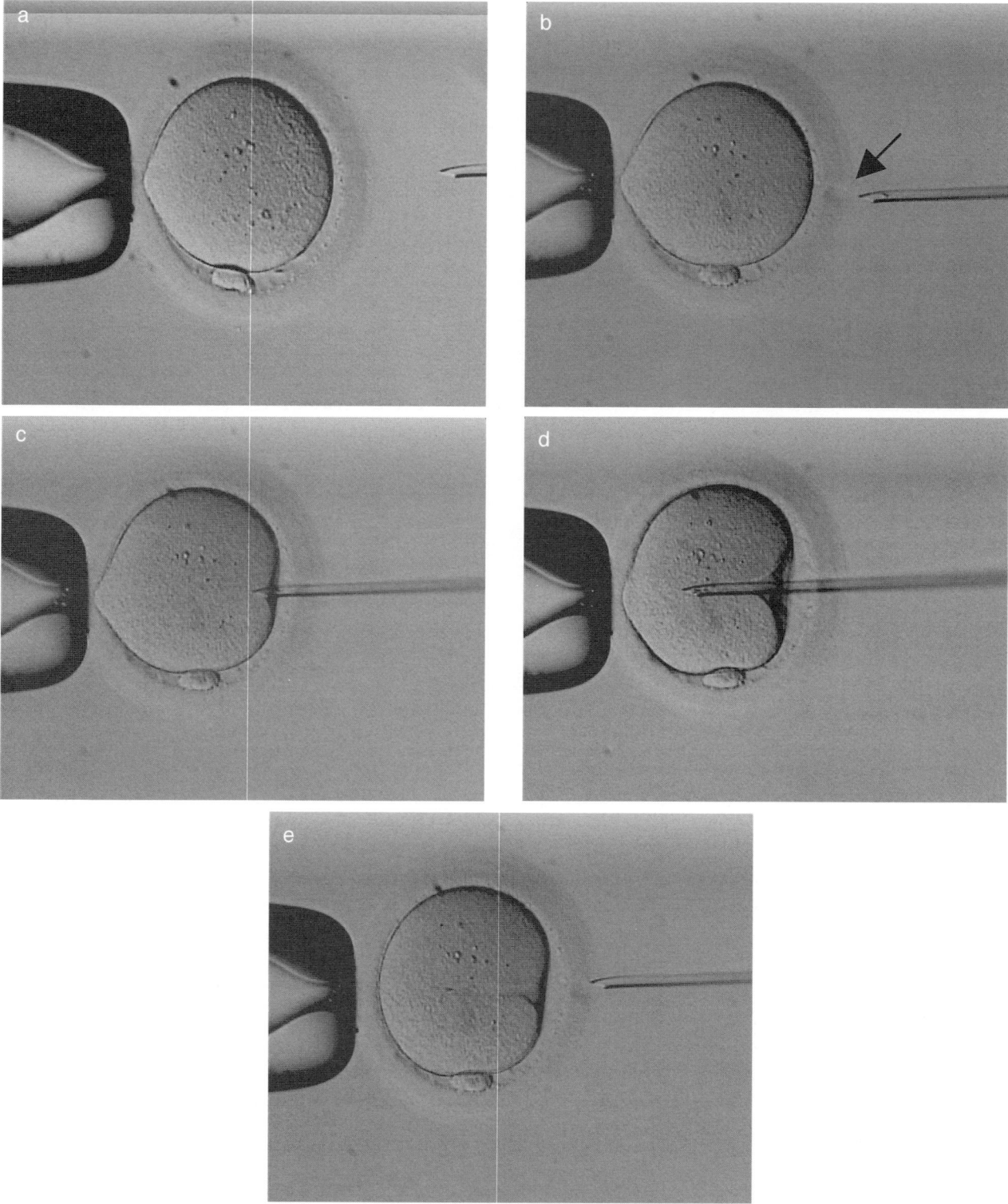

Figure 2 Laser-assisted intracytoplasmic sperm injection technique. The metaphase II oocyte is first held by the pipette (a) and then a hole (arrow) is created in the zona pellucida by the laser device (b). The injection pipette containing the immobilized spermatozoon is introduced through this hole directly in the perivitelline space (c) without deforming the oocyte and then deep inside the ooplasm (d) where the spermatozoon is released (e)

Table 2 Oocyte survival, fertilization and embryo development after conventional intracytoplasmic sperm injection (ICSI) and after laser-assisted ICSI in a selected population of patients displaying fragile oocyte membranes

	ICSI group (%)	Laser-assisted ICSI group (%)
Injected oocytes	140	140
Oocyte survival rate	130 (92.9)[a]	137 (97.9)[b]
Two pronuclei	97 (69.2)	110 (78.6)
One pronucleus	7 (5.0)	6 (4.3)
Three pronuclei	11 (7.8)[c]	4 (2.8)[d]
Embryos obtained	89 (91.7)	100 (90.9)

Data expressed as *n* (%). a–b, *p* = 0.04; c–d, *p* = 0.06

decondensation. Fertilization failure could be ascribed to either the unsuccessful release of the activation signal by the spermatozoon or the lack of a response of the oocyte to the activation signal. Some authors[127] have reported that the major cause of fertilization failure after ICSI is failure of the oocyte to initiate the biochemical processes necessary for activation. This inability of the oocyte to transduce the signal delivered by the injected spermatozoon could be ascribed to cytoplasmic immaturity of those gametes even if they have reached nuclear maturity.

The activation of a mature oocyte is characterized by release from MII arrest, cortical reaction and extrusion of the second polar body, followed by pronuclear formation. Both oocyte maturation and activation are apparently regulated by levels of intracellular free calcium ion (Ca^{2+})[128,129]. Sperm- and oocyte-borne oocyte activation failures can be overcome by artificial increases of Ca^{2+} concentration[130,131]. This increase has been obtained electrically or by using ionophores in assisted reproduction attempts in the case of globozoospermia, which is characterized by partial or complete failure of oocyte activation[15]. Despite the success in terms of oocyte activation and embryo development obtained, insufficient knowledge is available on the eventual cytotoxic, teratogenic or mutagenic effects on oocytes and embryos by the ionophores. Alternatively, oocyte activation can be obtained by the micromanipulation procedure (personal observations). Modification of the ICSI technique, based on repeated vigorous aspiration of the ooplasm into the injection needle before sperm deposition, enabled normal fertilization and the development of good-morphology embryos in cases of previous oocyte activation failures.

As in physiologic fertilization, oocyte activation is associated with the release of the cortical granules[132]. Nevertheless, ultrastructural studies have shown that some morphologic features of early fertilization are specific to ICSI-treated oocytes[133]. Membrane-bound vacuoles and oolemma inclusions are found at the injection site. Evidence of plasma-membrane damage and increased oocyte exchange processes with residual and multivesiculated bodies can be observed. Moreover, a sperm acrosome reaction is observed inside the oocyte cytoplasm; the sperm elements are then not incorporated into any oocyte vacuole but remain in the cytoplasm[133].

The timing of fertilization events after ICSI have been studied in detail[134,135]. Approximately half of the normally fertilized oocytes extrude their second polar body 2–3 h after ICSI, indicating that the oocytes have completed meiosis II by that time. The male pronucleus appears centrally, while the female pronucleus forms in close proximity to the second polar body. Pronuclei first appear as small bodies as soon as 3 h post-ICSI. Both pronuclei then increase in size and nucleoli are already present in each pronucleus[135]. Pronuclei end their formation at 6 h indicating that about 4 h are needed for complete pronucleus formation from the end of meiosis. At 8 h, 80% of the normally fertilized oocytes reveal two clear pronuclei, suggesting that in some oocytes, activation and pronucleus formation begins later. In some fertilized oocytes, the pronuclei appear asynchronously, while their disappearance is usually synchronous. The male pronucleus is larger and usually has more nucleoli[135,136].

Approximately 5% of the oocytes have three pronuclei after ICSI. The time sequence of pronucleus formation of these oocytes is similar to that described for normal oocyte fertilization[134,135]. Non-extrusion of the second polar body indicates that the development of the third pronucleus is linked to the second polar body which remains in the cytoplasm of the oocyte. The rate of parthenogenetic activation of the oocyte is normally between 3 and 5%. In these cases the oocytes develop a big female pronucleus and two polar bodies; however, no male pronucleus is developed in sperm[134,135]. Oocyte cleavage of normally fertilized oocytes starts approximately 20 h after microinjection. The subsequent development of the ICSI embryos is comparable to that observed in embryos obtained after conventional IVF[16].

RESULTS

Thanks to the efforts of the International Working Groups for Registers on Assisted Reproduction[137], the ESHRE Task Force on ICSI[138], the numerous data published by many IVF centers all over the world and the results generated from European registers by ESHRE in 1997 and 1998[139,140], we now have enough available data to evaluate the efficacy of this new micro-surgical procedure. These data however have biases due to their heterogeneity as there are important differences between centers with regard to experience, technique and number of cycles performed. From 1993 to 1995 the number of ICSI cycles using ejaculated epididymal and testicular spermatozoa increased significantly from 3.157 cycles in 1993 to 12.586 cycles in 1994 and to 47.654 cycles in 1995[137,138]. In 1997, the number of ICSI cycles rose to 62.253 and in 1998, 89.192 ICSI cycles were performed in 521 IVF clinics in 18 European countries[139,140]. This represents a significant increase in interest worldwide in the use of this technique. However, this significant increase in the number of ICSI cycles did not correspond to an improvement in the clinical results. The overall clinical pregnancy rate per cycle was 23.6% in 1993[137,138] and 24.8% in 1998. These data, as previously mentioned, are not homogeneous and huge differences exist between countries and centers.

In order to avoid heterogeneity due to differences in technique and experience between centers we will report the ICSI results obtained at the Centre for Reproductive Medicine of the European Hospital. From January 1997 to September 2001 a total of 1663 ICSI cycles with ejaculated spermatozoa were performed. Overall, 18 958 CCOCs were retrieved (mean 11.4 per oocyte retrieval). After removal of the surrounding cumulus and corona cells, 97% (18 358) of the CCOCs contained an oocyte with an intact zona pellucida. Of these oocytes, 84% (15 439) were at metaphase II stage, 5% at metaphase I and 11% at the germinal-vesicle stage. Therefore, 15 439 metaphase II oocytes were injected with a mean of 9.3 per cycle (Table 3). Out of 15 439 oocytes injected, 721 were damaged (degeneration rate of 4.7%). Overall, 10 715 oocytes showed two distinct pronuclei 16 to 18 h after ICSI (69.4% and 72.8% of the metaphase II oocytes injected and of the intact oocytes, respectively) (Table 3).

Of the 10 715 normally fertilized oocytes 7179 (67%) developed into good quality embryos (with less than 30% of anucleate fragments) suitable for transfer or for cryopreservation. The replacement of the embryos was performed in a total of 1554 cycles (93% of the ICSI cycles). An average of 2.8 embryos were replaced per cycle. Table 4 shows the outcome of embryo transfer after ICSI. Overall, 564 clinical pregnancies (defined as the presence of heart activity at ultrasound at 7 weeks of

Table 3 Oocyte maturity, fertilization and cleavage rates after intracytoplasmic sperm injection (ICSI)

| | *n* | *Percentage of* | | | | |
		A	*B*	*C*	*D*	*E*
CCOCs (A)	18 958					
Intact zona pellucida and oocyte (B)	18 358	97				
metaphase II oocytes (C)	15 439		84			
metaphase I oocytes	918		5			
germinal-vesicle stage	2 001		11			
Intact oocytes after ICSI (D)	14 718	78		96		
Oocytes with two pronuclei (E)	10 715	56		69	74	
Good quality embryos	7 179					67

CCOCs, cumulus–corona–oocyte complexes

Table 4 Outcome of the embryo transfer after intracytoplasmic sperm injection (ICSI) with ejaculated spermatozoa

	n	A	B	C	D
		Percentage of			
ICSI cycles (A)	1663				
Transfers (B)	1554	93			
Embryos transferred (C)	4339				
Embryos transferred/cycle (mean)	2.8				
Gestational sacs	732			16	
Clinical pregnancies (D)	564	34	36		
Miscarriages	87				15
Ectopic pregnancies	11				2
Deliveries/ongoing	466	28	30		

gestation) were recorded with a clinical pregnancy rate of 34% per started ICSI treatment and 36% per embryo transfer. Eighty-seven miscarriages (15.4%), 11 ectopic pregnancies (1.9%) and two eterotopic pregnancies (0.3%) were recorded. A total of 466 women delivered at least one child or are carrying an ongoing pregnancy (28% and 30% of the ICSI cycles and of the embryo replacements, respectively) (Table 4).

It has been shown that in assisted reproductive technology ongoing pregnancy rates decrease significantly with advancing age[141] and that decreased ovarian reserve and female age > 38–39 years are negative prognostic factors[142]. Most probably, the decrease in implantation rates with advancing age is related more to poor oocyte quality rather than to uterine quality[143]. Table 5 shows our clinical results according to female age. Higher implantation and clinical pregnancy rates have been recorded in patients < 39 years of age.

ICSI has been successfully used in azoospermic patients[22–25,28,29]. Recently in a case–control study we observed similar clinical pregnancy rates and delivery rates after ICSI in patients with extremely impaired semen parameters (controls) compared to patients with obstructive and non-obstructive azoospermia (cases)[79]. However, significantly lower implantation rates were recorded in patients with non-obstructive azoospermia.

Because of the novelty of the ICSI technique and because of the invasivity of the procedure there are major concerns about its safety. Recently a Belgian group reported the outcome of 2375 ICSI pregnancies leading

Table 5 Female age and outcome of embryo transfers after intracytoplasmic sperm injection (ICSI) with ejaculated spermatozoa

	≤ 39 years	≥ 39 years	Total
Number of ICSI cycles	1302	361	1663
Number of embryos transferred	1220	334	1554
Number of embryos transferred/cycle (mean)	2.7	3.0	2.8
Implantation rate (%)	20	8	
Clinical pregnancy rate per embryo transfer (%)	41	19	16
			36
Miscarriage rate (%)	11	28	
Ectopic pregnancy rate (%)	2	2	15
Delivery/ongoing pregnancy rate per embryo transfer (%)	36	13	2
			30

to the birth of 1987 children[104]. Fetal karyotypes were performed in 1082 fetuses (690 amniocenteses and 392 chorionic villus samplings) and in 28 cases an abnormal karyotype was observed. Eighteen of these (1.66%) were *de novo* chromosomal aberrations: nine (0.83%) showed *de novo* sex chromosomal aberrations and another nine cases (0.83%) were trisomies or structural aberrations. There was a statistically significant increase in sex chromosomal aberrations and an increase in structural *de novo* anomalies as compared to the incidence in the general population[144]. The causes of these phenomena need to be further investigated. Major malformations causing functional impairment or requiring surgical correction were found in 22 of 1063 (2.1%) singleton children, 22 of 805 (2.7%) twin children and two of 98 (2%) triplet children. These malformation rates were not different from those reported by other authors on IVF outcome (2.0–3.8%)[145,146].

With regard to the long-term outcome of children conceived by ICSI, Bowen and co-workers recently[147] reported the medical and developmental outcome of children conceived by ICSI at the age of 1 year. This developmental assessment was carried out using the Bayley Scales of Infant Development from which a mental index (MDI) was derived. In this prospective study, the authors compared the medical and developmental outcome at 1 year of 89 children conceived by ICSI with 84 children conceived by routine IVF, and with 80 children conceived naturally. The mean Bayley MDI was significantly lower for the children conceived by ICSI than for the children conceived by routine IVF or naturally[147]. A few months later Bonduelle and co-workers[103] reported the medical and developmental outcome of children conceived by ICSI at the age of 2 years, assessing the mental development with the same parameters (Bayley Scales of Infant Development) used by Bowen and co-workers[147]. Two hundred and one ICSI children and 131 IVF children were tested by the same pediatrician who was unaware of the type of assisted reproductive technology used for each couple. No differences were observed between ICSI and IVF children and the general population. Although the data from this latter study are reassuring, a detailed case–control study taking into account parental background is needed to be able to draw final conclusions.

CONCLUSIONS

The clinical approach to female and male infertility has changed since the introduction of IVF. This assisted reproductive technique has been used widely and successfully in treating female and male infertility. However when the cause of infertility is severe male-factor infertility, the fertilization rate after IVF remains disappointingly low. Corrective measures adopted in order to increase the fertilization rate such as higher numbers of spermatozoa for *in vitro* insemination[148] or various methods to improve the fertilizing potential of the spermatozoa[149] have been proposed although failing to improve the results[150]. Similarly, fertilization rates after new assisted reproductive micromanipulation techniques developed during the last 15 years (ZD, PZD and SUZI) failed to increase substantially in the case of severe male-factor infertility. It was only after the development of the ICSI technique that the fertilization rate and the clinical outcome improved dramatically. ICSI can be successfully performed in couples with quantitative and qualitative sperm anomalies since only one live, motile sperm is needed per oocyte and most steps of gamete interaction are by-passed.

Notwithstanding the brilliant results obtained from the beginning of the clinical application of this technique using ejaculated, epidydimal and testicular spermatozoa, during these years many technical aspects have been studied and improved with the aim of minimizing the invasivity of the technique. Moreover, new procedures such as piezo micromanipulation, laser-assisted ICSI and the use of the polarized microscope have been introduced to minimize oocyte damage during the insertion of the spermatozoon.

Despite reassuring data coming from the literature regarding the obstetric and the neonatal outcome of the ICSI pregnancies, couples should be informed of the risk of *de novo* sex chromosomal and structural aberrations and the risk of transmitting fertility problems to their offspring.

ACKNOWLEDGEMENTS

We are indebted to all our colleagues at the Centre for Reproductive Medicine: the clinicians, Dr. Greco, Dr. Ferrero, Dr. Anniballo, Dr. Cerulo, the clinical embryologist and the scientists, Dr. Iacobelli, Dr. Martinez, Dr. Tesarik, the nurses and the secretaries. Particular thanks to Dr. Minasi for her effort in preparing the manuscript.

References

1. Tournaye H, Devroey P, Camus M, *et al.* Comparison of *in vitro* fertilization in male and tubal infertility: a 3 year survey. *Hum Reprod* 1992;7:218–22
2. Chang MC. The meaning of sperm capacitation. A historical perspective. *J Androl* 1984;5:45–50
3. Tesarik J, Testart J. Human sperm–egg interactions and their disorders: implications in the management of infertility. *Hum Reprod* 1989;4:729–41
4. Metka M, Haromy T, Huber J, Schurz B. Artificial insemination using a micromanipulator. *Fertilitat* 1985;1:41–7
5. Laws-King A, Trounson A, Sathananthan H, Kola I. Fertilization of human oocytes by microinjection of a single spermatozoon under the zona pellucida. *Fertil Steril* 1987; 48:637–42
6. Lanzendorf SE, Maloney MK, Veeck LL, *et al.* A preclinical evaluation of pronuclear formation by microinjection of human spermatozoa into human oocytes. *Fertil Steril* 1988;49:835–42
7. Gordon JW, Grunfeld L, Garrisi GJ, *et al.* Fertilization of human oocytes by sperm from infertile males after zona pellucida drilling. *Fertil Steril* 1988;50:68–73
8. Cohen J, Malter H, Wright G, *et al.* Partial zona dissection of human oocytes when failure of zona pellucida penetration is anticipated. *Hum Reprod* 1989;4:435–42
9. Bongso TA, Sathananthan AH, Wong PC, *et al.* Human fertilization by micro-injection of immotile spermatozoa. *Hum Reprod* 1989;4:175–9
10. Palermo G, Joris H, Devroey P, Van Steirteghem AC. Pregnancies after intracytoplasmic injection of single spermatozoon into an oocyte. *Lancet* 1992;340:17–18
11. Van Steirteghem AC, Liu J, Joris H, *et al.* Higher success rate by intracytoplasmic sperm injection than by subzonal insemination. Report of a second series of 300 consecutive treatment cycles. *Hum Reprod* 1993;8:1055–60
12. Van Steirteghem AC, Nagy Z, Joris H, *et al.* High fertilization and implantation rates after intracytoplasmic sperm injection. *Hum Reprod* 1993;8:1061–6
13. Liu J, Nagy Z, Joris H, *et al.* Successful fertilization and establishment of pregnancies after intracytoplasmic sperm injection in patients with globozoospermia. *Hum Reprod* 1995;10:626–9
14. Bourne H, Richings N, Harari O, *et al.* The use of intracytoplasmic sperm injection for the treatment of severe and extreme male infertility. *Reprod Fertil Dev* 1995;7: 237–45
15. Rybouchkin AV, Van der Straeten F, Quatacker J, *et al.* Fertilization and pregnancy after assisted oocyte activation and intracytoplasmic sperm injection in a case of round-headed sperm associated with deficient oocyte activation capacity. *Fertil Steril* 1997;68:1144–7
16. Nagy ZP, Verheyen G, Tournaye H, Van Steirteghem AC. Special applications of intracytoplasmic sperm injection: the influence of sperm count, motility, morphology, source and sperm antibody on the outcome of ICSI. *Hum Reprod* 1998;13(Suppl.1):143–54
17. Kahraman S, Tasdemir M, Tasdemir I, *et al.* Pregnancies achieved with testicular and ejaculated spermatozoa in combination with intracytoplasmic sperm injection in men with totally or initially immotile spermatozoa in the ejaculate. *Hum Reprod* 1996;11:1343–6
18. Nijs M, Vanderzwalmen P, Vandamme B, *et al.* Fertilizing ability of immotile spermatozoa after intracytoplasmic sperm injection. *Hum Reprod* 1996;11:2180–5
19. Nagy ZP, Liu J, Joris H, *et al.* The result of intracytoplasmic sperm injection is not related to any of the three basic sperm parameters. *Hum Reprod* 1995;10: 1123–9
20. Casper RF, Meriano JS, Jarvi KA, *et al.* The hypo-osmotic swelling test for selection of viable sperm for intracytoplasmic sperm injection in men with complete asthenozoospermia. *Fertil Steril* 1996;65:972–6
21. Tournaye H, Liu J, Nagy Z, *et al.* The use of testicular sperm for intracytoplasmic sperm injection in patients with necrozoospermia. *Fertil Steril* 1996;66:331–4
22. Tournaye H, Devroey P, Liu J, *et al.* Microsurgical epididymal sperm aspiration and intracytoplasmic sperm injection: a new effective approach to infertility as a result of congenital bilateral absence of the vas deferens. *Fertil Steril* 1994;61:1045–51
23. Craft IL, Khalifa Y, Boulos A, *et al.* Factors influencing the outcome of *in vitro* fertilization with percutaneous aspirated epididymal spermatozoa and intracytoplasmic sperm injection in azoospermic men. *Hum Reprod* 1995;10:1791–4
24. Schoysman R, Vanderzwalmen P, Nijs M, *et al.* Pregnancy after fertilisation with human testicular spermatozoa. *Lancet* 1993;342:1237
25. Watkins W, Lim T, Bourne H, *et al.* Testicular aspiration of sperm for intracytoplasmic sperm injection: an alternative treatment to electro-emission: case report. *Spinal Cord* 1996;34:696–8
26. Schlegel PN. Testicular sperm extraction: microdissection improves sperm yield with minimal tissue excision. *Hum Reprod* 1999;14:131–5
27. Silber S. Evaluation and treatment of male infertility. *Clin Obstet Gynecol* 2000;43:854-8
28. Devroey P, Liu J, Nagy Z, *et al.* Normal fertilization of human oocytes after testicular sperm extraction and intracytoplasmic sperm injection. *Fertil Steril* 1994;62: 639–41
29. Devroey P, Liu J, Nagy Z, *et al.* Pregnancies after testicular sperm extraction and intracytoplasmic sperm injection in non-obstructive azoospermia. *Hum Reprod* 1995;10: 1457–60
30. Cohen J, Alikani M, Liu HC, Rosenwaks Z. Rescue of human embryos by micromanipulation. *Baillières Clin Obstet Gynaecol* 1994;8:95–116
31. Aboulghar MA, Mansour RT, Serour GI, *et al.* Prospective controlled randomized study of *in vitro* fertilization versus intracytoplasmic sperm injection in the treatment of tubal

factor infertility with normal semen parameters. *Fertil Steril* 1996;66:753–6

32. Aboulghar MA, Mansour RT, Serour GI, *et al.* Intracytoplasmic sperm injection and conventional *in vitro* fertilization for sibling oocytes in cases of unexplained infertility and borderline semen. *J Assist Reprod Genet* 1996;13:38–42

33. Ludwig M, al-Hasani S, Kupker W, *et al.* A new indication for an intracytoplasmic sperm injection procedure outside the cases of severe male factor infertility. *Eur J Obstet Gynecol Reprod Biol* 1997;75:207–10

34. Porcu E. Freezing of oocytes. *Curr Opin Obstet Gynecol* 1999;11:297–300

35. Kazem R, Thompson LA, Srikantharajah A, *et al.* Cryopreservation of human oocytes and fertilization by two techniques: *in vitro* fertilization and intracytoplasmic sperm injection. *Hum Reprod* 1995;10:2650–4

36. Nagy ZP, Cecile J, Liu J, *et al.* Pregnancy and birth after intracytoplasmic sperm injection of *in vitro* matured germinal-vesicle stage oocytes: case report. *Fertil Steril* 1996;65:1047–50

37. Ubaldi F, Nagy Z, Janssenwillen C, *et al.* Ovulation by repeated human chorionic gonadotrophin in 'empty follicle syndrome' yields a twin clinical pregnancy. *Hum Reprod* 1997;12:454–6

38. De Vos A, Van de Velde H, Joris H, Van Steirteghem A. *In vitro* matured metaphase-I oocytes have a lower fertilization rate but similar embryo quality as mature metaphase-II oocytes after intracytoplasmic sperm injection. *Hum Reprod* 1999;14: 1859–63

39. Edirisinghe WR, Junk SM, Matson PL, Yovich JL. Birth from cryopreserved embryos following *in vitro* maturation of oocytes and intracytoplasmic sperm injection. *Hum Reprod* 1997;12:1056–8

40. Hamberger L, Lundin K, Sjogren A, Soderlund B. Indications for intracytoplasmic sperm injection. *Hum Reprod* 1998;13 (Suppl.1):128–33

41. Liu DY, Bourne H, Baker HW. High fertilization and pregnancy rates after intracytoplasmic sperm injection in patients with disordered zona pellucida-induced acrosome reaction. *Fertil Steril* 1997;67:955–8

42. Garrisi GJ, Talansky BE, Grunfeld L, *et al.* Clinical evaluation of three approaches to micromanipulation-assisted fertilization. *Fertil Steril* 1990;54:671–7

43. Malter HE, Cohen J. Partial zona dissection of the human oocyte: a nontraumatic method using micromanipulation to assist zona pellucida penetration. *Fertil Steril* 1989;51:139–48

44. Feichtinger W, Strohmer H, Fuhrberg P, *et al.* Photoablation of oocyte zona pellucida by erbium-YAG laser for *in vitro* fertilisation in severe male infertility. *Lancet* 1992;339:811

45. Cohen J, Alikani M, Malter HE, *et al.* Partial zona dissection or subzonal sperm insertion: microsurgical fertilization alternatives based on evaluation of sperm and embryo morphology. *Fertil Steril* 1991;56:696–706

46. Ng SC, Bongso A, Chang SI, *et al.* Transfer of human sperm into the perivitelline space of human oocytes after zona-drilling or zona-puncture. *Fertil Steril* 1989;52:73–8

47. Fishel S, Antinori S, Jackson P, *et al.* Presentation of six pregnancies established by sub-zonal insemination (SUZI). *Hum Reprod* 1991;6:124–30

48. Joris H, Nagy Z, Van de Velde H, *et al.* Intracytoplasmic sperm injection: laboratory set-up and injection procedure. *Hum Reprod.* 1998;13(Suppl.1):76–86

49. Pickering SJ, Braude PR, Johnson MH, *et al.* Transient cooling to room temperature can cause irreversible disruption of the meiotic spindle in the human oocyte. *Fertil Steril* 1990;54:102–8

50. Van Steirteghem A, Tournaye H, Van der Elst J, *et al.* Intracytoplasmic sperm injection three years after the birth of the first ICSI child. *Hum Reprod* 1995;10:2527–8

51. Van de Velde H, Nagy ZP, Joris H, *et al.* Effects of different hyaluronidase concentrations and mechanical procedures for cumulus cell removal on the outcome of intracytoplasmic sperm injection. *Hum Reprod* 1997;12:2246–50

52. Iritani A. Micromanipulation of gametes for *in vitro* assisted fertilization. *Mol Reprod Dev* 1991;28:199–207

53. Flaherty SP, Payne D, Swann NJ, Matthews CD. Aetiology of failed and abnormal fertilization after intracytoplasmic sperm injection. *Hum Reprod* 1995;10:2623–9

54. Van de Velde H, De Vos A, Joris H, *et al.* Effect of timing of oocyte denudation and micro-injection on survival, fertilization and embryo quality after intracytoplasmic sperm injection. *Hum Reprod* 1998;13:3160–4

55. Rienzi L, Ubaldi F, Anniballo R, *et al.* Preincubation of human oocytes may improve fertilization and embryo quality after intracytoplasmic sperm injection. *Hum Reprod* 1998;13:1014–19

56. Junca AM, Mandelbaum J, Belaisch-Allart J, *et al.* Oocyte maturity and quality: value of intracytoplasmic sperm injection. Fertility of microinjected oocytes after *in vitro* maturation. *Contracept Fertil Sex* 1995;23:463–5

57. Nogueira D, Staessen C, Van de Velde H, Van Steirteghem A. Nuclear status and cytogenetics of embryos derived from *in vitro*-matured oocytes. *Fertil Steril* 2000;74:295–8

58. De Sutter P, Dozortsev D, Qian C, Dhont M. Oocyte morphology does not correlate with fertilization rate and embryo quality after intracytoplasmic sperm injection. *Hum Reprod* 1996;11:595–7

59. Balaban B, Urman B, Sertac A, *et al.* Oocyte morphology does not affect fertilization rate, embryo quality and implantation rate after intracytoplasmic sperm injection. *Hum Reprod* 1998;13:3431–3

60. Serhal PF, Ranieri DM, Kinis A, *et al.* Oocyte morphology predicts outcome of intracytoplasmic sperm injection. *Hum Reprod* 1997;12:1267–70

61. Xia P. Intracytoplasmic sperm injection: correlation of oocyte grade based on polar body, perivitelline space and cytoplasmic inclusions with fertilization rate and embryo quality. *Hum Reprod* 1997;12:1750–5

62. Veeck LL. Oocyte assessment and biological performance. *Ann N Y Acad Sci* 1988;541:259–74

63. Plachot M, Crozet N. Fertilization abnormalities in human *in vitro* fertilization. *Hum Reprod* 1992;7(Suppl.1):89–94

64. Fluker MR, Siu CK, Gunby J, Daya S. Cycle characteristics and outcome in relation to ovarian response during *in vitro* fertilization. *J Assist Reprod Genet* 1993;10:504–12

65. Devroey P, Godoy H, Smitz J, *et al.* Female age predicts embryonic implantation after ICSI: a case-controlled study. *Hum Reprod* 1996;11:1324–7

66. *World Health Organization Laboratory Manual for Examination of Human Semen.* Cambridge: Cambridge University Press, 1999

67. Kruger TF, Menkveld R, Stander FS, *et al.* Sperm morphologic features as a prognostic factor in *in vitro* fertilization. *Fertil Steril* 1986;46:1118–23

68. Palermo GD, Cohen J, Rosenwaks Z. Intracytoplasmic sperm injection: a powerful tool to overcome fertilization failure. *Fertil Steril* 1996;65:899–908

69. Liu J, Nagy Z, Joris H, *et al.* Intracytoplasmic sperm injection does not require special treatment of the spermatozoa. *Hum Reprod* 1994;9:1127–30

70. McClure RD, Nunes L, Tom R. Semen manipulation: improved sperm recovery and function with a two-layer Percoll gradient. *Fertil Steril* 1989;51:874–7

71. Van der Zwalmen P, Bertin-Segal G, Geerts L, *et al.* Sperm morphology and IVF pregnancy rate: comparison between Percoll gradient centrifugation and swim-up procedures. *Hum Reprod* 1991;6:581–8

72. De Vos A, Nagy ZP, Van de Velde H, *et al.* Percoll gradient centrifugation can be omitted in sperm preparation for intracytoplasmic sperm injection. *Hum Reprod* 1997;12:1980–4

73. Palermo GD, Cohen J, Alikani M, *et al.* Development and implementation of intracytoplasmic sperm injection (ICSI). *Reprod Fertil Dev* 1995;7:211–18

74. Nagy Z, Liu J, Cecile J, *et al.* Using ejaculated, fresh, and frozen–thawed epididymal and testicular spermatozoa gives rise to comparable results after intracytoplasmic sperm injection. *Fertil Steril* 1995;63:808–15

75. Barros A, Sousa M, Angelopoulos T, Tesarik J. Efficient modification of intracytoplasmic sperm injection technique for cases with total lack of sperm movement. *Hum Reprod* 1997;12:1227–9

76. Hull MGR, Glazener CMA, Kelly MJ, *et al.* Population study of causes treatment and outcome of infertility. *Br Med J* 1985;291:1693–7

77. Schlegel PN. How to do a workup for male infertility. *Med Aspect Hum Sex* 1991;25:28

78. Tournaye H, Verheyen G, Nagy P, *et al.* Are there any predictive factors for successful testicular sperm recovery in azoospermic patients? *Hum Reprod* 1997;12:80-6

79. Ubaldi F, Nagy ZP, Rienzi L, *et al.* Reproductive capacity of spermatozoa from men with testicular failure. *Hum Reprod* 1999;14:2796–800

80. Anniballo R, Ubaldi F, Cobellis L, *et al.* Criteria predicting the absence of spermatozoa in the Sertoli cell-only syndrome can be used to improve success rates of sperm retrieval. *Hum Reprod* 2000;15:2269–77

81. Tournaye H. Surgical sperm recovery for intracytoplasmic sperm injection: which method is to be preferred? *Hum Reprod* 1999;14(Suppl.1):71–81

82. Friedler S, Raziel A, Soffer Y, *et al.* Testicular sperm retrieval by percutaneous fine needle aspiration compared with testicular sperm extraction by open biopsy in men with non-obstructive azoospermia. *Hum Reprod* 1997;12:1488–93

83. Nagy ZP, Verheyen G, Tournaye H, *et al.* An improved treatment procedure for testicular biopsy specimens offers more efficient sperm recovery: case series. *Fertil Steril* 1997;68:376–9

84. Crabbe E, Verheyen G, Tournaye H, Van Steirteghem A. The use of enzymatic procedures to recover testicular germ cells. *Hum Reprod* 1997;12:1682–7

85. Schlegel PN, Su LM. Physiological consequences of testicular sperm extraction. *Hum Reprod* 1997;12:1688–92

86. Fishel S, Lisi F, Rinaldi L, *et al.* Systematic examination of immobilizing spermatozoa before intracytoplasmic sperm injection in the human. *Hum Reprod* 1995;10:497–500

87. Dozortsev D, Rybouchkin A, De Sutter P, Dhont M. Sperm plasma membrane damage prior to intracytoplasmic sperm injection: a necessary condition for sperm nucleus decondensation. *Hum Reprod* 1995;10:2960–4

88. Nagy ZP, Liu J, Joris H, *et al.* The influence of the site of sperm deposition and mode of oolemma breakage at intracytoplasmic sperm injection on fertilization and embryo development rates. *Hum Reprod* 1995;10:3171–7

89. Van Der Westerlaken LA, Helmerhorst FM, Hermans J, Naaktgeboren N. Intracytoplasmic sperm injection: position of the polar body affects pregnancy rate. *Hum Reprod* 1999; 14:2565–9

90. Blake M, Garrisi J, Tomkin G, Cohen J. Sperm deposition site during ICSI affects fertilization and development. *Fertil Steril* 2000;73:31–7

91. Stoddart NR, Fleming SD. Orientation of the first polar body of the oocyte at 6 or 12 o'clock during ICSI does not affect clinical outcome. *Hum Reprod* 2000;15:1580–5

92. Silva CP, Kommineni K, Oldenbourg R, Keefe DL. The first polar body does not predict accurately the location of the metaphase II meiotic spindle in mammalian oocytes. *Fertil Steril* 1999;71:719–21

93. Hardarson T, Lundin K, Hamberger L. The position of the metaphase II spindle cannot be predicted by the location of the first polar body in the human oocyte. *Hum Reprod* 2000;15:1372–6

94. Collas P, Robl JM. Factors affecting the efficiency of nuclear transplantation in the rabbit embryo. *Biol Reprod* 1990;43:877–84

95. Sato H, Ellis GW, Inoue S. Microtubular origin of mitotic spindle form birefringence. Demonstration of the applicability of Wiener's equation. *J Cell Biol* 1975;67:501–17

96. Oldenbourg R, Mei G. New polarized light microscope with precision universal compensator. *J Microsc* 1995;180:140–7

97. Wang WH, Meng L, Hackett RJ, *et al.* The spindle observation and its relationship with fertilization after intracytoplasmic sperm injection in living human oocytes. *Fertil Steril* 2001;75:348–53

98. Palermo GD, Alikani M, Bertoli M, *et al.* Oolemma characteristics in relation to survival and fertilization patterns of oocytes treated by intracytoplasmic sperm injection. *Hum Reprod* 1996;11:172–6

99. Hlinka D, Herman M, Vesela J, *et al.* A modified method of intracytoplasmic sperm injection without the use of polyvinylpyrrolidone. *Hum Reprod* 1998;13:1922–7

100. Cho YS, Traina V, Boyer P. Development of a successful ICSI programme without the use of PVP. *Hum Reprod* 1997;12:1116–17

101. Dozortsev D, De Sutter P, Rybouchkin A, Dhont M. Timing of sperm and oocyte nuclear progression after intracytoplasmic sperm injection. *Hum Reprod* 1995;10: 3012–17

102. Motoishi M, Goto K, Tomita K, et al. Examination of the safety of intracytoplasmic injection procedures by using bovine zygotes. *Hum Reprod* 1996;11:618–20

103. Bonduelle M, Joris H, Hofmans K, et al. Mental development of 201 ICSI children at 2 years of age. *Lancet* 1998;351:1553

104. Bonduelle M, Camus M, De Vos A, et al. Seven years of intracytoplasmic sperm injection and follow-up of 1987 subsequent children. *Hum Reprod* 1999;14(Suppl.1): 243–64

105. Ben-Rafael Z, Kopf GS, Blasco L, et al. Fertilization and cleavage after reinsemination of human oocytes *in vitro*. *Fertil Steril* 1986;45:58–62

106. Trounson A, Webb J. Fertilization of human oocytes following reinsemination *in vitro*. *Fertil Steril* 1984;41: 816–19

107. Imoedemhe DA, Sigue AB. Clinical experience with repeat subzonal microinsemination of oocytes failing to fertilize after an initial microinsemination. *Fertil Steril* 1994; 62:1072–4

108. Nagy ZP, Joris H, Liu J, et al. Intracytoplasmic single sperm injection of 1-day-old unfertilized human oocytes. *Hum Reprod* 1993;8:2180–4

109. Nagy ZP, Staessen C, Liu J, et al. Prospective, auto-controlled study on reinsemination of failed-fertilized oocytes by intracytoplasmic sperm injection. *Fertil Steril* 1995;64:1130–5

110. Edirisinghe WR, Murch A, Junk S, Yovich JL. Cytogenetic abnormalities of unfertilized oocytes generated from *in vitro* fertilization and intracytoplasmic sperm injection: a double-blind study. *Hum Reprod* 1997;12: 2784–91

111. Tsirigotis M, Redgment C, Craft I. Late intracytoplasmic sperm injection (ICSI) in *in vitro* fertilization (IVF) cycles. *Hum Reprod* 1994;9:1359

112. Lundin K, Sjogren A, Hamberger L. Reinsemination of one-day-old oocytes by use of intracytoplasmic sperm injection. *Fertil Steril* 1996;66:118–21

113. Morton PC, Yoder CS, Tucker MJ, et al. Reinsemination by intracytoplasmic sperm injection of 1-day-old oocytes after complete conventional fertilization failure. *Fertil Steril* 1997;68:488–91

114. Bussen S, Mulfinger L, Sutterlin M, et al. Dizygotic twin pregnancy after intracytoplasmic sperm injection of 1 day old unfertilized oocytes. *Hum Reprod* 1997;12:2560–2

115. Kimura Y, Yanagimachi R. Intracytoplasmic sperm injection in the mouse. *Biol Reprod* 1995;52:709–20

116. Huang T, Kimura Y, Yanagimachi R. The use of piezo micromanipulation for intracytoplasmic sperm injection of human oocytes. *J Assist Reprod Genet* 1996;13:320–8

117. Yanagida K, Katayose H, Yazawa H, et al. The usefulness of a piezo-micromanipulator in intracytoplasmic sperm injection in humans. *Hum Reprod* 1999;14:448–53

118. Takeuchi S, Minoura H, Shibahara T, et al. Comparison of piezo-assisted micromanipulation with conventional micromanipulation for intracytoplasmic sperm injection into human oocytes. *Gynecol Obstet Invest* 2001;52:158–62

119. Rienzi L, Greco E, Ubaldi F, et al. Laser-assisted intracytoplasmic sperm injection. *Fertil Steril* 2001;76: 1045–7

120. Rink K, Delacretaz G, Salathe RP, et al. Non-contact microdrilling of mouse zona pellucida with an objective-delivered 1.48-microns diode laser. *Lasers Surg Med* 1996;18:52–62

121. Germond M, Nocera D, Senn A, et al. Improved fertilization and implantation rates after non-touch zona pellucida microdrilling of mouse oocytes with a 1.48 micron diode laser beam. *Hum Reprod* 1996;11:1043–8

122. Obruca A, Strohmer H, Blaschitz A, et al. Ultrastructural observations in human oocytes and preimplantation embryos after zona opening using an erbium-yttrium-aluminium-garnet (Er:YAG) laser. *Hum Reprod* 1997; 12:2242–5

123. Van Blerkom J, Davis PW, Merriam J. The developmental ability of human oocytes penetrated at the germinal vesicle stage after insemination *in vitro*. *Hum Reprod* 1994;9: 697–708

124. Kim NH, Chung HM, Cha KY, Chung KS. Microtubule and microfilament organization in maturing human oocytes. *Hum Reprod* 1998;13:2217–22

125. Eppig JJ, Schultz RM, O'Brien M, Chesnel F. Relationship between the developmental programs controlling nuclear and cytoplasmic maturation of mouse oocytes. *Dev Biol* 1994;164:1–9

126. Kubiak JZ. Mouse oocytes gradually develop the capacity for activation during the metaphase II arrest. *Dev Biol* 1989;136:537–45

127. Tesarik J, Sousa M. Key elements of a highly efficient intracytoplasmic sperm injection technique: Ca^{2+} fluxes and oocyte cytoplasmic dislocation. *Fertil Steril* 1995;64: 770–6

128. Edwards RG, Van Steirteghem AC. Intracytoplasmic sperm injections (ICSI) and human fertilization: does calcium hold the key to success? *Hum Reprod* 1993;8: 988–9

129. Homa ST, Carroll J, Swann K. The role of calcium in mammalian oocyte maturation and egg activation. *Hum Reprod* 1993;8:1274–81

130. Tesarik J, Sousa M. More than 90% fertilization rates after intracytoplasmic sperm injection and artificial induction of oocyte activation with calcium ionophore. *Fertil Steril* 1995;63:343–9

131. Zhang J, Wang CW, Blaszcyzk A, et al. Electrical activation and *in vitro* development of human oocytes that fail to fertilize after intracytoplasmic sperm injection. *Fertil Steril* 1999;72:509–12

132. Ghetler Y, Raz T, Ben Nun I, Shalgi R. Cortical granules reaction after intracytoplasmic sperm injection. *Mol Hum Reprod* 1998;4:289–94

133. Bourgain C, Nagy ZP, De Zutter H, et al. Ultrastructure of gametes after intracytoplasmic sperm injection. *Hum Reprod* 1998;13(Suppl.1):107–16

134. Nagy ZP, Liu J, Joris H, *et al*. Time-course of oocyte activation, pronucleus formation and cleavage in human oocytes fertilized by intracytoplasmic sperm injection. *Hum Reprod* 1994;9:1743–8

135. Payne D, Flaherty SP, Barry MF, Matthews CD. Preliminary observations on polar body extrusion and pronuclear formation in human oocytes using time-lapse video cinematography. *Hum Reprod* 1997;12:532–41

136. Dieguez L, Soler C, Perez-Sanchez F, *et al*. Morphometric characterization of normal and abnormal human zygotes. *Hum Reprod* 1995;10:2339–42

137. De Mouzon J, Lancaster P (on behalf of the International Groups for Registers on Assisted Reproduction). World Collaborative Report. Presented at the *10th World Congress on In vitro Fertilization and Assisted Reproduction*, Vancouver, B.C., Canada, 1997

138. Tarlatzis BC, Bili H. Survey on intracytoplasmic sperm injection: report from the ESHRE ICSI Task Force. European Society of Human Reproduction and Embryology. *Hum Reprod* 1998;13(Suppl.1):165–77

139. Nygren KG, Andersen AN. Assisted reproductive technology in Europe, 1997. Results generated from European registers by ESHRE. European IVF-Monitoring Programme (EIM), for the European Society of Human Reproduction and Embryology (ESHRE). *Hum Reprod* 2001;16:384–91

140. Nygren KG, Andersen AN. Assisted reproductive technology in Europe, 1998. Results generated from European registers by ESHRE. *Hum Reprod* 2001;16: 2459–71

141. Lansac J, Thepot F, Mayaux MJ, *et al*. Pregnancy outcome after artificial insemination or IVF with frozen semen donor: a collaborative study of the French CECOS Federation on 21,597 pregnancies. *Eur J Obstet Gynecol Reprod Biol* 1997;74:223–8

142. Rosenwaks Z, Davis OK, Damario MA. The role of maternal age in assisted reproduction. *Hum Reprod* 1995;10(Suppl.1):165–73

143. Navot D, Bergh PA, Williams MA, *et al*. Poor oocyte quality rather than implantation failure as a cause of age-related decline in female fertility. *Lancet* 1991;337:1375–7

144. Van Steirteghem A, Devos A, Staessen C, *et al*. In Shoham Z, Howels CM, Jacobs HS, eds. *Should ICSI Apply to All IVF Cycles?* London: Blackwell Science, 1998:243–51

145. Rizk B, Doyle P, Tan SL, *et al*. Perinatal outcome and congenital malformations in *in vitro* fertilization babies from the Bourn-Hallam group. *Hum Reprod* 1991;6: 1259–64

146. Friedler S, Mashiach S, Laufer N. Births in Israel resulting from *in vitro* fertilization/embryo transfer, 1982–1989: National Registry of the Israeli Association for Fertility Research. *Hum Reprod* 1992;7:1159–63

147. Bowen JR, Gibson FL, Leslie GI, Saunders DM. Medical and developmental outcome at 1 year for children conceived by intracytoplasmic sperm injection. *Lancet* 1998;351: 1529–34

148. Ord T, Patrizio P, Balmaceda J, Asch R. Can severe male factor infertility be treated without micromanipulation? *Fertil Steril* 1993;60:110–15

149. Yovich JM, Edirisinghe WR, Cummins JM, Yovich JL. Influence of pentoxifylline in severe male male factor infertility. *Fertil Steril* 1990;53:715–22

150. Tournaye H, Janssens R, Camus M, *et al*. Pentoxifylline is not useful in enhancing sperm function in cases with previous *in vitro* fertilization failure. *Fertil Steril* 1993; 59:210–15

Techniques for embryo manipulation in human reproduction: assisted hatching, fragment removal, cytoplasmic transfer, nuclear transfer

24

Federica Moffa, John Zhang and Alberto Revelli

INTRODUCTION

During the last two decades, the introduction of specific techniques for the micromanipulation of gametes and embryos contributed to increase the pregnancy rates in human *in vitro* fertilization (IVF). Micromanipulation procedures such as intracytoplasmic sperm injection (ICSI) and embryo biopsy for prenatal genetic diagnosis (PGD) are now routinely performed in many embryology laboratories.

The aim to rescue suboptimal embryos by micromanipulation, thus improving their implantation potential, represents the rationale for the application of new micromanipulation techniques in human IVF. Despite several studies in the animal model, the techniques herein described remain to date essentially experimental, their effectiveness and safety in human IVF being still uncertain. Nevertheless, micromanipulation techniques offer the potential to create experimental models extremely useful for the comprehension of human gamete's and embryo physiology.

ASSISTED HATCHING

Since the early days of assisted reproductive technologies (ART), many steps of IVF have been modified in order to optimize results and therefore increase the pregnancy rate. Continuous progress in stimulation protocols, culture conditions, embryo transfer techniques and cryopreservation protocols has been achieved, but the overall chance of the transferred embryos implanting in the endometrium has not significantly improved, and implantation still represents a weak point in IVF, deeply affecting results.

Two main factors play a role in the implantation process: (1) the intrinsic embryonic quality, including the chromosomal asset[1-4]; and (2) the endometrial receptivity[5,6]. The ability of the embryo to escape the surrounding zona pellucida (ZP hatching) is one of the prerequisites for implantation, so that alterations in the ZP structure (e.g., increased thickness) may result in an impairment of the implantation potential. The ZP is a glycoprotein coat surrounding the mammalian oocyte and the embryo up to the blastocyst stage. In humans it is about 13–15 μm thick and it is produced by the oocyte itself and by the granulosa cells during the early phases of follicular growth. This extracellular matrix is mainly composed of three cross-linked glycoproteins (ZP1, ZP2 and ZP3), which form two layers of differing density and composition; the outer layer is thicker, but easy to dissolve with acidic Tyrode's solution in comparison with the inner layer[7]. The ZP has a structural and functional role during fertilization and early embryonic development. During fertilization the zona is able to bind the spermatozoon in a specific way and to induce the acrosome reaction, a prerequisite for sperm penetration[8]. Fertilization induces a sudden ZP hardening through the release of cortical granules located beneath the oolemma: this mechanism (cortical reaction) prevents penetration by additional spermatozoa and strongly contributes to avoid polyspermia[9]. During the preimplantation embryo development, the zona exerts a protective role: it prevents direct contact between the embryo and leukocytes, bacteria or substances (e.g., cytokines) that could damage the conceptus. Moreover, it prevents blastomere dispersion during the early developmental stages before embryo compaction, and it facilitates the journey of the embryo through the upper female reproductive tract, from the fertilization site to the uterine cavity[10]. At the blastocyst stage the embryo starts to hatch throughout the ZP. The increasing pressure of the expanding trophoblast, as well as the lytic action of both the uterine fluid and the lysins produced by the embryo itself, result in the progressive thinning of the ZP and in its rupture[11,12]. The embryo escaping the zona is able to implant within the endometrium.

Several studies have been performed both in the animal model and in humans to study the relationship between the hatching ability and the implantation potential of the embryo and the morphology of its ZP. The ZP thickness seems to be influenced by the hormonal environment during oogenesis, as it correlates with preovulatory serum estradiol and day 3 serum follicle stimulating hormone (FSH) levels; on average, it seems greater in patients with unexplained infertility compared to those with other infertility factors (endometriosis, tubal factor or male factor)[13]. Embryos showing zona thinning during their *in vitro* development appear to be more likely to implant after transfer[14]. Moreover, zygotes that stop cleaving during *in vitro* culture do not exhibit any change in zona thickness[15]. Palmstierna and co-workers[16] reported a pregnancy rate of 76.5% (26/34) when at least one of the transferred embryos showed ZP thinning of more than 20%, while the pregnancy rate dropped dramatically (1/22) when thinning was lower than 15% in all the transferred embryos. In contrast, other authors did not observe any correlation between ZP thickness and the implantation rate, and even the chemical removal of the outer layer of the zona in day 3 human embryos did not enhance implantation[17,18]. The hardness and resistance of the zona, rather than its overall thickness, could thus represent the most important factors for successful hatching and consequent implantation[17,19,20]. Both prolonged *in vitro* culture in suboptimal conditions and

cryopreservation procedures seem to induce ZP hardening, but the supplementation of culture and freezing media with serum, fetuin or other proteins such as albumin counteracts this phenomenon[21–24].

Overall, the proportion of *in vitro* produced embryos able to hatch is considerably lower than that observed in *in vivo* developing embryos, possibly because of the lack either of uterine fluid-derived lytic factors[11,12] or of the trophoectoderm-derived enzyme ZP lysin[20]. Since embryos transferred *in utero* after partial zona dissection (PZD) have been reported to possess a higher implantation potential[25], the artificial incision of the ZP by micromanipulation has been proposed as a tool to increase the implantation rate of *in vitro* produced embryos: this procedure is called 'assisted hatching'[26–28].

In vitro studies have demonstrated the safety of assisted hatching. Alikani and Cohen[29] cultured mouse embryos in a serum-free medium to induce ZP hardening and consequently hatching impairment; embryo hatching ability *in vitro* improved significantly with mechanical zona drilling, and there was no apparent embryo damage. An opening in the ZP was shown to be effective in increasing the *in vitro* hatching rate of human blastocysts, with no discernible blastomere loss or damage[30,31]. These observations led to the application of assisted hatching as a clinical procedure in IVF. Most IVF laboratories perform assisted hatching on the day of embryo replacement, usually on day 3 post-insemination or on day 5, when the procedure is applied to blastocyst-stage embryos.

Techniques of assisted hatching

Mechanically assisted hatching

This technique is similar to that used to assist ZP sperm penetration in PZD (see chapter 23). By using a micromanipulator, the zona of early cleaved embryos (two to eight cells) is pierced twice with a microneedle, leaving a small part of the zona trapped against the microneedle. If this part of the zona is rubbed against a holding pipette it is possible to create a slit opening in the ZP, but the artificial hole thus created must be big enough or abnormal hatching may result. In fact, the blastocyst may start to escape the zona before it is fully expanded and may then be trapped within the zona, causing the formation of trophoblastic vesicles or monozygotic twins when the inner cell mass splits as it passes through the narrow incision[32]. A three-dimensional PZD in the shape of a cross has been proposed to enlarge the artificial gap in the zona in order to facilitate hatching[33].

Chemically assisted hatching

The procedure is the same as that used to perform a blastomere biopsy for PGD. A micropipette is front-loaded with acidic Tyrode's solution (pH 2.35) and aligned against the embryo at the 3 o'clock position, facing either the empty perivitelline space or an area containing cytoplasmic fragments (Figure 1). With the tip of the micropipette gently touching the zona, the acidic solution is carefully expelled over a small area of the zona (about 30 μm). Depending on the time of exposure to the acidic solution, it is possible to obtain an area of thinning on the zona surface[34], or even to create a complete hole of about 20 μm diameter[28]. As soon as a gap is opened into the ZP, it is necessary to remove the excess acidic solution immediately and to rinse the embryo with fresh medium, as Tyrode's solution is toxic towards the embryo. The acidic Tyrode's solution, as well as the lytic enzyme pronase (10 IU/ml for 90 s), has been proposed to remove the zona pellucida totally and to obtain zona-free embryos for transfer[35–37].

Laser-assisted hatching

Different types of laser sources have been used for assisted hatching, including the Er:YAG laser at 2940 nm, the Nd:YAG laser at 1064 nm and the argon fluoride laser at 193 nm[38–42] (see also chapter 28). The microbeam can be delivered by either a 'contact' or 'non-contact' method. In contact mode, the laser beam is delivered through an optical fiber in direct contact with the ZP surface, but the need to manufacture and sterilize special glass pipettes and optical fibers every time is the main inconvenience of this method. The use of a non-contact technique overcomes this technical problem: the laser beam is passed through an optical lens and focused directly on the ZP surface, without the need to hold the embryo with a suction pipette. The most appropriate laser source for non-contact laser-assisted hatching appears to be the infrared diode laser at

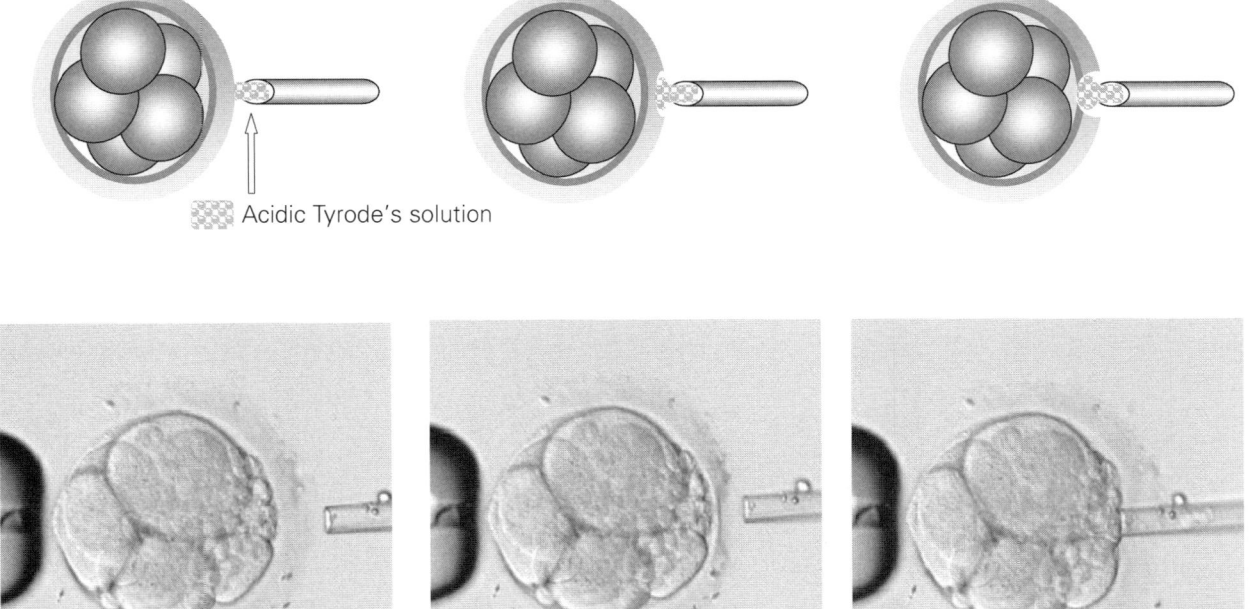

Acidic Tyrode's solution

Figure 1 Chemically assisted hatching. The acidic Tyrode's solution is front-loaded into the micropipette and gently expelled over the zona pellucida of a day 3 embryo. The lytic action of this solution opens a gap into the zona pellucida

1.48 μm[43,44]. Alternative laser sources have been tested; among them the krypton fluoride laser at 248 nm, the xenon chloride excimer laser at 308 nm and the PALM ultraviolet (UV) laser at 337 nm[45–47]. Using a laser, a very precise and localized lysis of the ZP can be achieved; it is possible either to create holes of a controlled size or just to thin a selected ZP area[48].

Piezo-assisted hatching

The piezoelectric pulse produces vibrations used to thin the ZP at a specific point; it is also possible to create a wider thinned ZP area by using repeated applications in adjacent points to obtain a hole of 20 μm diameter at the conjunction point of the blastomeres[49].

Clinical applications and effectiveness of assisted hatching

Is assisted hatching of help in improving embryo implantation rates? Which kind of patient could benefit from assisted hatching? Unfortunately, to date only a

few randomized studies are available to answer these questions (Table 1). Most of the studies on assisted hatching have been conducted in a non-randomized, retrospective way and their results are statistically inconclusive; while several of the contradictory findings reported are difficult to evaluate.

In the first prospective, randomized trial by Cohen[27], the application of assisted hatching to a non-selected group of embryos did seem to have an overall positive impact in terms of implantation of the manipulated embryos, the implantation rate in the study group (28%; 66/236) being significantly higher than that of the control embryos (22%; 50/226). However, in a subsequent similar study, this observation could not be confirmed as no statistically significant difference in implantation was observed between micromanipulated and control embryos[28]; nevertheless, embryos with a thick ZP were rescued by assisted hatching. Based on this finding, it was hypothesized that assisted hatching could be effective only in selected cases, and that it should be applied specifically to embryos with a poor prognosis due to a thickened ZP[28]. In a subsequent controlled,

Table 1 Prospective, randomized studies performed from 1996 to 2000 using different assisted hatching (AH) techniques

Author	No. of cycles	AH technique	Type of patients	PR and IR in AH group	PR and IR in control group	Conclusions
Tucker *et al.*, 1996[58]	100	Zona drilling	Unselected	30.0%/8.5%	32.0%/13.5%	No significant increase in PR and IR
Hellebaut *et al.*, 1996[57]	120	PZD	Unselected	42.1%/17.9%	38.1%/17.1%	No significant increase in PR and IR
Antinori *et al.*, 1996[47]	170	UV laser	Multiple IVF failures	44.4%/16.0%	19.3%/5.1%	Significant increase in PR and IR
Chao *et al.*, 1997[54]	64	PZD	Multiple IVF failures	42.4%/11.0%	16.1%/3.7%	Significant increase in PR and IR
Hurst *et al.*, 1998[59]	20	Acid Tyrode's	Good prognosis	23.0%/9.6%	43.0%/10.7%	No significant increase in PR and IR
Lazendorf *et al.*, 1998[55]	89	Acid Tyrode's	> 36 years old	39.0%/11.1%	41.7%/11.3%	No significant increase in PR and IR
Magli *et al.*, 1998[56]	248	Acid Tyrode's	> 38 years old and/or multiple IVF failures	33.0%/13.3%	12.0%/4.1%	Significant increase in PR and IR
Nakayama *et al.*, 1999[49]	248	Piezo micro-manipulator	Multiple IVF failures	19.4%/10.1%	5.9%/2.6%	Significant increase in PR and IR
Mansour *et al.*, 2000[37]	71	Acid Tyrode's zona removal	> 40 years old and/or multiple IVF failures	23.0%/6.8%	7.3%/1.9%	Significant increase in PR and IR

PR, pregnancy rate; IR, implantation rate; PZD, partial zone dissection; UV, ultraviolet; IVF, *in vitro* fertilization

randomized study, only embryos with a thick zona, a low developmental rate and fragmentation exceeding 20% were randomized and eventually submitted to zona drilling. The implantation rate here was significantly higher in the group undergoing selective assisted hatching (25%; 70/278) compared to the control group (18%; 51/258). Such a difference is emphasized when considering patients of over 38 years of age, in whom the implantation rate was 16% in the treated group against 3% in the control group. Moreover, in the same study assisted hatching was observed to be effective in selected patients with increased basal serum FSH (> 15 mIU/ml), the implantation rate reaching 26%.

Overall, however, among the numerous studies assessing the value of assisted hatching in a selected patient population[50–53], only a few prospective randomized trials demonstrated a real benefit in embryo implantation when either partial zona dissection[54], laser[47] or piezo zona drilling[49], or total zona removal[37] was performed in patients with repeated IVF failures. In other studies, when patient selection was based only on age, it was not possible to demonstrate an increased implantation rate for micromanipulated embryos in comparison with controls[55], but when the age factor was associated with repeated failed cycles as selection criteria[56] the difference was found to be statistically significant (Table 1). Finally, three prospective, randomized trials demonstrated that in either an unselected or a good-prognosis group of patients, assisted hatching has no effect on pregnancy or implantation rate[57–59] (Table 1).

In conclusion, assisted hatching is a micromanipulation technique that should not be applied routinely to all IVF patients, but when performed by expert operators it can be of help in rescuing poor-quality embryos and can be useful in selected groups of patients, resulting in increased rates of implantation and clinical pregnancy.

FRAGMENT REMOVAL

During early cleavage phases, a considerable proportion of *in vitro* produced human embryos show differing degrees of blastomere fragmentation, the significance and prognostic value of which are still controversial. Blastomere fragmentation has been claimed to be part of an apoptotic process leading to programed cell death in embryos with impaired developmental potential[60], and

on the basis of this claim the degree of fragmentation is cited as a morphologic parameter and is commonly used to select the best embryos for uterine transfer and to predict embryo implantation potential[61]. However, spontaneous fragmentation has been shown to be transient, and fragments may disappear by resorption or lysis in embryos with normal developmental competence[62]. The pattern, more than the degree of fragmentation, would appear to be related to an embryo's implantation potential[63]: the loss of large volumes of cytoplasm is probably detrimental in those embryos showing large fragments, while the presence of small, scattered fragments does not pose a serious threat to further development.

Cytoplasmic fragment removal has been proposed to rescue morphologically impaired embryos[63], based on the hypothesis that it could restore the spatial relationship of cells within the embryo, a prerequisite for normal compaction, cavitation and blastocyst formation. If fragments have a toxic impact on blastomeres, their removal could prevent the secondary degeneration of the remaining blastomeres immediately adjacent to fragments and cell degradation products.

The procedure of fragment removal is usually performed in conjunction with assisted hatching on the day of embryo transfer. A microneedle is introduced through the hole created in the ZP, and the fragments are carefully removed by gentle aspiration, constantly refocusing on the fragment membrane and on the tip of the micropipette (Figure 2). Fragment removal is a very delicate procedure, but performed by skillful operators it is safe; only on rare occasions are blastomeres accidentally lysed and therefore completely removed. Fragments that are in close contact with the blastomeres should not be removed because of the high risk of damage to the embryo. On the other hand, isolated and free fragments are easily aspirated and the injection of a small quantity of medium in the intercellular spaces facilitates the operation.

To date, no prospective, randomized studies have been conducted to assess the impact of fragment removal on pregnancy rates. Alikani and co-workers[63] retrospectively analyzed 2410 cycles in which more than one half of the embryos replaced for each cycle were homogeneous for the degree and/or the pattern of fragmentation; all the embryos showing more than 5% of fragmentation

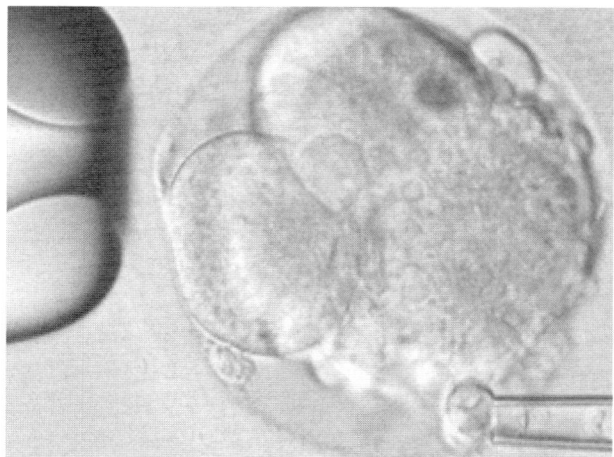

Figure 2 Fragment removal. Cytoplasmic fragments are gently removed from a day 3 embryo by a microneedle introduced through a gap created in the zona pellucida

were subjected to assisted hatching and fragment removal. This study demonstrated a significant decrease in implantation and pregnancy rates as fragmentation increased in degree and pattern, and showed a possible role of fragment removal in improving embryo implantation potential when applied to embryos with fragmentation not exceeding 35%.

CYTOPLASMIC TRANSFER

Among all patients undergoing ART treatments, a small but consistent proportion will experience poor embryo development and implantation failure in repeated IVF cycles despite the attempt to improve stimulation protocols and culture conditions, and to rescue impaired embryos by micromanipulation techniques such as assisted hatching and fragment removal. The reason (or reasons) for such impairment in embryo developmental potential are still under investigation, but there is probably more than one cause. The observation that embryo quality declines with increasing age in the mother has led several investigators to postulate that the metabolic capacity of the aging oocyte could be insufficient to sustain normal fertilization and regular embryo development[64–66]. The fact that mitochondria are clearly implicated in cell aging tends to substantiate the hypothesis that mitochondrial DNA (mtDNA) could play a role in female reproductive aging[67,68]. This

hypothesis is also well sustained by the observation that the level of mtDNA deletions is increased in the ovarian tissue and in the oocytes of older patients[65,69].

The metabolic competence of the fertilizing oocyte is strictly related to its mitochondrial content and to its capacity to generate ATP at levels sufficient to support normal chromosomal segregation and a biosynthetic process leading to regular mitotic divisions[70–72]. Around 200 000 mitochondria are contained in the human oocyte at the time of ovulation and their replication does not occur until the hatched blastocyst stage[66,70]; this mitochondrial complement is partitioned with each cell division in the early embryonic growth phases and any impairment in the original mitochondria number or function may be reflected in fertilization failure or embryo developmental abnormalities.

The critical role of mitochondria in normal embryo development and implantation led to the hypothesis that the transfer of cytoplasm derived from a fertile donor oocyte into a recipient oocyte may help in the rescue of metabolically impaired gametes by providing an infusion of normal mitochondria, together with other cytoplasmic organelles, stored mRNAs and proteins[73–77]. In 1997, after a few experiments in the animal model[78,79], Cohen[73] reported for the first time the birth of an infant conceived after the transfer of anucleate donor cytoplasm into a recipient egg subsequently fertilized by ICSI.

Cytoplasmic transfer may be performed using different techniques of ooplasm transfusion and different sources of donor cytoplasm. Most cases have been performed using fresh MII-stage oocytes retrieved simultaneously from the donor and from the recipient[74,80,81]. The use of cytoplasm derived from oocytes at the same developmental stage is probably advantageous, as it allows transfer into the recipient egg cellular components having a specific role at that particular cell stage. The disadvantage of fresh, synchronous cytoplasmic transfer is the need to synchronize both the patient's and the donor's ovarian stimulation protocols. To find a way to avoid this inconvenience, Lanzendorf and co-workers[77] performed cytoplasmic transfer using frozen MII oocytes as the source of the cytoplasm in four patients, all of whom had a history of repeated IVF failures and poor embryo quality, and a twin pregnancy resulted. Due to the limited number of reported cases, however, it is difficult to speculate on the effectiveness of cytoplasmic transfer and even more so on the possible use of cryopreserved eggs as a source of cytoplasm. Another possible source of cytoplasm are spared tripronucleate (3-PN) zygotes[76]: a few cases of asynchronous cytoplasmic transfer have been performed with encouraging results. In this case, though, it is difficult to understand the mechanism leading to the developmental improvement of the embryo after the transfer into an MII oocyte of a very small amount of cytoplasm derived from a zygote, donor and recipient cells being in a totally different stage of the cell cycle.

The most commonly used technique in both synchronous and asynchronous cytoplasmic transfer is the injection of a small amount of donor cytoplasm during the ICSI procedure[74,77,80–82]. Under an inverted microscope and using a micromanipulator, a standard ICSI needle is employed to immobilize and aspirate a spermatozoon; after a few seconds, this is inserted into the donor oocyte in an area opposite the polar body in order to avoid the spindle and minimize the risk of donor nuclear DNA contamination. Approximately 5–15% of the cytoplasm is removed from the donor oocyte by suction, the spermatozoon being always at the top of the column contained in the pipette. The needle is then introduced through the oolemma of the recipient oocyte, and the donor ooplasm together with the spermatozoon are injected into the recipient cytoplasm. Each donor oocyte can be used to provide cytoplasm for two or more patient's eggs.

A few cases of cytoplasmic transfer have been performed by electrofusion[74]. A small amount of cytoplasm is removed from the donor oocyte with a micropipette to form a cytoplast fragment surrounded by the oolemma. The donor cytoplast is inserted in the perivitelline space of the recipient oocyte under the ZP and electric pulses are applied in order to induce membrane fusion and mixing of the two cytoplasms. However, as no morphologic improvement of the embryo or pregnancy rate enhancement was noted after preliminary experiments with electrofusion-driven cytoplasmic transfer, the technique was abandoned in favor of microinjection.

So far, cytoplasmic transfer has been performed in patients whose history of failed IVF cycles was due to poor embryo quality. Despite the low number of such cases and the lack of prospective, controlled randomized studies, the pregnancy rates obtained (between 25 and 43% in the different studies) are satisfactory. It is claimed that the improvement in implantation rate is derived from the enhancement in morphology and the developmental potential of embryos after ooplasm transfusion, but further evidence is needed to substantiate this argument.

To date around 30 babies have been born after cytoplasmic transfer[74,76,80]. Despite the health of the babies being apparently normal according to reports, there is a growing concern about such a technique being applied without controlled studies in a suitable animal model being carried out first. At present, it is not known which mechanisms lead to embryo rescue after the infusion of such a small quantity of donor cytoplasm into the recipient egg, and it is therefore impossible to predict the short- and long-term consequences of this procedure for the newborn. Nevertheless, Barritt and co-workers[80] reported that two 45,XO karyotypes (Turner's syndrome) were conceived out of 18 fetuses after cytoplasmic transfer: one pregnancy resulted in a miscarriage and the other was terminated at 16 weeks. Compared to the overall incidence of a 45,XO karyotype in amniocentesis and chorionic villus sampling (0.2% and 0.4%, respectively), as well as to the rate of 1 to 6% of major congenital abnormalities in the natural population, the incidence of 11% (2/18) is higher than expected. Although no conclusion can be drawn from such a small sample, it is possible that the hybrid cytoplasm resulting from ooplasmic transfusion may improve the developmental potential in chromosomally normal as well as abnormal embryos.

The hybrid cytoplasm resulting from ooplasmic transfusion carries two different mtDNA populations: it has been demonstrated by confocal fluorescent imaging that donor mitochondria persist and replicate in the recipient cytoplasm, and that they may be found in embryo blastomeres[80]. In pregnancies occurring after cytoplasmic transfer, the presence of a small proportion of donor mtDNA was detected in amniotic fluid-derived cells, placental cells and fetal cord blood cells; two out of 15 children born still had traces of donor mtDNA in their blood cells at 1 year of age, as confirmed by mtDNA fingerprinting[80]. The consequences of such mtDNA heteroplasmy on maternal and paternal genome function, gene expression and epigenetic modification patterns are still unknown. It is generally thought that mitochondria exist in the body at a high level of homoplasmy as a result of strictly maternal inheritance. The paternal set of mitochondria are eliminated at fertilization in an attempt, supposedly, to minimize possible conflicts between nuclear and mitochondrial control elements[82]. Moreover, cytoplasmic factors exert epigenetic effects upon maternal and paternal genomes by mechanisms such as methylation on nuclear genes. In other words, much of the epigenetic information displayed in fetal and adult mammals is not completely elaborated in the genome, but is acquired onto-genetically[83]. In this scenario, the artificial infusion of an exogenous mitochondrial population may interfere with the regulation of imprinted genes, resulting in an aberrant expression of the parental genome possibly associated with diseases in the conceptus[83]. To this purpose, however, recent studies demonstrated that heteroplasmy in humans may be more common than suspected and may have a pathologic significance only when mitochondria contain rearranged mtDNA, with decreased functionality[84]. Also, the presence of mtDNA heteroplasmy in calves cloned from embryonic cells seems to indicate the existence of some compatibility between oocyte and embryo mito-chondria[85,86].

In conclusion, cytoplasmic transfer is a micromanipu-lation technique with some potential in human IVF, but its clinical application cannot be recommended until basic research has clarified the underlying biological mechanisms, and randomized studies have been conducted to prove its real effectiveness and safety.

NUCLEAR TRANSFER

The transfer of a cell nucleus into a recipient enucleated oocyte to obtain reconstructed female gametes by micro-manipulation is a fascinating and challenging step in the treatment of infertility, and is currently under investiga-tion in animal models as well as in humans. Essentially based on previous cloning studies, the technique of nuclear transfer has been touted as a potential remedy for various pathologies in human reproduction, including age-related female infertility, premature ovarian failure, mitochondrial diseases and ovarian failure induced by chemo- or radiotherapy.

The idea of using nuclear transfer to combat severe female infertility due to poor oocyte quality is based on the observation that the prevalence of IVF patients with poor embryo quality and a low implantation rate increases in parallel with the woman's age because of a higher incidence of embryo aneuploidy, linked to an abnormal state of the meiotic spindle[1–4]. Cytoplasmic factors appear to be responsible for such structural abnormalities of the meiotic spindle, resulting in chromosomal malsegregation[87]. The replacement of the entire cytoplasm of an 'old' oocyte with the cytoplasm derived from a 'young' oocyte (the oocyte of a younger patient) can be achieved by nuclear transfer with oocytes at the germinal-vesicle (GV) stage. Theoretically, oocytes containing specific cytoplasmic factors able to support a more normal meiotic spindle formation in older patients are obtained by use of the GV transfer procedure[88], thus reducing the aneuploidy rate in *in vitro* produced embryos in this population.

Furthermore, nuclear transfer has the potential to prevent the transmission of mitochondrial diseases. In fact, the replacement of a patient's ooplasm with a healthy donor's cytoplasm containing normal mito-chondria should prevent the transmission of the mitochondria-linked disease to the conceptus.

Technique of nuclear transfer

The nuclear transfer technique to reconstruct female gametes using the nucleus of a donor oocyte and the cytoplasm of a recipient oocyte may be performed at different stages of oocyte nuclear maturation (GV or MII). When the donor and the recipient oocyte have the same nuclear maturation, the nuclear transfer is termed 'synchronous', whereas if the stages are different it is termed 'asynchronous'.

Synchronous nuclear transfer seems to be the more effective of the two, and has been applied to human and other mammalian oocytes, more frequently in the case of GV-stage oocytes (GV transfer)[88–92]. The technique used is as follows (Figure 3). Immature oocytes at the GV stage (the donor oocyte and the recipient oocyte, belong-ing to different patients) are collected simultaneously and subsequently incubated for 30 min with substances able to increase the oolemma elasticity (e.g., cytochalasin B). Afterwards, under an inverted microscope and

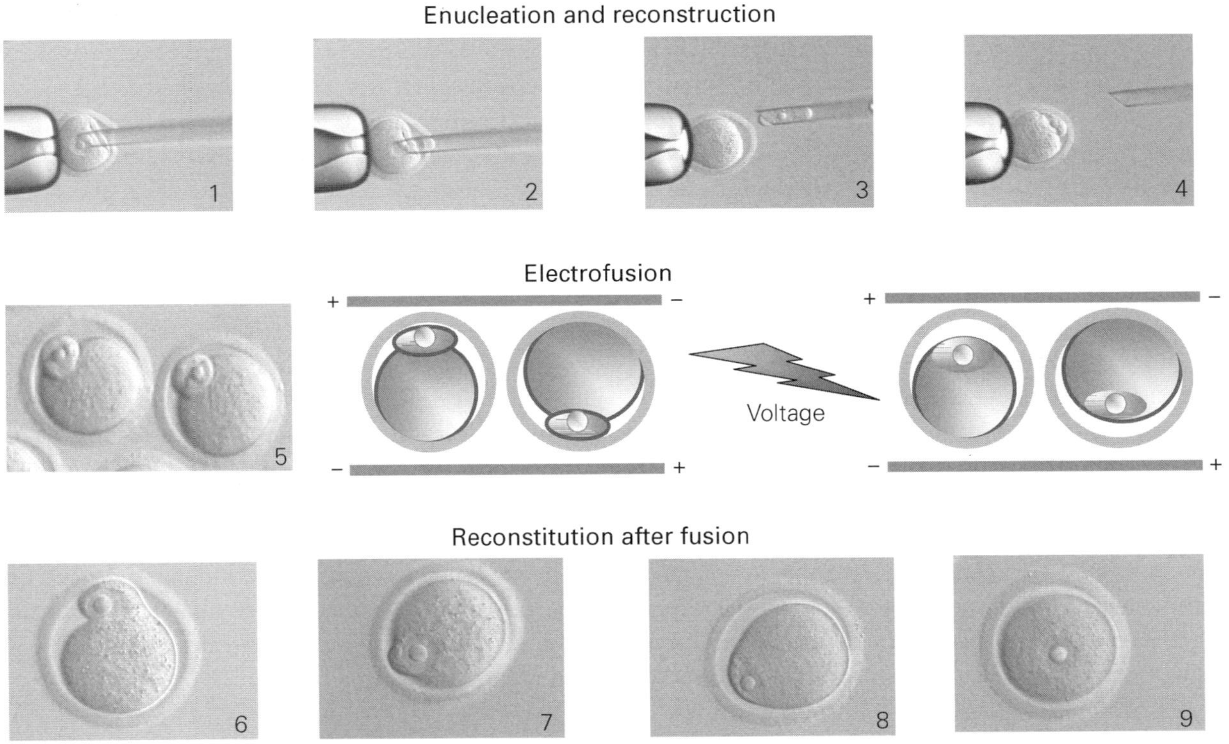

Figure 3 Nuclear transfer procedure with oocytes at the germinal vesicle (GV) stage. Reconstructed oocytes are obtained by micromanipulation and electrofusion. Stages 1–2, removal of the karyoplast containing the GV; stages 3–4, transfer into the perivitelline space of a previously prepared cytoplast. Stage 5, the karyoplast–cytoplast complexes are subjected to electrofusion. Stages 6–8, after the application of an electric pulse, the karyoplast progressively integrates into the cytoplasm. The reconstituted oocyte (stage 9) is morphologically identical to a non-manipulated oocyte, but contains cell components (cytoplast and karyoplast) derived from two different oocytes

utilizing a micromanipulator, a small opening in the ZP of both oocytes is created by either mechanical or chemical (Tyrode's acidic solution) PZD. A beveled glass micropipette with an inner diameter of 20 μm is introduced through the zona opening and the GV is gently aspirated and removed from both oocytes without breaking the oolemma. As a result of this procedure, two separate cell components are obtained from each oocyte: the karyoplast, containing the GV surrounded by a small amount of cytoplasm and encapsulated by a portion of the oolemma, and the cytoplast, which represents the enucleated oocyte cytoplasm, surrounded by the ZP. The karyoplast of the donor oocyte is then transferred into the perivitelline space of the cytoplast of the recipient oocyte, in direct contact with the oolemma. The next stage sees the fusion of the two cell components, achieved either by electrofusion[88] or by chemical or mechanical methods[93]. With fusion, the GV of

one oocyte and the enucleated cytoplasm of the other combine to form an oocyte. This is morphologically identical to non-manipulated eggs, but needs to be cultured *in vitro* for several hours to undergo GV breakdown and progress to the MII stage prior to fertilization. It has been shown, in fact, that the cytoplasm of GV-stage oocytes is able to aploidize diploid nuclei within a few hours of *in vitro* culture.

Synchronous nuclear transfer has also been performed using MII-stage oocytes, in both humans and animal models[93,94]. The procedure is similar to the previously described GV transfer, but in the case of MII oocytes the nucleus is not evident, and the oocyte DNA is removed by suction of the polar body and of the adjacent cytoplasm, presumably containing chromosomes aligned at the MII plate; this enucleation procedure may also be performed under UV light after spindle staining with fluorescent substances[90].

To date, the results of attempts to perform asynchronous nuclear transfer between oocytes having a different nuclear maturation have been poor: when the GVs were transferred into enucleated MII oocytes, only a few reconstructed oocytes underwent GV breakdown and none reached the MII stage[90]. This could be because of the fact that a more mature cytoplasm (coming from an MII oocyte) lacks the specific factors that promote nuclear aploidization – factors which are more abundant in the cytoplasm of GV-stage oocytes.

In order to overcome the problem generated by the need to have synchronous oocytes to perform a successful nuclear transfer, the use of frozen oocytes has been proposed. In the mouse model, the GV transfer has been proven to be effective even using frozen–thawed karyoplasts and cytoplasts, with no significant differences in terms of post-GV transfer oocyte maturation and chromosomal asset with respect to oocytes reconstructed using fresh material[92]. Speculating on this, the possibility of using frozen–thawed karyoplasts and cytoplasts could

mean it might be possible to cryopreserve the GV instead of the whole oocyte, thus enabling the egg to avoid the well-documented, low temperature-linked, cytoplasmic damage. Again, in the case of prepubertal girls at risk of sterilization because of oncostatic therapies, GVs could be taken from cryopreserved ovarian slices and oocytes could hypothetically be reconstructed using these GVs and fresh cytoplasts from donor patients.

More recently, oocytes have been reconstructed by the transfer of a patient's somatic cell nucleus to a recipient donor ooplast[93]. This technique, very similar to the previously described GV transfer and based on the ability of a GV oocyte cytoplasm to aploidize diploid nuclei, opens up the possibility of obtaining gametes from patients with no available oocytes starting from somatic cells.

Before the nuclear transfer techniques described here can be employed in the field of human reproduction, however, further studies about their feasibility, safety and effectiveness must be carried out.

References

1. Plachot M. Viability of preimplantation embryos. *Baillière's Clin Obst Gynaecol* 1992;6:327–38
2. van Kooij RJ, Looman CW, Habbema JD, *et al*. Age-dependent decrease in embryo implantation rate after *in vitro* fertilization. *Fertil Steril* 1996;66:769–75
3. Munne S, Alikani M, Tomkin G, *et al*. Embryo morphology, developmental rates, and maternal age are correlated with chromosome abnormalities. *Fertil Steril* 1995;64:382–91
4. Magli MC, Gianaroli L, Ferraretti AP. Chromosomal abnormality in embryos. *Mol Cell Endocrinol* 2001;22:183 (Suppl. 1):S29–34
5. Simon C, Dominguez F, Remohi J, *et al*. Embryo effects in human implantation: embryonic regulation of endometrial molecules in human implantation. *Ann N Y Acad Sci* 2001; 943:1–16
6. Damario MA, Lesnick TG, Lessey BA, *et al*. Endometrial markers of uterine receptivity utilizing the donor oocyte model. *Hum Reprod* 2001;16:1893–9
7. Harris HD, Hibler DW, Fontenot GK, *et al*. Cloning and characterization of zona pellucida genes and cDNAs from a variety of mammalian species: the ZPA, ZPB and ZPC gene families. *DNA Seq* 1994;4:361–93
8. Prasad SV, Skinner SM, Cario C, *et al*. Structure and function of the proteins of the mammalian zona pellucida. *Cell Tissues Organs* 2000;166:148–64
9. Wassarman PM. Early events in mammalian fertilization. *Annu Rev Cell Biol* 1987;3:109–42
10. Dunbar BS, Avery S, Lee V, *et al*. The mammalian zona pellucida: its biochemistry, immunochemistry, molecular biology, and developmental expression. *Reprod Fertil Dev* 1994;6:331–47
11. Lin SP, Lee RK, Tsai YJ. *In vivo* hatching phenomenon of mouse blastocysts during implantation. *J Assist Reprod Genet* 2001;18:341–5
12. Montag M, Koll B, Holmes P, van der Ven H. Significance of the number of embryonic cells and the state of the zona pellucida for hatching of mouse blastocysts *in vitro* versus *in vivo*. *Biol Reprod* 2000;62:1738–44
13. Loret de Mola JR, Garside WT, Bucci J, *et al*. Analysis of the human zona pellucida during culture: correlation with diagnosis and the preovulatory hormonal environment. *J Assist Reprod Genet* 1997;14:332–6
14. Cohen J, Inge KL, Suzman M, *et al*. Videocinematography of fresh, and cryopreserved embryos: a retrospective analysis of embryonic morphology and implantation. *Fertil Steril* 1989; 51:820–7
15. Chan PJ. Developmental potential of human oocytes according to zona pellucida thickness. *J In Vitro Fertil Embryo Transf* 1987;4:237–41
16. Palmstierna M, Murkes D, Csemiczdy G, *et al*. Zona pellucida thickness variation and occurrence of visible mononucleated blastomeres in preembryos are associated with a high pregnancy rate in IVF treatments. *J Assist Reprod Genet* 1998;15:70–5

17. Tucker MJ, Luecke NM, Wiker SR, Wright G. Chemical removal of the outside of the zona pellucida of day 3 human embryos has no impact on implantation rate. *J Assist Reprod Genet* 1993;10:187–91

18. Janssens RM, Carlé E, De Clerck P, *et al.* Can zona pellucida thickness predict the implantation rate? *Hum Reprod* 1994; 9(Suppl. 4):78–9

19. Tucker MJ, Wiker SR, Kort HI. Embryonal zona pellucida thinning and uterine transfer. *Assist Reprod Rev* 1993;3: 168–71

20. Schiewe MC, Araujo E Jr, Asch RH, Balmaceda JP. Enzymatic characterization of zona pellucida hardening in human eggs and embryos. *J Assist Reprod Genet* 1995;12: 2–7

21. Eppig JJ, Wigglesowrth K, O'Brien MJ. Comparison of embryonic developmental competence of mouse oocytes grown with and without serum. *Mol Reprod Dev* 1992;32: 33–40

22. Manna C, Rienzi L, Greco E, *et al.* Zona pellucida solubility and cortical granule complements in human oocytes following assisted reproductive techniques. *Zygote* 2001;9:201–10

23. Hoyer PE, Terkelsen OB, Grete Byskov A, *et al.* Fetuin and fetuin messenger RNA in granulosa cells of the rat ovary. *Biol Reprod* 2001;65:1655–62

24. George MA, Johnson MH. Use of fetal bovine serum substitutes for the protection of the mouse zona pellucida against hardening during cryoprotectant addition. *Hum Reprod* 1993;8:1898–900

25. Cohen J, Malter H, Wright G, *et al.* Partial zona dissection of human oocytes when failure of zona pellucida penetration is anticipated. *Hum Reprod* 1989;4:435–42

26. Cohen J, Elsner C, Kort H, *et al.* Impairment of the hatching process following IVF in the human and improvement of implantation by assisting hatching using micromanipulation. *Hum Reprod* 1990;5:7–13

27. Cohen J. Assisted hatching of human embryos. *J In Vitro Fertil Embryo Transf* 1991;8:179–90

28. Cohen J, Alikani M, Trowbridge J, Rosenwaks Z. Implantation enhancement by selective assisted hatching using zona drilling of human embryos with poor prognosis. *Hum Reprod* 1992;7:685–91

29. Alikani M, Cohen J. Micromanipulation of cleaved embryos cultured in protein-free medium: a mouse model for assisted hatching. *J Exp Zool* 1992;263:458–63

30. Mandelbaum J, Plachot M, Junca AM, *et al.* The effects of partial zona dissection on *in vitro* development hatching of human cryopreserved embryo. *Hum Reprod* 1994; 9(Suppl. 4):39

31. Dokras A, Ross C, Gosden B, *et al.* Micromanipulation of human embryos to assist hatching. *Fertil Steril* 1994;61: 514–20

32. Malter HE, Cohen J. Blastocyst formation and hatching *in vitro* following zona drilling of mouse and human embryos. *Gamete Res* 1989;24:67–80

33. Cieslak J, Ivakhnenko V, Wolf G, *et al.* Three-dimensional partial zona dissection for preimplantation genetic diagnosis and assisted hatching. *Fertil Steril* 1999;71:308–13

34. Khalifa EA, Tucker MJ, Hunt P. Cruciate thinning of the zona pellucida for more successful enhancement of blastocyst hatching in the mouse. *Hum Reprod* 1992;7:532–6

35. Fong CY, Bongso A, Sathananthan H, *et al.* Ultrastructural observations of enzymatically treated human blastocysts: zona-free blastocyst transfer and rescue of blastocysts with hatching difficulties. *Hum Reprod* 2001;16:540–6

36. Fong CY, Bongso A, Ng SC, *et al.* Blastocyst transfer after enzymatic treatment of the zona pellucida: improving *in-vitro* fertilization and understanding implantation. *Hum Reprod* 1998;13:2926–32

37. Mansour RT, Rhodes CA, Aboulghar MA, *et al.* Transfer of zona-free embryos improves outcome in poor prognosis patients: a prospective randomized controlled study. *Hum Reprod* 2000;15:1061–4

38. Tadir Y, Wright WH, Vafa O, *et al.* Micromanipulation of gametes using laser microbeams. *Hum Reprod* 1991;6: 1011–16

39. Feichtinger W, Strohmer H, Fuhrberg P, *et al.* Photoablation of oocyte zona pellucida by erbium-YAG laser for *in-vitro* fertilisation in severe male infertility. *Lancet* 1992;339:811

40. Antinori S, Panci C, Selman HA, *et al.* Zona thinning with the use of laser: a new approach to assisted hatching in humans. *Hum Reprod* 1996;11:590–4

41. Laufer N, Palanker D, Shufaro Y, *et al.* The efficacy and safety of zona pellucida drilling by a 193-nm excimer laser. *Fertil Steril* 1993;59:889–95

42. Coddington CC, Veeck LL, Swanson RJ, *et al.* The YAG laser used in micromanipulation to transect the zona pellucida of hamster oocytes. *J Assist Reprod Genet* 1992;9:557–63

43. Germond M, Nocera D, Senn A, *et al.* Microdissection of mouse and human zona pellucida using a 1.48-microns diode laser beam: efficacy and safety of the procedure. *Fertil Steril* 1995;64:604–11

44. Blake DA, Forsberg AS, Johansson BR, Wikland M. Laser zona pellucida thinning – an alternative approach to assisted hatching. *Hum Reprod* 2001;16:1959–64

45. Blanchet GB, Russell JB, Fincher CR Jr, Portmann M. Laser micromanipulation in the mouse embryo: a novel approach to zona drilling. *Fertil Steril* 1992;57:1337–41

46. Neer J, Tadir Y, Ho P, *et al.* Microscope-delivered ultraviolet laser zona dissection: principles and practices. *J Assist Reprod Genet* 1992;9:513–23

47. Antinori S, Selman HA, Caffa B, *et al.* Zona opening of human embryos using a non-contact UV laser for assisted hatching in patients with poor prognosis of pregnancy. *Hum Reprod* 1996;11:2488–92

48. Mantoudis E, Podsiadly BT, Gorgy A, *et al.* A comparison between quarter, partial and total laser assisted hatching in selected infertility patients. *Hum Reprod* 2001;16:2182–6

49. Nakayama T, Fujiwara H, Yamada S, *et al.* Clinical application of a new assisted hatching method using a piezo-micromanipulator for morphologically low-quality embryos in poor-prognosis infertile patients. *Fertil Steril* 1999;71: 1014–18

50. Schoolcraft WB, Schlenker T, Gee M, *et al.* Assisted hatching in the treatment of poor prognosis *in vitro* fertilization candidates. *Fertil Steril* 1994;62:551–4

51. Schoolcraft WB, Schlenker T, Jones GS, Jones HW Jr. *In vitro* fertilization in women age 40 and older: the impact of assisted hatching. *J Assist Reprod Genet* 1995;12:581–4

52. Stein A, Rufas O, Amit S, *et al.* Assisted hatching by partial zona dissection of human pre-embryos in patients with

recurrent implantation failure after *in vitro* fertilization. *Fertil Steril* 1995;63:838–41

53. Graham MC, Hoeger KM, Phipps WR. Initial IVF-ET experience with assisted hatching performed 3 days after retrieval followed by day 5 embryo transfer. *Fertil Steril* 2000; 74:668–71

54. Chao KH, Chen SU, Chen HF, *et al*. Assisted hatching increases the implantation and pregnancy rate of *in vitro* fertilization (IVF)-embryo transfer (ET), but not that of IVF-tubal ET in patients with repeated IVF failures. *Fertil Steril* 1997;67:904–8

55. Lanzendorf SE, Nehchiri F, Mayer JF, *et al*. A prospective, randomized, double-blind study for the evaluation of assisted hatching in patients with advanced maternal age. *Hum Reprod* 1998;13:409–13

56. Magli MC, Gianaroli L, Ferraretti AP, *et al*. Rescue of implantation potential in embryos with poor prognosis by assisted zona hatching. *Hum Reprod* 1998;13:1331–5

57. Hellebaut S, De Sutter P, Dozortsev D, *et al*. Does assisted hatching improve implantation rates after *in vitro* fertilization or intracytoplasmic sperm injection in all patients? A prospective randomized study. *J Assist Reprod Genet* 1996;13:19–22

58. Tucker MJ, Morton PC, Wright G, *et al*. Enhancement of outcome from intracytoplasmic sperm injection: does co-culture or assisted hatching improve implantation rates? *Hum Reprod* 1996;11:2434–7

59. Hurst BS, Tucker KE, Awoniyi CA, Schlaff WD. Assisted hatching does not enhance IVF success in good-prognosis patients. *J Assist Reprod Genet* 1998;15:62–4

60. Jurisicova A, Varmuza S, Casper RF. Programmed cell death and human embryo fragmentation. *Mol Hum Reprod* 1996;2: 93–8

61. Giorgetto C, Terrou P, Auquier P, *et al*. Embryo score to predict implantation after *in vitro* fertilization: based on 957 single embryo transfers. *Hum Reprod* 1995:2427–31

62. Van Blerkom J, Davis P, Alexander S. A microscopic and biochemical study of fragmentation phenotypes in stage-appropriate human embryos. *Hum Reprod* 2001;16:719–29

63. Alikani M, Cohen J, Tomkin G, *et al*. Human embryo fragmentation *in vitro* and its implications for pregnancy and implantation. *Fertil Steril* 1999;71:836–42

64. Gaulden ME. Maternal age effect: the enigma of Down syndrome and other trisomic conditions. *Mutat Res* 1992; 296:69–88

65. Keefe DL, Niven-Fairchild T, Powell S, Buradagunta S. Mitochondrial deoxyribonucleic acid deletions in oocytes and reproductive aging in women. *Fertil Steril* 1995;64:577–83

66. Van Blerkom J, Davis PW, Lee J. ATP content of human oocytes and developmental potential and outcome after *in-vitro* fertilization and embryo transfer. *Hum Reprod* 1995; 10:415–24

67. Jansen RP, de Boer K. The bottleneck: mitochondrial imperatives in oogenesis and ovarian follicular fate. *Mol Cell Endocrinol* 1998;145:81–8

68. Kirkwood TB, Kowald A. Network theory of aging. *Exp Gerontol* 1997;32:395–9

69. Kitagawa T, Suganuma N, Nawa A, *et al*. Rapid accumulation of deleted mitochondrial deoxyribonucleic acid in postmenopausal ovaries. *Biol Reprod* 1993;49:730–6

70. Reynier P, May-Panloup P, Chretien MF, *et al*. Mitochondrial DNA content affects the fertilizability of human oocytes. *Mol Hum Reprod* 2001;7:425–9

71. Cummins JM. Mitochondria: potential roles in embryogenesis and nucleocytoplasmic transfer. *Hum Reprod Update* 2001;7:217–28

72. Barnett DK, Bavister BD. What is the relationship between the metabolism of preimplantation embryos and their developmental competence? *Mol Reprod Dev* 1996;43:105–33

73. Cohen J, Scott R, Schimmel T, *et al*. Birth of infant after transfer of anucleate donor oocyte cytoplasm into recipient eggs. *Lancet* 1997;350:186–7

74. Cohen J, Scott R, Alikani M, *et al*. Ooplasmic transfer in mature human oocytes. *Mol Hum Reprod* 1998;4: 269–80

75. Van Blerkom J, Sinclair J, Davis P. Mitochondrial transfer between oocytes: potential applications of mitochondrial donation and the issue of heteroplasmy. *Hum Reprod* 1998; 13:2857–68

76. Huang CC, Cheng TC, Chang HH, *et al*. Birth after the injection of sperm and the cytoplasm of tripronucleate zygotes into metaphase II oocytes in patients with repeated implantation failure after assisted fertilization procedures. *Fertil Steril* 1999;72:702–6

77. Lanzendorf SE, Mayer JF, Toner J, *et al*. Pregnancy following transfer of ooplasm from cryopreserved–thawed donor oocytes into recipient oocytes. *Fertil Steril* 1999;71:575–7

78. Flood JT, Chillik CF, van Uem JF, *et al*. Ooplasmic transfusion: prophase germinal vesicle oocytes made developmentally competent by microinjection of metaphase II egg cytoplasm. *Fertil Steril* 1990;53:1049–54

79. Levron J, Willadsen S, Bertoli M, Cohen J. The development of mouse zygotes after fusion with synchronous and asynchronous cytoplasm. *Hum Reprod* 1996;11:1287–92

80. Barritt J, Willadsen S, Brenner C, Cohen J. Cytoplasmic transfer in assisted reproduction. *Hum Reprod Update* 2001; 7:428–35

81. Brenner CA, Barritt JA, Willadsen S, Cohen J. Mitochondrial DNA heteroplasmy after human ooplasmic transplantation. *Fertil Steril* 2000;74:573–8

82. Hurst LD, McVean GT. Clade selection, reversible evolution and the persistence of selfish elements – the evolutionary dynamics of cytoplasmic incompatibility. *Proc R Soc Lond* 1996;263:97–104

83. Latham KE. Epigenetic modification and imprinting of the mammalian genome during development. *Curr Top Dev Biol* 1999;43:1–49

84. Falls JG, Pulford DJ, Wylie AA, Jirtle RL. Genomic imprinting: implications for human disease. *Am J Pathol* 1999;154:635–47

85. Grzybowski T. Extremely high levels of human mitochondrial DNA heteroplasmy in single hair roots. *Electrophoresis* 2000; 21:548–53

86. Hiendleder S, Schmutz SM, Erhardt G, *et al*. Transmitochondrial differences and varying levels of heteroplasmy in nuclear transfer cloned cattle. *Mol Reprod Dev* 1999;54:24–31

87. Battaglia DE, Goodwin P, Klein NA, Soules MR. Influence of maternal age on meiotic spindle assembly in oocytes from naturally cycling women. *Hum Reprod* 1996;11:2217–22

88. Zhang J, Wang CW, Krey L, *et al. In vitro* maturation of human preovulatory oocytes reconstructed by germinal vesicle transfer. *Fertil Steril* 1999;71:726–31

89. Takeuchi T, Ergun B, Huang TH, *et al.* A reliable technique of nuclear transplantation for immature mammalian oocytes. *Hum Reprod* 1999;14:1312–17

90. Liu H, Wang CW, Grifo JA, *et al.* Reconstruction of mouse oocytes by germinal vesicle transfer: maturity of host oocyte cytoplasm determines meiosis. *Hum Reprod* 1999;14:2357–61

91. Takeuchi T, Gong J, Veeck LL, *et al.* Preliminary findings in germinal vesicle transplantation of immature human oocytes. *Hum Reprod* 2001;16:730–6

92. Moffa F, Comoglio F, Krey LC, *et al.* Germinal vesicle transfer between fresh and cryopreserved immature mouse oocytes. *Hum Reprod* 2002;17:178–83

93. Tesarik J, Nagy ZP, Mendoza C, Greco E. Chemically and mechanically induced membrane fusion: non-activating methods for nuclear transfer in mature human oocytes. *Hum Reprod* 2000;15:1149–54

94. Wang MK, Chen DY, Liu JL, *et al. In vitro* fertilisation of mouse oocytes reconstructed by transfer of metaphase II chromosomes results in live births. *Zygote* 2001;9:9–14

New technologies in human infertility

Modern transcervical techniques for the diagnosis and treatment of tubal obstruction

25

Ilan Tur-Kaspa

INTRODUCTION

Tubal obstruction is one of the main causes of female infertility. The introduction of assisted reproductive technologies (ART), especially *in vitro* fertilization (IVF), changed dramatically the management of women with mechanical infertility in the 1980s. With the increased popularity of IVF treatment, which basically bypasses the fallopian tubes, the need for better diagnosis and treatment to restore tubal function decreased. During the last decade, however, improved imaging techniques and miniature catheters, as well as the high cost of ART, led to a reversal of this trend.

This chapter will discuss modern transcervical techniques for the diagnosis and treatment of tubal obstruction. In 1985, pioneer researchers[1,2] were the first to report a successful fluoroscopically guided transcervical tubal catheterization (TTC) performed for proximal tubal obstruction (PTO), which resulted in an intrauterine pregnancy. A new era of the transcervical approach for the diagnosis and treatment of tubal infertility had begun. Meanwhile, ultrasound became a powerful tool in assessing female fertility. Its improved resolution in addition to new contrast media allows visualization of the normal tube. Thus, ultrasound examinations were introduced for the evaluation of tubal patency, and not only for studying the uterine cavity.

The advanced endoscopic techniques used for the investigation and treatment of infertile women is discussed in chapter 27. Thus, transcervical falloposcopy or tuboscopy, salpingoscopy and fertiloscopy or hydrolaparoscopy are not described here.

The evolving management of infertile women, from hysterosalpingography (HSG) to hystero-contrast sonography (HyCoSy) or three-dimensional Doppler tubal flow measurements, from laparoscopy to selective salpingography (SSG) and TTC, will be presented.

HYSTEROSALPINGOGRAPHY (HSG)

HSG is still one of the most commonly used screening tests for the investigation of mechanical infertility[3,4]. A survey of board-certified reproductive endocrinologists in the United States found that 96% of them rely on HSG for the initial infertility investigation[3]. The specificity, sensitivity, and positive and negative predictive values of HSG in predicting gynecologic pelvic pathology have been previously studied[4–8]. Its predictive value in determining tubal obstruction was recently evaluated by a meta-analysis of 20 studies which included 4179 patients[4]. They concluded that because of the test's high specificity, the presence of blocked tubes on HSG does not necessarily need confirmation by laparoscopy.

The main technical advances in recent years have aimed to reduce irradiation exposure to patients[9–11], to cause minimal pain and discomfort to women and to provide maximal technical ease in the performance of HSG. In order to achieve these goals, new less ionic contrast media were developed[12] and disposable balloon catheters were introduced to replace the traditional metal cannulae[13–16]. Tur-Kaspa and co-workers[13] recently performed a prospective, randomized, blinded study to compare HSG with a balloon catheter versus a metal cannula. They demonstrated that the use of a balloon catheter required significantly less fluoroscopic time, smaller amounts of contrast media, produced less pain and was easier for the physician to perform. Furthermore, when PTO was diagnosed (in 13% of patients), an immediate SSG (and/or TTC) was performed without the need to replace the cannula or to reschedule the patients.

The contraindications for HSG are pregnancy, pelvic infection (in the preceding 6 months), immediate pre- and postmenopausal period (increases risk of venous intravasation) and sensitivity to the contrast media[10–12]. Complications that may occur are pain, infection, hemorrhage, allergic reaction, vasovagal attacks and venous intravasation.

The level of irradiation was proved to be within safe limits – the absorbed dose for the ovary in the standard process is estimated to be 0.9 cGy[17,18]. Recently, Papaioannou and co-workers[19] specified that the radiation dose women are exposed to during HSG was only a fraction of the background annual radiation dose. The excess lifetime risk of developing cancer or any hereditary disorder were 4–13 and 2–6 per million procedures, respectively.

HSG technique

The optimal time to perform HSG lasts from the end of the menstrual period until day 12 of the menstrual cycle. This period encompasses the proliferative phase of the endometrium, ruling out a pregnancy, and occurs when the isthmus is at its most distensible state, so the tubes are readily filled with contrast medium[10–12].

Premedication with antiprostaglandins for pain relief may be used with prophylactic antibiotics. The bladder should be emptied immediately before the examination. The patient is placed in the lithotomy position on the screening table and the external os is visualized and swabbed with antiseptic solution. The anterior lip of the cervix may be grasped by a single-toothed tenaculum and a balloon catheter pre-filled with a water-soluble contrast dye or else a cannula is inserted into the cervical canal. Contrast medium is slowly and steadily injected under fluoroscopic control until peritoneal spillage or tubal occlusion is shown. Radiographs are taken during uterine filling and after visualization of the uterus, and the tubes delineated by the peritoneal spill. A further radiograph may be taken either immediately after 'rolling over the patient' (to enable the distribution of the dye), or 20 min later, in order to assess the peritoneal spillage pattern.

HYSTEROSONOGRAPHY (HSN) AND HYSTERO-CONTRAST SONOGRAPHY (HyCoSy)

Ultrasound has changed the role of HSG from being the most widespread source of information concerning the uterus to only a minor one. The use of contrast media, as simple as saline or more advanced as Echovist-200® (Schering AG, Germany), allows better evaluation of the uterine cavity and the fallopian tube patency.

HSN is a sonographic examination that uses saline as contrast medium and assesses only the uterine cavity. The procedure can be carried out with two- or three-dimensional ultrasound and may be comparable to diagnostic hysteroscopy in the evaluation of the uterine cavity for ART[20–22] (see three-dimensional ultrasound in ART in chapter 26).

HyCoSy is a new outpatient sonographic technique for the assessment of tubal patency using echogenic contrast media. A mixture of air and saline, in a ratio of 1:20, was used to visualize the fallopian tube for sonographic transcervical balloon tuboplasty[23]. A solution of galactose microparticle with air microbubble suspension was developed as an echogenic contrast medium for HyCoSy[24].

Several studies have investigated the reliability of HyCoSy in diagnosing tubal patency by comparing its performance to that of HSG and/or to laparoscopy and to dye studies[20,21,25–27]. The concordance rate regarding the uterine cavity was 90% and tubal patency was between 70 and 85.8%. The sensitivity, specificity, and positive and negative predictive values of HyCoSy were over 85%, 70%, 72% and 68% compared to those for laparoscopy, and 82%, 84%, 70% and 93% compared to those for HSG[20,21,25–27].

The performance of HyCoSy as a first-line outpatient investigation of tubal patency was examined recently by Hamilton and co-workers[25] in 500 consecutive infertile women. HyCoSy was completed in 92.6% of patients.

After the first 100 procedures only 4.8% of the tubes were not assessable. The concordance rate with previous laparoscopic findings was 85.8%. About 50% of patients described only mild discomfort and there were no significant post-procedure complications. Therefore, it was concluded that HyCoSy does indeed appear to be an acceptable first-line screening test for tubal infertility.

Tur-Kaspa and co-workers[26] investigated the implementation and learning curve capabilities of HyCoSy as a new medical technology in a tubal patency assessment trial taking in 215 women in 12 medical centers. HyCoSy achieved 83% and 87% of concordances compared to HSG and laparoscopy, respectively. In sequential cases 1–10, 26% of women reported pain compared to only 7% among cases 11–20 ($p = 0.0007$). Logistic regression analysis revealed that both the cases order and the significant decrease in the volume of Echovist used (13.1 ± 5.9 ml in cases 1–10 versus 8.9 ± 5.5 ml in cases 11–20; $p = 0.0001$) contributed significantly to the occurrence of pain ($p = 0.007$)[26].

The main drawback of HyCoSy is mild abdominal pain. Complications similar to HSG may occur, while galactosemia is an absolute contraindication. However, it may be concluded that HyCoSy is a reliable and safe procedure for assessing tubal patency and the uterine cavity. Because of the use of transvaginal ultrasound, patients are not exposed to radiation as they are with HSG. HyCoSy can be carried out early as an office procedure during the investigation of infertility and may replace HSG for most women.

HyCoSy technique

The patient should be in the dorsolithotomy position. After a speculum examination the vagina and cervix are disinfected, using a mild antiseptic solution, and a balloon catheter is inserted transcervically and filled with 1–2 ml of saline. Ultrasound using a vaginal probe (5 MHz transducer) is performed in order to exclude genital anomalies. Then 2–3 ml of Ecovist or saline are introduced into the uterine cavity to facilitate its assessment. The Ecovist contrast medium is applied intermittently and the maximal allowed dose is 16 ml per patient. The diagnosis is based on imaging of the moving medium in the proximal part of the tube

and on visualization of contrast media spillage into the pouch of Douglas. Sactosalpings should be ruled out. Doppler evaluation of tubal flow may also be used.

THREE-DIMENSIONAL DOPPLER IMAGING OF TUBAL PATENCY

Three-dimensional ultrasound Doppler imaging can be carried out in the color or power systems. The ultrasound images are digitized and then reconstructed in a three-dimensional volume (see chapter 26). The technique is performed in conjunction with HyCoSy or HSN and it is designated to enhance the visualization of contrast media passage through the fallopian tubes and of its peritoneal spillage[22,28]. The technique appears to achieve a better sensitivity than the conventional two-dimensional HyCoSy in terms of visualization of the tubal flow and spillage. It also uses a smaller amount of contrast medium[29]. Nevertheless, it has not gained popularity, mainly because of the high level of training required and the cost of the ultrasound machine.

RADIONUCLIDE HYSTEROSALPINGOGRAPHY (RHSG)

This technique is based on the spontaneous migration of radioactive tracer from the vagina into the uterine cavity, tubes and to the peritoneal cavity. The overall correlation rate when compared to laparoscopy or HSG is only 65%. In a recent study where RHSG was performed with radiolabeled spermatozoa, the sensitivity and specificity of the examination were 72% and 75%, respectively, compared to laparoscopic findings[30–32]. HSS is not used clinically.

TRANSCERVICAL TUBAL CATHETERIZATION (TTC)

Tubal obstruction is the cause of infertility in 25–35% of all infertile couples. PTO is the main finding in 10–15% of these cases[13,33,34]. In 1849, Smith[35] was the first to suggest catheterization of the tubal lumen through the vagina, cervix and the tubal ostium for the treatment of blocked tubes. He used whalebone. It took 136 years for new technology, accompanied with the growing experience of probing tubes and blood vessels with catheters, guidewires and fiberoptics, to overcome physicians' fears of damaging the narrow tortuous proximal part of the fallopian tube[36]. In 1985, Platia and Krudy[1] were the first to report a successful fluoroscopically guided TTC performed for PTO, which resulted in an intrauterine pregnancy. Confino, Friberg and Gleicher[2], other pioneer researchers, also attempted TTC while using balloon catheters (transcervical balloon tuboplasty, TBT).

TTC procedures, as less invasive and more cost-effective techniques, have since been implemented as an alternative to laparoscopy, tubal microsurgery and IVF in the management of infertile women with PTO. The use of SSG and TTC to recanalize the fallopian tubes has revolutionized the diagnosis and treatment of infertile women with PTO[10,11,23,37–56]. In a randomized, prospective, controlled study, SSG has recently been shown by Woolcott and co-workers[37] to be a better diagnostic test of PTO than laparoscopic dye studies. PTO was observed in 11.6% of patients (n = 135) who underwent laparoscopic dye studies first, and in 10% of tubes (n = 270). However, only 3.6% of the patients (n = 138) and 3.3% of the tubes were diagnosed with PTO when SSG was performed first (p = 0.02 and p = 0.003, respectively).

The fallopian tube recanalization procedure is simple for the interventional radiologist or gynecologist to perform and it is succssesfully completed in 70–92% of patients[10,11,38]. Tubal patency is restored in about 62–91% of tubes. Pregnancy rates after TTC average 40% (range from 7 to 67%). The complications of TTC are similar to those of HSG, with the addition of a low rate of tubal perforation which has no clinical sequelae. In patients with normal fallopian tubes following recanalization, the ectopic pregnancy rate is similar to that of the normal population. Some women, however, have peritubal adhesions and/or tubal mucosal abnormalities, which increase the risk of tubal pregnancy to an average of 3%. The radiation dose to the ovaries is approximately 1 rad (10 mGy), a dose that is within acceptable limits for women of reproductive age[9,11,18].

The presumed diagnosis of tubal spasm was used widely when a repeat test demonstrated recanalization of tubes diagnosed previously by an HSG and/or by laparoscopy to have PTO. If PTO persists, then it can be assumed that the tubal pathology will be more severe and that the pregnancy rates after recanalization by TTC will be lower. However, the results of the multicenter transcervical balloon tuboplasty (TBT) study showed the opposite[40,41]. After recanalization of the tubes by repeated HSG only, the pregnancy rates were just 17%, compared to 38% achieved by SSG or by TBT[41].

Other recent studies further support performing TTC for PTO caused by suspected tubal spasm. No effective medication for reversing tubal spasm has been identified[10,11]. Furthermore, it has been demonstrated that these cases might actually represent mechanical obstruction from a discrete plug or debris in the proximal part of the tube, which may be dislodged and missed by repeated diagnostic procedure or at the time of pathologic examination[34]. While no tubal pathology has been demonstrated in most clinically diagnosed proximally occluded resected tubal segments[34], the histology of tubal segments in cases of unsuccessful TTC for PTO revealed pathologic abnormalities in 93% of specimens[42].

High intrauterine pressure during HSG or laparoscopy can overcome an abnormal occluded corneal ostium, providing a false impression of tubal spasm and adequate tubal patency[44]. However, passing a guidewire beyond the site of the obstruction and/or by balloon tuboplasty results in a significant decrease in tubal opening pressures with a return to more normal tubal flow patterns and better pregnancy rates[40,43]. Thus, tubal spasm, if it exists, is a treatable disease using TTC.

Tur-Kaspa and co-workers[39] further investigated the efficacy and safety of TTC in a prospective study with a very large cohort of infertile women with PTO, and examined the impact of this new medical technology on patient counseling and management. A total of 625 infertile women with bilateral (61.6%) or unilateral (38.4%) PTO underwent TTC. Patients with a mean of 3.7 years (range of 1 to 22 years) of primary (23%) or secondary (77%) infertility were referred from elsewhere in Israel specifically for TTC. In 86% of the patients, at least one obstructed fallopian tube was successfully recanalized by TTC (bilateral recanalization was achieved in 71% of cases). Out of 1010 proximally obstructed fallopian tubes, 82% were successfully recanalized. Five cases (0.8%) were treated later for suspected pelvic inflammatory disease (PID), while one of them conceived spontaneously the following month.

Because of the TTC results, 86% of patients were recommended to try to conceive naturally. After the procedure, 330 consecutive patients were followed up by a telephone questionnaire, for a total follow-up of 1719 women's months. A 41% intrauterine pregnancy rate (with only a 1.4% ectopic pregnancy rate) was obtained. The cumulative pregnancy rates were 71% for women aged 20–29 years, 52% for those aged 30–39 and 18.5% for women aged $\geq$ 40 (p = 0.03; p = 0.0002 for trend)[39].

We further analyzed the data in order to identify prognostic factors for conception. A 53% pregnancy rate was achieved in 43% of the women who had no other causes for infertility but PTO, whereas a 34% pregnancy rate was obtained in 57% of the women who had additional causes for infertility (such as anovulation, a history of tuboplasty or other pelvic surgery or pelvic inflammatory disease, endometriosis, age $\geq$ 40 or male infertility) (p = 0.02). In addition, pregnancy rates were 54% for women complaining of infertility of < 2 years before they underwent TTC, 41% when the duration of infertility was 2–4 years and 23% for women with infertility of > 4 years (p = 0.02).

Tur-Kaspa and co-workers[39] clearly demonstrated that TTC has a major impact on the management and counseling of infertile women with PTO[39,57–61]. After TTC, 86% of patients are recommended to try to conceive naturally, instead of being referred for laparoscopy or ART. TTC is a safe and cost-effective procedure.

Thus, TTC should be recommended as first choice for further diagnosis and treatment to all infertile women with bilateral or unilateral PTO. This is also the recommendation of the Royal College of Obstetricians and Gynaecologists in the UK[38,45]. Evidence-based clinical guidelines on the management of infertility in secondary care recommend that when PTO is suspected, SSG and TTC should be attempted[45]. If TTC fails, then IVF or microsurgical anastomosis should be considered. Ransom and Garcia[59] clearly showed that irrespective of the etiology of the PTO, once successfully treated by TTC or microsurgery (after failed TTC), similar pregnancy rates and time to conception are expected. Furthermore, normal tubal function is

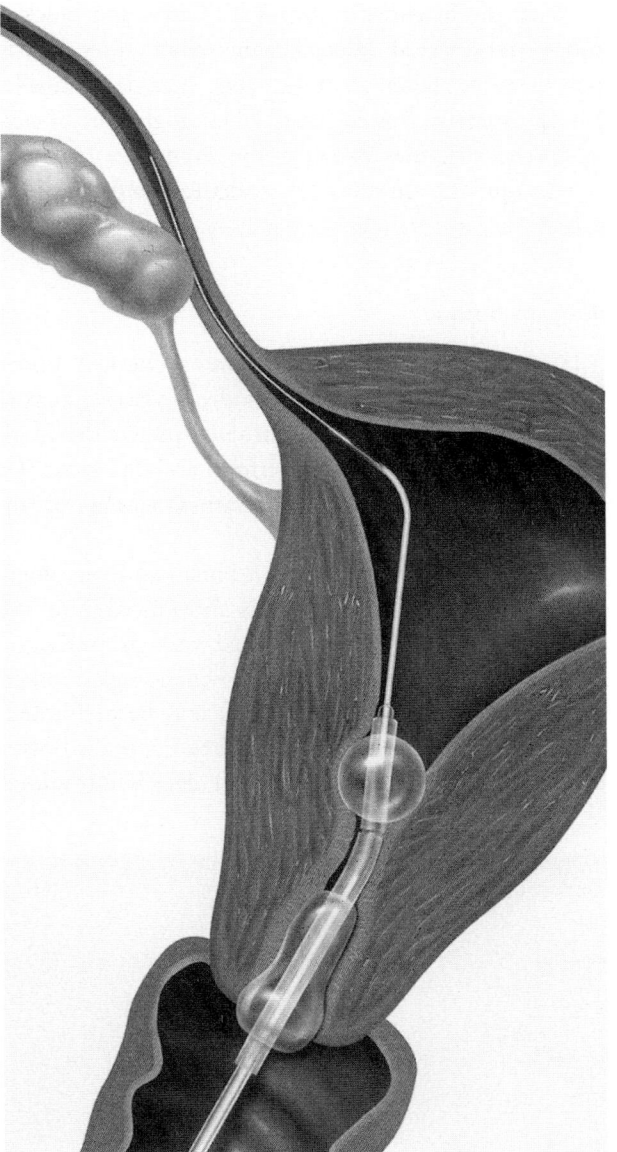

Figure 1 Specifically developed coaxial catheter systems, as seen here in the cervical canal and the uterine cavity, are used to perform a hysterosalpingography (HSG) and selective salpingography (SSG), before and after tubal catheterization with a selective catheter and/or soft-tip guidewire. The balloons of the HSG catheter are inflated with air and wedged at the cervix to seal the uterine cavity. A second curved catheter is then advanced through the HSG catheter into the cornual angle to perform SSG. Courtesy of Cook OB/GYN, Spencer, IN, USA

expected after TTC in most cases and even zygote intrafallopian transfer (ZIFT) may be successfully attempted[46].

With the increased usage of TTC to open blocked tubes, transcervical recanalization of strictures in the postoperative fallopian tube also became available. Several studies showed that TTC might be offered to correct strictures or occlusions even after chemical sterilization or at sites of reanastomosed fallopian tubes[62–67].

TTC technique

TTC is an ambulatory procedure performed under fluoroscopic observation. Specifically developed coaxial catheter systems (the author currently prefers the Cook Ob/Gyn system) are used to perform an HSG and SSG, before and after tubal catheterization with a soft-tip guidewire (Figure 1).

The procedure begins with a bimanual gynecologic examination and a speculum is introduced into the vagina. After the cervix is cleaned with an antiseptic solution, the anterior lip is grasped with a single-toothed tenaculum and a paracervical block may be performed. Thereafter, the balloon HSG catheter, prefilled with water-soluble radio-opaque contrast dye, is introduced into the cervical canal. The balloon is inflated with air and wedged at the level of the internal cervical os to seal the uterine cavity. The contrast medium is then injected into the uterus until visualization of the uterine cavity and the fallopian tubes or reconfirmation of PTO (Figure 2). A second curved catheter is then advanced through the HSG catheter into the cornual angle under fluoroscopic guidance to perform selective salpingography, before and after tubal catheterization with a soft-tip guidewire (Figure 2). After removal of the SSG catheter and the guidewire, more medium is injected in order to repeat the HSG and to confirm filling of the fallopian tubes and bilateral spillage after the TTC. SSG with TTC may be performed directly through the cervix without using the HSG catheter.

CONCLUSIONS

Modern transcervical techniques for the diagnosis and treatment of tubal obstruction revolutionized the management of mechanical infertility in the last 15 years. Preserving tubal function is important not only for natural conception but also for improving implantation

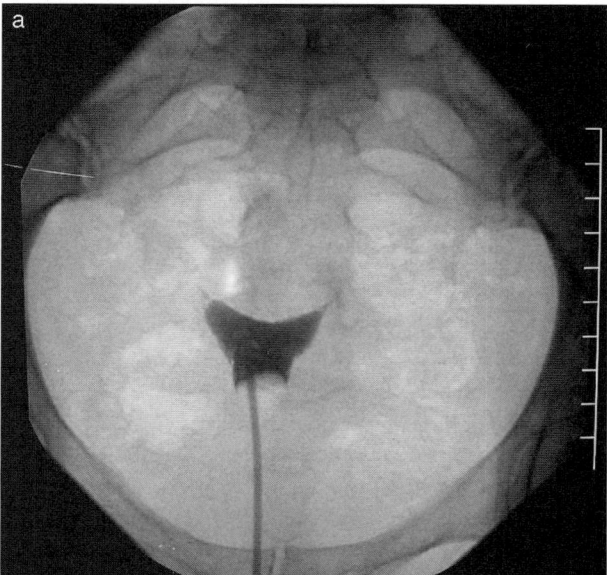
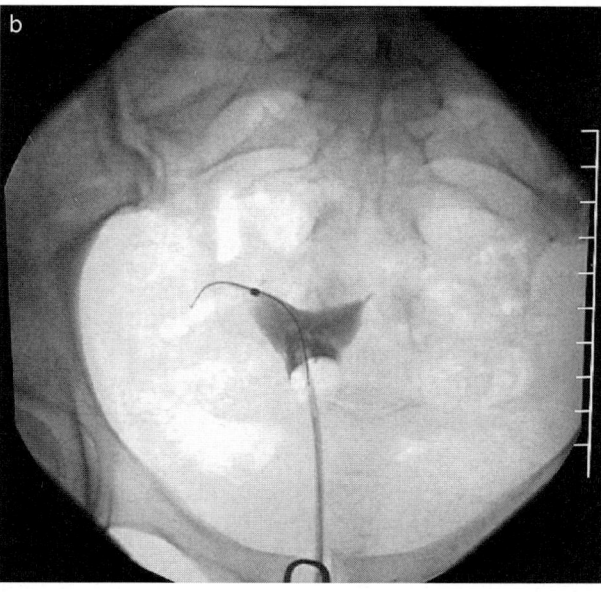

Figure 2 Transcervical tubal catheterization (TTC) performed under fluoroscopic observation. The anterior cervical lip is grasped with a single-toothed tenaculum. The balloon hysterosalpingography (HSG) catheter is introduced into the cervical canal. HSG confirmed the diagnosis of bilateral proximal tubal occlusion (PTO) (a). TTC was performed into the right fallopian tube (b) and selective salpingography (SSG) confirmed normal right tubal patency (c). TTC was then performed into the left tube (d), and left SSG and repeat HSG confirmed bilateral normal tubal flow and patency (e and f).

and pregnancy rates in ART. The application of advanced ultrasound and angiographic techniques to HSG provides excellent alternatives for better diagnosis and treatment of tubal obstruction and allows better counseling for infertile women. Evidence-based clinical guidelines currently recommend the use of SSG and TTC to recanalize blocked fallopian tubes. These procedures are ambulatory, minimally invasive and have a very low rate of complications. SSG has recently been demonstrated, in a randomized, prospective, controlled study, to be a better diagnostic test ('gold standard') of PTO than laparoscopic dye studies. TTC procedures, as a less invasive and more cost-effective technique, are being implemented as an alternative to laparoscopy, tubal microsurgery and IVF in the management of infertile women with PTO.

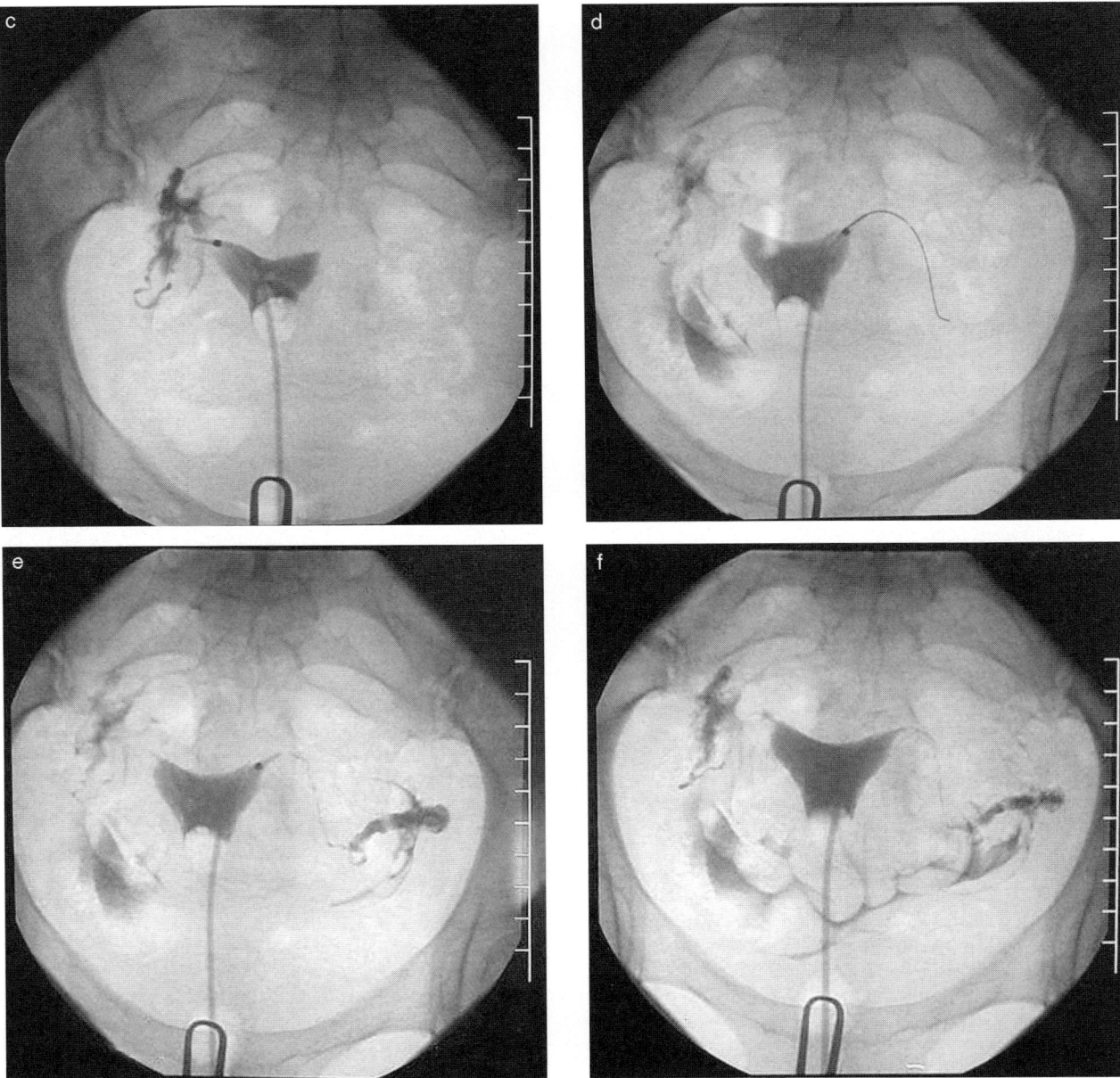

Figure 2 *Continued.*

References

1. Platia MP, Krudy AG. Transvaginal laparoscopic recanalization of a proximally occluded oviduct. *Fertil Steril* 1985;44:704–6

2. Confino E, Friberg J, Gleicher N. Transcervical balloon tuboplasty. *Fertil Steril* 1986;46:963–6

3. Glatstein IZ, Harlow B, Hornstein MD. Practice patterns among reproductive endocrinologists: the infertility evaluation. *Fertil Steril* 1997;67:443–51

4. Swart P, Mol BWJ, van der Veen F, et al. The accuracy of hysterosalpingography in the diagnosis of tubal pathology: a meta analysis. *Fertil Steril* 1995;64:486–91

5. Opsahl MS, Miller B, Klein TA. The predictive value of hysterosalpingography for tubal and peritoneal infertility factors. *Fertil Steril* 1993;60:444–8

6. Prevedourakis C, Loutradis D, Kalianidis C, et al. Hysterosalpingography and hysteroscopy in female inferility. *Hum Reprod* 1994;9:2353–5

7. Adelusi B, Al-Nuaim L, Makanjuola D, et al. Accuracy of hysterosalpingography and laparoscopic hydrotubation in diagnosis of tubal patency. *Fertil Steril* 1995;63:1016–20

8. Glatstein IZ, Sleeper LA, Lavy Y, Simon A. Observer variability in the diagnosis and management of the hysterosalpingogram. *Fertil Steril* 1997;67:233–7

9. Nakamura K, Ishiguchi T, Maekoshi H, et al. Selective fallopian tube catheterisation in female infertility: clinical results and absorbed radiation dose. *Eur Radiol* 1996;6:465–9

10. Maubon AJ, De Graef M, Boncoeur-Martel MP, Rouanet JP. Interventional radiology in female infertility: technique and role. *Eur Radiol* 2001;11:771–8

11. Thurmond AS, Machan LS, Maubon AJ, et al. A review of selective salpingography and fallopian tube catheterization. *Radiographics* 2000;20:1759–68

12. Grainger RG, Allison DJ. *Diagnostic Radiology*. Edinburgh: Churchill Livingstone 1999;1955–77

13. Tur-Kaspa I, Seidman DS, Soriano D, et al. Hysterosalpingography with a balloon catheter versus a metal cannula: a prospective, randomized, blinded comparative study. *Hum Reprod* 1998;13:75–7

14. Austin RM, Sacks BA, Nowell M, Feital C. Catheter hysterosalpingography. *Radiology* 1984;151:249

15. Sholkoff SD. Balloon hysterosalpingography catheter. *Am J Radiol* 1987;149:995–6

16. Varpula M. Hysterosalpingography with a balloon catheter versus a cannula: evaluation of patient pain. *Radiology* 1989;172:745–7

17. Murase E, Ishiguchi T, Ikeda M, Ishigaki T. Is lower-dose digital fluorography diagnostically adequate compared with higher-dose digital radiography for the diagnosis of fallopian tube stenosis? *Cardiovasc Intervent Radiol* 2000;23:126–30

18. Karande VC, Pratt DE, Balin MS, et al. What is the radiation exposure to patients during a gynecoradiologic procedure? *Fertil Steril* 1997;67:401–3

19. Papaioannou S, Afnan M, Coomarasamy A, et al. Long term safety of fluoroscopically guided selective salpingography and tubal catheterization. *Hum Reprod* 2002;17:370–2

20. Valenzano M, Ferraro F, Mansi M, Ferraiolo A. Use of a new ultrasonographic contrast medium (Echovist-200) in the study of the tubal factor of infertility. *Minerva Ginecol* 1996;48:445–50

21. Bloechle M, Schreiner T, Gouma E, Lisse K. Comparison between hysterosalpingo-contrast sonography and sonographically controlled selective tubal catheterization. *Hum Reprod* 1996;11:1423–6

22. Stern JJ, Peters AJ, Bustillo M, Coulam CB. Colour Doppler ultrasound guidance for transcervical wire tuboplasty. *Hum Reprod* 1993;8:1715–18

23. Confino E, Tur-Kaspa I, Gleicher N. Sonographic transcervical balloon tuboplasty. *Hum Reprod* 1992;7:1271–3

24. Deichart U. Transvaginal hysterosalpingo-contrast-sonography (Hy-Co-Sy) compared with conventional tubal diagnostics. *Hum Reprod* 1989;4:418–24

25. Hamilton JA, Larson AJ, Lower AM, et al. Evaluation of the performance of HSG in 500 consecutive, unselected, infertile women. *Hum Reprod* 1998;13:1519–26

26. Tur-Kaspa I, Achiron R, Aviram R, et al. Hysterosalpingo contrast sonography (HyCoSy) with Echovist®-200 for tubal patency assessment: a multicenter study. *Fertil Steril* 1998;701:S107

27. Session DR, Lerner JP, Tchen CK, Kelly AC. Ultrasound-guided fallopian tube cannulation using Albunex. *Fertil Steril* 1997;67:972–4

28. Wildt L, Kissler S, Licht P, Becker W. Sperm transport in the human female genital tract and its modulation by oxytocin as assessed by hysterosalpingoscintigraphy, hysterotonography, electrohysterography and Doppler sonography. *Hum Reprod* 1998;4:655–66

29. Sladkevicius P, Ojha K, Campbell S, Nargund G. Three-dimensional power Doppler imaging of the fallopian tube. *Ultrasound Obstet Gynecol* 2000;16:644–7

30. Ozgur K, Yildiz A, Uner M, et al. Radionuclide hysterosalpingography with radiolabeled spermatozoa. *Fertil Steril* 1997;67:751–5

31. Barrada M, Buxbaum P, Schatten C, et al. Hystero-salpingo scintigraphy: a routine investigation in sterile women? *Nucl Med Commun* 1995;16:447–51

32. Steck T, Wurfel W, Becker W, Albert PJ. Serial scintigraphic imaging for visualization of passive transport processes in the human fallopian tube. *Hum Reprod* 1991;6:1186–91

33. Tur-Kaspa I. Pathophysiology of the fallopian tube. In Gleicher N, ed. *Tubal Catheterization Procedures*. New York: Wiley-Liss Inc., 1992:5–14

34. Sulak PJ, Letterie GS, Coddington CC, et al. Histology of proximal tubal occlusion. *Fertil Steril* 1987;48:437–40

35. Smith WT. New method of treating sterility by the removal of obstructions of the fallopian tubes. *Lancet* 1849;1:529–30

36. DeCherney AH. Anything you can do I can do better . . . or differently! *Fertil Steril* 1987;48:374–6

37. Woolcott R, Fisher S, Thomas J, Kable W. A randomized, prospective, controlled study of laparoscopic dye studies and selective salpingography as diagnostic tests of fallopian tube patency. *Fertil Steril* 1999;72:879–84

38. Papaioannou S, Afnan M, Girling AJ, et al. The learning curve of selective salpingography and tubal catheterization. *Fertil Steril* 2002;77:1049–52

39. Tur-Kaspa I, Moscovici O, Meltzer S, *et al*. Transcervical tubal catheterization (TTC) is the treatment of choice for infertile women with proximal tubal obstruction – an experience with 1010 fallopian tubes. Presented at the *Annual Meeting of the American Society of Reproductive Medicine*, Seattle, Washington, October 2002

40. Confino E, Tur-Kaspa I, DeCherney A, *et al*. Transcervical balloon tuboplasty: a multicenter study. *J Am Med Assoc* 1990;264:2079–82

41. Gleicher N, Confino E, Corfman R, *et al*. The multicenter transcervical balloon tuboplasty study: conclusions and comparison to alternative technologies. *Hum Reprod* 1993; 8:1264–71

42. Letterie GS, Sakas EL. Histology of proximal tubal obstruction in cases of unsuccessful tubal canalization. *Fertil Steril* 1991;56:831–5

43. Gleicher N, Parrilli M, Redding L, *et al*. Standardization of hysterosalpingography and selective salpingography: a valuable adjunct to simple opacification studies. *Fertile Steril* 1992;58:1136–41

44. Dessole S, Meloni GB, Capobianco G, *et al*. A second hysterosalpingography reduces the use of selective technique for treatment of a proximal tubal obstruction. *Fertil Steril* 2000;73:1037–9

45. Royal College of Obstetricians and Gynecology. Evidence-based clinical guideline summary No. 3 – The management of infertility in secondary care. *Br J Gynaecol Int* 1999;83:641–5

46. Capitanio GL, Ferraiolo A, Croce S, *et al*. Transcervical selective salpingography: a diagnostic and therapeutic approach to cases of proximal tubal injection failure. *Fertil Steril* 1991;55:1045–50

47. Thurmond AS, Rosch J. Nonsurgical fallopian tube recanalization for treatment of infertility. *Radiology* 1990; 174:371–4

48. Thurmond AS. Selective salpingography and fallopian tube recanalization. *Am J Roentgenol* 1991;156:33–8

49. Eckstein N, Orron DE, Vagman I, *et al*. Digital road mapping image – a novel fluoroscopic real-time guide for selective transcervical catheterization in the treatment of proximal tubal obstruction. *Fertil Steril* 1992;58:850–3

50. Woolcott R, Petchpud A, O'Donnell P, Stanger J. Differential impact on pregnancy rate of selective salpingography, tubal catheterization and wire-guide recanalization in the treatment of proximal tube obstruction. *Hum Reprod* 1995;10:1423–6

51. Lang EK, Dunaway H, Roniger WE. Selective osteal salpingography and transvaginal catheter dilatation in the diagnosis and treatment of fallopian tube obstruction. *Am J Roentgenol* 1990;154:735–40

52. Lang EK, Dunaway HH. Recanalization of obstructed fallopian tube by selective salpingography and transvaginal bougie dilatation: outcome and cost analysis. *Fertil Steril* 1996;66:210–15

53. Golan A, Tur-Kaspa I. The management of the infertile patient with proximal tubal occlusion. *Hum Reprod* 1996; 11:1833–4

54. Gleicher N, Karande V. The diagnosis and treatment of proximal tubal disease. *Hum Reprod* 1996;11:1825–8

55. Wiedemann R, Montag M, Sterzik K. A modern approach to the diagnosis and treatment of proximal tubal occlusion. *Hum Reprod* 1996;11:1823–5

56. Hepp H, Korell M, Stowitzki T. Proximal tubal obstruction – is there a best way to treat it? *Hum Reprod* 1996;11: 1828–31

57. Woolcott R. Proximal tubal occlusion: a practical approach. *Hum Reprod* 1996;11:1831–3

58. Risquez F, Confino E. Transcervical tubal cannulation: past, present, and future. *Fertil Steril* 1993;60:211–26

59. Ransom MX, Garcia AJ. Surgical management of cornual-isthmic tubal obstruction. *Fertil Steril* 1997;68:887–91

60. Osada H, Kiyoshi Fujii T, Tsunoda I, *et al*. Outpatient evaluation and treatment of tubal obstruction with selective salpingography and balloon tuboplasty. *Fertil Steril* 2000;73: 1032–6

61. Lang EK, Dunaway HE Jr. Efficacy of salpingography and transcervical recanalization diagnosis, categorization, and treatment of fallopian tube obstruction. *Cardiovasc Intervent Radiol* 2000;23:417–22

62. Lang EK, Dunaway HH. Transcervical recanalization of strictures in the postoperative fallopian tube. *Radiology* 1994;191:507–12

63. Dunphy B, Greene C. Failed reversal of sterilization: transcervical transostial recannulation of occluded fallopian tube. *Am J Obstet Gynecol* 1994;171:274–5

64. Thurmond AS, Jones MK, Mullick B, Kessel E. Reversal of sterilization due to application of quinacrine by means of transcervical tubal catheterization. *J Vasc Interv Radiol* 1995; 6:147–9

65. Lang EK. The efficacy of transcervical recanalization of obstructed postoperative fallopian tubes. *Eur Radiol* 1998; 8:461–5

66. Thurmond AS, Brandt KR, Gorrill MJ. Tubal obstruction after ligation reversal surgery: results of catheter recanalization. *Radiology* 1999;210:747–50

67. Houston JG, Anderson D, Mills J, Harrold A. Fluoroscopically guided transcervical fallopian tube recanalization of post-sterilization reversal mid-tubal obstructions. *Cardiovasc Intervent Radiol* 2000;23:173–6

The art of imaging: three-dimensional ultrasound in assisted reproductive technologies

26

Ilan Tur-Kaspa, Shmuel Segal and Efraim Zohav

INTRODUCTION

The recent advances in ultrasound, and the new three-dimensional (3D) technology in particular, enable new insight for the evaluation of the uterus and ovaries before and during treatment with assisted reproductive technologies (ART). These advanced imaging technologies were introduced in the last decade in the field of infertility for basic research and clinical use. The new imaging technologies include 3D imaging and volume assessment, 3D color imaging and 3D power Doppler imaging. For the first time 3D ultrasound offers the possibility of evaluating a region of interest not only in the transverse and longitudinal planes but also in the coronal planes. This technique of 3D measurement and volume scanning amplifies the diagnostic potential of ultrasound in infertile women and may improve the management and ART outcome for these patients.

In this chapter we will address the advantages of 3D ultrasound in evaluating the uterus and the ovaries in ART. Mainly, diagnosing uterine cavity abnormalities, uterine and endometrial response to hormonal stimulation, and ovarian morphology and response to gonadotropin treatment. Furthermore, we will discuss the emerging role of endometrial and ovarian volume measurements in predicting ART outcome.

3D ULTRASOUND TECHNIQUE

Three-dimensional ultrasound is a new imaging technology, which uses a different scanning method than the conventional two-dimensional (2D) ultrasound. The understanding of the various terms and capabilities of 3D ultrasound is essential before describing the state of the art of this technology in infertility and ART. The 3D reconstruction of the zone of interest involves three steps[1].

Automatic volume acquisition Volume acquisition can be performed using special abdominal or vaginal volume probes. The latter is preferred for the more detailed resolution that can be obtained when the pelvic organs are targeted. The zone of interest is scanned first with conventional 2D ultrasound in order to define the area designated for 3D scan. When using a vaginal probe we define the area of interest exactly by opening the volume acquisition function and holding the probe by targeting to the center of the zone of interest. Automatic scanning of the defined zone is then performed. The collected volume data are saved and can be evaluated either immediately or after the patient has left the room.

Multiplanar volume analysis After the volume scan has been obtained, multiplanar ultrasonic information is shown immediately in three different views, saggital, transverse and coronal (Figure 1). The planes can be rotated around the center of interest to obtain the desired anatomic slices or standard uterine and ovarian orientation.

3D volume reconstruction (volume rendering) The third coronal plane is a virtual plane that is reconstructed from the volume data. Volume rendering is a visualization tool for a voxel-based 3D data set[1]. A voxel is the smallest information unit in 3D volume data, just as the pixel is the smallest information unit in 2D images. In surface mode reconstruction, surface surrounded by

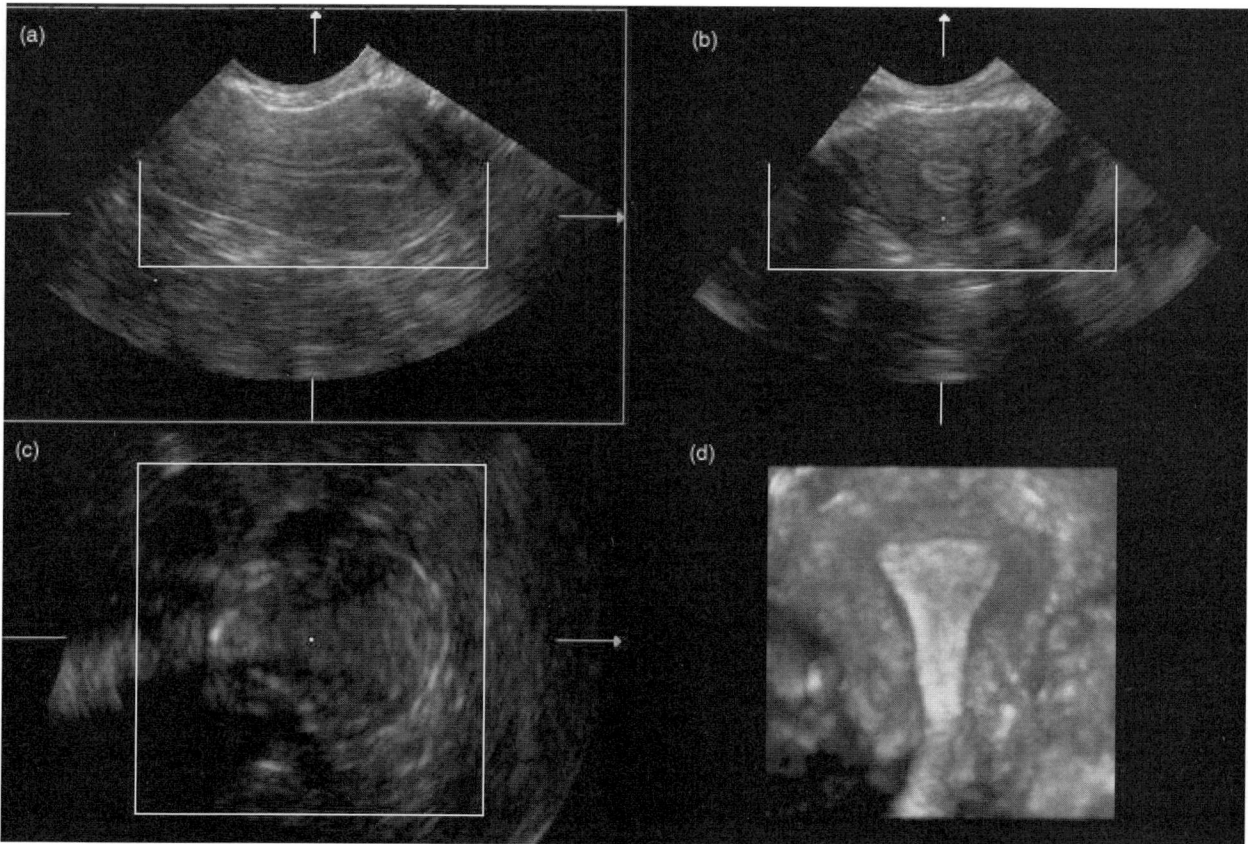

Figure 1 Multiplanar presentation and three-dimensional (3D) reconstruction of the uterus and endometrium under hormonal stimulation. (a) saggital plane; (b) transverse plane; (c) coronal plane; (d) 3D reconstruction of the endometrium in coronal plane

hypoechoic structures is visualized. In transparent mode reconstruction the object of interest is characterized by either hyper- or hypoechoic structures, such as bones or vessels or cystic structures.

3D volume measurements

Volume calculation may play an important role in clinical decisions. Three-dimensional ultrasound improves volume assessment due to its higher accuracy, low mean error and the possibility for proper documentation[2–4]. There are two methods for 3D volume calculation. In the three-distance method, volume calculation is obtained by multiplying the 3D-measured parameters, height, width and length, together with a factor of 0.526. The full planar (contour mode) is another method designated to calculate the volume of irregular shapes, like the endometrium and uterus. Tracing the contour of the zone of interest, slice by slice, is performed. The volume

is calculated by computerized software step by step until the total volume is obtained. Yaman and co-workers[5] performed 3D endometrial volume measurements in 57 women undergoing ovarian stimulation for *in vitro* fertilization (IVF) treatment. The volume measurements were done either with the full planar method or by the three-distance method. They concluded that both methods are reliable with no significant difference between them, but the full planar method seems to provide slightly better reproducibility. Further studies demonstrated also that 3D uterine and ovarian volume measurement was more accurate than volume calculation with 2D ultrasound[2–4].

THE UTERUS AND THE ENDOMETRIUM

Endometrial and uterine assessment is an important tool for the evaluation of uterine receptivity. Before ART treatment, routine evaluation of the uterine cavity is recommended. Assessing the contour and size of the uterus and the endometrial cavity is performed in order to rule out uterine and intracavitary structural abnormalities and anomalies. These investigations have usually been done by hysterosalpingography (HSG), 2D ultrasound with or without contrast media and hysteroscopy. The following will discuss the role of 3D ultrasound in the investigation of the uterus and endometrium of the infertile woman and its possible role in predicting ART outcome.

3D imaging of uterine anomalies

The diagnosis of congenital Müllerian anomalies is usually made in women with a history of infertility or recurrent abortion. The prevalence of these abnormalities is 0.9–4.5%[6,7]. The existence of a uterine anomaly may affect the management of infertile patients, either demanding further surgical intervention (i.e., hysteroscopic septectomy), or reducing the number of transferred embryos (i.e., embryo transfer of one embryo/blastocyst into the unicornuate uterus).

To classify uterine anomaly, according to the American Fertility Society classification, it is necessary to visualize the uterine fundus in the coronal plane. Since this plane is rarely obtained by 2D ultrasound, laparoscopy with HSG or hysteroscopy was used for diagnosing uterine anomalies. Computed tomography (CT) and magnetic resonance imaging (MRI) enable

visualization of the coronal plane and thus were recently introduced to replace the above invasive technique in these patients[8]. Three-dimensional ultrasound is similar to CT scan and MRI with the possibility of 3D reconstruction of the uterine anatomy. Because of the high cost of CT, MRI and laparoscopy with hysteroscopy compared to 3D ultrasound, this latest technique is potentially preferable and more cost-effective in diagnosing Müllerian anomalies.

A fundal external cleft of at least 1 cm depth is typical for a bicornuate uterus[6] (Figure 2a). In a septate uterus, the cavity is partially or completely divided by a septum and the fundal external contour is normal (Figure 2b). In an arcuate uterus the fundus is normal but the uterine cavity is concave (Figure 2c).

Jurkovic and co-workers[9,10] studied the potential value of 3D ultrasound for diagnosis of congenital uterine anomalies. It agreed with HSG in all cases of arcuate uterus and major congenital anomalies. The ability to visualize both the uterine cavity and the endometrium on the 3D scan facilitated the diagnosis of the anomaly and enabled easy differentiation between septate and bicornuate uteruses.

Raga and co-workers[11] confirmed the high correlation to HSG and in 91.6% of patients, the 3D ultrasound findings correlated with the external uterine configuration observed by laparoscopy. They further concluded that 3D ultrasound is the first technique that can be used reliably in an office setting to diagnose and classify Müllerian anomalies. The findings of the above

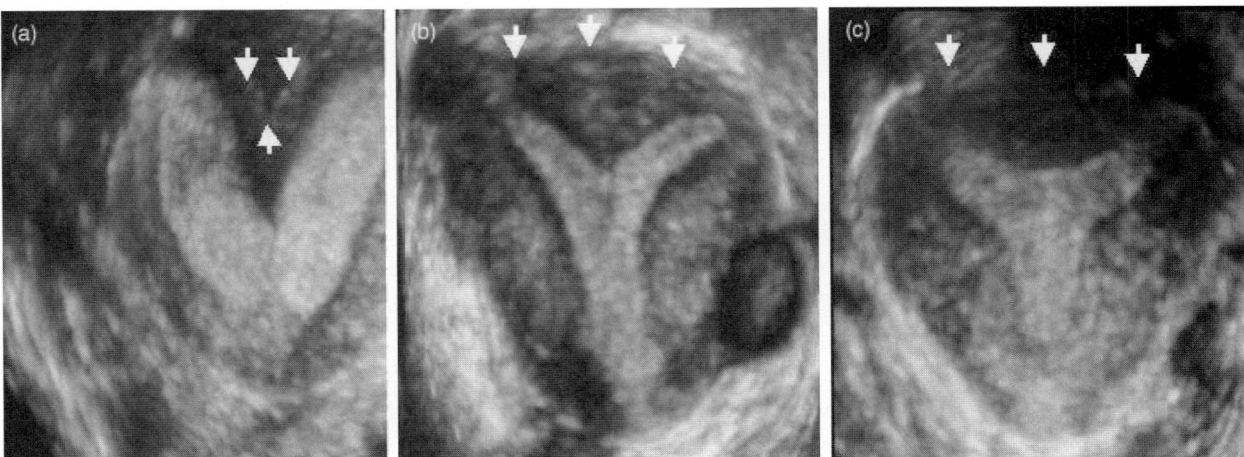

Figure 2 Three-dimensional (3D) reconstruction of a uterus with Müllerian anomalies, performed by transvaginal 3D ultrasound probe. (a) Bicornuate uterus; the diagnosis is based on the well-defined cleft uterine external contour (arrows); (b) septate uterus; the diagnosis is based on the well-defined external normal uterine contour (arrows); (c) arcuate uterus

studies were confirmed by Wu and co-workers[12], with 100% sensitivity and specificity of 3D ultrasound compared to laparoscopy and hysteroscopy.

Other investigators used 3D ultrasound with contrast media to evaluate uterine cavity anomalies[13,14]. However, because of the high accuracy of 3D ultrasound for the diagnosis of uterine malformations, the addition of contrast media had no advantages[15]. To conclude, 3D ultrasound is the procedure of choice for the diagnosis of congenital uterine anomalies. It is a non-invasive office procedure, involves no irradiation of patients (like HSG), and costs significantly less than laparoscopy with hysteroscopy, CT scan and MRI.

3D IMAGING OF THE ENDOMETRIAL CAVITY

A normal uterine cavity is a prerequisite for optimal implantation in ART. Intracavitary pathologies, such as polyps, adhesions and submucous fibroids, can be detected by transvaginal ultrasound. However, optimal visualization of the uterine cavity may require distension of the endometrial cavity with saline or contrast media in order to delineate the contour of the intracavitary lesion in relation to the endometrial lumen and walls. Two-dimensional ultrasound may miss small intracavitary pathologies. Therefore, hysteroscopy has become the 'gold standard' for the evaluation of the uterine cavity[8,14].

However, the addition of contrast media to 2D ultrasound examination, i.e. saline for sonohysterography (SHSG) or Echovist-200 (Schering, Germany), has increased the accuracy of these tests to that of diagnostic hysteroscopy. Weinraub and co-workers[13] were the first to describe 3D SHSG images of intrauterine pathologies, such as polyps, adhesions and submucous fibroids in a group of 32 women suspected previously of having these abnormalities by sonography or hysteroscopy. La Torre and co-workers[15] found 3D SHSG superior to 2D SHSG and equal to hysteroscopy, with a specificity of 100% in the detection of intracavitary polyps. Use of a contrast medium is mandatory in the case of a thin endometrium or in intracavitary pathology that precludes its visualization[16]. Lev-Toaff and co-workers[17] compared 3D SHSG to 2D SHSG and X-ray hysterosonography (HSG). In 11 out of 12 cases (92%),

3D SHSG was advantageous and provided additional and more specific information than HSG. In nine out of 13 patients (69%) 3D SHSG was superior to 2D SHSG, with a better detection of the intrauterine pathologies with regard to number and localization. Other researchers also investigated the value of 3D SHSG and reached similar conclusions[14,15].

We investigated a group of 32 infertile patients suspected previously by 2D transvaginal sonography of having intrauterine pathologies[18]. The 3D SHSG findings were compared with those from 2D ultrasound, 2D SHSG and hysteroscopy. The advantage of 3D SHSG versus 2D SHSG is more significant in cases of multiple intrauterine findings or combinations of two different intracavitary pathologies (Figure 3). In cases of suspected intrauterine pathologies, 3D SHSG may be recommended first, even before hysteroscopy.

3D ENDOMETRIAL RESPONSE TO HORMONAL STIMULATION

The endometrium is a very dynamic organ that undergoes significant, highly ordered, cyclical changes throughout the menstrual cycle. The importance of endometrial growth and its impact on female fertility has been widely investigated.

Endometrial thickness has been used for the last decade as a predictor of ART outcome. However, some investigators failed to demonstrate any relationship between endometrial thickness and pregnancy rate[19–25]. Retarded endometrium may be associated with primary infertility[26] and the role of endometrial thickness in unexplained infertility was also demonstrated[27].

Recently, endometrial volume, measured by 3D ultrasound, has also been suggested for endometrial assessment. Transvaginal 3D ultrasound was demonstrated to be highly reproducible in measuring endometrial volume[4,5]. Both methods for endometrial volume measurements available in 3D ultrasound, the full planar and three-distance methods, were found to be highly reliable and reproducible. The full planar method provided slightly better reproducibility with regard to endometrial volume measurement[5] and, therefore, we adapted this method for routine endometrial volume measurements[28,29].

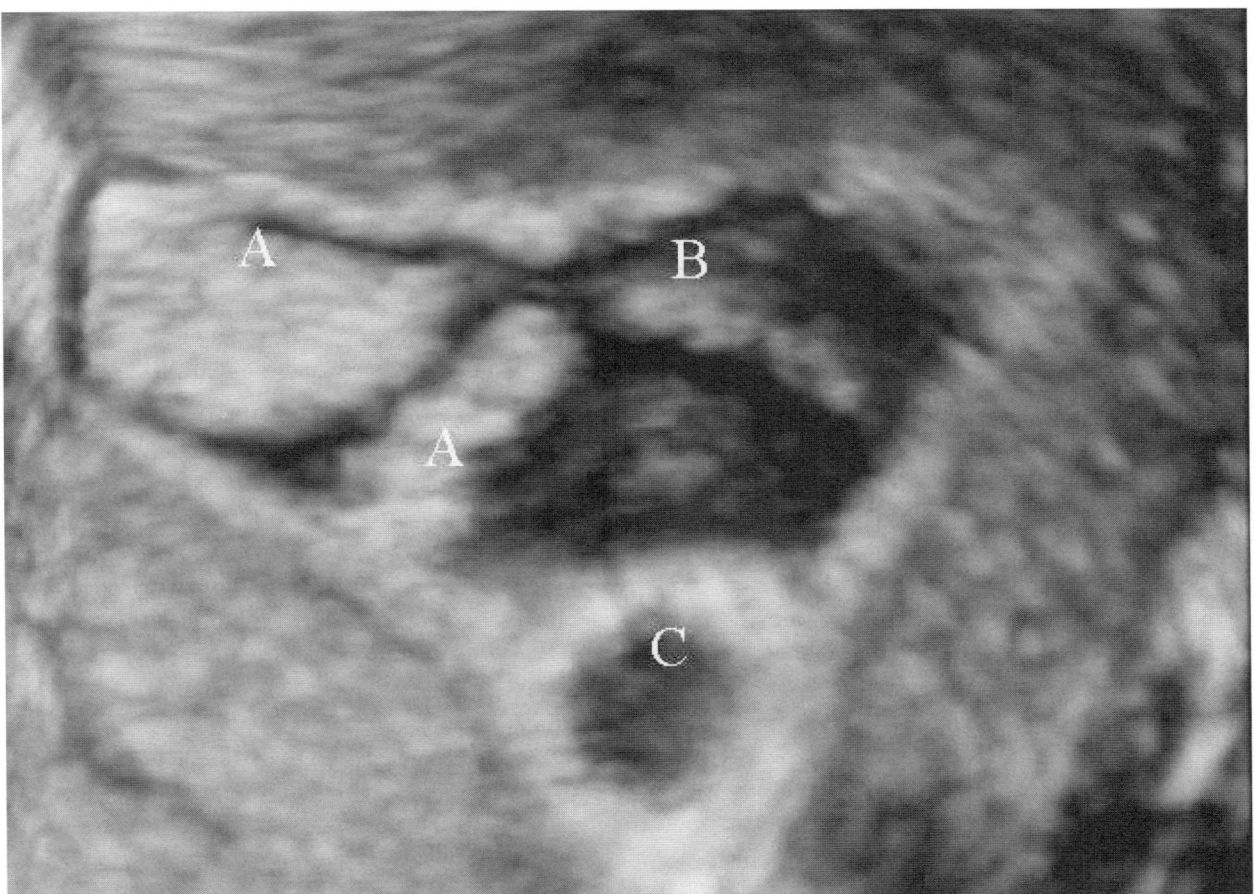

Figure 3 Three-dimensional (3D) reconstruction of a uterine cavity with multiple pathologies. (a) Endometrial polyp; (b) intrauterine adhesion; (c) the balloon of the catheter used for the procedure, filled with saline

Lee and co-workers[30] studied the endometrial volume in normally menstruating women. Significant variation between individuals was found. The ultrasound measurements were performed on various days of the cycle. Mean endometrial volume was 1.23 ml and mean uterine volume was 48.93 ml. The uterus:endometrial volume ratio showed a better correlation with the day of menstrual cycle than endometrial volume alone.

Several authors have studied the role of endometrial volume as a predictor for pregnancy in ART treatment, with conflicting results[31–34]. The endometrial volume measurements played different roles in successful ART outcome, mainly because they were taken on different days of the stimulated cycles. Yaman and co-workers[31] measured the endometrium on the day of human chorionic gonadotropin (hCG) administration. While Schild and co-workers[32] performed it on the day of

oocyte retrieval. In contrast, the groups of Raga[33] and Kupesic[34] measured the endometrial volume on the day of embryo transfer, which was performed 2 days after oocyte retrieval[33] or 5 days after the retrieval[34].

Raga and co-workers[33] analyzed the endometrial volume measured on the day of embryo transfer into three quantitative groups: < 2 ml, 2–4 ml, and > 4 ml. Pregnancy and implantation rates were significantly lower in the group of patients with an endometrial volume < 2 ml. No pregnancy was achieved with an endometrial volume < 1 ml. The role of endometrial volume on the day of hCG administration was evaluated by Yaman and co-workers[31]. They found that endometrial volume > 2.5 ml appeared to favor pregnancy.

We prospectively investigated the possible role of endometrial volume on the day of embryo transfer, measured by new 3D ultrasound technology, on IVF outcome[35]. Endometrial volume, thickness, surface and morphology were measured on the day of hCG administration and the day of embryo transfer. Women were divided by the endometrial volume on the day of embryo transfer (3 days after retrieval) to < 2 ml, 2–4 ml and > 4 ml subgroups. The clinical pregnancy and the implantation rates were significantly lower in the endometrial volume < 2 ml group compared to the > 4 ml group, 17.6% versus 53.3% (p = 0.04) and 7.1% versus 37% (p < 0.001), respectively. The pregnancy and implantation rates of the endometrial volume 2–4 ml subgroup were 37.5% and 22.5%, respectively (p = not significant and p = 0.04 compared to the < 2 ml endometrial volume group, respectively). On the day of hCG administration, however, there were no significant differences between the endometrial volume and endometrial thickness of the < 2 ml and 2–4 ml subgroups, 2.8 ± 1.3 ml versus 3.4 ± 1.6 ml and 9.3 ± 1.3 mm versus 10.5 ± 2.2 mm, respectively. Interestingly, only within the < 2 ml subgroup was there a significant decrease of the endometrial volume and endometrial thickness from the day of hCG administration to the day of embryo transfer (p = 0.02).

Endometrial thickness, as well as endometrial volume, on the day of hCG failed to identify this unique group of poor responders[35].

We may conclude that endometrial volume on the day of embryo transfer of < 2 ml is an important new negative prognostic factor for IVF outcome. Evaluation of the endometrium by 3D ultrasound on the day of embryo transfer may become an important tool for better management of infertile patients undergoing ART.

Recently, the new color histogram mode in 3D power Doppler (3DPD) ultrasonography was used to evaluate subendometrial blood flow in addition to the endometrial volume and thickness. Schild and co-workers[36], using the 3DPD mode of the spiral artery, investigated the role of subendometrial blood flow, endometrial volume and thickness on the first day of ovarian stimulation in downregulated cycles. The 3D indices were significantly lower in conception cycles compared with non-conception cycles. This new observation suggests that a better functional downregulation of the endo-metrium with less intrauterine vascularization before the initiation of ovarian stimulation indicates a more favorable endometrial milieu.

A coordinated endometrial growth during the stimulation phase increases the chances of successful implantation. Using the same 3DPD technology Kupesic and co-workers[34] assessed endometrial receptivity on the day of embryo transfer. A significantly higher flow index of the subendometrial vessels was found in conception versus non-conception cycles (13.2 ± 2.2 versus 11.9 ± 2.4). Neither endometrial volume nor endometrial thickness had a predictive value for successful outcome. A high degree of endometrial perfusion on the day of embryo transfer can indicate a more favorable endometrial milieu for successful implantation.

The heterogeneity of the studies that were published using the 3D technology, including different stimulation protocols, parameters studied and timing of sonography, make it almost impossible to compare the results and make definitive conclusions about the role of endometrial volume and thickness in favoring successful implantation. We investigated the 3D endometrial volume and uterine changes during IVF treatment with gonadotropin releasing hormone (GnRH) antagonists (two different doses of cetrorelix (Cetrotide®)) versus GnRH agonist[28,29,37].

Mean endometrial and uterine volumes, and mean uterine artery volume flow increased gradually and similarly in all groups until the day of hCG administration. Only on the day of embryo transfer were significant differences between the three groups found. The mean endometrial volume of the GnRH agonist group was significantly greater than that of the cetrorelix 0.25 mg group while that of the cetrorelix 3 mg group was in between (4.6 ± 2.8 ml, 2.8 ± 1.2 ml (p = 0.016) and 3.3 ± 2.9 ml, respectively).

These are the first studies to report 3D endometrial volume and uterine changes during ART treatment with GnRH antagonists versus GnRH agonist. After the addition of the GnRH antagonist to the gonadotropin stimulation a change in the rate of endometrial growth was observed, irrespective of the continuous rise of estradiol and uterine volume blood flow. Moreover, different endometrial responses were found with different doses of the antagonist.

PRETREATMENT EVALUATION AND PREDICTION OF OVARIAN RESPONSE

Early prediction of ovarian response is essential for significant improvement in the individual ovarian stimulation protocol and successful ART outcome. The 3D technology enables new insight and better assessment of the ovaries before and during the ART treatment (Figure 4). The main and well-documented parameters that are still useful for prediction of ovarian response are follicle stimulating hormone (FSH) levels and patient age. Elevated early follicular FSH level and/or advanced maternal age predict poor ovarian response. Low FSH levels and young age predict good ovarian response. Nevertheless these traditional parameters cannot predict the ovarian response when they are in the normal range during the reproductive age. The use of direct ultra-sonographic ovarian markers allows better individual evaluation of ovarian function. Three-dimensional ultrasound increases the capability of evaluating the ovary. Stroma volume, ovarian volume and follicular volume were obtained more accurately by 3D ultrasound than by 2D ultrasound[4,38]. The third reconstructed plane of the ovary is available only in this technique. Controlled rotation of the desired ovarian plane in multiplanar mode enables more accurate volume measurements of the entire ovary and its stroma. Spatial localization, size and number of follicles are more easily detected and measured with this technique. Clinical studies based on this technology define the ovarian parameters of poor responders as ovaries with small volume with low number of antral follicles[39,40]. The role of antral follicle number and size was also investigated with 3D technology[41]. The number of antral follicles

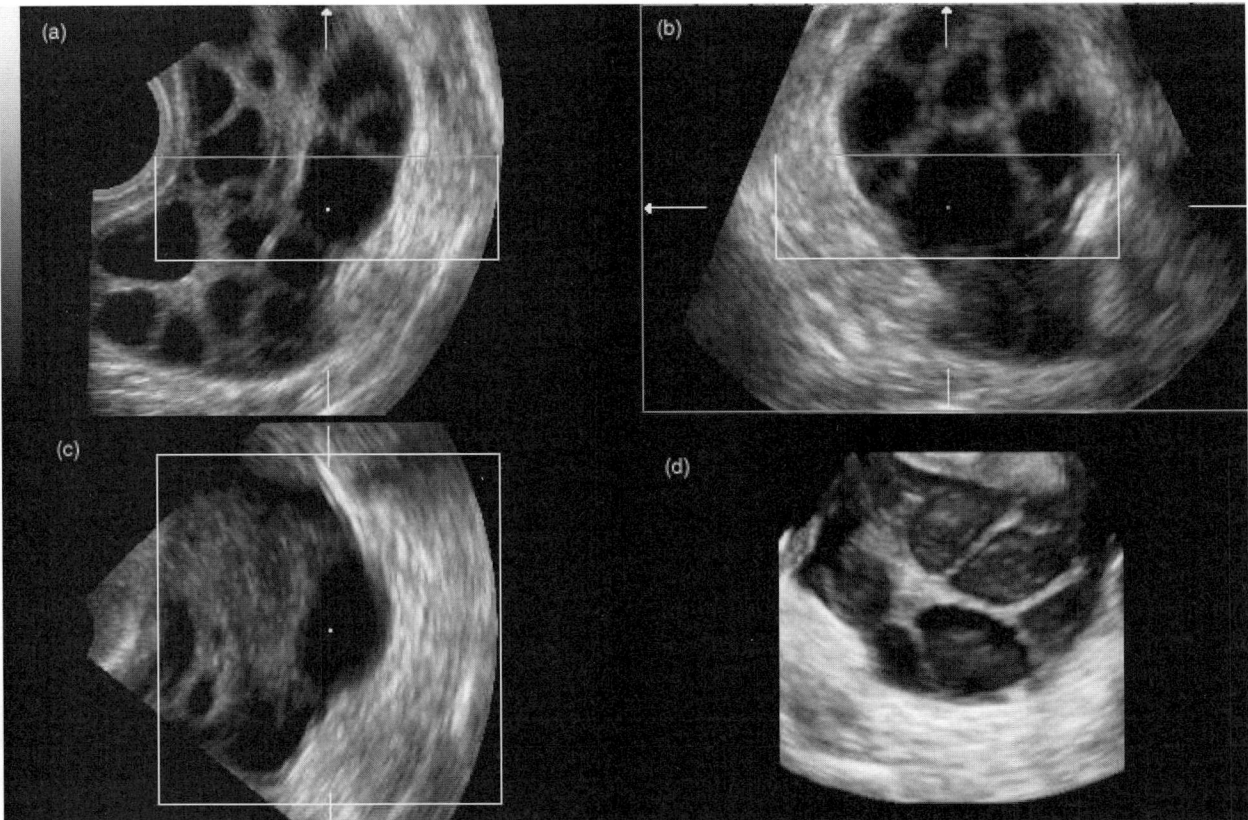

Figure 4 Multiplanar images and three-dimensional (3D) reconstruction of a stimulated ovary demonstrating follicular growth after 7 days of gonadotropin treatment. (a) saggital plane; (b) transverse plane; (c) coronal plane; (d) 3D reconstruction of the ovary

5–10 mm in size has been found to be a predictor for successful outcome in ART. A significantly higher number of antral follicles was found in this category in pregnant patients versus non-pregnant patients. This category of antral follicles decreased significantly with age. The presence of follicles in early follicular phase, with mean diameter > 11 mm was found to be associated with higher cancellation rate[41]. Recently, some authors[42,43] investigated the role of hormonal and 3D ovarian parameters in early follicular phase as predictors of oocyte recruitment. The number of antral follicles was found to be the best positive predictor for the number of retrieved oocytes. Other hormonal and ovarian markers, such as estradiol, inhibin A and B, FSH, luteinizing hormone (LH), testosterone, androstenedione, ovarian stromal volume, ovarian volume and the woman's age, were found to have lower predictive value.

We investigated the effect of age on the ovary by using 3D parameters of ovarian volume and number of antral follicles, in infertile women attending an ART program[44]. One hundred and seventy-five infertile women, aged 20–45 years old, undergoing IVF treatment, were enrolled in this prospective clinical observational study. The women underwent 3D transvaginal ultrasound examination on days 1–3 of the cycle. Ovarian volume and the number of antral follicles and estradiol levels were measured. The study group was divided into 5-year age subgroups. Mean ovarian volume decreased steadily and significantly from 12.3 ± 3.8 ml at age 20–24 years to 6.9 ± 4.4 ml at age 41–45 years ($p = 0.03$), while mean menstrual estradiol levels remained constant during the study period. Similarly, significant decrease with age in the mean number of antral follicles was observed, from 6.9 ± 4.4 to 2.1 ± 0.7 ($p = 0.04$), respectively. Mean ovarian volume and mean number of antral follicles were negatively correlated with age ($r = -0.200$, $p = 0.008$; $r = -0.298$, $p = 0.001$, respectively). Mean number of oocytes collected were negatively correlated with age ($r = -0.214$, $p = 0.04$) and positively correlated with the number of antral follicles detected on days 1–3 of the cycle ($r = 0.494$, $p = 0.001$). This is the largest study to investigate human ovarian volume changes during the reproductive life, in women 20–45 years of age, by 3D ultrasound technology. The ovarian volume and the number of antral follicles are

maximal at age of 20–25 years, and then start to decline. Menstrual estradiol levels, however, remained constant. The number of antral follicles was found to be a better predictor for the number of oocytes retrieved than age or ovarian volume. We suggest that these characteristics of ovarian changes may explain response in natural and ART cycles.

Ovarian perfusion is an important parameter of ovarian function and morphology[45,46]. Transvaginal color and pulse Doppler ultrasonography have become important tools for the evaluation of ovarian perfusion[47–49]. Engmann and co-workers[48] found peak systolic velocity of ovarian stromal vessels to be the representative index of ovarian stromal blood flow. Low stromal blood flow velocities were found to be predictors of low ovarian response and low pregnancy rate as well[47–49]. In stimulated polycystic ovaries, proliferation of stromal blood flow delivers more gonadotropins to the granulosa layer cells and may result in increased incidence of hyperstimulated ovaries[50]. A risk of poor ovarian response and cancellation rate higher than 50% is expected in patients with normal FSH levels with low ovarian volume (< 3 ml) and low number of antral follicles (< 3)[51].

Polycystic ovary syndrome

Polycystic ovary syndrome (PCOS) is a clinical representation of approximately 17% of women during reproductive age[52]. The classical signs of this syndrome result from excessive androgen secretion, inappropriate gonadotropin secretion and chronic anovulation with oligomenorrhea. The clinical signs are associated with hormonal changes, i.e., elevated levels of LH, testosterone and the LH:FSH ratio. These hormonal changes are associated with larger ovarian volume and high number of follicles. The most common clinical signs of this syndrome are oligomenorrhea, obesity, hirsutism and infertility[52]. Polycystic ovarian morphology alone as demonstrated by ultrasonography does not predict the severity of hormonal changes[53]. Recently, Dolz and co-workers[54] conducted extensive 3D ultrasound, color Doppler flow and color Doppler amplitude studies on PCOS ovaries, demonstrating an increased impedance in the uterine arteries, decreased resistant index and pulsatility index in ovarian stromal vessels and increased stromal vascularity during menstrual and follicular phase. Patients with PCOS were found to have an average of 13.7 follicles per ovary. The lower cut-off was 10 follicles (−2 SD). PCOS ovaries differ from multifollicular ovaries by having increased stromal area.

Tulandi and co-workers[55] measured ovarian volume by 3D ultrasound in PCOS patients undergoing laparoscopic ovarian drilling and with clomiphene-resistant PCOS. Preoperative mean ovarian volume was 12.2 ml. One week after the operation the mean ovarian volume increased to 13.6 ml and 3 weeks thereafter it decreased to a mean of 6.9 ml. In this group of patients ovulation rate was 88.2% and cumulative pregnancy rate was 70% at 12 months.

The combination of 3D ultrasound with special software for 3D reconstruction imaging improves further the ability to investigate small intrafollicular changes during superovulation induction for IVF treatment. It has been possible to demonstrate the intrafollicular cumulus during IVF treatment for many years[56] and it was a sign of follicular maturation, but mature oocytes were retrieved also from follicles in which a cumulus was not identified. Recently, using 3D technology, Poehl and co-workers[57] demonstrated significant correlation between the number of detected cumuli and the number of retrieved mature fertilized oocytes. Using 3D technology, cumulus visualization appears to be an indicator for mature oocytes and successful fertilization.

CONCLUSIONS

Three-dimensional ultrasound emerges as a valuable new technology for ART. It may be used for diagnosing uterine cavity abnormalities, uterine and endometrial responses to hormonal stimulation, and ovarian morphology and responses to gonadotropin treatment. The role of endometrial and ovarian volume measurements in predicting ART outcome emphasizes the advantages of 3D ultrasound in evaluating the uterus and the ovaries of women undergoing infertility and ART treatment.

The accuracy of diagnosing Müllerian anomalies by 3D ultrasound is preferable and more cost-effective than CT scan or MRI. In cases with multiple intrauterine findings, 3D SHSG was demonstrated to be superior to 2D SHSG and HSG. Calculating uterine, endometrial and ovarian volume is undoubtedly more accurate with the full plane method available only with 3D technology. Counting antral follicles, the main predictor for ovarian response, is also superior with 3D ultrasound compared with 2D.

Endometrial volume on the day of ET has been recently shown also to be a predictor for ART outcome. Low endometrial volume (< 2 ml) on the day of ET is a bad prognostic factor, and a volume of < 1 ml may require the cancellation of ET and freezing of all embryos. On the other hand, endometrial volume of > 4 ml suggests better endometrial receptivity. A new entity of 'low endometrial responder' which is different from 'low ovarian responder' may explain some repeated implantation failure.

References

1. Gritzki A, Brandl H. The Voluson (Kretz) technique. In Merz E, ed. *3-D Ultrasound in Obstetrics and Gynecology*. Philadelphia: Lippincott Williams & Wilkins, 1998:9–15
2. Hosli I, Tercanli S, Surbek D, Holzgreve W. Comparison between transvaginal *in vitro* and *in vivo* volume measurements. In Merz E, ed. *3-D Ultrasound in Obstetrics and Gynecology*. Philadelphia: Lippincott Williams & Wilkins, 1998:49–55
3. Riccabona M. Nelson TR, Pretorius DH. Three-dimensional ultrasound: accuracy of distance and volume measurements. *Ultrasound Obstet Gynecol* 1996;7:429–34
4. Kyei-Mensah A, Maconochie N, Zaidi J, *et al.* Transvaginal three-dimensional ultrasound: reproducibility of ovarian and endometrial volume measurements. *Fertil Steril* 1996;66: 718–22
5. Yaman C, Sommergruber M, Ebner T, *et al.* Reproducibility of transvaginal three-dimensional endometrial volume measurements during ovarian stimulation. *Hum Reprod* 1999;14:2604–8
6. Speroff L, Glass RH, Kase NG. The uterus. Clinical gynecologic endocrinology and infertility. In Speroff L, Glass RH, Kase NG, eds. *Clinical Gynecologic Endocrinology and Infertility*. Baltimore: Williams & Wilkins, 1994; 109–34
7. Nussbaum-Blask A,Taylor WS, Rubin A, *et al.* Prevalence of mullerian duct anomalies detected at ultrasound. *Am J Med Genet* 2000;94:9–12

8. Pellerito JS, McCarthy SM, Doyle MB, *et al*. Diagnosis of uterine anomalies: relative accuracy of MR imaging, endovaginal sonography, and hysterosalpingography. *Radiology* 1992;183:795–800

9. Jurkovic D, Geipel A, Gruboeck K, *et al*. Three-dimensional ultrasound for the assessment of uterine anatomy and detection of congenital anomalies: a comparison with hysterosalpingography and two-dimensional sonography. *Ultrasound Obstet Gynecol* 1995;5:233–7

10. Jurkovic D, Aslan N. Three-dimensional ultrasound for diagnosis of congenital uterine anomalies. In Merz E, ed. *3-D Ultrasound in Obstetrics and Gynecology*. Philadelphia: Lippincott Williams & Wilkins, 1998:27–9

11. Raga F, Bonilla-Musoles F, Blanes J, Osborne NG. Congenital Müllerian anomalies: diagnostic accuracy of three-dimensional ultrasound. *Fertil Steril* 1996;65:523–8

12. Wu MH, Hsu CC, Huang KE. Detection of congenital mullerian duct anomalies using three-dimensional ultrasound. *J Clin Ultrasound* 1997;25:487–92

13. Weinraub Z, Maymon R, Shulman A, *et al*. Three-dimensional saline contrast hysterosonography and surface rendering of uterine cavity pathology. *Ultrasound Obstet Gynecol* 1996;8:277–82

14. Ayida G, Kennedy S, Barlow D, Chamberlain P. Contrast sonography for uterine cavity assessment: a comparison of conventional two-dimensional with three-dimensional transvaginal ultrasound: a pilot study. *Fertil Steril* 1996;66: 848–50

15. La Torre R, De Felice C, De Angelis C, *et al*. Transvaginal sonographic evaluation of endometrial polyp: comparison with two dimensional and three-dimensional contrast sonography. *Clin Exp Obstet Gynecol* 1999;26:171–3

16. Dubinsky TJ, Parvey HR, Gormaz G, Makland N. Transvaginal hysterosonography in the evaluation of small endoluminal masses. *J Ultrasound Med* 1995;14:1–6

17. Lev-Toaff AS, Pinheiro LW, Bega G, *et al*. Three-dimensional multiplanar sonohysterography: comparison with conventional two-dimensional sonohysterography and X-ray hysterosalpingography. *J Ultrasound Med* 2001;20: 295–306

18. Zohav E, Segal O, Meltzer S, *et al*. Three-dimensional hysterosonography (3D-HSG) is comparable to hysteroscopy for diagnosing abnormal intrauterine findings in infertile women. *Fertil Steril* 2001;Suppl.76:3

19. Gonen Y, Casper RF. Prediction of implantation by the sonographic appearance of the endometrium during controlled ovarian stimulation for *in vitro* fertilization (IVF). *J In Vitro Fertil Embryo Transf* 1990;7:146–52

20. Noyes N, Liu HC, Sultan K, *et al*. Endometrial thickness appears to be a significant factor in embryo implantation in *in-vitro* fertilization. *Hum Reprod* 1995;10:919–22

21. Oliveira JB, Baruffi RL, Mauri AL, *et al*. Endometrial ultrasonography as a predictor of pregnancy in an *in-vitro* fertilization programme after ovarian stimulation and gonadotrophin-releasing hormone and gonadotrophins. *Hum Reprod* 1997;12:2515–18

22. Ohno Y, Fujimoto Y. Endometrial oestrogen and progesterone receptors and their relationship to sonographic appearance of the endometrium. *Hum Reprod Update* 1998; 4:560–4

23. Randall JM, Fisk NM, McTavish A, Templeton AA. Transvaginal ultrasonic assessement of endometrial growth in spontaneous and hyperstimulated menstrual cycles. *Br J Obstet Gynaecol* 1989;96:954–9

24. Soham Z, DiCarlo C, Patel A. Is it possible to run a successful ovulation induction program based solely on ultrasound monitoring? The importance of endometrial measurements. *Fertil Steril* 1991;56:836–41

25. Rabinowitz R, Laufer N, Lewin A, *et al*. The value of ultrasonographic endometrial measurement in the prediction of pregnancy following *in vitro* fertilization. *Fertil Steril* 1986;45:824–8

26. Sachiko N, Tsutomu D, Toshimichi O, *et al*. Relationship between sonographic endometrial thickness and progestin induced withdrawal bleeding. *Obstet Gynecol* 1996:87:722–5

27. Li TC, Lenton EA, Dockery P, Cooke ID. A comparison of some clinical and endocrinological features between cycles with normal and defective luteal phases in women with unexplained infertility. *Hum Reprod* 1990;5:805–10

28. Zohav E, Segal O, Meltzer S, *et al*. 3-D endometrial and uterine volume changes during IVF treatment with GnRH antagonist, 17th annual meeting of the European Society of Human Reproduction. *Hum Reprod* 2001;16:74

29. Tur-Kaspa I, Zohav E, Segal O, *et al*. Three-dimensional endometrial and uterine volumes changes during induction of ovulation for IVF, GnRH antagonist vs GnRH agonist protocols. *Fertil Steril* 2001;Suppl.76:54

30. Lee A, Sator M, Kratochwil A, *et al*. Endometrial volume change during spontaneous menstrual cycles: volumetry by transvaginal three-dimensional ultrasound. *Fertil Steril* 1997; 68:831–5

31. Yaman C, Ebner T, Sommergruber M, *et al*. Role of three-dimensional ultrasonographic measurement of endometrium volume as a predictor of pregnancy outcome in an IVF-ET program: a preliminary study. *Fertil Steril* 2000;74:797–801

32. Schild RL, Indefrei D, Eschweiler S, *et al*. Three-dimensional endometrial volume calculation and pregnancy rate in an *in-vitro* fertilization programme. *Hum Reprod* 1999;14: 1255–8

33. Raga F, Bonilla-Musoles F, Casan EM, *et al*. Assessment of endometrial volume by three-dimensional ultrasound prior to embryo transfer: clues to endometrial receptivity. *Hum Reprod* 1999;14:2851–4

34. Kupesic S, Bekavac I, Bjelos D, Kurjak A. Assessment of endometrial receptivity by transvaginal color Doppler and three-dimensional power Doppler ultrasonography in patients undergoing *in vitro* fertilization procedures. *J Ultrasound Med* 2001;20:125–34

35. Tur-Kaspa I, Segal O, Meltzer S, *et al*. 3-D endometrial measurement on the day of ET and IVF outcome. *Fertil Steril* 2002;Suppl.0–243

36. Schild RL, Knobloch C, Dorn C, *et al*. Endometrial receptivity in an *in vitro* fertilization program as assessed by spiral artery blood flow, endometrial thickness, endometrial volume, and uterine artery blood flow. *Fertil Steril* 2001; 75:361–6

37. Tur-Kaspa I, Meltzer S, Segal O, *et al*. 3-D endometrial volume and uterine changes during ART treatment with GnRH antagonists vs GnRH agonist. *Hum Reprod* 2002; Suppl.17:19

38. Wu MH, Tang HH, Hsu CC, *et al.* The role of three-dimensional ultrasonographic images in ovarian measurement. *Fertil Steril* 1998;69:1152–5

39. Pellicer A, Ardilles G, Neuspiller F, *et al.* Evaluation of the ovarian reserve in young low responders with normal basal levels of follicle stimulating hormone using three-dimensional ultrasonography. *Fertil Steril* 1998:70:671–5

40. Chang MY, Chiang CH, Hsieh TT, *et al.* Use of antral follicle count to predict the outcome of assisted reproduction follicle technologies. *Fertil Steril* 1998;69:505–10

41. Pohl M, Hohlagschwandtner M, Obscura A, *et al.* Number and size of antral follicles as predictive factors in *in vitro* fertilization and embryo transfer. *J Assist Reprod Genet* 2000;17:315–18

42. Kupesic S, Kurjak A. Predictors of IVF outcome by three-dimensional ultrasound. *Hum Reprod* 2002;17:950–5

43. Dumesic DA, Damario MA, Session DR, *et al.* Ovarian morphology and serum hormone markers as predictors of ovarian follicle recruitment by gonadotropins for *in vitro* fertilization. *J Clin Endocrinol Metab* 2001;86: 2538–43

44. Zohav E, Segal O, Meltzer S, *et al.* 3-D ovarian changes during the fertility period: from age 20 to 45 years old. *Fertil Steril* 2002;Suppl.0–241

45. Dada T, Salha O, Allgar V, Sharma V. Utero-ovarian blood flow characteristics of pituitary desensitization. *Hum Reprod* 2001;16:1663–70

46. Tan SL. Clinical applications of Doppler and three-dimensional ultrasound in assisted reproductive technology. *Ultrasound Obstet Gynecol* 1999;13:153–6

47. Zaidi J, Campbell S, Pittrof R, *et al.* Ovarian stromal blood flow in women with polycystic ovaries – a possible new marker for diagnosis? *Hum Reprod* 1995;10:1992–6

48. Engmann L, Sladkevicius P, Agrawal R, *et al.* The pattern of changes in ovarian stromal and uterine artery blood flow velocities during *in vitro* fertilization treatment and its relationship with outcome of the cycle. *Ultrasound Obstet Gynecol* 1999;13:26–33

49. Zaidi J, Jacobs H, Campbell S, Tan SL. Blood flow changes in the ovarian and uterine arteries in women with polycystic ovary syndrome who respond to clomiphene citrate: correlation with serum hormone concentrations. *Ultrasound Obstet Gynecol* 1998;12:188–96

50. Pinkas H, Mashiach R, Rabinerson D, *et al.* Doppler parameters of uterine and ovarian stromal blood flow in women with polycystic ovary syndrome and normally ovulating women undergoing controlled ovarian stimulation. *Ultrasound Obstet Gynecol* 1998;2:197–200

51. Kupesic S. The present and future role of three-dimensional ultrasound in assisted conception. *Ultrasound Obstet Gynecol* 2001;18:191–4

52. Botesis D, Kassanos D, Pyrgiotis E, *et al.* Sonographic incidence of polycystic ovaries in a gynecological population. *Ultrasound Obstet Gynecol* 1995;6:182–5

53. van der Westhuizen S, van der Spuy ZM. Ovarian morphology as a predictor of hormonal values in polycystic ovarian syndrome. *Ultrasound Obstet Gynecol* 1996;17:335–41

54. Dolz M, Osborne NG, Blanes J, *et al.* Polycystic ovarian syndrome: assessment with color Doppler and three-dimensional ultrasonography. *J Ultrasound Med* 1999;18: 303–13

55. Tulandi T, Watkin K, Tan SL. Reproductive performance and three-dimensional ultrasound volume measurements of polycystic ovaries following laparoscopic ovarian drilling. *Int J Fertil Womens Med* 1997;42:436–40

56. Cacciatore B, Liukkonen S, Koskimies A, Yelostalo P. Ultrasound detection of cumulus oophorus in patients undergoing *in vitro* fertilization. *J In Vitro Fertil Embryo Transf* 1985;2:224–8

57. Poehl M, Hohlagschwandtner M, Doerner V, *et al.* Cumulus assessment by three-dimensional ultrasound for *in vitro* fertilization. *Ultrasound Obstet Gynecol* 2000;16:251–3

Endoscopic technologies in human infertility 27

Giovanni B. La Sala, Maria T. Villani and Sandro Gambino

INTRODUCTION

Gynecologic endoscopy has an important role in the diagnosis and therapy of female infertility. It has been repeatedly claimed that endoscopic technologies can be substituted for other diagnostic tools used in the past and can represent effective therapies in some cases. Nevertheless, after about 20 years of clinical use on a large scale, there is now an unavoidable need to determine the criteria applicable to their use in the diagnosis and therapy of human infertility. It is necessary to define precisely the indications for the extensive use of endoscopy and to evaluate the advantages its use can bring in comparison with other available techniques. When endoscopic technology is used as a therapy, the results, limits and complications must be critically assessed as objectively as possible. The aim of this chapter is not to plead the cause of endoscopy, but to contribute to the addressing of the above-mentioned issues.

DIAGNOSTIC HYSTEROSCOPY

Indications

Sterility and infertility

The study of the uterine, tubal and pelvic factors in an infertile woman has to be personalized according to: (1) woman's age; (2) sterility duration; (3) history of pelvic surgery and/or pelvic inflammatory disease (PID); and (4) the partner's semen quality.

Figures 1–4 summarize our approach to the study of the uterine, tubal and pelvic factors. These diagrams do not include hysterosalpingography (HSG), and the reason for this will be explained later.

According to our experience, ultrasound hysterosalpingography (USHSG) is indicated as the first diagnostic approach only in young women with no history of pelvic surgery and/or PID and whose partner has a normal semen examination (Figure 1). Otherwise, we think it

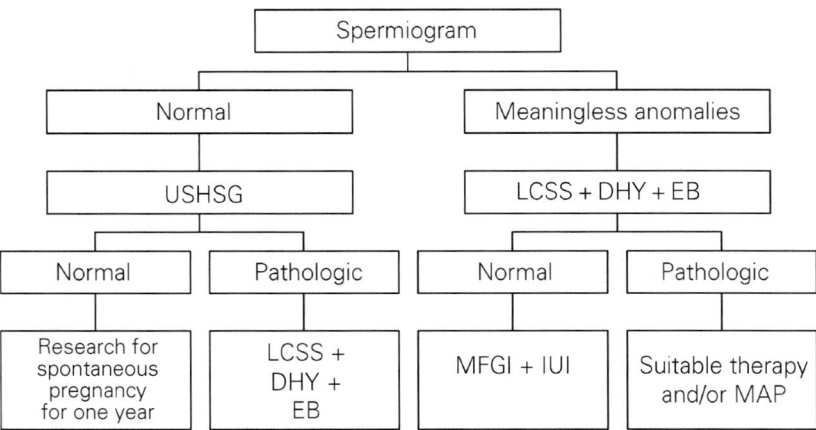

Figure 1 Diagnostic hysteroscopy (DHY). The sequence illustrated here pertains to a woman ≤ 30 years of age with no history of a pelvic surgical operation and/or pelvic inflammatory disease, but with regular ovulatory menstrual cycles and a sterility duration of ≥ 3 years. USHSG, ultrasound hysterosalpingography; LCSS, laparoscopy plus chromosalpingography; EB, endometrial biopsy; MFGI + IUI, multifollicular growth induction + intrauterine insemination; MAP, medically assisted procreation

more appropriate to perform a diagnostic hysteroscopy (DHY), immediately, coupled with a laparoscopy plus chromosalpingoscopy (LCSS) (Figures 2 and 3).

If the patient is 35 years old or older, she will be considered for *in vitro* fertilization and embryo transfer (IVF-ET), in which case only a DHY will be necessary (Figure 4). In our opinion, DHY is the most reliable investigation to evaluate the cervix and the uterine cavity, in addition to which the study of the tubal and pelvic factors is useless in IVF-ET patients.

Artificial insemination with donor semen

Where couples undergo artificial insemination with donor semen (AID), study of the uterine and tubal factors must be always performed before proceeding to

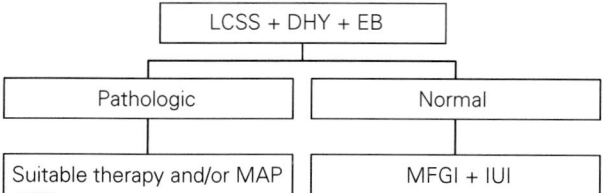

Figure 2 Diagnostic hysteroscopy (DHY). The sequence illustrated here pertains to a woman ≤ 30 years of age, whose history includes a pelvic surgical operation and/or pelvic inflammatory disease, but who has regular ovulatory menstrual cycles and a sterility duration of ≥ 3 years, and whose parner's spermiogram is either normal or shows only meaningless anomalies. LCSS, laparoscopy plus chromosalpingography; EB, endometrial biopsy; MFGI + IUI, multifollicular growth induction + intrauterine insemination; MAP, medically assisted procreation

the insemination. Missing this point represents, in our opinion, a methodological mistake.

If PID and/or pelvic surgery are not included in the patient's history, we suggest that the USHSG be performed first, whereas in cases of a positive history, we think it more appropriate to perform a DHY coupled with an LCSS (Figure 5).

Medically assisted procreation (MAP)

Women candidates for IVF-ET Some studies have reported the presence of undiagnosed uterine pathologies in 18–50% of the women candidates for IVF-ET and have suggested that DHY should be completed before IVF-ET is commenced[1–5]. Until now, we have acted in accordance with other authors[6–8], proceeding to a DHY only after two or more unsuccessful IVF-ET cycles, even when a DHY could conceivably have been performed as a preliminary investigation prior to IVF-ET.

Women candidates for intracytoplasmic sperm injection Even when semen quality is the major factor responsible for infertility, there is a concrete risk attached to intracytoplasmic sperm injection (ICSI) unless the female factor is studied first, and a DHY is indicated as a preliminary investigation.

Results and controversies

Results

The DHY can be easily performed in 95% of patients. With an experienced operator, the incidence of

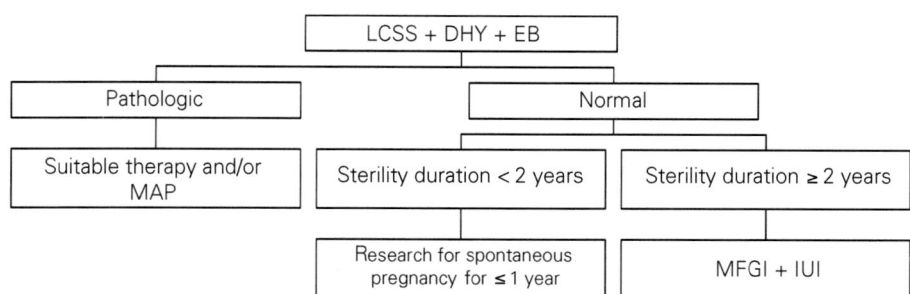

Figure 3 Diagnostic hysteroscopy (DHY). The sequence illustrated here pertains to a woman > 30 and < 35 years of age, whose sterility has lasted for 3 years, who has regular ovulatory cycles and whose partner's spermiogram is either normal or shows only meaningless anomalies. LCSS, laparoscopy plus chromosalpingography; EB, endometrial biopsy; MFGI + IUI, multifollicular growth induction + intrauterine insemination; MAP, medically assisted procreation

complications is very low[9] and diagnostic reliability is almost 100%. Table 1 lists the results of 1133 DHYs performed on infertile women in our Reproductive Medicine Unit.

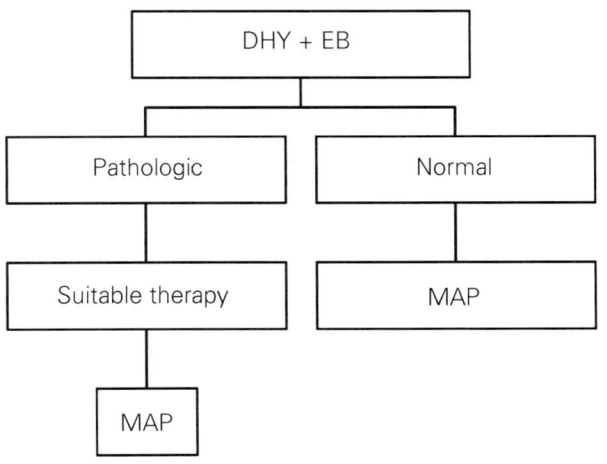

Figure 4 Diagnostic hysteroscopy (DHY). The sequence illustrated here pertains to a woman ≥ 35 years of age. EB, endometrial biopsy; MAP, medically assisted procreation

Controversies

There is some controversy over the respective roles of HSG, DHY and USHSG in the study of uterine, tubal and pelvic factors in infertile women. Currently, in the guidelines of the World Health Organization (WHO) and of the European Society of Human Reproduction and Embryology (ESHRE), HSG is considered the first-level investigation[10,11], and is therefore the most widely employed in the study of uterotubal factors in infertile patients. DHY is still underestimated, and the use of USHSG has been proposed only recently; its clinical value in the study of uterotubal factors is under investigation.

The data presented in Table 2 confirm those of previous studies; in particular, they show that the overall sensibility and specificity are significantly lower for HSG than for DHY.

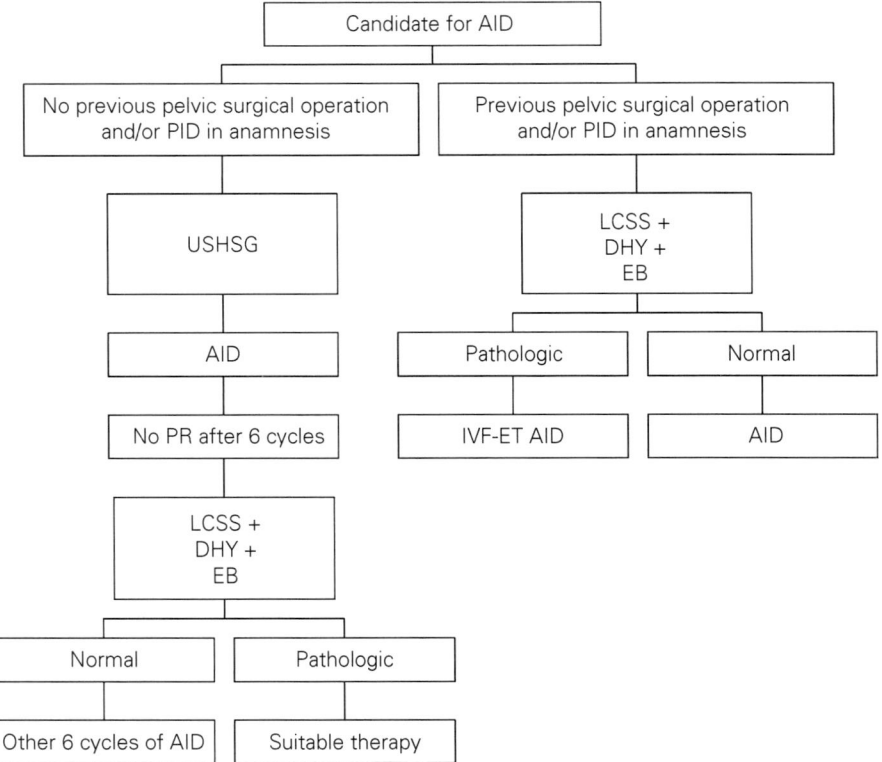

Figure 5 Diagnostic hysteroscopy (DHY). The sequence illustrated here pertains to a woman proposed as candidate for artificial insemination with donor semen (AID). PID, pelvic inflammatory disease; USHSG, ultrasound hysterosalpingography; LCSS, laparoscopy plus chromosalpingography; EB, endometrial biopsy; PR, pregnancy; IVF-ET, *in vitro* fertilization and embryo transfer

The data in Table 3, although still preliminary, induce a cautious optimism about the diagnostic reliability of USHSG.

Current trends lean towards the use of USHSG and DHY (even in association) rather than HSG: the rationale of this choice is summarized in Table 4.

Table 1 Diagnostic hysteroscopy (DHY) in 1133 infertile women, performed at the Reggio Emilia Reproductive Medicine Unit: results

	Patients	
Diagnosis at DHY	n	%
Normal	841	74.2
Uterine malformation	132	11.7
Submucous myoma	41	3.6
Endometrial polyp	38	3.4
Uterine sinechiae	35	3.1
Tubal occlusion	32	2.8
Aspecific endometritis	6	0.5
Hyoperplastic endometrium	3	0.3
Cervical sinechiae	3	0.3
Specific endometritis	2	0.2
Overall	1133	100

Table 2 Hysterosalpingography (HSG) versus diagnostic hysteroscopy (DHY) in the study of the uterine factor

References	Studied women (n)	Specificity HSG versus DHY (%)	Sensibility HSG versus DHY (%)
Golan *et al.*, 1996[12]	464	15	98
Gaglione *et al.*, 1996[13]	70	82	79
Wang *et al.*, 1996[14]	214	70	80

Table 3 Ultrasound hysterosalpingography (USHSG) versus diagnostic hysteroscopy (DHY) in the study of the uterine factor

References	Studied women (n)	Specificity USHSG versus DHY (%)	Sensibility USHSG versus DHY (%)
Cicinelli *et al.*, 1995[15]	52	94	100
Gronlund *et al.*, 1999[16]	20	100	100
Roudigoz *et al.*, 1995[17]	104	94	98

We think that DHY is the best choice to study the uterine factor, while HSG should no longer be used in this context. We consider USHSG to be a promising investigation, but still at the experimental stage. It is conceivable to choose USHSG as a first-level technique and to keep DHY as a complementary investigation for use only in those women in whom endouterine pathology is suspected.

OPERATIVE HYSTEROSCOPY

Polypectomy

A few controversial data[18,19] exist in the literature about the incidence and role of endometrial polyps in infertility. According to our experience, an endometrial polyp bigger than 1 cm represents an indication for hysteroscopic polypectomy, because it could interfere with the embryo's implantation and/or its further development, while if it is smaller than 1 cm it is unlikely to be a cause of infertility, and polypectomy is not indicated.

In infertile women, hysteroscopic polypectomy, like every other hysteroscopic operation, must be performed following these criteria:

(1) Absolute respect is to be shown for endometrial and myometrial integrity; and

(2) Hysteroscopic control is indicated 2–3 months after the operation.

Table 4 Comparative advantages and disadvantages of diagnostic hysteroscopy (DHY), hysterosalpingography (HSG) and ultrasound hysterosalpingography (USHSG)

Method	DHY Direct	HSG Indirect	USHSG Indirect
Localization of pathology	Careful	Presumptive	Careful
Diagnostic care	High	Medium	Medium/high
Cost	Low	High	Low
Exposure to radiations	No	Yes	No
Repetivity	Yes	No	Yes
Discomfort	Low/medium	Medium/high	Low
Complications	Low	Low/medium	Low

Myomectomy

Neither the real incidence, nor the etiopathogenetic role, of submucous myomas in infertile women is as yet well defined[20–22]. The most suitable therapy for their treatment is hysteroscopic myomectomy[23–25], which can be recommended for any infertile woman with submucous myomas because of its effectiveness coupled with its low invasiveness. Reliable data[22,26,27] show that submucous myomas reduce implantation and pregnancy rates after IVF-ET or ICSI (Table 5).

The results of hysteroscopic myomectomy in infertile women are reported in Table 6; our results are listed in Table 7.

The most serious complication of hysteroscopic myomectomy in infertile women is a uterine rupture during the subsequent pregnancy or labor, but the incidence of this is not well known[28].

The most controversial points are the following:

(1) Which technique should be employed in G1 and G2 myomas to preserve myometrial integrity? Some authors perform myomectomy in two operations, others in one single operation by removing the endocavitary portion of the myoma by electroresection ('warm loop'), and then enucleating the intramural portion mechanically ('cold loop'). In our opinion, the 'warm loop' + 'cold loop' technique is the best choice.

(2) Which kind of delivery (spontaneous or Cesarean section, CS) should be chosen after a hysteroscopic myomectomy in the absence of obstetric

complications? Several variables should be considered, among them the woman's age; the number and size of myomas; their location and the depth of myometrial invasion; the feasibility of myomectomy and the possible myometrial damage or perforation during an operation. The systematic use of

Table 5 Submucous myomas: results of medically assisted procreation

IVF-ET or ICSI	No myomas	Subserous myomas	Intramural myomas	Submucous myomas
Number of cycles	318	41	55	10
Pregnancies	98	14	9	1
Pregnancy rate (%)	30.1	34.1	16.4*	10*
Implantation rate (%)	15.5	15.1	6.4**	4.3**

IVF-ET, *in vitro* fertilization and embryo transfer; ICSI, intracytoplasmic sperm injection

*$p < 0.005$ versus women without myomas and women with subserous myomas; **$p < 0.005$ versus women without myomas and women with subserous myomas

Table 6 Hysteroscopic myomectomy in infertile women

References	Year of study	Treated women (n)	Pregnancies (n)	Childbirth (n)
Hallez et al.[29]	1987	11	7	5
Brooks et al.[30]	1989	15	5	—
Valle[31]	1990	16	10	8
Loffer[32]	1990	12	9	7
Corson and Brooks[23]	1991	13	11	9
Hucke et al.[33]	1992	14	4	2
Mergui et al.[34]	1993	15	6	4
Donnez et al.[24]	1994	24	16	—
Cravello et al.[35]	1995	16	4	—
Kuhulmann et al.[36]	1997	61	41	—
La Sala et al.[37]	1999	54	35	30
Bernard et al.[38]	2000	31	11	—
Fernandez et al.[39]	2001	59	16	6

Table 7 Hysteroscopic myomectomy in sterile and infertile women: results obtained in the Reggio Emilia Sterility Centre and by the Division of Gynaecology of the San Carlo Nancy Hospital of Rome

	Women with sterility		Women with infertility		Overall	
	n	%	n	%	n	%
Treated	40	—	14	—	54	—
Pregnant	21	52.5*	12	85.7*	33	61.1*
Pregnancies	23	—	12	—	35	—
Miscarriages	2	8.6**	3	25**	5	14.2**
Total childbirths	21	91.4**	9	75**	30	85.8**
Spontaneous deliveries	8	38.1§	0	0§	8	26.6§
Caesarian sections	13	61.9§	9	100§	22	73.4§

*Refers to the 40, 14 and 54 treated women, respectively; **refers to the 23, 12 and 35 obtained pregnancies, respectively; §refers to the 21, 9 and 30 childbirths, respectively

ultrasound in the follow-up to myomectomy could be helpful for estimating myometrial integrity, and would also be of use when choosing between vaginal delivery and CS. In addition, ultrasound can assist the identification of women who have a higher risk of uterine rupture during pregnancy[40].

Metroplasty

The real incidence of uterine septa in infertile women is controversial[41–45]. Sterility and infertility in women with a uterine septum or with other uterine malformations are relative, the clinical history of women with uterine septa being very variable.

Hysteroscopic metroplasty has definitively replaced the abdominal metroplasty. In the endoscopic technique, the section of the septum is stopped before reaching the myometrium, so that finally a uterine cavity similar to an arcuate uterus is obtained; in this way, the risk of myometrial perforation is highly reduced, and even partial lesions of the myometrium at the base of the septum are avoided. This minimizes the risk of uterine rupture during a subsequent pregnancy and labor[46–48].

The most important questions about hysteroscopic metroplasty concern the diagnosis and indications. Regarding diagnosis, some authors think that transvaginal ultrasound is able to differentiate between a septate uterus and a bicornuate uterus, and that a diagnostic laparoscopy is therefore not necessary[59–65]; we, on the contrary, think that a laparoscopy is essential[66].

The debate over the indications of metroplasty is still in progress[46–69]. Must a nulliparous woman with a uterine septum undergo a metroplasty before trying for a child? There is not a univocal answer to this question at present, but we generally advise such women to undergo a metroplasty before attempting pregnancy[65].

Must a sterile woman with a uterine septum undergo a metroplasty? The answer is a matter of opinion[49–66,68]; we believe metroplasty is opportune because the disadvantages of such an operation are much lower that those of spontaneous abortion or premature delivery. Table 8 contains the results obtained by hysteroscopic metroplasty in infertile women. Table 9 reports our results in 47 women with repeated miscarriage and 24 women with recurrent miscarriage who underwent

hysteroscopic metroplasty; it must be underlined that none of the 71 women had a living child before the operation and that 63 of the 71 women were followed-up for at least a year.

Synechiolysis

The incidence of uterine synechiae in infertile women is unknown, but according to one report 90% of women with the condition had undergone a uterine cavity curettage at least once, and 50% of women submitted for three curettages had uterine synechiae[70]. Other

Table 8 Results of hysteroscopic metroplasty in infertile women

References	Year of study	Treated women (n)	Pregnancies (n)	Term deliveries n	Term deliveries %*
Corson and Batzer[50]	1986	18	17	14	82.3
De Cherney and Polan[51]	1983	72	72	64	88.9
Fayez[52]	1986	19	16	14	87.5
March and Israel[53]	1987	66	63	55	87.3
Blanc et al.[54]	1994	45	31	25	80.6
Grimbizis et al.[55]	1998	57	44	32	72.7
La Sala et al.[37]	1999	71	46	34	74.1
Perino et al.[56]	1999	297	234	189	80.7
Hickok[69]	2000	21	22	17	77.3
Porcu et al.[71]	2000	63	45	28	62.2

*Refers to pregnancies

Table 9 Metroplasty in women with repeated or recurrent miscarriage; results of the Reggio Emilia Reproductive Medicine Unit

	Repeated miscarriage n	Repeated miscarriage %	Recurrent miscarriage n	Recurrent miscarriage %	Overall n	Overall %
Treated women	47	—	24	—	71	—
Pregnant women	29	—	15	—	44	—
Total pregnancies	30	—	16	—	46	—
Spontaneous abortions	6	20	3	18.7	9	19.5
Preterm deliveries	2*	6.7	1*	6.2	3	6.5
Term deliveries	22	73.3	12	75.1	34	74

*No mortality – survival of three babies

important etiopathogenetic causes are endometritis, hysteroscopic operations, abdominal myomectomy, and metroplasty with opening of the uterine cavity.

In our opinion, hysteroscopic lysis of uterine synechiae is technically the most difficult and dangerous hysteroscopic operation[72-76]. Infertility and repeated or recurrent miscarriage are correct indications for the synechiolysis.

Table 10 reports the results after synechiolysis, but note that these are not subdivided according to the severity of the synechiae. The results after lysis of severe synechiae are listed in Table 11.

The most frequent and immediate complication of the synechiolysis operation is uterine perforation, the risk of which is very high when there is a total obliteration of the uterine cavity. The most important backward complications are synechiae recurrence, uterine rupture during a subsequent pregnancy, and the abnormal placental site in a subsequent pregnancy[70,77,78].

How can the incidence of uterine synechiae be reduced? The only way is to lower the frequency and the aggressiveness of the uterine curettage as much as possible. How can the incidence of uterine synechiae recurrence be reduced? The application of intrauterine devices, as well as the medical therapies tried thus far, has proved unsuccessful, and to date there is no answer. What is the effectiveness of hysteroscopic synechiolysis? It is certainly good for little synechiae, but unsatisfactory for severe cases – the best therapy is prevention.

Resection of the 'ossificated' endometrium

The ossificated endometrium, previously called 'bone metaplasia', is a rare pathologic occurrence. We have seen four cases, and about 100 have been described in the literature[79-90]. If a woman has fertility problems, then a hysteroscopic exeresis of the ossified endometrial areas is indicated.

Since this pathology is very rare, there aren't reliable data about results in these cases. A patient of ours underwent exeresis after two failed IVF-ET cycles, and conceived in the subsequent IVF cycle; another woman with unexplained infertility became spontaneously pregnant after exeresis.

Table 10 Results of hysteroscopic synechiolysis in infertile women

References	Year of study	Treated women (n)	Pregnancies (n)	Abortions n	Abortions %*	Term delivery n	Term delivery %*
Sugimoto[91]	1978	113	79	29	36.8	50	63.2
Valle[92]	1988	187	143	29	20.1	114	79.9
Roge et al.[93]	1996	52	34	10	19.5	24	70.5
Colacurci et al.[94]	1997	53	27	5	18.5	21	81.5
Capella-Allouc et al.[95]	1999	28	12	3	25	9	75

*Refers to the pregnancies

Table 11 Results of hysteroscopic synechiolysis in infertile women (severe synechiae only)

References	Year of study	Patients (n)	Pregnancies n	Pregnancies %*	Abortions n	Abortions %**	Term deliveries n	Term deliveries %**
Colacurci et al.[94]	1997	23	5	21.7	3	60	2	40
Capella-Allouc et al.[95]	1999	28	12	42.8	3	25	9	75

*Refers to the 23 and 28 treated women, respectively; **refers to the five and 12 pregnancies, respectively

Transcervical tubal transfer of gametes or embryos

The first ultrasound-guided intratubal insemination was reported in 1988[96]. Since then, a few studies about transcervical intratubal insemination and transcervical intratubal embryo transfer have been published[97-103], but claims that homolog or eterolog intratubal insemination is more effective than intrauterine insemination are not justified in our experience.

Transferring gametes (gamete intrafallopian transfer, GIFT; zygote intrafallopian transfer, ZIFT) or embryos (tubal embryo transfer, TET) in the salpinx using the transcervical method is more interesting because an invasive operation like laparoscopy may be avoided. Unfortunately, though, this technique has not been proven to be as effective as the laparoscopic version. This could be due

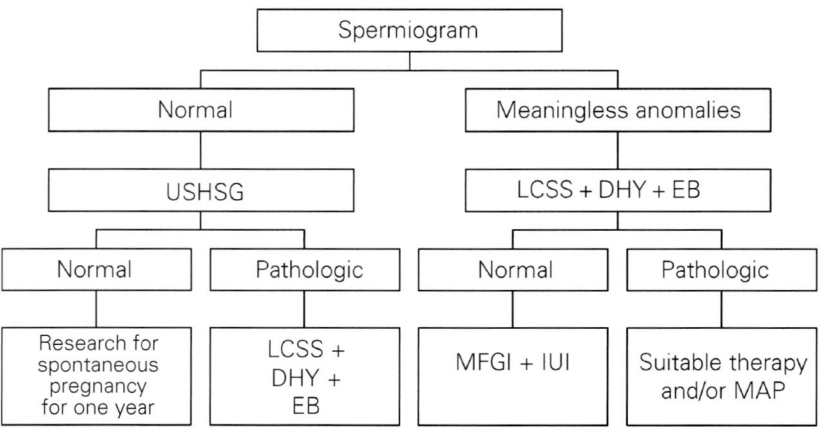

Figure 6 Diagnostic laparoscopy. The sequence illustrated here pertains to a woman ≤ 30 years of age, with no history of a pelvic surgical operation and/or pelvic inflammatory disease, but with regular ovulatory menstrual cycles and a sterility duration of ≥ 3 years. USHSG, ultrasound hysterosalpingography; LCSS, laparoscopy plus chromosalpingography; DHY, diagnostic hysteroscopy; EB, endometrial biopsy; MFGI + IUI, multifollicular growth induction + intrauterine insemination; MAP, medically assisted procreation

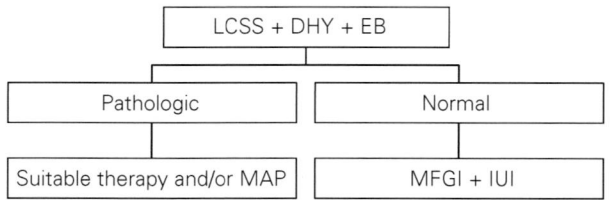

Figure 7 Diagnostic laparoscopy. The sequence illustrated here pertains to a woman ≤ 30 years of age, whose history includes a pelvic operation and/or pelvic inflammatory disease, who has regular ovulatory menstrual cycles and a sterility duration of ≥ 3 years, and whose partner's spermiogram is either normal or has only meaningless anomalies. LCSS, laparoscopy plus chromosalpingography; DHY, diagnostic hysteroscopy; EB, endometrial biopsy; MFGI + IUI, multifollicular growth induction + intrauterine insemination; MAP, medically assisted procreation

(1) The study of 'mechanical' sterility is not only a matter of checking the presence or absence of tubal occlusion, but also includes the examination of the anatomy and functionality of the entire female pelvis. We therefore think that LCSS is the only reliable investigative technique allowing the thorough study of the role of mechanical factors in female sterility.

(2) LCSS is nevertheless an invasive and expensive investigation and may have very serious, sometimes even fatal, complications. The choice to perform LCSS in an infertile woman must therefore be carefully considered in every case, all available data being evaluated, together with the cost–benefit rate.

(3) The woman's age and the sterility duration are two very important variables. As a general rule, LCSS may be postponed if a woman is younger than 30, whereas if older than 30 it must be done sooner in the diagnostic schedule. The possibility of a spontaneous pregnancy is generally good when sterility has lasted less than 3 years, but drops when duration exceeds 3 years, and it is in the latter cases that we consider LCSS to be opportune.

(4) A history of pelvic surgical operations and/or PID is an absolute indication for LCSS, and conversely a contraindication for HSG and/or USHSG; in these

to the lack of adequate technical equipment, however, and hopefully results will improve.

DIAGNOSTIC LAPAROSCOPY

Indications

Infertility

Figures 6–9 below illustrate the way we personalize the study of the uterine, tubal and pelvic factors in the treatment of 'mechanical' female infertility. Our approach embodies the following precepts:

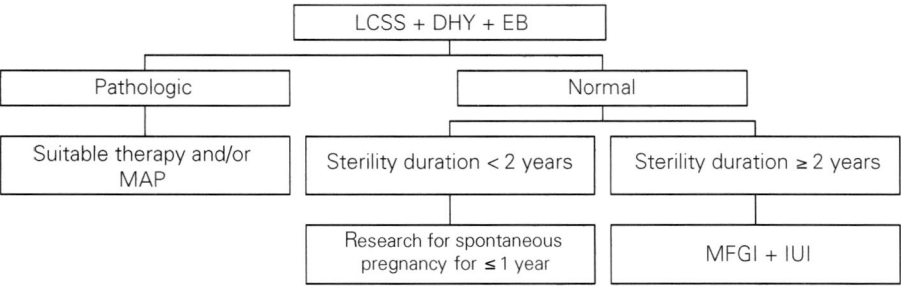

Figure 8 Diagnostic laparoscopy. The sequence illustrated here pertains to a woman > 30 and < 35 years of age, who has regular ovulatory menstrual cycles, and whose partner's spermiogram is either normal or has only meaningless anomalies. LCSS, laparoscopy plus chromosalpingography; DHY, diagnostic hysteroscopy; EB, endometrial biopsy; MFGI + IUI, multifollicular growth induction + intrauterine insemination; MAP, medically assisted procreation; PR, pregnancy

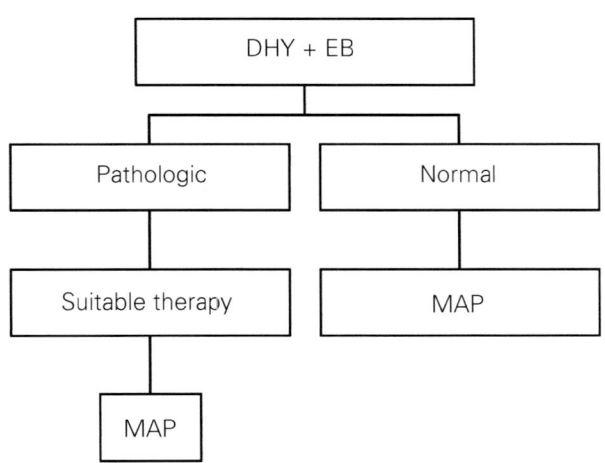

Figure 9 Diagnostic laparoscopy. The sequence illustrated here pertains to a woman ≥ 35 years of age. DHY, diagnostic hysteroscopy; EB, endometrial biopsy; MAP, medically assisted procreation

women, the direct study of tubal anatomy and of the whole pelvic anatomy is very important.

(5) LCSS should be performed in all women candidates for ovulation induction (multifollicular growth induction, MFGI) with or without intrauterine insemination (IUI). In fact, with this therapy there are low but significant risks, such as multiple pregnancies and ovarian hyperstimulation syndrome, and it should be undertaken only when a potential patient has been thoroughly studied with respect to the 'mechanical' factor.

Artificial insemination by donor

In women candidates for AID without a history of pelvic surgery and/or PID, the evaluation of tubal patency by USHSG is probably sufficient (Figure 10). Executing a LCSS before AID in a young and healthy woman would be, in our opinion, an 'overtreatment'.

However, if pelvic surgery and/or PID have been previously performed or if at least six AID cycles have been unsuccessful, then we feel that LCSS should be undertaken, and if tubal damage is documented, the couple should be shifted to etherologous IVF-ET.

Results and controversies

Results The diagnostic reliability of LCSS is approximately 100%. Tables 12 and 13 report our results with the technique in 1011 sterile or infertile women between January 1984 and April 1998. The mean age of the women in this group was 35.6 years, and the mean sterility duration 4.1 years.

It is noteworthy that 37.3% of the women had endometriosis (Table 12) and that 50.8% had open tubae, but with an associated pathology of the uterus, the ovaries and/or the pelvis (Table 13).

Controversies The current debate concerns the respective roles of HSG, USHSG and laparoscopic salpingoscopy in the study of the 'mechanical' factor in sterile women.

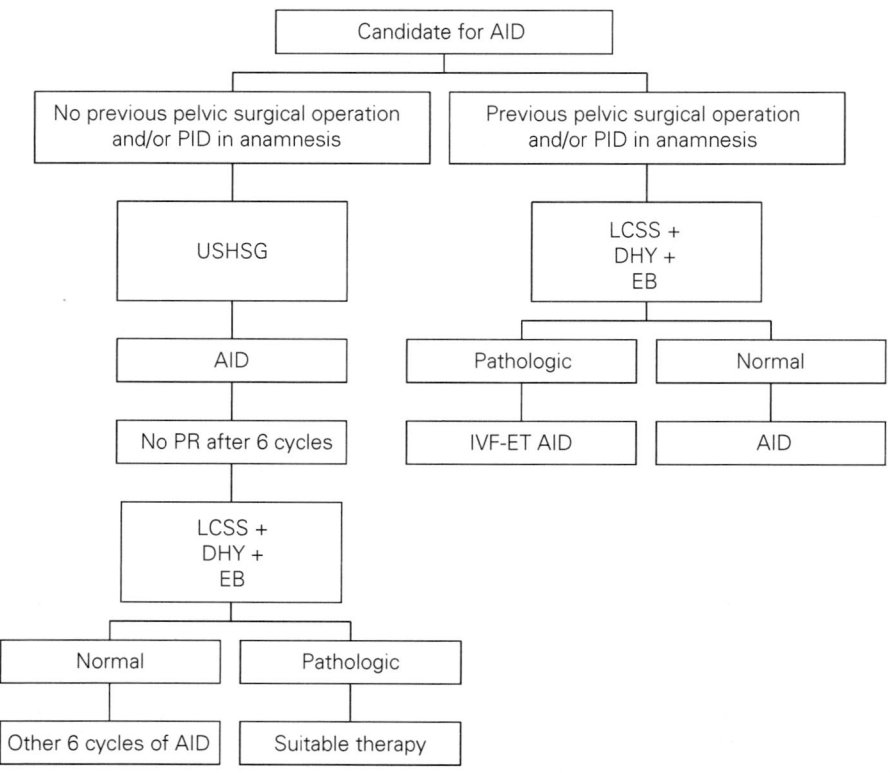

Figure 10 Diagnostic laparoscopy. The sequence illustrated here pertains to a woman proposed as a candidate for artificial insemination with donor semen (AID). PID, pelvic inflammatory disease; USHSG, ultrasound hysterosalpingography; LCSS, laparoscopy plus chromosalpingography; DHY, diagnostic hysteroscopy; EB, endometrial biopsy; PR, pregnancy; IVF-ET, *in vitro* fertilization and embryo transfer

Table 12 Laparoscopy plus chromosalpingography in the evaluation of sterility or infertility*: results of the Reggio Emilia Sterility Centre study of internal genitals and pelvis in 1011 women

Internal genitals and pelvis	Women (n)	Women (%**)
Normal	164	16.2
Endometriosis I-IV St***	377	37.3
Adhesion***	275	27.2
Fibromyomas***	203	20.1
Uterine anomalies***	102	10.1

*Infertile women suspected of uterus septum or bicurnate; **refers to the 1011 women who have undergone laparoscopy plus chromosalpingography; ***110 women presented with two pathologies, have been counted twice

HSG cannot be used to study the whole pelvis; in the study of the tubae, its results are in agreement with LCSS in about 65% of patients, at least in our experience.

USHSG is a promising investigation. Although we consider it still experimental, we do perform it as a

Table 13 Laparoscopy plus chromosalpingography in the evaluation of sterility or infertility*: results of the Reggio Emilia Sterility Centre study of internal genitals, pelvis and tubal opening/obstruction

Internal genitals, pelvis and tubal opening/obstruction	Women, n (%)
Normality + BTO	164 (16.3)
Pathology + BTO	515 (50.8)
Pathology + MTO	210 (20.8)
Normality or pathology + BTOB	122 (12.1)
Total	1011

*Infertile women suspected of uterus septum or bicurnate. Normality, normal internal genitals and pelvis; pathology, pathologic internal genitals and pelvis; BTO, bilateral tubal opening; MTO, monolateral tubal opening; BTOB, bilateral tubal obstruction

first-level procedure in young women without a history of pelvic surgery and/or PID and with a partner having a normal semen examination (Figure 6). However, it too cannot be used to study the whole pelvis, nor can it replace LCSS.

Laparoscopic salpingoscopy is still regarded as an experimental procedure; we nonetheless consider it to be a safe, reliable and useful technique that may be associated with the classic LCSS in the evaluation of the tubal factor.

Table 14 Literature results of laparoscopic adhesiolysis in sterile woman

References	Treated patients (n)	Intrauterine pregnancies		Extrauterine pregnancies	
		n	%*	n	%*
Bruhat et al., 1992[104]	92	48	90.5	5	(9.5
Fayez, 1983[105]	49	28	93.3	2	6.7
Chew et al., 1998[106]	21	7	87.5	1	12.5
Donnez and Casanas-Roux, 1986[107]	719	412	—	—	—
Gomel, 1983[108]	92	57	91.9	5	8.1

*Refers to total pregnancies (intrauterine + extrauterine) obtained by each author

Table 15 Literature results of laparoscopic fimbrioplasty and distal salpingostomy

References	Treated patients (n)	Intrauterine pregnancies		Extrauterine pregnancies	
		n	%*	n	%*
Mettler et al., 1979[109]	38	10	100	0	0
Fayez 1983[105]	14	3	60	2	40
Reich et al., 1987[110]	7	2	100	0	0
Donnez and Casanas-Roux, 1986[107]	25	5	100	0	0
Bruhat et al., 1983[111]	68	17	77.2	5	22.8
Dubuisson et al., 1994[112]	112	34	79	9	21
Bracco et al., 1995[113]	44	15	93	1	7
Lavergne et al., 1996[114]	46	16	88.8	2	11.2

*Refers to total pregnancies (intrauterine + extrauterine) obtained by each author

OPERATIVE LAPAROSCOPY

Introduction

In our opinion, operative laparoscopy and IVF should be considered as complementary and not as antagonistic options. Operative laparoscopy is probably the best therapy when performed with the right indication (e.g., endometriosis), whereas IVF is a good choice when laparoscopy is contraindicated or when it has failed.

Deciding the choice between the two requires a global evaluation of the pelvis and a careful examination of all data regarding a prospective patient's situation. For example, age is very important: if a woman is older than 35, we suggest skipping operative laparoscopy and going directly to IVF-ET.

Adhesiolysis, fimbrioplasty and salpingostomy

Currently, only adhesiolysis, fimbroplasty and distal salpingostomy are performed laparoscopically. Laparotomy has been abandoned because of its higher invasiveness and the higher risk of postoperative adhesions[108,115–118].

The American Fertility Society (AFS) classification of periadnexal adhesions and of distal tubal pathologies is the most widely used[119]. The various scores recently proposed consider not only the presence of adhesions and the type of distal tubal pathology, but also the tubal diameter and thickness and the endosalpingeal anatomic conditions[107,112,119–126]. It has been shown that there is a positive correlation between these scores and the results of laparoscopic surgery. In fact, if the score is favorable the results in terms of intrauterine pregnancies are good, while if the score is unfavorable, the results are also unsatisfactory.

Tables 14 and 15 report the results of adhesiolysis, fimbrioplasty and laparoscopic salpingostomy. It can be argued that not all women with pelvic adhesions and/or phymosis and/or distal tubal occlusion should be treated with laparoscopic surgery. Those with serious pelvic adhesions and/or with an unfavorable score should be shifted to IVF. The results of adhesiolysis are very good if pelvic adhesions are light or moderate, but insufficient if adhesions are serious[104,126]. Fimbrioplasty and distal salpingostomy achieve good results with lower scores, while if the score is high, IVF is the best option.

Table 16 Laparotomic salpingotomy versus laparoscopic salpingotomy: results of three randomized prospective studies

	Laparotomy*	Laparoscopy**
Treated women (*n*)	123	108
Days in hospital (*n*)	3.8	1.6
Days before working resumption (*n*)	43	14
Persistence of EUP (%)	1.7	12.2
Women with IUP after salpingostomy (%)	52	61
Women with EUP after salpingostomy (%)	14.4	8

EUP, extrauterine pregnancy; IUP, intrauterine pregnancy
*Laparotomic salpingotomy; **laparoscopic salpingotomy
Data derived from references 165–167

Table 17 Laparotomic salpingotomy versus laparoscopic salpingotomy: results of ten comparative studies of the persistence of extrauterine pregnancy

Technique	Treated women (*n*)	Women with persistent EUPs *n*	%
Laparotomic salpingotomy	230	9	3.9
Laparoscopic salpingotomy	699	58	8.3

EUP, extrauterine pregnancy. Data derived from references 165–174

Salpingotomy or salpingectomy after a tubal pregnancy

In recent decades, despite the growing incidence of extrauterine pregnancy (EUP), there has been a significant reduction in mortality resulting from it[127,128]. Yet EUP remains one of the most important causes of mortality in pregnant women[129–131]. The incidence is higher in sterile women, where EUP can be the first sign of a pre-existing tubal–pelvic sterility. It can also be the cause of *de novo* tubal–pelvic sterility.

In a woman wishing to have a child, the less invasive and more conservative the EUP treatment, the better. Three therapeutic options meet this criterion[132].

Table 18 Laparotomic salpingotomy versus laparoscopic salpingotomy: results of 30 non-comparative studies of intrauterine pregnancy and extrauterine pregnancy

Technique	Treated women* (*n*)	Women with IUP *n*	%	Women with EUP *n*	%
Laparotomic salpingostomy	811	498	61.4	125	15.4
Laparoscopic salpingostomy	703	429	61	109	15.5

EUP, extrauterine pregnancy; IUP, intrauterine pregnancy.
*Women wishing to conceive
Reproduced with permission from Yao, M, Tulandi T. Current status of surgical and non-surgical management of ecotopic pregnancy. *Fertil Steril* 1997;67:421–33

Table 19 Laparoscopic salpingotomy versus laparoscopic salpingectomy: results of a meta-analysis of nine comparative studies

Technique	Treated women* (*n*)	Treated women with IUP *n*	%	Treated women with EUP *n*	%
Laparoscopic salpingotomy	528	290	53	78	14.8
Laparoscopic salpingectomy	1246	614	49.3	128	9.9

IUP, intrauterine pregnancy; EUP, extrauterine pregnancy
*Women wishing to conceive
Reproduced with permission from Yao, M, Tulandi T. Current status of surgical and non-surgical management of ecotopic pregnancy. *Fertil Steril* 1997;67:421–33

Observational therapy

The outcome of 69.2% of EUPs is a spontaneous tubal abortion[133–142]. Therefore, careful observation of the patient may be justified, even if some cases of tubal rupture may occur during the observational therapy.

Medical therapy

Methotrexate (MTX) treatment is successful in 59–91% of cases[143–148,149–152], with complications arising in between 2 and 21%[143–148]. MTX therapy failure is more frequent when concentrations of beta human chorionic gonadotropin are high, when the tubal diameter is more than 2 cm and when the embryonic heartbeat can be seen at ultrasound[145,153,154].

Table 20 Endometriosis incidence in sterility and infertility: results obtained by the Reggio Emilia Sterility Centre in 1011 women who received laparoscopy due to sterility or infertility*

Endometriosis stage	Women (n)	Percentage of total 1011 who received laparoscopy	Percentage of 377 with endometriosis
I	158	15.6	41.9
II	103	10.2	27.4
III	60	5.9	15.9
IV	56	5.5	14.8

*Infertile women suspected of uterum septum or bicurnate

Table 21 Laparoscopic operation versus danazol and laparoscopic operation versus waiting behavior: literature studies *pro** laparoscopic therapy

References	Year of study	Type of study
Nowroozi et al.[155]	1987	Meta-analysis
Fayez et al.[156]	1988	Cohort
Paulson et al.[157]	1991	Cohort
Hughes et al.[158]	1993	Meta-analysis
Marcoux et al.[159]	1997	Randomized

*Laparoscopic therapy is more efficacious than danazol and waiting behavior

Table 22 Laparoscopic operation versus danazol and laparoscopic operation versus waiting behavior: literature studies *against** laparoscopic therapy

References	Year of study	Type of study
Seiler et al.[160]	1986	Meta-analysis
Levinson[161]	1989	Retrospective
Chong et al.[162]	1990	Cohort
Adamson and Pasta[163]	1994	Meta-analysis
Gruppo Italiano[164]	1999	Randomized

*Danazol and waiting behavior are as efficacious as laparoscopic therapy

Surgical therapy

Laparotomic surgical therapy has been almost completely abandoned, and is justified only when there is a high risk of mortality linked to massive hemorrhage. Tables 16–18[143,165–174] show that laparotomic and laparoscopic surgical approaches give the same results, but that there is a higher incidence of postoperative

Table 23 Literature results of laparoscopic therapy in sterile women with endometriosis grade III–IV

References	Pregnancies in women with III St* (%)	Pregnancies in women with IV St* (%)
Damewood and Rock, 1988[175]	25	16
Nezhat et al., 1989[176]	67	69
Donnez et al., 1992[177]	52	40
Bruhat et al., 1992[178]	44.4	36.8
Adamson et al., 1993[179]	82	82
Cook and Rock, 1991[180]	36–58	50–64
Busacca et al., 1999[181]	51	16.7

*AFS classification

adhesions with the former[182]. Nowadays, the laparoscopic approach is the most widely used in the treatment of EUP[150,151,183].

Salpingotomy or salpingectomy?

Most authors prefer salpingotomy, if possible. According to a 40-study meta-analysis, the percentage of intrauterine pregnancies (IUP) is 46% after salpingotomy and 44% after salpingectomy, whereas the incidence of a new EUP is 15 and 10%, respectively[184]. Another meta-analysis of nine studies shows similar results (Table 19)[145].

In favor of salpingotomy it has been observed that the percentages of IUP and recurrent EUP in women with only one salpinx are 54 and 20.5%, respectively (results derived from 18 studies)[143]. In favor of salpingectomy, it has reported that the rate of EUP persistence is 4.8–11% after salpingotomy, in comparison with 0% after salpingectomy[174,185–188].

Laparoscopic therapy of endometriosis (grade I–II)

Endometriosis represents the most frequent indication for operative laparoscopy in the USA[189,190], with the incidence being higher in sterile women than in fertile women. The AFS classification is the most widely used[119]. Table 20 reports the incidence of endometriosis in 1011 sterile women submitted to laparoscopy in our Centre.

Is endometriosis grade I–II a cause of female sterility? If this is the case, which is the best therapy? Different opinions have been reported[191,192,193–197]. Tables 21 and 22 show data of some of the most reliable studies about the efficiency of laparoscopic therapy, danazol therapy and observation with no therapy.

In our judgment, endometriosis grade I–II is not a cause of sterility and we therefore use no therapy except observation. We wait for 1–2 years, after which, if no pregnancy has occurred, we perform IUI cycles and then eventually IVF.

Laparoscopic therapy of endometriosis grade III–IV

Endometriosis grade III–IV is a well-known cause of sterility whose elective therapy is laparoscopic surgery. Table 23 reports some results published in the literature[198–203].

In our experience, post-surgery medical therapy does not improve results and indeed often reduces the probability of a spontaneous pregnancy in the first months after the operation.

Although laparoscopy is the elective treatment, we still use laparotomic surgery for the most difficult cases. However, neither the incidence of endometriosis recurrence, nor that of new adhesions after surgery, can be reduced to zero even with the laparoscopic approach.

After the surgical therapy, we suggest looking for a spontaneous pregnancy for 1 year if the woman is younger than 35, and for 6 months if the woman is between 35 and 37. In older women, we prefer to perform IVF immediately after surgery.

Myomectomy

Uterine subserous and pedunculate myomas are not a cause of sterility or infertility and, therefore, do not represent an indication for myomectomy[209–212]. Small (less than 5 cm diameter) intramural myomas that do not affect the uterine cavity are also unlikely to cause infertility and, similarly, do not represent a clear indication for myomectomy. However, bigger intramural myomas can affect fertility negatively[209–212]. In the latter case, myomectomy is indicated, as it is when sterility duration exceeds 3 years, or when other causes of sterility are absent, or there is deformation of the uterine cavity caused by the myoma(s).

Myomectomy can be performed laparotomically or laparoscopically. Laparoscopic myomectomy appears to obtain results similar to those of laparotomic myomectomy in sterile women (Tables 24 and 25)[213–215]. The most important complications of both laparoscopic and laparotomic myomectomy are adhesion formation, uterine rupture during the subsequent pregnancy, and the recurrence of myomas[206,216–220].

We perform myomectomy in the case of intramural myomas bigger than 5 cm in diameter, and we usually do it laparoscopically. We try to respect myometrial integrity as much as possible, in order to minimize the risk of uterine rupture during future pregnancies. A laparoscopic myomectomy does not necessarily require the performance of a Cesarean section, the type of delivery selected being dependent on the individual case, and based on data derived from a careful ultrasound follow-up after myomectomy[221].

Table 24 Literature results of laparotomic myomectomy in sterile women

References	Year of study	Pregnant women after treatment (%)
Smith and Uhlir[204]	1990	50
Verkauf[205]	1992	58
Gehlbach et al.[206]	1993	57
Acien and Quereda[207]	1996	62.7
Li et al.[208]	1999	57

Table 25 Literature results of laparoscopic myomectomy in sterile women

References	Year of study	Pregnant women after treatment (%)
Nezhat et al.[222]	1991	15.7
Hasson et al.[223]	1992	70.1
Dubuisson et al.[224]	1996	44.4
Darai et al.[216]	1997	36.3
Ribeiro et al.[217]	1999	64.3
Dubuisson et al.[225]	2000	53.1
Rossetti et al.[226]	2001	65.5

OTHER APPLICATIONS OF ENDOSCOPY IN REPRODUCTIVE MEDICINE

The anatomic and functional integrity of endosalpinges is one of the principal prerequisites for female fertility, yet even now, their anatomic and functional study plays only a marginal role in the diagnostic protocols. The most probable causes of this apparent failing are:

(1) The objective technical limit of the procedures used until now (HSG, LCSS and USHSG) allows endosalpinges to be studied only indirectly.

(2) The fact that tubal patency is usually considered the most important prognostic factor, rather than tubal 'good or bad functionality', creates a subjective conceptual limit.

Transcervical salpingoscopy and laparoscopic salpingoscopy represent two ways of overcoming these limits, and make the study of tubal factor complete and more reliable.

Transcervical salpingoscopy

The literature on transcervical salpingoscopy[227–241] concludes that:

(1) It is a safe, reliable and reproducible procedure;

(2) It allows false positives and false negatives of both HSG and LCSS to be identified;

(3) It is helpful in making a more reliable diagnosis of the tubal factor and, therefore, in choosing the best therapy, be that no therapy, surgical therapy or IVF-ET;

(4) It can be successfully employed even for intra-fallopian gametes/embryo transfer; and

(5) The incidence of complications such as tubal perforation is very low.

Our opinions on transcervical salpingoscopy are the following:

(1) We recognize the value of this technique, as theoretically it is both useful and non-invasive.

(2) We nevertheless think it is technically difficult to perform and that the interpretation and classification of the observed images need further experience and revision.

(3) The cost of the procedure remains high.

(4) Finally, we think that it can no longer be used as a routine clinical test, even though its use as an experimental procedure should be encouraged.

Laparoscopic salpingoscopy

Laparoscopic salpingoscopy was first proposed as a useful technique in the study of female infertility more than 15 years ago, but it has not been taken up widely[123,242,243]. The rationale for its use is the same as that for transcervical salpingoscopy. Even the synthesis of the literature results is similar to that reported for transcervical salpingoscopy[122,123,242–251]. Our opinions on laparoscopic salpingoscopy are the following:

(1) Even if some authorities still place it in the category of experimental and/or controversial techniques, we think that it has already got over its experimental phase and, therefore, can be proposed as a useful, simple, safe, and reproducible clinical procedure.

(2) Laparoscopic salpingoscopy is technically easier to perform than transcervical salpingoscopy.

(3) The quality of images is higher and the classification of the different endotubal findings more consistent than the results achieved by transcervical salpingoscopy.

(4) Its cost is limited.

(5) Laparoscopic salpingoscopy is a necessary procedure that can be employed systematically in all sterile women undergoing LCSS.

Laparoscopic salpingoscopy could also be indicated in the case of EUP, the study of the endosalpinx of the controlateral tube playing a possibly fundamental role not only when the right intraoperative choice between salpingotomy and salpingectomy has to be made, but above all, when a reliable reproductive prognosis is to be reached and a therapeutic option offered.

Minilaparoscopy

Minilaparoscopy, otherwise called microlaparoscopy, has been proposed as an alternative to classic laparoscopy, being less costly, less invasive and suitable for office use. However, studies and experiences so far reported in the literature are scanty[252–259], although they are only preliminary.

Female sterility is the most important indication for minilaparoscopy, but the technique has also been mooted for the performance of GIFT and for TET.

Our opinions on minilaparoscopy are as follows:

(1) At the moment, it remains an experimental procedure.

(2) Pain and the consequent low compliance rate in women seem to be the most important problems.

(3) It is necessary to deepen the study concerning the physiopathology of respiratory and cardiocirculatory problems connected with minilaparoscopy.

(4) Many technical aspects of minilaparoscopy can improve, among them local anesthesia, analgesia, pneumoperitoneum induction and the grade of Trendelenburg to be applied.

(5) Probably, in the future, tool diameters will be further reduced and optic resolution will increase.

(6) In conclusion, we foresee that minilaparoscopy will soon have an important role in the diagnosis of female sterility.

Transvaginal hydrolaparoscopy

Transvaginal hydrolaparoscopy has been recently proposed[260] and, theoretically, it can be considered as a modern culdoscopy. The rationale of transvaginal hydrolaparoscopy use is similar to that of minilaparoscopy and its most important indication is for female sterility. Our opinions are:

(1) At the moment, transvaginal hydrolaparoscopy must be considered an experimental procedure as only few authors have reported their experience of it in the literature[260–269].

(2) We believe randomized clinical trials are needed to evaluate its diagnostic accuracy, safety and feasibility.

(3) We are strongly against the widespread clinical use of this procedure at present.

(4) We think that the procedure has two major limitations: the impossibility of giving complete vision of the pelvic organs and of the whole pelvis, and the technical difficulties and/or risks of complications in the case of serious pelvic pathologies (adhesions, excluded pelvis, stage III–IV endometriosis, etc.).

Isthmus-ampullar/isthmus-isthmic/isthmus-cornual anastomosis

Only one study in the literature has given the results of isthmus-ampullar/isthmus-isthmic/isthmus-cornual anastomosis performed laparoscopically[270]. In the case of isthmic or cornual obstructive pathology, laparotomic microsurgery remains the first-choice therapy[2,3], while the use of both operative laparoscopy and radiologic or hysteroscopic techniques for tubal disobstruction must be considered experimental[270,271].

Our opinions on laparoscopic isthmus-ampullar/isthmus-isthmic/isthmus-cornual anastomosis are:

(1) After technical optimization, these procedures should give the same results as those obtained by

Table 26 Literature results of laparoscopic tubal sterilization reversal

Author	Year of study	Treated women (n)	Women with IUP (%)	Women with EUP (%)	Non-pregnant women (%)
Koh[272]	1995	31	71	3.2	25.8
Kim et al.[273]	1997	922	54.8	Not reported	Not reported
Dubuisson and Chapron[274]	1998	32	58.1	6.2	35.7
Bissonette et al.[275]	1999	102	70	7.2	22.8
Yoon et al.[276]	1999	202	83.3	3.2	13.5

IUP, intrauterine pregnancy; EUP, extrauterine pregnancy

laparotomic microsurgey but with lower costs and invasiveness.

(2) Their correct application requires an operator with uncommon technical ability as well as a 'reconstructive and not disruptive' mind.

(3) Currently, therefore, there are few gynecologic laparoscopists who are able to perform these techniques in a satisfactory way.

(4) Tools can be improved: it is not difficult to hypothesize the future construction of laparoscopic optics with a magnification system connected to a microscope, and it is reasonable to think that a good control of precise movements could also be achieved by laparoscopy.

(5) It seems therefore that the application and improvement of these techniques will increase in the future.

Tubal sterilization reversal

Microsurgical tubal sterilization reversal has, according to reports, led to 60–90% of all IUPs[277–280], results that are comparable to those obtained by IVF-ET[280–282]. The feasibility of laparoscopic tubal sterilization reversal had already been demonstrated with more than 100 pregnancies (70% of them IUPs) reported in the literature, when, in 1995, a technique for *microsurgical* laparoscopic 'tubal sterilization reversal' was described[272]. The technique is the same as isthmus-ampullar/isthmus-isthmic/isthmus-cornual anastomosis. Table 26 shows the literature data.

Our opinions on tubal sterilization reversal are almost the same as those previously stated about isthmus-ampullar/isthmus-isthmic/isthmus-cornual anastomosis.

GIFT/ZIFT/TET

Most MAP Centres around the world no longer use tubal gametes/embryos transfer, preferring to perform IVF with intrauterine transfer[283]. Initially, GIFT/ZIFT/TET aroused great interest on the basis of results obtained by retrospective studies[284–291]. Yet these results were not confirmed by following prospective studies and interest dropped[292–297]. Some years ago, while the GIFT/ZIFT/TET decline seemed unceasing, an

unexpected debate about their role re-emerged[298–305,306]. Our opinions on GIFT/ZIFT/TET and our comments about their old splendor and current misery follow:

(1) This case is an example of how to make a blunder in medicine, by drawing conclusions that are not based on confirmed data.

(2) The results reported by the first retrospective studies, although promising, did not in fact justify the conclusion that tubal transfer was more successful than uterine transfer. Besides, it must be borne in mind

Table 27 Laparoscopic therapy of PCO: literature results

Author	Year of study	Patients (n)	Ovulation (%)	Pregnancy (%)
Campo	1983	12	69	54
Aakvaag	1984	58	91	42
Greenblatt	1987	6	83	67
Vander-Weiden	1987	11	82	42
Sumioki	1988	7	86	57
Daniell	1989	85	71	56
Yanagibori	1989	6	—	50
Kojima	1989	12	83	58
Armar	1990	21	81	52
Abdel Gadir	1990	29	87	48
Gjonnaess	1990	113	> 90	65
Keckstein	1990	27	70	37
Sakata	1990	9	89	33
Tasaka	1990	11	91	36
Utsunomiva	1990	16	94	50
Gurgen	1991	17	71	47
Kovacs	1991	10	90	40
Rossmanith	1991	11	72	36
Gurgen	1992	40	70	50
Ostrzenski	1992	12	100	75
Armar	1993	50	92	62
Naether	1993	104	100	70
Campo	1993	23	61	56
Verhelst	1993	17	82	65
Tiitinen	1993	10	70	20
Greenblatt	1993	8	100	88
Szilagyi	1993	4	75	25
Total		729	614 (84.2%)	406 (55.7%)

Reproduced with permission from Donesky BW, Adashi EY. Surgically induced ovulation in the polycystic ovary disease: wedge resection revisited in the age of laparoscopy. *Fertil Steril* 1995;63: 439–63

that this conclusion was drawn despite the fact that, for tubal transfer, an invasive technique like laparoscopy, which has complications, was necessary.

(3) It is not possible to draw any conclusions about the uselessness of these techniques, however, as the results available so far came from only a few prospective studies based on a small number of cases.

(4) Therefore, our idea is that a double mistake was made.

(5) We began operating as a MAP Centre in 1987, performing GIFT as an assisted reproductive technology, and continued to practice GIFT and TET until 1999, even in the face of current opinions[307,308].

(6) In our experience, the results of tubal embryo transfer are better than those of uterine embryo transfer[307,308], but we are aware that this is a subjective opinion and that it is easily contested because it is based on retrospective data.

(7) Although we have not performed GIFT or TET for 3 years, we continue to think that the results obtained are better than those from IVF-ET.

(8) We think the truth sides neither with supporters nor with detractors of GIFT/ZIFT/TET.

(9) In the future, we would like to see suitable prospective studies undertaken, the results of which would show whether or not GIFT/ZIFT/TET are effectively useful.

Laparoscopic therapy of a micropolycystic ovary

Currently, ovulation induction in women with a micropolycystic ovary (PCO) can be obtained either by administering clomiphene citrate (CC), follicular stimulating hormone (FSH) or gonadotrophin releasing hormone (GnRH), or else by laparoscopic surgery. The laparoscopic surgery of choice for ovulation induction is, theoretically, the classic cuneiform ovarian resection[309], which is an empiric therapy[310].

It is agreed that CC is the best therapy to induce ovulation in women with PCO; only in CC-resistant patients is the use of FSH or GnRH, or laparoscopic surgery, justified[310–314]. Table 27 shows the literature results of laparoscopic treatment; periovarian adhesion

and anatomic–functional ovarian damage are the complications of this therapy.

Our opinions about ovulation induction in women with PCO are:

(1) In overweight or obese women, a weight loss program designed to achieve a body mass index (BMI) of not more than 25 is the best therapeutic option. In normal weight PCO women the best choice is CC.

(2) In CC-resistant patients, our second therapeutic option is FSH.

(3) In patients who respond to CC or FSH but who do not become pregnant, we perform LCSS in association with multiple ovarian electrocoagulations.

(4) We consider laparoscopic surgery for PCO patients as an experimental therapy, because neither its real therapeutic efficacy nor the incidence of its complications have yet been defined. Today, we do have efficacious therapeutic alternatives such as FSH and GnRH; it therefore is not justifiable to treat a woman with laparoscopy, a procedure with complications, just to induce ovulation.

Laparoscopic exeresis of rudimentary uterine horns

A bicornuate uterus with a rudimentary horn and its different forms can cause cryptomenorrhea with dysmenorrhea, and even acute abdomen, hematometra, endometriosis or hematosalpinges. The rudimentary horn can be the site of an EUP, when the risk of uterine rupture is very high. For these reasons, most authors suggest the exeresis of rudimentary uterine horns.

We know that, in the last 10 years, only 17 cases of exeresis of rudimentary uterine horns were made laparoscopically and that in four of these a pregnancy was already implanted in the rudimentary uterine horn at the time of laparoscopy[315–317].

Results reported in the literature are promising, and in the future laparotomic exeresis will probably be replaced by laparoscopic exeresis in all cases. Yet, in the meanwhile, we think it is opportune to underline the point that laparoscopic exeresis of rudimentary uterine horns can face technical difficulties and that, therefore, it must be performed by an experienced laparoscopist.

References

1. Seinera P, Maccario S, Visentin L, DiGregorio A. Hysteroscopy in an IVF-ET program. Clinical experience with 360 infertile patients. *Acta Obstet Gynecol Scand* 1988; 67:135–7

2. Dicker D, Goldman JA, Ashkenazi J, *et al.* The value of hysteroscopy in elderly women prior to *in vitro* fertilization-embryo transfer (IVF-ET): a comparative study. *J IVF Embryo Transf* 1990;7:267–70

3. Goldenberg M, Bider D, Ben-Rafael Z, *et al.* Hysteroscopy in a program of *in vitro* fertilization. *J IVF Embryo Transf* 1991;8:336–8

4. Shamma FN, Lee G, Gutmann JN, Lavy G. The role of office hysteroscopy in *in vitro* fertilization. *Fertil Steril* 1992;58:1237–9

5. Golan A, Ron-El R, Herman A, *et al.* Diagnostic hysteroscopy: its value in an *in vitro* fertilization/embryo transfer unit. *Hum Reprod* 1992;7:1433–4

6. Balmaceda JP, Ciuffardi I. Hysteroscopy and assisted reproductive technology. *Obstet Gynecol Clin N Am* 1995; 22:507–18

7. La Sala GB, Montanari R, Dessanti L, *et al.* The role of diagnostic hysteroscopy and endometrial biopsy in assisted reproductive technologies. *Fertil Steril* 1998;70:378–80

8. Schiano A, Jourdain O, Papaxanthos A, *et al.* The value of hysteroscopy after repeated implantation failures with *in vitro* fertilization. *Contracept Fertil Sex* 1999;27: 129–32

9. Hulka JF, Petrson HB, Phillips JM, *et al.* Operative hysteroscopy: American Association of Gynecologic Laparoscopists. Membership survey. *J Am Assoc Gynecol Laparosc* 1993;4:39–45

10. Hrowe PJ, Comhaire FH, Hargreave TB, Mellows HJ, eds. *WHO Manual for the Standardized Investigation of the Infertile Couple.* Cambridge, UK: Press Syndicate of University of Cambridge, 1993

11. ESHRE guidelines to the prevalence, diagnosis, treatment and management of infertility. *ESHRE Capri Workshop Hum Reprod* 1996;11:1775–807

12. Golan A, Eilat E, Ron-El R, *et al.* Hysteroscopy is superior to hysterosalpingography in infertility investigation. *Acta Obstet Gynecol Scand* 1996;75:654–6

13. Gaglione R, Valentini AL, Pistilli E, Nuzzi NP. A comparison of hysteroscopy and hysterosalpingography. *Int J Gynecol Obstet* 1996;52:151–3

14. Wang CW, Lee CL, Lai YM, *et al.* Comparison of hysterosalpingography and hysteroscopy in female infertility. *J Am Assoc Gynecol Laparosc* 1996;33:581–4

15. Cicinelli E, Romano F, Anastasio PS, *et al.* Transabdominal sonohysterography, transvaginal sonography, and hysteroscopy in the evaluation of submucous myomas. *Obstet Gynecol* 1995;85:42–7

16. Gronlund L, Hertz J, Helm P, Colov NP. Transvaginal sonohysterography and hysteroscopy in the evaluation of female infertility, habitual abortion or metrorrhagia. A comparative study. *Acta Obstet Gynecol Scand* 1999;78: 415–18

17. Roudigoz R, Gaucherand P, Piacenza JM, Salle B. Sonohysterography of uterine cavity: preliminary investigation. *J Clin Ultrasound* 1995;23:339–48

18. Lass A, William G, Aushevkhe N, Brindsen R. Effect of endometrial polyps on outcomes in IVF cycles. *J Assist Reprod Genet* 1999;16:410–5

19. Mastrominas M, Pistofidis GA, Dimitropoulos K. Fertility outcome after outpatient hysteroscopic removal of endometrial polyps and submucous fibroids. *J Am Assoc Gynecol Laparosc* 1996;3(Suppl)

20. Garcia CR, Tureck RW. Submucosal leiomyomas and infertility. *Fertil Steril* 1984;42:16–19

21. Forssman L. Distribution of blood flow in a myometus uteri as measured by locally injected 133 xenon. *Acta Obstet Gynecol Scand* 1996;55:101–6

22. Wood C, *et al.* Effect of intramural, subserosal and submucosal uterine fibroids on ART. Presented at the *14th Annual Meeting of ESHRE*, Goteborg, Sweden, 1998, 0-055; *Hum Reprod* 13:abstr.

23. Corson SL, Brooks PG. Resectoscopic myomectomy. *Fertil Steril* 1991;55:1041

24. Donnez J, Nisolle M, Casanas-Roux F, *et al.* CO$_2$ laser laparoscopic surgery: adhesiolysis salpingostomy and fimbrioplsty. In Donnez J, Nisolle M, eds. *Atlas of Laser Operative Laparoscopy and Hysteroscopy.* New York: Parthenon Publishing Group, 1994;97–112

25. Tulandi T, al-Took S. Endoscopic myomectomy. Laparoscopy and hysteroscopy. *Obstet Gynecol Clin N Am* 1999;26:135–48

26. Stovall DW, Parrish SB, Van Voorhis BJ, *et al.* Uterine leiomyomas reduce the efficacy of assisted reproduction cycles: results of a matched follow-up study. *Hum Reprod* 1998;13:192–7

27. Ramzy AM, Sattar M, Amin Y, *et al.* Uterine myomata and outcome of assisted reproduction. *Hum Reprod* 1998;13: 198–202

28. Abbas A, Montgomery IL. Uterine rupture during labor myomectomy. *Gynecol Endosc* 1997;6:245–6

29. Hallez JP, Netter A, Carter R. Methodical intra-uterine resection. *Am J Obstet Gynecol* 1987;156:1080–4

30. Brooks PG, Loffer FD, Serden SP. Resectoscopic removal of symptomatic intrauterine lesions. *J Reprod Med* 1989;34: 435–7

31. Valle RF. Hysteroscopic removal of submucous leiomyomas. *J Gynecol Surg* 1990;6:89–96

32. Loffer FD. Removal of large symptomatic intrauterine growths by hysteroscopic resectoscope. *Obstet Gynecol* 1990;76:836–40

33. Hucke J, Campo RL, Debbruyne F, *et al.* Hysteroscopic resection of submucous myoma. *Geburt Frauen* 1992;52: 214–18

34. Mergui JL, Renolleau C, Salat-Baroux J. Hystéroscopie opératoire et fibromes. *Gynecologie* 1993;1:325–37

35. Cravello L, d'Ercole C, Azoulay P, *et al.* Le traitement hystéroscopique des fibromes utérines. *Eur J Obstet Gynecol Reprod Biol* 1995;24:374–80

36. Kuhulmann M, Gartner A, Schindler EM. Uterine leiomyomata and sterility: therapy with gonadotrophin releasing hormone agonists and myomectomy. *Gynecol Endocrinol* 1997;11:169–74

37. La Sala GB, Montanari R. La terapia medica e chirurgica della sterilità. Presented at the *1999 Atti del congresso SIGO Viareggio*, 288–93

38. Bernard G, Darai E, Poncelet C, *et al.* Fertility after hysteroscopic myomectomy: effect of intramural myomas associated. *Eur J Obstet Gynecol Reprod Biol* 2000;88:85–90

39. Fernandez H, Sefrioui O, Virelizier C, *et al.* Hysteroscopic resection of submucosal myomas in patients with infertility. *Hum Reprod* 2001;16:1489–92

40. Seinera P, Gagliati P, Volpi E, *et al.* Ultrasound evaluation of uterine wound healing following laparoscopic myomectomy: preliminary results. *Hum Reprod* 1999;14:2460–3

41. Green LK, Harris RE. Frequency of diagnosis and associated obstetric complications. *Obstet Gynecol* 1976;47:427–9

42. Heinonen PK, Saarikoski S, Pystynen P. Reproductive performance of women with uterine anomalies. *Acta Obstet Gynecol Scand* 1982;61:157–62

43. Golan A, Langer R, Bukovsky I, Caspi E. Congenital anomalies of the Mullerian system. *Fertil Steril* 1989;51:747–55

44. Simòn C, Martinez L, Pardo P, *et al.* Mullerian defects in women with normal reproductive outcome. *Fertil Steril* 1991;56:1192–93

45. Acien P. Incidence of Mullerian defects in fertile and infertile women. *Hum Reprod* 1997;12:1372–6

46. Halvorson LM, Aserkoff RD, Oskowitz SP. Spontaneous uterine rupture after hysteroscopic metroplasty with uterine perforation. A case report. *J Reprod Med* 1993;38:236–8

47. Lobaugh ML, Bammel BM, Duke D, Webster BW. Uterine rupture during pregnancy in a patient with a history of hysteroscopic metroplasty. *Obstet Gynecol* 1994;83:838–40

48. Gabriele A, Zanetta G, Pasta F, Colombo M. Uterine rupture after hysteroscopic metroplasty and labor induction. A case report. *J Reprod Med* 1999;44:642–4

49. Pellicer A. Shall we operate on Mullerian defects? *Hum Reprod* 1997;12:1371

50. Corson SL, Batzer FR. CO_2 uterine distension for hysteroscopic septal incision. *J Reprod Med* 1986;31:710

51. De Cherney AH, Polan ML. Hysteroscopic management of intrauterine lesions and intractable uterine bleeding. *Obstet Gynecol* 1983;61:392–7

52. Fayez JA. Comparison between abdominal and hysteroscopic metroplasty. *Obstet Gynecol* 1986;68:399–403

53. March CM, Israel R. Hysteroscopic management of recurrent abortion caused by the septate uterus. *Am J Obstet Gynecol* 1987;156:834–42

54. Blanc B, d'Ercole C, Gaiato ML, Boubli L. Le traitement endoscopique des cloisons utérines. *J Gynecol Obstet Biol Reprod* 1994;23:596–601

55. Grimbizis G, Camus M, Clasen K, *et al.* Hysteroscopic septum resection in patients with recurrent abortions or infertility. *Hum Reprod* 1998;13:1188–93

56. Perino A, Biondo A, Castelli A, *et al.* Isteroscopia operatoria: valutazione decca tecnica operatoria e deirisulta ti in S69 casi. Presented at the *1999 Atti del 75° Congresso SIGO Viareggio*, 115–18

57. Querleu D, Brasme TL, Parmentier D. Ultrasound-guided transcervical metroplasty. *Fertil Steril* 1990;54:995–8

58. Raga F, Bonilla-Musoles F, Blanes J, Osborne NG. Congenital Mullerian anomalies: diagnostic accuracy of three-dimensional ultrasound. *Fertil Steril* 1996;65:523–8

59. Clifford K, Rai R, Watson H, Regan L. An informative protocol for the investigation of recurrent miscarriage: preliminary experience of 500 consecutive cases. *Hum Reprod* 1994;9:1328–32

60. Exalto N, Eskes TKAB, Hein PR. Ultrasonographic examination of the congenitally malformed uterus. *Eur J Obstet Gynecol Reprod Biol* 1978;8:369–75

61. Jurkovic D, Geipel A, Gruboeck K, *et al.* Three-dimensional ultrasound for the assessment of uterine anatomy and detection of congenital anomalies: a comparison with hysterosalpingography and two-dimensional sonography. *Ultrasound Obstet Gynecol* 1995;5:233–7

62. Nasry MN, Setchell ME, Chard T. Transvaginal ultrasound for diagnosis of uterine malformations. *Br J Obstet Gynaecol* 1990;97:1043–5

63. Pelosi MA III, Pelosi MA. Laparoscopic-assisted metroplasty for the treatment of bicornate uterus: a case study. *Fertil Steril* 1996;65:886–90

64. Homer HA, Li T-C, Cooke ID. The septate uterus: a review of management and reproductive outcome. *Fertil Steril* 2000;73:1–14

65. Cohen LS, Valle RF. Role of vaginal sonography and hysterosonography in the endoscopic treatment of uterine myomas. *Fertil Steril* 2000;73:197–204

66. Zabak K, Benifla JL, Uzan S. Septate uterus and reproduction disorders: current results of hysteroscopic septoplasty. *Gynecol Obstet Fertil* 2001;29:829–40

67. Wai CY, Zekam N, Sanz LE. Septate uterus with double cervix and longitudinal vaginal septum. A case report. *J Reprod Med* 2001;46:613–17

68. Grimbizis GF, Camus M, Tarlatzis BC, *et al.* Clinical implications of uterine malformations and hysteroscopic treatment results. *Hum Reprod Update* 2001;7:161–74

69. Hickok LR. Hysteroscopic treatment of the uterine septum: a clinician's experience. *Am J Obstet Gynecol* 2000;182:1414–20

70. Shenker JG, Margalioth EJ. Intrauterine adhesions: an update appraisal. *Fertil Steril* 1982;37:593

71. Porcu G, Cravello L, d'Ercole C, *et al.* Hysteroscopic metroplasty for septate uterus and repetitive abortions: reproductive outcome. *Eur J Obstet Gynecol Reprod Biol* 2000;88:81–4

72. March CM, Israel R, March AD. Hysteroscopic management of intrauterine adhesions. *Am J Obstet Gynecol* 1978;130:653–7

73. Siegler AM, Valle RF. Therapeutic hysteroscopic procedures. *Fertil Steril* 1988;50:685–701

74. American Fertility Society. The American Fertility Society classifications of intrauterine adhesions. *Fertil Steril* 1988;49:944–55

75. Valle RF, Sciarra JJ. Intrauterine adhesions: hysteroscopic diagnosis, classification, treatment, and reproductive outcome. *Am J Obstet Gynecol* 1988;158:1459–70

76. Donnez J, Nisolle M. Operative laser hysteroscopy in Mullerian defects and uterine adhesions. In Donnez J, ed. *Operative Laser Laparoscopy and Hysteroscopy*. Louvain, Belgium: Nauweraerts Printing, 1989:249–61

77. Friedman A, de Fazio J, de Cherney A. Severe obstetric complications after aggressive treatment of Asherman syndrome. *Obstet Gynecol* 1986;67:864–7

78. Deaton JL, Maier D, Andreoli J. Spontaneous uterine rupture during pregnancy after treatmente of Asherman's syndrome. *Am J Obstet Gynecol* 1989;160:1053–4

79. Chan NS. Intrauterine retention of fetal bone. *Aust N Z J Obstet Gynaecol* 1996;36:368–71

80. Coccia ME, Becattini C, Bracco GL, Scarselli G. Ultrasound-guided hysteroscopic management of endometrial osseous metaplasia. *Ultrasound Obstet Gynecol* 1996;8:134–6

81. Torne A, Jou P, Pagano R, *et al*. Endometrial ossification successfully treated by hysteroscopic resection. *Eur J Obstet Gynecol Reprod Biol* 1996;66:75–7

82. Hoang-Ngoc Minh, Lemay B, Smadja A. Reflections on endometrial osteogenesis. Apropos of 3 cases. *Rev Fr Gynecol Obstet* 1984;79:471–4

83. Verstraete JP, Adnet JJ, Wahl P. Osteogenic metaplasia or residual embryonic endometrial ossification. *J Gynecol Obstet Biol Reprod (Paris)* 1984;13:425–31

84. Torne A, Jou P, Pagano R, *et al*. Endometrial ossification successfully treated by hysteroscopic resection. *Eur J Obstet Gynecol Reprod Biol* 1996;66:75–7

85. Bahceci M, Demirel LC. Osseous metaplasia of the endometrium: a rare cause of infertility and its hysteroscopic management. *Hum Reprod* 1996;11:2537–9

86. Camus M, Ropert JF, Iloki LH, *et al*. Endometrial ossification. Apropos of five recent cases. *J Gynecol Obstet Biol Reprod (Paris)* 1990;19:295–300

87. Ombelet W. Endometrial ossification, an unusual finding in an infertility clinic. A case report. *J Reprod Med* 1989;34:303–6

88. Melius FA, Julian TM, Nagel TC. Prolonged retention of intrauterine bones. *Obstet Gynecol* 1991;78:919–21

89. Pace S, Torcia F, Palazzetti PL, *et al*. Successful diagnostic and surgical hysteroscopy for endometrial ossification. *Clin Exp Obstet Gynecol* 2001;28:24–5

90. Van de Basch T, Dubin M, Cornelis A. Favorable pregnancy outcome in a woman with osseus metaplasia of the uterus. *Ultrasound Obstet Gynecol* 2000;15:445–6

91. Sugimoto O. Diagnostic and therapeutic hysteroscopy for traumatic intrauterine adhesions. *Am J Obstet Gynecol* 1978;131:539–47

92. Valle RF. Future growth and development of hysteroscopy. *Obstet Gynecol Clin N Am* 1988;15:111–26

93. Roge P, d'Ercole C, Cravello L, *et al*. Hysteroscopic management of uterine synechiae: a series of 102 observations. *Eur J Obstet Gynecol Reprod Biol* 1996;65:189–93

94. Colacurci N, Fortunato N, Nasto R, *et al*. Reproductive outcome of hysteroscopic lysis of intrauterine adhesions. *Minerva Ginecol* 1997;49:325–7

95. Capella-Allouc S, Morsad F, Rongieres-Bertrand C, *et al*. Hysteroscopic treatment of severe Asherman's syndrome and subsequent fertility. *Hum Reprod* 1999; 14:1230–3

96. Bustillo M, Munali AK, Shulman JD. Pregnancy after non surgical ultrasound-guided gamete intra-fallopian transfer. *N Engl J Med* 1988;319:313

97. Seracchioli R, Possati G, Affora P, *et al*. Hysteroscopic gamete intra-fallopian transfer: a good alternative in selected cases to laparoscopic intra-fallopian transfer. *Hum Reprod* 1991;67:1388–90

98. Ferraiolo A, Croce S, Anserini P, *et al*. 'Blind' transfer of gamete in the fallopian tube: a preliminary study. *Hum Reprod* 1991;6:537–40

99. Possati G, Seracchioli R, Melega C, *et al*. Gamete intra-fallopian transfer by hysteroscopy as an alternative treatment for infertility. *Fertil Steril* 1991;56:845–57

100. Diedrich K, Bauer O, Werner A. Transvaginal intratubal embryo transfer: a new treatment of male infertility. *Hum Reprod* 1991;6:672–5

101. Jansen R, Anderesen J. Transvaginal versus laparoscopic gamete intra-fallopian transfer: a case controlled retrospective comparison. *N Engl J Med* 1991;319:288–91

102. Sholtes M, Roozenburg B, Albreda A, Zeilmaker GH. Transcervical transfer of zygotes. *Fertil Steril* 1990;54:283–6

103. Sholtes M, Roozenburg B, Verhoeff A, Zeilmaker GH. A randomized study of transcervical intrafallopian transfer of pronuclear embryos controlled by ultrasound vs intra-uterine transfer of four- to eight-cell embryos. *Fertil Steril* 1994;61:102–4

104. Bruhat MA, Mage G, Pouly J-L. *Operative Laparoscopy*. New York: McGraw Hill, 1992

105. Fayez JA. An assessment of the role of operative laparoscopy in tuboplasty. *Fertil Steril* 1983;39:476–9

106. Chew S, Chang C, Ng SC, Ratnam SS. Laparoscopic adhesiolysis for subfertility. *Singapore Med J* 1998;39:491–5

107. Donnez J, Casanas-Roux F. Prognostic factors of fimbrial microsurgery. *Fertil Steril* 1986;46:200–4

108. Gomel V. *Microsurgery in Female Infertility*. Boston: Little Brown & Co., 1983

109. Mettler L, Giesel H, Semm K. Treatment of female infertility due to tubal obstruction by operative laparoscopy. *Fertil Steril* 1979;32:384–8

110. Reich H, Freifeld ML, McGlynn F, Reich E. Laparoscopic treatment of tubal pregnancy. *Obstet Gynecol* 1987;69:275–9

111. Bruhat MA, Mage G, Manhes H, *et al*. Laparoscopy procedures to promote fertility ovariolysis and salpingolysis. Results of 93 selected cases. *Acta Eur Fertil* 1983;14:113–15

112. Dubuisson JB, Chapron C, Morice P, *et al*. Laparoscopic salpingostomy: fertility results according to the tubal mucosa appearance. *Hum Reprod* 1994;9:334–9

113. Bracco GL, Coccia E, Barciulli F, Scarselli G. Le salpingo-plastiche: tecniche a confronto. Presented at the *1995 Atti Congresso Nazionale SIELG Firenze Editi da E.Cittadini-G.Scarselli*. Sevizi Editoriali-Firenze, 141

114. Lavergne N, Krimly A, Roge P, Erny R. Resultats et indications de la coeliochirurgie tubaire distale. *Contracept Fertil Sex* 1996;24:41–8

115. Luciano AA, Withman G, Maier DB, *et al*. A comparative study of postoperative adhesions: following laser surgery by laparoscopy versus laparotomy in the rabbit model. *Obstet Gynecol* 1989;74:220–4

116. Lundorff P, Halhin M, Kallfelt B, *et al*. Adhesions: formation after laparoscopic surgery in tubal pregnancy: a randomized trial versus laparotomy. *Fertil Steril* 1991; 55:911–15

117. Diamond MP, Daniell JF, Johns DA, *et al*. Postoperative adhesions after operative laparoscopy: evaluation at early second-look procedures. Operative Laparoscopy Study Group. *Fertil Steril* 1991;55:700–4

118. Surrey ES, Schoolcraft WB. Laparoscopic management of hydrosalpinges before *in vitro* fertilization-embryo transfer: salpingectomy versus proximal tubal occlusion. *Fertil Steril* 2001;75:612–17

119. American Fertility Society. The American Fertility Society classifications of adnexal adhesions, distal tubal occlusion secondary to tubal ligation, tubal pregnancies, Mullerian anomalies and intrauterine adhesions. *Fertil Steril* 1988; 49:944–6

120. Mage G, Pouly JL, Bouquet de Jolinerie J, *et al*. A preoperative classification to predict the intrauterine and ectopic pregnancy rates after distal tubal microsurgery. *Fertil Steril* 1986;46:807–10

121. Boer-Meisel ME, Te Velde ER, Habbema JD, *et al*. Predicting the pregnancy outcome in patients treated for hydrosalpinx: a prospective study. *Fertil Steril* 1986;45: 23–9

122. Puttemans P, Brosens IA, Delattin PH, *et al*. Salpingoscopy versus hysterosalpingography in hydrosalpinges. *Hum Reprod* 1987;2:535–40

123. Brosens IA, Boeckx W, Delattin PH, *et al*. Salpingoscopy: a new preoperative diagnostic tool in tubal infertility. *J Obstet Gynecol* 1987;94:768–73

124. Vasquez G, Boeckx W, Brosens IA. Prospective study of tubal mucosal lesions and fertility in hydrosalpinges. *Hum Reprod* 1995;10:1075–8

125. Marana R, Rizzi M, Muzzii L, *et al*. Correlation between the American Fertility Society classifications of adnexal adhesions, distal tubal occlusion, salpingoscopy and reproductive outcome in tubal surgery. *Fertil Steril* 1995; 64:924–9

126. Hulka JS, Reich H. *Textbook of Laparoscopy*, 3rd edn. Philadelphia: WB Saunders, 1998

127. Goldner TE, Lawson HW, Xia Z, Atrash HLK. Surveillance of ectopic pregnancy: United States 1970–89. *MMWR CDC Surveillance Summaries* 1993;42:73–85

128. Department of Health. *Report on Confidential Enquiries into Maternal Deaths in the United Kingdom 1991–1993*. London: HMSO, 1996:68–73

129. Department of Health. *Reports on Confidential Enquiries into Maternal Deaths in the United Kingdom 1973–75, 1988–90*. London: HMSO, 1994

130. Ling FW, Stovall TG. Update on the diagnosis and management of ectopic pregnancy. *Adv Obstet Gynecol* 1994;16:55–83

131. Kim DS, Chung SR, Park MI, Kim YP. Comparative review of diagnostic accuracy in tubal pregnancy: a 14 year survey of 1040 cases. *Obstet Gynecol* 1987;70: 547–52

132. Lehner R, Kucera E, Jirecek S, *et al*. Ectopic pregnancy. *Arch Gynecol Obstet* 2000;263:87–92

133. Shalev E, Romano S, Peleg D, *et al*. Spontaneous resolution of ectopic tubal pregnancy. Natural history. *Fertil Steril* 1995;63:1519

134. Trio D, Strobelt N, Picciolo C, *et al*. Prognostic factors for successful expectant management of ectopic pregnancy. *Fertil Steril* 1995;63:469–72

135. Korhonen J, Stenman UH, Ylostalo P. Serum human chorionic gonadotrophin diagnosis during spontaneous resolution of ectopic pregnancy. *Fertil Steril* 1994;61: 632–6

136. Makinen JI, Kivijarvi AK, Idala KMA. Success of non-surgical management of ectopic pregnancy. *Lancet* 1990;335:1099

137. Fernandez H, Rainborn JL, Papiernik E, *et al*. Spontaneous resolution of ectopic pregnancy. *Obstet Gynecol* 1988;7: 171–4

138. Derricks-Tan JSE, Scholz C, Taubert HD. Spontaneous recovery of ectopic pregnancy: a preliminary report. *Eur J Obstet Gynecol Reprod Biol* 1997;25:181–5

139. Garcia AJ, Aubert JM, Sama J, Josimovich JB. Expectant management of presumed ectopic pregnancies. *Fertil Steril* 1987;48:395–400

140. Sauer MV, Gorril MJ, Rodi A, Yeko TR. Non-surgical management of unruptured ectopic pregnancy: an extended clinical trial. *Fertil Steril* 1987;48:752–5

141. Adoni A, Milwidsky A, Hurwitz A, Palti Z. Declining beta-hCG levels: an indicator for expectant approach in ectopic pregnancy. *Int J Fertil* 1986;31:40–2

142. Carp HJA, Oelsner G, Serr DM, Mashiach S. Fertility after non surgical treatment of ectopic pregnancy. *J Reprod Med* 1986;31:119–22

143. Yao M, Tulandi T. Current status of surgical and non-surgical management of ectopic pregnancy. *Fertil Steril* 1997;67:421–33

144. Clasen K, Camus H, Tounaye H, Devroey I. Ectopic pregnancy: let's cut! Strict laparoscopic approach to 194 consecutive cases and review of the literature on alternatives. *Hum Reprod* 1997;12:596–601

145. Kooi S, Kock HCLV. A review of the literature on non-surgical treatment in tubal pregnancies. *Obstet Gynecol Surv* 1992;47:739–49

146. Slaughter JL, Grimes DA. Methotrexate therapy: non-surgical management of ectopic pregnancy. *West J Med* 1995;162:225–8

147. Horrigan TJ, Fanning J, Marcotte MP. Methotrexate pneumonitis after systemic treatment for ectopic pregnancy. *Am J Obstet Gynecol* 1997;176:714–15

148. Isaacs SD, McGehee RP, Covan BD. Life-threatening neutropenia following methotrexate treatment of ectopic pregnancy: a report of two cases. *Obstet Gynecol* 1996;8: 694–6

149. Lipscomb GH, McCord ML, Stovall TG, *et al*. Predictors of success of methotrexate treatment in women with tubal ectopic pregnancies. *N Engl J Med* 1999;341:1974

150. Varma R, Vindla S, Mascarenhas L. Fertility following radical, conservative-surgical or medical treatment for tubal pregnancy: a population-based study. *Br J Obstet Gynaecol* 2001;108:130–1

151. Bouyer J, Job-Spira N, Pouly JL, *et al.* Fertility following radical, conservative-surgical or medical treatment for tubal pregnancy: a population-based study. *Br J Obstet Gynaecol* 2000;107: 714–21

152. Debby A, Golan A, Sadan O, *et al.* Fertility outcome following combined methotrexate treatment of unruptured extrauterine pregnancy. *Br J Obstet Gynaecol* 2000;107: 626–30

153. Stovall TG, Ling M. Single-dose methotrexate: an expanded clinical trial. *Am J Obstet Gynecol* 1993;168:1759–65

154. Sagiv R, Golan A, Arbel-Alon S, Glezerman M. Three conservative approaches to treatment of interstitial pregnancy. *J Am Assoc Gynecol Laparosc* 2001;8:154–8

155. Nowroozi K,Chase JS, Check JH, *et al.* The importance of laparoscopic coagulation of mild endometriosis in infertile women. *Int J Fertil* 1987;32:442–4

156. Fayez JA, Collazo LM, Vernon C. Comparison of different modalities of treatment for minimal and mild endometriosis. *Am J Obstet Gynecol* 1988;159:927–32

157. Paulson JD, Asmar P, Saffan DS. Mild and moderate endometriosis: compaarison of treatment modalities (laser excluded). *J Reprod Med* 1991;36:151–5

158. Hughes EG, Fedorkow DM, Collins JA. A quantitative overview of controlled trials in endometriosis-associated infertility. *Fertil Steril* 1993;59:963–70

159. Marcoux S, Maheux R, Berube S, *et al.* Laparoscopic surgery in infertile women with minimal or mild endometriosis. *N Engl J Med* 1997;337:217–22

160. Seiler JC, Gidwani G, Ballard L. Laparoscopic cauterization of endometriosis for infertility: a controlled study. *Fertil Steril* 1986;46:1098–100

161. Levinson CJ. Endometriosis therapy: rationale for expectant or minimal therapy in minimal/mild cases (AFSI). Presented at the *Second World Congress on Gynecology and Endoscopy*, Clermont-Ferrand, France, 1989;abstr.

162. Chong AP, Keene ME, Thorton NL. Comparison of 3 modes of treatment for infertility patients with minimal pelvic endometriosis. *Fertil Steril* 1990;53:407–10

163. Adamson GD, Pasta DJ. Surgical treatment of endometriosis-associated infertility: meta-analysis compared with survival analysis. *Am J Obstet Gynecol* 1994; 171:1488–505

164. Gruppo Italiano per lo studio dell'Endometriosi. Ablation of lesions or no treatment in minimal-mild endometriosis in infertile women: a randomized trial. *Hum Reprod* 1999;14:1332–4

165. Murphy AA, Nager CW, Wujek JJ, *et al.* Operative laparoscopy versus laparotomy for the management of ectopic pregnancy: a prospective trial. *Fertil Steril* 1992;57:1180–5

166. Vermesh M, Silva PD, Rosen GF, *et al.* Management of ruptured ectopic gestation by linear salpingostomy: a prospective randomised clinical trial of laparoscopy versus laparotomy. *Obstet Gynecol* 1989;73:400–4

167. Lundorff P, Thorburn J, Hahlin M, *et al.* Laparoscopic surgery in ectopic pregnancy: a randomised trial versus laparotomy. *Acta Obstet Gynecol Scand* 1991;70:343–88

168. Seifer DB, Gutmann J, Grant WD, *et al.* Comparison of persistent ectopic pregnancy after laparoscopic salpingostomy versus salpingostomy at laparotomy for ectopic pregnancy. *Obstet Gynecol* 1993;81:370–82

169. Henderson SR. Ectopic tubal pregnancy treated by operative laparoscopy. *Am J Obstet Gynecol* 1989;160: 1462–9

170. Hoppe DE, Bekkar BE, Nager CW. Single-dose systemic methotrexate for the treatment of persistent ectopic pregnancy after consecutive surgery. *Obstet Gynecol* 1994; 83:51–4

171. Keckstein G, Wolf AS, Hepp S, *et al.* Tubenerhaltende endoskopische Operationsverfahren bei nicht rupturierter Tubagravidität. Welche Bedeutung hat dabei der Lasereinsatz? *Geburt Frauen* 1990;50:207–11

172. Brumsted J, Kessler C, Gibson C, *et al.* A comparison of laparoscopy and laparotomy for the treatment of ectopic pregnancy. *Obstet Gynecol* 1988;71:889–92

173. Silva PD. A laparoscopic approach can be applied to most cases of ectopic pregnancy. *Obstet Gynecol* 1988;72:944–7

174. Pouly JL, Mahnes H, Mage G, *et al.* Conservative laparoscopic treatment of 321 ectopic pregnancies. *Fertil Steril* 1986;46:1093–7

175. Damewood MD, Rock JA. Treatment independent pregnancy with operative laparoscopy for endometriosis. *Am J Obstet Gynecol* 1988;159:927

176. Nezhat C, Winer W, Cooper J, *et al.* Endoscopic infertility surgery. *J Reprod Med* 1989;34:127–34

177. Donnez J, Nisolle M, Casanas F. Co$_2$ laser laparoscopy in peritoneal endometriosis and in ovarian endometrial cyst. Presented at the *Atti del XV Congresso Nazionale della SIFES Ed.CO.FE.SE*, 1992

178. Bruhat MA, Mage G, Pouly J-L, *et al.* Endometriosis: results. In *Operative Laparoscopy*. New York: McGraw Hill, 1992:125

179. Adamson GD, Hurth SJ, Pasta DJ, *et al.* Laparoscopic endometriosis treatment: is it better? *Fertil Steril* 1993;59: 35–44

180. Cook AS, Rock JA. The role of laparoscopy in the treatment of endometriosis. *Fertil Steril* 1991;56:628–68

181. Busacca MD, Bianchi S, Agnoli B, *et al.* Follow-up of laparoscopic treatment of stage III–IV endometriosis. *J Am Assoc Gynecol Laparosc* 1999;6:55–8

182. Lundorff P, Thorburn J, Hahlin M, Kallfelt B. Adhesion formation after laparoscopic surgery in tubal pregnancy: a randomised trial versus laparotomy. *Fertil Steril* 1998; 55:911–15

183. Sagiv R, Debby A, Sadan O, *et al.* Laparoscopic surgery for extrauterine pregnancy in hemodynamically unstable patients. *J Am Assoc Gynecol Laparosc* 2001;8:529–32

184. Clausen I. Conservative versus radical surgery for tubal pregnancy. *Acta Obstet Gynecol Scand* 1996;75:8–12

185. Dwarakanath LS, Mascarenhas L, Penketh RJA, Newton JR. Persistent ectopic pregnancy following conservative surgery for tubal pregnancy. *Br J Obstet Gynaecol* 1996; 103:1021–4

186. Stromme WB. Conservative surgery for ectopic pregnancy. *Obstet Gynecol* 1973;41:215–23

187. Maymon R, Shulman A, Halperin R, *et al.* Ectopic pregnancy and laparoscopy: a review of 1197 patients

treated by salpingectomy or salpingostomy. *Eur J Obstet Gynecol Reprod Biol* 1995;62:61–7

188. Hajenius PJ, Mol BW, Bossuyt PM, *et al.* Interventions for tubal ectopic pregnancy. *Cochrane Database Syst Rev* 2000;2:CD000324

189. Peterson HB, Hulka JF, Phillips JM. American Association of Laparoscopists' 1988 membership survey on operative laparoscopy. *J Reprod Med* 1990;36:587–9

190. Rawson JM. Prevalence of endometriosis in asymptomatic women. *J Reprod Med* 1991;36:513–15

191. Audebert A. Medical treatment of endometriosis. *Rev Prat* 1999;49:269–75

192. Ledger WL. Endometriosis and infertility: an integrated approach. *Int J Gynecol Obstet* 1999;64(Suppl.):33–4

193. Meden-Vrtovec H, Tomazevic T, Verdenik I. Infertility treatment by *in vitro* fertilization in patients with minimal or mild endometriosis. *Clin Exp Obstet Gynecol* 2000;27:191–3

194. Paulson JD, Borremeo R, Speck G. The success of laser laparoscopy in the treatment of endometriosis: a two-step analysis. *JSLS* 2001;5:21–7

195. Al-Inany H. Laparoscopic ablation is not necessary for minimal or mild lesions in endometriosis associated subfertility. *Acta Obstet Gynecol Scand* 2001;80:593–5

196. Omland AK, Fedorcsak P, Storeng R, *et al.* Natural cycle IVF in unexplained, endometriosis-associated and tubal factor infertility. *Hum Reprod* 2001;16:2587–92

197. Osuga Y, Koga K, Tsutsumi O, *et al.* Role of laparoscopy in the treatment of endometriosis-associated infertility. *Gynecol Obstet Invest* 2002;53(Suppl.1):33–9

198. Candiani GB, Fedele L, Vercellini P. Conservative surgical treatment of endometriosis. *Acta Eur Fertil* 1986;17:173–80

199. Jones HW, Rock JA. Others factors associated with infertility, endometriosis externa, fibromyomata uteri. In Pepperell RJ, Hudson B, Wood C, eds. *The Infertile Couple*. Edinburgh, UK: Churchill Livingstone, 1987:181

200. Buttram VC Jr. Endometriosis. In Beherman SJ, Kistner RW, Patton GW Jr, eds. *Progress in Infertility*, 3rd edn. Boston: Little, Brown, 1987

201. Crosignani PG, Vecellini P, Biffignandi F, *et al.* Laparoscopy versus laparotomy in conservative surgical treatment of severe endometriosis. *Fertil Steril* 1996;66:706–11

202. Henry-Suchet J. Severe endometriosis. Current microsurgical strategies. Information for endometriotic women. *Gynecol Obstet Fertil* 2000;28:839–43

203. El Amrani R, Henry-Suchet J, Cornier E, *et al.* Comparison of 2 therapeutic strategies in severe endometriosis, in young women consulting for sterility or pain. II. In the case of infertility, value of ovarian stimulation with intrauterine insemination after surgery. *Gynecol Obstet Fertil* 2001;29:192–9

204. Smith DC, Uhlir JK. Myomectomy as a reproductive procedure. *Am J Obstet Gynecol* 1990;162:1476–82

205. Verkauf BS. Myomectomy for fertility enhancement and preservation. *Fertil Steril* 1992;58:1–15

206. Gehlbach DL, Sousa RC, Carpenter SE, Rock JA. Abdominal myomectomy in the treatment of infertility. *Int J Gynaecol Obstet* 1993;40:45–50

207. Acien P, Quereda F. Abdominal myomectomy: results of a simple operative technique. *Fertil Steril* 1996;65:41–51

208. Li TC, Mortimer R, Cooke ID. Myomectomy a retrospective study to examine reproductive performance before and after surgery. *Hum Reprod* 1999;14:1735–40

209. Buttram VC, Reiter RC. Uterine leiomyomata: etiology, symptomatology and management. *Fertil Steril* 1981;36:433–45

210. De Cherney AH. The effect of leiomyomata on ferility. *Obstet Gynecol Forum* 1990;4:3–5.

211. Amiel C, Mollard J, Cravello L, *et al.* Le traitement des fibromes utérines. *Ann Chir* 1996;50:40–50

212. Jourdan O, Descamps P, Abusada N, *et al.* Treatment of fibromas. *Eur J Obstet Gynecol Reprod Biol* 1996;66:99–107

213. Fauconnier A, Dubuisson JB, Ancel Py, Chapron C. Prognostic factors of reproduction outcome after myomectomy in infertile patients. *Hum Reprod* 2000;15:1751–7

214. Nezhat CH, Kane L, Abolfathian P, *et al.* Successful pregnancy in a series of patients with adenomyosis. *Obstet Gynecol* 2001;4(Suppl. 1):S22

215. Dubuisson JB, Chapron C, Fauconnier A, Babaki-Fard K. Laparoscopic myomectomy fertility results. *Ann N Y Acad Sci* 2001;943:269–75

216. Darai E, Dechaud H, Benifla JL, *et al.* Fertility after laparoscopic myomectomy: preliminary results. *Hum Reprod* 1997;12:1931–4

217. Ribeiro SC, Reich H, Rosenberg J, *et al.* Laparoscopic myomectomy and pregnancy outcome in infertile patients. *Fertil Steril* 1999;71:571–4

218. Ingersoll FM. Fertility following myomectomy. *Fertil Steril* 1963;14:596–602

219. Babaknia A, Rock JA, Jones HW. Pregnancy success following abdominal myomectomy for fertility. *Fertil Steril* 1978;30:644–7

220. Lumsden MA, Gartner A, Scindler EM. Uterine fibroids. *Baillière's Clin Obstet Gynaecol* 1998;12:181–4

221. Seinera P, Gagliati P, Volpi E, *et al.* Ultrasound evaluation of uterine wound healing following laparoscopic myomectomy: preliminary results. *Hum Reprod* 1999;14:2460–3

222. Nezhat C, Nezhat F, Silfen SL, *et al.* Laparoscopic myomectomy. *Int J Fertil* 1991;36:75–80

223. Hasson HM, Rotman C, Rana N, *et al.* Laparoscopic myomectomy. *Obstet Gynecol* 1992;80:545–7

224. Dubuisson JB, Chapron C, Chavet X, *et al.* Fertility after laparoscopic myomectomy of large intramural myomas: preliminary results. *Hum Reprod* 1996;11:518–22

225. Dubuisson JB, Fauconnier A, Chapron C, *et al.* Reproductive outcome after laparoscopic myomectomy in infertile women. *J Reprod Med* 2000;45:23–30

226. Rossetti A, Sizzi O, Soranna L, *et al.* Fertility outcome: long-term results after laparoscopic myomectomy. *Gynecol Endocrinol* 2001;15:129–34

227. Kerin J, Daykhovsky L, Segalowitz J, *et al.* Falloposcopy: a microendoscopic technique for visual exploration of the human fallopian tube from the uterotubal ostium to the fimbria using a transvaginal approach. *Fertil Steril* 1990;54:390–400

228. Kerin JF. Nonhysteroscopic falloposcopy: a proposed method for visual guidance and verification of tubal

cannula placement for endotuboplasty, gamete, and embryo transfer procedures. *Fertil Steril* 1992;57:1133–5

229. Kerin J, Williams D, San Roman G, *et al*. Falloposcopic classification and treatment of fallopian tube lumen disease. *Fertil Steril* 1992;57:731–41

230. David A, *et al*. Human hydrosalpinx. Histologic study and chemical composition of fluid. *Am J Obstet Gynecol* 1969;105:400–11

231. Kerin JF, Pearlstone AC, Surrey ES. Cannulation of the fallopian tube and falloposcopy: difficulties and complications. In Corfman RS, Diamond MP, DeCherney A, eds. *Complications of Laparoscopy and Hysteroscopy*. Boston: Blackwell Scientific, 1993:223–32

232. Dunphy B, Tawzer P, Bultz B, *et al*. A comparison of pain experienced during hysterosalpingography and in-office falloposcopy. *Fertil Steril* 1994;62:62–70

233. Grow DR, Coddington CC, Flood JF. Proximal tubal occlusion by hysterosalpingogram: a role for falloposcopy. *Fertil Steril* 1993;60:170–4

234. Venezia R, Zangara C, Knight C, *et al*. Initial experience of a new linear everting falloposcopy system in comparison with hysterosalpingography. *Fertil Steril* 1993;60:771–5

235. Dechaud H, Hedon B. Fallopian tube endoscopy in 1996. *Contracept Fertil Sex* 1996;24:543–8

236. Kerin JF. Falloposcopy: antegrade imaging in the management of oviductal disease. *J Am Assoc Gynecol Laparosc* 1996;3:S21

237. Porcu E, Dal Prato L, Seracchioli R, *et al*. Births after transcervical gamete intrafallopian transfer with a falloposcopic delivery system. *Fertil Steril* 1997;67:1175–7

238. Surrey ES, Adamson GD, Nagel TC, *et al*. Multicenter feasibility study of a new coaxial falloposcopy system. *J Am Assoc Gynecol Laparosc* 1997;4:473–8.

239. Surrey ES, Adamson GD, Surrey ME, *et al*. Introduction of a new coaxial falloposcopy system: a multicenter feasibility study. *J Am Assoc Gynecol Laparosc* 1997;4:473–8

240. Rimbach S, Bastert G, Wallwiener D. Technical results of falloposcopy for infertility diagnosis in a large multicentre study. *Hum Reprod* 2001;16:925–30

241. Hucke J, de Bruyne F, Balan P. Hysteroscopy in infertility – diagnosis and treatment including falloposcopy. *Contrib Gynecol Obstet* 2000;20:13–20

242. Cornier E, Feintuch MJ, Bouccara L. Ampulla fibrotuboscopy. *J Gynecol Obstet Biol Reprod (Paris)* 1984;13:49–53

243. Henry-Suchet J, Loffredo V, Tesquier L, Pez J. Endoscopy of the tube (= tuboscopy): its prognostic value for tuboplasties. *Acta Eur Fertil* 1985;16:139–45

244. De Bruyne F, Puttemans P, Boeckx W, Brosens IA. The clinical value of salpingoscopy in tubal infertility. *Fertil Steril* 1989;51:339–40

245. Marana R, Muscatello P, Muzii L, *et al*. Pelaparoscopic salpingoscopy in the evaluation of the tubal factor in infertile women. *Int J Fertil* 1990;35:211

246. Hershalag A, Seifer DB, Carcangiu ML, *et al*. Salpingoscopy: light microscopic and electron microscopic correlations. *Obstet Gynecol* 1991;77:399–405

247. Heylen SM, Brosens IA, Puttemans PJ. Clinical value and cumulative pregnancy rates following rigid salpingoscopy during laparoscopy for infertility. *Hum Reprod* 1995;10:2913–6

248. Marana R, Rizzi M, Muzii L, *et al*. Correlation between the American Fertility Society classifications of adnexal adhesions and distal tubal occlusion, salpingoscopy, and reproductive outcome in tubal surgery. *Fertil Steril* 1995;64:924–9

249. Surrey ES, Surrey MW. Correlation between salpingoscopic and laparoscopic staging in the assessment of the distal fallopian tube. *Fertil Steril* 1996;65:267–71

250. Marana R, Catalano GF, Muzii L, *et al*. The prognostic role of salpingoscopy in laparoscopic tubal surgery. *Hum Reprod* 1999;14:2991–5

251. Marchino GL, Gigante V, Gennarelli G, *et al*. Salpingoscopic and laparoscopic investigations in relation to fertility outcome. *Am Assoc Gynecol Laparosc* 2001;8:218–21

252. Dorsey JM, Tabb CR. Minilaparoscopy and fiber optic lasers. *Obstet Gynecol Clin N Am* 1991;18:613–7

253. Risquez F, Dubuisson JB, Madelenat P, *et al*. Transcervical tubal cannulation and falloposcopy for the management of tubal pregnancy. *Hum Reprod* 1992;7:375–6

254. Bauer O, Devoroey P, Wisanto A, *et al*. Small diameter laparoscopy using a microlaparoscope. *Hum Reprod* 1995;10:1461–4

255. Fernanadez E, Fernandez C, Zegers P, Balmaceda JP. Second look: microlaparoscopy. *J Am Assoc Gynecol Laparosc* 1996;4(Suppl.):S13

256. Zupi E, Marcon D, Salina E, *et al*. Microlaparoscopy. *J Am Assoc Gynecol Laparosc* 1996;4(Suppl.):S13

257. Eden B, Bryam I. Microlaparoscopy under local anaesthesia: a simplification of infertility investigation. Presented at the *14th Annual Meeting of ESHRE*, Goteborg, Sweden, 1998;abstr.

258. Palter S. Office microlaparoscopy under local anaesthesia. *Obstet Gynecol Clin N America* 1999;26:109–20

259. Faber BM, Coddington CC III. Microlaparoscopy: a comparative study of diagnostic accuracy. *Fertil Steril* 1997;67:952–4

260. Gordts S, Campo R, Rombauts L, Brosens I. Transvaginal hydrolaparoscopy as an outpatient procedure for infertility investigation. *Hum Reprod* 1998;13:99–103

261. Gordts S, Campo R, Rombauts L, Brosens I. Endoscopic visualization of the process of fimbrial ovum retrieval in the human. *Hum Reprod* 1998;13:1425–8

262. Gordts S, Campo R, Rombauts L, Brosens I. Transvaginal salpingoscopy: an office procedure for infertility investigation. *Fertil Steril* 1998;70:523–6

263. Campo R, Gordts S, Rombauts L, Brosens I. Diagnostic accuracy of transvaginal hydrolaparoscopy in infertility. *Fertil Steril* 1999;71:1157–60

264. Brosens I, Campo R, Gordts S. Office hydrolaparoscopy for the diagnosis of endometriosis and tubal infertility. *Curr Opin Obstet Gynecol* 1999;11:371–7

265. Gordts S, Campo R, Brosens I. Office transvaginal hydrolaparoscopy for early diagnosis of pelvic endometriosis and adhesions. *J Am Assoc Gynecol Laparosc* 2000;7:45–9

266. Brosens I, Gordts S, Campo R. Transvaginal hydrolaparoscopy but not standard laparoscopy reveals subtle endometriotic adhesions of the ovary. *Fertil Steril* 2001;75:1009–12

267. Dechaud H, Ali Ahmed SA, Aligier N, *et al*. Does transvaginal hydrolaparoscopy render standard diagnostic laparoscopy obsolete for unexplained infertility investigation? *Eur J Obstet Gynecol Reprod Biol* 2001; 94:97–102

268. Darai E, Dessolle L, Lecuru F, Soriano D. Transvaginal hydrolaparoscopy compared with laparoscopy for the evaluation of infertile women: a prospective comparative blind study. *Hum Reprod* 2000;15:2379–82

269. Moore ML, Cohen M. Diagnostic and operative transvaginal hydrolaparoscopy for infertility and pelvic pain. *J Am Assoc Gynecol Laparosc* 2001;8:393–7

270. Bissonette F, Lapensee L, Bouzayen R. Outpatient laparoscopic tubal anastomosis and subsequent fertility. *Fertil Steril* 1999;72:549–52

271. Honorè GM, Holden AEC, Shenken RS. Pathophysiology and management of proximal tubal blockage. *Fertil Steril* 1999;71:785–95

272. Koh CH. Microsurgical laparoscopic tubal resection and anastomosis: techniques and results. Références en gynécologie obstétrique. *Review Congrès Vichy* 1995; IFS:102–4

273. Kim JD, Kim KS, Doo JK, Rhyou CH. A report on 387 cases of microsurgical tubal reversals. *Fertil Steril* 1997;68:875–80

274. Dubuisson JB, Chapron CL. Single suture laparoscopic tubal reanastomosis. *Curr Opin Obstet Gynecol* 1998;10. 307–13

275. Bissonette F, Lapensee L, Bouzayen R. Outpatient laparoscopic tubal anastomosis and subsequent fertility. *Fertil Steril* 1999;72:549–52

276. Yoon TK, Sung HR, Kang HG, *et al*. Laparoscopic tubal anastomosis: fertility outcomes in 202 cases. *Fertil Steril* 1999;72:1121–6

277. Winston RM. Microsurgical tubocornual anastomosis for reversal of sterilization. *Lancet* 1977;1:284–5

278. Gomel V. Microsurgical reversal of female sterilization: a reappraisal. *Fertil Steril* 1980;33:587–9

279. Hulka JF, Halme J. Sterilization reversal: results of 101 attempts. *Am J Obstet Gynecol* 1988;159:767–74

280. Dubuisson JB, Chapron C, Nos C, *et al*. Sterilization reversal: fertility results. *Hum Reprod* 1995;10:1145–51

281. Istre O, Olsboe F, Trolle B. Laparoscopic tubal anastomosis: reversal sterilization. *Acta Obstet Gynecol Scand* 1993;68:680–1

282. Glock JL, Kim AH, Hulka JF, *et al*. Reproductive outcome after tubal reversal in women 40 years of age or older. *Fertil Steril* 1996;65:863–5

283. Tournaye H, Camus M, Ubaldi F, *et al*. Is there still an important role for tubal transfer procedures? *Hum Reprod* 1996;11:1815–18

284. Asch R, Balmaceda J, Ellsworth L, Wong P. Pregnancy after translaparoscopic gamete intra-fallopian transfer (GIFT). *Lancet* 1984;2:1034–5

285. Devroey P, Braeckmans P, Smitz J, *et al*. Pregnancy after translaparoscopic zygote intra-fallopian transfer in a patient with sperm antibodies. *Lancet* 1986;1:1329

286. Yovich JL, Yovich JM, Edrisinghe WR. The relative chance of pregnancy following tubal or uterine transfer procedures. *Fert Steril* 1988;49:858–64

287. Hammitt D, Syropo C, Hahn S, *et al*. Comparison of concurrent pregnancy rates for *in vitro* fertilization-embryo transfer, pronuclear stage embryo and gamete intra-fallopian transfer. *Hum Reprod* 1990;5:947–54

288. Tournaye H, Camus M, Kahan I, *et al*. *In vitro* fertilization, gamete or zygote intra-fallopian transfer for female infertility. *Hum Reprod* 1991;6:263–6

289. Bollen N, Camus M, Staessen C, *et al*. The incidence of multiple pregnancy after *in vitro* fertilization and embryo transfer, gamete or zygote intra-fallopian transfer. *Fertil Steril* 1991;55:314–18

290. Asch R. Uterine versus tubal embryo transfer in the human: comparative analysis of implantation, pregnancy and live birth rate. *Ann N Y Acad Sci* 1991;626:461–6

291. Mills M, Eddowas H, Cahill D, *et al*. A prospective control study of *in vitro* fertilization, gamete intra-fallopian transfer and intrauterine insemination combined with super-ovulation. *Hum Reprod* 1992;7:490–4

292. Leeton J, Rogers P, Caro C, *et al*. A control study between the use of gamete intra-fallopian transfer (GIFT) and *in vitro* fertilization and embryo transfer in the management of idiopathic and male infertility. *Fertil Steril* 1987;48:605–7

293. Tanbo T, Daele P, Abyholm T. Assisted fertilization in infertile women with patent fallopian tubes: a comparison of *in vitro* fertilization, gamete intra-fallopian transfer and tubal embryo stage transfer. *Hum Reprod* 1990;5:266–70

294. Crosignani PG, Walters DE, Soliani A. The ESHRE multicentre trial on the treatment of unexplained infertility: a preliminary report. *Hum Reprod* 1991;6:953–8

295. Tournaye H, Devroey P, Camus M, *et al*. Zygote intra-fallopian transfer or *in vitro* fertilization and embryo transfer for the treatment of male-factor infertility: a prospective randomized trial. *Fertil Steril* 1992;58:344–50

296. Balmaceda J, Allam V, Roszjteim D, *et al*. Embryo implantation rates in oocyte donation: a prospective comparison of tubal versus uterine transfers. *Fertil Steril* 1992;57:362–5

297. Fluker M, Zouves C, Bebbington M. A prospective randomised comparison of zygote intra-fallopian transfer and *in vitro* fertilization – embryo transfer for non-tubal factor infertility. *Fertil Steril* 1993;60:515–9

298. Kenny DT. *In vitro* fertilization and gamete intra-fallopian transfer: an integrative analysis of research. *Br J Obstet Gynaecol* 1995;102:317–25

299. Ranieri M, Beckett VA, Marchant S, *et al*. Gamete intra-fallopian transfer or *in vitro* fertilization after failed ovarian stimulation and intrauterine insemination in unexplained infertility. *Hum Reprod* 1995;10:2023–6

300. Van Voorhis BJ, Shyropo CH, Vincent RD Jr, *et al*. Tubal versus uterine transfer of cryopreserved embryos: a prospective randomized trial. *Fertil Steril* 1995;63:578–83

301. Tournaye H, Camus M, Ubaldi F, *et al*. Tubal transfer: a forgotten ART? *Hum Reprod* 1996;11:1815–8

302. Menezo YJR, Janny L. Is there a rationale for tubal transfer in human ART? *Hum Reprod* 1996;11:1818–20

303. Bulletti C. Debating tubal transfer in assisted reproductive technologies. *Hum Reprod* 1996;11:1820–2

304. Chen C-D, Ho H-H, Yang Y-S. Tubal embryo transfer improves pregnancy rate (Letter). *Hum Reprod* 1997;12: 630

305. Tournaye H. Tubal embryo transfer improves pregnancy rate (Letter). *Hum Reprod* 1997;12:630–1

306. Farhi J, Weissman A, Nahum H, Levran D. Zygote intrafallopian transfer in patients with tubal factor infertility after repeated failure of implantation with *in vitro* fertilization-embryo transfer. *Fertil Steril* 2000;74:390–3

307. La Sala GB, Campari C, Montanari R, *et al*. A retrospective comparison of 151 tubal versus 548 uterine embryo transfer cycles. *Isr J Obstet Gynecol* 1999;10:47–54

308. La Sala GB, Montanari R, Cantarelli M, *et al*. PMA: Casistica e risultati nel periodo 21-2-1987/30-6-1997 da Bambini e genitori speciali? Dal bambino desiderato al bambino reale. Presented at the *Atti del Convegno Internazionale*, Reggio Emilia, Rome, 30–31 October 1998; *Percorsi Editoriali Carrocci* 1999

309. Stein IF, Leventhal ML. Amenorrhea associated with bilateral polycystic ovaries. *Am J Obstet Gynecol* 1935;29:181

310. Donesky BW, Adashi EY. Surgically induced ovulation in the polycystic ovary disease: wedge resection revisited in the age of laparoscopy. *Fertil Steril* 1995;63:439–63

311. Wang CF, Gemzell C. The use of human gonadotrophins for induction of ovulation in women with polycystic ovarian disease. *Fertil Steril* 1980;33:479–86

312. Shoam Z, Patel A, Jacobs HS. Polycystic ovary syndrome: safety and effectiveness of stepwise and low-dose administration of purified follicle stimulating hormone. *Fertil Steril* 1991;55:1051–6

313. Filicori M, Flamigni C, Sellai P, *et al*. Treatment of anovulation with pulsatile GnRH: prognostic factors and clinical results in 600 cycles. *J Clin Endocr Metab* 1994;79:1215–20

314. Donesky BW, Adashi EY. Surgical ovulation induction: the role of ovarian diathermy in polycystic ovary syndrome. *Baillière's Clin Obstet Gynaecol* 1996;10:195–202

315. Canis M, Wattiez A, Pouly JL, *et al*. Laparoscopic management of unicornuate uterus with rudimentary horn and unilateral extensive endometriosis: case report. *Hum Reprod* 1990;5:819–20

316. Nezhat CR, Smith KS. Laparoscopic management of a unicornuate uterus with two cavitated, non-communicating rudimentary horns: case report. *Hum Reprod* 1999;14:1965–8

317. Morgans D, Scott F. Twin pregnancy in a rudimentary uterine horn diagnosed by ultrasound and managed laparoscopically before rupture. *Gynecol Endosc* 1999; 8:293–5

Laser techniques in assisted reproductive technologies

28

Yona Tadir, Bruce J. Tromberg and Michael W. Berns

INTRODUCTION

During the summer of 1988 a team of scientists and clinicians gathered to evaluate the potential of laser microbeams in the fast moving area of assisted reproductive technologies (ART). At this time the only clinical experience with lasers was in reproductive surgery, where laser beams were delivered through operative microscopes and laparoscopes and the effective beam spot size was 300–800 µm[1]. Conventional micro-

manipulation on gametes was at its early stage and the availability of a large variety of laser equipment delivered through inverted microscopes to sub-micron spot sizes (Figure 1) enabled testing its potential on gametes and embryos[2]. The two main applications tested during the past 15 years are: (1) sperm manipulations with optical tweezers to improve fertilization *in vitro* and to study basic sperm physiology; and (2) drilling of oocytes and

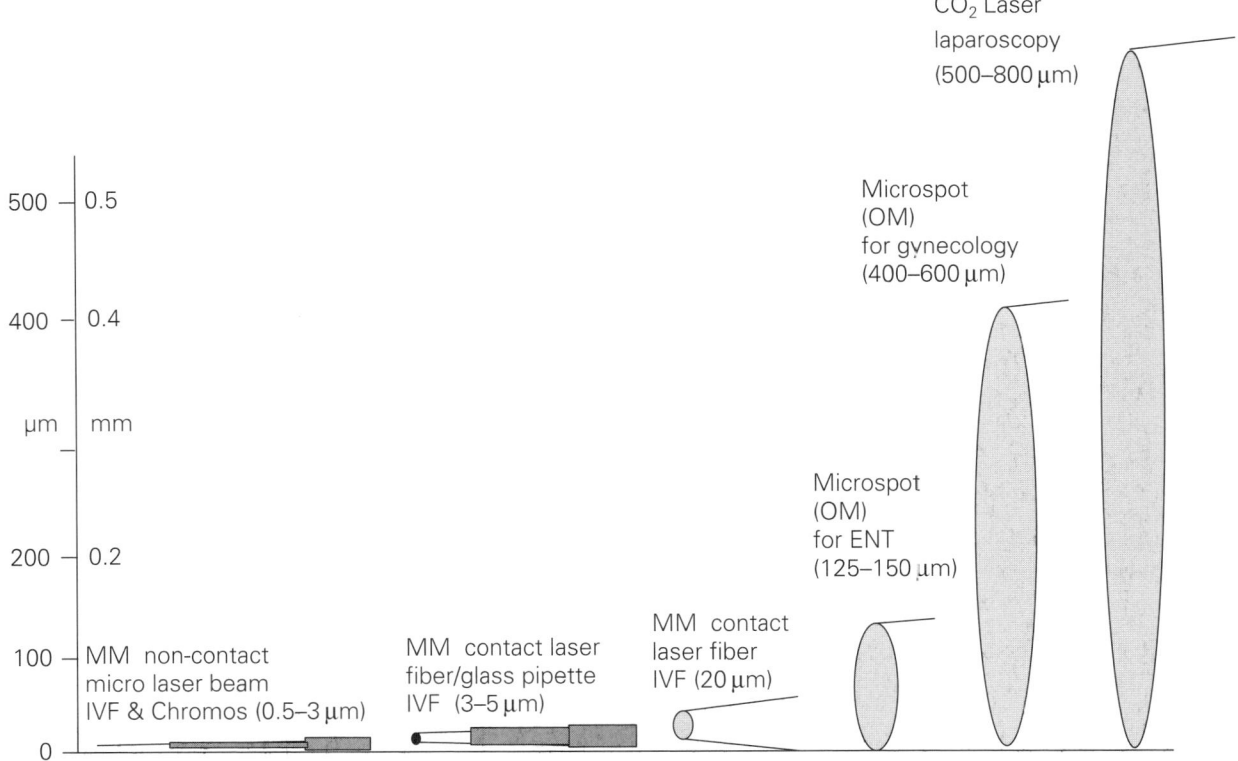

Figure 1 Laser beam spot sizes delivered through various systems for clinical use. IVF, *in vitro* fertilization; MM, micromanipulation; ENT, ear/nose/throat; OM, operating microscope

embryos to improve fertilization, assist hatching, remove blastomeres for pre-embryo genetic diagnosis and assess zona pellucida properties. The purpose of this chapter is to review progress and evaluate the current status of laser microbeams vis-à-vis other technologies available in the ART laboratory.

LASERS AND DELIVERY SYSTEMS AVAILABLE FOR MICROMANIPULATIONS

Lasers (light amplification by stimulated emission of radiation) are electromagnetic waves with unique properties. The beam is collimated, monochromatic and coherent. Lasers differ from each other by their wavelengths, which are in the visible (red, green or blue), invisible (ultraviolet, UV) or infra-red. Effects on gametes may also vary as a result of different parameters and application modes. Some heat may be generated in the micromanipulated object if exposure time is long enough. Conversely, heat formation may be minimized by short exposure in the order of micro- or nanoseconds.

Laser beams for gamete manipulation are typically reduced to a spot size of $1–5\ \mu m^3$. In principle, lasers can be delivered to the target as a free beam or via flexible quartz fibers. This is dependent on the wavelengths and the absorption by the nurturing liquid medium, which is relatively low between 200 and 2000 nm[4]. As such, wavelengths that are shorter or longer than this range require fiber delivery. Light absorption in proteins and DNA is also wavelength-dependent and this should be considered as an important factor when selecting the optimal laser for gamete manipulation (Figure 2). The advantage of using light as an accurate cutting tool is the ability to eliminate the need for disposable or reusable tools. As such, the non-contact free beam delivered through the microscope objective is a preferred approach. Moreover, availability of solid-state compact diode lasers inserted into the body of the microscope makes this combination very practical (Figure 3a).

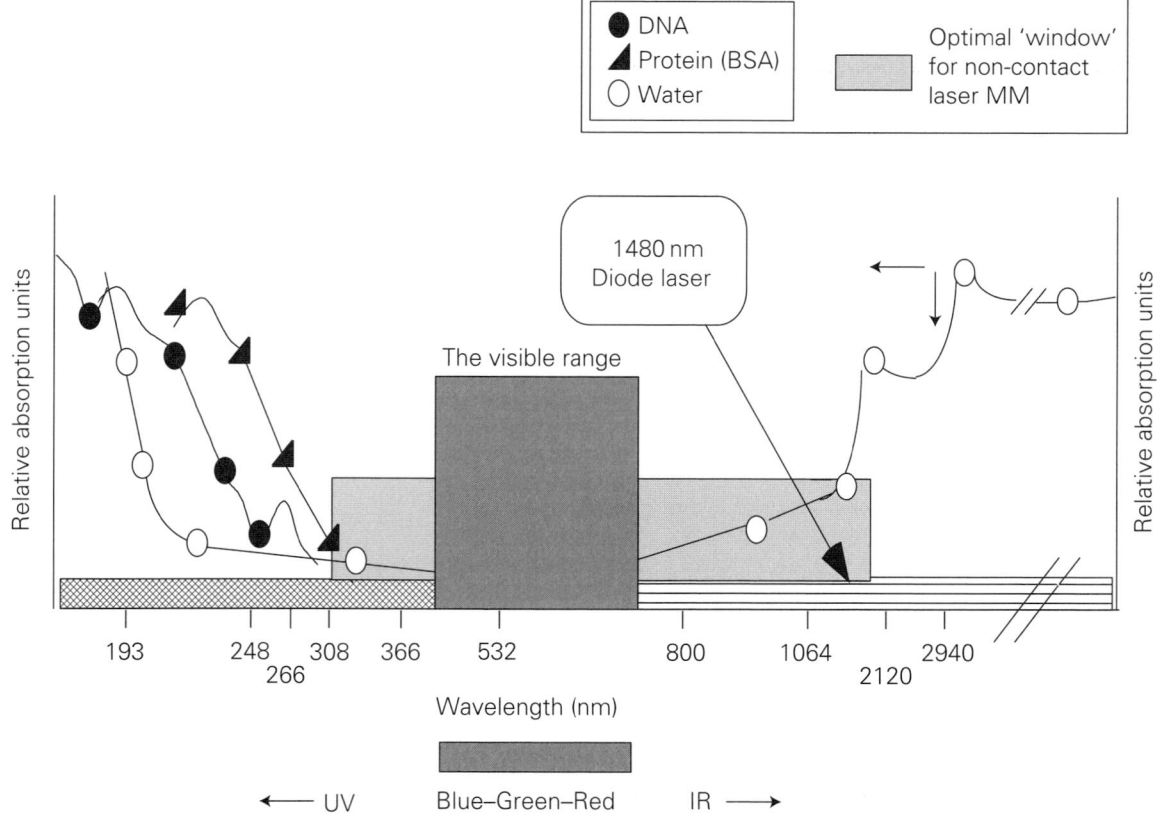

Figure 2 The electromagnetic spectrum and relative light absorption of DNA, protein (bovine serum albumin, BSA) and water at various wavelengths used for micromanipulations (MM). UV, ultraviolet; IR, infra-red

(a)

(b)

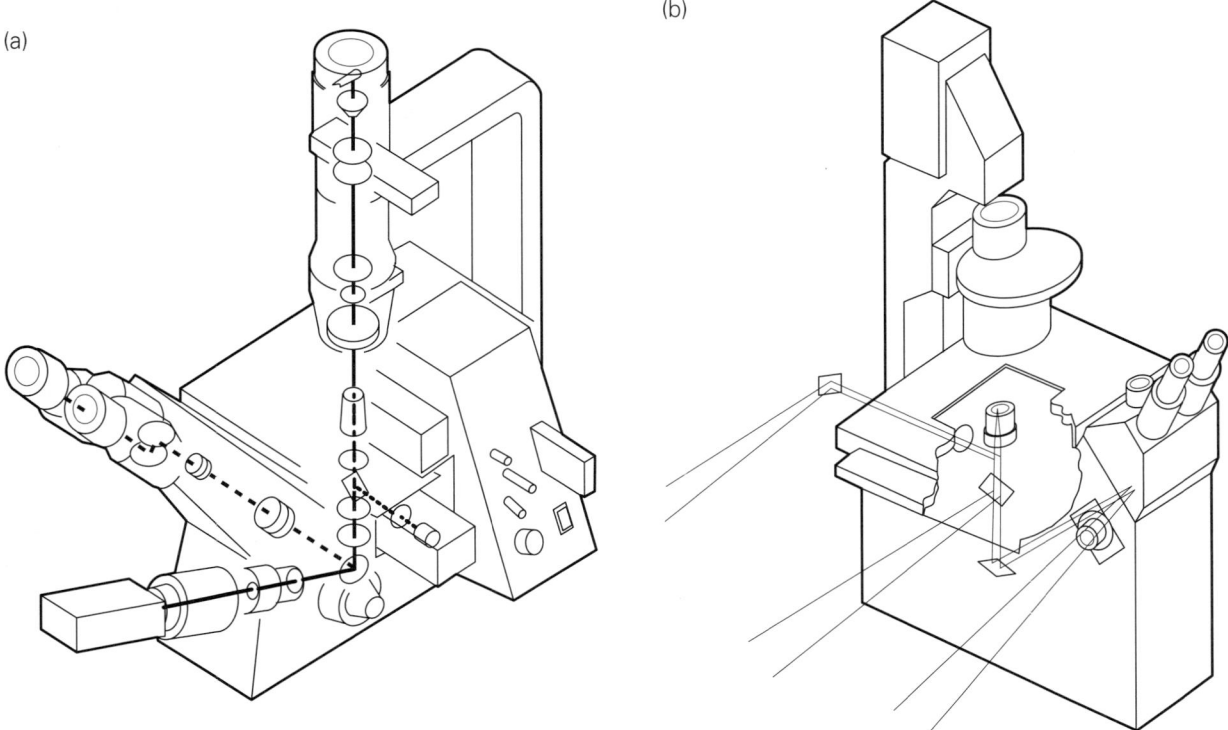

Figure 3 (a) Inverted microscope with a single laser beam delivered through the objective for gamete manipulation; (b) optional three beams for sperm trapping, tail cutting and/or zona pellucida cutting. Adapted courtesy of Cell Robotics/ZILOS, Hamilton Thorne Research

SPERM MANIPULATIONS

The ability to trap and immobilize cells with optical tweezers initially described by Ashkin[5] has opened new possibilities for manipulating sperm[6,7]. The principles of cell trapping are based on mechanical force, which is exerted on a microscopic particle by light. A single beam gradient force trap consists of a laser beam with a Gaussian intensity profile, focused to a spot smaller than the particle being trapped. This trap confines the particle to a location just below the focal point of the laser beam in the axial direction and centered in the beam in the transverse direction. The magnitude and direction of the net force on the particle is determined by the scattering of the laser light through the object. The force generated by the light is greater than all other forces acting on the particle and as such creates a trapping effect.

In a set of studies we demonstrated how single motile sperm could be trapped and subsequently released by reducing the trapping power[8-10]. Single sperm can be guided from one location to another without any mechanical tool. Although sperm can be optically trapped and guided through a hole in the zona pellucida to fertilize an egg[11] it is probably not the preferred approach to improve the fertilization potential in modern assisted reproduction, since other technologies such as the intracytoplasmic sperm injection (ICSI) have already proven to be very successful. However, recent articles demonstrate potential advantages of laser-assisted immobilization of sperm prior to ICSI[12], and laser-assisted ICSI in which non-contact drilling was followed by sperm injection in patients with inherent oocyte fragility[13] (Figure 3b).

Using these principles, a laser generated optical trap was applied to manipulate sperm in two[6] and three dimensions[7]. Initially, the continuous-wave neodymium:yttrium–aluminum garnet laser (Nd:YAG) operating at 1064 nm was used to determine the relative force generated by a single sperm[8]. The results demonstrated that zigzag motile sperm swam with more force than straight-swimming sperm. Other experiments revealed that similar effects could be achieved with a tunable, continuous-wave Titanium Sapphinfrarede laser (700–800 nm wavelength)[9,10]. Several studies were performed to explore relative and absolute sperm force under various physiologic conditions. This demonstrated a significant increase in swimming force following interaction with the cumulus mass[14], and a significant force increase following exposure to pentoxifylline, a motility-enhancing agent[15]. Relative force of human sperm before and after cryopreservation demonstrated that there was no significant difference when a yolk buffer freezing medium was used as cryoprotectant[10]. In another study, relative escape force of human epididymal sperm (aspirated microsurgically for *in vitro* fertilization (IVF)) was tested and compared to normal sperm. Data suggested that the relative swimming force of the epididymal sperm was significantly lower (60%) than ejaculated sperm[9]. ATP-driven motility forces were calculated from calibrated trapping forces generated during the interaction of an 800 nm laser beam with single sperm cells[16]. Sperm heads were obtained by microsurgically removing the flagellum with a pulsed laser beam ('laser scissors'). A trapping efficiency of 0.12 ± 0.02 and a mean intrinsic motility force of 44 ± 20 pN (pico Newton) were determined for motile spermatozoa from healthy donors.

OOCYTE MANIPULATIONS

Laser zona drilling

Non-contact mode

Laser–zona interaction and the fate of oocytes exposed to light beams delivered through the microscope objective has been studied by several groups. Initially, a tunable dye laser at various wavelengths (266–532 nm) was tested on mouse, hamster and discarded human oocytes[2,3]. The beam was delivered through the microscope objective and the depth of incision was observed on a television monitor, and adjusted by a joystick-operated motorized stage. This method is simple and accurate when compared to conventional micromanipulations. Subsequently, other groups used a krypton fluoride laser (operating at 248 nm)[17], nitrogen laser (337 nm)[11,18] and a nitrogen-pumped dye laser (440 nm)[19] to test simplicity, accuracy and local effects. The investigators concluded that from a technical point of view these lasers could be used for zona incision.

In order to further elucidate the effect on the zona pellucida, we designed a set of experiments with various laser parameters. Oocytes were exposed to two different XeCl excimer laser systems (both operating at 308 nm) that offer a large variety of parameters such as pulse duration or pulse repetition rate. High quality images were video-recorded and analyzed by computerized image processing and the oocytes were further processed for scanning electron microscopy[20,21]. Ablation holes smaller than 1 μm were obtained in a reproducible fashion without causing any apparent damage to neighboring cells. Pulse energy and the beam focal plane position were shown to be the most critical parameters in defining the ablated spot diameters. It was concluded that excimer lasers of 308 nm operating in a short pulse duration (15–250 ns) are effective microsurgical tools for achieving 'clean' zona pellucida removal in a non-contact mode. At this particular wavelength, the optical absorption is strong enough to cause selective interaction with the zona pellucida, yet weak enough to induce heat or explosive ablation. In addition, the 308 nm radiation can be delivered through glass slides, microscope objectives and liquid medium or oil. It can facilitate easy, accurate and highly reproducible material removal without the need for handling and maintaining a contact delivery system. However, the known potential damage of UV irradiation still raises some concerns and the sensitivity of a gamete's genetic material deserves extra caution.

Ng and co-workers[22] studied the potential use of nitrogen laser (337 nm) delivered through an inverted microscope to provide a spot of less than 1 μm in non-contact mode. The laser was used at 2.5 μl/pulse with a repetition rate of 10 pulses/s. A 10 μm opening was made in each zona pellucida of mouse oocytes. The drilled oocytes were then inseminated in micro-droplets with murine sperm at 2×10^5 sperm/ml. There was a significant improvement in fertilization and blastocyst formation at day 5 following laser zona drilling (LZD) (89/158 (65.2%) compared to 46/127 (36.2%) $p < 0.001$).

Lasers used in these studies were in the UV or the visible range. A more advanced system that can selectively disrupt the zona pellucida and be delivered as a 'free beam' is the compact diode laser operating in the near infra-red range (1480 nm). This option will be discussed in the laser-assisted hatching and the blastomere biopsy sections where it has more clinical relevance.

Contact mode

A different approach to zona drilling that uses a glass pipette or laser fibers in a contact mode has been suggested by several investigators. In these studies, the argon fluoride excimer laser (ArFl) at 193 nm[23], Nd:YAG laser (1640 nm)[24], holmium:YAG laser (2100 nm)[25] and Er:YAG laser (2940 nm)[26,27] were applied to oocytes. The 193 nm short wavelength was delivered to mouse oocyte zona pellucida[23] through a series of mirrors and a long focal length lens connected to an alumina silicate pipette. The glass pipette was pulled from capillaries with a 1 mm outer diameter to a tip of about 3–5 μm and filled with positive air pressure. Insemination at low sperm densities led to fertilization and further development to the blastocyst stage[28].

Successful fertilization and pregnancy in humans following Er:YAG and LZD indicated feasibility of the technique[26,27]. Laufer and co-workers[28] examined the safety and efficacy of the 193 nm laser by drilling the zona pellucida of mouse oocytes to improve fertilization rate. The LZD significantly enhanced fertilization rate over controls, and the rate of hatching was also enhanced. Normal litters were born following the transfer of the embryos into the uteri of pseudo-pregnant recipients. Improved fertilization rate following ICSI and the ability to use sperm of very poor quality in IVF suggest that zona drilling for this application is obsolete.

LASER-ASSISTED HATCHING

Laser-assisted hatching (LAH) was introduced into the clinical practice in 1990 by Cohen and co-workers[29,30] to improve implantation rate in patients with thick zona pellucida (> 15 μm) or in patients over the age of 38 years. Eight years later assisted hatching following IVF is still controversial (see also chapter 24). It is being offered to selective groups of patients as a standard procedure in some institutions. A literature search reveals conflicting data regarding the clinical use of assisted hatching. Some articles provide statistical data suggesting that assisted hatching improved pregnancy rates for selective groups of randomized patients, mainly previous failures over the age of 39 years[29,31–38], and other reports revealed non-significant improvement or no improvement at all[38–41]. A recent study offers firm statistical evidence that pregnancy rates arising from quarter LAH were higher in comparison with partial and total LAH[42]. Micromanipulation techniques used in these studies were: mechanical slitting, microinjection of acid Tyrode's solution, chemical zona thinning and laser

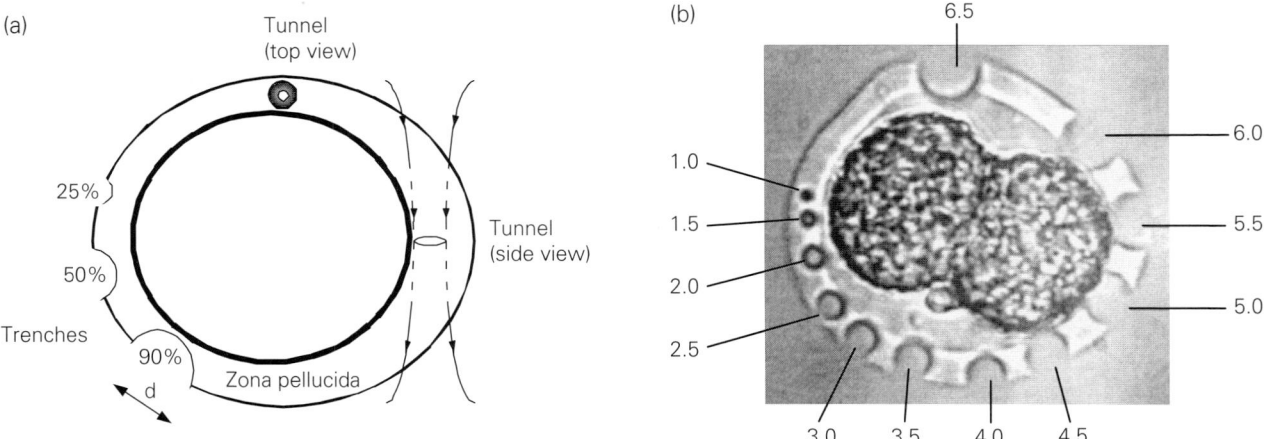

Figure 4 Various hole sizes performed by laser drilling. (a) Schematic representation; (b) microscopic view in a mouse oocyte (power: 100 mW; diameter: 82 μm; duration: variable). The laser procedure can be controlled and accurately repeated in multiple locations. d, diameter. Panel b courtesy of Cell Robotics/ZILOS, Hamilton Thorne Research

drilling. More information is needed to define the common denominators for patients and embryos that may benefit from assisted hatching. However, from a technical point of view, it appears that the laser is the most accurate technique. The procedure can be easily performed, controlled, monitored and repeated in multiple locations in the same embryo or in multiple embryos (Figures 4a and 4b). Moreover, a computer-controlled IVF workstation can pre-define the size and location of the crater in the zona pellucida, document various parameters and automatically transfer the data to a spreadsheet or to the patient's records. Wavelengths tested for LAH range from 308 nm (UV) to 2940 nm (infra-red) and in the contact or free beam modes.

XeCl laser (306 nm)

Several investigators[21,42] studied topical effects of this UV laser on mouse blastomeres using a commercially available system coupled to an inverted microscope. Effects were determined by microinjection of a vital fluorescence dye (fluoresein isothiocyanate (FITC) dextran) into the cell immediately adjacent to the site of zona photo-ablation. This dye is only passed onto daughter blastomeres and therefore permits study of specific cell lines. Embryonic growth was assessed following cell separation at the morula and blastocyst stage. Four-cell-stage embryos treated with this laser had significantly fewer cells 12 h after zona photo-ablation than control embryos. This information suggests that the 308 nm UV excimer laser has some detrimental effect on pre-compacted mouse embryos. However, a different set of laser parameters at the same wavelength may eliminate this problem.

Nitrogen laser (337 nm)

Antinori and co-workers[37] in a randomized trial evaluated pregnancy and implantation rates in three groups of women with repeated IVF failures. One hundred and seven patients received mixed embryos (with or without LAH), 72 patients received only laser-treated embryos and a control group of 98 patients were treated by regular IVF. The resulting clinical pregnancies were 39 (36.4%) in the mixed embryo group, 32 (44.4%) in the LAH group and 19 (19.3%) in the IVF controls. The implantation rates per embryo were 9.3%, 16% and 5.1% in the three groups respectively. In total, 17 normal babies have been delivered (10 in the mixed group and seven in the pure LAH group). These results demonstrate that LAH increased the pregnancy and implantation rates. The increase was slight but significant in the mixed embryo group ($p < 0.01$ and $p < 0.02$); it was even higher in the LAH group ($p < 0.05$). The laser parameters used did not cause any visible damage to the embryos as assessed immediately after birth.

Ho:YAG and Ho:YSGG lasers (2100 nm)

Light absorption by fluid at a wavelength of 2100 nm (infra-red) (Ho:YAG or holmium:yttrium–scandium–gallium garnet, Ho:YSGG) is significant and several factors may affect the amount of energy deposited in the zona pellucida: (1) quality and thickness of the Petri dish; (2) protein content of the culture medium; or (3) distance that the beam has to pass in fluid before it hits the zona pellucida. For this reason some investigators used this wavelength in contact mode and some used it as a free beam. It is not the intent of this review to discuss technical details such as pulse duration, energy or pulse repetition rate; however, it is important to realize that such details will further determine the laser effects[44].

Contact mode

Reshef and co-workers[25] applied the Ho:YAG laser delivered via fibers on the zona pellucida of two- to eight-cell-stage mouse embryos to assist hatching. The rate of development to blastocyst stage and the rate of hatching between the laser-treated and control embryos were compared. Further development was assessed 72 h post-lasing. Thirty-three out of 49 laser-drilled embryos (67%) progressed to hatching blastocysts as compared to 36 of 82 (44%) untreated controls ($p < 0.01$).

Non-contact mode

Our group[43–45] used the Ho:YSGG as a free beam delivered through an inverted microscope and quartz glass dish to perform LAH in two-cell mouse embryos. Control embryos were treated with human tubal fluid

(HTF) culture with or without serum (HTF-s, HTF-o), or with late-serum supplementation (HTF-o/s). Fewer ($p < 0.05$) embryos developed to the blastocyst stage in the HTF-s group (81%) in contrast to the LAH (90%), HTF-o (94%) and HTF-o/s (92%) treatments. The level of hatching was significantly increased ($p < 0.01$) in the LAH treatment (57%) compared to HTF-o/s (32%), HTF-s (18%) or HTF-o (5%). Implantation rates were not impaired following the LAH treatment (21%). These data suggested that LAH using the Ho:YSGG laser is accurate and effective; however, in view of technical limitations of light delivery through fluids near the 2000 nm range there might be better wavelengths to perform LAH in non-contact mode.

Er:YAG laser (2940 nm)

A wavelength of 2940 nm (infra-red) has a high absorption peak in water and can be delivered to the zona pellucida only via fibers in contact mode. Embryos must be kept stable with a holding pipette during the procedure[4]. Strohmer and Feichtinger[46] tested it in mouse embryos, and subsequently in human embryos. Groups of 10–15 mouse embryos were placed under oil on two slides. A control slide was maintained on a warming stage while embryos on the other slide were subjected to the laser to produce 20–30 μm holes in the zona pellucida. Subsequently, embryos were assessed up to the blastocyst stage. There was no difference between the laser-treated mouse embryos and the untreated controls on days 1 and 2 of culture. On day 3, however, complete hatching was significantly enhanced in the laser-treated group (44/55 (80%) in the laser group, 17/58 (29.3%) for controls, $p = 0.0001$).

The same laser was used in a multicenter study for human assisted hatching[32]. Embryos obtained from 129 patients who previously experienced repeated IVF failures were exposed to similar laser effects for assisted hatching. Ablation was performed by applying some pressure with the laser fiber to deposit approximately 10 μl in the zona pellucida. Five to eight pulses were employed to penetrate the zona pellucida creating a 20–30 μm opening. A pregnancy rate of 36% (30/84) and 29% (13/45) in the two centers was achieved, which was encouraging considering the selective groups of patients studied. Preliminary results of a prospective randomized study in patients undergoing first IVF attempts exhibited a pregnancy rate of 50% (10/20 patients) in the LAH group compared to 44% (10/23) in patients without assisted hatching. Differences in the implantation rate per embryo in this preliminary study were also not significant (23.8% versus 21%, LAH versus control). Improved pregnancy rate following LAH using the 2940 nm laser in patients with repeated IVF failures and first IVF attempts as compared to controls was reported by Antinori and co-workers[36]. Embryo implantation rate was 7.3% to 12.2% in patients with previous repeated failures.

Diode laser (1480 nm)

Near infra-red solid-state lasers are small and can emit light at power levels sufficient to cause selective damage to the zona pellucida. The laser module that contains the diode, the electronic board and the collimated lens are all small and can be inserted into the inverted microscope. The 1480 nm wavelength is ideal since it is just minimally absorbed by water, but is highly absorbed by zona pellucida glycoproteins (Figure 2). This system may serve as an optimal cutting tool for the IVF laboratory as suggested by Rink and co-workers[47]. In this study, the beam was delivered through a 45 × objective of an inverted microscope (2–4 μm spot diameter) to produce zona dissection in mouse and human oocytes and zygotes. One laser exposure was sufficient to drill openings in the zona pellucida ranging from 5 to 20 μm depending on laser power and exposure time. The same group[48] demonstrated that the energy needed to drill a hole of a given diameter is greater for mouse and human zygotes than for oocytes. This confirmed a previous observation with the XeCl laser that ethanol-induced zona hardening can be verified and quantified with a non-contact laser[49]. Zona hardening with regard to assisted hatching is summarized by De Vos and Van Steirteghem[50].

The safety of microdrilling the zona pellucida of mouse oocytes with a 1.48 μm diode laser has been investigated by determining the ability of mouse oocytes to develop *in vivo*[51]. Mice born after transfer of control and zona pellucida-microdrilled embryos into foster mothers were submitted to anatomic and immuno-histochemical investigations, and their aptitude to breed was assessed in two subsequent generations. Decoronization of the oocytes with hyaluronidase induced a reduction of the fertilization and implantation rates, which was attributed to a zona-hardening phenomenon. After laser microdrilling of the zona pellucida, these rates were restored to those obtained with embryos derived from untreated oocyte–cumulus complexes. Pups derived from zona pellucida-microdrilled embryos were comparable with those obtained from control embryos, confirming the lack of deleterious effects of the laser treatment.

Another group of mouse zygotes were microdrilled by exposing their zona pellucida to a short pulse of the same 1.4 μm diode laser and allowed to develop *in vitro*[52]. Various sharp-edged holes could be generated and sizes varied by changing infra-red radiation time (3–100 ms) or laser power (22–55 mW). Drilled zygotes presented no signs of thermal damage under light and scanning electron microscopy. Embryos were allowed to develop *in vitro* and showed no sign of abnormality. In a crossed-beam experiment a HeNe laser probe was used to detect the temperature-induced change in the refractive index of an aqueous solution, and estimate the local thermal gradient. The authors found that the 1480 nm laser beam produced superheated water approaching 200°C on the beam axis. Thermal histories during and following the laser pulse were given for regions in the neighborhood of the beam. They concluded that an optimum regime exists with pulse duration ≤ 5 ms and laser power approximately 100 mW[53]. Recent human studies using the same diode laser for zona drilling[38] and zona thinning[54] demonstrated improved pregnancy rate.

BLASTOMERE BIOPSY AND PREIMPLANTATION GENETIC DIAGNOSIS

Preimplantation genetic diagnosis (PGD) is offered worldwide for an expanded range of genetic defects causing disease. This very early form of prenatal diagnosis involves the detection of affected embryos by fluorescent *in situ* hybridization (FISH; for sex determination or chromosomal defects) or by polymerase chain reaction (PCR; for monogenic diseases) prior to implantation. Genetic analysis of the embryos involves the

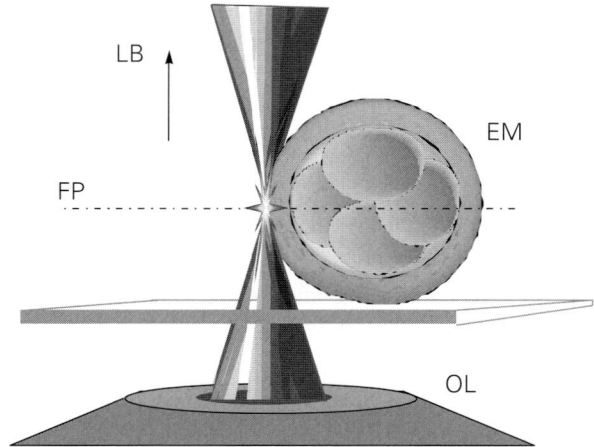

Figure 5 Non-contact laser manipulation of the zona pellucida. The laser beam is delivered through the microscope objective. FP, focal plane; LB, laser beam; OL, optic lense; EM, embryo (at four-cell stage)

removal of some cellular mass from the embryos by means of an embryo biopsy procedure. Genetic analysis can also be performed preconceptionally by removal of the first polar body. Removal of polar bodies or cellular material from embryos requires an opening in the zona pellucida, which can be created in a mechanical (partial zona dissection) or chemical way (acidic Tyrode's solution). In several articles[55–57] it is stated that the introduction of laser technology has facilitated this step enormously.

CONCLUSIONS

Competitive technologies should be tested by non-prejudicial investigators in clinical studies to avoid bias regarding the role and outcome of any new technique. This is especially true when expensive tools are introduced into clinical practice. Advances in gamete manipulations have changed indications for ART and opened new avenues for research and clinical applications. Though lasers may not be beneficial to the process of fertilization they may play a role in assisted hatching. The combined dissecting and the diode laser microscope may be the ultimate approach for non-contact gamete manipulations (Figure 5) since no disposable tools are needed. The computerized workstation offers significant benefits since no extra handling is needed to assess various parameters, and data can automatically be stored and loaded in a patient's records. Other delicate procedures such as blastomere or polar body biopsy can be assisted by the non-contact effects laser method. Systems that combine mechanical tools for gamete manipulations with diode lasers or even systems that provide more than one laser beam delivered through the same optical system may be useful for cutting and trapping. This may be used in sperm tail cutting prior to

ICSI or for targeting subcellular organelles that should be removed or inactivated during the IVF.

ACKNOWLEDGEMENTS

The authors wish to acknowledge the Laser Microbeam and Medical Program (P41-RR01192) at the Ramat Marpe Hospital, Israel and the Beckman Laser Institute, Department of Obstetrics and Gynecology, University of California, Irvine, USA.

References

1. Tadir Y, Kaplan I, Zukerman Z, Ovadia J. Laparoscopic CO2 laser sterilization. In Semm K, Mettler L, eds. *Human Reproduction*. Amsterdam: Exerpta Medica 1981;551:423–31

2. Tadir Y, Wright WH, Berns MW. Cell micromanipulation with laser beam. In Capitanio GL, Asch RH, De Cecco L, Croce S, eds. *G.I.F.T.: From Basics to Clinics*. New York: Raven Press, 1989:359–68

3. Tadir Y, Wright WH, Vafa O, *et al.* Micromanipulation of gametes using laser microbeams. *Hum Reprod* 1991; 6:1011–16

4. Tadir Y, Neev J, Ho P, Berns MW. Lasers for gamete micromanipulation: basic concepts. *J Assist Reprod Genet* 1993;10:121–5

5. Ashkin A, Dziedzic JM, Bjorkholm JE, Chu S. Observation of a single beam gradient force optical trap for dielectric particles. *Opt Lett* 1986;11:288–90

6. Tadir Y, Wright WH, Vafa O, *et al.* Micromanipulation of sperm by a laser generated optical trap. *Fertil Steril* 1989;52:870–3

7. Colon JM, Sarosi P, McGovern PG, *et al.* Controlled micromanipulation of human spermatozoa in three dimensions with an infrared laser optical trap: effect on sperm velocity. *Fertil Steril* 1992;57:695–8

8. Tadir Y, Wright WH, Vafa O, *et al.* Force generated by human sperm correlated to velocity and determined using a laser generated optical trap. *Fertil Steril* 1990;53: 944–6

9. Araujo E, Tadir Y, Patrizio P, *et al.* Relative force of human epididymal sperm correlated to the fertilizing capacity *in vitro*. *Fertil Steril* 1994;62;585–90

10. Zoentania ND, Araujo E, Tadir Y, *et al.* Effect of freezing on the relative escape force of sperm as measured by laser optical trap. *Fertil Steril* 1995;63:185–8

11. Schutze K, Clement-Sengewald A, Ashkin A. Zona drilling and sperm insertion with combined laser microbeam and optical tweezers. *Fertil Steril* 1994;61:783–6

12. Ebner T, Yaman C, Moser M, *et al.* Laser assisted immobilization of spermatozoa prior to intracytoplasmic sperm injection in humans. *Hum Reprod* 2001;16:2628–31

13. Rienzi L, Greco E, Ubaldi F, *et al.* Laser-assisted intracytoplasmic sperm injection. *Fertil Steril* 2001;76: 1045–7

14. Westphal L, El-Danasouri IE, Shimizu S, *et al.* Exposure of human sperm to the cumulus oophorus results in increased relative force as measured by a 760 nm laser optical trap. *Hum Reprod* 1993;8:1083–6

15. Patrizio P, Liu Y, Sonek JG, *et al.* Effect of pentoxifylline on the intrinsic swimming forces of human sperm assessed by optical tweezers. *Int J Androl* 2000;21:753–6

16. Konig K, Tadir Y, Patrizio P, *et al.* Effects of ultraviolet exposure and near infrared laser tweezers on human spermatozoa. *Hum Reprod* 1996;11:2161–4

17. Blanchet BB, Russel JB, Fincher CR, Portman M. Laser micromanipulation in the mouse embryo: a novel approach to zona drilling. *Fertil Steril* 1992;57:1337–47

18. Schutze K, Clement-Sengewald A. Catch and move-cut or fuse. *Nature* 1994;6472:667–9

19. Godke RA, Beetem DD, Burleigh DW. A method for zona pellucida drilling using a compact nitrogen laser. Presented at the *VII World Congress on Human Reproduction*, Helsinky, Finland, June–July 1990;abstr.258

20. Neev J, Tadir Y, Ho P, *et al.* Microscope-delivered UV laser zona dissection: principles and practices. *J Assist Reprod Genet* 1992;9:513–23

21. Li L, Munne S, Licciardi F, *et al.* Microinjection of FITC-dextran into mouse blastomeres to assess topical effects of zona penetration. *Zygote* 1993;1:43–8

22. Ng SC, Liow SL, Schutze K, *et al.* The use of ultra violet microbeam laser zona dissection in the mouse. VIII World Congress of *In Vitro* Fertilization and Assisted Reproductive Technologies, Japan, September 12–15. *J Assist Reprod Prog Suppl* 1993;abstr.273

23. Palanker D, Ohad S, Lewis A, *et al.* Technique for cellular microsurgery using the 193 nm excimer laser. *Laser Surg Med* 1991;11:580–6

24. Coddington CC, Veeck LL, Swanson RJ, *et al.* The YAG laser used in micromanipulation to transect the zona pellucida of hamster oocytes. *J Assist Reprod Genet* 1992;9:557–63

25. Reshef E, Haaksma CJ, Bettinger TL, *et al*. Gamete and embryo micromanipulation using the holmium:YAG laser. Presented at the 49th American Fertility Society Meeting, Montreal, Canada, October 11–14. *Fertil Steril* 1993; S88(Suppl.):P-016

26. Feichtinger W, Strohmer H, Fuhrberg P, *et al*. Photoablation of oocyte zona pellucida by erbium:YAG laser for *in-vitro* fertilization in severe male infertility. *Lancet* 1992;339:811

27. Antinori S, Versaci C, Fuhrberg P, *et al*. Seventeen births after the use of an erbium:YAG laser in the treatment of male factor infertility. *Hum Reprod* 1994;9:1891–6

28. Laufer N, Palanker D, Shufaro Y, *et al*. The efficacy and safety of zona pellucida drilling by a 193-nm excimer laser. *Fertil Steril* 1993;59:889–95

29. Cohen J, Elsner C, Kort H, *et al*. Impairment of the hatching process following IVF in the human and improvement of implantation by assisting hatching using micromanipulation. *Hum Reprod* 1990;5:7–13

30. Cohen J. Assisted hatching of human embryos. *J In Vitro Fertil Embryo Transf* 1991;8:179–90

31. Cohen J, Alikani M, Trowbridge J, Rosenwaks Z. Implantation enhancement by selective assisted hatching using zona drilling of human embryos with poor prognosis. *Hum Reprod* 1992;5:685–91

32. Obruca A, Strohmer H, Sakkas D, *et al*. Use of lasers in assisted fertilization and hatching. *Hum Reprod* 1994;9: 1723–6

33. Schoolcraft WB, Schlenker T, Jones GS, Jones HW Jr. *In vitro* fertilization in women age 40 and older: the impact of assisted hatching. *J Assist Reprod Genet* 1995;9:581–4

34. Takahashi K, Takenaka M, Ishizuka B. The effect of assisted hatching on patients repeatedly failed to conceive by *in vitro* fertilization. Nippon Sanka Fujinka Gakkai Zasshi. *Acta Obstet Gynaecol Japonica* 1994;10:1009–12

35. Stein A, Rufas O, Amit S, *et al*. Assisted hatching by partial zona dissection of human pre-embryos in patients with recurrent implantation failure after *in vitro* fertilization. *Fertil Steril* 1995;4:838–41

36. Antinori S, Panci C, Selman HA, *et al*. Zona thinning with the use of laser: a new approach to assisted hatching in humans. *Hum Reprod* 1996;11:590–4

37. Antinori S, Selman HA, Caffa B, *et al*. Zona opening of human embryos using a non-contact UV laser for assisted hatching in patients with poor prognosis of pregnancy. *Hum Reprod* 1996;11:2488–92

38. Montag M, van der Ven H. Laser-assisted hatching in assisted reproduction. *Croat Med J* 1999;40:398–403

39. Hellebaut S, De Sutter P, Dozortsev D, *et al*. Does assisted hatching improve implantation rates after *in vitro* fertilization or intracytoplasmic sperm injection in all patients? A prospective randomized study. *J Assist Reprod Genet* 1996; 13:19–22

40. Tucker MJ, Cohen J, Massey JB, *et al*. Partial dissection of the zona pellucida of frozen–thawed human embryos may enhance blastocyst hatching, implantation, and pregnancy rates. *Am J Obstet Gynecol* 1991;165:341–4

41. Tucker MJ, Luecke NM, Wicker SR, Wright G. Chemical removal of the zona pellucida of day 3 human embryo has no impact on implantation rate. *J Assist Reprod Genet* 1993;10:187–91

42. Mantoudis E, Podsiadly BT, Gorgy A, *et al*. A comparison between quarter, partial and total laser assisted hatching in selected infertility patients. *Hum Reprod* 2001;16:2182–6

43. Neev J, Gonzales A, Licciardi F, *et al*. A contact-free microscope delivered laser ablation system for assisted hatching of the mouse embryo without the use of a micromanipulator. *Hum Reprod* 1993;8:939–44

44. Neev Y, Schiewe MC, Sung WV, *et al*. Assisted hatching in mouse embryos using a non-contact Ho:YSSG laser system. *J Assist Reprod Genet* 1995;12:228–93

45. Schiewe MC, Neev J, Hazeleger NL, *et al*. Developmental competence of mouse embryos following zona drilling using a non-contact holmium:yttrium scandium gallium garnet (Ho:YSGG) laser system. *Hum Reprod* 1995;10:1821–4

46. Strohmer H, Feichtinger W. Successful clinical application of laser for micromanipulation in an *in vitro* fertilization program. *Fertil Steril* 1992;58:212–14

47. Rink K, Delacretaz G, Salathe RP, *et al*. Non-contact microdrilling of mouse zona pellucida with an objective-delivered 1.48-microns diode laser. *Lasers Surg Med* 1996;18: 52–62

48. Germond M, Nocera D, Senn A, *et al*. Microdissection of mouse and human zona pellucida using a 1.48-microns diode laser beam: efficacy and safety of the procedure. *Fertil Steril* 1995;64:604–11

49. Tadir Y, Neev Y, Schiewe M, *et al*. Spontaneous and induced zona pellucida hardness: measurements using enzyme assay and a non-contact laser micromanipulation. Pacific Coast Fertility Society, Palm Springs, CA, April 14–16. *Fertil Steril Prog* 1993 (Suppl.):O–21

50. De Vos A, Van Steirteghem A. Zona hardening, zona drilling and assisted hatching: new achievements in assisted reproduction. *Cells Tissues Organs* 2000;166:220–7

51. Germond M, Nocera D, Senn A, *et al*. Improved fertilization and implantation rates after non-touch zona pellucida microdrilling of mouse oocytes with a 1.48 microm diode laser beam. *Hum Reprod* 1996;11:1043–8

52. Rink K, Delacretaz G, Salathe RP, *et al*. Non-contact microdrilling of mouse zona pellucida with an objective-delivered 1.48-microns diode laser. *Lasers Surg Med* 1996; 18:52–62

53. Douglas-Hamilton DH, Conia J. Thermal effects in laser-assisted pre-embryo zona drilling. *J Biomed Opt* 2001;6: 205–13

54. Blake DA, Forsberg AS, Johansson BR, Wikland M. Laser zona pellucida thinning – an alternative approach to assisted hatching. *Hum Reprod* 2001;16:1959–64

55. De Vos A, Van Steirteghem A. Aspects of biopsy procedures prior to preimplantation genetic diagnosis. *Prenat Diag* 2001;21:767–80

56. Montag M, van der Ven K, Delacretaz G, *et al*. Laser-assisted microdissection of the zona pellucida facilitates polar body biopsy. *Fertil Steril* 1998;69:539–42

57. Licciardi F, Gonzalez A, Tang YX, *et al*. Laser ablation of the mouse zona pellucida for blastomere biopsy. *J Assist Reprod Genet* 1995;12:462–6

Transgenic models for the study of human reproduction

29

Kathleen H. Burns and Martin M. Matzuk

INTRODUCTION

Our ability to alter the mammalian genome using transgenic technology has forever changed our means of inventing and testing medical hypotheses. Transgenic mouse models are compelling demonstrations of the functions of gene products and genetic control elements *in vivo*. Despite the relatively recent advent of the technology, many phenotypes relevant to the field of reproductive biology have been described. These models identify factors critical to fertility and herein rests their potential to reveal both genetic and non-genetic causes of infertility in men and women.

Fertility in both sexes relies on complex physiologic and molecular processes with many levels of regulation. Defects intrinsic to the germ cells, their surrounding and supporting gonadal somatic cells or hormone-elaborating cells in distant organs can cause failure of prenatal or postnatal developmental programs and lead to infertility. Among the most commonly recognized genetic causes of human infertility today are chromosomal aneuploidies and, in the case of the male, deletions of regions of the Y chromosome[1–4]. However, many single-gene mutations in patients have been described that cause defined defects in the gonadotropin signaling pathways, the progression of germ cell development and the enzymatic steps of steroidogenesis[3–5]. These cases provide the most straightforward examples for the utility of transgenic mice to model human pathophysiologies. We begin this chapter by examining mouse models of human gonadotropin signaling defects as an illustration of this principle. We will then consider the technologies that underlie the making of these mouse models and present examples of murine mutations that cause reproductive phenotypes.

MOUSE MODELS OF HUMAN GONADOTROPIN SIGNALING PATHOLOGIES

The hypothalamic–pituitary–gonadal (HPG) axis is fundamental to the endocrine control of gametogenesis in mammals, and because of its long-recognized physiologic importance, many human mutations disrupting the normal function of the axis have already been characterized and several relevant mouse models have been engineered. The pituitary gonadotropins, follicle stimulating hormone (FSH) and luteinizing hormone (LH), are central to this endocrine communication. FSH and LH are heterodimeric glycoproteins each comprised of a common α-subunit and a unique β-subunit; functional dimers are synthesized and secreted into the circulation in response to hypothalamic gonadotropin releasing hormone (GnRH)[6]. FSH and LH elicit intracellular signaling pathways by binding to their respective G-protein-coupled transmembrane receptors, FSHR and LHR, in somatic gonadal cells to regulate follicular development, ovulation and steroidogenesis in females and spermatogenesis, growth and steroidogenesis in males[7,8]. Examples of loss-of-function and gain-of-function human mutations in components of the HPG axis have been described[9–11] (Table 1).

Hypothalamic production of GnRH and signaling through pituitary GnRH receptors (GnRHR) appear to be essential to maintaining serum gonadotropins and ultimately fertility in man. Mutations in genes mediating the developmental migration of GnRH-releasing neurons (as seen in the *kal* gene causing Kallman's syndrome), or aspects of GnRH processing (as seen in mutations in the *pc1* protein processing enzyme), lead to hypogonadotropic hypogonadism (HHG)[12,13].

Table 1 Examples of human mutations in the gonadotropin signaling pathways

Gene	Type of mutation	References
GnRH (deletions)	Deletion	Weiss *et al.*, 1991[14]
GnRHR	Point mutations	de Roux and Milgram, 2001[11]; Chanson *et al.*, 1998[55]
α-subunit	No human mutations	Themmen and Huhtaniemi, 2000[9]
FSHβ	Frameshift mutations, point mutations	Matthews *et al.*, 1993[56]; Layman *et al.*, 1997[57]
LHβ	Point mutations	Weiss *et al.*, 1992[25]
FSHR	Point mutations, activating mutations	Tapanainen *et al.*, 1997[22]; Aitomaki *et al.*, 1995[58]; Gromoll *et al.*, 1996[59]
LHR	Point mutations, activating mutations	Latronica *et al.*, 1996[60]; Kremer *et al.*, 1995[61]; Toleda *et al.*, 1996[62]; Shenker *et al.*, 1993[27]

However, to date, no defined loss-of-function mutations in the GnRH gene itself have been described in patients with HHG[14]. Mutations in GnRHR resulting in GnRH resistance have been described in HHG patients that are homozygous or compound heterozygous for missense mutations[11].

In vitro studies of these mutant GnRHR products indicate that they are hypomorphic mutations in which deficiencies in GnRH binding or intracellular signal initiation are observed. Clinically, a majority of these patients respond to administration of exogenous GnRH, and in one reported case, pulsatile provision of GnRH was able to induce ovulation and restore female fertility[15].

A naturally occurring 33.5 kb deletion truncating the *gnrh* gene results in a mouse model of hereditary hypogonadism (*hpg*), which resembles the HHG seen in patients with defects in GnRH production or responsiveness[16]. Hypogonadism is also a feature of a knockout mouse model harboring a null allele at the glycoprotein hormone common α-subunit locus; besides reproductive defects, the knockout mouse is hypothyroid owing to a loss of thyroid-stimulating hormone function[17]. No mutations altering the amino acid sequence of the common α-subunit have been described in humans. It has been hypothesized that a deleterious mutation would result in embryonic lethality because in humans (and not in mice) the α-subunit is also shared with chorionic gonadotropin (hCG). Substantiating this, the most highly expressed hCGβ subunit-encoding genes are highly conserved, with no individuals homozygous for null alleles at these loci[18].

Loss of FSH signaling causes infertility in women[19], and this condition can be modeled by disrupting either the *fshβ* or *fshr* loci in mice. Women who are homozygous or compound heterozygous for inactivating ligand (frameshift/truncation or missense) or receptor (missense) mutations exhibit normal preantral follicle development, but no antral-stage follicles capable of ovulation form. These women typically present clinically as cases of primary amenorrhea and sexual infantilism. Both *fshb* and *fshr* knockouts in mice phenocopy the human mutations, displaying female infertility, uterine hypoplasia and folliculogenesis blocks prior to antrum formation[20,21].

In men, *fshb* and *fshr* mutations have been associated with variable degrees of impairment of spermatogenesis and sometimes with delayed puberty[9]. However, men with *fshr* null mutations can father children[22]. Both *fshb* and *fshr* knockout male mouse models are fertile, but have reduced sperm counts and decreased testicular size. Therefore, it seems that FSH activity, while necessary for optimal sexual development and spermatogenesis, is dispensable for male, but not female, fertility in both humans and mice. A transgenic mouse model over-expressing high levels of human FSH has been developed[23]. The transgenic males demonstrate stimulated Leydig cell function, elevated serum testosterone and infertility; females exhibit infertility with hemorrhagic and cystic ovaries. These mice may prove valuable in modeling gonadal and more global physiologic effects of FSH hyperstimulation. Consistent with the high conservation of FSH in mice and humans, synthesis of human FSHβ in the gonadotrophs of mice lacking the endogenous FSHβ subunit results in normal fertility in male and female mice[24].

Loss of isolated LH signaling function can arise from defects in the unique LHβ subunit or in the LHR. One loss-of-function missense mutation in the hormone was identified in an infertile man presenting with low testosterone levels, Leydig cell hypoplasia and elevated immunoreactive but functionally ineffective LH[25]. Several LHR mutations have been discovered to cause defects in receptor activity of variable severity. Phenotypes range from micropenis and hypospadias in the case of hypomorphic alleles to male pseudohermaphroditism and female infertility associated with a barrier to preovulatory follicle development, ovulation and luteinization[9]. These conditions are partially modeled by targeted deletion of the *lhr* locus in mice; homozygotes demonstrate normal prenatal sexual development but are infertile[26]. Male *lhr* knockouts have defects in testicular growth and descent, Leydig cell hypoplasia and a block in spermatogenesis at the round spermatid stage. Female knockouts have underdeveloped ovaries and uteri and their ovaries do not contain preovulatory stage follicles or corpora lutea; thus the LHR knockout in female mice very closely models the pathology of women with LH resistance. Gain-of-function mutations of LHR have also been described in families with autosomal dominant male precocious puberty[27]. Although there are no mouse models that phenocopy this disorder, transgenics have been engineered to overexpress the bovine LHβ subunit with a C-terminal extension of the hCGβ subunit, thereby extending the serum halflife of the chimeric LH hormone[28]. Interestingly, female transgenics have a 10-fold increase in circulating immunoreactive LH and exhibit impaired ovulation, a prolonged luteal phase, ovarian cysts and granulosa/theca cell tumors on some genetic backgrounds[29]. Studies of these latter transgenic female mice may shed light on the roles of LH in polycystic ovarian syndrome and in the development of postmenopausal ovarian stromal tumors as well as identify genetic modifiers of these phenotypes.

TRANSGENIC TECHNOLOGY AND MANIPULATION OF THE MOUSE GENOME

A transgenic mouse has heritable engineered DNA sequences incorporated into its genome. Transgenic technology first became a reality with the work of Gordon and co-workers[30,31] who microinjected foreign DNA into male pronuclei of one-cell zygotes (Figure 1). Transgene concatemers thus injected generally integrate during the one-cell stage at a single, random site within

Preparation of the transgene construct

Day 1: PMSG injection

Day 3: hCG injection and mating
       pseudopregnant foster mothers mated with vasectomized males

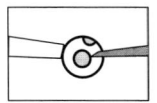

Day 4: embryo collection and DNA microinjection
       embryo transfer to foster mothers

Day 23: F_0 generation is born

Day 37: tail DNA taken for Southern blot or PCR analysis

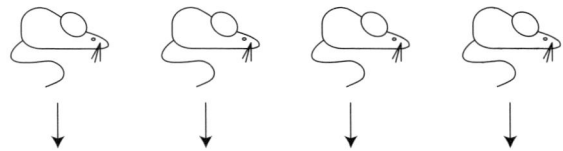

Each founder is crossed to wild-type mice to begin a unique transgenic line. Fifty percent of the F_1 generation is expected to inherit the transgene

Figure 1 Generation of transgenic mice by DNA microinjection. Immature, adolescent female mice are superovulated using a pregnant mare serum gonadotropin (PMSG)/human chorionic gonadotropin (hCG) regime and are then mated. Fertilized eggs are recovered from these females and the transgene construct DNA is directly injected into male pronuclei. The microinjection depends on the use of contrast optics and micromanipulators. Many factors must be considered in the choice of mouse strains, including the desire for genetic uniformity, the responsiveness of female mice to gonadotropins, the reproductive capacities of the strain and strain-specific characteristics of the zygote that may facilitate microinjection. The embryos are transferred to pseudopregnant recipient females that have been mated with vasectomized males. The F_0 transgenic mice each have a unique transgene integration and become founders for a line of offspring which are expected to inherit the transgene with Mendelian frequency. Intercrossing members of a transgenic line allows for the establishment of homozygosity at the transgene locus. Homozygotes may exhibit a phenotype that is distinct from heterozygotes, reflecting the effects of the increased dosage of the introduced DNA or the effects of insertional mutations. PCR, polymerase chain reaction

Table 2 Transgenic mouse models with reproductive phenotypes

(a) Overexpresser/misexpresser models

Construct	Reproductive phenotype	References
mMT promoter and activin/inhibin βA sequence	Males with high expression in testes are infertile; seminiferous tubules show patchy lesions and vacuoles	Tanimoto *et al.*, 1999[63]
Mifeprisone-inducible transgene overexpressing inhibin A from liver	Females have a block in folliculogenesis at the early antral stage; males have decreased testis size	Pierson *et al.*, 2000[45]
Rat androgen-binding protein (*abpa*) promoter and coding sequence	Increased Sertoli cell number and compromised male fertility	Larriba *et al.*, 1995[64]
Inhibin α promoter and *bcl2* coding sequence	Increased folliculogenesis, decreased follicle apoptosis and germ cell tumors	Hsu *et al.*, 1996[65]
EF-1α promoter and *bcl2* coding sequence	Compromised testicular germ cell apoptosis/differentiation	Furuchi *et al.*, 1996[66]
hMT promoter and rat c-*myc* coding sequence	Early arrest in spermatogenesis and apoptosis	Suzuki *et al.*, 1996[67]
dax1 overexpresser (*dax* regulatory region)	XY mice with a weak *sry* allele develop as females	Swain *et al.*, 1998[68]
Bovine glycoprotein hormone-α promoter and diphtheria toxin coding sequence	Pituitary gonadotrophs lost; mice are hypogonadal	Kendall *et al.*, 1991[69]
Rat histone H1t promoter and diphtheria toxin coding sequence	Male infertility; small testes with loss of germ cells; females fertile	Bartell *et al.*, 2000[70]
Hydroxymethyl glutaryl CoA reductase promoter and E2F-1 transcription factor sequence	Males are infertile; testicular atrophy and apoptosis in the germinal epithelium	Holmberg *et al.*, 1998[71]
β-actin promoter and epidermal growth factor (EGF) sequence	Males are sterile; few post-meiosis II gametes	Wong *et al.*, 2000[72]
mMT-1 promoter and follistatin (*fst*) coding sequence	Progressive infertility in both sexes; males have somatic cell and germ cell defects; females have folliculogenesis defects	Guo *et al.*, 1998[73]
mMT-1 promoter and β1,4-galactosyltransferase sequence	Altered sperm–egg binding	Youakim *et al.*, 1994[74]
Growth hormone overexpressers (multiple constructs studied)	Compromised reproduction in both sexes	Bartke *et al.*, 1994[75]
β-actin promotor and active human heat shock transcription factor (*hsf1*) coding sequence	Males are infertile; arrest during meiosis and germ cell apoptosis	Nakai *et al.*, 2000[76]
mMT promoter and interferon-α1 coding sequence	Degeneration of spermatogenic cells and seminiferous tubule atrophy	Hekman *et al.*, 1988[77]
mMT promoter and interferon-β coding sequence	Male sterility; testes involuted with degeneration of spermatocytes and spermatids	Iwakura *et al.*, 1988[78]
mMT promoter and human interleukin-2 sequence	Males exhibit atrophic testes and defects in spermatogenesis	Ohta *et al.*, 1990[79]
Liver promoter and leptin coding sequence (skinny mice)	Females have accelerated puberty and late-onset hypothalamic hypogonadism	Yura *et al.*, 2000[80]
Bovine α promoter and bLHβ-hCG fusion	Female infertility, polycystic ovaries, granulosa cell tumors	Risma *et al.*, 1995[28]
mMMT promoter and matrilysin (MAT) (*mmp7*) coding sequence	Progressive male infertility associated with degeneration in testes and loss of interstitial cells	Rudolph-Owen *et al.*, 1998[81]
mMT promoter and human Müllerian inhibiting substance	Both sexes have defects in development of reproductive structures	Behringer *et al.*, 1990[82]
mMTV promoter and N-*ras* oncogene sequence	Male infertility; sperm motility defects	Mangues *et al.*, 1990[83]

Table 2 *Continued*

Construct	Reproductive phenotype	References
mMMT promoter and *neu* oncogene sequence	Male infertility; epithelial cell hyperplasia within the epididymis	Guy *et al.*, 1996[84]
Human ornithine decarboxylase overexpresser (human ODC regulatory region)	Alterations in spermatogenic DNA synthesis and compromised male fertility	Hakovirta *et al.*, 1993[85]
P450 aromatase (*cyp19*) overexpression	Subfertility and Leydig tumor development in males	Fowler *et al.*, 2000[86]
mMT promoter and *p53* coding sequence	Males exhibit variable subfertility; increased spermatid apoptosis and teratozoospermia	Allemand *et al.*, 1999[87]
Rat proenkephalin cDNA; human promoter and flanking sequence	Male subfertility or infertility; abnormal testicular morphology	O'Hara *et al.*, 1994[88]
Mouse protamine 1 promoter and avian protamine coding sequence	Disrupted sperm chromatin condensation	Rhim *et al.*, 1995[85]
Protamine 1 Δ3'UTR transgene	Males are infertile; premature protamine accumulation and arrested spermatid differentiation	Lee *et al.*, 1995[90]
Mouse mammary tumor virus promoter and dominant negative RARα coding sequence	Males are infertile or subfertile; spermatogenesis intact; squamous metaplasia of epididymis and vas deferens	Costa *et al.*, 1997[91]
Spermidine/spermine N1-acetyltransferase (*sat*) promoter and coding sequence	Female infertility; ovarian hypofunction and hypoplastic uteri	Pietila *et al.*, 1997[92]
Wilms tumor homolog (*wt1*) promoter and *sox9* coding sequence	XX transgenics develop as sterile, sex-reversed males	Vidal *et al.*, 2001[93]
Sex determining region of chromosome Y (*sry*) promoter and coding sequence	Transgenic females develop as males	Koopman *et al.*, 1991[94]
Human GnRH promoter and SV40 T antigen sequence	GnRH neurons do not migrate appropriately; mice are hypogonadal	Radovick *et al.*, 1991[95]
Bovine FSHβ promoter and HSV-tk coding sequence	Thymidine kinase expression in the pituitary and testis	Markkula *et al.*, 1993[96]
Group 1 Mup promoter and HSV-tk coding sequence	Male infertility	Al-Shawi *et al.*, 1988[97]
mMT promoter and hTGF-α coding sequence	Delayed implantation and parturition	Das *et al.*, 1997[98]
Mouse MTV LTR promoter and vascular endothelial growth factor (*vegf*) sequence	Males are infertile; defects in spermatogenesis and aberrant blood vessel formation in the testes and epididymis	Korpelainen *et al.*, 1998[99]
Phosphoglycerate kinase-2 promoter and v-*mos* coding sequence	Male infertility; germ cell development arrest at metaphase I of meiosis	Rosenberg *et al.*, 1995[100]

(b) Random insertional mutations

Construct	Reproductive phenotype	References
ELKL motif kinase (*emk*)	β-gal gene trap insertion creates a null allele; homozygotes intercrossed are not fertile	Bessone *et al.*, 1999[101]
gcd (germ-cell deficient)	Infertility in homozygotes of both sexes; germ cell migration/proliferation failure	Pellas *et al.*, 1991[102]
Histone 3.3A gene	β-gal gene trap insertion creates a hypomorphic allele; homozygotes have multiple defects including male subfertility	Couldrey *et al.*, 1999[103]

Table 2 *Continued*

Gene	Reproductive phenotype	References
ho (hotfoot)	Male homozygotes are infertile; sperm cannot penetrate the zona pellucida	Gordon *et al.*, 1990[104]
Kisimo mouse (*theg*)	Infertility in homozygous males; asthenospermia	Yanaka *et al.*, 2000[105]
lvs (lacking vigorous sperm)	Infertility in hemizygous males; abnormal nuclear condensation during spermatogenesis	Magram and Bishop, 1991[106]
Microtubule-associated protein (*mtap7*) (E-MAP-115)	Gene trap insertion generates a null allele; male sterility with abnormal microtubules in germ cells and Sertoli cells	Komada *et al.*, 2000[107]
morc (microrchidia)	Infertility in homozygous males; early arrest in meiosis and germ cell apoptosis	Watson *et al.*, 1998[108]
ods (*odsex*, ocular degeneration with sex reversal)	XX develop as infertile males; dominant mutation associated with 150 kb deletion and increased *sox9* expression	Bishop *et al.*, 2000[109]
pcd (Purkinje cell degeneration)	Male infertility; abnormal spermatozoa	Krulewski *et al.*, 1989[110]
Spermatid perinuclear RNA-binding protein (*spnr*)	β-geo gene trap insertion; subfertility in homozygote males; defects in seminiferous epithelium and spermatogenesis	Pires-daSilva *et al.*, 2001[111]
sys (symplastic spermatids)	Infertility in homozygous males; Sertoli cell and spermatid abnormalities	MacGregor *et al.*, 1990[112]
2:12 translocation associated with transgene insertion	Male infertility; defects in chromosome synapsis and spermatogenesis	Gordon *et al.*, 1989[113]

the mouse genome[32]. Embryos with DNA thus introduced are expected to develop into mice that are hemizygous for a unique transgene integration to be passed on to 50% of their F$_1$ progeny. A number of reported transgenic mouse models exhibit reproductive findings (Table 2).

Because transgene integration takes place randomly within the recipient genome, there is a chance that the transgene will disrupt important endogenous sequences[33]. Such random insertional mutations can be fortuitous for researchers as they identify and mark loci with important functions. Indeed, the goal of 'gene-trapping' experiments is to generate insertional mutations at loci with an interesting pattern of expression; the strategy typically introduces a β-galactosidase reporter construct for random genome integration[34–36]. Insertional mutations having reproductive phenotypes are listed (Table 2).

BITRANSGENIC SYSTEMS: CONTROLLING THE EXPRESSION OF INSERTED TRANSGENES

Bitransgenic regulatory systems allow researchers to direct the timing of transgene expression. These systems require two transgene constructs that function in sequence. The first transgene is the regulator construct, which encodes a transcription factor expressed under the control of a constitutive or tissue-specific promoter. The transcription factor protein product is activated or inactivated by a pharmacologic agent administered by the researcher. The second transgene contains a responder operon, which directs the expression of the gene of interest only in the presence of the active transcription factor. Thus, by giving or withholding a drug, biologists can control the expression of a gene of interest. The most commonly used bitransgenic systems are based on the provision of tetracycline or tetracycline

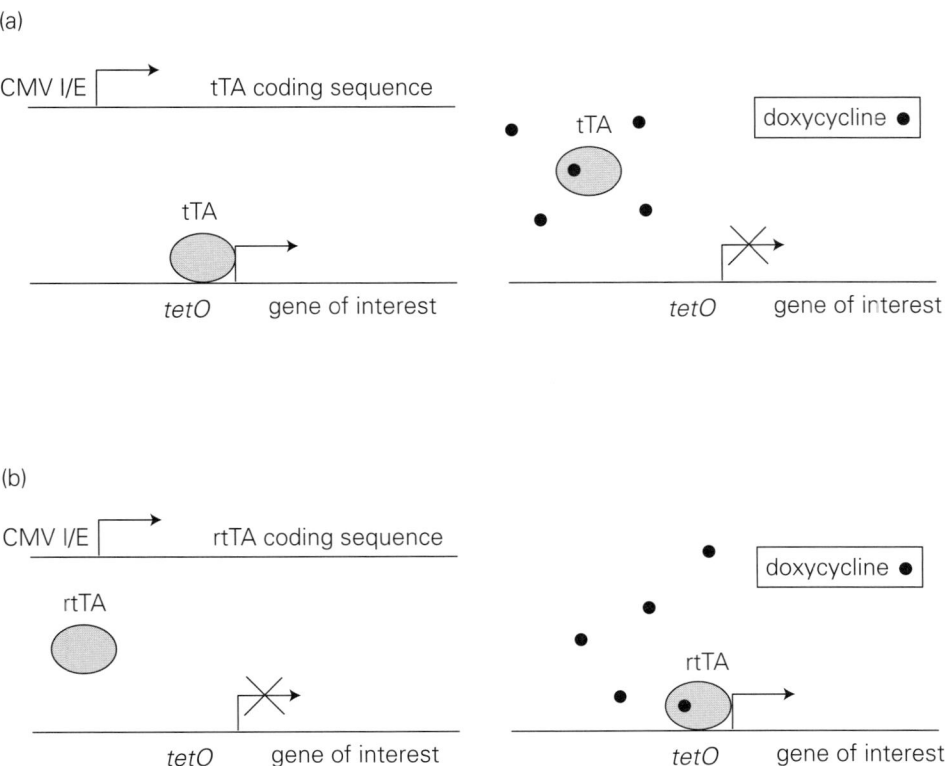

(a)

CMV I/E tTA coding sequence

tTA

tetO   gene of interest

tTA   doxycycline

tetO   gene of interest

(b)

CMV I/E rtTA coding sequence

rtTA

tetO   gene of interest

doxycycline

rtTA

tetO   gene of interest

Figure 2 Indicible systems to regulate transgene expression. (a) The tet-off system as designed by Gossen and Bujard[53]. The cytomegalovirus immediate early (CMV I/E) promoter/enhancer directs the constitutive expression of the tet transactivator protein (tTA). The tTA is a fusion of the *Escherichia coli* tet repressor and the transcriptional activation domain of the herpes virus VP16 protein. The tTA protein binds to the Tn10 tetracycline resistance operator (tetO) sequence in the absence of tetracycline and recruits transcriptional machinery. Seven tandem tetO sequences and a minimal promoter element precede the coding sequence of a gene of interest and expression takes place in the absence of doxycycline (the commonly used tetracycline-like drug). Administration of doxycycline results in drug binding and conformational changes to the tTA which abrogate expression of the responder construct – hence the nomenclature 'tet-off'. (b) The tet-on system[54]. The tet-on bitransgenic system depends on a reverse transactivator (rtTA) which has affinity for tetO only when tetracycline derivatives are provided. The gene of interest in the responder construct is expressed with the administration of these drugs – hence 'tet-on'

derivatives (doxycycline or anhydrotetracycline); these include the tet-off and tet-on regulators (Figure 2).

TARGETING ENDOGENOUS LOCI IN EMBRYONIC STEM CELLS AND STUDIES OF NECESSITY

Embryonic stem (ES) cells, which are derived from mouse blastocysts, can be maintained for long periods in pluripotent states *in vitro*, mutagenized or provided with DNA for selected integration. Deleting or altering a particular site in the genome depends on a cell's homo-

logous recombination machinery to mediate a crossover exchange between the endogenous locus and an introduced targeting vector. Correctly targeted exchanges are identified by a two-step process wherein researchers isolate cells that carry the DNA integration (positive selection), while excluding those with random vector incorporation (negative selection)[37] (Figure 3). ES cells may be re-introduced into blastocysts to give rise to somatic and germ cell lineages in the resulting chimeras (Figure 4)[38,39].

The goal of many targeting experiments is to disrupt a given gene in ES cells and study the resultant phenotype

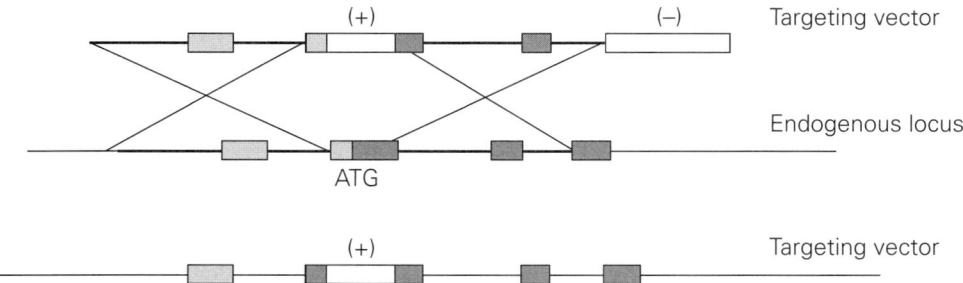

Figure 3 A replacement (Ω) targeting vector. The two regions representing sites for homologous recombination are indicated by the thick lines. The positive selection cassette (+) replaces a portion of the coding sequence and confers antibiotic resistance allowing for the selection of embryonic stem (ES) cell clones that have incorporated the transgene. Neo and Puro are commonly used. Here the (+) cassette is also positioned to replace the endogenous initiation codon in an attempt to ensure a null allele is left at the targeted locus. The (−) cassette, if retained because of random integration, will make cells susceptible to an administered compound (herpes simplex virus thymidine kinase is commonly used and makes cells sensitive to gancyclovir treatment). Together, this (+)/(−) strategy allows for the selective survival of ES cells that have undergone incorporation at the desired locus, which have the (+) cassette and have excluded the (−) cassette

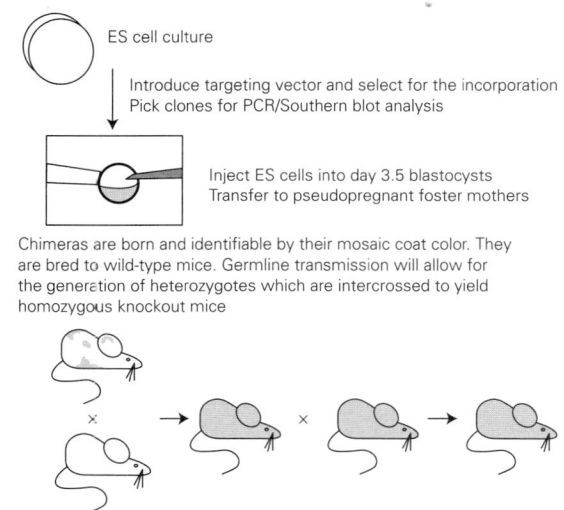

Figure 4 Generation of knockout mice by embryonic stem (ES) cell injection. Under optimal culture conditions with fibroblast-derived feeder cells, ES cells retain pluripotency *in vitro* for multiple passages, allowing selection and characterization of a targeted disruption. ES cells having undergone the desired recombination are injected into day 3.5 blastocysts. A coat color difference between the ES cell donor strain and the blastocyst donor strain allow for the ready assessment of chimerism. Chimeras with a germline component can generate uniform heterozygote offspring when mated with wild-type mice. Finally, barring effects on development, crosses between heterozygotes are expected to yield knockout mice at the Mendelian frequency of 1:3. PCR, polymerase chain reaction

in mice bred to heterozygosity and homozygosity for the alteration. When one completely abrogates the expression of a gene, these null or knockout mice display the direct, indirect and/or compensatory *in vivo* effects of the loss-of-function. Knockout models are leading to major conceptual breakthroughs in many fields of interest; some notable examples relevant to reproductive biology are listed in Table 3.

Transgenic experimentation led to the interesting observation that male ES cells injected into female blastocysts can be transmitted to the germ line within female chimeras. Furthermore, when these chimeras are bred to wild-type males, male offspring have a high incidence of sex chromosome aneuploidies (i.e., XXY and XYY)[40]. Such mice may prove valuable models of human sex chromosome disorders. XXY male mice demonstrate hypoplastic testes and progressive germ cell loss similar to Klinefelter's syndrome patients[41].

RESCUE EXPERIMENTS AND KNOCK-IN MODELS: STUDIES OF SUFFICIENCY

Knockout models are used not only to study the *in vivo* function of a gene but also the potential for other related sequences to compensate for its absence. Underscoring the relevance of mouse models for understanding human

Table 3   Knockout mouse models with reproductive findings

Gene	Reproductive phenotype (in homozygotes unless specified)	References
A-*myb* (*myb11*)	Male infertility with germ cell arrest; females have mammary gland defects	Toscani *et al.*, 1997[114]
Acrosin	Males exhibit delayed fertility, though sperm are capable of binding and penetrating the zona pellucida	Adham *et al.*, 1997[115]
Activin receptor-type IIA (*actrII*)	Infertility in females and delayed fertility in males; small gonads	Matzuk *et al.*, 1995[116]
Activin/inhibin βB subunit (*inhbb*)	Females are subfertile and litters do not survive postnatally	Vasalli *et al.*, 1994[117]
Ahch (*dax1* or DSS-AHC region on the human X chromosome)	Males are infertile with progressive degeneration of the germinal epithelium	Yu *et al.*, 1999[118]
Angiotensin-converting enzyme (*ace*)	Male subfertility; compromised ability of sperm to fertilize ova	Krege *et al.*, 1995[119]
Apaf 1 (Apoptotic protease activating factor 1)	Few homozygotes survive to reproduce; males that do are infertile with spermatogonial degeneration	Honarpour *et al.*, 2000[120]
Apolipoprotein B (*apob*)	Heterozygote males infertile; decreased sperm count, motility, survival time and ability to fertilize ova	Huang *et al.*, 1996[121]
Aryl-hydrocarbon receptor (*ahr*)	Female subfertility; early development of primordial follicles; decreased numbers of antral follicles	Benedict *et al.*, 2000[122]; Abbot *et al.*, 1999[123]
Ataxia telangiectasia (*atm*)	Germ cells of both sexes degenerate; disruptions evident early in meiosis I	Barlow *et al.*, 1998[124]
ATP-binding cassette transporter 1 (*abca1*)	Placental malformations leading to impaired embryo growth, embryo loss and neonatal death	Christiansen-Weber *et al.*, 2000[125]
Basigin (*bsg*)	Males are infertile with a block in spermatogenesis; females are infertile with defects in fertility and implantation	Kune *et al.*, 1998[126]; Igakura *et al.*, 1998[127]
bax	Males are infertile with a spermatogenesis block; females have an extended reproductive lifespan and increased primordial follicle reserve	Knudson *et al.*, 1995[128]; Perez *et al.*, 1999[129]
bcl2	Females have fewer oocytes/primordial follicles in the postnatal ovary	Ratts *et al.*, 1995[130]
bcl6	Compromised male fertility; apoptosis in metaphase I spermatocytes	Kojima *et al.*, 2001[131]
BCLW (*bcl2l2*)	Male infertility; spermatogenesis block with eventual loss of germ cells and Sertoli cells	Ross *et al.*, 1998[132]
Bone morphogenetic protein 8A (*bmp8A*)	Males exhibit progressive infertility; degeneration of germ cells and epididymis	Zhao *et al.*, 1998[133]
Bone morphogenetic protein 8B (*bmp8B*)	Males infertile; germ cell proliferation/depletion defects	Zhao *et al.*, 1996[134]
Bone morphogenetic protein 15 (GDF-9B) (*bmp15*)	Female subfertility; normal ovarian histology	Yan *et al.*, 2001[135]
BMP receptor, type IB (*bmpr1b*)	Female infertility; defects in estrus cyclicity, cumulus expansion and endometrial gland development	Yi *et al.*, 2001[136]
c-*mos*	Female subfertility; parthogenetic activation, cysts and teratomas	Colledge *et al.*, 1994[137]; Hashimoto *et al.*, 1994[138]
c-*ros*	Male infertility due to sperm motility defects	Yeung *et al.*, 1999[139]; Yeung *et al.*, 2000[140]
Calmegin (*clgn*)	Males are infertile; defect in sperm–zona pellucida binding	Ikawa *et al.*, 1997[141]
camk4	Males are infertile; impaired chromatin packaging during spermiogenesis	Wu *et al.*, 2000[142]
Casein kinase IIα	Males are infertile; exhibit globozoospermia (no acrosomal cap)	Xu *et al.*, 1999[143]

Table 3 *Continued*

Gene	Reproductive phenotype (in homozygotes unless specified)	References
Caspase-2 (*casp2*)	Decreased apoptosis of female germ cells	Bergeron *et al.*, 1998[144]
Centromere protein B	Males are hypogonadal and have low sperm counts; females have strain-dependent uterine epithelium defects	Hudson *et al.*, 1998[145]; Fowler *et al.*, 2000[146]
C/EPBβ (CCAAT/enhancer-binding protein β)	Female infertility; reduced ovulation and block in corpus luteum (CL) differentiation	Sterneck *et al.*, 1997[147]
Colony stimulating factor (*csf1*)	Subfertility in both sexes; males have reduced testosterone; females have implantation and lactation defects	Cohen *et al.*, 1997[148]
Connexin 37 (*gja4*)	Female infertility; defects in late folliculogenesis and oocyte meiosis	Simon *et al.*, 1997[149]
Connexin 43 (*gja1*)	Neonatal lethality; small ovaries and testes; decreased numbers of germ cells from E11.5	Juneja *et al.*, 1999[150]
crem (cAMP-responsive element modulator)	Males are infertile due to defective spermatogenesis	Blendy *et al.*, 1996[151]; Nantel *et al.*, 1996[152]
Cyclin A1 (*ccna1*)	Male infertility; block in spermatogenesis before the first meiotic division	Liu *et al.*, 1998[153]
Cyritestin (*adam3*)	Male infertility; altered sperm protein expression and adhesion defects during fertilization	Shamsadin *et al.*, 1999[154]; Nishimura *et al.*, 2001[155]
Cyclin D2 (*ccnd2*)	Female infertility due to a failure of granulosa cell proliferation; males fertile with decreased testis size	Sicinski *et al.*, 1996[156]
Cyclin dependent kinase 4 (*cdk4*)	Female infertility; defects in the hypothalamic–pituitary–gonadal axis	Rane *et al.*, 1999[157]
Cyclooxygenase 2 (*ptgs2*)	Females are mostly infertile; defects in ovulation and implantation	Dinchuk *et al.*, 1995[158]; Lim *et al.*, 1997[159]
dazl (deleted in azoospermia-like autosomal)	Differentiation failure and degeneration of both male and female germ cells	Ruggiu *et al.*, 1997[160]
Desert hedgehog (*dhh*)	Males are infertile; likely roles for *dhh* in early and late stages of spermatogenesis	Bitgood *et al.*, 1996[161]
Dmc1 (*dmc1h*)	Defects in chromosome synapsis in meiosis; both males and females infertile	Pittman *et al.*, 1999[162]; Yoshida *et al.*, 1998[163]
Dynein heavy chain 7 (*dnahc1*)	Male sterility due to defects in sperm flagellar motility	Nishimura *et al.*, 2001[155]
Early growth response 4 (*erg4*)	Infertility in males; most germ cells undergo apoptosis during pachytene stage	Tourtellotte *et al.*, 1999[164]
emx2	Defective development of gonads and genital tracts	Miyamoto *et al.*, 1997[165]
EP$_2$ prostaglandin E2 receptor (*ptger2*)	Female subfertility; decreased fertilization and defects in cumulus expansion	Kennedy *et al.*, 1999[166]; Tilley *et al.*, 1999[167]; Hizaki *et al.*, 1999[168]
Estrogen receptor α (ERα) (*esr1*)	Females are infertile with hemorrhagic ovarian cysts and uterine defects; males develop disruptions of the seminiferous epithelium	Lubahn *et al.*, 1993[169]; Hess *et al.*, 1999[170]
Estrogen receptor β (ERβ) (*esr2*)	Females are subfertile; males are fertile, but develop prostate hyperplasia	Krege *et al.*, 1998[171]
Fanconi anemia complementation group C (*fac*)	Compromised gametogenesis in males and females; impaired fertility	Chen *et al.*, 1996[172]; Whitney *et al.*, 1996[173]
Fertilin (*adam2*)	Male infertility; altered sperm protein expression and adhesion defects during fertilization	Nishimura *et al.*, 2001[155]; Cho *et al.*, 1998[174]
figla or FIGα (factor in the germline α)	Female infertility; no primordial follicles develop at birth and oocytes die	Soyal *et al.*, 2000[175]

Table 3 *Continued*

Gene	Reproductive phenotype (in homozygotes unless specified)	References
Fragile-X (*fmr1*)	Males exhibit macroorchidism	Kooy *et al.*, 1996[176]
FSH hormone β-subunit (*fshb*)	Female infertility; pre-antral block in folliculogenesis; males fertile with decreased testis size	Kumar *et al.*, 1997[20]
FSH receptor (*fshr*)	Female infertility; block in folliculogenesis prior to antral formation	Dierich *et al.*, 1998[21]
β1,4-Galactosyltransferase	Male infertility; defects in sperm–egg interaction; females exhibit dystocia and agalactosis	Lu and Shur, 1997[177]; Cho *et al.*, 1998[178]
γ-Glutamyl transpeptidase (*ggtp*)	Both males and females are hypogonadal and infertile; phenotype corrected by feeding mice *N*-acetylcysteine	Kumar *et al.*, 2000[179]
Glycoprotein hormone α-subunit	Males and females are infertile; hypogonadal due to FSH and LH deficiency	Kendall *et al.*, 1995[17]
Growth differentiation factor-7 (*gdf7*)	Males demonstrate infertility owing to defects in seminal vesicle development	Settle *et al.*, 2001[180]
Growth differentiation factor-9 (*gdf9*)	Female infertility; folliculogenesis arrest at the one-layer follicle stage	Dong *et al.*, 1996[181]
Growth hormone receptor (*ghr*)	Females show delayed puberty and prolonged pregnancy	Zhou *et al.*, 1997[182]
Heat shock protein 70-2 (*hsp70-2*)	Male infertility; meiosis defects and germ cell apoptosis	Dix *et al.*, 1996[183]
Heat shock transcription factor 1 (*hsf1*)	Female infertility; pre- and post-implantation defects	Xiao *et al.*, 1999[184]
High mobility group box 2 (*hmgb2*)	Male subfertility; Sertoli and germ cell degeneration and immotile spermatozoa	Ronfani *et al.*, 2001[185]
hoxa10	Variable infertility; males have cryptorchidism and females have frequent embryo loss prior to implantation	Satokata *et al.*, 1995[186]
hoxa11	Infertility in both sexes; females have uterine defects; males have malformed vas deferens and undescended testes	Hsieh-Li *et al.*, 1995[187]
Inhibin α (*inha*)	Female infertility; male secondary infertility; granulosa/Sertoli tumors	Matzuk *et al.*, 1992[188]
Insulin-like growth factor 1 (*igf1*)	Females are hypogonadal and infertile; impaired antral follicle formation	Baker *et al.*, 1996[189]
Insulin receptor substrate 2 (*irs2*)	Females infertile; small, anovulatory ovaries with reduced numbers of follicles	Burks *et al.*, 2000[190]
Interleukin 11 (*il11*)	Female infertility; compromised implantation and decidualization	Robb *et al.*, 1998[191]
JunD (*jund1*)	Males are infertile; anomalous hormone levels and sperm structure defects	Thepot *et al.*, 2000[192]
Leukemia inhibitory factor (*lif*)	Females infertile; failed implantation	Stewart *et al.*, 1992[193]
Lipase, hormone sensitive (HSL) (*lipe*)	Male infertility; multiple abnormalities in spermatogenesis	Chung *et al.*, 2001[194]
Luteinizing hormone receptor (*lhcgr*)	Underdeveloped sex organs and infertility in both males and females; spermatogenesis arrested at round spermatid stage; folliculogenesis block prior to antral stage	Zhang *et al.*, 2001[26]; Lei *et al.*, 2001[195]
Mismatch repair gene homolog (*pms2*)	Males are infertile; abnormal chromosome synapsis in meiosis	Baker *et al.*, 1995[196]
mlh1 (MutL homolog 1)	Infertility in both sexes; defects in meiosis and genome instability	Edelmann *et al.*, 1996[197]; Baker *et al.*, 1996[198]
msh5 (MutS homolog 5)	Infertility in both sexes; prophase I meiotic defects with aberrant chromosome synapsis and apoptosis	de Vries *et al.*, 1999[199]; Edelmann *et al.*, 1999[200]

Table 3 *Continued*

Gene	Reproductive phenotype (in homozygotes unless specified)	References
Müllerian inhibiting substance (*amh*)	Uteri development in males causes obstruction and secondary infertility; females exhibit early delpetion of primordial follicles	Behringer *et al.*, 1994[201]; Durlinger *et al.*, 1999[202]
Müllerian inhibiting substance receptor	Male subfertility; Müllerian duct causes physical blockage	Mishina *et al.*, 1996[203]
Na$^{(+)}$-K$^{(+)}$-2Cl$^{(-)}$ cotransporter	Males are infertile; low spermatid counts and compromised sperm transport	Pace *et al.*, 2000[204]
NGFI-A transcription factor (*egr-1*)	Female infertility; LH deficiency	Lee *et al.*, 1996[205]; Topilko *et al.*, 1998[206]
Neuronal helix-loop-helix 2 (*nhlh2*)	Males are infertile and hypogonadal; females are fertile when reared with males	Good *et al.*, 1997[207]
NIRKO mice (neuronal insulin receptor knockout)	Mice exhibit hypothalamic hypogonadism; impaired spermatogenesis and follicle maturation	Bruning *et al.*, 2000[208]
Nuclear receptor co-repressor RIP40 (*nrip1*)	Female infertility due to an ovulation defect; ovaries accumulate luteinized, unruptured follicles	White *et al.*, 2000[209]
Osp-11/Claudin-11	Males are infertile; no tight junctions between Sertoli cells	Gow *et al.*, 1999[210]
Ovo	Males have reduced fertility and hypogenitalism	Dai *et al.*, 1998[211]
Oxytocin (*oxt*)	Females unable to nurse offspring	Nishimori *et al.*, 1996[212]
P450 aromatase (*cyp19*)	Females are infertile; ovaries do not form CL; males develop progressive infertility with defects in spermatogenesis	Robertson *et al.*, 1999[213]; Fisher *et al.*, 1998[214]
P450 25-hydroxyvitamin D-1αhydroxylase (*cyp40*)	Female infertility with uterine hypoplasia and absence of CL	Panda *et al.*, 2001[215]
P2X1 receptor (*p2rx1*)	Male infertility; oligospermia and defective vas deferens contraction	Mulryan *et al.*, 2000[216]
p19^{Ink4d} (*cdkn2d*)	Males are fertile despite testicular atrophy and germ cell apoptosis	Zindy *et al.*, 2000[217]
p27^{Kip1} (*cdkn1b*)	Female infertility with CL differentiation failure and granulosa cell hyperplasia; males fertile with testicular hyperplasia	Nakayama *et al.*, 1996[218]; Fera *et al.*, 1996[219]
p53 (*trp53*)	Males are compromised in recovering spermatogenesis after irradiation	Hendry *et al.*, 1996[220]
PC4 (testicular germ cell protease)	Males are infertile; sperm have impaired fertilization ability	Mbikay *et al.*, 1997[221]
Phosphatidylinositol 3'-kinase	Males are infertile; defects in proliferation and increased apoptosis of spermatogonia	Blume-Jensen *et al.*, 2000[222]
Phosphatidylinositol glycan, class A (*piga*)	Chimeric males have abnormal testes, epididymis and seminal vesicles	Lin *et al.*, 2000[223]
Polyomavirus enhancer activator 3 (*pea3*)	Male infertility; males exhibit normal mating behavior but do not set plugs or release sperm	Pandka *et al.*, 2001[215]
Progesterone receptor (*pgr*)	Female infertility; ovulation failure, uterine implantation defects and mammary defects	Lydon *et al.*, 1995[224]
Prolactin (*prl*)	Females are infertile with irregular estrus cycles	Horseman *et al.*, 1997[225]
Prolatin receptor (*prlr*)	Males infertile; females infertile related to compromised ovulation, fertilization and preimplantation development	Ormandy *et al.*, 1997[226]
Prostaglandin F receptor (*ptgfr*)	Females do not undergo parturition; failed luteolysis	Sugimoto *et al.*, 1997[227]
Protamine 1 (*prm1*)	Male chimeras exhibit protamine haploinsufficiency; abnormal spermatogenesis and infertility	Cho *et al.*, 2001[228]
Protamine 2 (*prm2*)	Male chimeras exhibit protamine haploinsufficiency; abnormal spermatogenesis and infertility	Cho *et al.*, 2001[228]

Table 3   *Continued*

Gene	Reproductive phenotype (in homozygotes unless specified)	References
Protein C inhibitor (*serpina5*)	Male infertility; Sertoli cell destruction	Uhrin *et al.*, 2000[229]
Protein phosphatase 1 catalytic subunit γ (*pp1cc*)	Male infertility; defects in spermiogenesis	Varmuza *et al.*, 1999[230]
Retinoic acid receptor α (*rara*)	Male infertility with seminiferous tubule degeneration	Lufkin *et al.*, 1993[231]
Retinoic acid receptor γ (*rarg*)	Male infertility; squamous metaplasia of the seminal vesicles and prostate	Lohnes *et al.*, 1993[232]
Retinoid X receptors (*rxrb*)	Male infertility; germ cell maturation defects and tubular degeneration	Kastner *et al.*, 1996[233]
Rho GDIα (GDP dissociation inhibitor)	Both sexes infertile; impaired spermatogenesis in males; post-implantation pregnancy defects in females	Togana *et al.*, 1999[234]
Scavenger receptor, class B1 (*srb1*)	Female infertility; defects in oocyte maturation and early embryo development	Rigotti *et al.*, 1997[235]; Trigatti *et al.*, 1999[236]
SCP3 (synaptonemal complex protein 3) (*sycp3*)	Males are infertile; defects in chromosome synapsis during meiosis; germ cell apoptosis	Yuan *et al.*, 2000[237]
Sp4 transcription factor (*sp4*)	Males are infertile; defects in reproductive behavior	Supp *et al.*, 1996[238]
Sperm-1	Males are subfertile; defect in haploid sperm function	Pearse *et al.*, 1997[239]
SPO11 homolog (*spo11*)	Infertility in both sexes; defects in meiosis; spermatocytes undergo apoptosis in early prophase; oocytes are lost soon after birth	Baudat *et al.*, 2000[240]; Romanienko and Camarini-Otero, 2000[241]
Steroid 5α-reductase type 1 (*srd5a1*)	Female infertility; defects in parturition	Mahendroo *et al.*, 1996[242]; Mahendroo *et al.*, 1997[243]
Steroidogenic acute regulatory protein (*star*)	Males have female external genitalia; both sexes die of adrenocortical insufficiency	Caron *et al.*, 1997[244]
Steroidogenic factor-1 (SF-1) (*nr5a1*)	Gonadal agenesis in both sexes	Luo *et al.*, 1994[245]
Superoxide dismutase 1 (*sod1*) (copper–zinc superoxide dismutase)	Females are subfertile; folliculogenesis defect; failure to maintain pregnancy	Matzuk *et al.*, 1998[246]; Ho *et al.*, 1998[247]
TATA-binding protein-like protein (TRF2) (*tlp*)	Males are sterile due to defects in post-meiotic spermatogenesis	Zhang *et al.*, 2001[248]; Martianov *et al.*, 2001[249]
TBP-associated factor TAF$_{II}$105	Female sterility due to defects in follicular development, oocyte maturation/fertilization	Freiman *et al.*, 2001[250]
Telomerase	Progressive infertility in both sexes; males show germ cell apoptosis; females have few oocytes and uterine abnormalities	Lee *et al.*, 1998[251]
TIAR RNA binding protein	PGCs are lost by E13.5; no spermatogonia or oogonia develop	Beck *et al.*, 1998[252]
TLS (translocated in liposarcoma) (*fus1*)	Males are infertile; defects in spermatocyte chromosome pairing	Kuroda *et al.*, 2000[253]
tnp1 (transition protein 1)	Male subfertility; abnormal chromosome condensation and reduced sperm motility	Yu *et al.*, 2000[254]
Tumor necrosis factor type I receptor (*tnfrsf1a*)	Enhanced prepubertal response to gonadotropins; early ovarian senescence	Roby *et al.*, 1999[255]
Ubiquitin-conjugating enzyme E2B (HR6B) (*ube2b*)	Males are sterile with alterations in chromatin structure	Roest *et al.*, 1996[256]
VASA (*ddx4*)	Infertile males; defective proliferation/differentiation of primordial germ cells	Tanaka *et al.*, 2000[257]
Wilms tumor homolog (*wt1*)	Embryonic lethality with gonadal agenesis	Kreidberg *et al.*, 1993[258]

Table 3 *Continued*

Gene	Reproductive phenotype (in homozygotes unless specified)	References
wnt4	Female infertility; ovaries are depleted of oocytes and contain Leydig cells; Müllerian ducts do not form	Vainio *et al.*, 1999[259]
wnt7a	Females show abnormal development of oviducts and uterus; males do not have Müllerian duct regression	Parr and McMahon, 1998[260]
zfx	Reduced germ cell numbers in both sexes; defects in mitotic proliferation	Luoh *et al.*, 1997[261]
Zona pellucida protein (*zp1*)	Reduced female fertility; defects in fertilization	Rankin *et al.*, 1999[262]
Zona pellucida protein (*zp2*)	Female infertility; fragile oocytes with defects in developmental competence	Rankin *et al.*, 2001[263]
Zona pellucida protein (*zp3*)	Female infertility; fragile oocytes	Rankin *et al.*, 1996[264]; Liu *et al.*, 1996[265]

biology, human FSHβ, zona pellucida protein 3 (ZP3) and deleted in azoospermia (*daz*) transgenes have been shown to rescue or partially rescue fertility defects in their respective mouse knockout models[24,42,43].

Targeting in ES cells may also be employed to replace one coding sequence with another, creating a 'knock-in' model. This is a powerful technique for examining differences in the potential functions of related genes. For example, mice null for activin/inhibin βA subunit gene (*inhba*[−/−]) die neonatally due to craniofacial defects that prevent suckling. This phenotype can be rescued by replacing the activin/inhibin βA coding sequence with that of the activin/inhibin βB gene, conferring the activin/inhibin βA expression pattern on this related sequence (63% amino acid identity). Interestingly the homozygous knock-in mice demonstrate enlarged external genitalia, hypogonadism and diminished female fertility, indicating unique and previously unrecognized functions of the activin/inhibin βA protein product in reproduction[44].

Combinations of knockout and bitransgenic models are also being used to investigate whether a specific spatiotemporal expression pattern is key to a gene function. For example, a mifeprisone-inducible system regulating ectopic expression of inhibin A subunits can avert testicular tumorigenesis in inhibin α null mice as long as the transgene expression is maintained[45].

SITE-SPECIFIC RECOMBINATION

Recombinases from P1 bacteriophage (Cre) and yeast (FLP) have been shown to mediate double-stranded exchanges between defined DNA sequences when these targets are introduced into mammalian cells[46,47]. Recombinases and their target sites are used in transgenic systems in the design of tissue-specific knockouts[48]. An endogenous gene to be deleted in the Cre/loxP systems is first flanked by tandem loxP sites, 34 bp target sequences for Cre recombinase. Expression of the Cre protein in a tissue-specific and/or inducible manner will mediate deletion of the sequence between the loxP sites, whereas in tissues without Cre activity, the intervening sequence remains intact. The introduction of subtle mutations or replacement alleles at a targeted locus are important applications of the Cre/loxP system (Figure 5). This approach was used recently to introduce a missense mutation within the Kit receptor (Y[719]F), revealing roles for a specific aspect of Kit signaling (i.e., the phosphatidylinositol-3′-kinase pathway) in spermatogenesis and oogenesis[49]. A missense mutation similarly engineered in fibroblast growth factor 3 (G[374]R) shortens the female reproductive lifespan and is associated with defects in the maintenance of serum gonadotropins[50,51].

LoxP sites can be recognized by Cre recombinase over large distances within the genome allowing the recombination system to be harnessed to engineer large

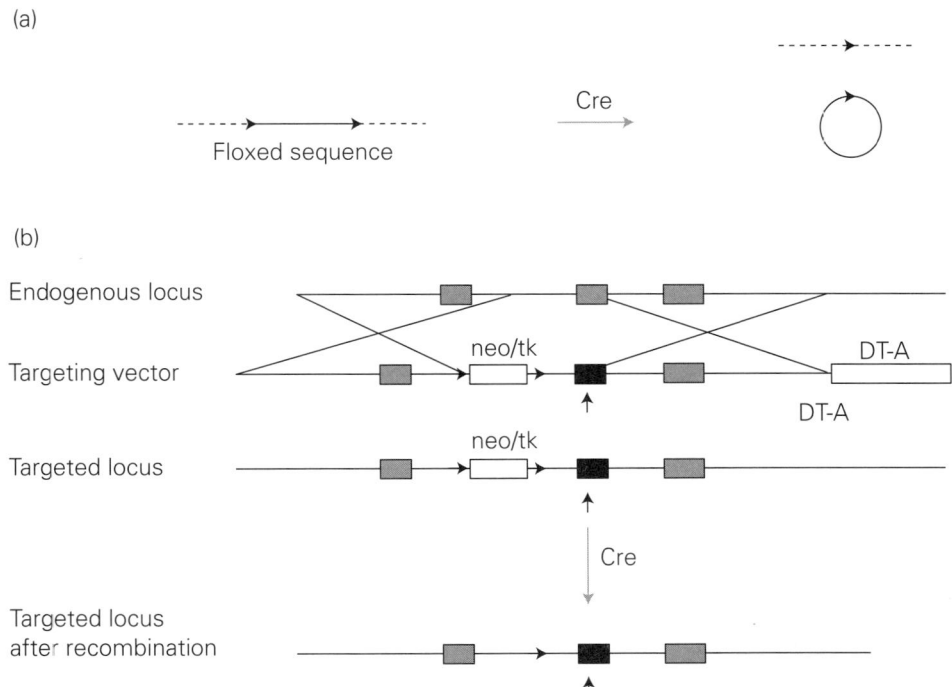

Figure 5 Cre/loxP-mediated recombination. (a) Tissue-specific knockout. Two locus of crossover x in P1 sites (loxP sites) have been introduced in tandem orientations flanking a sequence to be deleted. This is the 'floxed' allele. A promoter driving the Cre recombinase transgene determines the temporal and tissue specificity of the deletion. When the Cre transgene is expressed, active recombinase enzyme will promote a recombination between the loxP sequences. The result is an excision and circularization of the intervening sequence. In the absence of Cre the functional floxed allele remains intact. (b) Introduction of a subtle mutation using Cre/loxP. A targeting vector replaces a portion of coding sequence with the desired allele (here the second exon; arrow) and introduces a floxed positive/negative dual selection cassette. Two-part selection for the incorporation of the dual cassette and for the loss of a second cassette (here, the diphtheria toxin (DT) A gene fragment) ensures a homologous and not random introduction. Finally, Cre recombinase electroporation and selection for loss of the positive/negative cassette promotes the excision of the markers and the locus is left with only the subtle change of codon replacement. neo, neomycin; tk, thymidine kinase

chromosomal rearrangement[52]. Long-range recombination can thus be used to generate chromosome deletions, duplications and inversions. Transgenic models with large rearrangements will be used to study effects of aneuploidies, regional deletions and for balancing and tracking chromosomes during random mutagenesis experiments.

CONCLUSIONS

Transgenic technology has gripped the medical scientific community as a means of producing mammalian genetic models that phenocopy human pathologies. There are numerous commonalities in human and mouse physiology and the list of mouse models recapitulating recognized or idiopathic human reproductive defects is growing at an ever-increasing rate. These have been discovered by deliberate investigation of gene function, as well as by serendipity – researchers can hardly fail to recognize an overt reproductive phenotype while generating a line of transgenic mice. What we have to learn from these models will continue to revise our conceptions of physiology, identify new targets for contraception and improve our means of understanding, diagnosing and treating cases of human infertility.

ACKNOWLEDGEMENTS

We thank Drs. Julia A. Elvin, T. Rajendra Kumar and Francesco J. DeMayo for their assistance in preparing the tables and figures presented. We also thank Ms. Shirley Baker and Ms. Jennifer Newton for their expert assistance in the preparation of the manuscript. Transgenic mouse research in the Matzuk lab has been supported by Wyeth Ayerst Research, National Institutes of Health grants CA60651, HD32067 and HD33438, and the Specialized Cooperative Centers Program in Reproduction Research (HD07495). KHB is a student in the Medical Scientist Training Program at Baylor College of Medicine and is supported in part by NIH grant T32GM07330 and National Eye Institute grant T32EY07102.

References

1. Fauser BCJM, Conway GS, Franks S. Genetics of female reproductive dysfunction. In Fauser BCJM, ed. *Molecular Biology in Reproductive Medicine*. New York: Parthenon Publishing, 1999:449–79

2. Elliott DJ, Cooke HJ. The molecular genetics of male infertility. *Bioessays* 1997;19:801–9

3. Thielemans BF, Spiessens C, D'Hooghe T, *et al*. Genetic abnormalities and male infertility. A comprehensive review. *Eur J Obstet Gynecol Reprod Biol* 1998;81:217–25

4. Hargreave TB. Genetic basis of male fertility. *Br Med Bull* 2000;56:650–71

5. Adashi EY, Hennebold JD. Single-gene mutations resulting in reproductive dysfunction in women. *N Engl J Med* 1999;340:709–18

6. Bousfield GR, Perry WM, Ward DN. Gonadotropins: chemistry and biosynthesis. In: Knobil E, Neill JD, eds. *The Physiology of Reproduction*, 2nd edn. New York: Raven Press, 1994:1749–92

7. Greenwald GS, Roy SK. Follicular development and its control. In Knobil E, Neill J, eds. *The Physiology of Reproduction*, 2nd edn. New York: Raven Press, 1994: 629–724

8. Sharpe RM. Regulation of spermatogenesis. In Knobil E, Neill J, eds. *The Physiology of Reproduction* 2nd edn. New York: Raven Press, 1994:1363–1434

9. Themmen APN, Huhtaniemi IT. Mutations of gonadotropins and gonadotropin receptors: elucidating the physiology and pathophysiology of pituitary–gonadal function. *Endocr Rev* 2000;21:551–83

10. Achermann JC, Weiss J, Lee EJ, Jameson JL. Inherited disorders of the gonadotropin hormones. *Mol Cell Endocrinol* 2001; 179:89–96

11. de Roux N, Milgrom E. Inherited disorders of GnRH and gonadotropin receptors. *Mol Cell Endocrinol* 2001;179: 83–7

12. Franco B, Guioli S, Pragliola A, *et al*. A gene deleted in Kallmann's syndrome shares homology with neural cell adhesion and axonal path-finding molecules. *Nature* 1991;353:529–36

13. Jackson RS, Creemers JW, Ohagi S, *et al*. Obesity and impaired prohormone processing associated with mutations in the human prohormone convertase 1 gene. *Nat Genet* 1997;16:303–6

14. Weiss J, Adams E, Whitcomb RW, *et al*. Normal sequence of the gonadotropin-releasing hormone gene in patients with idiopathic hypogonadotropic hypogonadism. *Biol Reprod* 1991;45:743–7

15. Seminara SB, Beranova M, Oliveira LM, *et al*. Successful use of pulsatile gonadotropin-releasing hormone (GnRH) for ovulation induction and pregnancy in a patient with GnRH receptor mutations. *J Clin Endocrinol Metab* 2000;85:556–62

16. Mason AJ, Hayflick JS, Zoeller RT, *et al*. A deletion truncating the gonadotropin-releasing hormone gene is responsible for hypogonadism in the hpg mouse. *Science* 1986;234:1366–71

17. Kendall SK, Samuelson LC, Saunders TL, *et al*. Targeted disruption of the pituitary glycoprotein hormone α-subunit produces hypogonadal and hypothyroid mice. *Genes Dev* 1995;9:2007–19

18. Miller-Lindholm AK, Bedows E, Bartels CF, *et al*. A naturally occurring genetic variant in the human chorionic gonadotropin-beta gene 5 is assembly inefficient. *Endocrinology* 1999;140:3496–506

19. Layman LC, McDonough PG. Mutations of follicle stimulating hormone-beta and its receptor in human and mouse: genotype/phenotype. *Mol Cell Endocrinol* 2000; 161:9–17

20. Kumar TR, Wang Y, Lu N, Matzuk MM. Follicle stimulating hormone is required for ovarian follicle maturation but not male fertility. *Nat Genet* 1997;15:201–4

21. Dierich A, Sairam MR, Monaco L, *et al*. Impairing follicle-stimulating hormone (FSH) signaling *in vivo*: targeted disruption of the FSH receptor leads to aberrant gametogenesis and hormonal imbalance. *Proc Natl Acad Sci USA* 1998;95:13612–17

22. Tapanainen JS, Aittomaki K, Min J, *et al*. Men homozygous for an inactivating mutation of the follicle-stimulating hormone (FSH) receptor gene present variable suppression of spermatogenesis and fertility. *Nat Genet* 1997;15:205–6

23. Kumar TR, Palapattu G, Wang P, *et al*. Transgenic models to study gonadotropin function: the role of follicle-stimulating hormone in gonadal growth and tumorigenesis. *Mol Endocrinol* 1999;13:851–65

24. Kumar TR, Low MJ, Matzuk MM. Genetic rescue of follicle-stimulating hormone β-deficient mice. *Endocrinology* 1998;139:3289–95

25. Weiss J, Axelrod L, Whitcomb RW, *et al*. Hypogonadism caused by a single amino acid substitution in the beta

subunit of luteinizing hormone. *N Engl J Med* 1992; 326:179–83

26. Zhang FP, Poutanen M, Wilbertz J, Huhtaniemi I. Normal prenatal but arrested postnatal sexual development of luteinizing hormone receptor knockout (LuRKO) mice. *Mol Endocrinol* 2001;15:172–83

27. Shenker A, Laue L, Kosugi S, *et al*. A constitutively activating mutation of the luteinizing hormone receptor in familial male precocious puberty. *Nature* 1993;365:652–4

28. Risma KA, Clay CM, Nett TM, *et al*. Targeted over-expression of luteinizing hormone in transgenic mice leads to infertility, polycystic ovaries, and ovarian tumors. *Proc Natl Acad Sci USA* 1995;92:1322–6

29. Keri RA, Lozada KL, Abdul-Karim FW, *et al*. Luteinizing hormone induction of ovarian tumors: oligogenic differences between mouse strains dictates tumor disposition. *Proc Natl Acad Sci USA* 2000;97:383–7

30. Gordon JW, Scangos GA, Plotkin DJ, *et al*. Genetic transformation of mouse embryos by microinjection of purified DNA. *Proc Natl Acad Sci USA* 1980;77:7380–4

31. Gordon JW, Ruddle FH. Integration and stable germ line transmission of genes injected into mouse pronuclei. *Science* 1981;214:1244–6

32. Brinster RL, Chen HY, Trumbauer M, *et al*. Somatic expression of herpes thymidine kinase in mice following injection of a fusion gene into eggs. *Cell* 1981;27:223–31

33. Costantini F, Radice G, Lee JL, *et al*. Insertional mutations in transgenic mice. *Prog Nucleic Acid Res Mol Biol* 1989;36:159–69

34. Wurst W, Rossant J, Prideaux V, *et al*. A large-scale gene-trap screen for insertional mutations in developmentally regulated genes in mice. *Genetics* 1995;139: 889–99

35. Gossler A, Joyner AL, Rossant J, Skarnes WC. Mouse embryonic stem cells and reporter constructs to detect developmentally regulated genes. *Science* 1989;244:463–5

36. Friedrich G, Soriano P. Promoter traps in embryonic stem cells: a genetic screen to identify and mutate developmental genes in mice. *Genes Dev* 1991;5:1513–23

37. Mansour SL, Thomas KR, Capecchi MR. Disruption of the proto-oncogene *int-2* in mouse embryo-derived stem cells: a general strategy for targeting mutations to non-selectable genes. *Nature* 1988;336:348–52

38. Robertson E, Bradley A, Kuehn M, Evans M. Germ-line transmission of genes introduced into cultured pluripotential cells by retroviral vector. *Nature* 1986;323:445–8

39. Kuehn MR, Bradley A, Robertson EJ, Evans M. A potential animal model for Lesch–Nyhan syndrome through introduction of HPRT mutations into mice. *Nature* 1987; 326:295–8

40. Bronson SK, Smithies O, Mascarello JT. High incidence of XXY and XYY males among the offspring of female chimeras from embryonic stem cells. *Proc Natl Acad Sci USA* 1995;92:3120–3

41. Lue Y, Rao PN, Sinha Hikim AP, *et al*. XXY male mice: an experimental model for Klinefelter syndrome. *Endocrinology* 2001;142:1461–70

42. Rankin TL, Tong ZB, Castle PE, *et al*. Human ZP3 restores fertility in *Zp3* null mice without affecting order-specific sperm binding. *Development* 1998;125: 2415–24

43. Slee R, Grimes B, Speed RM, *et al*. A human DAZ transgene confers partial rescue of the mouse *Dazl* null phenotype. *Proc Natl Acad Sci USA* 1999;96:8040–5

44. Brown CW, Houston-Hawkins DE, Woodruff TK, Matzuk MM. Insertion of *Inhbb* into the *Inhba* locus rescues the *Inhba*-null phenotype and reveals new activin functions. *Nat Genet* 2000;25:453–7

45. Pierson MP, DeMayo FJ, Matzuk MM, *et al*. Regulable expression of inhibin A in wild-type and inhibin α null mice. *Mol Endocrinol* 2000;14:1075–85

46. Lakso M, Sauer B, Mosinger B, *et al*. Targeted oncogene activation by site-specific recombination in transgenic mice. *Proc Natl Acad Sci USA* 1992;89:6232–6

47. O'Gorman S, Fox DT, Wahl GM. Recombinase-mediated gene activation and site-specific integration in mammalian cells. *Science* 1991;251:1351–5

48. Gu H, Marth JD, Orban PC, *et al*. Deletion of a DNA polymerase beta gene segment in T cells using cell type-specific gene targeting. *Science* 1994;265:103–6

49. Kissel H, Timokhina I, Hardy MP, *et al*. Point mutation in kit receptor tyrosine kinase reveals essential roles for kit signaling in spermatogenesis and oogenesis without affecting other kit responses. *EMBO J* 2000;19: 1312–26

50. Wang Y, Spatz MK, Kannan K, *et al*. A mouse model for achondroplasia produced by targeting fibroblast growth factor receptor 3. *Proc Natl Acad Sci USA* 1999;96: 4455–60

51. Amsterdam A, Kannan K, Givol D, *et al*. Apoptosis of granulosa cells and female infertility in achondroplastic mice expressing mutant fibroblast growth factor receptor 3(g374r). *Mol Endocrinol* 2001;15:1610–23

52. Zheng B, Mills AA, Bradley A. A system for rapid generation of coat color-tagged knockouts and defined chromosomal rearrangements in mice. *Nucleic Acids Res* 1999;27:2354–60

53. Gossen M, Bujard H. Tight control of gene expression in mammalian cells by tetracycline-responsive promoters. *Proc Natl Acad Sci USA* 1992;89:5547–51

54. Gossen M, Freundlieb S, Bender G, *et al*. Transcriptional activation by tetracyclines in mammalian cells. *Science* 1995;268:1766–9

55. Chanson P, De Roux N, Young J, *et al*. Absence of activating mutations in the GnRH receptor gene in human pituitary gonadotroph adenomas. *Eur J Endocrinol* 1998; 139:157–60

56. Matthews CH, Borgato S, Beck-Peccoz P, *et al*. Primary amenorrhoea and infertility due to a mutation in the beta-subunit of follicle-stimulating hormone. *Nat Genet* 1993;5:83–6

57. Layman LC, Lee EJ, Peak DB, *et al*. Delayed puberty and hypogonadism caused by mutations in the follicle-stimulating hormone beta-subunit gene. *N Engl J Med* 1997;337:607–11

58. Aittomaki K, Lucena JLD, Pakarinen P, *et al*. Mutation in the follicle stimulating hormone receptor gene causes hereditary hypergonadotropic ovarian failure. *Cell* 1995; 82:959–68

59. Gromoll J, Simoni M, Nieschlag E. An activating mutation of the follicle-stimulating hormone receptor autonomously sustains spermatogenesis in a hypophysectomized man. *J Clin Endocrinol Metab* 1996;81:1367–70

60. Latronico AC, Anasti J, Arnhold IJ, *et al*. Brief report: testicular and ovarian resistance to luteinizing hormone caused by inactivating mutations of the luteinizing hormone-receptor gene. *N Engl J Med* 1996;334:507–12

61. Kremer H, Kraaij R, Toledo SP, *et al*. Male pseudohermaphroditism due to a homozygous missense mutation of the luteinizing hormone receptor gene. *Nat Genet* 1995;9:160–4

62. Toledo SP, Brunner HG, Kraaij R, *et al*. An inactivating mutation of the luteinizing hormone receptor causes amenorrhea in a 46,XX female. *J Clin Endocrinol Metab* 1996;81:3850–4

63. Tanimoto Y, Tanimoto K, Sugiyama F, *et al*. Male sterility in transgenic mice expressing activin betaA subunit gene in testis. *Biochem Biophys Res Commun* 1999;259:699–705

64. Larriba S, Esteban C, Toran N, *et al*. Androgen binding protein is tissue-specifically expressed and biologically active in transgenic mice. *J Steroid Biochem Mol Biol* 1995;53:573–8

65. Hsu SY, Lai RJ, Finegold M, Hsueh AJ. Targeted overexpression of Bcl-2 in ovaries of transgenic mice leads to decreased follicle apoptosis, enhanced folliculogenesis, and increased germ cell tumorigenesis. *Endocrinology* 1996;137:4837–43

66. Furuchi T, Masuko K, Nishimune Y, *et al*. Inhibition of testicular germ cell apoptosis and differentiation in mice misexpressing Bcl-2 in spermatogonia. *Development* 1996; 122:1703–9

67. Suzuki M, Abe K, Yoshinaga K, *et al*. Specific arrest of spermatogenesis caused by apoptotic cell death in transgenic mice. *Genes Cells* 1996;1:1077–86

68. Swain A, Narvaez V, Burgoyne P, *et al*. Dax1 antagonizes Sry action in mammalian sex determination. *Nature* 1998;391:761–7

69. Kendall SK, Saunders TL, Jin L, *et al*. Targeted ablation of pituitary gonadotropes in transgenic mice. *Mol Endocrinol* 1991;5:2025–36

70. Bartell JG, Fantz DA, Davis T, *et al*. Elimination of male germ cells in transgenic mice by the diphtheria toxin A chain gene directed by the histone H1t promoter. *Biol Reprod* 2000;63:409–16

71. Holmberg C, Helin K, Sehested M, Karlstrom O. E2F-1-induced p53-independent apoptosis in transgenic mice. *Oncogene* 1998;17:143–55

72. Wong RW, Kwan RW, Mak PH, *et al*. Overexpression of epidermal growth factor induced hypospermatogenesis in transgenic mice. *J Biol Chem* 2000;275):18297–301

73. Guo Q, Kumar TR, Woodruff T, *et al*. Overexpression of mouse follistatin causes reproductive defects in transgenic mice. *Mol Endocrinol* 1998;12:96–106

74. Youakim A, Hathaway HJ, Miller DJ, *et al*. Overexpressing sperm surface beta 1,4-galactosyltransferase in transgenic mice affects multiple aspects of sperm–egg interactions. *J Cell Biol* 1994;126:1573–83

75. Bartke A, Cecim M, Tang K, *et al*. Neuroendocrine and reproductive consequences of overexpression of growth hormone in transgenic mice. *Proc Soc Exp Biol Med* 1994;206:345–59

76. Nakai A, Suzuki M, Tanabe M. Arrest of spermatogenesis in mice expressing an active heat shock transcription factor 1. *EMBO J* 2000;19:1545–54

77. Hekman AC, Trapman J, Mulder AH, *et al*. Interferon expression in the testes of transgenic mice leads to sterility. *J Biol Chem* 1988;263:12151–5

78. Iwakura Y, Asano M, Nishimune Y, Kawade Y. Male sterility of transgenic mice carrying exogenous mouse interferon-beta gene under the control of the metallothionein enhancer-promoter. *EMBO J* 1988;7:3757–62

79. Ohta M, Mitomi T, Kimura M, *et al*. Anomalies in transgenic mice carrying the human interleukin-2 gene. *Tokai J Exp Clin Med* 1990;15:307–15

80. Yura S, Ogawa Y, Sagawa N, *et al*. Accelerated puberty and late-onset hypothalamic hypogonadism in female transgenic skinny mice overexpressing leptin. *J Clin Invest* 2000;105:749–55

81. Rudolph-Owen LA, Cannon P, Matrisian LM. Overexpression of the matrix metalloproteinase matrilysin results in premature mammary gland differentiation and male infertility. *Mol Biol Cell* 1998;9:421–35

82. Behringer RR, Cate RL, Froelick GJ, *et al*. Abnormal sexual development in transgenic mice chronically expressing mullerian inhibiting substance. *Nature* 1990; 345:167–70

83. Mangues R, Seidman I, Pellicer A, Gordon JW. Tumorigenesis and male sterility in transgenic mice expressing a MMTV/N-ras oncogene. *Oncogene* 1990;5:1491–7

84. Guy CT, Cardiff RD, Muller WJ. Activated *neu* induces rapid tumor progression. *J Biol Chem* 1996;271:7673–8

85. Hakovirta H, Keiski A, Toppari J, *et al*. Polyamines and regulation of spermatogenesis: selective stimulation of late spermatogonia in transgenic mice overexpressing the human ornithine decarboxylase gene. *Mol Endocrinol* 1993;7:1430–6

86. Fowler KA, Gill K, Kirma N, *et al*. Overexpression of aromatase leads to development of testicular Leydig cell tumors: an *in vivo* model for hormone-mediated testicular cancer. *Am J Pathol* 2000;156:347–53

87. Allemand I, Anglo A, Jeantet AY, *et al*. Testicular wild-type p53 expression in transgenic mice induces spermiogenesis alterations ranging from differentiation defects to apoptosis. *Oncogene* 1999;18:6521–30

88. O'Hara BF, Donovan DM, Lindberg I, *et al*. Proenkephalin transgenic mice: a short promoter confers high testis expression and reduced fertility. *Mol Reprod Dev* 1994;38:275–84

89. Rhim JA, Connor W, Dixon GH, *et al*. Expression of an avian protamine in transgenic mice disrupts chromatin structure in spermatozoa. *Biol Reprod* 1995;52:20–32

90. Lee K, Haugen HS, Clegg CH, Braun RE. Premature translation of protamine 1 mRNA causes precocious nuclear condensation and arrests spermatid differentiation in mice. *Proc Natl Acad Sci USA* 1995;92:12451–5

91. Costa SL, Boekelheide K, Vanderhyden BC, *et al*. Male infertility caused by epididymal dysfunction in transgenic mice expressing a dominant negative mutation

of retinoic acid receptor alpha 1. *Biol Reprod* 1997; 56:985–90

92. Pietila M, Alhonen L, Halmekyto M, *et al*. Activation of polyamine catabolism profoundly alters tissue polyamine pools and affects hair growth and female fertility in transgenic mice overexpressing spermidine/spermine N1-acetyltransferase. *J Biol Chem* 1997;272:18746–51

93. Vidal VP, Chaboissier MC, de Rooij DG, Schedl A. Sox9 induces testis development in XX transgenic mice. *Nat Genet* 2001;28:216–17

94. Koopman P, Gubbay J, Vivian N, *et al*. Male development of chromosomally female mice transgenic for Sry. *Nature* 1991;351:117–21

95. Radovick S, Wray S, Lee E, *et al*. Migratory arrest of gonadotropin-releasing hormone neurons in transgenic mice. *Proc Nat Acad Sci USA* 1991;88:3402–6

96. Markkula MA, Hamalainen TM, Zhang F, *et al*. The FSH beta-subunit promoter directs the expression of herpes simplex virus type 1 thymidine kinase to the testis of transgenic mice. *Mol Cell Endocrinol* 1993;96:25–36

97. Al-Shawi R, Burke J, Jones CT, *et al*. A Mup promoter-thymidine kinase reporter gene shows relaxed tissue-specific expression and confers male sterility upon transgenic mice. *Mol Cell Biol* 1988;8:4821–8

98. Das SK, Lim H, Wang J, *et al*. Inappropriate expression of human transforming growth factor (TGF)-alpha in the uterus of transgenic mouse causes downregulation of TGF-beta receptors and delays the blastocyst-attachment reaction. *J Mol Endocrinol* 1997;18:243–57

99. Korpelainen EI, Karkkainen MJ, Tenhunen A, *et al*. Overexpression of VEGF in testis and epididymis causes infertility in transgenic mice: evidence for nonendothelial targets for VEGF. *J Cell Biol* 1998;143:1705–12

100. Rosenberg MP, Aversa CR, Wallace R, Propst F. Expression of the v-*Mos* oncogene in male meiotic germ cells of transgenic mice results in metaphase arrest. *Cell Growth Differ* 1995;6:325–36

101. Bessone S, Vidal F, Le Bouc Y, *et al*. EMK protein kinase-null mice: dwarfism and hypofertility associated with alterations in the somatotrope and prolactin pathways. *Dev Biol* 1999;214:87–101

102. Pellas TC, Ramachandran B, Duncan M, *et al*. Germ-cell deficient (*gcd*), an insertional mutation manifested as infertility in transgenic mice. *Proc Natl Acad Sci USA* 1991;88:8787–91

103. Couldrey C, Carlton MB, Nolan PM, *et al*. A retroviral gene trap insertion into the histone 3.3A gene causes partial neonatal lethality, stunted growth, neuromuscular deficits and male sub-fertility in transgenic mice. *Hum Mol Genet* 1999;8:2489–95

104. Gordon JW, Uehlinger J, Dayani N, *et al*. Analysis of the hotfoot (*ho*) locus by creation of an insertional mutation in a transgenic mouse. *Dev Biol* 1990;137:349–58

105. Yanaka N, Kobayashi K, Wakimoto K, *et al*. Insertional mutation of the murine *kisimo* locus caused a defect in spermatogenesis. *J Biol Chem* 2000;275:14791–4

106. Magram J, Bishop JM. Dominant male sterility in mice caused by insertion of a transgene. *Proc Natl Acad Sci USA* 1991;88:10327–31

107. Komada M, McLean DJ, Griswold MD, *et al*. E-MAP-115, encoding a microtubule-associated protein, is a retinoic acid-inducible gene required for spermatogenesis. *Genes Dev* 2000;14:1332–42

108. Watson ML, Zinn AR, Inoue N, *et al*. Identification of morc (microrchidia), a mutation that results in arrest of spermatogenesis at an early meiotic stage in the mouse. *Proc Natl Acad Sci USA* 1998;95:14361–6

109. Bishop CE, Whitworth DJ, Qin Y, *et al*. A transgenic insertion upstream of *sox9* is associated with dominant XX sex reversal in the mouse. *Nat Genet* 2000;26:490–4

110. Krulewski TF, Neumann PE, Gordon JW. Insertional mutation in a transgenic mouse allelic with Purkinje cell degeneration. *Proc Natl Acad Sci USA* 1989;86: 3709–12

111. Pires-daSilva A, Nayernia K, Engel W, *et al*. Mice deficient for spermatid perinuclear RNA-binding protein show neurologic, spermatogenic, and sperm morphological abnormalities. *Dev Biol* 2001;233:319–28

112. MacGregor GR, Russell LD, Van Beek ME, *et al*. Symplastic spermatids (*sys*): a recessive insertional mutation in mice causing a defect in spermatogenesis. *Proc Natl Acad Sci USA* 1990;87:5016–20

113. Gordon JW, Pravtcheva D, Poorman PA, *et al*. Association of foreign DNA sequence with male sterility and translocation in a line of transgenic mice. *Somat Cell Mol Genet* 1989;15:569–78

114. Toscani A, Mettus RV, Coupland R, *et al*. Arrest of spermatogenesis and defective breast development in mice lacking A-*myb*. *Nature* 1997;386:713–17

115. Adham IM, Nayernia K, Engel W. Spermatozoa lacking acrosin protein show delayed fertilization. *Mol Reprod Dev* 1997;46:370–6

116. Matzuk MM, Kumar TR, Bradley A. Different phenotypes for mice deficient in either activins or activin receptor type II. *Nature* 1995;374:356–60

117. Vassalli A, Matzuk MM, Gardner HAR, *et al*. Activin/inhibin βB subunit gene disruption leads to defects in eyelid development and female reproduction. *Genes Dev* 1994;8:414–27

118. Yu RN, Ito M, Saunders TL, *et al*. Role of Ahch in gonadal development and gametogenesis. *Nat Genet* 1998;20: 353–7

119. Krege JH, John SW, Langenbach LL, *et al*. Male–female differences in fertility and blood pressure in ACE-deficient mice. *Nature* 1995;375:146–8

120. Honarpour N, Du C, Richardson JA, *et al*. Adult Apaf-1-deficient mice exhibit male infertility. *Dev Biol* 2000;218:248–58

121. Huang LS, Voyiaziakis E, Chen HL, *et al*. A novel functional role for apolipoprotein B in male infertility in heterozygous apolipoprotein B knockout mice. *Proc Natl Acad Sci USA* 1996;93:10903–7

122. Benedict JC, Lin TM, Loeffler IK, *et al*. Physiological role of the aryl hydrocarbon receptor in mouse ovary development. *Toxicol Sci* 2000;56:382–8

123. Abbott BD, Schmid JE, Pitt JA, *et al*. Adverse reproductive outcomes in the transgenic Ah receptor-deficient mouse. *Toxicol Appl Pharmacol* 1999;155:62–70

124. Barlow C, Liyanage M, Moens PB, *et al*. Atm deficiency results in severe meiotic disruption as early as leptonema of prophase I. *Development* 1998;125:4007–17

125. Christiansen-Weber TA, Voland JR, Wu Y, *et al*. Functional loss of ABCA1 in mice causes severe placental malformation, aberrant lipid distribution, and kidney glomerulonephritis as well as high-density lipoprotein cholesterol deficiency. *Am J Pathol* 2000;157:1017–29

126. Kuno N, Kadomatsu K, Fan Q-W, *et al*. Female sterility in mice lacking the *basigin* gene, which encodes a transmembrane glycoprotein belonging to the immunoglobulin superfamily. *FEBS Lett* 1998;425:191–4

127. Igakura T, Kadomatsu K, Kaname T, *et al*. A null mutation in *basigin*, an immunoglobulin superfamily member, indicates its important roles in peri-implantation development and spermatogenesis. *Dev Biol* 1998;194:152–65

128. Knudson CM, Tung KSK, Tourtellotte WG, *et al*. Bax-deficient mice with lymphoid hyperplasia and male germ cell death. *Science* 1995;270:96–9

129. Perez GI, Robles R, Knudson CM, *et al*. Prolongation of ovarian lifespan into advanced chronological age by *Bax*-deficiency. *Nat Genet* 1999;21:200–3

130. Ratts VS, Flaws JA, Kolp R, *et al*. Ablation of *bcl-2* gene expression decreases the numbers of oocytes and primordial follicles established in the post-natal female mouse gonad. *Endocrinology* 1995;136:3665–8

131. Kojima S, Hatano M, Okada S, *et al*. Testicular germ cell apoptosis in *Bcl6*-deficient mice. *Development* 2001;128: 57–65

132. Ross AJ, Waymire KG, Moss JE, *et al*. Testicular degeneration in *Bclw*-deficient mice. *Nat Genet* 1998;18:251–6

133. Zhao G-Q, Liaw L, Hogan BLM. Bone morphogenetic protein 8A plays a role in the maintenance of spermatogenesis and the integrity of the epididymis. *Development* 1998;125:1103–12

134. Zhao G-Q, Deng K, Labosky PA, *et al*. The gene encoding bone morphogeneetic protein 8B is required for the initiation and maintenance of spermatogenesis in the mouse. *Genes Dev* 1996;10:1657–69

135. Yan C, Wang P, DeMayo J, *et al*. Synergistic roles of bone morphogenetic protein 15 and growth differentiation factor 9 in ovarian function. *Mol Endocrinol* 2001;15: 854–66

136. Yi SE, LaPolt PS, Yoon BS, *et al*. The type I BMP receptor BmprIB is essential for female reproductive function. *Proc Natl Acad Sci USA* 2001;98:7994–9

137. Colledge WH, Carlton MB, Udy GB, *et al*. Disruption of c-*mos* causes parthenogenetic development of unfertilized mouse eggs. *Nature* 1994;370:65–8

138. Hashimoto N, Watanabe N, Furuta Y, *et al*. Parthenogenetic activation of oocytes in c-*mos*-deficient mice. *Nature* 1994;370:68–71

139. Yeung CH, Sonnenberg-Riethmacher E, Cooper TG. Infertile spermatozoa of c-*ros* tyrosine kinase receptor knockout mice show flagellar angulation and maturational defects in cell volume regulatory mechanisms. *Biol Reprod* 1999;61:1062–9

140. Yeung CH, Wagenfeld A, Nieschlag E, Cooper TC. The cause of infertility of male c-*ros* tyrosine kinase receptor knockout mice. *Biol Reprod* 2000;63:612–18

141. Ikawa M, Wada I, Kominami K, *et al*. The putative chaperone calmegin is required for sperm fertility. *Nature* 1997;387:607–11

142. Wu JY, Ribar TJ, Cummings DE, *et al*. Spermiogenesis and exchange of basic nuclear proteins are impaired in male germ cells lacking Camk4. *Nat Genet* 2000;25:448–52

143. Xu X, Toselli PA, Russell LD, Seldin DC. Globozoospermia in mice lacking the casein kinase II alpha' catalytic subunit. *Nat Genet* 1999;23:118–21

144. Bergeron L, Perez GI, Macdonald G, *et al*. Defects in regulation of apoptosis in caspase-2-deficient mice. *Genes Dev* 1998;12:1304–14

145. Hudson DF, Fowler KJ, Earle E, *et al*. Centromere protein B null mice are mitotically and meiotically normal but have lower body and testis weights. *J Cell Biol* 1998;141: 309–19

146. Fowler KJ, Hudson DF, Salamonsen LA, *et al*. Uterine dysfunction and genetic modifiers in centromere protein B-deficient mice. *Genome Res* 2000;10:30–41

147. Sterneck E, Tessarollo L, Johnson PF. An essential role for C/EBPβ in female reproduction. *Genes Dev* 1997;11: 2153–62

148. Cohen PE, Zhu L, Pollard JW. Absence of colony stimulating factor-1 in osteopetrotic (*csfm*^{op}/*csfm*^{op}) mice disrupts estrous cycles and ovulation. *Biol Reprod* 1997; 56:110–18

149. Simon AM, Goodenough DA, Li E, Paul DL. Female infertility in mice lacking connexin 37. *Nature* 1997; 385:525–9

150. Juneja SC, Barr KJ, Enders GC, Kidder GM. Defects in the germ line and gonads of mice lacking connexin43. *Biol Reprod* 1999;60:1263–70

151. Blendy JA, Kaestner KH, Weinbauer GF, *et al*. Severe impairment of spermatogenesis in mice lacking the *CREM* gene. *Nature* 1996;380:162–5

152. Nantel F, Monaco L, Foulkes NS, *et al*. Spermiogenesis deficiency and germ-cell apoptosis in *CREM*-mutant mice. *Nature* 1996;380:159–62

153. Liu D, Matzuk MM, Sung WK, *et al*. Cyclin A1 is required for meiosis in the male mouse. *Nat Genet* 1998; 20:377–88

154. Shamsadin R, Adham IM, Nayernia K, *et al*. Male mice deficient for germ-cell cyritestin are infertile. *Biol Reprod* 1999;61:1445–51

155. Nishimura H, Cho C, Branciforte DR, *et al*. Analysis of loss of adhesive function in sperm lacking cyritestin or fertilin beta. *Dev Biol* 2001;233:204–13

156. Sicinski P, Donaher JL, Gene Y, *et al*. Cyclin D2 is an FSH-responsive gene involved in gonadal cell proliferation and oncogenesis. *Nature* 1996;384:470–4

157. Rane SG, Dubus P, Mettus RV, *et al*. Loss of Cdk4 expression causes insulin-deficient diabetes and Cdk4 activation results in β-islet cell hyperplasia. *Nat Genet* 1999;22:44–52

158. Dinchuk JE, Car BD, Focht RJ, *et al*. Renal abnormalities and an altered inflammatory response in mice lacking cyclooxygenase II. *Nature* 1995;378:406–9

159. Lim H, Paria BC, Das SK, *et al*. Multiple female reproductive failures in cyclooxygenase 2-deficient mice. *Cell* 1997;91:197–208

160. Ruggiu M, Speed R, Taggart M, *et al.* The mouse *Dazla* gene encodes a cytoplasmic protein essential for gametogenesis. *Nature* 1997;389:73–7

161. Bitgood MJ, Shen L, McMahon AP. Sertoli cell signaling by Desert hedgehog regulates the male germline. *Curr Biol* 1996;6:298–304

162. Pittman DL, Cobb J, Schimenti KJ, *et al.* Meiotic prophase arrest with failure of chromosome synapsis in mice deficient for *Dmc1*, a germline-specific RecA homolog. *Mol Cell* 1998;1:697–705

163. Yoshida K, Kondoh G, Matsuda Y, *et al.* The mouse RecA-like gene *Dmc1* is required for homologous chromosome synapsis during meiosis. *Mol Cell* 1998;1:707–18

164. Tourtellotte WG, Nagarajan R, Auyeung A, *et al.* Infertility associated with incomplete spermatogenic arrest and oligozoospermia in *Egr4*-deficient mice. *Development* 1999; 126:5061–71

165. Miyamoto N, Yoshida M, Kuratani S, *et al.* Defects of urogenital development in mice lacking *Emx2*. *Development* 1997;124:1653–64

166. Kennedy CRJ, Zhang Y, Brandon S, *et al.* Salt-sensitive hypertension and reduced fertility in mice lacking the prostaglandin EP$_2$ receptor. *Nat Med* 1999;5:217–20

167. Tilley SL, Audoly LP, Hicks EH, *et al.* Reproductive failure and reduced blood pressure in mice lacking the EP2 prostaglandin E2 receptor. *J Clin Invest* 1999;103:1539–45

168. Hizaki H, Segi E, Sugimoto Y, *et al.* Abortive expansion of the cumulus and impaired fertility in mice lacking the prostaglandin E receptor subtype EP(2). *Proc Natl Acad Sci USA* 1999;96:10501–6

169. Lubahn DB, Moyer JS, Golding TS, *et al.* Alteration of reproductive function but not prenatal sexual development after insertional disruption of the mouse estrogen receptor gene. *Proc Nat Acad Sci USA* 1993;90:11162–6

170. Hess RA, Bunick D, Lee KH, *et al.* A role for oestrogens in the male reproductive system. *Nature* 1997;390:509–12

171. Krege JH, Hodgin JB, Couse JF, *et al.* Generation and reproductive phenotypes of mice lacking estrogen receptor β. *Proc Natl Acad Sci USA* 1998;95:15677–82

172. Chen M, Tomkins DJ, Auerbach W, *et al.* Inactivation of *Fac* in mice produces inducible chromosomal instability and reduced fertility reminiscent of Fanconi anaemia. *Nat Genet* 1996;12:448–51

173. Whitney MA, Royle G, Low MJ, *et al.* Germ cell defects and hematopoietic hypersensitivity to gamma-interferon in mice with a targeted disruption of the Fanconi anemia C gene. *Blood* 1996;88:49–58

174. Cho C, Bunch DO, Faure JE, *et al.* Fertilization defects in sperm from mice lacking fertilin beta. *Science* 1998;281: 1857–9

175. Soyal SM, Amleh A, Dean J. FIG(alpha), a germ cell-specific transcription factor required for ovarian follicle formation. *Development* 2000;127:4645–54

176. Kooy RF, D'Hooge R, Reyniers E, *et al.* Transgenic mouse model for the fragile X syndrome. *Am J Med Genet* 1996; 64:241–5

177. Lu Q, Shur BD. Sperm from beta 1,4-galactosyl-transferase-null mice are refractory to ZP3-induced acrosome reactions and penetrate the zona pellucida poorly. *Development* 1997;124:4121–31

178. Lu Q, Hasty P, Shur BD. Targeted mutation in beta1,4-galactosyltransferase leads to pituitary insufficiency and neonatal lethality. *Dev Biol* 1997;181:257–67

179. Kumar TR, Wiseman AL, Kala G, *et al.* Reproductive defects in gamma-glutamyl transpeptidase-deficient mice. *Endocrinology* 2000;141:4270–7

180. Settle S, Marker P, Gurley K, *et al.* The BMP family member Gdf7 is required for seminal vesicle growth, branching morphogenesis, and cytodifferentiation. *Dev Biol* 2001;234:138–50

181. Dong J, Albertini DF, Nishimori K, *et al.* Growth differentiation factor-9 is required during early ovarian folliculogenesis. *Nature* 1996;383:531–5

182. Zhou Y, Xu BC, Maheshwari HG, *et al.* A mammalian model for Laron syndrome produced by targeted disruption of the mouse growth hormone receptor/binding protein gene (the Laron mouse). *Proc Natl Acad Sci USA* 1997;94:13215–20

183. Dix DJ, Allen JW, Collins BW, *et al.* Targeted gene disruption of Hsp70-2 results in failed meiosis, germ cell apoptosis, and male infertility. *Proc Natl Acad Sci USA* 1996;93:3264–8

184. Xiao X, Zuo X, Davis AA, *et al.* HSF1 is required for extra-embryonic development, postnatal growth and protection during inflammatory responses in mice. *EMBO J* 1999;18:5943–52

185. Ronfani L, Ferraguti M, Croci L, *et al.* Reduced fertility and spermatogenesis defects in mice lacking chromosomal protein Hmgb2. *Development* 2001;128:1265–73

186. Satokata I, Benson G, Maas R. Sexually dimorphic sterility phenotypes in *Hoxa10*-deficient mice. *Nature* 1995;374: 460–3

187. Hsieh-Li HM, Witte DP, Weinstein M, *et al.* *Hoxa 11* structure, extensive antisense transcription, and function in male and female fertility. *Development* 1995;121:1373–85

188. Matzuk MM, Finegold MJ, Su J-GJ, *et al.* α-Inhibin is a tumor-suppressor gene with gonadal specificity in mice. *Nature* 1992;360:313–19

189. Baker J, Hardy MP, Zhou J, *et al.* Effects of an *Igf1* gene null mutation on mouse reproduction. *Mol Endocrinol* 1996;10:903–18

190. Burks DJ, de Mora JF, Schubert M, *et al.* IRS-2 pathways integrate female reproduction and energy homeostasis. *Nature* 2000;407:377–82

191. Robb L, Li R, Hartley L, *et al.* Infertility in female mice lacking the receptor for interleukin II is due to a defective uterine response to implantation. *Nat Med* 1998;4: 303–8

192. Thepot D, Weitzman JB, Barra J, *et al.* Targeted disruption of the murine *junD* gene results in multiple defects in male reproductive function. *Development* 2000;127:143–53

193. Stewart CL, Kaspar P, Brunet LJ, *et al.* Blastocyst implantation depends on maternal expression of leukemia inhibitory-function. *Nature* 1992;359:76–9

194. Chung S, Wang SP, Pan L, *et al.* Infertility and testicular defects in hormone-sensitive lipase-deficient mice. *Endocrinology* 2001;142:4272–81

195. Lei ZM, Mishra S, Zou W, *et al.* Targeted disruption of luteinizing hormone/human chorionic gonadotropin receptor gene. *Mol Endocrinol* 2001;15:184–200

196. Baker SM, Bronner CE, Zhang L, *et al.* Male mice defective in the DNA mismatch repair gene *PMS2* exhibit abnormal chromosome synapsis in meiosis. *Cell* 1995;82: 309–19

197. Edelmann W, Cohen PE, Kane M, *et al.* Meitoic pachytene arrest in *MLH1*-deficient mice. *Cell* 1996;85:1125–34

198. Baker SM, Plug AW, Prolla TA, *et al.* Involvement of mouse *Mlh1* in DNA mismatch repair and meiotic crossing over. *Nat Genet* 1996;13:336–41

199. de Vries SS, Baart EB, Dekker M, *et al.* Mouse MutS-like protein Msh5 is required for proper chromosome synapsis in male and female meiosis. *Genes Dev* 1999;13:523–31

200. Edelmann W, Cohen PE, Kneitz B, *et al.* Mammalian MutS homologue 5 is required for chromosome pairing in meiosis. *Nat Genet* 1999;21:123–7

201. Behringer RR, Finegold MJ, Cate RL. Mullerian-inhibiting substance function during mammalian sexual development. *Cell* 1994;79:415–25

202. Durlinger AL, Kramer P, Karels B, *et al.* Control of primordial follicle recruitment by anti-Mullerian hormone in the mouse ovary. *Endocrinology* 1999;140:5789–96

203. Mishina Y, Rey R, Finegold MJ, *et al.* Genetic analysis of the Mullerian-inhibiting substance signal transduction pathway in mammalian sexual differentiation. *Genes Dev* 1996;10:2577–87

204. Pace AJ, Lee E, Athirakui K, *et al.* Failure of spermatogenesis in mouse lines deficient in the Na(+)-K(+)-2Cl(−) cotransporter. *J Clin Invest* 2000;105(4):441–50

205. Lee SL, Sadovsky Y, Swirnoff AH, *et al.* Luteinizing hormone deficiency and female infertility in mice lacking the transcription factor NGFI-A (Egr-1). *Science* 1996; 273:1219–21

206. Topilko P, Schneider-Maunoury S, Levi G, *et al.* Multiple pituitary and ovarian defects in Krox-24 (NGFI-A, Egr-1)-targeted mice. *Mol Endocrinol* 1998;12:107–22

207. Good DJ, Porter FD, Mahon KA, *et al.* Hypogonadism and obesity in mice with a targeted deletion of the *Nhlh2* gene. *Nat Genet* 1997;15:397–401

208. Bruning JC, Gautam D, Burks DJ, *et al.* Role of brain insulin receptor in control of body weight and reproduction. *Science* 2000;289:2122–5

209. White R, Leonardsson G, Rosewell I, *et al.* The nuclear receptor co-repressor *nrip1* (RIP140) is essential for female fertility. *Nat Med* 2000;6:1368–74

210. Gow A, Southwood CM, Li JS, *et al.* CNS myelin and sertoli cell tight junction strands are absent in *Osp/claudin-11* null mice. *Cell* 1999;99:649–59

211. Dai X, Schonbaum C, Degenstein L, *et al.* The *ovo* gene required for cuticle formation and oogenesis in flies is involved in hair formation and spermatogenesis in mice. *Genes Dev* 1998;12:3452–63

212. Nishimori K, Young LJ, Guo Q, *et al.* Oxytocin is required for nursing but is not essential for parturition or reproductive behavior. *Proc Nat Acad Sci USA* 1996;93: 11699–704

213. Robertson KM, O'Donnell L, Jones ME, *et al.* Impairment of spermatogenesis in mice lacking a functional aromatase (*cyp 19*) gene. *Proc Natl Acad Sci USA* 1999;96):7986–91

214. Fisher CR, Graves KH, Parlow AF, Simpson ER. Characterization of mice deficient in aromatase (ArKO) because of targeted disruption of the *cyp19* gene. *Proc Natl Acad Sci USA* 1998;95:6965–70

215. Panda DK, Miao D, Tremblay ML, *et al.* Targeted ablation of the 25-hydroxyvitamin D 1alpha-hydroxylase enzyme: evidence for skeletal, reproductive, and immune dysfunction. *Proc Natl Acad Sci USA* 2001;98:7498–503

216. Mulryan K, Gitterman DP, Lewis CJ, *et al.* Reduced vas deferens contraction and male infertility in mice lacking P2X1 receptors. *Nature* 2000;403:86–9

217. Zindy F, van Deursen J, Grosveld G, *et al.* INK4d-deficient mice are fertile despite testicular atrophy. *Mol Cell Biol* 2000;20:372–8

218. Nakayama K, Ishida N, Shirane M, *et al.* Mice lacking p27(Kip1) display increased body size, multiple organ hyperplasia, retinal dysplasia, and pituitary tumors. *Cell* 1996;85:707–20

219. Fero ML, Rivkin M, Tasch M, *et al.* A syndrome of multiorgan hyperplasia with features of gigantism, tumorigenesis, and female sterility in p27[Kip1]-deficient mice. *Cell* 1996;85:733–44

220. Hendry JH, Adeeko A, Potten CS, Morris ID. P53 deficiency produces fewer regenerating spermatogenic tubules after irradiation. *Int J Radiat Biol* 1996;70:677–82

221. Mbikay M, Tadros H, Ishida N, *et al.* Impaired fertility in mice deficient for the testicular germ-cell protease PC4. *Proc Natl Acad Sci USA* 1997;94:6842–6

222. Blume-Jensen P, Jiang G, Hyman R, *et al.* Kit/stem cell factor receptor-induced activation of phosphatidylinositol 3′-kinase is essential for male fertility. *Nat Genet* 2000; 24:157–62

223. Lin SR, Yu IS, Huang PH, *et al.* Chimaeric mice with disruption of the gene coding for phosphatidylinositol glycan class A (*Pig-a*) were defective in embryogenesis and spermatogenesis. *Br J Haematol* 2000;110:682–93

224. Lydon JP, DeMayo FJ, Funk CR, *et al.* Mice lacking progesterone receptor exhibit pleiotropic reproductive abnormalities. *Genes Dev* 1995;9:2266–78

225. Horseman ND, Zhao W, Montecino-Rodriguez E, *et al.* Defective mammopoiesis, but normal hematopoiesis, in mice with a targeted disruption of the prolactin gene. *EMBO J* 1997;16:6926–35

226. Ormandy CJ, Camus A, Barra J, *et al.* Null mutation of the prolactin receptor gene produces multiple reproductive defects in the mouse. *Genes Dev* 1997;11:167–78

227. Sugimoto Y, Yamasaki A, Segi E, *et al.* Failure of parturition in mice lacking the prostaglandin F receptor. *Science* 1997;277:681–3

228. Cho C, Willis WD, Goulding EH, *et al.* Haplo-insufficiency of protamine-1 or -2 causes infertility in mice. *Nat Genet* 2001;28:82–6

229. Uhrin P, Dewerchin M, Hilpert M, *et al.* Disruption of the protein C inhibitor gene results in impaired spermatogenesis and male infertility. *J Clin Invest* 2000;106:1531–9

230. Varmuza S, Jurisicova A, Okano K, *et al.* Spermiogenesis is impaired in mice bearing a targeted mutation in the protein phosphatase 1cgamma gene. *Dev Biol* 1999;205:98–110

231. Lufkin T, Lohnes D, Mark M, *et al.* High postnatal lethality and testis degeneration in retinoic acid receptor alpha mutant mice. *Proc Natl Acad Sci USA* 1993;90: 7225–9

232. Lohnes D, Kastner P, Dierich A, *et al*. Function of retinoic acid receptor gamma in the mouse. *Cell* 1993;73:643–58

233. Kastner P, Mark M, Leid M, *et al*. Abnormal spermatogenesis in RXR beta mutant mice. *Genes Dev* 1996;10:80–92

234. Togawa A, Miyoshi J, Ishizaki K, *et al*. Progressive impairment of kidneys and reproductive organs in mice lacking Rho GDIalpha. *Oncogene* 1999;18:5373–80

235. Rigotti A, Trigatti BL, Penman M, *et al*. A targeted mutation in the murine gene encoding the high density lipoprotein (HDL) receptor scavenger receptor class B type I reveals its key role in HDL metabolism. *Proc Natl Acad Sci USA* 1997;94:12610–15

236. Trigatti B, Rayburn H, Vinals M, *et al*. Influence of the high density lipoprotein receptor SR-BI on reproductive and cardiovascular pathophysiology. *Proc Natl Acad Sci USA* 1999;96:9322–7

237. Yuan L, Liu JG, Zhao J, *et al*. The murine *SCP3* gene is required for synaptonemal complex assembly, chromosome synapsis, and male fertility. *Mol Cell* 2000;5:73–83

238. Supp DM, Witte DP, Branford WW, *et al*. Sp4, a member of the Sp1-family of zinc finger transcription factors, is required for normal murine growth, viability, and male fertility. *Dev Biol* 1996;176:284–99

239. Pearse RV, Drolet DW, Kalla KA, *et al*. Reduced fertility in mice deficient for the POU protein sperm-1. *Proc Natl Acad Sci USA* 1997;94:7555–60

240. Baudat F, Manova K, Yuen JP, *et al*. Chromosome synapsis defects and sexually dimorphic meiotic progression in mice lacking spo11. *Mol Cell* 2000;6:989–98

241. Romanienko PJ, Camerini-Otero RD. The mouse *spo11* gene is required for meiotic chromosome synapsis. *Mol Cell* 2000;6:975–87

242. Mahendroo MS, Cala KM, Russell DW. 5α-Reduced androgens play a key role in murine parturition. *Mol Endocrinol* 1996;10:380–92

243. Mahendroo MS, Cala KM, Landrum CP, Russell DW. Fetal death in mice lacking 5α-reductase type I caused by estrogen excess. *Mol Endocrinol* 1997;11:1–11

244. Caron KM, Soo SC, Wetsel WC, *et al*. Targeted disruption of the mouse gene encoding steroidogenic acute regulatory protein provides insights into congenital lipoid adrenal hyperplasia. *Proc Natl Acad Sci USA* 1997;94:11540–5

245. Luo X, Ikeda Y, Parker KL. A cell-specific nuclear receptor is essential for adrenal and gonadal development and sexual differentiation. *Cell* 1994;77:481–90

246. Matzuk MM, Dionne L, Guo Q, *et al*. Ovarian function in superoxide dismutase 1 and 2 knockout mice. *Endocrinology* 1998;139:4008–11

247. Ho YS, Gargano M, Cao J, *et al*. Reduced fertility in female mice lacking copper–zinc superoxide dismutase. *J Biol Chem* 1998;273:7765–9

248. Zhang D, Penttila TL, Morris PL, *et al*. Spermiogenesis deficiency in mice lacking the *Trf2* gene. *Science* 2001;292:1153–5

249. Martianov I, Fimia GM, Dierich A, *et al*. Late arrest of spermiogenesis and germ cell apoptosis in mice lacking the TBP-like *TLF/TRF2* gene. *Mol Cell* 2001;7:509–15

250. Freiman RN, Albright SR, Zheng S, *et al*. Requirement of tissue-selective TBP-associated factor TAFII105 in ovarian development. *Science* 2001;293:2084–7

251. Lee H-W, Blasco MA, Gottlieb GJ, *et al*. Essential role of mouse telomerase in highly proliferative organs. *Nature* 1998;392:569–77

252. Beck AR, Miller IJ, Anderson P, *et al*. RNA-binding protein TIAR is essential for primordial germ cell development. *Proc Natl Acad Sci USA* 1998;95:2331–6

253. Kuroda M, Sok J, Webb L, *et al*. Male sterility and enhanced radiation sensitivity in TSL (–/–) mice. *EMBO J* 2000;19:453–62

254. Yu YE, Zhang Y, Unni E, *et al*. Abnormal spermatogenesis and reduced fertility in transition nuclear protein 1-deficient mice. *Proc Natl Acad Sci USA* 2000;97:4683–8

255. Roby KF, Son DS, Terranova PF. Alterations of events related to ovarian function in tumor necrosis factor receptor type I knockout mice. *Biol Reprod* 1999;61:1616–21

256. Roest HP, van Klaveren J, de Wit J, *et al*. Inactivation of the HR6B ubiquitin-conjugating DNA repair enzyme in mice causes male sterility associated with chromatin modification. *Cell* 1996;86:799–810

257. Tanaka SS, Toyooka Y, Akasu R, *et al*. The mouse homolog of *Drosophila vasa* is required for the development of male germ cells. *Genes Dev* 2000;14:841–53

258. Kreidberg JA, Sariola H, Loring JM, *et al*. WT-1 is required for early kidney development. *Cell* 1993;74:679–91

259. Vainio S, Heikkila M, Kispert A, *et al*. Female development in mammals is regulated by Wnt-4 signalling. *Nature* 1999;397:405–9

260. Parr BA, McMahon AP. Sexually dimorphic development of the mammalian reproductive tract requires *Wnt-7a*. *Nature* 1998;395:707–710

261. Luoh S-W, Bain PA, Polakiewicz RD, *et al*. Zfx mutation results in small animal size and reduced germ cell number in male and female mice. *Development* 1997;124:2275–84

262. Rankin T, Talbot P, Lee E, Dean J. Abnormal zonae pellucidae in mice lacking ZP1 result in early embryonic loss. *Development* 1999;126:3847–55

263. Rankin TL, O'Brien M, Lee E, *et al*. Defective zonae pellucidae in *Zp2*-null mice disrupt folliculogenesis, fertility and development. *Development* 2001;128:1119–26

264. Rankin T, Familari M, Lee E, *et al*. Mice homozygous for an insertional mutation in the *Zp3* gene lack a zona pellucida and are infertile. *Development* 1996;122:2903–10

265. Liu C, Litscher ES, Mortillo S, *et al*. Targeted disruption of the *mZP3* gene results in production of eggs lacking a zona pellucida and infertility in female mice. *Proc Nat Acad Sci USA* 1996;93:5431–6

Stem cells biotechnology in human reproduction

Human embryonic stem cells and embryo cloning

30

Michal Amit, Edith Suss-Toby, Dorit Manor and Joseph Itskovitz-Eldor

INTRODUCTION

Embryonic stem (ES) cells are cells derived from the inner cell mass (ICM) of the mammalian blastocyst whose uniqueness lies in their pluripotency. Usually, during early embryonic mammalian development, there is only a short period when each cell of the developing embryo possesses the capacity to differentiate into every cell type of the adult body. This capacity, known as pluripotency, takes place between fertilization and the blastocyst stage, and lasts until the early post-implantation period. However, as the embryo reaches the gastrulation stage, it consists of specialized precursors of all three germ layers; thus, the pluripotency of each individual cell is lost.

ES cells possess additional unique characteristics. As well as being pluripotent, ES cells are also immortal cell lines, capable of unlimited undifferentiated proliferation, which does not affect their remarkable developmental potential. During prolonged culture, ES cell lines maintain normal karyotypes. Some of the existing mouse cell lines have been shown to be able to contribute to all fetal tissues, including the germ line. The main characteristics of ES cells are summarized in Table 1.

Initially, pluripotent cell lines were derived from the stem cells of mouse teratocarcinomas[1]. Some of these tumor-derived cell lines were found to be able to differentiate *in vitro* into a variety of cell types, including muscle and nerve cells[2]; to aggregate into cell clusters known as embryoid bodies (EBs), in which part of the cells differentiate spontaneously; and to form teratocarcinomas after their injection into recipient mice[3]. Unlike ES cells, embryonal carcinoma (EC) cells may undergo minor karyotypic changes during prolonged culture and exhibit a lower tendency to differentiate both *in vitro* and *in vivo*[4]. The methods developed for EC cell line derivation and EC cell culture led to the ability to derive ES cell lines from blastocysts[5,6].

In 1981 the first mammalian ES cell lines were derived from mouse blastocysts[5,6]. Since then many research studies have been published that demonstrated the enormous developmental potential of these cells. Following their injection into blastocysts, mouse ES cells were reported to integrate into all fetal germ layers, including the germ line[7], and, in some cases, to develop into mature chimeric animals. A few mouse ES lines can form entire viable newborns when injected into tetraploid embryos or heat-treated blastocysts[8,9].

Since the first publication on ES cell line derivation in 1981, much effort throughout the world has been invested in the derivation of additional ES cell lines from mice and other species. ES cell lines and ES cell-like lines have also been successfully derived from other species, including other rodents such as golden hamsters[10], rats[11], and rabbits[12,13], several domestic animal species[14–17], and

Table 1 Main characteristics of embryonic stem cells

Derived from the preimplantation embryo

Pluripotent, capable of differentiating into representative cells from all three germ layers of the embryo

Immortal, with long-term proliferation at the undifferentiated stage (self-maintenance)

Maintaining normal karyotype after prolonged culture

Capable, after injection into blastocysts, of contributing to all three embryonic germ layers, including the germ line

Expressing unique markers like transcription factor Oct-4 or cell surface markers like stage-specific embryonic antigen

Clonogenic, i.e., each individual cell possessing the above characteristics

two non-human primate species (rhesus monkey and marmoset)[18,19]. Only mouse ES cells have been shown to possess all the ES cell characteristics listed in Table 1.

Many years of research have led to the development of a number of *in vitro* differentiation systems, based on both directed and spontaneous differentiation. One example is neural differentiation systems. Under strict conditions, mouse ES cells can be directed to differentiate into various neural precursors, while in other conditions they can be induced to differentiate into glial and neural precursors that can contribute to brain development after transplantation into rat embryo brains[20,21].

Another example is directed differentiation of ES cells into hematopoietic cells and stem cells. Since there is no efficient *in vitro* model for maintaining hematopoietic stem cells (HSC) in an undifferentiated stage, an alternative ES cell-derived HSC model may offer a unique research tool. Mouse ES cells have been shown to be capable of *in vitro* differentiation into most hematopoietic cell types[22,23]. Hematopoiesis in differentiating mouse ES cell cultures parallels embryonic events[24,25]. In recent years a number of groups have reported on the generation of relatively reproducible models of *in vitro* directed differentiation of mouse ES cells into HSC-like cells. For example, Kennedy and co-workers[25] demonstrated that primitive erythrocytes and other hematopoietic lineages arise from a common multipotential precursor that develops within the EBs. A single publication by Palacios and colleagues[26] claimed successful production of HSC-like cells from mouse ES cells by co-culturing these cells with stromal cells and a combination of various cytokines. These HSC-like cells repopulated the lymphoid, myeloid and erythroid lineages of recipient mice[26]. However, in most of the existing ES cell models there is no convincing evidence, as yet, for the emergence of self-renewing HSC.

Some of the existing directed-differentiation mouse models include the genetic manipulation of ES cells in order to generate relatively pure populations of specific cells. One of the most convincing models is the directed-differentiation model into insulin-secreting cells[27,28].

In this chapter we will discuss the derivation, culture and spontaneous and non-spontaneous differentiation of human ES cells, together with the possible role of cloning in the therapeutic use of ES cells.

DERIVATION OF HUMAN EMBRYONIC STEM CELL LINES

Since the derivation of the first non-human primate ES cell line in 1995[18], a silent competition was raging between several groups of scientists throughout the world on the derivation of the first human ES cell line. In 1998 the contest ended with the publication resulting from the collaboration of Thomson (University of Wisconsin, Madison) and Itskovitz-Eldor and co-workers[29] reporting the derivation of the first human ES cell line. Although all five human ES cell lines described by these authors met most of the criteria listed in Table 1, the clonality of the new human ES cell lines was not tested in that study, and for ethical reasons, the ability of the cells to contribute to all three embryonic germ layers during embryo development, including germ cells, after injection into blastocysts was not examined. Two years later, Reubinoff and co-workers[30] reported the derivation of two additional human ES cell lines. At present, there are more than 70 human ES cell lines in several laboratories around the world, according to a list published by the National Institutes of Health (NIH; www.nih.gov/news/stemcell/index.htm). Although this list does not offer full information on all the lines fulfilling all the ES cell criteria listed in Table 1, it suggests that the derivation of human ES cells is a reproducible procedure with reasonable success rates.

The most common procedure used to isolate the ICM is immunosurgery. This procedure (Figure 1) was developed in the early 1970s by Solter and Knowles[31] and was used for the derivation of some of the EC lines and for early embryonic development research. The tremendous effort described above and the development of the basic methodology employed in the wide-ranging studies using the EC cell lines, laid the groundwork for the first ES cell line derivation in 1981[5,6]. The aim of immunosurgery is to selectively isolate the ICM of the blastocyst from the outer layer of the trophoectoderm. Initially, the zona pellucida of the embryo is removed using traditional methods like Tyrode's solution, and then the embryo is exposed to anti-human whole antiserum (Figure 1b). The antibodies recognize and attach to all human cells, thereby marking all the trophoblast cells. ICM cells, however, remain untouched due to cell-to-cell connections between the trophoblast cells, preventing antibody penetration of the embryo. Then

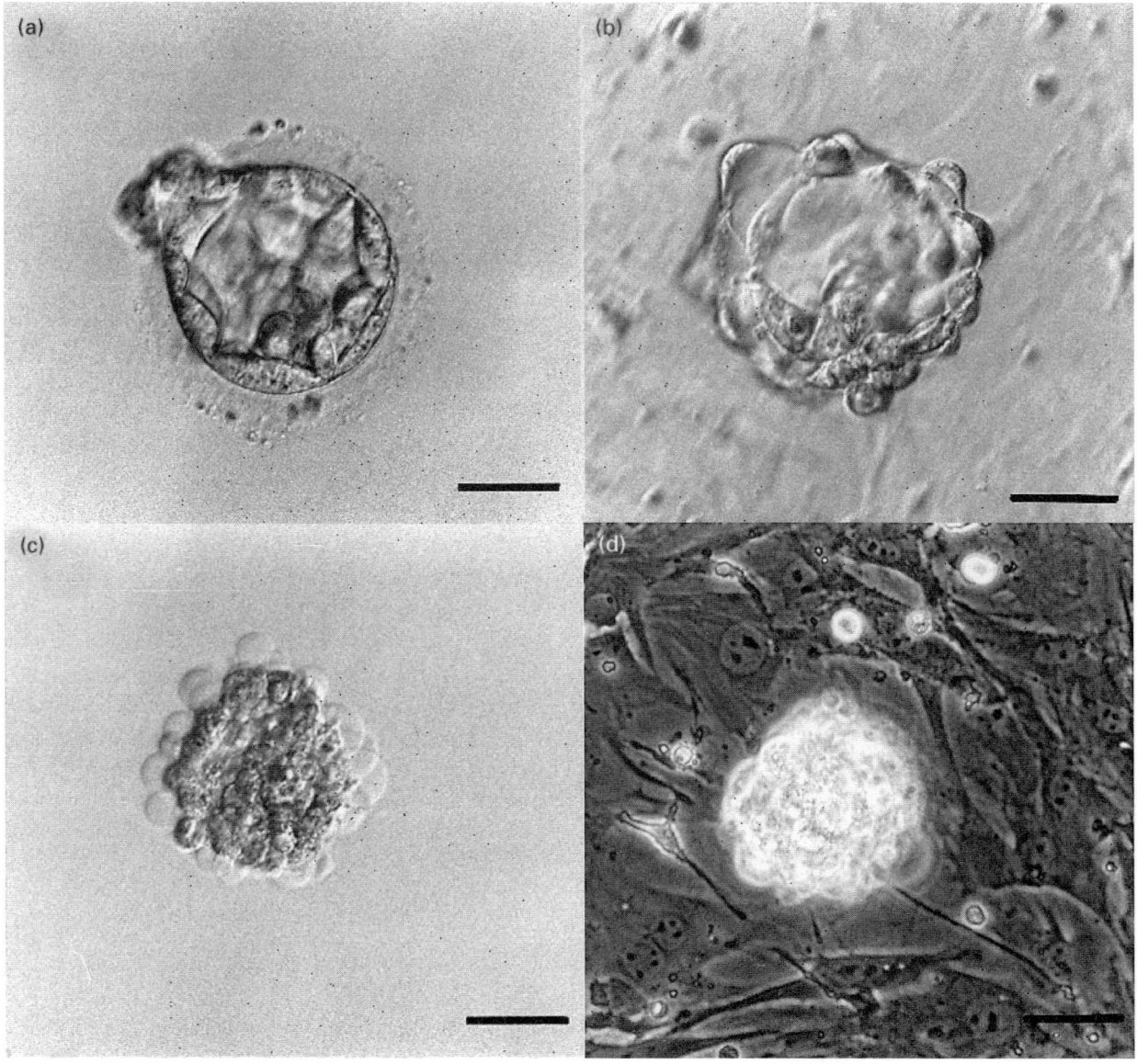

Figure 1 Immunosurgery of a human blastocyst. (a) Donated human embryo produced by *in vitro* fertilization at the blastocyst stage. (b) Human blastocyst after zona pellucida removal by Tyrode's solution, during exposure to rabbit anti-human whole antiserum. (c) The same embryo after exposure to guinea-pig complement. (d) The intact inner cell mass immediately after immunosurgery on mitotically inactivated mouse embryonic fibroblast feeder layer. Bar = 50 μm. Reproduced with permission from Amit M, Itskovitz-Eldor J. Derivation and spontaneous differentiation of human embryonic stem cells. *J Anat* 2002;200;255–32

the embryo is exposed to guinea-pig complement, which lyses all cells marked with the antibody, i.e., the trophoectoderm cells (Figure 1c). The intact ICM is further cultured on a mitotically inactivated mouse embryonic fibroblast (MEF) feeder layer (Figure 1d). The feeder layer has a dual role: firstly, as the term implies, the

MEFs support ES cell growth; secondly, the MEFs prevent spontaneous differentiation of the ES cells during culture. The mechanism by which the MEFs prevent differentiation is still not completely understood.

Four human ES cell lines (I-3, I-4, I-6 and J-3) have been derived in our laboratory. These lines fulfill the characteristics reported regarding the existing human ES cell lines[32]. Two lines (I-3 and I-6) were derived using immunosurgery, and one line (I-4) was derived by gentle removal of the trophoblast with 27-gauge needles. J-3 was derived using a different method. The overall success rate was 60%, which is consistent with other reports on human ES cell line derivation[29,30]. An example of a human ES cell colony growing on MEF feeder layers can be seen in Figure 2. Note the high nucleus-to-cytoplasm ratio and the presence of two to three nucleoli, typical of ES cells.

Primate ES cells have been found to express different surface markers compared with mouse ES cells. While mouse ES cells highly express surface marker stage-specific embryonic antigen-1 (SSEA-1), non-human primate ES cells and human ES cells do not express this marker at all[33]. In addition, non-human primate ES cells and human ES cells strongly express SSEA-3, tumor rejecting antigen (TRA)-1-60 and TRA-1-81 and weakly express SSEA-4, while mouse ES cells never express these markers[29,33]. Another remarkable difference between mouse ES cells and primate ES cells is the latter's ability to differentiate into trophoblasts *in vitro*, a characteristic clearly absent in mouse ES cells[18,19,29]. The implication of these differences on embryonic development requires further investigation.

BASIC METHODS FOR CULTURING EMBRYONIC STEM CELLS

All methods used for culturing human ES cells have been developed according to the knowledge accumulated through the derivation and culture of EC cells and mouse ES cells[34,35]. While mouse ES cells can be grown directly on gelatin-coated plates with the addition of leukemia inhibitory factor (LIF), human ES cells require a feeder layer in order to continuously grow in an undifferentiated stage in culture. As of today, there is no efficient alternative to the traditional MEFs as feeder layers to human ES cells. All reported human ES cell lines have been derived using 80% Dulbecco's modified Eagle's medium supplemented with 20% fetal bovine serum, 1 mmol/l L-glutamine, 0.1 mmol/l β-mercaptoethanol and 1% non-essential amino acid stock. Under these conditions, no additional growth factors are needed either to derive or to grow the human ES cells in culture[29,30].

ES cells are known for their reluctance to grow. They usually require daily medium change, passage every 5 to 7 days depending on the culture conditions and simultaneous growth of the feeder layers. Even though they require rigorous care, ES cells can be cultured in large numbers, frozen and thawed with high survival rates and continuously cultured for over a year without

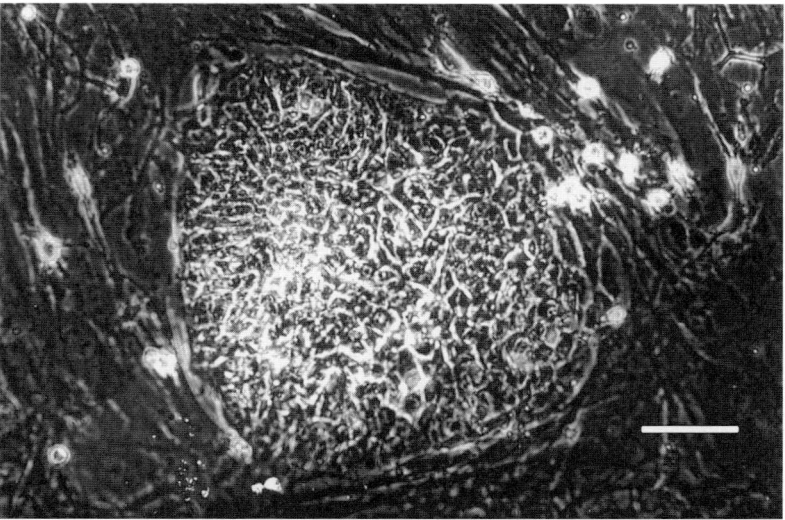

Figure 2 Human embryonic stem cell colony forming I-3 normal XX line, growing on mitotically inactivated mouse embryonic fibroblast feeder layer. Bar = 50 μm

losing their unique characteristics. Overall, ES cells may serve as a powerful scientific tool and be available for cell-consuming research.

Two major improvements in the basic culture methods of human ES cells are their ability to grow under feeder-free culture conditions and the use of a serum-free medium. As noted, the main role of the MEF feeder layer is to prevent the spontaneous differentiation of ES cells. However, this preventive mechanism is not entirely clear. Studies on murine ES cells have shown that LIF, simply by its addition to the culture medium, can prevent the differentiation of ES cells and maintain their potential for self-renewal[36,37]. Therefore, provided that LIF is added to the medium, mouse ES cells can be cultured directly on gelatin-coated plates, without a MEF feeder layer.

Unfortunately, LIF has little or no effect on the prevention of differentiation or on the self-renewal of human ES cells[29,30]. These data raise the possibility that the self-renewal mechanism differs between mouse and primate ES cells. Since it has become clear that the traditional methods of feeder-free culture of mouse ES cells cannot be applied to the human ES cell system, the search for alternatives has begun.

In the first method reported, human ES cells were grown successfully on Matrigel matrix (Becton, Dickinson & Co., Bedford, MA) with 100% MEF-conditioned medium supplemented with basic fibroblast growth factor (bFGF)[38]. This system still requires massive growth of MEFs for the production of conditioned medium. Therefore, although it may prove unsuitable for large-scale growth of human ES cells, it may serve as the basis for completely MEF-free culture systems for human ES cells in the future. The second major improvement in human ES cell culture is the ability to grow these cells under serum-free conditions, using a serum replacer supplemented with bFGF[39]. Serum-free growth of human ES cells provides better defined culture conditions, which will be crucial for further development of controlled directed-differentiation systems based on human ES cell technology.

Another advantage of the serum-free condition is that it has been found to be suitable for the derivation of single-cell human ES cell clones[39]. Originally, human ES cell lines were derived from the clump of cells in the ICM, which might not represent a homologous cell population. To date, eight single-cell clones from five different parental ES cell lines have been derived in our laboratory using the serum-free condition (unpublished data). All eight clones fulfilled the main criteria described for human ES cell lines (Table 1). In our experience, single-cell clones are easier to grow and manipulate. They represent homogeneous populations of cells, which may be important for the development of research models based on gene knockout. Some of our preliminary results indicate that there may be a difference between various single-cell clones and the tendency of parental lines to differentiate in a specific direction. Further studies are being carried out in order to clarify the molecular basis of these differences.

DIFFERENTIATION

Most of the existing methods for directed differentiation of mouse ES cells include the formation of EBs as one of the initial steps, e.g., the *in vitro* differentiation of mouse ES cells into hematopoietic cells[24] and cardiomyocytes[28]. Apparently this step encourages the ES cells to differentiate and consequently increase the rate and efficiency of differentiation. Human ES cells, like mouse ES cells, spontaneously create EBs, including cystic EBs, when they are cultured in suspension. These EBs contain derivatives of the three embryonic germ layers. Figure 3 shows EBs formed from human ES cells in suspension, demonstrating differentiation into various tissues including epithelium, blood vessels and connective tissue. The increasing data on human ES cells suggest that they can create EBs with the same efficiency as mouse ES cells, although the EBs created from human ES cells have been found to be somewhat less organized than mouse ES cell-derived EBs[40]. One report, however, indicates that some of the existing human ES cell lines do not create EBs in suspension[30]. It is still unclear whether this failure reflects differences in the developmental potential between different human ES cell lines or some technical problems.

In addition to the EB mechanism, ES cells can differentiate *in vivo*. Following injection into the hind muscle of SCID mice, ES cells spontaneously create teratomas, in which they differentiate into tissues of either of the three embryonic germ layers. This is the classic and simplest way to examine ES cell pluripotency. Human

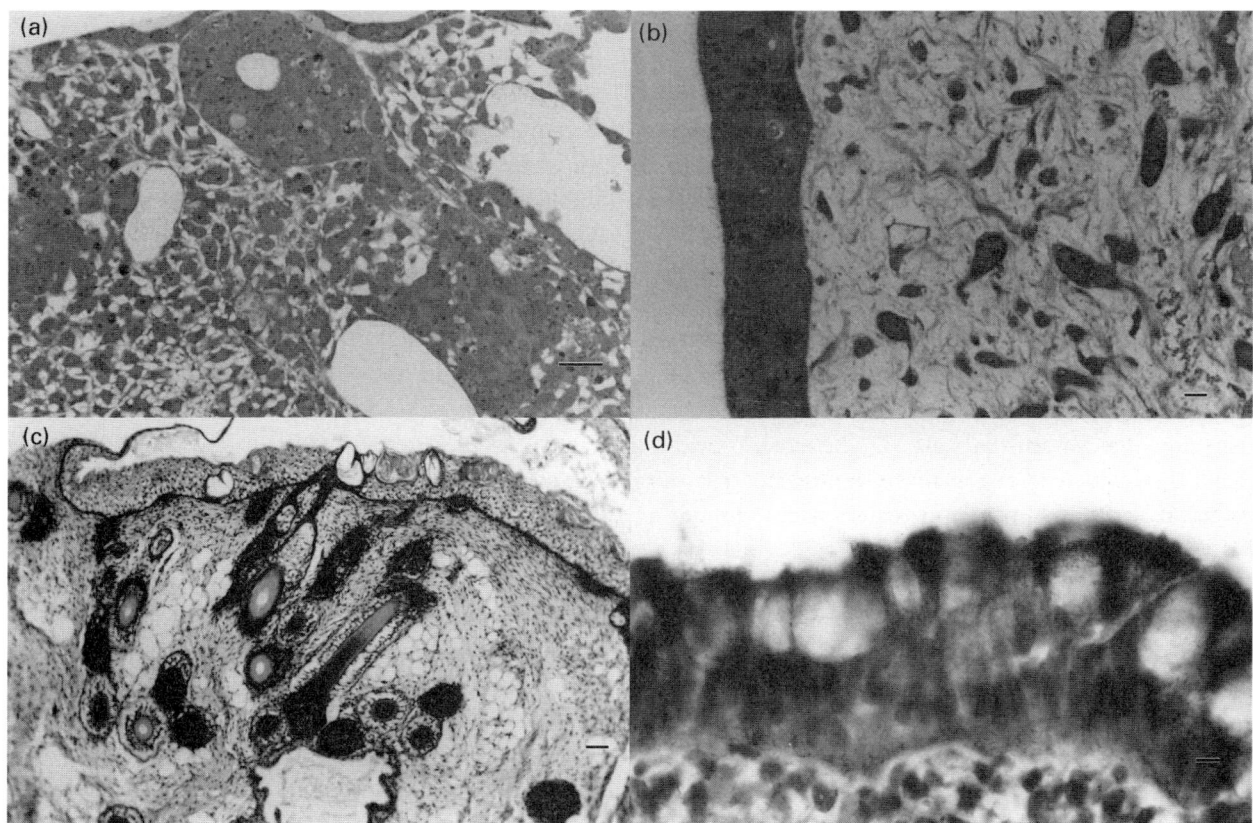

Figure 3 Human embryonic stem (ES) cells undergoing *in vitro* and *in vivo* differentiation. (a, b) Embryoid bodies formed from human ES cells in suspension show differentiation into various tissues. (a) Differentiation into connective tissue including blood vessels. Bar = 25 μm. (b) Differentiation into connective tissue including epithelium. Bar = 5 μm. (c, d) Differentiation of human embryonic stem cells from stably transfected subclone H-9.2.4, expressing *lacz* gene, under the control of the phosphoglycerate kinase promoter, in teratomas. (c) Group of developing hair follicles (hematoxylin and eosin staining). Bar = 50 μm. (d) *lacz*-positive staining in respiratory epithelium containing goblet cells. Bar = 5 μm

ES cells have been shown to differentiate into bone tissue, cartilage tissue, striated muscle, gut-like structures, structures resembling fetal glomeruli, neural rosettes, etc.[29,30] Whereas in EBs, ES cells differentiate mainly into simple structures or unorganized groups of cells, in teratomas, human ES cells can also create more complex and well-organized organ-like structures, a process which requires co-operation between cells and tissues derived from different germ layers. An example of such an organ is seen in Figure 3c, which shows a group of hair follicles. The teratoma model cannot serve as a controlled system for ES cell differentiation because, although it is usually a benign tumor, it also has the potential for malignancy, and incidences of abnormal karyotype among the differentiating cells may appear.

The next step in developing differentiation methods for human ES cells was to examine the abilities of different growth factors to influence human ES cell differentiation *in vitro*. Eight different growth factors have been explored[41]. In that study, ES cells were induced to differentiate as EBs in suspension, and then the EBs were dissociated into single cells and allowed to differentiate further as monolayers on gelatin-coated plates. Under each of the various conditions used, a distinct differentiation direction was observed. None of the growth factors examined was able to direct homogeneous differentiation into a specific cell type. Nevertheless, the study demonstrated the possibility of manipulating human ES cell differentiation.

One of the most important steps in creating directed-differentiation methods with mouse ES cells has been the ability to genetically manipulate these cells. Efficient stable transfection methods are useful for creating pure populations of cells and for inducing directed differentiation. These stable transfection methods can also be used both to mark the cells so that they can be recognized in histologic slides and to examine the role of specific genes during specific differentiation steps or embryonic development.

Judging by the first publications dealing with transfection protocols, it would seem that human ES cells are relatively easy to transfect. In one report, several promoters and transfection agents were shown to be useful in transfecting human ES cells, with different success rates[42]. That study offered a unique method to separate undifferentiated human ES cells from differentiating ones, using a construct with green fluorescent protein under the control of the *rex-1* promoter. In undifferentiated cells expressing *rex-1*, the green protein would be translated and the green cells could be separated using a fluorescence-activated cell sorter.

In another innovative work, lentiviruses were used as vectors to transfect mouse and human ES cells. Human ES cells were shown to hold and express the lentiviral vector for several passages without being silenced[43]. Therefore, lentiviral vector may offer an alternative method for transfection. Finally, our own data have shown that stably transfected subclones can hold the transfected plasmids for more than a year in continuous culture, after freeze–thaw cycles, and after long differentiation periods. The genes inserted were expressed both in undifferentiated cells and in cells differentiated both in an EB system and within teratomas (Figure 3d). Taken together, these findings suggest that genetic manipulation may be performed in human ES cells.

At present, several studies have been published on specific differentiation of human ES cells in both spontaneous and directed-differentiation models. Most of these differentiation systems are summarized in Table 2. One of the first reports was on a reproducible method for differentiation of human ES cells into cardiomyocytes, based on spontaneous differentiation[44]. The cardiomyocytes described demonstrated many characteristic features of cardiomyocytes, including typical myofibrillar organization consistent with early-stage cardiomyocytes, positive staining with specific cardiomyocytic markers including anti-cardiac myosin

heavy chain, anti-α-actinin and anti-desmin, and also expressed specific cardiac genes. In addition, the cardiomyocytes derived from human ES cells demonstrated physiologic features of cardiac cells. Due to the lack of early human cardiomyocyte differentiation models, this innovative model can provide an insight into early human cardiac differentiation.

Levenberg and co-workers[45] demonstrated spontaneous differentiation of human ES cells into endothelial cells which can form vascular-like structures. Human ES cell-derived endothelial cells express endothelial markers like platelet endothelial cell adhesion molecule (PECAM-1), vascular endothelial cadherin (VECAD) and CD34. Following transplantation of these cells into SCID mice, they form vascular-like structures containing mouse blood cells. Human embryonic endothelial cells may be used in the future for the tissue engineering of blood vessels, for transplantation purposes and for the repair of ischemic tissues.

Assady and co-workers[46] demonstrated the ability of human ES cells to differentiate into insulin-secreting cells in growing EBs. These cells immunostained positively for insulin, secreted insulin into the growth medium and expressed β-cell-specific genes. These insulin-secreting cells cannot be considered β-cells, since they did not respond to increasing concentrations of glucose and since the ability of these cells to normalize glycemia in diabetic mice was not examined. Nonetheless, the conditions presented may provide the direction

Table 2 Cell types developed in spontaneous and induced human embryonic stem cell differentiation

Cell type	Reference
Spontaneous differentiation	
cardiomyocytes	Kehat *et al.*, 2001[44]
endothelial cells	Levenberg *et al.*, 2002[45]
insulin-secreting cells	Assady *et al.*, 2001[46]
Directed differentiation	
neural cells	Reubinoff *et al.*, 2000[30]
	Carpenter *et al.*, 2001[47]
	Reubinoff *et al.*, 2001[48]
	Zhang *et al.*, 2001[49]
hematopoietic cells	Kaufman *et al.*, 2001[50]

for further studies on the capability of human ES cells to form insulin-secreting cells and, in the far future, pancreatic β-cells.

The first reported directed-differentiation model for human ES cells showed differentiation of the ES cells into neural precursors in serum-free conditions[30]. Under these conditions, human ES cells formed a unique sphere of neural progenitor cells that gave rise to neural cells expressing specific mature neural surface markers, including neurofilament protein and β-tubulin, and cells that synthesized glutamate and contained glutamic acid decarboxylase and the $GABA_A$ receptor α_2 subunit. Another study on the differentiation of human ES cells into neural precursors demonstrated through electrophysiologic analysis that these resultant neural progenitors possess voltage-dependent channels that can be triggered by neurotransmitters[47]. Further investigation showed the incorporation of these precursors into brain tissue during brain development in a rat model[48,49]. Overall, the increasing data on neural differentiation by human ES cells strongly suggest that an ES cell-based model for neural differentiation could be used in the future for research and for therapeutic uses.

Another directed-differentiation system reported for human ES cells demonstrated their ability to differentiate into hematopoietic precursors[50]. That study demonstrated that only 1–2% of human ES cells examined showed the CD34[+]/CD38[−] phenotype consistent with early hematopoietic cells. In addition, the CD34[+]/CD38[−] cells expressed specific hematopoietic genetic markers. As in the mouse ES cell model, the small pool of human ES cell-derived hematopoietic progenitors tends to differentiate and disappear after 28 days in culture. Much work remains to be done in order to establish an efficient hematopoietic differentiation model.

THE NUCLEAR TRANSFER TECHNIQUE AND ITS LINK TO STEM CELLS

The nuclear transfer (NT) technique is the most commonly used method for animal cloning. In the process of NT, metaphase II (MII) oocytes are enucleated, the genetic material is removed and a donor nucleus is inserted by either fusion or injection. The reconstructed oocyte is activated electrically and/or chemically, and the embryo produced is transferred to a foster mother.

The first successful nuclear transfer was described in 1952[51], and the first somatically cloned frogs were described a decade later[52]. Since then a number of sheep[53], calves[54], mice[55,56], rabbits[57], pigs[58] and rhesus monkeys[59] have been produced, using blastomeres of preimplantation embryos at the two-cell to blastocyst stages as nuclear donors and MII oocytes as recipient cytoplasts. Enucleated zygotes have also been used as recipient cytoplasts; however, their developmental potential is limited.

The next stage in animal cloning was achieved using cell lines of fetal origin[60,61] as a source for nuclear donors. Soon after, Dolly the sheep was reported as the first animal clone produced from an adult somatic cell[61]. This novel finding of adult cells that can be reprogramed in MII oocyte cytoplasm to produce an offspring was confirmed shortly after in mice[62] and calves[63], using cumulus cells. Over the past few years, a large number of cloned animals have been produced using various somatic cells. Sheep[64], mice[65,66], calves[67], goats[68], pigs[69], and recently rabbits[70] have been produced from nuclear donors derived from skin, ear, muscle, liver, tail and Sertoli cells. ES cells have also been shown to be potential nuclear donors[65], unlike fetal germ cells, which exhibit limited developmental potential[71].

The NT technique, combined with genetic manipulations applied to the cells prior to their use as nuclear donors, has many uses in the fields of human medicine, agriculture and research, such as increasing livestock products, production of human therapeutic proteins and the use of tissues and organs for xenotransplantation.

NT research today is mainly focused on the reasons for its low production rate. Although the rate of NT-derived blastocysts is high, the implantation rate after embryo transfer to a foster mother is low. Abortion occurs at various stages of the pregnancy, and only a low percentage of live clones is obtained[72]. Moreover, many of the clones developed to term are overgrown, a phenomenon known as 'large offspring syndrome'[73]. Many die shortly after birth from respiratory distress and circulatory problems[61] or, at later stages, suffer from immune dysfunction or kidney or brain malformations[64,74]. This variety of abnormalities is probably a result of insufficient epigenetic reprograming of the inserted nucleus[75]. During NT, a short time window is available for the inserted nuclear donor to undergo reprograming to support full-term development of the embryo. Little is known about this process, and major efforts are currently being put forth to understand the mechanisms underlying nuclear reprograming.

The low production rate of NT and the abnormalities accompanying the clones, as well as the ethical issues involved, make human reproductive cloning unfeasible. One should clearly distinguish between human reproductive cloning and human therapeutic cloning. Human reproductive cloning refers to the use of NT to produce a live human newborn. Human therapeutic cloning refers to the use of the NT technique to produce an embryo for human ES cell line derivation and for future therapeutic uses. Figure 4 clarifies the steps during somatic NT and the link to stem cell derivation. These results were established in our laboratory using an animal (bovine) model. Figures 4a, 4b and 4c, respectively, show a mature bovine oocyte, oocyte enucleation and a somatic NT-derived blastocyst. Figure 4d shows a Hoechst-stained blastocyst under ultraviolet illumination following parthenogenetic activation. The ICM (arrow) was plated on a mitotically inactivated feeder layer for the derivation of a parthenogenetic ES cell-like line (Figure 4e; unpublished data).

Human therapeutic cloning can serve to create ES cell lines for a specific patient or for pooling ES cell lines. It is one of the options available to overcome the rejection problem that arises when differentially induced ES cells are clinically applied. However, it is associated with two major obstacles. NT is a multi-stage process, and therefore requires a large number of human oocytes which are rather difficult to obtain. In addition, the NT technique has not yet been officially established using human oocytes. To date, only one group has published results related to NT-derived human embryos[76]. They demonstrated poorly developed parthenogenetic embryos and one arrested six-cell embryo following somatic NT. Another newspaper report described 5% blastocyst formation following somatic NT, achieved by a Chinese group[77]. A successful parthenogenetic protocol, i.e., a protocol leading to development until the blastocyst stage following chemical and/or electrical activation,

should be demonstrated on human eggs, since, in order to establish an NT-derived blastocyst, the reconstructed oocytes are exposed to such activation.

Overcoming the problematic issue of the human egg source may be achieved in different ways. It may be feasible to use parthenogenetically activated human embryos once a more defined protocol has been established. This process is simple and requires only a small number of oocytes. Using parthenogenetic embryos for ES cell derivation was recently shown in monkeys[78], in which establishing a normal karyotype of an ES cell line from parthenogenetically activated embryos was demonstrated. ES cell lines may be established using a different approach that would obviate the need for oocytes. The latter approach, which is based on work previously performed in mouse EC cells[79], refers to the possibility of fusing enucleated ES cells with karyoplasts derived from differentiated cells. If the ES cell cytoplasm forces the differentiated karyoplasts to reprogram, a new karyotypic ES cell line can be achieved and redifferentiated for therapeutic uses. Moreover, electrical or spontaneous fusion was demonstrated to occur between mouse ES cells and differentiated cells *in vitro*[80,81]. The resultant tetraploid cells were dominated by the stem cells[81]. Thus, it is possible to assume that somatic cells will be successfully reprogramed to the stem cell stage and redifferentiated for therapeutic uses.

The use of non-human donor–recipient cytoplasm has also been discussed as an optional way to overcome the low number of human oocytes available. Lanza and co-workers[82] reported on one embryo which developed beyond the 16 cells, following NT of human somatic cells into tens of enucleated bovine cytoplasts. This embryo was plated onto a fibroblast feeder layer, and a few ES cell-like colonies were detected. No further description of these ES cell-like cells has been published since. The use of non-human recipient cytoplasm seems to be problematic. First, the number of embryos developed to the morula and blastocyst stages is very small[81] (and unpublished data). Second, and more importantly, the human nuclear donor was actively exposed to an animal cytoplasmic environment. Therefore, if ES cells are derived using a non-human recipient cytoplasm, they will probably not be suitable for clinical applications. However, this model is an interesting research model, as the reprogramed nucleus differs genetically from its cytoplasmic environment, thereby opening a unique situation for studying nuclear– cytoplasmic interactions and for genetic analysis of the factors involved in nuclear reprograming.

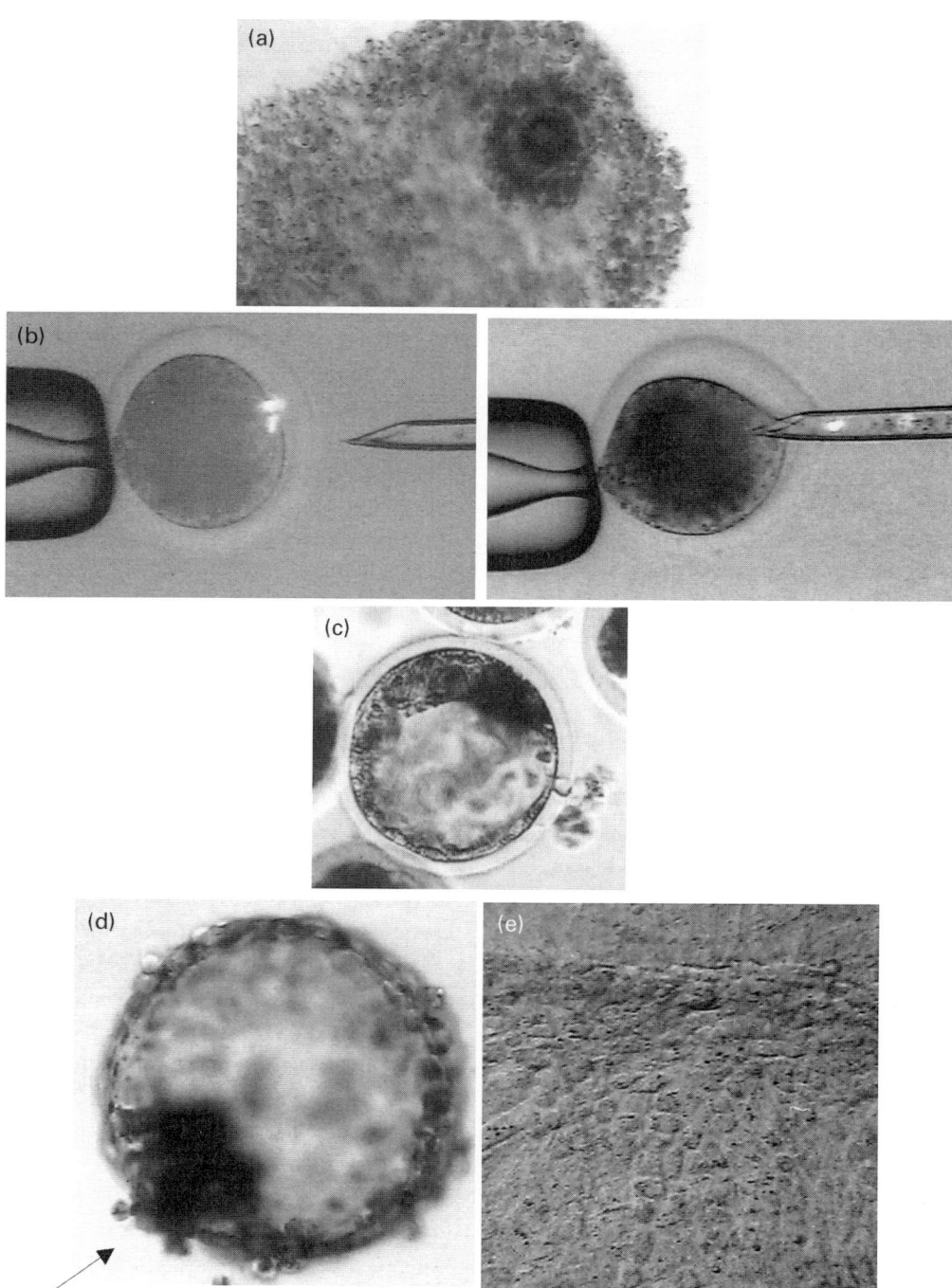

Figure 4 Steps during somatic nuclear transfer (NT) and the link to stem cell derivation. These results were established in our laboratory in the cow oocyte. (a) Mature bovine oocyte. (b) Oocyte enucleation. (c) Somatic NT-derived blastocyst. (d) Hoechst-stained blastocyst under ultraviolet illumination following parthenogenetic activation. (e) Inner cell mass (arrow, d) plated on mitotically inactivated feeder layer for derivation of parthenogenetic embryonic stem cell-like line

In summary, human therapeutic cloning appears to be an achievable goal, which may be accomplished over the next decade. Still, careful genetic and functional characterization of the derived ES cells is needed, as well as the resolution of its pertinent ethical issues.

CONCLUSIONS

Due to their unique characteristics, ES cells are a powerful scientific tool. They are pluripotent, which means they can differentiate into any mature cell type of the adult body; they are immortal cell lines, capable of unlimited undifferentiated proliferation without any influence on their enormous developmental potential; and they maintain normal karyotypes during prolonged culture. Taken together, these three characteristics make human ES cells highly suitable for developmental research, drug testing, tissue engineering and cell-based therapy.

Since the first publication on the derivation of human ES cell lines, several additional derivations of human ES cell lines have been reported, indicating that human ES cell line development is a reproducible procedure with reasonable success rates.

ES cells require meticulous culture, including daily medium exchange and co-culture with a feeder layer. The basic culture methods used for human ES cells have been greatly improved by the ability to grow these cells under feeder-free culture conditions with the use of serum-free medium instead of fetal bovine serum. These methods will offer better-defined culture conditions and, in the future, may help to avoid the risks associated with the transplantation of cells exposed to murine retroviruses.

Another obstacle to using ES cells for cell-based therapy is the rejection of the donor cells by the recipient's immune system. One solution to this problem may be offered by human therapeutic cloning. NT-derived ES cell lines may be used to create ES cell lines for a specific patient or, alternatively, for the development of a pool of human ES cell lines for clinical uses.

As previously demonstrated for mouse ES cells, human ES cells have been shown to differentiate into all three embryonic germ layers both in EBs and in teratomas. Since their derivation, several *in vitro* methods have been developed for spontaneous and induced human ES cell differentiation. Human ES cells have been shown to differentiate into insulin-secreting cells, cardiomyocytes, neural progenitors, hematopoietic precursors and endothelial cells.

From the growing knowledge about human ES cell differentiation and culture conditions, it appears that these cells will fulfill their tremendous potential and expectations.

ACKNOWLEDGEMENTS

The research done in our laboratory was partly supported by the Fund for Medical Research and Development of Infrastructure and Health Services, Rambam Medical Center. We thank Ruth Singer and Hadas Perry for editing the manuscript.

References

1. Kahan BW, Ephrussi B. Developmental potentialities of clonal *in vitro* cultures of mouse testicular teratoma. *J Natl Cancer Inst* 1970;44:1015–36
2. Martin GR, Evans MJ. The morphology and growth of a pluripotent teratocarcinoma cell line and its derivatives in tissue culture. *Cell* 1974;2:163–72
3. Evans MJ. The isolation and properties of a clonal tissue culture strain of pluripotent mouse teratoma cells. *J Embryol Exp Morphol* 1972;28:163–76
4. Andrews PW, Przyborski SA, Thomson JA. Embryonal carcinoma cells as embryonic stem cells. In Marshak DR, Gardner RL, Gottlieb D, eds. *Stem Cell Biology*. Cold Spring Harbor, NY: Cold Spring Harbor Laboratory Press, 2001: 231–65
5. Evans MJ, Kaufman MH. Establishment in culture of pluripotential cells from mouse embryos. *Nature* 1981;292: 154–6
6. Martin GR. Isolation of a pluripotent cell line from early mouse embryos cultured in medium conditioned by teratocarcinoma stem cells. *Proc Natl Acad Sci USA* 1981; 78:7634–8
7. Bradley A, Evans M, Kaufman MH, Robertson E. Formation of germ-line chimaeras from embryo-derived teratocarcinoma cell lines. *Nature* 1984;309:255–6

8. Nagy A, Rossant J, Nagy R, *et al*. Derivation of completely cell culture-derived mice from early-passage embryonic stem cells. *Proc Natl Acad Sci USA* 1993;90:8424–8

9. Amano T, Nakamura K, Tani T, *et al*. Production of mice derived entirely from embryonic stem cells after injecting the cells into heat treated blastocysts. *Theriogenology* 2000;53:1449–58

10. Doetschman T, Williams P, Maeda N. Establishment of hamster blastocyst-derived embryonic stem (ES) cells. *Dev Biol* 1988;127:224–7

11. Iannaccone PM, Taborn GU, Garton RL, *et al*. Pluripotent embryonic stem cells from the rat are capable of producing chimeras. *Dev Biol* 1994;163:288–92 [erratum in *Dev Biol* 1997;185:124–5]

12. Giles JR, Yang X, Mark W, Foote RH. Pluripotency of cultured rabbit inner cell mass cells detected by isozyme analysis and eye pigmentation of fetuses following injection into blastocysts or morulae. *Mol Reprod Dev* 1993;36:130–8

13. Graves KH, Moreadith RW. Derivation and characterization of putative pluripotential embryonic stem cells from preimplantation rabbit embryos. *Mol Reprod Dev* 1993;36:424–33

14. Notarianni E, Galli C, Laurie S, *et al*. Derivation of pluripotent, embryonic cell lines from the pig and sheep. *J Reprod Fertil Suppl* 1991;43:255–60

15. Sims M, First NL. Production of calves by transfer of nuclei from cultured inner cell mass cells. *Proc Natl Acad Sci USA* 1994;91:6143–7

16. Wheeler MB. Development and validation of swine embryonic stem cells: a review. *Reprod Fertil Dev* 1994;6:563–8

17. Mitalipova M, Beyhan Z, First NL. Pluripotency of bovine embryonic stem cell line derived from precompacting embryos. *Cloning* 2001;3:59–67

18. Thomson JA, Kalishman J, Golos TG, *et al*. Isolation of a primate embryonic stem cell line. *Proc Natl Acad Sci USA* 1995;92:7844–8

19. Thomson JA, Kalishman J, Golos TG, *et al*. Pluripotent cell lines derived from common marmoset (*Callithrix jacchus*) blastocysts. *Biol Reprod* 1996; 55:254–9

20. Brüstle O, Spiro AC, Karram K, *et al*. In vitro-generated neural precursors participate in mammalian brain development. *Proc Natl Acad Sci USA* 1997;94:14809–14

21. Brüstle O, Jones KN, Learish RD, *et al*. Embryonic stem cell-derived glial precursors: a source of myelinating transplants. *Science* 1999;285:754–6

22. Doetschman TC, Eistetter H, Katz M, *et al*. The *in vitro* development of blastocyst-derived embryonic stem cell lines: formation of visceral yolk sac, blood islands and myocardium. *J Embryol Exp Morphol* 1985;87:27–45

23. Hole N. Embryonic stem cell-derived haematopoiesis. *Cells Tissues Organs* 1999;165:181–9

24. Keller GM. *In vitro* differentiation of embryonic stem cells. *Curr Opin Cell Biol* 1995;7:862–9

25. Kennedy M, Firpo M, Choi K, *et al*. A common precursor for primitive erythropoiesis and definitive haematopoiesis. *Nature* 1997;386:488–93

26. Palacios R, Golunski E, Samaridis J. *In vitro* generation of hematopoietic stem cells from an embryonic stem cell line. *Proc Natl Acad Sci USA* 1995;92:7530–4

27. Soria B, Roche E, Berná G, *et al*. Insulin-secreting cells derived from embryonic stem cells normalize glycemia in streptozotocin-induced diabetic mice. *Diabetes* 2000;49:157–62

28. Klug MG, Soonpaa MH, Koh GY, Field LJ. Genetically selected cardiomyocytes from differentiating embryonic stem cells form stable intracardiac grafts. *J Clin Invest* 1996;98:216–24

29. Thomson JA, Itskovitz-Eldor J, Shapiro SS, *et al*. Embryonic stem cell lines derived from human blastocysts. *Science* 1998;282:1145–7 [erratum in *Science* 1998;282:1827]

30. Reubinoff BE, Pera MF, Fong C, *et al*. Embryonic stem cell lines from human blastocysts: somatic differentiation *in vitro*. *Nat Biotechnol* 2000;18: 399–404

31. Solter D, Knowles BB. Immunosurgery of mouse blastocyst. *Proc Natl Acad Sci USA* 1975;72:5099–102

32. Amit M, Itskovitz-Eldor J. Derivation and spontaneous differentiation of human embryonic stem cells. *J Anat* 2002;200:225–32

33. Thomson JA, Marshall VS. Primate embryonic stem cells. *Curr Top Dev Biol* 1998;38:133–65

34. Robertson EJ. Embryo-derived stem cell lines. In Robertson EJ, ed. *Teratocarcinomas and Embryonic Stem Cells: A Practical Approach*. Oxford: IRL Press, 1987:71–112

35. Marshall VS, Waknitz MA, Thomson JA. Isolation and maintenance of primate embryonic stem cells. *Methods Mol Biol* 2001;158:11–18

36. Smith AG, Heath JK, Donaldson DD, *et al*. Inhibition of pluripotential embryonic stem cell differentiation by purified polypeptides. *Nature* 1988;336:688–90

37. Williams RL, Hilton DJ, Pease S, *et al*. Myeloid leukaemia inhibitory factor maintains the developmental potential of embryonic stem cells. *Nature* 1988;336:684–7

38. Xu C, Inokuma MS, Denham J, *et al*. Feeder-free growth of undifferentiated human embryonic stem cells. *Nat Biotechnol* 2001;19:971–4

39. Amit M, Carpenter MK, Inokuma MS, *et al*. Clonally derived human embryonic stem cell lines maintain pluripotency and proliferative potential for prolonged periods of culture. *Dev Biol* 2000;227:271–8

40. Itskovitz-Eldor J, Schuldiner M, Karsenti D, *et al*. Differentiation of human embryonic stem cells into embryoid bodies comprising the three embryonic germ layers. *Mol Med* 2000;6:88–95

41. Schuldiner M, Yanuka O, Itskovitz-Eldor J, *et al*. Effects of eight growth factors on the differentiation of cells derived from human embryonic stem cells. *Proc Natl Acad Sci USA* 2000;97:11307–12

42. Eiges R, Schuldiner M, Drukker M, *et al*. Establishment of human embryonic stem cell-transfected clones carrying a marker for undifferentiated cells. *Curr Biol* 2001;11:514–18

43. Pfeifer A, Okawa M, Dayn Y, Verma IM. Transgenesis by lentiviral vectors: lack of gene silencing in mammalian embryonic stem cells and preimplantation embryos. *Proc Natl Acad Sci USA* 2002;99:2140–5

44. Kehat I, Kenyagin-Karsenti D, Snir M, *et al*. Human embryonic stem cells can differentiate into myocytes with structural and functional properties of cardiomyocytes. *J Clin Invest* 2001;108:407–14

45. Levenberg S, Golub JS, Amit M, *et al.* Endothelial cells derived from human embryonic stem cells. *Proc Natl Acad Sci USA* 2002;99:4391–6

46. Assady S, Maor G, Amit M, *et al.* Insulin production by human embryonic stem cells. *Diabetes* 2001;50:1691–7

47. Carpenter MK, Inokuma MS, Denham J, *et al.* Enrichment of neurons and neural precursors from human embryonic stem cells. *Exp Neurol* 2001;172: 383–97

48. Reubinoff BE, Itsykson P, Turetsky T, *et al.* Neural progenitors from human embryonic stem cells. *Nat Biotechnol* 2001;19:1134–40

49. Zhang S-C, Wernig M, Duncan ID, *et al. In vitro* differentiation of transplantable neural precursors from human embryonic stem cells. *Nat Biotechnol* 2001; 19:1129–33

50. Kaufman DS, Hanson ET, Lewis RL, *et al.* Hematopoietic colony-forming cells derived from human embryonic stem cells. *Proc Natl Acad Sci USA* 2001;98: 10716–21

51. Briggs R, King TJ. Transplantation of living nuclei from blastula cells into enucleated frogs' eggs. *Proc Natl Acad Sci USA* 1952;38:455–63

52. Gurdon JB. Adult frogs derived from the nuclei of single somatic cells. *Dev Biol* 1962;4:256–73

53. Wiladsen SM. Nuclear transplantation in sheep embryos. *Nature* 1986;320:63–5

54. Chesné P, Heyman Y, Peynot N, Renard JP. Nuclear transfer in cattle: birth of cloned calves and estimation of blastomere totipotency in morula used as a source of nuclei. *C R Acad Sci III* 1993;316:487–91

55. McGrath J, Solter D. Nuclear transplantation in mouse embryo by microsurgery and cell fusion. *Science* 1983;220: 1300–2

56. Kono T, Kwon OY, Nakahara T. Development of enucleated mouse oocytes reconstituted with embryonic nuclei. *J Reprod Fertil* 1991;93:165–72

57. Stice SL, Robl JM. Nuclear reprogramming in nuclear transplant rabbit embryos. *Biol Reprod* 1988;39:657–64

58. Prather RS, Sims MM, First NL. Nuclear transplantation in early pig embryos. *Biol Reprod* 1989;41:414–18

59. Meng L, Ely JJ, Stouffer RL, Wolf DP. Rhesus monkeys produced by nuclear transfer. *Biol Reprod* 1997;57:454–9

60. Campbell KHS, McWhir J, Ritchie WA, Wilmut I. Sheep cloned by nuclear transfer from a cultured cell line. *Nature* 1996;380:64–6

61. Wilmut I, Schnieke AE, McWhir J, *et al.* Viable offspring derived from fetal and adult mammalian cells. *Nature* 1997;385:810–13

62. Wakayama T, Perry ACF, Zuccotti M, *et al.* Full-term development of mice from enucleated oocytes injected with cumulus cell nuclei. *Nature* 1998;394:369–74

63. Kato Y, Tani T, Sotomaru Y, *et al.* Eight calves cloned from somatic cells of a single adult. *Science* 1998;282:2095–8

64. McCreath KJ, Howcroft J, Campbell KHS, *et al.* Production of gene-targeted sheep by nuclear transfer from cultured somatic cells. *Nature* 2000; 405:1066–9

65. Wakayama T, Rodriguez I, Perry AC, *et al.* Mice cloned from embryonic stem cells. *Proc Natl Acad Sci USA* 1999;96: 14984–9

66. Oruga A, Inouce K, Ogonuki N, *et al.* Production of male cloned mice from fresh, cultured and cryopreserved immature Sertoli cells. *Biol Reprod* 2000;62:1579–84

67. Wells DN, Misica PM, Tervit HR. Production of cloned calves following nuclear transfer with cultured mural granulosa cells. *Biol Reprod* 1999;60:996–1005

68. Baguisi A, Behboodi E, Melican DT, *et al.* Production of goats by somatic cell nuclear transfer. *Nat Biotechnol* 1999;17: 456–6

69. Polejaeva IA, Chen SH, Vaught TD, *et al.* Cloned pigs produced by nuclear transfer from adult somatic cloning. *Nature* 2000;407:86–90

70. Chesné P, Adenot PG, Viglietta C, *et al.* Cloned rabbits produced by nuclear transfer from adult somatic cells. *Nat Biotechnol* 2002;20:366–9

71. Kato Y, Rideout WM III, Hilton K, *et al.* Developmental potential of mouse primordial germ cells. *Development* 1999; 126:1823–32

72. Renard JP, Zhou Q, LeBourhis D, *et al.* Nuclear transfer technologies: between successes and doubts. *Theriogenology* 2002;57:203–22

73. Young LE, Sinclair KD, Wilmut I. Large offspring syndrome in cattle and sheep. *Rev Reprod* 1998:3:155–63

74. Lanza RP, Cibelli JB, Blackwell C, *et al.* Extension of cell life-span and telomere length in animals cloned from senescent somatic cells. *Science* 2000;288:665–9

75. Rideout WM III, Eggan K, Jaenisch R. Nuclear cloning and epigenetic reprogramming of the genome. *Science* 2001;293: 1093–8

76. Cibelli JB, Kiessling AA, Cunniff K, *et al.* Somatic cell nuclear transfer in humans: pronuclear and early embryonic development. *J Regen Med* 2001;2:25–31

77. Leggett K, Regalado A. Fertile ground: as West mulls ethics, China forges ahead in stem-cell research – Prof. Lu's lab says it cloned a human embryo in '99 and is getting better at it – seeking stamp of credibility. *The Wall Street Journal* 2002 Mar 6;Sect. A:1

78. Cibelli JB, Grant KA, Chapman KB, *et al.* Parthenogenetic stem cells in nonhuman primates. *Science* 2002;295: 819

79. McBurney MW, Strutt B. Fusion of embryonal carcinoma cells to fibroblast cells, cytoplasts, and karyoplasts. Developmental properties of viable fusion products. *Exp Cell Res* 1979;124:171–80

80. Tada M, Takahama Y, Abe K, *et al.* Nuclear reprogramming of somatic cells by *in-vitro* hybridization with ES cells. *Curr Biol* 2001;11:1553–8

81. Ying Q-L, Nichols J, Evans EP, Smith AG. Changing potency by spontaneous fusion. *Nature* 2002;416: 545–8

82. Lanza RP, Cibelli JB, West MD. Human therapeutic cloning. *Nat Med* 1999;9:975–7

Transplantation of male germ line stem cells: a technique for man?

31

Stefan Schlatt

INTRODUCTION

Fascinating breakthroughs have been achieved in recent years with regard to our knowledge and to novel applications on spermatogonia, the male germ-line stem cell in the testis. The ability to transplant these cells and manipulate them *in vitro* opened new pathways for research in the testis, preservation and genetic engineering of the male germ line in livestock, and gonadal protection in oncological patients (Figure 1). This chapter gives a brief introduction to the status of research on spermatogonia and future clinical applications.

ORIGIN OF SPERMATOGONIA

Spermatogonia represent a heterogeneous subset of cells consisting of undifferentiated (stem cells) and differentiating diploid germ cells[1]. An easy nomenclature for the various spermatogonial subtypes was recently presented[2].

Testicular stem cells are defined, from a functional point of view, as the self-renewing precursor cells constantly generating differentiating germ cell populations to replenish the loss of cells occuring during spermatogenesis and due to the release of mature gametes. Spermatogonial precursor cells differentiate in the extraembryonic tissue early during embryogenesis. Few cells are now recognized as primordial germ cells. After migration into the undifferentiated gonads, these cells develop into female (oogonia) or male (gonocytes, spermatogonia) germ cells depending on the sexual gonadal differentiation. The number of spermatogonial stem cells is low. A mouse testis contains about 35 000 and a rat testis about 350 000 spermatogonial stem cells[3].

Sertoli cells support a limited number of spermatogonia through the provision of niches allowing a defined

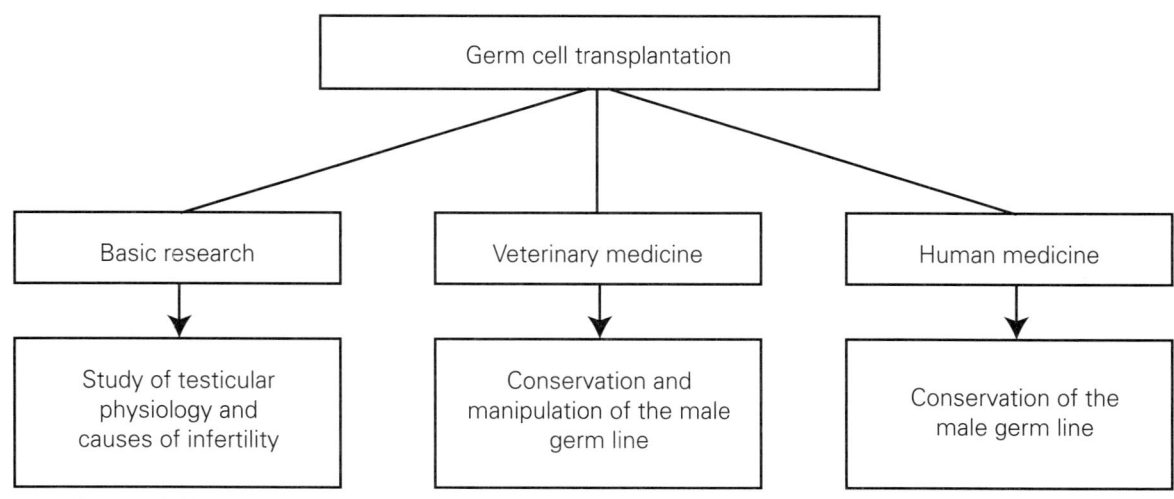

Figure 1 The main fields and the specific applications for germ cell transplantation

number of stem cells to reside. The existence of such niches for germ cell precursors was reported in the fly ovary[5]. Spermatogonial recolonization after transplantation supports the presence of a male germ cell niche[5,6] and the regulation of such niches during postnatal testicular development[7]. Transplanted germ cells from dog, rabbit, and other large domestic species are able to recolonize the mouse testis[6,8,9], although they do not differentiate. Hormones do not influence the expansion of stem cells and the size of the stem cell population is not dependent on the ability of stem cells to express c-Kit or respond to stem cell factor[10]. This indicates that fundamental mechanisms of cell recognition, but not specific hormonal or growth factor-mediated interactions, are regulating the process of testicular stem cell settlement and colonisation.

IN VITRO APPROACHES ON SPERMATOGONIA

Various approaches have been described for spermatogonial isolation, depending on the cell size and shape for the discrimination of testicular cell types. Since spermatogonia differ only marginally from differentiating germ cells such as spermatocytes or round spermatids, this approach is only partially successful using adult testes when the testis is populated by an abundance of more mature germ cell types, and spermatogonia represent only a small subset of germ cells[11–13]. Specific immunologically detectable markers of spermatogonia are rare but can be used for the isolation of these cells even from adult testes: magnetic cell sorting has been described as a tool for the isolation of spermatogonia from testicular cell suspensions of various species using c-Kit antibodies[14]. However, since differentiating but not spermatogonial stem cells express c-Kit, it is not the 'true' stem cell that is enriched using such an approach[15].

Spermatogonial culture and *in vitro* expansion of spermatogonial stem cells appears to be difficult. In single-cell suspensions of spermatogonia or in co-cultures of spermatogonia and testicular cell types, the initiation of spermatogonial differentiation from stem cells and the entry of differentiating spermatogonia into meiosis is blocked. On the other hand, spermatogonial stem cells survive for several months in culture and their

subsequent transplantation allows reinitiation of spermatogenesis[16]. Initiation of spermatogonial proliferation, however, is observed in organ culture[17–19]. A culture method for clonal expansion of gonocytes has been described[20,21] and provides evidence that testicular cell underlays inhibit the proliferation of male germ-line stem cells, suggesting that the testicular environment may act to suppress stem cell expansion.

SPERMATOGONIA: EASY ACCESS TO THE MALE GERM LINE

Spermatogonia play important roles both in the individual and in the genetic stability of populations, as they are the most vulnerable cell type with respect to the integrity of the gene pool of any given species. The integrity of the male genome relies on continuous spermatogonial proliferation without the spontaneous occurrence of changes to the DNA. On the other hand, spermatogonia open a pathway for changes to the male germ line. Recently, the combination of transfection and transplantation of spermatogonia has revealed that it is possible to use such a strategy for the generation of transgenic animals[22,23]. If this approach is optimized it might allow more simple access to the germ line than through embryonic stem cells and might be easily adaptable to many species.

SPERMATOGONIAL TRANSPLANTATION IN MICE AND RODENTS

The first description of germ cell transplantation in mice was presented in 1994[24,25]. Donor spermatogenesis, recognized by developing germ cells carrying the *lac-z* gene, was restored from spermatogonial stem cells microinjected into the seminiferous tubules of host animals. The transmission of the germ cell donor haplotype was proven in mating experiments[25,26]. Later studies showed that most transplanted cells degenerate and disappear from the testis before the first meiotic donor germ cells appear 1 month after germ cell transfer[27]. As an easy alternative to intratubular infusion of germ cells by microinjection, injections into the rete testis of the recipients have been successfully employed[28]. Xenogeneic transfer of rat germ cells into mouse testes prompted rat germ cells to associate with mouse Sertoli cells[29] and vice versa[30]. The differentiation of rat germ cells in mouse testes occurred according to the kinetics and topography typical for the rat[31]. Transplanted spermatogonia from hamsters, rabbits, dogs, monkeys and large domestic species were not able to fully restore spermatogenesis in the immunodeficient mouse testis[6,8,9,32]. The hamster-to-mouse transfer resulted in the production of abnormal hamster sperm. Rabbit or dog spermatogonia recolonized the basal compartment of the host seminiferous tubules but were unable to undergo differentiation.

Germ cell transplantation has enabled the study of several aspects of testicular physiology. These studies have shown that the blood–testis barrier did not block the migration of stem spermatogonia from the adluminal to the basal compartment of the seminiferous tubules, and that absence of c-Kit allowed stem cell colonization and expansion, but not differentiation of germ cells[5,26]. Meanwhile, germ cell transplantation has become an assay with which to assess the potential of germ cell development and the site of action in transgenic animals with disturbed fertility. For example, transplanted estrogen receptor α-deficient germ cells induced qualitatively normal spermatogenesis after transplantation into wild-type testes[33]. In contrast, after transplantation of germ cells from mice carrying the *jsd* mutation, no donor-derived spermatogenesis was established[34].

It was important to develop additional uses for spermatogonia. A combination of different approaches widens the spectrum of applications of testicular stem cells (Figure 2). For example, it is possible to cryo-preserve[35] or culture[16] spermatogonia prior to transplantation and the sorting of germ cells allowed the characterization of marker genes on spermatogonial stem cells and improved transplantation efficiency[14,15].

SPERMATOGONIAL TRANSPLANTATION IN MAN

Optimized treatment regimens have led to high survival rates of oncological patients. However, since the rapidly dividing diploid spermatogonia are the testicular cells that are most sensitive to the cytotoxic effects of radiotherapy[36] and chemotherapy[37], severe damage of the gonads often leads to temporary or permanent infertility as a side-effect of therapy. In men, all hormonal attempts to protect the seminiferous epithelium by reducing the rate of spermatogenesis failed and it appears necessary to develop new approaches to protect or preserve fertility in oncological patients[38–40]. A most obvious subgroup are prepubertal patients who have a poor fertility prognosis and cannot supply a semen sample for cryopreservation. Autologous germ cell transplantation might become a clinically relevant technique by which germ cells can be removed from and later reintroduced into the male gonad[41] (Figure 3).

Modifications of the transfer technique originally described for the mouse were necessary for germ cell transfer into the testes of primates. The easiest, least invasive, and most efficient approach to fill the seminiferous tubules in the monkey testis was ultrasound-guided injection into the rete testis[42]. This technique has also been applied to surgically removed

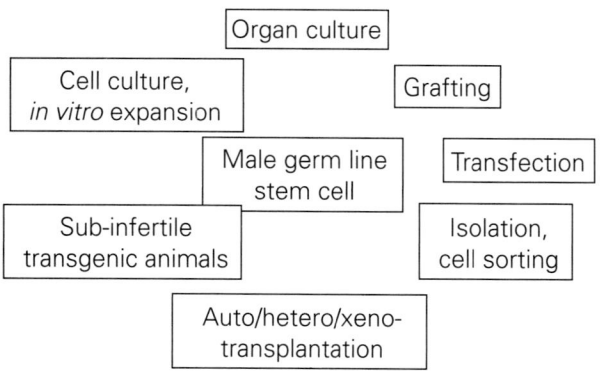

Figure 2 Techniques and research tools influencing future applications of male germ line stem cells

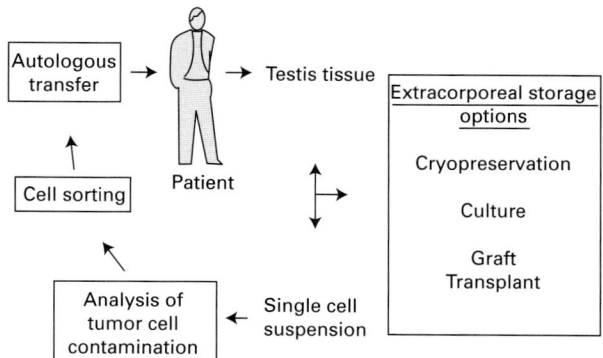

Figure 3 Scheme for the application of germ cell transplantation as a strategy to preserve fertility in oncological patients

human testes. Immaturity or involution of the testis appeared to be a prerequisite for achieving good results. The detection of differentiating spermatogonia 4 weeks after autologous germ cell transfer revealed that germ cell transplantation can, in principle, be applied to primates[42].

In studies using the adult macaque monkey as a preclinical model, attempts were made to mimic the damage occurring in oncological patients to spermatogenesis and to test the application of germ cell transplantation as a gonadal protection strategy[43]. Testicular X-irradiation was used to induce complete azoospermia for more than 200 days, after which spontaneous regrowth of the testes and recovery of spermatogenesis were observed. Unilateral germ cell transfer with the comparison of testicular regrowth enabled the study of the efficacy of germ cell transplantation as a future tool for gonadal protection. Although four out of five monkeys responded with a better regrowth of the germ cell infused testis compared to the contralateral control, the data do not unequivocally show a beneficial effect of germ cell transplantation on the recovery of spermatogenesis. As the transfer procedure, exposure to irradiation, organ culture, preparation of cell suspensions, and cryoprotection procedure represent sources of variation and individual differences, the complexity of such a study renders the results difficult to interpret. However, the approach to cryopreserve the germ cell suspension for the time of extracorporeal storage appears to be successful. A comparison of various cryoprotectants for preservation of testicular cells showed that several of the commonly used cryoprotectants are similarly effective in achieving good survival of germ cells after a freezing and thawing cycle[44].

A major concern is the safety of the procedure since the testis is an organ for settlement of metastases. A study in rats has shown that testicular infusion of only a few malignant cells induces transmission of the disease[45]. Additional strategies combined with the germ cell transplantation procedure have to ensure that the risk for transfer of malignant cells back to the testis of a patient is minimal. In addition to practical implications regarding the safety and efficacy of the procedure it appears noteworthy to consider ethical issues for the further development of the procedure.

OUTLOOK

Germ cell transplantation is a new technique. It opens interesting scenarios for basic research and has immediate implications for veterinary and human medicine. However, as these new applications consist of combinations of a variety of techniques that incorporate cell isolation and cell sorting approaches, *in vitro* systems for stem cell expansion, strategies for cryopreservation, and other tools of extracorporeal storage of cells, the range and the importance of future applications is difficult to judge and remains to be elucidated. Clinical applications in men will be centered around the conservation of testicular tissue as an option for fertility preservation. The efficacy and safety of such applications have to be optimized and proven by future studies.

ACKNOWLEDGEMENTS

I am grateful for the support by grants and fellowships from the Deutsche Forschungsgemeinschaft and I am indebted to Prof. E. Nieschlag, Prof. G. F. Weinbauer, Dr. C. Rolf, Dr. A. Kamischke, Dr. L. Foppiani, Dr. G. Rosiepen, Dr. A. Schepers, Dr. J. Wistuba, C. Cantauw and V. von Schönfeldt who were involved in the work presented.

References

1. Meistrich ML, van Beek MEAB. Spermatogonial stem cells. In Desjardins C, Ewing LL, eds. *Cell and Molecular Biology of the Testis*. New York: Oxford University Press, 1993:266–95
2. de Rooij DG, Russell LD. All you wanted to know about spermatogonia but were afraid to ask. *J Androl* 2000;21: 776–98
3. Tegelenbosch RA, de Rooij DG. A quantitative study of spermatogonial multiplication and stem cell renewal in the C3H/101 F1 hybrid mouse. *Mutat Res* 1993;290: 193–200
4. Xie T, Spradling AC. A niche maintaining germ line stem cells in the Drosophila ovary. *Science* 2000;290:328–30

5. Ohta H, Yomogida K, Yamada S, *et al*. Real-time observation of transplanted 'green germ cells': proliferation and differentiation of stem cells. *Dev Growth Differ* 2000;42:105–12

6. Nagano M, McCarrey JR, Brinster RL. Primate spermatogonial stem cells colonize mouse testes. *Biol Reprod* 2001; 64:1409–16

7. Shinohara T, Orwig KE, Avarbock MR, Brinster RL. Remodeling of the postnatal mouse testis is accompanied by dramatic changes in stem cell number and niche accessibility. *Proc Natl Acad Sci USA* 2001;98:6186–91

8. Dobrinski I, Avarbock MR, Brinster RL. Transplantation of germ cells from rabbits and dogs into mouse testes. *Biol Reprod* 1999;61:1331–9

9. Dobrinski I, Avarbock MR, Brinster RL. Germ cell transplantation from large domestic animals into mouse testes. *Mol Reprod Dev* 2000;57:270–9

10. Ohta H, Yomogida K, Dohmae K, Nishimune Y. Regulation of proliferation and differentiation in spermatogonial stem cells: the role of c-kit and its ligand SCF. *Development* 2000; 127:2125–31

11. Bucci LR, Brock WA, Johnson TS, Meistrich ML. Isolation and biochemical studies of enriched populations of spermatogonia and early primary spermatocytes from rat testes. *Biol Reprod* 1986;34:195–206

12. Dirami G, Ravindranath N, Jia MC, Dym M. Isolation and culture of immature rat type A spermatogonial stem cells in signal transduction in testicular cells. In Hansson FO, Levy K, Tasken V, eds. *Signal Transduction in Testicular Cells*. Berlin: Springer-Verlag, 1997:141–65

13. Bellvé AR, Cavicchia JC, Millette CF, *et al*. Spermatogenic cells of the prepuberal mouse. Isolation and morphological characterization. *J Cell Biol* 1977;74:68–85

14. von Schönfeldt V, Krishnamurthy H, Foppiani L, Schlatt S. Magnetic cell sorting is a fast and efficient method of enriching viable spermatogonia from rodent and primate testes. *Biol Reprod* 1999;61:582–89

15. Shinohara T, Avarbock MR, Brinster RL. Beta1- and alpha6-integrin are surface markers on mouse spermatogonial stem cells. *Proc Natl Acad Sci USA* 1999;96:5504–9

16. Nagano M, Avarbock MR, Leonida EB, *et al*. Culture of mouse spermatogonial stem cells. *Tissue Cell* 1998;30:389–97

17. Boitani C, Politi MG, Menna T. Spermatogonial cell proliferation in organ culture of immature rat testis. *Biol Reprod* 1993;48:761–7

18. Schlatt S, Zhengwei Y, Meehan T, *et al*. Application of morphometric techniques to postnatal rat testes in organ culture: insights into testis growth. *Cell Tissue Res* 1999; 298:335–43

19. Meehan T, Schlatt S, O'Bryan MK, *et al*. Regulation of germ cell and Sertoli cell development by activin, follistatin, and FSH. *Dev Biol* 2000;220:225–37

20. Hasthorpe S, Barbic S, Farmer PJ, Hutson JM. Neonatal mouse gonocyte proliferation assayed by an *in vitro* clonogenic method. *J Reprod Fertil* 1999;116:335–44

21. Hasthorpe S, Barbic S, Farmer PJ, Hutson JM. Growth factor and somatic cell regulation of mouse gonocyte-derived colony formation *in vitro*. *J Reprod Fertil* 2000;119:85–91

22. Nagano M, Shinohara T, Avarbock MR, Brinster RL. Retrovirus-mediated gene delivery into male germ line stem cells. *FEBS Lett* 2000;475:7–10

23. Nagano M, Brinster CJ, Orwig KE, *et al*. Transgenic mice produced by retroviral transduction of male germ-line stem cells. *Proc Natl Acad Sci USA* 2001;98:13090–5

24. Brinster RL, Zimmermann JW. Spermatogenesis following male germ-cell transplantation. *Proc Natl Acad Sci USA* 1994;91:11289–302

25. Brinster RL, Avarbock MR. Germline transmission of donor haplotype following spermatogonial transplantation. *Proc Natl Acad Sci USA* 1994;91:11303–7

26. Ogawa T, Dobrinski I, Avarbock, MR, Brinster RL. Transplantation of male germ line stem cells restores fertility in infertile mice. *Nat Med* 2000;6:29–34

27. Parreira CG, Ogawa T, Avarbock MR, Brinster RL. Development of germ cell transplants in mice. *Biol Reprod* 1998; 59:1360–70

28. Ogawa T, Arechaga JM, Avarbock MR, Brinster RL. Transplantation of testis germinal cells into mouse seminiferous tubules. *Int J Dev Biol* 1997;41:111–22

29. Clouthier DE, Avarbock MR, Maika SD, *et al*. Rat spermatogenesis in mouse testis. *Nature* 1996;381:418–21

30. Ogawa T, Dobrinski I, Brinster RL. Recipient preparation is critical for spermatogonial transplantation in the rat. *Tissue Cell* 1999;31:461–72

31. Franca LR, Ogawa T, Avarbock MR, *et al*. Germ cell genotype controls cell cycle during spermatogenesis in the rat. *Biol Reprod* 1998;59:1371–7

32. Ogawa T, Dobrinski I, Avarbock MR, Brinster RL. Xenogeneic spermatogenesis following transplantation of hamster germ cells to mouse testes. *Biol Reprod* 1999;60:515–21

33. Mahato D, Goulding EH, Korach KS, Eddy EM. Spermatogenic cells do not require estrogen receptor-alpha for development or function. *Endocrinology* 2000;141:1273–6

34. Boettger-Tong HL, Johnston DS, Russell LD, *et al*. Juvenile spermatogonial depletion (*jsd*) mutant seminiferous tubules are capable of supporting tranplanted spermatogenesis. *Biol Reprod* 2000;63:1185–91

35. Avarbock MR, Brinster CJ, Brinster RL. Reconstitution of spermatogenesis from frozen spermatogonial stem cells. *Nat Med* 1996;2:693–6

36. Meistrich ML. Effects of chemotherapy and radiotherapy on spermatogenesis. *Eur Urol* 1993;23:136–41

37. Meistrich ML, Finch M, da Cunha MF, *et al*. Damaging effects of fourteen chemotherapeutic drugs on mouse testis cells. *Cancer Res* 1982;42:122–31

38. Naysmith TE, Blake DA, Harvey VJ, Johnson NP. Do men undergoing sterilizing cancer treatments have a fertile future? *Hum Reprod* 1998;13:3250–5

39. Radford J, Shalet S, Lieberman B. Fertility after treatment for cancer. Questions remaining over ways of preserving ovarian and testicular tissue. *Br Med J* 1999;319:935–6

40. Aslam I, Fishel S, Moore H, *et al*. Fertility preservation of boys undergoing anti-cancer therapy: a review of the existing situation and prospects for the future. *Hum Reprod* 2000; 15:2154–9

41. Schlatt S, von Schönfeldt V, Schepers AG. Male germ cell transplantation: an experimental approach with a clinical perspective. *Br Med Bull* 2000;56:824–836

42. Schlatt S, Rosiepen G, Weinbauer GF, *et al*. Germ cell transfer into rat, bovine, monkey and human testes. *Hum Reprod* 1999;14:144–50

43. Schlatt S, Foppiani L, Rolf C, *et al.* Germ cell transplantation into X-irradiated monkey testes. *Hum Reprod* 2002;17: 55–62

44. Brook PF, Radford JA, Shalet SM, *et al.* Isolation of germ cells from human testicular tissue for low temperature storage and autotransplantation. *Fertil Steril* 2001;75: 269–74

45. Jahnukainen K, Hou M, Petersen C, *et al.* Intratesticular transplantation of testicular cells from leukemic rats causes transmission of leukemia. *Cancer Res* 2001;61:706–10

Index